A Note from the Author

You are about to embark on an amazing journey of discovery. The study of life spans from the inner workings of cells to the complex interactions of entire ecosystems, through the information stored in DNA to the ways genetic information evolves over time. At the same time that our understanding of biology is growing in leaps and bounds, so too are great insights into how learners acquire new knowledge and skills. We are thrilled to join Scott Freeman on *Biological Science*, a book dedicated to active, research-based learning and to exploring the experimental evidence that informs what we know about biology. The next few pages highlight the features in this book and in MasteringBiology® that will help you succeed.

From left to right: Michael Black, Emily Taylor, Jon Monroe, Lizabeth Allison, Greg Podgorski, Kim Quillin

New chapter-opening Roadmaps visually group and organize information to help you anticipate key ideas as well as recognize meaningful relationships and connections between them.

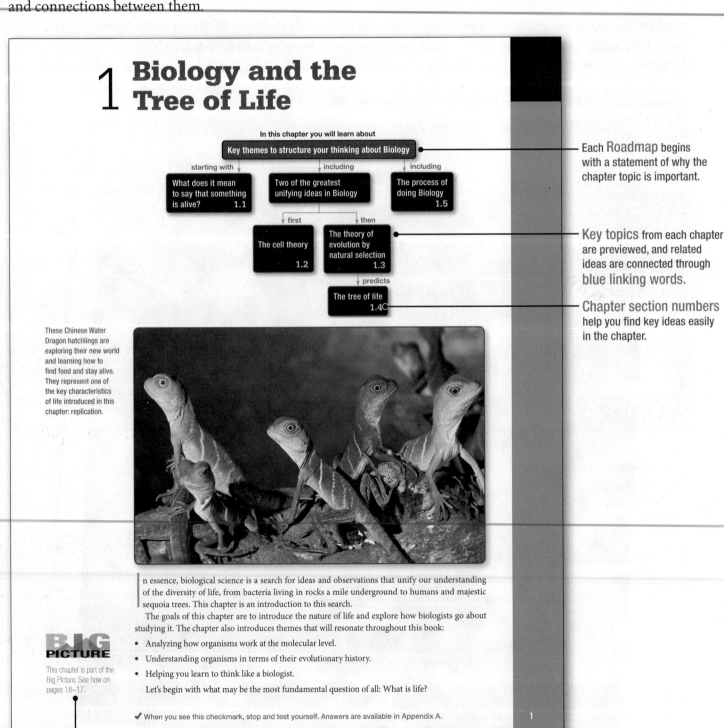

1 Biology and the Tree of Life

In this chapter you will learn about

Key themes to structure your thinking about Biology

starting with → What does it mean to say that something is alive? **1.1**

including → Two of the greatest unifying ideas in Biology

including → The process of doing Biology **1.5**

first → The cell theory **1.2**

then → The theory of evolution by natural selection **1.3**

predicts → The tree of life **1.4**

Each Roadmap begins with a statement of why the chapter topic is important.

Key topics from each chapter are previewed, and related ideas are connected through **blue linking words.**

Chapter section numbers help you find key ideas easily in the chapter.

These Chinese Water Dragon hatchlings are exploring their new world and learning how to find food and stay alive. They represent one of the key characteristics of life introduced in this chapter: replication.

BIG PICTURE

This chapter is part of the Big Picture. See how on pages 16–17.

In essence, biological science is a search for ideas and observations that unify our understanding of the diversity of life, from bacteria living in rocks a mile underground to humans and majestic sequoia trees. This chapter is an introduction to this search.

The goals of this chapter are to introduce the nature of life and explore how biologists go about studying it. The chapter also introduces themes that will resonate throughout this book:

- Analyzing how organisms work at the molecular level.
- Understanding organisms in terms of their evolutionary history.
- Helping you learn to think like a biologist.

Let's begin with what may be the most fundamental question of all: What is life?

✔ When you see this checkmark, stop and test yourself. Answers are available in Appendix A.

1

Big Picture Concept Maps are referenced on the opening page of related chapters, pointing you to summary pages that help you synthesize challenging topics.

Big Picture Concept Maps integrate visuals and words to help you synthesize information about challenging topics in biology that span multiple chapters and units.

Three New Big Picture topics have been added to the Fifth Edition:

- NEW! Doing Biology
- NEW! The Chemistry of Life
- Energy for Life
- Genetic Information
- Evolution
- NEW! Plant and Animal Form and Function
- Ecology

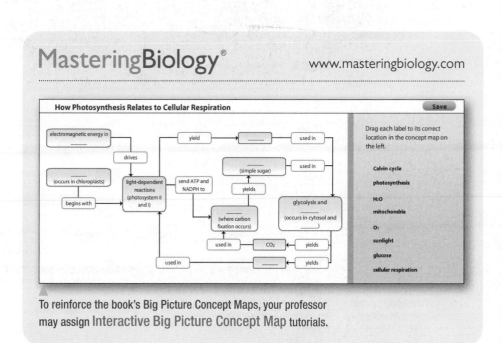

To reinforce the book's Big Picture Concept Maps, your professor may assign Interactive Big Picture Concept Map tutorials.

Practice for success
on tests and exams

Intertwined color-coded "active learning threads" are embedded in the text. The gold thread helps you to identify important ideas, and the blue thread helps you to test your understanding.

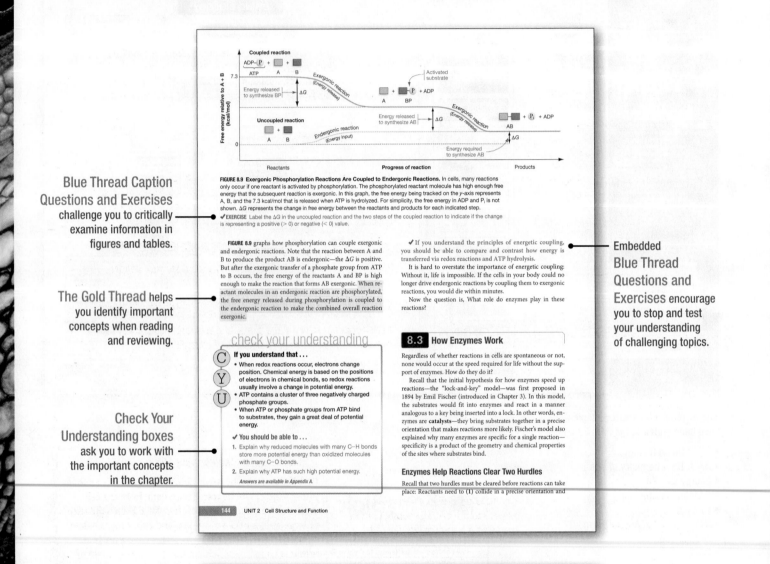

FIGURE 8.9 Exergonic Phosphorylation Reactions Are Coupled to Endergonic Reactions. In cells, many reactions only occur if one reactant is activated by phosphorylation. The phosphorylated reactant molecule has high enough free energy that the subsequent reaction is exergonic. In this graph, the free energy being tracked on the y-axis represents A, B, and the 7.3 kcal/mol that is released when ATP is hydrolyzed. For simplicity, the free energy in ADP and P_i is not shown. ΔG represents the change in free energy between the reactants and products for each indicated step.

✔ **EXERCISE** Label the ΔG in the uncoupled reaction and the two steps of the coupled reaction to indicate if the change is representing a positive (> 0) or negative (< 0) value.

Blue Thread Caption Questions and Exercises challenge you to critically examine information in figures and tables.

FIGURE 8.9 graphs how phosphorylation can couple exergonic and endergonic reactions. Note that the reaction between A and B to produce the product AB is endergonic—the ΔG is positive. But after the exergonic transfer of a phosphate group from ATP to B occurs, the free energy of the reactants A and BP is high enough to make the reaction that forms AB exergonic. When reactant molecules in an endergonic reaction are phosphorylated, the free energy released during phosphorylation is coupled to the endergonic reaction to make the combined overall reaction exergonic.

The Gold Thread helps you identify important concepts when reading and reviewing.

check your understanding

Check Your Understanding boxes ask you to work with the important concepts in the chapter.

If you understand that . . .

- When redox reactions occur, electrons change position. Chemical energy is based on the positions of electrons in chemical bonds, so redox reactions usually involve a change in potential energy.
- ATP contains a cluster of three negatively charged phosphate groups.
- When ATP or phosphate groups from ATP bind to substrates, they gain a great deal of potential energy.

✔ You should be able to . . .

1. Explain why reduced molecules with many C–H bonds store more potential energy than oxidized molecules with many C–O bonds.

2. Explain why ATP has such high potential energy.

Answers are available in Appendix A.

✔ If you understand the principles of energetic coupling, you should be able to compare and contrast how energy is transferred via redox reactions and ATP hydrolysis.

It is hard to overstate the importance of energetic coupling: Without it, life is impossible. If the cells in your body could no longer drive endergonic reactions by coupling them to exergonic reactions, you would die within minutes.

Now the question is, What role do enzymes play in these reactions?

Embedded Blue Thread Questions and Exercises encourage you to stop and test your understanding of challenging topics.

8.3 How Enzymes Work

Regardless of whether reactions in cells are spontaneous or not, none would occur at the speed required for life without the support of enzymes. How do they do it?

Recall that the initial hypothesis for how enzymes speed up reactions—the "lock-and-key" model—was first proposed in 1894 by Emil Fischer (introduced in Chapter 3). In this model, the substrates would fit into enzymes and react in a manner analogous to a key being inserted into a lock. In other words, enzymes are **catalysts**—they bring substrates together in a precise orientation that makes reactions more likely. Fischer's model also explained why many enzymes are specific for a single reaction—specificity is a product of the geometry and chemical properties of the sites where substrates bind.

Enzymes Help Reactions Clear Two Hurdles

Recall that two hurdles must be cleared before reactions can take place: Reactants need to (1) collide in a precise orientation and

8.2 Nonspontaneous Reactions May Be Driven Using Chemical Energy

- Redox reactions transfer energy by coupling exergonic oxidation reactions to endergonic reduction reactions.
- High-energy C–H bonds may be formed during the reduction step of a redox reaction when an H^+ is combined with a transferred electron.
- The hydrolysis of ATP is an exergonic reaction and may be used to drive a variety of cellular processes.
- When a phosphate group from ATP is added to a substrate that participates in an endergonic reaction, the potential energy of the substrate is raised enough to make the reaction exergonic and thus spontaneous.

✔ You should be able to explain what energetic coupling means, and why life would not exist without it.

End-of-Chapter Blue Thread Exercises, integrated in the chapter summary, help you review the major themes of the chapter and synthesize information.

8.3 How Enzymes Work

- Enzymes are protein catalysts. They speed reaction rates but do not affect the change in free energy of the reaction.
- The structure of an enzyme has an active site that brings sub-

- Protein cleavage and phosphorylation are examples of how enzymes may be regulated by modifying their primary structure.

✔ You should be able to compare and contrast the effect of allosteric regulation versus phosphorylation on enzyme function.

8.5 Enzymes Can Work Together in Metabolic Pathways

- In cells, enzymes often work together in metabolic pathways that sequentially modify a substrate to make a product.
- A pathway may be regulated by controlling the activity of one enzyme, often the first in the series of reactions. Feedback inhibition results from the accumulation of a product that binds to an enzyme in the pathway and inactivates it.
- Metabolic pathways were vital to the evolution of life, and new pathways continue to evolve in cells.

✔ You should be able to predict how the removal of the intermediate in a two-step metabolic pathway would affect the enzymatic rates of the first and last.

 www.masteringbiology.com

1. MasteringBiology Assignments

Identify gaps in your understanding, then fill them

The Fifth Edition provides many opportunities for you to test yourself and offers **helpful learning strategies.**

Analyze: Can I recognize underlying patterns and structure?

Evaluate: Can I make judgments on the relative value of ideas and information?

Create: Can I put ideas and information together to generate something new?

Apply: Can I use these ideas in the same way or in a new situation?

Understand: Can I explain this concept in my own words?

Remember: Can I recall the key terms and ideas?

◄ **Bloom's Taxonomy** describes six learning levels: Remember, Understand, Apply, Analyze, Evaluate, and Create. Questions in the book span all levels, including self-testing at the higher levels to help you develop higher-order thinking skills that will prepare you for exams.

Steps to Building Understanding
Each chapter ends with three groups of questions that build in difficulty:

✔ TEST YOUR KNOWLEDGE
Begin by testing your basic knowledge of new information.

✔ TEST YOUR UNDERSTANDING
Once you're confident with the basics, demonstrate your deeper understanding of the material.

✔ TEST YOUR PROBLEM-SOLVING SKILLS
Work towards mastery of the content by answering questions that challenge you at the highest level of competency.

BIOSKILL 16 using Bloom's taxonomy

Most students have at one time or another wondered why a particular question on an exam seemed so hard, while others seemed easy. The explanation lies in the type of cognitive skills required to answer the question. Let's take a closer look.

NEW! BioSkill Covering Bloom's Taxonomy helps you to recognize question types using the Bloom's cognitive hierarchy, and it provides specific strategies to help you study for questions at all six levels.

Answer Appendix Includes Bloom's Taxonomy Information
Answers to all questions in the text now include the Bloom's level being tested. You can simultaneously practice assessing your understanding of content and recognizing Bloom's levels. Combining this information with the guidance in the BioSkill on Bloom's Taxonomy will help you form a plan to improve your study skills.

► **✔ Test Your Problem-Solving Skills**

13. `analyze` A scientific theory is not a guess—it is an idea whose validity can be tested with data. Both the cell theory and the theory of evolution have been validated by large bodies of observational and experimental data.
14. `apply` If all eukaryotes living today have a nucleus, then it is logical to conclude that the nucleus arose in a common ancestor of all eukaryotes, indicated by the arrow you should have added to the figure. See **FIGURE A1.2**. If it had arisen in a common ancestor of Bacteria or Archaea, then species in those groups would have had to lose the trait—an unlikely event.
15. `evaluate` The data set was so large and diverse that it was no longer reasonable to argue that noncellular life-forms would be discovered. **16.** `apply` b

Mastering**Biology**®
www.masteringbiology.com

NEW! End-of-chapter questions from the book are now available for your professor to assign as homework in MasteringBiology.

Practice scientific thinking and scientific skills

A unique emphasis on the process of scientific discovery and experimental design teaches you how to think like a scientist as you learn fundamental biology concepts.

RESEARCH

QUESTION: Do horses minimize the cost of locomotion?

HYPOTHESIS: Horses choose gaits that minimize energy use at different speeds.

NULL HYPOTHESIS: Horses do not choose gaits based on cost of locomotion.

EXPERIMENTAL SETUP:

1. Measure oxygen consumption of horses trained to walk, trot, and gallop at a range of speeds on a treadmill. Calculate energy used per distance travelled at different speeds.

5.6 m/s

Oxygen mask

Treadmill

2. Videotape the same horses locomoting freely in the paddock, and measure the gaits and speeds they choose to use naturally.

PREDICTION: For each gait, there is a range of speeds where energy use is minimized. Horses will favor these gaits and speeds.

PREDICTION OF NULL HYPOTHESIS: There will be no correlation between chosen gaits and energy consumption.

RESULTS:

Energy/distance (joules per meter)

500
400
300
200
100
0

Walking   Trotting   Galloping

Speeds and gaits chosen by free horses

1 2 3 4 5 6 7
Speed (m/s)

©1981 Nature Publishing Group

CONCLUSION: Horses choose gaits that minimize energy use at different speeds and avoid speeds with high energy consumption.

FIGURE 48.16 Horses Minimize the Cost of Locomotion by Choosing Appropriate Gaits.
SOURCE: Hoyt, D. F., and C. R. Taylor. 1981. Gait and the energetics of locomotion in horses. *Nature* 292: 239–240.

✔ **QUANTITATIVE** Use the graph to estimate the relative energy expense of galloping rather than trotting at 3.5 meters/second (m/s).

All of the Research Boxes cite the original research paper and include a question that asks you to analyze the design of the experiment or study.

◄ Research Boxes explain how research studies are designed and give you additional practice interpreting data. Each Research Box consistently models the scientific method, presenting the research question, hypotheses, experimental setup, predictions, results, and conclusion. 15 Research Boxes are new to the Fifth Edition.

MasteringBiology®
www.masteringbiology.com

► **NEW! Solve It Tutorials** are available for homework assignments in MasteringBiology and give you an opportunity to work like a scientist through a simulated investigation that requires you to analyze and interpret data.

Experimental Inquiry Tutorials based on some of biology's most seminal experiments give you a chance to analyze data and the reasoning that led scientists from the data to their conclusions.

Experimental Inquiry tutorial topics include:

- What Can You Learn About the Process of Science from Investigating a Cricket's Chirp?
- Which Wavelengths of Light Drive Photosynthesis?
- What Is the Inheritance Pattern of Sex-Linked Traits?
- Does DNA Replication Follow the Conservative, Semiconservative, or Dispersive Model?
- How Do Calcium Ions Help to Prevent Polyspermy During Egg Fertilization?
- Did Natural Selection of Ground Finches Occur When the Environment Changed?
- What Effect Does Auxin Have on Coleoptile Growth?
- What Role Do Genes Play in Appetite Regulation?
- Can a Species' Niche Be Influenced by Interspecific Competition?
- What Factors Influence the Loss of Nutrients from a Forest Ecosystem?

Build important skills scientists use to perform, evaluate, and communicate scientific research.

FIGURE 28.14 **The Big Five Mass Extinction Events.** This graph shows the percentage of lineages called families that went extinct over each interval in the fossil record since the Cambrian explosion. Over 50 percent of families and 90 percent of species went extinct during the end-Permian extinction.

DATA: Benton, M. J., 1995. *Science* 268: 52–58.

✔**QUANTITATIVE** Which extinction event ended the era of the dinosaurs 65 million years ago? About what percentage of families went extinct?

NEW! Graphs and tables now include their data sources, emphasizing the research process that leads to our understanding of biological ideas.

NEW! Quantitative questions are identified throughout the text, helping you practice computational problem solving and data analysis.

Expanded BioSkills Appendix helps you build skills that will be important to your success in biology. At relevant points in the text, you'll find references to the BioSkills appendix that will help you learn and practice foundational skills.

BioSkills Topics include:

- The Metric System and Significant Figures
- Some Common Latin and Greek Roots Used in Biology
- Reading Graphs
- Using Statistical Tests and Interpreting Standard Error Bars
- Combining Probabilities

- Using Logarithms
- Reading a Phylogenetic Tree
- Reading Chemical Structures
- Separating and Visualizing Molecules
- Separating Cell Components by Centrifugation
- Biological Imaging: Microscopy and X-ray Crystallography

- Cell and Tissue Culture Methods
- Model Organisms
- NEW! Primary Literature and Peer Review
- Making Concept Maps
- NEW! Using Bloom's Taxonomy

MasteringBiology®
www.masteringbiology.com

You can access self-paced BioSkills activities in the Study Area, and your instructor can assign additional activities in MasteringBiology.

Visualize biology processes and structures

A carefully crafted visual program helps you gain a better understanding of biology through accurate, appropriately detailed figures.

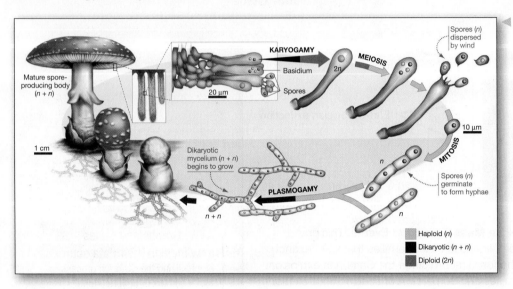

◄ **NEW! Redesigned Life Cycle diagrams** in Unit 6 and 7 help you compare and contrast processes among different organisms.

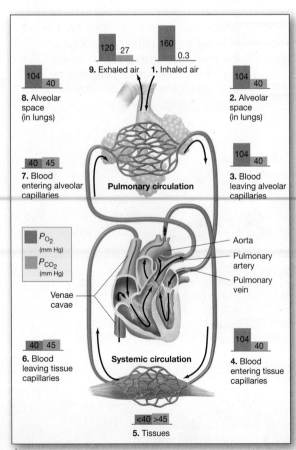

Informative figures help you think through complex biological processes in manageable steps.

SUMMARY TABLE 28.3	Branch Lengths in Phylogenetic Trees
Graphical Symbol	**Meaning of Branch Lengths**
	Branch lengths are *arbitrary*. Emphasis is on the branching *pattern*, which estimates evolutionary relationships among populations. This is the type of tree used in this book.
0.1 Number of base substitutions per site	Horizontal branch lengths show the extent of *genetic difference* among populations. A scale bar is included.
4 3 2 1 0 Millions of years ago	Horizontal branch lengths show the extent of *evolutionary time* between nodes. A scale bar is included.

Visual Summary Tables pull together important information in a format that allows for easy comparison and review.

Instructor and Student Resources

For Instructors

Instructor Resource DVD-ROM
978-0-321-86112-2 • 0-321-86112-4

Everything you need for lectures in one place, including video segments that demonstrate how to incorporate active-learning techniques into your own classroom. Enhanced menus make locating and assessing the digital resources for each chapter easy. The Instructor Resource CD/DVD-ROM includes PowerPoint® Lecture Outlines that integrate figures and animations for classroom presentations. All textbook figures, art, and photos are in JPEG format, and all PowerPoint slides and JPEGs have editable labels. Over 300 Instructor Animations accurately depict complex topics and dynamic processes described in the book.

Instructor Guide (Download only)
Available in the instructor resource area of MasteringBiology.®

TestGen® (Download only)

All of the exam questions in the Test Bank have been peer reviewed and student tested, providing questions that set the standard for quality and accuracy. To improve the Test Bank, Metadata from MasteringBiology users has been incorporated directly into the software. Test questions that are ranked according to Bloom's taxonomy and improved TestGen® software makes assembling tests that much easier.

For Students

Study Guide
978-0-321-85832-0 • 0-321-85832-8

The Study Guide presents a breakdown of key biological concepts, difficult topics, and quizzes to help students prepare for exams. Unique to this study guide are four introductory, stand-alone chapters that introduce students to foundational ideas and skills necessary for classroom success: Introduction to Experimentation and Research in the Biological Sciences, Presenting Biological Data, Understanding Patterns in Biology and Improving Study Techniques, and Reading and Writing to Understand Biology. "Looking Forward" and "Looking Back" sections help students make connections across the chapters instead of viewing them as discrete entities.

Practicing Biology: A Student Workbook
978-0-321-88647-7 • 0-321-88647-X

This workbook focuses on key ideas, principles, and concepts that are fundamental to understanding biology. A variety of hands-on activities such as mapping and modeling suit different learning styles and help students discover which topics they need more help on. Students learn biology by doing biology. An instructors guide can be downloaded from the Instructor Area of MasteringBiology.

MasteringBiology®

www.masteringbiology.com

MasteringBiology is an online homework, tutorial, and assessment system that delivers self-paced tutorials that provide individualized coaching, focus on your course objectives, and respond to each student's progress. The Mastering system helps instructors maximize class time with customizable, easy-to-assign, and automatically graded assessments that motivate students to learn outside of class and arrive prepared for lecture. MasteringBiology is also available with a complete Pearson eText edition of *Biological Science*.

Highlights of the Fifth Edition Item Library include:

* **NEW! assignment options** include Solve It activities, end-of-chapter problems, and questions that accompany new BioSkills and new Big Picture Interactive Concept Maps.
* **NEW! "best of" homework pre-built assignments** help professors assign popular, key content quickly, including a blend of tutorials, end-of-chapter problems, and test bank questions.

* ***Get Ready for Biology* and Chemistry Review assignment options** help students get up to speed with activities that review chemistry, mathematics, and basic biology.

MasteringBiology® Virtual Labs

978-0-321-88644-6 • 0-321-88644-5

MasteringBiology: Virtual Labs is an online environment that promotes critical-thinking skills using virtual experiments and explorations that might be difficult to perform in a wet-lab environment due to time, cost, or safety concerns. MasteringBiology: Virtual Labs offers unique learning experiences in the areas of microscopy, molecular biology, genetics, ecology, and systematics.

For more information, please visit www.pearsonhighered.com/virtualbiologylabs

BIOLOGICAL SCIENCE

VOLUME 1 The Cell, Genetics, & Development

**After you explore this book . . .
you should be able to . . .**

• Pose an evolutionary hypothesis to explain why meter-long male water dragons are larger and have more colorful throats than the females.

• Propose how DNA sequences could be used to determine the relationships among populations of these lizards in China, India, and Southeast Asia.

• Design an experiment to study the relative importance of swimming, tree climbing, and running to the ability of these semi-aquatic lizards to find food and escape from predators.

• Create questions of your own and suggest methods for finding the answers!

Chinese Water Dragon,
Physignathus cocincinus

BIOLOGICAL SCIENCE

FIFTH EDITION

SCOTT FREEMAN
University of Washington

LIZABETH ALLISON
College of William & Mary

MICHAEL BLACK
California Polytechnic State University in San Luis Obispo

GREG PODGORSKI
Utah State University

KIM QUILLIN
Salisbury University

JON MONROE
James Madison University

EMILY TAYLOR
California Polytechnic State University in San Luis Obispo

PEARSON

Boston   Columbus   Indianapolis   New York   San Francisco   Upper Saddle River
Amsterdam   Cape Town   Dubai   London   Madrid   Milan   Munich   Paris   Montréal   Toronto
Delhi   Mexico City   São Paulo   Sydney   Hong Kong   Seoul   Singapore   Taipei   Tokyo

Editor-in-Chief: Beth Wilbur
Senior Acquisitions Editor: Michael Gillespie
Executive Director of Development: Deborah Gale
Senior Development Editor: Sonia DiVittorio
Project Editor: Anna Amato
Development Editors: Mary Catherine Hager, Moira Lerner Nelson, Bill O'Neal
Art Development Editors: Fernanda Oyarzun, Adam Steinberg
Associate Editor: Brady Golden
Assistant Editor: Leslie Allen
Editorial Assistant: Eddie Lee
Executive Media Producer: Laura Tommasi
Media Producer: Joseph Mochnick
Associate Media Producer: Daniel Ross
Associate Media Project Manager: David Chavez
Text Permissions Project Manager: Joseph Croscup
Text Permissions Specialist: Sheri Gilbert
Director of Production: Erin Gregg
Managing Editor: Michael Early

Production Project Manager: Lori Newman
Production Management: S4Carlisle Publishing Services
Copyeditor: Christianne Thillen
Compositor: S4Carlisle Publishing Services
Design Manager: Mark Ong
Interior Designer: Integra Software Services
Cover Designer: tt eye
Illustrators: Imagineering Media Services
Photo Permissions Management: Phutu
Photo Researchers: Kristin Piljay, Eric Schrader, Maureen Spuhler
Senior Photo Editor: Travis Amos
Manufacturing Buyer: Michael Penne
Director of Marketing: Christy Lesko
Executive Marketing Manager: Lauren Harp
Sales Director for Key Markets: David Theisen

Cover Photo Credit: *Physignathus cocincinus* Eric Isselée/Fotolia

Credits and acknowledgments for materials borrowed from other sources and reproduced, with permission, in this textbook appear on the appropriate page within the text or beginning on page C:1 of the backmatter.

Library of Congress Cataloging-in-Publication Data is available upon request.

1 2 3 4 5 6 7 8 9 10—CRK—16 15 14 13 12

ISBN 10: 0-321-74367-9; ISBN 13: 978-0-321-74367-1 (Student Edition)
ISBN 10: 0-321-84159-X; ISBN 13: 978-0-321-84159-9 (Instructor's Review Copy)
ISBN 10: 0-321-86216-3; ISBN 13: 978-0-321-86216-7 (Books a la Carte Edition)
ISBN 10: 0-321-84180-8; ISBN 13: 978-0-321-84180-3 (Volume 1)
ISBN 10: 0-321-84181-6; ISBN 13: 978-0-321-84181-0 (Volume 2)
ISBN 10: 0-321-84182-4; ISBN 13: 978-0-321-84182-7 (Volume 3)

www.pearsonhighered.com

Detailed Contents

40 Plant Sensory Systems, Signals, and Responses 793

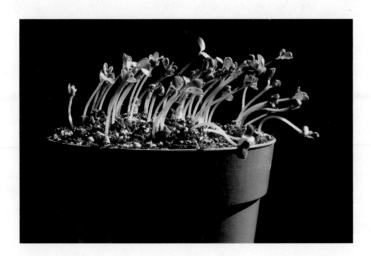

41 Plant Reproduction 822

UNIT 8 HOW ANIMALS WORK 842

42 Animal Form and Function 842

About the Authors

A Letter from Scott:

I started working on *Biological Science* in 1997 with a simple goal: To help change the way biology is taught. After just shy of 20,000 hours of work on four editions of this text, that goal still gets me out of bed in the morning. But instead of focusing my energies on textbook writing, I've decided to devote myself full-time to research on student learning and developing new courses for undergraduate and graduate students at the University of Washington.

So with this edition I am passing the torch—to an all-star cast of leading scientists and educators who have enthusiastically taught from, and contributed to, previous editions of *Biological Science*. Working with them, I have seen the new team bring their passion, talent, and creativity to the book, with expertise that spans the breadth of the life sciences. Just as important, they work beautifully together because they think alike. They are driven by a shared concern for student learning, a commitment to the craft of writing, and a background in evidence-based teaching.

These pages provide a brief introduction to Liz Allison, Michael Black, Greg Podgorski, Kim Quillin, Jon Monroe, and Emily Taylor. As a group, they've built on the book's existing strengths and infused this edition with fresh energy, perspective, and ideas. I'm full of admiration for what they have accomplished, and excited about the impact this edition will have on biology students from all over the world.—*Scott Freeman*

Lizabeth A. Allison is professor and chair of the Biology Department at the College of William & Mary. She received her Ph.D. in Zoology from the University of Washington, specializing in molecular and cellular biology. Before coming to William & Mary, she spent eight years as a faculty member at the University of Canterbury in New Zealand. Liz teaches introductory biology for majors and upper-division molecular biology courses. She has mentored graduate students and more than 80 undergraduate research students, many of them coauthoring papers with her on intracellular trafficking of the thyroid hormone receptor in normal and cancer cells. The recipient of numerous awards, including a State Council for Higher Education in Virginia (SCHEV) Outstanding Faculty Award in 2009, Liz received one of the three inaugural Arts & Sciences Faculty Awards for Teaching Excellence in 2011, and a Plumeri Award for Faculty Excellence in 2012. In addition to her work on this text, she is author of *Fundamental Molecular Biology*, now in its second edition.
Lead Author; Chapter 1 and BioSkills
laalli@wm.edu

Scott Freeman received a Ph.D. in Zoology from the University of Washington and was subsequently awarded an Alfred P. Sloan Postdoctoral Fellowship in Molecular Evolution at Princeton University. He has done research in evolutionary biology on topics ranging from nest parasitism to the molecular systematics of the blackbird family and is coauthor, with Jon Herron, of the standard-setting undergraduate text *Evolutionary Analysis*. Scott is the recipient of a Distinguished Teaching Award from the University of Washington and is currently a Senior Lecturer in the UW Department of Biology, where he teaches introductory biology for majors, a writing-intensive course for majors called The Tree of Life, and a graduate seminar in college science teaching. Scott's current research focuses on how active learning affects student learning and academic performance.

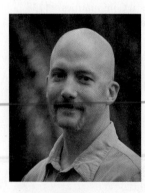

Michael Black received his Ph.D. in Microbiology & Immunology from Stanford University School of Medicine as a Howard Hughes Predoctoral Fellow. After graduation, he studied cell biology as a Burroughs Wellcome Postdoctoral Fellow at the MRC Laboratory of Molecular Biology in Cambridge, England. His current research focuses on the use of molecules to identify and track the transmission of microbes in the environment. Michael is a professor of Cell & Molecular Biology at California Polytechnic State University in San Luis Obispo, where he teaches introductory and advanced classes for majors in cell biology and microbiology. In addition to his teaching and research activities, Michael serves as the director of the Undergraduate Biotechnology Lab, where he works alongside undergraduate technicians to integrate research projects and inquiry-based activities into undergraduate classes.
Chapters 2–12, 36, and 51
mblack@calpoly.edu

Greg Podgorski received his Ph.D. in Molecular and Cellular Biology from Penn State University and has been a postdoctoral fellow at the Max Plank Institute for Biochemistry and Columbia University. His research interests are in biology education, developmental genetics, and computational biology. Greg's most recent work has been in mathematical modeling of how patterns of different cell types emerge during development and how tumors recruit new blood vessels in cancer. Greg has been teaching at Utah State University for more than 20 years in courses that include introductory biology for majors and for nonmajors, genetics, cell biology, developmental biology, and microbiology, and he has offered courses in nonmajors biology in Beijing and Hong Kong. He's won teaching awards at Utah State University and has been recognized by the National Academies as a Teaching Fellow and a Teaching Mentor.

Chapters 13–24

greg.podgorski@usu.edu

Jon Monroe is professor of Biology at James Madison University in Harrisonburg, Virginia. Jon completed his undergraduate work in Botany at the University of Michigan and his graduate work in Plant Physiology at Cornell University. He began his current position after a postdoc in biochemistry at Michigan State University. He currently teaches Plant Biology, and Cell and Molecular Biology. Jon's interest in plants is broad, ranging from systematics and taxonomy to physiology and biochemistry. His research, mostly with undergraduates, uses Arabidopsis thaliana to study the functions of a family of β-amylase genes in starch metabolism. Jon has been active in promoting undergraduate research through his work with the American Society of Plant Biologists (ASPB) and the Council on Undergraduate Research. He has received ASPB's Excellence in Teaching award and James Madison University Alumni Association's Distinguished Faculty Award.

Chapters 29–32; 37–41

monroejd@jmu.edu

Kim Quillin received her B.A. in Biology at Oberlin College *summa cum laude* and her Ph.D. in Integrative Biology from the University of California, Berkeley (as a National Science Foundation Graduate Fellow). Kim has worked in the trenches with Scott Freeman on every edition of *Biological Science*, starting with the ground-up development of the illustrations in the first edition in 1999 and expanding her role in each edition, always with the focus of helping students to think like biologists. Kim currently teaches introductory biology at Salisbury University, a member of the University System of Maryland, where she is actively involved in the ongoing student-centered reform of the concepts-and-methods course for biology majors. Her current research focuses on the scholarship of teaching and learning with an emphasis on measuring science process skills and the advantages and pitfalls of active multimedia learning.

Chapters 25–28; 33–35; 48; 52–57

kxquillin@salisbury.edu

Emily Taylor earned a B.A. in English at the University of California, Berkeley followed by a Ph.D. in Biological Sciences from Arizona State University, where she conducted research in the field of environmental physiology as a National Science Foundation Graduate Research Fellow. She is currently an associate professor of Biological Sciences at the California Polytechnic State University in San Luis Obispo, California. Her student-centered research program focuses on the endocrine and reproductive physiology of free-ranging reptiles, especially rattlesnakes. She teaches numerous undergraduate and graduate courses, including introductory biology, anatomy and physiology, and herpetology, and received the California Faculty Association's Distinguished Educator Award in 2010 and Cal Poly's Distinguished Teaching Award in 2012. Her revision of Unit 8 is her first foray into textbook writing.

Chapters 42–50

etaylor@calpoly.edu

Preface to Instructors

The first edition of *Biological Science* was visionary in its unique emphasis on the process of scientific discovery and experimental design—teaching how we know what we know. The goal was for students not only to learn the language of biology and understand fundamental concepts but also to begin to apply those concepts in new situations, analyze experimental design, synthesize results, and evaluate hypotheses and data—to learn how to think like biologists. Each edition since has proudly expanded on this vision. The Fifth Edition is no exception.

A team of six dedicated teacher-scholars has joined Scott to build on and refine the original vision, and by so doing, make the book an even better teaching and learning tool. The pace of biological discovery is rapid, and with each novel breakthrough it becomes even more challenging to decide what is essential to include in an introductory biology text. Pulling together an author team with firsthand expertise from molecules to ecosystems has ensured that the content of the Fifth Edition reflects cutting-edge biology that is pitched at the right level for introductory students and is as accurate and as exciting as ever for instructors and students alike.

New findings from education research continue to inform and inspire the team's thinking about *Biological Science*—we know more today than ever before about how students learn. These findings demand that we constantly look for new ways to increase student engagement in the learning process, and to help instructors align course activities and learning goals with testing strategies.

The New Coauthors

The new coauthor team brings a broad set of talents and interests to the project, motivated by a deep commitment to undergraduate teaching, whether at a small liberal arts college or a large university. Kim Quillin has been a partner in this textbook in every edition. For the Fifth Edition, she revised chapters across three units in addition to spearheading the continued effort to enhance the visual-teaching program. Michael Black, Greg Podgorski, Jon Monroe, and Emily Taylor, who served as unit advisors on the Fourth Edition, were already familiar with the book. And most of the authorial team have been avid users of previous editions for many years.

Core Values

Together, the coauthor team has worked to extend the vision and maintain the core values of *Biological Science*—to provide a book for instructors who embrace the challenge of boosting students to higher levels of learning, and to provide a book for students that helps them each step of the way in learning to think like scientists. Dedicated instructors have high expectations of their students—the Fifth Edition provides scaffolding to help students learn at the level called for by the National Academy of Sciences, the Howard Hughes Medical Institute, the American Association of Medical Academies, and the National Science Foundation.

What's New in This Edition

The Fifth Edition contains many new or expanded features, all of them targeted at ways to help students learn to construct their own knowledge and think like biologists.

- **Road Maps** The new Road Maps at the beginning of each chapter pair with the Big Picture concept maps introduced in the Fourth Edition. Together they help students navigate chapter content and see the forest for the trees. Each Road Map starts with a purpose statement that tells students what they can expect to learn from each chapter. It then goes on to visually group and organize information to help students anticipate key ideas as well as recognize meaningful relationships and connections between the ideas.

- **The Big Picture** Introduced in the Fourth Edition, Big Picture concept maps integrate words and visuals to help students synthesize information about challenging topics that span multiple chapters and units. In response to requests from instructors and students, three new Big Pictures focused on additional tough topics have been added: Doing Biology, The Chemistry of Life, and Plant and Animal Form and Function. In addition, the Ecology Big Picture is completely revised to reflect changes to that unit.

- **New Chapters** Two new chapters are added to better serve instructors and students. Unit 2 now contains a new Chapter 8, Energy and Enzymes: An Introduction to Metabolic Pathways. This chapter consolidates these critical topics in a place where students and instructors need it most—right before the chapters on cellular respiration and photosynthesis. In the Fourth Edition, animal movement was discussed in a chapter largely focused on animal sensory systems. In the Fifth Edition, this important topic is treated in depth in a new Chapter 48, Animal Movement, that explores how muscle and skeletal systems work together to produce locomotion.

- **New BioSkills** Instructors recognize that biology students need to develop foundational science skills in addition to content knowledge. While these skills are emphasized throughout the book, *Biological Science,* beginning with the Third

Edition, has provided a robust set of materials and activities to guide students who need extra help. To promote even fuller use of this resource, the BioSkills are now updated, expanded, and reorganized. New in this edition are a discussion of significant figures within the BioSkills on the Metric System, and two new BioSkills on Primary Literature and Peer Review and Using Bloom's Taxonomy. BioSkills are located in Appendix B, and practice activities can be assigned online in MasteringBiology®.

- **Promotion of Quantitative Skills** Reports like *Biology 2010, Scientific Foundations for Future Physicians*, and *Vision and Change* all place a premium on quantitative skills. To infuse a quantitative component throughout the text, new and existing quantitative questions are flagged in each chapter to encourage students to work on developing their ability to read or create a graph, perform or interpret a calculation, or use other forms of quantitative reasoning.

- **Bloom's Taxonomy** In the Fifth Edition, all questions in the text are assigned a Bloom's Taxonomy level to help both students and instructors understand whether a question requires higher-order or lower-order cognitive skills. Questions span all six Bloom's levels. (Bloom's levels are identified in Appendix A: Answers.) The coauthors were trained by experts Mary Pat Wenderoth and Clarissa Dirks[1] to ensure we followed a process that would result in high inter-rater reliability—or agreement among raters—in assigning Bloom's levels to questions. The new BioSkill, Using Bloom's Taxonomy, explains the six Bloom's levels to students and offers a practical guide to the kinds of study activities best suited for answering questions at each level.

- **Expanded Emphasis on "Doing Biology"** A constant hallmark of this text is its emphasis on experimental evidence—on teaching how we know what we know. To reflect the progress of science, in the Fifth Edition, the coauthor team replaced many experiments with fresh examples and added new Research Boxes. And as noted earlier, they added a new Big Picture on Doing Biology, focusing on the process of science and the organizational levels of biology. Data sources are now cited for all graphs and data tables to model the importance of citing data sources to students. Updated Research Box questions continue to encourage students to analyze some aspect of experimental design. Also new to this edition is a BioSkill on Primary Literature and Peer Review.

- **Art Program** The art program is further enhanced in this edition by the addition of more illustrated summary tables. These tables make subject areas more accessible to visual learners and reinforce key concepts of the chapter. Many of the life-cycle figures in Unit 6 are significantly overhauled.

[1] Crowe, A., C. Dirks, and M. P. Wenderoth. 2008. Biology in Bloom: Implementing Bloom's Taxonomy to enhance student learning in biology. *CBE–Life Sciences Education* 7: 368–381.

Updated Blue Thread Scaffolding

In the Third and Fourth editions of *Biological Science*, a metacognitive tool was formulated as the now popular feature known as "Blue Thread"—sets of questions designed to help students identify what they do and don't understand. The fundamental idea is that if students really understand a piece of information or a concept, they should be able to do something with it.

In the Fifth Edition, the Blue Thread is revised to reflect changes in chapter content, and to incorporate user feedback. Blue-Thread questions appear in the following locations:

- **In-text "You should be able to's"** offer exercises on topics that professors and students have identified as the most difficult concepts in each chapter.

- **Caption questions and exercises** challenge students to examine the information in a figure or table critically—not just absorb it.

- **Check Your Understanding boxes** present two to three tasks that students should be able to complete in order to demonstrate a mastery of summarized key ideas.

- **Chapter summaries** include "You should be able to" problems or exercises related to each key concept.

- **End-of-chapter** questions are organized in three levels of increasing difficulty so students can build from lower to higher-order cognitive questions.

Integration of Media

The textbook continues to be supported by MasteringBiology®, the most powerful online homework, tutorial, and assessment system available. Tutorials follow the Socratic method, coaching students to the correct answer by offering feedback specific to a student's misconceptions as well as providing hints students can access if they get stuck. Instructors can associate content with publisher-provided learning outcomes or create their own. Content highlights include the following:

- **NEW! Solve It Tutorials** These activities allow students to act like scientists in simulated investigations. Each tutorial presents an interesting, real-world question that students will answer by analyzing and interpreting data.

- **Experimental Inquiry Tutorials** The call to teach students about the process of science has never been louder. To support such teaching, there are 10 interactive tutorials on classic scientific experiments—ranging from Meselson–Stahl on DNA replication to the Grants' work on Galápagos finches and Connell's work on competition. Students who use these tutorials should be better prepared to think critically about experimental design and evaluate the wider implications of the data—preparing them to do the work of real scientists in the future.

- **BioFlix® Animations and Tutorials** BioFlix are movie-quality, 3-D animations that focus on the most difficult core topics and are accompanied by in-depth, online tutorials that provide hints and feedback to guide student learning. Eighteen BioFlix animations and tutorials tackle topics such as meiosis, mitosis, DNA replication, photosynthesis, homeostasis, and the carbon cycle.

- **NEW! End-of-Chapter Questions** Multiple choice end-of-chapter questions are now available to assign in MasteringBiology.

- **Blue-Thread Questions** Over 500 questions based on the Blue-Thread Questions in the textbook are assignable in MasteringBiology.

- **Big Picture Tutorials** Interactive concept map activities based on the Big Picture figures in the textbook are assignable in MasteringBiology, including tutorials to support the three new Big Pictures: Doing Biology, The Chemistry of Life, and Plant and Animal Form and Function.

- **BioSkills Activities** Activities based on the BioSkills content in the textbook are assignable in MasteringBiology, including activities to support the new BioSkills on Primary Literature and Peer Review and Using Bloom's Taxonomy.

- **Reading Quiz Questions** Every chapter includes reading quiz questions you can assign to ensure students read the textbook and understand the basics. These quizzes are perfect as a pre-lecture assignment to get students into the content before class, allowing you to use class time more effectively.

Serving a Community of Teachers

All of us on the coauthor team are deeply committed to students and to supporting the efforts of dedicated teachers. Doing biology is what we love. At various points along our diverse paths, we have been inspired by our own teachers when we were students, and now are inspired by our colleagues as we strive to become even better teacher-scholars. In the tradition of all previous editions of *Biological Science*, we have tried to infuse this textbook with the spirit and practice of evidence-based teaching. We welcome your comments, suggestions, and questions.

Thank you for your work on behalf of your students.

Content Highlights
of the Fifth Edition

As discussed in the preface, a major focus of this revision is to enhance the pedagogical utility of *Biological Science*. Another major goal is to ensure that the content reflects the current state of science and is accurate. The expanded author team has scrutinized every chapter to add new, relevant content, update descriptions when appropriate, and adjust the approach to certain topics to enhance student comprehension. In this section, some of the key content improvements to the textbook are highlighted.

Chapter 1 Biology and the Tree of Life   A concept map summarizing the defining characteristics of life is added. The process of doing biology coverage is expanded to include discussion of both experimental and descriptive studies, and more rigorous definitions of the terms hypothesis and theory.

Chapter 2 Water and Carbon: The Chemical Basis of Life   A stronger emphasis on chemical evolution is threaded throughout the chapter to bring chemistry to life for the student reader. Two prominent models for chemical evolution are introduced; the historic Miller prebiotic soup experiment was moved here. Advanced discussion of energy and chemical reactions was moved to a new chapter (see Chapter 8).

Chapter 3 Protein Structure and Function   The chapter is reorganized to emphasize the link between structure and function, from amino acids to folded proteins. Updated content illustrates that protein shapes are flexible and dynamic, and may remain incompletely folded until the protein interacts with other molecules or ions. Details of how enzymes work were moved to Chapter 8.

Chapter 4 Nucleic Acids and the RNA World   New experimental results concerning the synthesis of nucleotides and nucleic acids in a prebiotic environment are discussed. The section on the RNA world is expanded to include the artificial evolution of a novel ribozyme involved in nucleotide synthesis.

Chapter 5 An Introduction to Carbohydrates   The molecular basis for resistance of structural polymers, such as cellulose, to degradation is clarified. A new research box illustrates the role of carbohydrates in cellular recognition and attachment using the egg and sperm of mice as a model system.

Chapter 6 Lipids, Membranes, and the First Cells   New content on lipid and membrane evolution and the proposed characteristics of the first protocell is introduced. The aquaporin and potassium channel figures are updated; how key amino acids serve as selectivity filters is now highlighted.

Chapter 7 Inside the Cell   Several new electron micrographs were selected to more clearly illustrate cell component structure and function. A new figure is added to better depict the pulse–chase assay used to identify the secretory pathway. Coverage of nuclear transport is expanded to differentiate between passive diffusion and active nuclear import. Updated content emphasizes the role of the cytoskeleton in localizing organelles, and how polarity of microtubules and microfilaments influences their growth rate.

Chapter 8 Energy and Enzymes: An Introduction to Pathways   This new chapter pulls together concepts in energy, chemical reactions, and enzymes that previously were covered in three different chapters. Oxidation and reduction reactions are emphasized to prepare students for Chapters 9 and 10. The energetics behind ATP hydrolysis and its role in driving endergonic reactions is discussed, and figures are revised to better illustrate the process. Updated content on enzyme regulation and a new process figure show a model for how metabolic pathways may have evolved.

Chapter 9 Cellular Respiration and Fermentation   Two new summary tables for glycolysis and the citric acid cycle are added that provide the names of the enzymes and the reaction each catalyzes. New content is introduced to propose a connection between the universal nature of the proton motive force and the story of the chemical evolution of life.

Chapter 10 Photosynthesis   More extensive comparison between the chemical reactions in mitochondria and chloroplasts is added. A new figure is introduced to illustrate noncyclic electron flow in the context of the thylakoid membrane. Greater emphasis is placed on the number of ATPs and NADPHs required for each cycle of carbon fixation and reduction.

Chapter 11 Cell–Cell Interactions   Coverage of extracellular matrix structure and function is expanded, including its role in intercellular adhesions and cell signaling. The plant apoplast and symplast are now introduced as key terms in the text and illustrated in a new figure. New content and a new figure on unicellular models for intercellular communication via pheromone sensing (yeast) and quorum sensing (slime mold) are added.

Chapter 12 The Cell Cycle   A new figure helps explain the pulse–chase assay for identifying phases of the cell cycle. Content is added to the text and to a figure that illustrates the similarities between chromosome segregation in eukaryotes and prokaryotes. A revised description of anaphase emphasizes how microtubule fraying at the kinetochore can drive chromosome movement. The explanation of how phosphorylation and dephosphorylation turns on MPF activity is updated to reflect current research.

Chapter 13 Meiosis   To improve the flow of the chapter, the section on advantages of sexual reproduction was moved before mistakes in meiosis. The discussion of the role and timing of crossing over during meiosis I is updated. A new study that supports the hypothesis that sex evolved in response to the selective pressure of pathogens is introduced.

Chapter 14 Mendel and the Gene Material on gene linkage is revised to emphasize the importance of genetic mapping. A new matched set of figures on pedigree analysis brings together the various modes of transmission that were previously shown in four individual figures. A new summary table on characteristics of different patterns of inheritance is added.

Chapter 15 DNA and the Gene: Synthesis and Repair A new research figure is added that focuses on the relationship between telomere length and senescence in cultured somatic cells.

Chapter 16 How Genes Work Coverage of the evolving concept of the gene and of different types of RNA is expanded. A figure showing the karyotype of a cancer cell is revised to improve clarity.

Chapter 17 Transcription, RNA Processing, and Translation The sections on transcription in bacteria and eukaryotes are now separated, and content on charging tRNAs was moved to a new section. The discussion of translation is reorganized, first to emphasize the process in bacteria and then to highlight differences in eukaryotes.

Chapter 18 Control of Gene Expression in Bacteria Coverage of *lac* operon positive regulation is updated to reflect current research. A new section and new process figure on global gene regulation are added, using the *lexA* regulon as an example.

Chapter 19 Control of Gene Expression in Eukaryotes Extensive updates to the discussion of epigenetics include a new research box and a section on DNA methylation. Coverage of transcription initiation is updated to reflect current science. A new figure illustrates the role of p53 in the cell cycle in normal and cancerous cells.

Chapter 20 Analyzing and Engineering Genes The material on sequencing the Neanderthal genome is updated, including evidence of limited Neanderthal genetic material in some modern human populations. New information on current generation sequencing technologies and massive parallelism is added. Recent advances in gene therapy are highlighted.

Chapter 21 Genomics and Beyond Extensive updates throughout reflect recent advances in genomics. Changes include sequence database statistics, genomes that have been sequenced to study evolutionary relationships, and new figures illustrating gene count versus genome size in prokaryotes and eukaryotes and functional classes of human DNA sequences. A new section on systems biology is added. Also included are notes on the discovery of widespread transcription of eukaryotic genomes, deep sequencing, and the spectrum of mutations in human tumors.

Chapter 22 Principles of Development New information is added on dedifferentiation in induced pluripotent stem cells, maternal genes in *Drosophila* development, how morphogens work, and tool-kit genes. The order of topics in the discussion

of developmental principles is reorganized. The figure on *Hox* genes in *Drosophila* and the mouse is updated.

Chapter 23 An Introduction to Animal Development The chapter is streamlined by focusing on principles of animal development. The discussion of gametogenesis was moved to Chapter 50 (Animal Reproduction). The presentation of fertilization is simplified, and a new figure summarizing steps of fertilization is added. The figure on gastrulation is modified to better depict the arrangement of the germ layers and their movement.

Chapter 24 An Introduction to Plant Development The chapter is modified to impart an evolutionary perspective on the similarities and differences in plant and animal development. The chapter also was streamlined by removing material such as details of gametogenesis, which now appears in Chapter 41 (Plant Reproduction).

Chapter 25 Evolution by Natural Selection Several new key passages are included, among them the use of the Grand Canyon as a context for understanding relative dating of fossils, Darwin's artificial selection experiments with fancy pigeons, and Malthus's concept of struggle for existence. A new example of people living at high altitude in Tibet clarifies the difference between acclimatization and adaptation. An illustrated summary table of common misconceptions is added.

Chapter 26 Evolutionary Processes Discussion of sexual selection now falls within the section on natural selection, and the terms intersexual and intrasexual selection are added. Several new examples replace those in the Fourth Edition, including inbreeding depression in Florida panthers, gene flow in Oregon steelhead trout, and lateral gene transfer in aphids. An illustrated summary table is added on modes of selection.

Chapter 27 Speciation Several points are clarified, such as the gradient-like (rather than all-or-nothing) nature of reproductive isolation. The section on polyploidy is reorganized and the figures revised, including a side-by-side comparison of autopolyploidy and allopolyploidy.

Chapter 28 Phylogenies and the History of Life The phylogenetics section is reorganized and expanded to include three illustrated summary tables and updated life-history timelines. New content is added, including the concept of the Anthropocene, the calendar analogy to the history of the Earth, the Chengjiang fossils, and a Life-in-the-Cambrian illustration. Evidence for the impact hypothesis is combined into an illustrated summary table.

Chapter 29 Bacteria and Archaea The chapter is updated to include a description of metagenomic experiments with an emphasis on the role of gut bacteria in digestion. A newly recognized phylum of Archaea, the Thaumarchaeota, is included, and the table comparing key characteristics of the Bacteria, Archaea, and Eukarya is streamlined.

Chapter 30 Protists For simplicity, protist lineages are now referred to throughout the chapter by their more familiar common names. Also, some key lineage boxes were consolidated to trim the number to one box per major lineage. Discussion of the origin of the nuclear envelope and mitochondria is expanded to reflect new thinking on the evolution of eukaryotic cells. Protist life cycle figures are significantly overhauled.

Chapter 31 Green Algae and Land Plants Coverage of the evolution of land plants is expanded to include the importance of UV light and UV-absorbing molecules on the colonization of land. The "Redwood group" is now referred to as the Cupressophyta. Updates emphasize the role(s) of each stage of a life cycle in dispersal and in increasing genetic variation and individual numbers. A new research box is added, showing the importance of flower color to pollinator preference. Plant life-cycle figures are significantly overhauled.

Chapter 32 Fungi Coenocytic fungal hyphae are illustrated with a new image showing GFP-labeled nuclei in *Neurospora crassa*. The chapter now points out the similarity between fungal and animal modes of nutrition, in terms of extracellular digestion and absorption of small molecules. The discussion of lignin degradation is updated, and new descriptions of mutualisms of fungi with animals are included. Fungal life cycle figures are significantly overhauled.

Chapter 33 An Introduction to Animals The chapter is extensively revised to streamline and modernize the presentation, including emphasis on genetic tool kits and symmetry in the phylogeny of animals. The "Themes of Diversification" section is reorganized around five illustrated summary tables. The discussion of life cycles is revised to be more general. Insect metamorphosis has moved to Chapter 34, and a sea urchin life cycle replaces *Obelia*.

Chapter 34 Protostome Animals Two themes are threaded throughout the chapter: the water-to-land transition and modular body plans. The section on lophotrochozoans emphasizes spiral cleavage, indirect versus direct development, hemocoels, and radulas. The section on ecdysozoans highlights segmentation and *Hox* genes, including discussion of the origin of the wing and metamorphosis. Key lineage boxes include new phylogenies for annelids, crustaceans, and chelicerates.

Chapter 35 Deuterostome Animals Updates to reflect current research include revised phylogenies, evolution of flight and feathers, *Australopithecus sediba*, human migration out of Africa, and genetic evidence for interbreeding of *Homo neanderthalensis* and *Homo sapiens*.

Chapter 36 Viruses New content focuses on how viruses contribute to evolution via lateral gene transfer and direct addition of genes to cellular genomes. Content is updated and expanded on viral structure and function, and on lytic and latent infections. Three new figures are added, including a comparison of replication of viruses and cells, how pandemic strains of influenza arise via reassortment, and the devastating impact of the 1918 influenza pandemic.

Chapter 37 Plant Form and Function The use of terminology is streamlined for consistency and clarity. For example,

"lateral meristem" is replaced with "vascular cambium" to avoid confusion with lateral buds, and the description of bark is clarified to avoid using the term phelloderm. Several complex figures were converted to illustrated summary tables.

Chapter 38 Water and Sugar Transport in Plants The chapter is revised to improve accuracy, and points out that water loss is also a means for transporting minerals from roots to shoots. Updated content clarifies the role of energy expenditure in moving water across roots, and the Casparian strip as a barrier to the back diffusion of ions and water out of the root. A new research figure shows the importance of the sucrose proton symporter in long-distance transport in *Arabidopsis*.

Chapter 39 Plant Nutrition In this chapter the coverage of mycorrhizae is modified to emphasize their overall role in nutrient acquisition. In the section on nitrogen fixation, a description of the worldwide practice of crop rotation involving legumes and grains is added.

Chapter 40 Plant Sensory Systems, Signals, and Responses A new research box is added that reveals the essential role of PHOT1 phosphorylation in phototropism. A section on the effect of day length on flowering was moved here from Chapter 41 and is integrated with the discussion of phytochromes. New content on the role of plasmodesmata in plant action potentials is added. The section on how plants respond to pathogens is simplified and updated with an example of control of stomata during a bacterial infection.

Chapter 41 Plant Reproduction To provide a clearer example of a gametophyte-dominant life cycle, the liverwort life cycle has been replaced with a moss life cycle. The term "pollination syndrome" is clarified. Also, a research box on how capsaicin prevents seed predation and facilitates dispersal was reinstated from an earlier edition of *Biological Science*.

Chapter 42 Animal Form and Function The chapter includes a new experiment illustrating physiological trade-offs, along with improved examples of thermoregulatory strategies in animals. Several complex figures were converted to illustrated summary tables.

Chapter 43 Water and Electrolyte Balance in Animals The chapter is reorganized to better integrate the relationship between excretion and water and electrolyte balance. Osmoregulatory strategies are now organized according to the challenges presented by marine, freshwater, and terrestrial habitats. Coverage of osmoregulation in bony fishes versus cartilaginous fishes, mammalian kidney function, and how nonmammalian vertebrates concentrate their urine is expanded and clarified.

Chapter 44 Animal Nutrition This chapter contains new information on nutritional imbalances, including diabetes and obesity.

Chapter 45 Gas Exchange and Circulation Discussion of the insect tracheal system is expanded, including new content on how respiration restricts upper limits of body size of insects.

Details regarding the lymphatic system and heart anatomy in vertebrates are updated, and a new section on cardiovascular disease is added.

Chapter 46 Electrical Signals in Animals This chapter is greatly expanded to reflect recent research and growing interest in neuroscience. New information includes comparative anatomy of vertebrate brains, more case studies of brain injuries or dysfunctions that have led to major discoveries in neuroscience, and the concept of neuroplasticity—especially neurogenesis.

Chapter 47 Animal Sensory Systems Content from the Fourth Edition has been split into two chapters (47, Animal Sensory Systems; 48, Animal Movement). The chapter on sensory systems is now organized by type of sensory reception: mechanoreception (with new coverage on the lateral line system of fishes), photoreception, chemoreception (with new coverage of pheromones), and a new section introducing thermoreception, electroreception, and magnetoreception.

Chapter 48 Animal Movement This new chapter introduces the importance of movement in animals, building from small to large scale. The mechanism of muscle contraction (with revised figures) is covered, followed by discussions of types of muscle tissue (with new content on skeletal-muscle fiber types and parallel- versus pennate-muscle fiber orientation), and skeletal systems (hydrostatic skeletons, exoskeletons, endoskeletons). A completely new final section discusses how biologists study locomotion on land, in the air, and in the water.

Chapter 49 Chemical Signals in Animals Figures and content are updated for clarity. The chapter includes a new section on endocrine disruptors.

Chapter 50 Animal Reproduction New content includes details of sperm and egg structure and function, reproduction in the spotted hyena, and human contraceptive methods.

Chapter 51 The Immune System in Animals Coverage of the innate immune response is expanded to include more detail on Toll-like receptors and how they transmit signals. The section on adaptive immunity is reorganized to improve flow. Updated content on inappropriate immune responses (autoimmunity and allergies) and inadequate responses (immunodeficiency) is grouped together in one section. The hygiene hypothesis is introduced to explain the growing trend of inappropriate immune responses in populations that have reduced exposure to common pathogens and parasites.

Chapter 52 An Introduction to Ecology The first section on levels of ecological study is expanded to include global ecology. The rest of the chapter is reorganized, beginning with the factors that determine the distribution and abundance of organisms (including a new Argentine ant case study) and ending with biomes. The biome section is streamlined with a new emphasis on human impacts, including an introduction to anthropogenic biomes and the Anthropocene.

Chapter 53 Behavioral Ecology The introduction is revised to provide a clearer framework for types of behavior. Sections are now organized around proximate versus ultimate causation. These new examples replace those in the Fourth Edition: Argentine ant territorial behavior (replacing spiny lobsters), optimal foraging in gerbils (replacing white-fronted bee-eaters), sexual selection in *Anolis* lizards (replacing barn swallows), and map orientation in green sea turtles.

Chapter 54 Population Ecology A new introductory section focuses on the distribution of organisms in populations, including dispersion patterns and a consolidated discussion of measurement methods. The human population content is updated and separated into a new section. The quantitative methods boxes and life table are now more student friendly.

Chapter 55 Community Ecology Several changes to content are made, such as a clarification of competitive exclusion versus niche differentiation. New content includes a summary table on constitutive defenses, a discovery story on mimicry (including Bates and Müller), an introduction to food webs, and the process of soil formation in primary succession.

Chapter 56 Ecosystems The title and scope of the chapter are updated to include global ecology. Extensive revisions include many content updates and new figures on the relationship between GPP and NPP, the one-way flow of energy and cycling of nutrients, the food web, open versus closed aquifers, the High Plains Aquifer, the biomagnification of DDT, and the greenhouse effect. The climate change section is expanded and updated, including an illustrated summary table.

Chapter 57 Biodiversity and Conservation Biology Throughout the chapter, there is more emphasis on conserving ecosystem function. Many new examples are added, including Smits's restoration project in Borneo, the Census of Marine Life, the IUCN Red List, the Sinervo lizard extinction experiment, and Florida panther genetic restoration. New summary tables highlight ecosystem services and threats to biodiversity.

Acknowledgments

Reviewers

The peer review system is the key to quality and clarity in science publishing. In addition to providing a filter, the investment that respected individuals make in vetting the material—catching errors or inconsistencies and making suggestions to improve the presentation—gives authors, editors, and readers confidence that the text meets rigorous professional standards.

Peer review plays the same role in textbook publishing. The time and care that this book's reviewers have invested is a tribute to their professional integrity, their scholarship, and their concern for the quality of teaching. Virtually every paragraph in this edition has been revised and improved based on insights from the following individuals.

Tamarah Adair, *Baylor University*
Sandra D. Adams, *Montclair State University*
Marc Albrecht, *University of Nebraska at Kearney*
Larry Alice, *Western Kentucky University*
Leo M. Alves, *Manhattan College*
David R. Angelini, *American University*
Dan Ardia, *Franklin & Marshall College*
Paul Arriola, *Elmhurst College*
Davinderjit K. Bagga, *University of Montevallo*
Susan Barrett, *Wheaton College*
Donald Baud, *University of Memphis*
Vernon W. Bauer, *Francis Marion University*
Robert Bauman, *Amarillo College*
Christopher Beck, *Emory University*
Vagner Benedito, *West Virginia University*
Scott Bingham, *Arizona State University*
Stephanie Bingham, *Barry University*
Wendy Birky, *California State University, Northridge*
Jason Blank, *California Polytechnic State University*
Kristopher A. Blee, *California State University, Chico*
Margaret Bloch-Qazi, *Gustavus Adolphus College*
Lanh Bloodworth, *Florida State College at Jacksonville*
Catherine H. Borer, *Berry College*
James Bottesch, *Brevard Community College*
Jacqueline K. Bowman, *Arkansas Tech University*
John Bowman, *University of California, Davis*
Chris Brochu, *University of Iowa*
Matthew Brown, *Dalhousie University*
Mark Browning, *Purdue University*
Carolyn J. W. Bunde, *Idaho State University*
David Byres, *Florida State College at Jacksonville*
Michael Campbell, *Penn State Erie*
Manel Camps, *University of California, Santa Cruz*
Geralyn M. Caplan, *Owensboro Community and Technical College*
Richard Cardullo, *University of California, Riverside*

David Carlini, *American University*
Dale Casamatta, *University of North Florida*
Deborah Chapman, *University of Pittsburgh*
Joe Coelho, *Quincy University*
Allen Collins, *Smithsonian Museum of Natural History*
Robert A. Colvin, *Ohio University*
Kimberly L. Conner, *Florida State College at Jacksonville*
Karen Curto, *University of Pittsburgh*
Clarissa Dirks, *Evergreen State College*
Peter Ducey, *SUNY Cortland*
Erastus Dudley, *Huntingdon College*
Jeffrey P. Duguay, *Delta State University*
Tod Duncan, *University of Colorado, Denver*
Joseph Esdin, *University of California, Los Angeles*
Brent Ewers, *University of Wyoming*
Amy Farris, *Ivy Tech Community College*
Bruce Fisher, *Roane State Community College*
Ryan Fisher, *Salem State University*
David Fitch, *New York University*
Elizabeth Fitch, *Motlow State Community College*
Michael P. Franklin, *California State University, Northridge*
Susannah French, *Utah State University*
Caitlin Gabor, *Texas State University*
Matthew Gilg, *University of North Florida*
Kendra Greenlee, *North Dakota State University*
Patricia A. Grove, *College of Mount Saint Vincent*
Nancy Guild, *University of Colorado, Boulder*
Cynthia Hemenway, *North Carolina State University*
Christopher R. Herlihy, *Middle Tennessee State University*
Kendra Hill, *South Dakota State University*
Sara Hoot, *University of Wisconsin, Milwaukee*
Kelly Howe, *University of New Mexico*
Robin Hulbert, *California Polytechnic State University*
Rick Jellen, *Brigham Young University*
Russell Johnson, *Colby College*
William Jira Katembe, *Delta State University*
Elena K. Keeling, *California Polytechnic State University*
Jill B. Keeney, *Juniata College*
Greg Kelly, *University of Western Ontario*
Scott L. Kight, *Montclair State University*
Charles Knight, *California Polytechnic State University*
Jenny Knight, *University of Colorado, Boulder*
William Kroll, *Loyola University Chicago*
Dominic Lannutti, *El Paso Community College*
Brenda Leady, *University of Toledo*
David Lindberg, *University of California, Berkeley*
Barbara Lom, *Davidson College*
Robert Maxwell, *Georgia State University*
Marshall D. McCue, *St. Mary's University*
Kurt A. McKean, *SUNY Albany*
Michael Meighan, *University of California, Berkeley*
John Merrill, *Michigan State University*

Richard Merritt, *Houston Community College*
Alan Molumby, *University of Illinois at Chicago*
Jeremy Montague, *Barry University*
Chad E. Montgomery, *Truman State University*
Kimberly D. Moore, *Lone Star College System, North Harris*
Michael Morgan, *Berry College*
James Mulrooney, *Central Connecticut State University*
John D. Nagy, *Scottsdale Community College*
Margaret Olney, *St. Martin's University*
Nathan Okia, *Auburn University at Montgomery*
Robert Osuna, *SUNY Albany*
Daniel Panaccione, *West Virginia University*
Stephanie Pandolfi, *Michigan State University*
Michael Rockwell Parker, *Monell Chemical Senses Center*
Lisa Parks, *North Carolina State University*
Nancy Pelaez, *Purdue University*
Shelley W. Penrod, *Lone Star College System, North Harris*
Andrea Pesce, *James Madison University*
Raymond Pierotti, *University of Kansas*
Melissa Ann Pilgrim, *University of South Carolina Upstate*
Paul Pillitteri, *Southern Utah University*
Debra Pires, *University of California, Los Angeles*
P. David Polly, *Indiana University, Bloomington*
Vanessa Quinn, *Purdue University North Central*
Stacey L. Raimondi, *Elmhurst College*
Stephanie Randell, *McLennan Community College*
Marceau Ratard, *Delgado Community College*
Flona Redway, *Barry University*
Srebrenka Robic, *Agnes Scott College*
Dave Robinson, *Bellarmine University*
George Robinson, *SUNY Albany*
Adam W. Rollins, *Lincoln Memorial University*
Amanda Rosenzweig, *Delgado Community College*
Leonard C. Salvatori, *Indian River State College*
Dee Ann Sato, *Cypress College*
Leena Sawant, *Houston Community College*
Jon Scales, *Midwestern State University*
Oswald Schmitz, *Yale University*
Joan Sharp, *Simon Fraser University*
Julie Schroer, *North Dakota State University*
Timothy E. Shannon, *Francis Marion University*
Lynnette Sievert, *Emporia State University*
Susan Skambis, *Valencia College*
Ann E. Stapleton, *University of North Carolina, Wilmington*
Mary-Pat Stein, *California State University, Northridge*
Christine Strand, *California Polytechnic State University*
Denise Strickland, *Midlands Technical College*
Jackie Swanik, *Wake Technical Community College*
Billie J. Swalla, *University of Washington*
Zuzana Swigonova, *University of Pittsburgh*
Briana Timmerman, *University of South Carolina*
Catherine Ueckert, *Northern Arizona University*
Sara Via, *University of Maryland, College Park*
Thomas J. Volk, *University of Wisconsin–La Crosse*
Jeffrey Walck, *Middle Tennessee State University*
Andrea Weeks, *George Mason University*
Margaret S. White, *Scottsdale Community College*
Steven D. Wilt, *Bellarmine University*
Candace Winstead, *California Polytechnic State University*
James A. Wise, *Hampton University*

Correspondents

One of the most enjoyable interactions we have as textbook authors is correspondence or conversations with researchers and teachers who take the time and trouble to contact us to discuss an issue with the book, or who respond to our queries about a particular data set or study. We are always amazed and heartened by the generosity of these individuals. They care, deeply.

Lawrence Alice, *Western Kentucky University*
David Baum, *University of Wisconsin–Madison*
Meredith Blackwell, *Louisiana State University*
Nancy Burley, *University of California, Irvine*
Thomas Breithaupt, *University of Hull*
Philip Cantino, *Ohio University*
Allen Collins, *Smithsonian Museum of Natural History*
Robert Full, *University of California, Berkeley*
Arundhati Ghosh, *University of Pittsburgh*
Jennifer Gottwald, *University of Wisconsin–Madison*
Jon Harrison, *Arizona State University*
David Hawksworth, *Natural History Museum, London*
Jim Herrick, *James Madison University*
John Hunt, *University of Exeter*
Doug Jensen, *Converse College*
Scott Kight, *Montclair State University*
Scott Kirkton, *Union College*
Mimi Koehl, *University of California, Berkeley*
Rodger Kram, *University of Colorado*
Matthew McHenry, *University of California, Irvine*
Alison Miyamoto, *California State University, Fullerton*
Sean Menke, *Lake Forest College*
Rich Mooi, *California Academy of Sciences*
Michael Oliver, *MalawiCichlids.com*
M. Rockwell Parker, *Monell Chemical Senses Center*
Andrea Pesce, *James Madison University*
Chris Preston, *Monterey Bay Aquarium Research Institute*
Scott Sakaluk, *Illinois State University*
Kyle Seifert, *James Madison University*
Jos Snoeks, *Royal Museum for Central Africa*
Jeffrey Spring, *University of Louisiana*
Christy Strand, *California Polytechnic State University, San Luis Obispo*
Torsten Struck, *University of Osnabrueck, Germany*
Oswald Schmitz, *Yale University*
Ian Tattersal, *American Museum of Natural History*
Robert Turgeon, *Cornell University*
Tom Volk, *University of Wisconsin–La Crosse*
Naomi Wernick, *University of Massachusetts, Lowell*

Supplements Contributors

Instructors depend on an impressive array of support materials—in print and online—to design and deliver their courses. The student experience would be much weaker without the study guide, test bank, activities, animations, quizzes, and tutorials written by the following individuals.

Brian Bagatto, *University of Akron*
Scott Bingham, *Arizona State University*
Jay L. Brewster, *Pepperdine University*

Mirjana Brockett, *Georgia Institute of Technology*
Warren Burggren, *University of North Texas*
Jeff Carmichael, *University of North Dakota*
Tim Christensen, *East Carolina University*
Erica Cline, *University of Washington—Tacoma*
Patricia Colberg, *University of Wyoming*
Elia Crisucci, *University of Pittsburgh*
Elizabeth Cowles, *Eastern Connecticut State University*
Clarissa Dirks, *Evergreen State College*
Lisa Elfring, *University of Arizona, Tucson*
Brent Ewers, *University of Wyoming*
Rebecca Ferrell, *Metropolitan State University of Denver*
Miriam Ferzli, *North Carolina State University*
Cheryl Frederick, *University of Washington*
Cindee Giffen, *University of Wisconsin–Madison*
Kathy M. Gillen, *Kenyon College*
Linda Green, *Georgia Institute of Technology*
Christopher Harendza, *Montgomery County Community College*
Cynthia Hemenway, *North Carolina State University*
Laurel Hester, *University of South Carolina*
Jean Heitz, *University of Wisconsin–Madison*
Tracey Hickox, *University of Illinois, Urbana–Champaign*
Jacob Kerby, *University of South Dakota*
David Kooyman, *Brigham Young University*
Barbara Lom, *Davidson College*
Cindy Malone, *California State University, Northridge*
Jim Manser, retired, *Harvey Mudd College*
Jeanette McGuire, *Michigan State University*
Mark Music, *Indian River State College*
Jennifer Nauen, *University of Delaware*
Chris Pagliarulo, *University of California, Davis*
Stephanie Scher Pandolfi, *Michigan State University*
Lisa Parks, *North Carolina State University*
Debra Pires, *University of California, Los Angeles*
Carol Pollock, *University of British Columbia*
Jessica Poulin, *University at Buffalo, the State University of New York*
Vanessa Quinn, *Purdue University North Central*
Eric Ribbens, *Western Illinois University*
Christina T. Russin, *Northwestern University*
Leonard Salvatori, *Indian River State College*
Joan Sharp, *Simon Fraser University*
Chrissy Spencer, *Georgia Institute of Technology*
Mary-Pat Stein, *California State University, Northridge*
Suzanne Simon-Westendorf, *Ohio University*
Fred Wasserman, *Boston University*
Cindy White, *University of Northern Colorado*
Edward Zalisko, *Blackburn College*

Book Team

Anyone who has been involved in producing a textbook knows that many people work behind the scenes to make it all happen. The coauthor team is indebted to the many talented individuals who have made this book possible.

Development editors Mary Catherine Hager, Moira Lerner-Nelson, and Bill O'Neal provided incisive comments on the revised manuscript. Fernanda Oyarzun and Adam Steinberg used their artistic sense, science skills, and love of teaching to hone the figures for many chapters.

The final version of the text was copyedited by Chris Thillen and expertly proofread by Pete Shanks. The final figure designs were rendered by Imagineering Media Services and carefully proofread by Frank Purcell. Maureen Spuhler, Eric Schrader, and Kristen Piljay researched images for the Fifth Edition.

The book's clean, innovative design was developed by Mark Ong and Emily Friel. Text and art were skillfully set in the design by S4Carlisle Publishing Services. The book's production was supervised by Lori Newman and Mike Early.

The extensive supplements program was managed by Brady Golden and Katie Cook. All of the individuals mentioned—and more—were supported with cheerful, dedicated efficiency by Editorial Assistant Leslie Allen for the first half of the project; Eddie Lee has since stepped in to skillfully fill this role.

Creating MasteringBiology® tutorials and activities also requires a team. Media content development was overseen by Tania Mlawer and Sarah Jensen, who benefited from the program expertise of Caroline Power and Caroline Ross. Joseph Mochnick and Daniel Ross worked together as media producers. Lauren Fogel (VP, Director, Media Development), Stacy Treco (VP, Director, Media Product Strategy), and Laura ensured that the complete media program that accompanies the Fifth Edition, including MasteringBiology, will meet the needs of the students and professors who use our offerings.

Pearson's talented sales reps, who listen to professors, advise the editorial staff, and get the book in students' hands, are supported by tireless Executive Marketing Manager Lauren Harp and Director of Marketing Christy Lesko. The marketing materials that support the outreach effort were produced by Lillian Carr and her colleagues in Pearson's Marketing Comunications group. David Theisen, national director for Key Markets, tirelessly visits countless professors each year, enthusiastically discussing their courses and providing us with meaningful editorial guidance.

The vision and resources required to run this entire enterprise are the responsibility of Vice President and Editor-in-Chief Beth Wilbur, who provided inspirational and focused leadership, and President of Pearson Science Paul Corey, who displays unwavering commitment to high-quality science publishing.

Becky Ruden recruited the coauthor team, drawing us to the project with her energy and belief in this book. The editorial team was skillfully directed by Executive Director of Development Deborah Gale. Finally, we are deeply grateful for three key drivers of the Fifth Edition. Project Editor Anna Amato's superb organizational skills and calm demeanor assured that all the wheels and cogs of the process ran smoothly to keep the mammoth project steadily rolling forward. Supervising Development Editor Sonia DiVittorio's deep expertise, creative vision, keen attention to detail, level, and clarity, and inspiring insistence on excellence kept the bar high for everyone on every aspect of the project. Lastly, Senior Acquisitions Editor Michael Gillespie's boundless energy and enthusiasm, positive attitude, and sharp intellect have fueled and united the team and also guided the book through the hurdles to existence. The coauthor team thanks these exceptional people for making the art and science of book writing a productive and exhilarating process.

1 Biology and the Tree of Life

In this chapter you will learn about

Key themes to structure your thinking about Biology

starting with → **What does it mean to say that something is alive?** **1.1**

including → **Two of the greatest unifying ideas in Biology**

including → **The process of doing Biology** **1.5**

first → **The cell theory** **1.2**

then → **The theory of evolution by natural selection** **1.3**

predicts → **The tree of life** **1.4**

These Chinese water dragon hatchlings are exploring their new world and learning how to find food and stay alive. They represent one of the key characteristics of life introduced in this chapter—replication.

n essence, biological science is a search for ideas and observations that unify our understanding of the diversity of life, from bacteria living in rocks a mile underground to humans and majestic sequoia trees. This chapter is an introduction to this search.

The goals of this chapter are to introduce the nature of life and explore how biologists go about studying it. The chapter also introduces themes that will resonate throughout this book:

- Analyzing how organisms work at the molecular level.

- Understanding organisms in terms of their evolutionary history.

- Helping you learn to think like a biologist.

Let's begin with what may be the most fundamental question of all: What is life?

This chapter is part of the Big Picture. See how on pages 16–17.

✔ When you see this checkmark, stop and test yourself. Answers are available in Appendix A.

1.1 What Does It Mean to Say That Something Is Alive?

An **organism** is a life-form—a living entity made up of one or more cells. Although there is no simple definition of life that is endorsed by all biologists, most agree that organisms share a suite of five fundamental characteristics.

- *Energy* To stay alive and reproduce, organisms have to acquire and use energy. To give just two examples: plants absorb sunlight; animals ingest food.

- *Cells* Organisms are made up of membrane-bound units called cells. A cell's membrane regulates the passage of materials between exterior and interior spaces.

- *Information* Organisms process hereditary, or genetic, information encoded in units called genes. Organisms also respond to information from the environment and adjust to maintain stable internal conditions. Right now, cells throughout your body are using information to make the molecules that keep you alive; your eyes and brain are decoding information on this page that will help you learn some biology, and if your room is too hot you might be sweating to cool off.

- *Replication* One of the great biologists of the twentieth century, François Jacob, said that the "dream of a bacterium is to become two bacteria." Almost everything an organism does contributes to one goal: replicating itself.

- *Evolution* Organisms are the product of evolution, and their populations continue to evolve.

You can think of this text as one long exploration of these five traits. Here's to life!

1.2 The Cell Theory

Two of the greatest unifying ideas in all of science laid the groundwork for modern biology: the cell theory and the theory of evolution by natural selection. Formally, scientists define a **theory** as an explanation for a very general class of phenomena or observations that are supported by a wide body of evidence. The cell theory and theory of evolution address fundamental questions: What are organisms made of? Where do they come from?

When these concepts emerged in the mid-1800s, they revolutionized the way biologists think about the world. They established

two of the five attributes of life: Organisms are cellular, and their populations change over time.

Neither insight came easily, however. The cell theory, for example, emerged after some 200 years of work. In 1665 the Englishman Robert Hooke devised a crude microscope to examine the structure of cork (a bark tissue) from an oak tree. The instrument magnified objects to just 30× (30 times) their normal size, but it allowed Hooke to see something extraordinary. In the cork he observed small, pore-like compartments that were invisible to the naked eye. Hooke coined the term "cells" for these structures because of their resemblance to the cells inhabited by monks in a monastery.

Soon after Hooke published his results, a Dutch scientist named Anton van Leeuwenhoek succeeded in developing much more powerful microscopes, some capable of magnifications up to 300×. With these instruments, van Leeuwenhoek inspected samples of pond water and made the first observations of a dazzling collection of single-celled organisms that he called "animalcules." He also observed and described human blood cells and sperm cells, shown in **FIGURE 1.1**.

In the 1670s an Italian researcher who was studying the leaves and stems of plants with a microscope concluded that plant tissues were composed of many individual cells. By the early 1800s, enough data had accumulated for a German biologist to claim that *all* organisms consist of cells. Did this claim hold up?

All Organisms Are Made of Cells

Advances in microscopy have made it possible to examine the amazing diversity and complexity of cells at higher and higher magnifications. Biologists have developed microscopes that are tens of thousands of times more powerful than van Leeuwenhoek's and have described over a million new species. The basic conclusion made in the 1800s remains intact, however: All organisms are made of cells.

The smallest organisms known today are bacteria that are barely 200 nanometers wide, or 200 *billionths* of a meter. (See **BioSkills 1** in Appendix B to review the metric system and its prefixes.[1]) It would take 5000 of these organisms lined up side by side to span a millimeter. This is the distance between the smallest hash marks on a metric ruler. In contrast, sequoia trees can be over 100 meters tall. This is the equivalent of a 20-story building. Bacteria and sequoias are composed of the same fundamental building block, however—the cell. Bacteria consist of a single cell; sequoias are made up of many cells.

Today a **cell** is defined as a highly organized compartment that is bounded by a thin, flexible structure called a plasma membrane and that contains concentrated chemicals in an aqueous (watery) solution. The chemical reactions that sustain life take place inside cells. Most cells are also capable of reproducing by dividing—in effect, by making a copy of themselves.

The realization that all organisms are made of cells was fundamentally important, but it formed only the first part of the cell

---

[1]BioSkills are located in the second appendix at the back of the book. They focus on general skills that you'll use throughout this course. More than a few students have found them to be a life-saver. Please use them!

(a) van Leeuwenhoek built his own microscopes—which, while small, were powerful. They allowed him to see, for example . . .

Lens

(b) . . . human blood cells (this modern photo was shot through one of van Leeuwenhoek's original microscopes) . . .

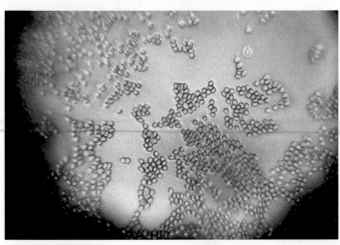

(c) . . . and animal sperm (drawing by van Leeuwenhoek of canine sperm cells on left, human on right).

FIGURE 1.1 Van Leeuwenhoek's Microscope Made Cells Visible.

theory. In addition to understanding what organisms are made of, scientists wanted to understand how cells come to be.

Where Do Cells Come From?

Most scientific theories have two components: The first describes a pattern in the natural world; the second identifies a mechanism or process that is responsible for creating that pattern. Hooke and his fellow scientists articulated the pattern component of the cell theory. In 1858, a German scientist named Rudolph Virchow added the process component by stating that all cells arise from preexisting cells.

The complete **cell theory** can be stated as follows: All organisms are made of cells, and all cells come from preexisting cells.

Two Hypotheses   The cell theory was a direct challenge to the prevailing explanation of where cells come from, called spontaneous generation. In the mid-1800s, most biologists believed that organisms could arise spontaneously under certain conditions. For example, the bacteria and fungi that spoil foods such as milk and wine were thought to appear in these nutrient-rich media of their own accord—springing to life from nonliving materials. In contrast, the cell theory maintained that cells do not spring to life spontaneously but are produced only when preexisting cells grow and divide. The all-cells-from-cells explanation was a **hypothesis:** a testable statement to explain a phenomenon or a set of observations.

Biologists usually use the word theory to refer to proposed explanations for broad patterns in nature and prefer hypothesis to refer to explanations for more tightly focused questions. A theory serves as a framework for the development of new hypotheses.

An Experiment to Settle the Question   Soon after Virchow's all-cells-from-cells hypothesis appeared in print, a French scientist named Louis Pasteur set out to test its predictions experimentally. An experimental **prediction** describes a measurable or observable result that must be correct if a hypothesis is valid.

Pasteur wanted to determine whether microorganisms could arise spontaneously in a nutrient broth or whether they appear only when a broth is exposed to a source of preexisting cells. To address the question, he created two treatment groups: a broth that was not exposed to a source of preexisting cells and a broth that was.

The spontaneous generation hypothesis predicted that cells would appear in both treatment groups. The all-cells-from-cells hypothesis predicted that cells would appear only in the treatment exposed to a source of preexisting cells.

FIGURE 1.2 (on page 4) shows Pasteur's experimental setup. Note that the two treatments are identical in every respect but one. Both used glass flasks filled with the same amount of the same nutrient broth. Both were boiled for the same amount of time to kill any existing organisms such as bacteria or fungi. But because the flask pictured in Figure 1.2a had a straight neck, it was exposed to preexisting cells after sterilization by the heat treatment. These preexisting cells are the bacteria and fungi that cling to dust particles in the air. They could drop into the nutrient broth because the neck of the flask was straight.

In contrast, the flask drawn in Figure 1.2b had a long swan neck. Pasteur knew that water would condense in the crook of the swan neck after the boiling treatment and that this pool of water

QUESTION: Do cells arise spontaneously or from other cells?

SPONTANEOUS GENERATION HYPOTHESIS: Cells arise spontaneously from nonliving materials.

ALL-CELLS-FROM-CELLS HYPOTHESIS: Cells are produced only when preexisting cells grow and divide.

(a) Pasteur experiment with straight-necked flask:	(b) Pasteur experiment with swan-necked flask:

(a) Pasteur experiment with straight-necked flask:

1. Place nutrient broth in straight-necked flask.

Cells

2. Boil to sterilize the flask (killing any living cells that were in the broth).

No cells

Cells

3. Preexisting cells enter flask from air.

(b) Pasteur experiment with swan-necked flask:

1. Place nutrient broth in swan-necked flask.

Cells

2. Boil to sterilize the flask (killing any living cells that were in the broth).

No cells

Condensation settles in neck

Cells

3. Preexisting cells from air are trapped in swan neck.

PREDICTION OF SPONTANEOUS GENERATION HYPOTHESIS:
Cells will appear in broth.

PREDICTION OF ALL-CELLS-FROM-CELLS HYPOTHESIS:
Cells will appear in broth.

PREDICTION OF SPONTANEOUS GENERATION HYPOTHESIS:
Cells will appear in broth.

PREDICTION OF ALL-CELLS-FROM-CELLS HYPOTHESIS:
Cells will not appear in broth.

RESULTS:

Cells

Both hypotheses supported

No cells

Spontaneous generation hypothesis rejected

CONCLUSION: Cells arise from preexisting cells, not spontaneously from nonliving material.

FIGURE 1.2 The Spontaneous Generation and All-Cells-from-Cells Hypotheses Were Tested Experimentally.

✓**QUESTION** What problem would arise in interpreting the results of this experiment if Pasteur had (1) put different types of broth in the two treatments, (2) heated them for different lengths of time, or (3) used a ceramic flask for one treatment and a glass flask for the other?

would trap any bacteria or fungi that entered on dust particles. Thus, the contents of the swan-necked flask were isolated from any source of preexisting cells even though still open to the air.

Pasteur's experimental setup was effective because there was only one difference between the two treatments and because that difference was the factor being tested—in this case, a broth's exposure to preexisting cells.

One Hypothesis Supported   And Pasteur's results? As Figure 1.2 shows, the treatment exposed to preexisting cells quickly filled with bacteria and fungi. This observation was important because it showed that the heat sterilization step had not altered the nutrient broth's capacity to support growth.

The broth in the swan-necked flask remained sterile, however. Even when the flask was left standing for months, no organisms appeared in it. This result was inconsistent with the hypothesis of spontaneous generation.

Because Pasteur's data were so conclusive—meaning that there was no other reasonable explanation for them—the results persuaded most biologists that the all-cells-from-cells hypothesis was correct. However, you will see that biologists now have evidence that life did arise from nonlife early in Earth's history, through a process called chemical evolution (Chapters 2–6).

The success of the cell theory's process component had an important implication: If all cells come from preexisting cells, it follows that all individuals in an isolated population of single-celled organisms are related by common ancestry. Similarly, in you and most other multicellular individuals, all the cells present are descended from preexisting cells, tracing back to a fertilized egg. A fertilized egg is a cell created by the fusion of sperm and egg—cells that formed in individuals of the previous generation. In this way, all the cells in a multicellular organism are connected by common ancestry.

The second great founding idea in biology is similar, in spirit, to the cell theory. It also happened to be published the same year as the all-cells-from-cells hypothesis. This was the realization, made independently by the English scientists Charles Darwin and Alfred Russel Wallace, that all species—all distinct, identifiable types of organisms—are connected by common ancestry.

1.3 The Theory of Evolution by Natural Selection

In 1858 short papers written separately by Darwin and Wallace were read to a small group of scientists attending a meeting of the Linnean Society of London. A year later, Darwin published a book that expanded on the idea summarized in those brief papers. The book was called *The Origin of Species*. The first edition sold out in a day.

What Is Evolution?

Like the cell theory, the theory of evolution by natural selection has a pattern and a process component. Darwin and Wallace's theory made two important claims concerning patterns that exist in the natural world.

1. Species are related by common ancestry. This contrasted with the prevailing view in science at the time, which was that species represent independent entities created separately by a divine being.

2. In contrast to the accepted view that species remain unchanged through time, Darwin and Wallace proposed that the characteristics of species can be modified from generation to generation. Darwin called this process descent with modification.

Evolution is a change in the characteristics of a population over time. It means that species are not independent and unchanging entities, but are related to one another and can change through time.

What Is Natural Selection?

This pattern component of the theory of evolution was actually not original to Darwin and Wallace. Several scientists had already come to the same conclusions about the relationships between species. The great insight by Darwin and Wallace was in proposing a process, called **natural selection,** that explains *how* evolution occurs.

Two Conditions of Natural Selection Natural selection occurs whenever two conditions are met.

1. Individuals within a population vary in characteristics that are **heritable**—meaning, traits that can be passed on to offspring.

A **population** is defined as a group of individuals of the same species living in the same area at the same time.

2. In a particular environment, certain versions of these heritable traits help individuals survive better or reproduce more than do other versions.

If certain heritable traits lead to increased success in producing offspring, then those traits become more common in the population over time. In this way, the population's characteristics change as a result of natural selection acting on individuals. This is a key insight: Natural selection acts on individuals, but evolutionary change occurs in populations.

Selection on Maize as an Example To clarify how selection works, consider an example of **artificial selection**—changes in populations that occur when *humans* select certain individuals to produce the most offspring. Beginning in 1896, researchers began a long-term selection experiment on maize (corn).

1. In the original population, the percentage of protein in maize kernels was variable among individuals. Kernel protein content is a heritable trait—parents tend to pass the trait on to their offspring.

2. Each year for many years, researchers chose individuals with the highest kernel protein content to be the parents of the next generation. In this environment, individuals with high kernel protein content produced more offspring than individuals with low kernel protein content.

FIGURE 1.3 shows the results. Note that this graph plots generation number on the *x*-axis, starting from the first generation (0 on the graph) and continuing for 100 generations. The average percentage of protein in a kernel among individuals in this population is plotted on the *y*-axis.

FIGURE 1.3 Response to Selection for High Kernel Protein Content in Maize.

DATA: Moose, S. P., J. W. Dudley, and T. R. Rocheford. 2004. *Trends in Plant Sciences* 9: 358–364; and the Illinois long-term selection experiment for oil and protein in corn (University of Illinois at Urbana–Champaign).

To read this graph, put your finger on the *x*-axis at generation 0. Then read up the *y*-axis, and note that kernels averaged about 11 percent protein at the start of the experiment. Now read the graph to the right. Each dot is a data point, representing the average kernel protein concentration in a particular generation. (A generation in maize is one year.) The lines on this graph simply connect the dots, to make the pattern in the data easier to see. During a few years the average protein content goes down, because of poor growing conditions or chance changes in how the many genes responsible for this trait interact. However, at the end of the graph, after 100 generations of selection, average kernel protein content is about 29 percent. (For more help with reading graphs, see **BioSkills 3** in Appendix B.)

This sort of change in the characteristics of a population, over time, is evolution. Humans have been practicing artificial selection for thousands of years, and biologists have now documented evolution by *natural* selection—where humans don't do the selecting—occurring in thousands of different populations, including humans. Evolution occurs when heritable variation leads to differential success in reproduction.

✔ QUANTITATIVE If you understand the concepts of selection and evolution, you should be able to describe how protein content in maize kernels changed over time, using the same *x*-axis and *y*-axis as in Figure 1.3, when researchers selected individuals with the *lowest* kernel protein content to be the parents of the next generation. (This experiment was actually done, starting with the same population at the same time as selection for high protein content.)

Fitness and Adaptation Darwin also introduced some new terminology to identify what is happening during natural selection.

- In everyday English, fitness means health and well-being. But in biology, **fitness** means the ability of an individual to produce viable offspring. Individuals with high fitness produce many surviving offspring.

- In everyday English, adaptation means that an individual is adjusting and changing to function in new circumstances. But in biology, an **adaptation** is a trait that increases the fitness of an individual in a particular environment.

Once again, consider kernel protein content in maize: In the environment of the experiment graphed in Figure 1.3, individuals with high kernel protein content produced more offspring and had higher fitness than individuals with lower kernel protein content. In this population and this environment, high kernel protein content was an adaptation that allowed certain individuals to thrive.

Note that during this process, the amount of protein in the kernels of any individual maize plant did not change within its lifetime—the change occurred in the characteristics of the population over time.

Together, the cell theory and the theory of evolution provided the young science of biology with two central, unifying ideas:

1. The cell is the fundamental structural unit in all organisms.

2. All species are related by common ancestry and have changed over time in response to natural selection.

If you understand that . . .

- Natural selection occurs when heritable variation in certain traits leads to improved success in reproduction. Because individuals with these traits produce many offspring with the same traits, the traits increase in frequency and evolution occurs.
- Evolution is a change in the characteristics of a population over time.

✔ **You should be able to . . .**

Using the graph you just analyzed in Figure 1.3, describe the average kernel protein content over time in a maize population where *no* selection occurred.

Answers are available in Appendix A.

1.4 The Tree of Life

Section 1.3 focuses on how individual populations change through time in response to natural selection. But over the past several decades, biologists have also documented dozens of cases in which natural selection has caused populations of one species to diverge and form new species. This divergence process is called **speciation.**

Research on speciation has two important implications: All species come from preexisting species, and all species, past and present, trace their ancestry back to a single common ancestor.

The theory of evolution by natural selection predicts that biologists should be able to construct a **tree of life**—a family tree of organisms. If life on Earth arose just once, then such a diagram would describe the genealogical relationships between species with a single, ancestral species at its base.

Has this task been accomplished? If the tree of life exists, what does it look like?

Using Molecules to Understand the Tree of Life

One of the great breakthroughs in research on the tree of life occurred when American biologist Carl Woese (pronounced *woze*) and colleagues began analyzing the chemical components of organisms as a way to understand their evolutionary relationships. Their goal was to understand the **phylogeny** of all organisms—their actual genealogical relationships. Translated literally, phylogeny means "tribe-source."

To understand which organisms are closely versus distantly related, Woese and co-workers needed to study a molecule that is found in all organisms. The molecule they selected is called small subunit ribosomal RNA (rRNA). It is an essential part of the machinery that all cells use to grow and reproduce.

Although rRNA is a large and complex molecule, its underlying structure is simple. The rRNA molecule is made up of sequences of four smaller chemical components called ribonucleotides. These ribonucleotides are symbolized by the letters A, U, C, and G. In rRNA, ribonucleotides are connected to one another linearly, like the boxcars of a freight train.

Analyzing rRNA Why might rRNA be useful for understanding the relationships between organisms? The answer is that the ribonucleotide sequence in rRNA is a trait that can change during the course of evolution. Although rRNA performs the same function in all organisms, the sequence of ribonucleotide building blocks in this molecule is not identical among species.

In land plants, for example, the molecule might start with the sequence A-U-A-U-C-G-A-G (**FIGURE 1.4**). In green algae, which are closely related to land plants, the same section of the molecule might contain A-U-A-U-G-G-A-G. But in brown algae, which are not closely related to green algae or to land plants, the same part of the molecule might consist of A-A-A-U-G-G-A-C.

The research that Woese and co-workers pursued was based on a simple premise: If the theory of evolution is correct, then rRNA sequences should be very similar in closely related organisms but less similar in organisms that are less closely related. Species that are part of the same evolutionary lineage, like the plants, should share certain changes in rRNA that no other species have.

To test this premise, the researchers determined the sequence of ribonucleotides in the rRNA of a wide array of species. Then they considered what the similarities and differences in the sequences implied about relationships between the species. The goal was to produce a diagram that described the phylogeny of the organisms in the study.

A diagram that depicts evolutionary history in this way is called a phylogenetic tree. Just as a family tree shows relationships between individuals, a phylogenetic tree shows relationships between species. On a phylogenetic tree, branches that share a recent common ancestor represent species that are closely related; branches that don't share recent common ancestors represent species that are more distantly related.

The Tree of Life Estimated from Genetic Data To construct a phylogenetic tree, researchers use a computer to find the arrangement of branches that is most consistent with the similarities and differences observed in the data.

Although the initial work was based only on the sequences of ribonucleotides observed in rRNA, biologists now use data sets that include sequences from a wide array of genetic material. **FIGURE 1.5** shows a recent tree produced by comparing these sequences. Because this tree includes such a diverse array of

Compare the rRNA nucleotide sequence observed in land plants ...

Land plant rRNA

Green algae rRNA

... with the nucleotide sequence found at the same location in the rRNA molecule of green algae

FIGURE 1.4 RNA Molecules Are Made Up of Smaller Molecules. A complete small subunit rRNA molecule contains about 2000 ribonucleotides; just 8 are shown in this comparison.

✔**QUESTION** Suppose that in the same section of rRNA, molds and other fungi have the sequence A-U-A-U-G-G-A-C. Are fungi more closely related to green algae or to land plants? Explain your logic.

FIGURE 1.5 The Tree of Life. A phylogenetic tree estimated from a large amount of genetic sequence data. The three domains of life revealed by the analysis are labeled. Common names are given for lineages in the domains Bacteria and Eukarya. Phyla names are given for members of the domain Archaea, because most of these organisms have no common names.

species, it is often called the universal tree, or the tree of life. (For help in learning how to read a phylogenetic tree, see **BioSkills 7** in Appendix B.) Notice that the tree's main node is the common ancestor of all living organisms. Researchers who study the origin of life propose that the tree's root extends even further back to the "*last universal common ancestor*" of cells, or **LUCA.**

The tree of life implied by rRNA and other genetic data established that there are three fundamental groups or lineages of organisms: **(1)** the Bacteria, **(2)** the Archaea, and **(3)** the Eukarya. In all **eukaryotes,** cells have a prominent component called the nucleus (**FIGURE 1.6a**). Translated literally, the word eukaryotes means "true kernel." Because the vast majority of bacterial and archaeal cells lack a nucleus, they are referred to as **prokaryotes** (literally, "before kernel"; see **FIGURE 1.6b**). The vast majority of bacteria and archaea are unicellular ("one-celled"); many eukaryotes are multicellular ("many-celled").

When results based on genetic data were first published, biologists were astonished. For example:

- Prior to Woese's work and follow-up studies, biologists thought that the most fundamental division among organisms was between prokaryotes and eukaryotes. The Archaea were virtually unknown—much less recognized as a major and highly distinctive branch on the tree of life.

- Fungi were thought to be closely related to plants. Instead, they are actually much more closely related to animals.

(a) Eukaryotic cells have a membrane-bound nucleus.

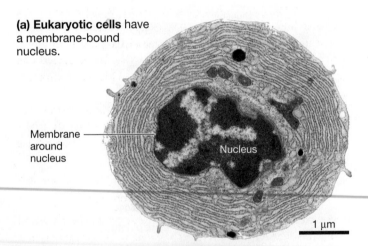

Membrane around nucleus

Nucleus

1 μm

(b) Prokaryotic cells do *not* have a membrane-bound nucleus.

No nucleus

0.1 μm

FIGURE 1.6 Eukaryotes and Prokaryotes.

✔**QUANTITATIVE** How many times larger is the eukaryotic cell in this figure than the prokaryotic cell? (Hint: Study the scale bars.)

- Traditional approaches for classifying organisms—including the system of five kingdoms divided into various classes, orders, and families that you may have learned in high school—are inaccurate in many cases, because they do not reflect the actual evolutionary history of the organisms involved.

The Tree of Life Is a Work in Progress Just as researching your family tree can help you understand who you are and where you came from, so the tree of life helps biologists understand the relationships between organisms and the history of species. The discovery of the Archaea and the accurate placement of lineages such as the fungi qualify as exciting breakthroughs in our understanding of evolutionary history and life's diversity.

Work on the tree of life continues at a furious pace, however, and the location of certain branches on the tree is hotly debated. As databases expand and as techniques for analyzing data improve, the shape of the tree of life presented in Figure 1.5 will undoubtedly change. Our understanding of the tree of life, like our understanding of every other topic in biological science, is dynamic.

How Should We Name Branches on the Tree of Life?

In science, the effort to name and classify organisms is called **taxonomy.** Any named group is called a **taxon** (plural: **taxa**). Currently, biologists are working to create a taxonomy, or naming system, that accurately reflects the phylogeny of organisms.

Based on the tree of life implied by genetic data, Woese proposed a new taxonomic category called the **domain.** The three domains of life are the Bacteria, Archaea, and Eukarya.

Biologists often use the term **phylum** (plural: **phyla**) to refer to major lineages within each domain. Although the designation is somewhat arbitrary, each phylum is considered a major branch on the tree of life. Within the lineage called animals, biologists currently name 30–35 phyla—each of which is distinguished by distinctive aspects of its body structure as well as by distinctive gene sequences. For example, the mollusks (clams, squid, octopuses) constitute a phylum, as do chordates (the vertebrates and their close relatives).

Because the tree of life is so new, though, naming systems are still being worked out. One thing that hasn't changed for centuries, however, is the naming system for individual species.

Scientific (Latin) Names In 1735, a Swedish botanist named Carolus Linnaeus established a system for naming species that is still in use today. Linnaeus created a two-part name unique to each type of organism.

- *Genus* The first part indicates the organism's **genus** (plural: **genera**). A genus is made up of a closely related group of species. For example, Linnaeus put humans in the genus *Homo*. Although humans are the only living species in this genus, at least six extinct organisms, all of which walked upright and made extensive use of tools, were later also assigned to *Homo*.

- *Species* The second term in the two-part name identifies the organism's species. Linnaeus gave humans the species name *sapiens*.

An organism's genus and species designation is called its **scientific name** or Latin name. Scientific names are always italicized. Genus names are always capitalized, but species names are not—as in *Homo sapiens*.

Scientific names are based on Latin or Greek word roots or on words "Latinized" from other languages. Linnaeus gave a scientific name to every species then known, and also Latinized his own name—from Karl von Linné to Carolus Linnaeus.

Linnaeus maintained that different types of organisms should not be given the same genus and species names. Other species may be assigned to the genus *Homo*, and members of other genera may be named *sapiens*, but only humans are named *Homo sapiens*. Each scientific name is unique.

Scientific Names Are Often Descriptive Scientific names and terms are often based on Latin or Greek word roots that are descriptive. For example, *Homo sapiens* is derived from the Latin *homo* for "man" and *sapiens* for "wise" or "knowing." The yeast that bakers use to produce bread and that brewers use to brew beer is called *Saccharomyces cerevisiae*. The Greek root *saccharo* means "sugar," and *myces* refers to a fungus. *Saccharomyces* is aptly named "sugar fungus" because yeast is a fungus and because the domesticated strains of yeast used in commercial baking and brewing are often fed sugar. The species name of this organism, *cerevisiae*, is Latin for "beer." Loosely translated, then, the scientific name of brewer's yeast means "sugar-fungus for beer."

Scientific names and terms often seem daunting at first glance. So, most biologists find it extremely helpful to memorize some of the common Latin and Greek roots. To aid you in this process, new terms in this text are often accompanied by a translation of their Latin or Greek word roots in parentheses. (A glossary of common root words with translations and examples is also provided in **BioSkills 2** in Appendix B.)

check your understanding

If you understand that . . .
- A phylogenetic tree shows the evolutionary relationships between species.
- To infer where species belong on a phylogenetic tree, biologists examine genetic and other characteristics of the species involved. Closely related species should have similar characteristics, while less closely related species should be less similar.

✔ **You should be able to . . .**
Examine the following rRNA ribonucleotide sequences and draw a phylogenetic tree showing the relationships between species A, B, and C that these data imply:

Species A: A A C T A G C G C G A T
Species B: A A C T A G C G C C A T
Species C: T T C T A G C G G T A T

Answers are available in Appendix A.

1.5 Doing Biology

This chapter has introduced some of the great ideas in biology. The development of the cell theory and the theory of evolution by natural selection provided cornerstones when the science was young; the tree of life is a relatively recent insight that has revolutionized our understanding of life's diversity.

These theories are considered great because they explain fundamental aspects of nature, and because they have consistently been shown to be correct. They are considered correct because they have withstood extensive testing.

How do biologists go about testing their ideas? Before answering this question, let's step back a bit and consider the types of questions that researchers can and cannot ask.

The Nature of Science

Biologists ask questions about organisms, just as physicists and chemists ask questions about the physical world or geologists ask questions about Earth's history and the ongoing processes that shape landforms.

No matter what their field, all scientists ask questions that can be answered by observing or measuring things—by collecting data. Conversely, scientists cannot address questions that can't be answered by observing or measuring things.

This distinction is important. It is at the root of continuing controversies about teaching evolution in publicly funded schools. In the United States and in Turkey, in particular, some Christian and Islamic leaders have been particularly successful in pushing their claim that evolution and religious faith are in conflict. Even though the theory of evolution is considered one of the most successful and best-substantiated ideas in the history of science, they object to teaching it.

The vast majority of biologists and many religious leaders reject this claim; they see no conflict between evolution and religious faith. Their view is that science and religion are compatible because they address different types of questions.

- Science is about formulating hypotheses and finding evidence that supports or conflicts with those hypotheses.

- Religious faith addresses questions that cannot be answered by data. The questions addressed by the world's great religions focus on why we exist and how we should live.

Both types of questions are seen as legitimate and important.

So how do biologists go about answering questions? After formulating hypotheses, biologists perform experimental studies, or studies that yield descriptive data, such as observing a behavior, characterizing a structure within a cell by microscopy, or sequencing rRNA. Let's consider two recent examples of this process.

Why Do Giraffes Have Long Necks?
An Introduction to Hypothesis Testing

If you were asked why giraffes have long necks, you might say based on your observations that long necks enable giraffes to reach food that is unavailable to other mammals. This hypothesis is expressed in African folktales and has traditionally been accepted by many biologists. The food competition hypothesis is so plausible, in fact, that for decades no one thought to test it.

In the mid-1990s, however, Robert Simmons and Lue Scheepers assembled data suggesting that the food competition hypothesis is only part of the story. Their analysis supports an alternative hypothesis—that long necks allow giraffes to use their heads as effective weapons for battering their opponents, and that longer-necked giraffes would have a competitive advantage in fights.

Before exploring these alternative explanations, it's important to recognize that hypothesis testing is a two-step process:

Step 1 State the hypothesis as precisely as possible and list the predictions it makes.

Step 2 Design an observational or experimental study that is capable of testing those predictions.

If the predictions are accurate, the hypothesis is supported. If the predictions are not met, then researchers do further tests, modify the original hypothesis, or search for alternative explanations. But the process does not end here. Biologists also talk to other researchers. Over coffee, at scientific meetings, or through publications, biologists communicate their results to the scientific community and beyond. (You can see the Big Picture of the process of doing biology on pages 16–17.)

Now that you understand more about hypothesis testing, let's return to the giraffes. How did biologists test the food competition hypothesis? What data support their alternative explanation?

The Food Competition Hypothesis: Predictions and Tests The food competition hypothesis claims that giraffes compete for food with other species of mammals. When food is scarce, as it is during the dry season, giraffes with longer necks can reach food that is unavailable to other species and to giraffes with shorter necks. As a result, the longest-necked individuals in a giraffe population survive better and produce more young than do shorter-necked individuals, and average neck length of the population increases with each generation.

To use the terms introduced earlier, long necks are adaptations that increase the fitness of individual giraffes during competition for food. This type of natural selection has gone on so long that the population has become extremely long necked.

The food competition hypothesis makes several explicit predictions. For example, the food competition hypothesis predicts that:

• neck length is variable among giraffes;

• neck length in giraffes is heritable; and

• giraffes feed high in trees, especially during the dry season, when food is scarce and the threat of starvation is high.

The first prediction is correct. Studies in zoos and natural populations confirm that neck length is variable among individuals.

The researchers were unable to test the second prediction, however, because they studied giraffes in a natural population and were unable to do breeding experiments. As a result, they simply had to accept this prediction as an assumption. In

(a) Most feeding is done at about shoulder height.

(b) Typical feeding posture in giraffes

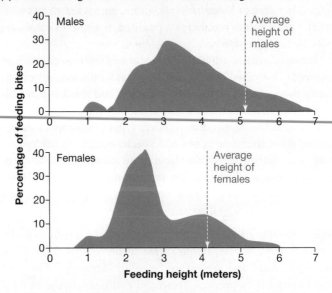

FIGURE 1.7 Giraffes Do Not Usually Extend Their Necks Upward to Feed.

DATA: Young, T. P., L. A. Isbell. 1991. *Ethology* 87: 79–89.

general, though, biologists prefer to test every assumption behind a hypothesis.

What about the prediction regarding feeding high in trees? According to Simmons and Scheepers, this is where the food competition hypothesis breaks down.

Consider, for example, data collected by a different research team on the amount of time that giraffes spend feeding in vegetation of different heights. **FIGURE 1.7a** plots the height of vegetation versus the percentage of bites taken by a giraffe, for males and for females from the same population in Kenya. The dashed line on each graph indicates the average height of a male or female in this population.

Note that the average height of a giraffe in this population is much greater than the height where most feeding takes place. In this population, both male and female giraffes spend most of their feeding time eating vegetation that averages just 60 percent of their full height. Studies on other populations of giraffes, during both the wet and dry seasons, are consistent with these data. Giraffes usually feed with their necks bent (**FIGURE 1.7b**).

These data cast doubt on the food competition hypothesis, because one of its predictions does not appear to hold. Biologists have not abandoned this hypothesis completely, though, because feeding high in trees may be particularly valuable during extreme droughts, when a giraffe's ability to reach leaves far above the ground could mean the difference between life and death. Still, Simmons and Scheepers have offered an alternative explanation for why giraffes have long necks. The new hypothesis is based on the mating system of giraffes.

The Sexual Competition Hypothesis: Predictions and Tests
Giraffes have an unusual mating system. Breeding occurs year round rather than seasonally. To determine when females are coming into estrus or "heat" and are thus receptive to mating, the males nuzzle the rumps of females. In response, the females urinate into the males' mouths. The males then tip their heads back and pull their lips to and fro, as if tasting the liquid. Biologists who have witnessed this behavior have proposed that the males taste the females' urine to detect whether estrus has begun.

Once a female giraffe enters estrus, males fight among themselves for the opportunity to mate. Combat is spectacular. The bulls stand next to one another, swing their necks, and strike thunderous blows with their heads. Researchers have seen males knocked unconscious for 20 minutes after being hit and have cataloged numerous instances in which the loser died. Giraffes are not the only animals known to fight in this way—male giraffe weevils also use enormously long necks to fight for mating rights.

These observations inspired a new explanation for why giraffes have long necks. The sexual competition hypothesis is based on the idea that longer-necked giraffes are able to strike harder blows during combat than can shorter-necked giraffes. In engineering terms, longer necks provide a longer "moment arm." A long moment arm increases the force of an impact. (Think about the type of sledgehammer you'd use to bash down a concrete wall—one with a short handle or one with a long handle?)

The idea here is that longer-necked males should win more fights and, as a result, father more offspring than shorter-necked males do. If neck length in giraffes is inherited, then the average neck length in the population should increase over time. Under the sexual competition hypothesis, long necks are adaptations that increase the fitness of males during competition for females.

Although several studies have shown that long-necked males are more successful in fighting and that the winners of fights gain access to estrous females, the question of why giraffes have long necks is not closed. With the data collected to date, most biologists would probably concede that the food competition hypothesis needs further testing and refinement and that the sexual competition hypothesis appears promising. It could also be true that both hypotheses are correct. For our purposes, the important take-home message is that all hypotheses must be tested rigorously.

In many cases in biological science, testing hypotheses rigorously involves experimentation. Experimenting on giraffes is difficult. But in the case study considered next, biologists were able to test an interesting hypothesis experimentally.

How Do Ants Navigate? An Introduction to Experimental Design

Experiments are a powerful scientific tool because they allow researchers to test the effect of a single, well-defined factor on a particular phenomenon. Because experiments testing the effect of neck length on food and sexual competition in giraffes haven't been done yet, let's consider a different question: When ants leave their nest to search for food, how do they find their way back?

The Saharan desert ant lives in colonies and makes a living by scavenging the dead carcasses of insects. Individuals leave the burrow and wander about searching for food at midday, when temperatures at the surface can reach 60°C (140°F) and predators are hiding from the heat.

Foraging trips can take the ants hundreds of meters—an impressive distance when you consider that these animals are only about a centimeter long. But when an ant returns, it doesn't follow the same long, wandering route it took on its way away from the nest. Instead, individuals return in a straight line (**FIGURE 1.8**).

Once individuals are close to the nest, they engage in a characteristic set of back-and-forth U-turns until they find their nest hole. How do they do know how far they are from the nest?

The Pedometer Hypothesis Early work on navigation in desert ants showed that they use the Sun's position as a compass—meaning that they always know the approximate direction of the nest relative to the Sun. But how do they know how far to go?

After experiments had shown that the ants do not use landmarks to navigate, Matthias Wittlinger and co-workers set out to test a novel idea. The biologists proposed that Saharan desert ants know how far they are from the nest by integrating information from leg movements.

According to this pedometer hypothesis, the ants always know how far they are from the nest because they track the number

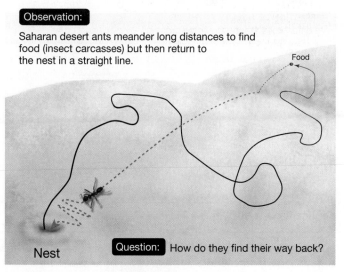

Observation:

Saharan desert ants meander long distances to find food (insect carcasses) but then return to the nest in a straight line.

Food

Nest

Question: How do they find their way back?

FIGURE 1.8 Navigation in Foraging Desert Ants.

of steps they have taken and their stride length. The idea is that they can make a beeline back toward the burrow because they integrate information on the angles they have traveled *and* the distance they have gone—based on step number and stride length.

If the pedometer hypothesis is wrong, however, then stride length and step number should have no effect on the ability of an ant to get back to its nest. This latter possibility is called a **null hypothesis**. A null hypothesis specifies what should be observed when the hypothesis being tested isn't correct.

Testing the Hypothesis   To test their idea, Wittlinger's group allowed ants to walk from a nest to a feeder through a channel—a distance of 10 m. Then they caught ants at the feeder and created three test groups, each with 25 individuals (**FIGURES 1.9** and **1.10**):

- *Stumps*   By cutting the lower legs of some individuals off, they created ants with shorter-than-normal legs.

- *Normal*   Some individuals were left alone, meaning that they had normal leg length.

FIGURE 1.9 An Experimental Test: Do Desert Ants Use a "Pedometer"?

SOURCE: Wittlinger, M., R. Wehner, and H. Wolf. 2006. The ant odometer: Stepping on stilts and stumps. *Science* 312: 1965–1967.

✔ **QUESTION** What is the advantage of using 25 ants in each group instead of just one?

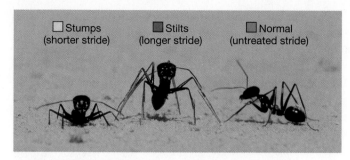

Stumps (shorter stride) Stilts (longer stride) Normal (untreated stride)

FIGURE 1.10 Manipulation of Desert Ant Stride Length.

- *Stilts* By gluing pig bristles onto each leg, the biologists created ants with longer-than-normal legs.

Next they put the ants in a different channel and recorded how far they traveled in a direct line before starting their nest-searching behavior. To see the data they collected, look at the graph on the left side of the "Results" section in Figure 1.9.

- *Stumps* The ants with stumps stopped short, by about 5 m, before starting to search for the nest opening.
- *Normal* The normal ants walked the correct distance—about 10 m.
- *Stilts* The ants with stilts walked about 5 m too far before starting to search for the nest opening.

To check the validity of this result, the researchers put the test ants back in the nest and recaptured them one to several days later, when they had walked to the feeder on their stumps, normal legs, or stilts. Now when the ants were put into the other channel to "walk back," they all traveled the correct distance—10 m—before starting to search for the nest (see the graph on the right side of the "Results" section in Figure 1.9).

The graphs in the "Results" display "box-and-whisker" plots that allow you to easily see where most of the data fall. Each box indicates the range of distances where 50 percent of the ants stopped to search for the nest. The whiskers indicate the lower extreme (stopping short of the nest location) and the upper extreme (going too far) of where the ants stopped to search. The vertical line inside each box indicates the median—meaning that half the ants stopped above this distance and half below. (For more details on how biologists report medians and indicate the variability and uncertainty in data, see **BioSkills 4** in Appendix B.)

Interpreting the Results The pedometer hypothesis predicts that an ant's ability to walk home depends on the number and length of steps taken on its outbound trip. Recall that a prediction specifies what we should observe if a hypothesis is correct. Good scientific hypotheses make testable predictions—predictions that can be supported or rejected by collecting and analyzing data. In this case, the researchers tested the prediction by altering stride length and recording the distance traveled on the return trip. Under the null hypothesis in this experiment, all the ants—altered and unaltered—should have walked 10 m in the first test before they started looking for their nest.

Important Characteristics of Good Experimental Design In relation to designing effective experiments, this study illustrates several important points:

- It is critical to include **control** groups. A control checks for factors, other than the one being tested, that might influence the experiment's outcome. In this case, there were two controls. Including a normal, unmanipulated individual controlled for the possibility that switching the individuals to a new channel altered their behavior. In addition, the researchers had to control for the possibility that the manipulation itself—and not the change in leg length—affected the behavior of the stilts and stumps ants. This is why they did the second test, where the outbound and return runs were done with the same legs.

- The experimental conditions must be as constant or equivalent as possible. The investigators used ants of the same species, from the same nest, at the same time of day, under the same humidity and temperature conditions, at the same feeders, in the same channels. Controlling all the variables except one—leg length in this case—is crucial because it eliminates alternative explanations for the results.

- Repeating the test is essential. It is almost universally true that larger sample sizes in experiments are better. By testing many individuals, the amount of distortion or "noise" in the data caused by unusual individuals or circumstances is reduced.

✔ If you understand these points, you should be able to explain: (1) What you would conclude if in the first test, the normal individual had not walked 10 m on the return trip before

check your understanding

If you understand that . . .

- Hypotheses are proposed explanations that make testable predictions.
- Predictions describe observable outcomes of particular conditions.
- Well-designed experiments alter just one condition—a condition relevant to the hypothesis being tested.

✔ **You should be able to . . .**

Design an experiment to test the hypothesis that desert ants feed during the hottest part of the day because it allows them to avoid being eaten by lizards. Then answer the following questions about your experimental design:

1. How does the presence of a control group in your experiment allow you to test the hypothesis?

2. How are experimental conditions controlled or standardized in a way that precludes alternative explanations of the data?

Answers are available in Appendix A.

looking for the nest; and (2) What you would conclude if the stilts and stumps ants had not navigated normally during the second test.

From the outcomes of these experiments, the researchers concluded that desert ants use stride length and number to measure how far they are from the nest. They interpreted their results as strong support for the pedometer hypothesis.

The giraffe and ant studies demonstrate a vital point: Biologists practice evidence-based decision making. They ask questions about how organisms work, pose hypotheses to answer those questions, and use experimental or observational evidence to decide which hypotheses are correct.

The data on giraffes and ants are a taste of things to come. In this text you will encounter hypotheses and research on questions ranging from how water gets to the top of 100-meter-tall sequoia trees to how the bacterium that causes tuberculosis has become resistant to antibiotics. As you work through this book, you'll get lots of practice thinking about hypotheses and predictions, analyzing the nature of control treatments, and interpreting graphs.

A commitment to tough-minded hypothesis testing and sound experimental design is a hallmark of biological science. Understanding their value is an important first step in becoming a biologist.

CHAPTER 1 REVIEW

For media, go to MasteringBiology

If you understand . . .

1.1 What Does It Mean to Say That Something Is Alive?

- There is no single, well-accepted definition of life. Instead, biologists point to five characteristics that organisms share.

✔ You should be able to explain why the cells in a dead organism are different from the cells in a live organism.

1.2 The Cell Theory

- The cell theory identified the fundamental structural unit common to all life.

✔ You should be able to describe the evidence that supported the pattern and the process components of the cell theory.

1.3 The Theory of Evolution by Natural Selection

- The theory of evolution states that all organisms are related by common ancestry.

- Natural selection is a well-tested explanation for why species change through time and why they are so well adapted to their habitats.

✔ You should be able to explain why the average protein content of seeds in a natural population of a grass species would increase over time, if seeds with higher protein content survive better and grow into individuals that produce many seeds with high protein content when they mature.

1.4 The Tree of Life

- The theory of evolution predicts that all organisms are part of a genealogy of species, and that all species trace their ancestry back to a single common ancestor.

- To construct this phylogeny, biologists have analyzed the sequences in rRNA and in an array of genetic material found in all cells.

- A tree of life, based on similarities and differences in these molecules, has three fundamental lineages, or domains: the Bacteria, the Archaea, and the Eukarya.

✔ You should be able to explain how biologists can determine which of the three domains a newly discovered species belongs to by analyzing its rRNA.

1.5 Doing Biology

- Biology is a hypothesis-driven, experimental science.

✔ You should be able to explain (1) the relationship between a hypothesis and a prediction and (2) why experiments are convincing ways to test predictions.

MasteringBiology

1. MasteringBiology Assignments

Tutorials and Activities An Introduction to Graphing; Experimental Inquiry: What Can You Learn about the Process of Science from Investigating a Cricket's Chirp?; Introduction to Experimental Design; Levels of Life Card Game; Metric System Review; The Scientific Method

Questions Reading Quizzes, Blue-Thread Questions, Test Bank

2. eText Read your book online, search, take notes, highlight text, and more.

3. The Study Area Practice Test, Cumulative Test, BioFlix® 3-D Animations, Videos, Activities, Audio Glossary, Word Study Tools, Art

You should be able to . . .

1. Anton van Leeuwenhoek made an important contribution to the development of the cell theory. How?
   a. He articulated the pattern component of the theory—that all organisms are made of cells.
   b. He articulated the process component of the theory—that all cells come from preexisting cells.
   c. He invented the first microscope and saw the first cell.
   d. He invented more powerful microscopes and was the first to describe the diversity of cells.

2. What does it mean to say that experimental conditions are controlled?
   a. The test groups consist of the same individuals.
   b. The null hypothesis is correct.
   c. There is no difference in outcome between the control and experimental treatment.
   d. All physical conditions except for one are identical for all groups tested.

3. The term *evolution* means that _____ change through time.

4. What does it mean to say that a characteristic of an organism is heritable?
   a. The characteristic evolves.
   b. The characteristic can be passed on to offspring.
   c. The characteristic is advantageous to the organism.
   d. The characteristic does not vary in the population.

5. In biology, to what does the term *fitness* refer?

6. Could *both* the food competition hypothesis and the sexual competition hypothesis explain why giraffes have long necks? Why or why not?
   a. No. In science, only one hypothesis can be correct.
   b. No. Observations have shown that the food competition hypothesis cannot be correct.
   c. Yes. Long necks could be advantageous for more than one reason.
   d. Yes. All giraffes have been shown to feed at the highest possible height and fight for mates.

7. What would researchers have to demonstrate to convince you that they had discovered life on another planet?

8. What did Linnaeus's system of naming organisms ensure?
   a. Two different organisms never end up with the same genus and species name.
   b. Two different organisms have the same genus and species name if they are closely related.
   c. The genus name is different for closely related species.
   d. The species name is the same for each organism in a genus.

9. What does it mean to say that a species is adapted to a particular habitat?

10. Explain how selection occurs during natural selection. What is selected, and why?

11. The following two statements explain the logic behind the use of molecular sequence data to estimate evolutionary relationships:

    "If the theory of evolution is true, then rRNA sequences should be very similar in closely related organisms but less similar in organisms that are less closely related."

    "On a phylogenetic tree, branches that share a recent common ancestor represent species that are closely related; branches that don't share recent common ancestors represent species that are more distantly related."

    Is the logic of these statements sound? Why or why not?

12. Explain why researchers formulate a null hypothesis in addition to a hypothesis when designing an experimental study.

13. A scientific theory is a set of propositions that defines and explains some aspect of the world. This definition contrasts sharply with the everyday usage of the word theory, which often carries meanings such as "speculation" or "guess." Explain the difference between the two definitions, using the cell theory and the theory of evolution by natural selection as examples.

14. Turn back to the tree of life shown in Figure 1.5. Note that Bacteria and Archaea are prokaryotes, while Eukarya are eukaryotes. On the simplified tree below, draw an arrow that points to the branch where the structure called the nucleus originated. Explain your reasoning.

BACTERIA

ARCHAEA

EUKARYA

15. The proponents of the cell theory could not "prove" that it was correct in the sense of providing incontrovertible evidence that all organisms are made up of cells. They could state only that all organisms examined to date were made of cells. Why was it reasonable for them to conclude that the theory was valid?

16. Some humans have heritable traits that make them resistant to infection by HIV. In areas of the world where HIV infection rates are high, are human populations evolving? Explain your logic.
    a. No. HIV infection rates would not affect human evolution.
    b. Yes. The heritable traits that confer resistance to HIV should increase over time.
    c. No. The heritable traits that confer resistance to HIV should decrease over time.
    d. Yes. The heritable traits that confer resistance to HIV should decrease over time.

The Big Picture

Biologists study the characteristics of life. The cell theory, the theory of evolution by natural selection, and the tree of life are some of the great ideas in biology that came about by biologists asking questions that can be answered by observing or measuring things—that is, by collecting data.

Notice that the study of life is not a series of linear steps with a beginning and an end. Instead, the process of doing biology is dynamic and ongoing. The answer to one question may lay the foundation for twenty more questions. Working together, biologists from different disciplines integrate data across many levels, from atoms to the biosphere.

Note that the gray numbers in boxes tell you where to go for more information. Also, be sure to do the blue exercises in the Check Your Understanding box.

Characteristics of living things
- Energy
- Cells
- Information
- Replication
- Evolution

1.1

Text section where you can find more information

focuses on

Scientists regularly integrate across many of these levels

Levels of biological organization

BIOSPHERE

ECOSYSTEM

COMMUNITY

POPULATION

MULTICELLULAR ORGANISM

ORGAN SYSTEM

ORGAN

TISSUE

CELL

ORGANELLE

MOLECULE

ATOM

check your understanding

(C) (Y) (U)

If you understand the big picture . . .

✔ **You should be able to . . .**

1. Describe how biologists go about testing their ideas.
2. Provide an example of how an experimental study could span more than one level of biological organization.
3. Compare and contrast a hypothesis with a theory.
4. Propose the next step to take if data support the hypothesis you are testing.

Answers are available in Appendix A.

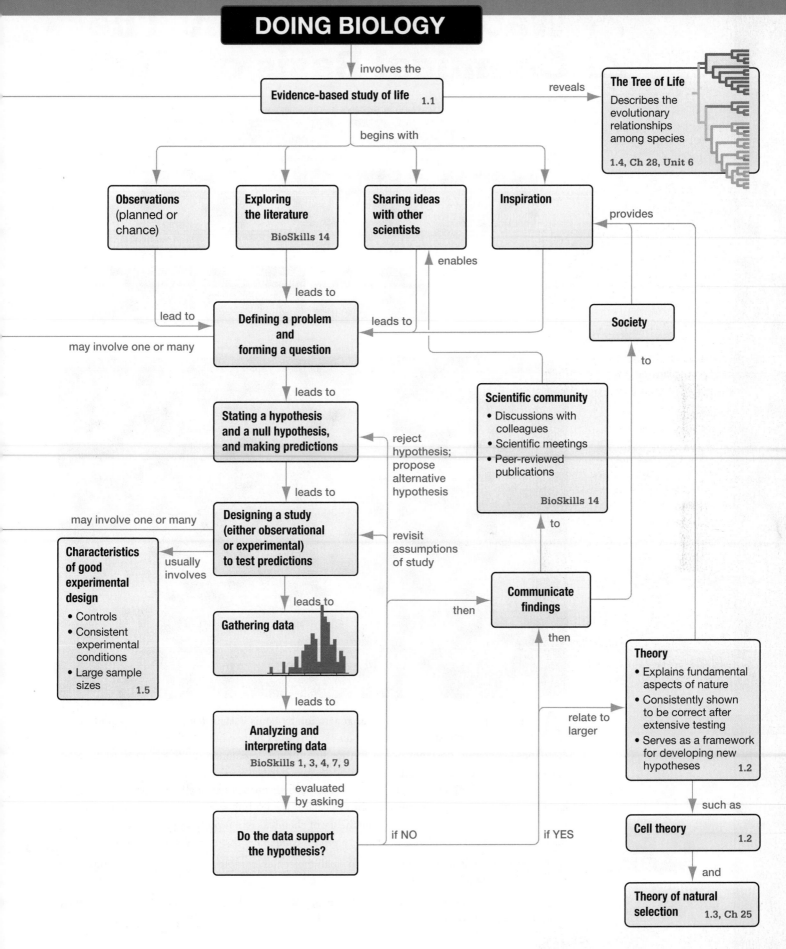

DOING BIOLOGY

involves the

Evidence-based study of life 1.1

— reveals → **The Tree of Life** Describes the evolutionary relationships among species 1.4, Ch 28, Unit 6

begins with

Observations (planned or chance)

Exploring the literature BioSkills 14

Sharing ideas with other scientists

Inspiration

provides

lead to

leads to

enables

Defining a problem and forming a question

leads to

may involve one or many

Society

to

leads to

Stating a hypothesis and a null hypothesis, and making predictions

reject hypothesis; propose alternative hypothesis

Scientific community
- Discussions with colleagues
- Scientific meetings
- Peer-reviewed publications

BioSkills 14

leads to

to

may involve one or many

Designing a study (either observational or experimental) to test predictions

revisit assumptions of study

Characteristics of good experimental design
- Controls
- Consistent experimental conditions
- Large sample sizes 1.5

usually involves

Communicate findings

then

leads to

Gathering data

then

Theory
- Explains fundamental aspects of nature
- Consistently shown to be correct after extensive testing
- Serves as a framework for developing new hypotheses 1.2

leads to

relate to larger

Analyzing and interpreting data BioSkills 1, 3, 4, 7, 9

evaluated by asking

such as

Do the data support the hypothesis?

if NO

if YES

Cell theory 1.2

and

Theory of natural selection 1.3, Ch 25

2 Water and Carbon: The Chemical Basis of Life

In this chapter you will learn that

Chemistry is intimately linked to the evolution of life

by reviewing

How are atoms bonded together in molecules? ←and asking— The structure of atoms, ions, and molecules **2.1** —and asking→ How do chemical reactions occur? **2.3**

then examining

The unique properties of water **2.2**

then examining

Current models for chemical evolution **2.4**

Carbon: The most versatile atom on Earth **2.5**

These deep-sea hydrothermal vents produce hydrogen-rich, highly basic fluids at temperatures that range from 40° to 90°C. It has been proposed that life emerged from similar seafloor chimneys early in Earth's history via chemical evolution.

A classic experiment on spontaneous generation by Louis Pasteur tested the idea that life arises from nonliving materials (see Chapter 1). This work helped build a consensus that spontaneous generation does not occur. But for life to exist, spontaneous generation must have occurred at least once, early in Earth's history.

How did life begin? This simple query has been called "the mother of all questions." This chapter examines a theory, called **chemical evolution,** that is the leading scientific explanation for the origin of life. Like all scientific theories, the theory of chemical evolution has a *pattern component* that makes a claim about the natural world and a *process component* that explains that pattern.

- *The pattern component* In addition to small molecules, complex carbon-containing substances exist and are required for life.

- *The process component* Early in Earth's history, simple chemical compounds combined to form more complex carbon-containing substances before the evolution of life.

This chapter is part of the Big Picture. See how on pages 104–105.

✔ When you see this checkmark, stop and test yourself. Answers are available in Appendix A.

The theory maintains that inputs of energy led to the formation of increasingly complex carbon-containing substances, culminating in a compound that could replicate itself. At this point, there was a switch from chemical evolution to biological evolution.

As the original molecule multiplied, the process of evolution by natural selection took over. Eventually a descendant of the original molecule became metabolically active and acquired a membrane. When this occurred, the five attributes of life (discussed in Chapter 1) were fulfilled. Life had begun.

At first glance, the theory of chemical evolution may seem implausible. But is it? What evidence do biologists have that chemical evolution occurred? What approaches do they take to gathering this evidence?

Let's start with the fundamentals—the atoms and molecules that would have combined to get chemical evolution started.

2.1 Atoms, Ions, and Molecules: The Building Blocks of Chemical Evolution

Just four types of atoms—hydrogen, carbon, nitrogen, and oxygen—make up 96 percent of all matter found in organisms today. Many of the molecules found in your cells contain thousands, or even millions, of these atoms bonded together. But early in Earth's history, these elements existed only in simple substances such as water and carbon dioxide, which contain just three atoms apiece.

Two questions are fundamental to understanding how elements could have evolved into the more complex substances found in living cells:

1. What is the physical structure of the hydrogen, carbon, nitrogen, and oxygen atoms found in living cells?

2. What is the structure of the simple molecules—water, carbon dioxide, and others—that served as the building blocks of chemical evolution?

The focus on structure follows from one of the most central themes in biology: *Structure affects function*. To understand how a molecule affects your body or the role it played in chemical evolution, you have to understand how it is put together.

Basic Atomic Structure

FIGURE 2.1a shows a simple way of depicting the structure of an atom, using hydrogen and carbon as examples. Extremely small particles called electrons orbit an atomic nucleus made up of larger particles called protons and neutrons. **FIGURE 2.1b** provides a sense of scale at the atomic level.

Protons have a positive electric charge ($+1$), neutrons are electrically neutral, and electrons have a negative electric charge (-1). When the number of protons and the number of electrons in an atom are the same, the charges balance and the atom is electrically neutral.

(a) Diagrams of atoms

Hydrogen

Carbon

(b) Most of an atom's volume is empty space.

If an atom occupied the same volume as this stadium, the nucleus would be about the size of a pea

FIGURE 2.1 Parts of an Atom. The atomic nucleus, made up of protons and neutrons, is surrounded by orbiting electrons. In reality, electrons do not orbit the nucleus in circles; their actual orbits are complex.

FIGURE 2.2 shows a segment of the periodic table of the elements. Notice that each atom of a given **element** contains a characteristic number of protons, called its **atomic number.** The atomic number is written as a subscript to the left of an element's symbol in Figure 2.2. The sum of the protons and neutrons in an atom is called its **mass number** and is written as a superscript to the left of its symbol.

Mass number
(number of protons
+ neutrons)

Atomic number
(number of protons)

$^{1}_{1}H$							$^{4}_{2}He$
$^{7}_{3}Li$	$^{9}_{4}Be$	$^{11}_{5}B$	$^{12}_{6}C$	$^{14}_{7}N$	$^{16}_{8}O$	$^{19}_{9}F$	$^{20}_{10}Ne$
$^{23}_{11}Na$	$^{24}_{12}Mg$	$^{27}_{13}Al$	$^{28}_{14}Si$	$^{31}_{15}P$	$^{32}_{16}S$	$^{35}_{17}Cl$	$^{40}_{18}Ar$

FIGURE 2.2 A Portion of the Periodic Table. Each element has a unique atomic number and is represented by a unique one- or two-letter symbol. The mass numbers given here are the most common for each element. (Appendix C provides a complete periodic table of elements.)

The number of protons in an element does not vary—if the atomic number of an atom changes, then it is no longer the same element. The number of neutrons present in an element can vary, however. Forms of an element with different numbers of neutrons are known as **isotopes** (literally, "equal-places" in regard to position in the periodic table).

Different isotopes have different masses, yet are the same element. For example, all atoms of the element carbon have 6 protons. But naturally occurring isotopes of carbon can have 6, 7, or even 8 neutrons, giving them a mass number of 12, 13, or 14, respectively. The **atomic weight** of an element is an average of all the mass numbers of the naturally occurring isotopes based on their abundance. This is why the atomic weights for elements are often slightly different from the mass numbers—the atomic weight of carbon, for example, is 12.01.

Most isotopes are stable, but not all. For example, ^{14}C, with 6 protons and 8 neutrons, represents an unstable **radioactive isotope.** Its nucleus will eventually decay and release energy (radiation). When ^{14}C decays, one of its neutrons changes into a proton, converting ^{14}C to the stable ^{14}N isotope of nitrogen, with 7 protons and 7 neutrons. Timing of decay is specific to each radioisotope, a fact that has been very useful in estimating the dates of key events in the fossil record (see Chapter 25).

Although the masses of protons, neutrons, and electrons can be measured in grams, the numbers involved are so small that biologists prefer to use a special unit called the **dalton.** The masses of protons and neutrons are virtually identical and are routinely rounded to 1 dalton. A carbon atom that contains 6 protons and 6 neutrons has a mass of 12 daltons, while a carbon atom with 6 protons and 7 neutrons would have a mass of 13 daltons. These isotopes would be written as ^{12}C and ^{13}C,

respectively. The mass of an electron is so small that it is normally ignored.

To understand how the atoms involved in chemical evolution behave, focus on how electrons are arranged around the nucleus:

- Electrons move around atomic nuclei in specific regions called **orbitals.**

- Each orbital can hold up to two electrons.

- Orbitals are grouped into levels called **electron shells.**

- Electron shells are numbered 1, 2, 3, and so on, to indicate their relative distance from the nucleus. Smaller numbers are closer to the nucleus.

- Each electron shell contains a specific number of orbitals. An electron shell comprising a single orbital can hold up to two electrons; a shell with four orbitals can contain up to eight electrons.

- The electrons of an atom fill the innermost shells first, before filling outer shells.

To understand how the structures of atoms differ, take a moment to study **FIGURE 2.3**. This chart highlights the elements that are most abundant in living cells. The gray ball in the center of each box represents an atomic nucleus, and the orange circle or circles represent the electron shells around that nucleus. The small orange balls on the circles indicate the number of electrons that are distributed in the shells of each element. Electrons shown as pairs share the same orbital within a shell.

Now focus on the outermost shell of each atom. This is the element's **valence shell.** The electrons found in this shell are referred to as **valence electrons.** Two observations are important:

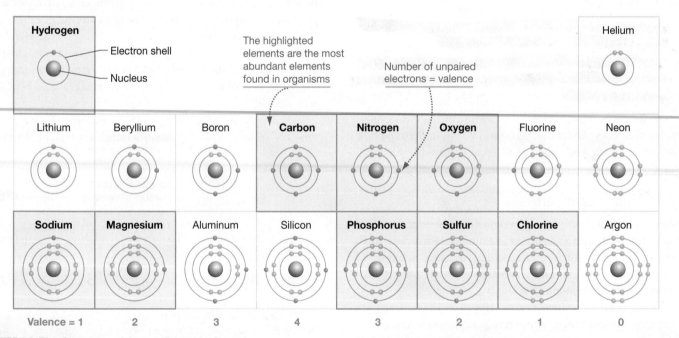

FIGURE 2.3 The Structure of Atoms Found in Organisms.

✔**QUESTION** If the mass number of phosphorus is 31, how many neutrons exist in the most common isotope of phosphorus?

1. In each of the highlighted elements, the outermost electron shell is not full—not all orbitals in the valence shell have two electrons. The highlighted elements have at least one unpaired valence electron—meaning at least one unfilled valence shell orbital.

2. The number of unpaired valence electrons varies among elements. Carbon, for example, has four valence electrons, all unpaired. Oxygen has six valence electrons; four are paired, two are not. The number of unpaired electrons found in an atom is called its **valence.** Carbon's valence is four, oxygen's is two.

These observations are significant because an atom is most stable when its valence shell is filled. One way that shells can be filled is through the formation of strong **chemical bonds**—attractions that bind atoms together. A strong attraction where two atoms share one or more pairs of electrons is called a **covalent bond.**

How Does Covalent Bonding Hold Molecules Together?

To understand how atoms can become more stable by making covalent bonds, consider hydrogen. The hydrogen atom has just one electron, which resides in a shell that can hold two electrons.

Because it has an unpaired valence electron, the hydrogen atom is not very stable. But when two atoms of hydrogen come into contact, the two electrons become shared by the two nuclei (**FIGURE 2.4**). Both atoms now have a completely filled outer shell. Together, the hydrogen atoms are more stable than the two individual hydrogen atoms.

Shared electrons "glue" two hydrogen atoms together. Substances held together by covalent bonds are called **molecules.** In the case of two hydrogen atoms, the bonded atoms form a single molecule of hydrogen, written as H—H or H_2.

It can also be helpful to think about covalent bonding as electrical attraction and repulsion. Opposite charges attract; like charges repel. As two hydrogen atoms move closer together, their positively charged nuclei repel each other and their negatively charged electrons repel each other. But each proton attracts both electrons, and each electron attracts both protons. Covalent bonds form when the attractive forces overcome the repulsive forces. This is the case when hydrogen atoms interact to form the hydrogen molecule (H_2).

Nonpolar and Polar Bonds In **FIGURE 2.5a**, the covalent bond between hydrogen atoms is represented by a dash and the electrons are drawn as dots halfway between the two nuclei. This depiction shows that the electrons are shared equally between the two hydrogen atoms, resulting in a covalent bond that is symmetrical.

It's important to note, though, that the electrons participating in a covalent bond are not always shared equally between the atoms involved. This happens because some atoms hold the electrons in covalent bonds much more tightly than do other atoms. Chemists call this property **electronegativity.**

What is responsible for an atom's electronegativity? It's a combination of two things—the number of protons in the nucleus and the distance between the nucleus and the valence shell. If

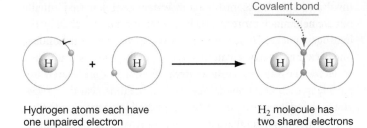

FIGURE 2.4 Covalent Bonds Result from Electron Sharing. When two hydrogen atoms come into contact, their electrons are attracted to the positive charge in each nucleus. As a result, their orbitals overlap, the electrons are shared by each nucleus, and a covalent bond forms.

you return to Figure 2.3 and move your finger along a row from left to right, you will be moving toward elements that increase in protons and in electronegativity (ignoring the elements in the far right column, which have full outer shells). Each row, however, represents shells of electrons, so if your finger moved down the table, the elements would decrease in electronegativity.

Oxygen, which has eight protons and only two electron shells, is among the most electronegative of all elements. It attracts covalently bonded electrons more strongly than does any other atom commonly found in organisms. Nitrogen's electronegativity is somewhat lower than oxygen's. Carbon and hydrogen, in turn, have relatively low and approximately equal electronegativities. Thus, the electronegativities of the four most abundant elements in organisms are related as follows: $O > N > C \cong H$.

Because carbon and hydrogen have approximately equal electronegativity, the electrons in a C—H bond are shared equally or symmetrically. The result is a **nonpolar covalent bond.** In contrast, asymmetric sharing of electrons results in a **polar covalent bond.** The electrons in a polar covalent bond spend most of their time close to the nucleus of the more electronegative atom. Why is this important?

Polar Bonds Produce Partial Charges on Atoms To understand the consequences of differences in electronegativity and the formation of polar covalent bonds, consider the water molecule.

Water consists of an oxygen bonded to two hydrogen atoms, and is written H_2O. As **FIGURE 2.5b** illustrates, the electrons

(a) Nonpolar covalent bond in hydrogen molecule

Electrons are halfway between the two atoms, shared equally

(b) Polar covalent bonds in water molecule

Electrons are not shared equally (O is more electronegative than H), so partial charges exist on the O and H atoms

FIGURE 2.5 Electron Sharing and Bond Polarity. Delta (δ) symbols in polar covalent bonds refer to partial positive and negative charges that arise owing to unequal electron sharing.

involved in the covalent bonds in water are not shared equally but are held much more tightly by the oxygen nucleus than by the hydrogen nuclei. Hence, water has two polar covalent bonds—one between the oxygen atom and each of the hydrogen atoms.

Here's the key observation: Because electrons are shared unequally in each O–H bond, they spend more time near the oxygen atom, giving it a partial negative charge, and less time near the hydrogen atoms, giving them a partial positive charge. These partial charges are symbolized by the lowercase Greek letter delta, δ.

As Section 2.2 shows, the partial charges on water molecules—due simply to the difference in electronegativity between oxygen and hydrogen—are one of the primary reasons that life exists.

Ionic Bonding, Ions, and the Electron-Sharing Continuum

Ionic bonds are similar in principle to covalent bonds, but instead of being shared between two atoms, the electrons in ionic bonds are completely transferred from one atom to the other. The electron transfer occurs because it gives the resulting atoms a full outermost shell.

Sodium atoms (Na), for example, tend to lose an electron, leaving them with a full second shell. This is a much more stable arrangement, energetically, than having a lone electron in their third shell (**FIGURE 2.6a**). The atom that results has a net electric charge of +1, because it has one more proton than it has electrons.

An atom or molecule that carries a full charge, rather than the partial charges that arise from polar covalent bonds, is called an **ion**. The sodium ion is written Na^+ and, like other positively charged ions, is called a **cation** (pronounced *KAT-eye-un*).

Chlorine atoms (Cl), in contrast, tend to gain an electron, filling their outermost shell (**FIGURE 2.6b**). The ion has a net charge of −1, because it has one more electron than protons. This

negatively charged ion, or **anion** (pronounced *AN-eye-un*), is written Cl^- and is called chloride.

When sodium and chlorine combine to form sodium chloride (NaCl, common table salt), they pack into a crystal structure consisting of sodium cations and chloride anions (**FIGURE 2.6c**). The electrical attraction between the ions is so strong that salt crystals are difficult to break apart.

This discussion of covalent and ionic bonding supports an important general observation: The degree to which electrons are shared in chemical bonds forms a continuum from equal sharing in nonpolar covalent bonds to unequal sharing in polar covalent bonds to the transfer of electrons in ionic bonds.

As the left-hand side of **FIGURE 2.7** shows, covalent bonds between atoms with exactly the same electronegativity—for example, between the atoms of hydrogen in H_2—represent one end of the continuum. The electrons in these nonpolar bonds are shared equally.

In the middle of the continuum are bonds where one atom is much more electronegative than the other. In these asymmetric bonds, substantial partial charges exist on each of the atoms. These types of polar covalent bonds occur when a highly electronegative atom such as oxygen or nitrogen is bonded to an atom with a lower affinity for electrons, such as carbon or hydrogen. Ammonia (NH_3) and water (H_2O) are examples of molecules with polar covalent bonds.

At the right-hand side of the continuum are molecules made up of atoms with extreme differences in their electronegativities. In this case, electrons are transferred rather than shared, the atoms have full charges, and the bonding is ionic. Sodium chloride (NaCl) is a familiar example of a molecule formed by ionic bonds.

Most chemical bonds that occur in biological molecules are on the left-hand side and the middle of the continuum; in the molecules found in organisms, ionic bonding is less common.

(a) A sodium ion being formed

(c) Table salt (NaCl) is a crystal composed of two ions.

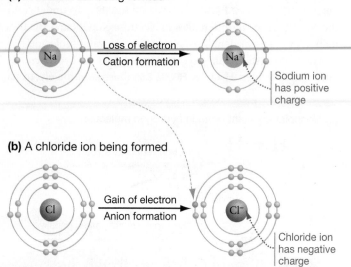

Loss of electron
Cation formation

Sodium ion has positive charge

(b) A chloride ion being formed

Gain of electron
Anion formation

Chloride ion has negative charge

Cl⁻

Na⁺

FIGURE 2.6 Ion Formation and Ionic Bonding. The sodium ion (Na^+) and the chloride ion (Cl^-) are stable because they have full valence shells. In table salt (NaCl), sodium and chloride ions pack into a crystal structure held together by electrical attraction between their positive and negative charges.

Equal sharing of electrons ◄─────────────────────────────► Transfer of electrons

| Nonpolar covalent bonds (atoms have no charge) | Polar covalent bonds (atoms have partial charge) | Ionic bonds (atoms have full charge) |

Hydrogen   Methane   Ammonia   Water   Sodium chloride

FIGURE 2.7 The Electron-Sharing Continuum. The degree of electron sharing in chemical bonds can be thought of as a continuum, from equal sharing in nonpolar covalent bonds to no sharing in ionic bonds.

✔**QUESTION** Why do most polar covalent bonds involve nitrogen or oxygen?

Some Simple Molecules Formed from C, H, N, and O

Look back at Figure 2.3 and count the number of unpaired electrons in the valence shells of carbon, nitrogen, oxygen, and hydrogen atoms. Each unpaired electron in a valence shell can make up half of a covalent bond. It should make sense to you that a carbon atom can form a total of four covalent bonds; nitrogen can form three; oxygen can form two; and hydrogen, one.

When each of the four unpaired electrons of a carbon atom covalently bonds with a hydrogen atom, the molecule that results is written CH_4 and is called methane (**FIGURE 2.8a**). Methane is the most common molecule found in natural gas. When a nitrogen atom's three unpaired electrons bond with three hydrogen atoms, the result is NH_3, or ammonia. Similarly, an atom of oxygen can form covalent bonds with two atoms of hydrogen, resulting in a water molecule (H_2O). As Figure 2.4 showed, a hydrogen atom can bond with another hydrogen atom to form hydrogen gas (H_2).

In addition to forming more than one single bond, atoms with more than one unpaired electron in the valence shell can form double bonds or triple bonds. **FIGURE 2.8b** shows how carbon forms double bonds with oxygen atoms to produce carbon dioxide (CO_2). Triple bonds result when three pairs of electrons are shared. **FIGURE 2.8c** shows the structure of molecular nitrogen (N_2), which forms when two nitrogen atoms establish a triple bond.

✔ If you understand how electronegativity affects covalent bonds, you should be able to draw arrows between the atoms in each molecule shown in Figure 2.8 to indicate the relative position of the shared electrons. If they are equally shared, then draw a double-headed arrow.

The Geometry of Simple Molecules

In many cases, the overall shape of a molecule dictates how it behaves. In chemistry and in biology, function is based on structure.

The shapes of the simple molecules you've just learned about are governed by the geometry of their bonds. Nitrogen (N_2) and

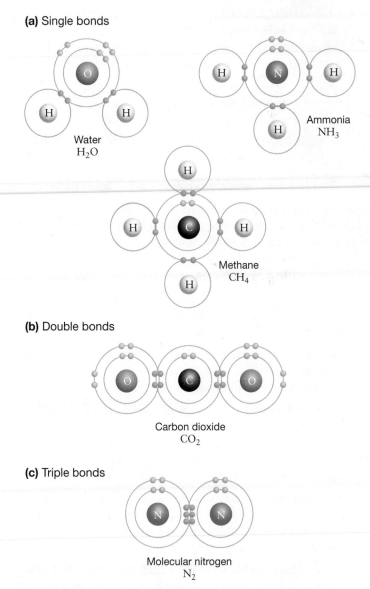

(a) Single bonds

Water H_2O

Ammonia NH_3

Methane CH_4

(b) Double bonds

Carbon dioxide CO_2

(c) Triple bonds

Molecular nitrogen N_2

FIGURE 2.8 Unpaired Electrons in the Valence Shell Participate in Covalent Bonds. Covalent bonding is based on sharing of electrons in the outermost shell. Covalent bonds can be **(a)** single, **(b)** double, or **(c)** triple.

(a) Methane (CH_4)   **(b)** Water (H_2O)

Unshared electrons

Shared electrons

FIGURE 2.9 The Geometry of Methane and Water.

carbon dioxide (CO_2), for example, have linear structures (see Figure 2.8). Molecules with more complex geometries include

- Methane (CH_4)—which is tetrahedral, a structure with four triangular faces like a pyramid (**FIGURE 2.9a**). The tetrahedron forms because the electrons in the four C—H bonds repulse each other equally. The electron pairs are as far apart as they can get.

- Water (H_2O)—which is bent and two-dimensional, or planar (**FIGURE 2.9b**). Why? The electrons in the four orbitals of oxygen's valence shell repulse each other, just as they do in methane. But in water, two of the orbitals are filled with electron pairs from the oxygen atom, and two are filled with electron pairs from covalent bonds between oxygen and hydrogen. The shared electrons form a molecule that is V-shaped and flat.

Section 2.2 explores how water's shape, in combination with the partial charges on the oxygen and hydrogen atoms, makes it the most important molecule on Earth.

Representing Molecules

Molecules can be represented in a variety of increasingly complex ways—only some of which reflect their actual shape. Each method has advantages and disadvantages.

- **Molecular formulas** are compact, but don't contain a great deal of information—they indicate only the numbers and types of atoms in a molecule (**FIGURE 2.10a**).

- **Structural formulas** indicate which atoms in a molecule are bonded together. Single, double, and triple bonds are represented by single, double, and triple dashes, respectively. Structural formulas also indicate geometry in two dimensions (**FIGURE 2.10b**). This method is useful for planar molecules such as water and O_2.

- **Ball-and-stick models** take up more space than structural formulas, but provide information on the three-dimensional shape of molecules and indicate the relative sizes of the atoms involved (**FIGURE 2.10c**).

- **Space-filling models** are more difficult to read than ball-and-stick models but more accurately depict the spatial relationships between atoms. (**FIGURE 2.10d**).

In both ball-and-stick and space-filling models, biologists use certain colors to represent certain atoms. A black ball, for

	Methane	Ammonia	Water	Oxygen
(a) Molecular formulas:	CH_4	NH_3	H_2O	O_2

(b) Structural formulas:

$$H-\overset{\displaystyle H}{\underset{\displaystyle H}{C}}-H \qquad H-\overset{\displaystyle H}{\underset{\displaystyle H}{N}} \qquad \overset{O}{H \quad H} \qquad O=O$$

(c) Ball-and-stick models:

(d) Space-filling models:

FIGURE 2.10 Molecules Can Be Represented Several Ways. Each method of representing a molecule has particular advantages.

example, always symbolizes carbon. (For more information on interpreting chemical structures, see **BioSkills 8** in Appendix B.)

Some of the small molecules you've just learned about are found in volcanic gases, the atmospheres of nearby planets, and in deep-sea hydrothermal vents, like those shown in the photograph at the start of this chapter. Based on these observations, researchers claim that they were important components of Earth's ancient atmosphere and oceans. If so, then they provided the building blocks for chemical evolution. The question is: How did these simple building blocks combine to form more complex products, early in Earth's history?

Researchers postulate that most of the critical reactions in chemical evolution occurred in an aqueous, or water-based, environment. To understand what happened and why, let's delve into the properties of water and then turn to analyzing the reactions that triggered chemical evolution.

check your understanding

If you understand that . . .

- Covalent bonds are based on electron sharing. Electron sharing allows atoms to fill all the orbitals in their valence shell, making them more stable.
- Covalent bonds can be polar or nonpolar, depending on whether the electronegativities of the two atoms involved are the same or different.

✔ **You should be able to . . .**

Draw the structural formula of formaldehyde (CH_2O) and add dots to indicate the relative locations of the electrons being shared in each covalent bond, based on the relative electronegativities of C, H, and O.

Answers are available in Appendix A.

2.2 Properties of Water and the Early Oceans

Life is based on water. It arose in an aqueous environment and remains dependent on water today. In fact, 75 percent of the volume in a typical cell is water; water is the most abundant molecule in organisms (**FIGURE 2.11**). You can survive for weeks without eating, but you aren't likely to live more than 3 or 4 days without drinking.

Water is vital for a simple reason: It is an excellent **solvent**—that is, an agent for dissolving substances and getting them into **solution.** The reactions that were responsible for chemical evolution some 3.5 billion years ago, like those occurring inside your body right now, depend on direct, physical interaction between molecules. Substances are most likely to come into contact with one another and react as **solutes**—meaning, when they are dissolved in a solvent like water. The formation of Earth's first ocean, about 3.8 billion years ago, was a turning point in chemical evolution because it gave the process a place to happen.

Why Is Water Such an Efficient Solvent?

To understand why water is such an effective solvent, recall that

1. Both of the O—H bonds in a water molecule are polar, owing to the difference in the electronegativities of hydrogen and oxygen. As a result, the oxygen atom has a partial negative charge and each hydrogen atom has a partial positive charge.

2. The molecule is bent. Consequently, the partial negative charge on the oxygen atom sticks out, away from the partial positive charges on the hydrogen atoms, giving a water molecule an overall polarity (**FIGURE 2.12a**).

FIGURE 2.12b illustrates how water's structure affects its interactions with other water molecules. When two water molecules approach each other, the partial positive charge on hydrogen attracts the partial negative charge on oxygen. This weak electrical attraction forms a **hydrogen bond** between the molecules.

✔ If you understand how water's structure makes hydrogen bonding possible, you should be able to (1) draw a fictional version of Figure 2.12b that shows water as a linear (not bent)

(a) Water is polar.

Electrons are pulled toward oxygen

(b) Hydrogen bonds form between water molecules.

FIGURE 2.12 Water Is Polar and Participates in Hydrogen Bonds. (a) Because of oxygen's high electronegativity, the electrons that are shared between hydrogen and oxygen spend more time close to the oxygen nucleus, giving the oxygen atom a partial negative charge and the hydrogen atom a partial positive charge. **(b)** The electrical attraction that occurs between the partial positive and negative charges on water molecules forms a hydrogen bond.

molecule with partial charges on the oxygen and hydrogen atoms; and (2) explain why electrostatic attractions between such water molecules would be much weaker as a result.

In an aqueous solution, hydrogen bonds also form between water molecules and other polar molecules. Similar interactions occur between water and ions. Ions and polar molecules stay in solution because of their interactions with water's partial charges (**FIGURE 2.13**). Substances that interact with water in this way are said to be **hydrophilic** ("water-loving"). Hydrogen bonding makes it possible for almost any charged or polar molecule to dissolve in water.

In contrast, compounds that are uncharged and nonpolar do not interact with water through hydrogen bonding and do not dissolve in water. Substances that do not interact with water are said to be **hydrophobic** ("water-fearing"). Because their interactions with water are minimal or nonexistent, they are forced to interact with each other (**FIGURE 2.14**, see page 26). The water molecules surrounding nonpolar molecules form hydrogen bonds with one another and increase the stability of these **hydrophobic interactions.**

Although individual hydrogen bonds are not as strong as covalent or ionic bonds, many of them occur in a solution. Hydrogen bonding is extremely important in biology owing to the

FIGURE 2.11 Fruits Shrink When They Are Dried Because They Consist Primarily of Water.

Salt in absence of water

Salt dissolved in water

FIGURE 2.13 Polar Molecules and Ions Dissolve Readily in Water. Water's polarity makes it a superb solvent for polar molecules and ions.

FIGURE 2.14 Nonpolar Molecules Do Not Dissolve in Water.
In aqueous solution, nonpolar molecules and compounds are forced to interact with each other. This occurs because water is much more stable when it interacts with itself rather than with the nonpolar molecules.

✔ **QUESTION** What is the physical basis of the expression, "Oil and water don't mix"?

(a) A meniscus forms where water meets a solid surface, as a result of two forces.

Adhesion: Water molecules that adhere to the glass pull upward at the perimeter.

Cohesion: Water molecules at the surface form hydrogen bonds with nearby water molecules and resist the upward pull of adhesion.

(b) Water has high surface tension.

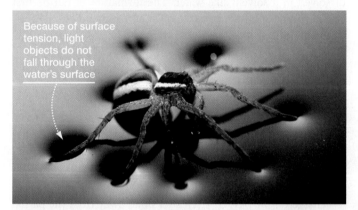

Because of surface tension, light objects do not fall through the water's surface

FIGURE 2.15 Cohesion, Adhesion, and Surface Tension.
(a) Meniscus formation is based on hydrogen bonding. **(b)** Water resists forces—like the weight of a spider—that increase its surface area. The resistance is great enough that light objects do not break the surface.

sheer number of hydrogen bonds that form between water and hydrophilic molecules.

What Properties Are Correlated with Water's Structure?

Water's small size, highly polar covalent bonds, and bent shape resulting in overall polarity are unique among molecules. Because the structure of molecules routinely correlates with their function, it's not surprising that water has some remarkable properties, in addition to its extraordinary capacity to act as a solvent.

Cohesion, Adhesion, and Surface Tension Attraction between like molecules is called **cohesion.** Water is cohesive—meaning that it stays together—because of the hydrogen bonds that form between individual molecules.

Attraction between unlike molecules, in contrast, is called **adhesion.** Adhesion is usually analyzed in regard to interactions between a liquid and a solid surface. Water adheres to surfaces that have any polar or charged components.

Cohesion and adhesion are important in explaining how water can move from the roots of plants to their leaves against the force of gravity (see Chapter 38). But you can also see them in action in the concave surface, or meniscus, that forms in a glass tube (**FIGURE 2.15a**). A meniscus forms as a result of two forces:

1. Water molecules at the perimeter of the surface adhere to the glass, resulting in an upward pull.

2. Water molecules at the surface hydrogen-bond with water molecules next to them and below them, resulting in a net lateral and downward pull that resists the upward pull of adhesion.

Cohesion is also instrumental in the phenomenon known as **surface tension.** When water molecules are at the surface, there

are no water molecules above them for hydrogen bonding. As a result, they exhibit stronger attractive forces between their nearest neighboring molecules. This enhanced attraction between the surface water molecules results in tension that minimizes the total surface area.

This fact has an important consequence: Water resists any force that increases its surface area. More specifically, any force that depresses a water surface meets with resistance. This resistance makes a water surface act like an elastic membrane (**FIGURE 2.15b**).

In water, the "elastic membrane" is stronger than it is in other liquids. Water's surface tension is extraordinarily high because of the stronger hydrogen bonding that occurs between molecules at the surface. This explains why it is better to cut the water's surface with your fingertips when you dive into a pool, instead of doing a belly flop.

Water Is Denser as a Liquid than as a Solid When factory workers pour molten metal or plastic into a mold and allow it to cool to the solid state, the material shrinks. When molten lava pours out of a volcano and cools to solid rock, it shrinks. But when you fill an ice tray with water and put it in the freezer to make ice, the water expands.

Unlike most substances, water is denser as a liquid than it is as a solid. In other words, there are more molecules of water in a given volume of liquid water than there are in the same volume of solid water, or ice. **FIGURE 2.16** illustrates why this is so.

Note that in ice, each water molecule participates in four hydrogen bonds. These hydrogen bonds cause the water molecules to form a regular and repeating structure, or crystal (Figure 2.16a). The crystal structure of ice is fairly open, meaning that there is a relatively large amount of space between molecules.

Now compare the extent of hydrogen bonding and the density of ice with that of liquid water, illustrated in Figure 2.16b. Note that the extent of hydrogen bonding in liquid water is much less than that found in ice, and that the hydrogen bonds in liquid water are constantly being formed and broken. As a result, molecules in the liquid phase are packed much more closely together than in the solid phase.

Normally, heating a substance causes it to expand because molecules begin moving faster and colliding more often and with greater force. But heating ice causes hydrogen bonds to break and the open crystal structure to collapse. In this way, hydrogen bonding explains why water is denser as a liquid than as a solid.

This property of water has an important result: Ice floats (Figure 2.16c). If it didn't, ice would sink to the bottom of lakes, ponds, and oceans soon after it formed. The ice would stay frozen in the cold depths. Instead, ice serves as a blanket, insulating the liquid below from the cold air above. If water weren't so unusual, it is almost certain that Earth's oceans would have frozen solid before life had a chance to start.

Water Has a High Capacity for Absorbing Energy Hydrogen bonding is also responsible for another of water's remarkable physical properties: Water has a high capacity for absorbing energy.

Specific heat, for example, is the amount of energy required to raise the temperature of 1 gram of a substance by 1°C. Water has a high specific heat because when a source of energy hits it, hydrogen bonds must be broken before heat can be transferred and the water molecules begin moving faster. As **TABLE 2.1** indicates, as molecules increase in overall polarity, and thus in their ability to form hydrogen bonds, it takes an extraordinarily large amount of energy to change their temperature.

TABLE 2.1 Specific Heats of Some Liquids

The specific heats reported in this table were measured at 25°C and are given in units of joules per gram of substance per degree Celsius. (The joule is a unit of energy.)

With extensive hydrogen bonding	Specific Heat
Water (H_2O)	4.18
With some hydrogen bonding	
Ethanol (C_2H_6O)	2.44
Glycerol ($C_3H_8O_3$)	2.38
With little or no hydrogen bonding	
Benzene (C_6H_6)	1.74
Xylene (C_8H_{10})	1.72

DATA: D. R. Lide (editor). 2008. Standard Thermodynamic Properties of Chemical Substances, in *CRC Handbook of Physics and Chemistry.* 89th ed. Boca Raton, FL: CRC Press.

(a) In ice, water molecules form a crystal lattice.

(b) In liquid water, no crystal lattice forms.

(c) Liquid water is denser than ice. As a result, ice floats.

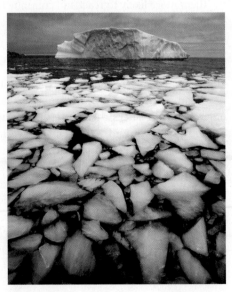

FIGURE 2.16 Hydrogen Bonding Forms the Crystal Structure of Ice. In ice, each molecule can form four hydrogen bonds at one time. Each oxygen atom can form two; each hydrogen atom can form one.

Property	Cause	Biological Consequences
Solvent for charged or polar compounds		Most chemical reactions important for life take place in aqueous solution.
Denser as a liquid than a solid	As water freezes, each molecule forms a total of four hydrogen bonds, leading to the formation of the low-density crystal structure called ice.	
High specific heat	Water molecules must absorb lots of heat energy to break hydrogen bonds and experience increased movement (and thus temperature).	Oceans absorb and release heat slowly, moderating coastal climates.
High heat of vaporization		Evaporation of water from an organism cools the body.

✔ **EXERCISE** You should be able to fill in the missing cells in this table.

Similarly, it takes a large amount of energy to break the hydrogen bonds in liquid water and change the molecules from the liquid phase to the gas phase. Water's **heat of vaporization**—the energy required to change 1 gram of it from a liquid to gas—is higher than that of most molecules that are liquid at room temperature. As a result, water has to absorb a great deal of energy to evaporate. Water's high heat of vaporization is the reason that sweating or dousing yourself with water is an effective way to cool off on a hot day. Water molecules absorb a great deal of energy from your body before they evaporate, so you lose heat.

Water's ability to absorb energy is critical to the theory of chemical evolution. Molecules that were formed in the ocean were well protected from sources of energy that could break them apart, such as intense sunlight. As a result, they would have persisted and slowly increased in concentration over time, making them more likely to react and continue the process.

TABLE 2.2 summarizes some of the key properties of water.

The Role of Water in Acid–Base Reactions

You've seen that water's high specific heat and heat of vaporization tend to keep its temperature and liquid form stable. One other aspect of water's chemistry is important for understanding chemical evolution and how organisms work: Water is not a completely stable molecule. In reality, water molecules continually undergo a chemical reaction with themselves. When a **chemical reaction** occurs, one substance is combined with others or broken down into another substance. Atoms may also be rearranged; in most cases, chemical bonds are broken and new bonds form. The chemical reaction that takes place between water molecules is called a "dissociation" reaction. It can be written as follows:

$$H_2O \rightleftharpoons H^+ + OH^-$$

The double arrow indicates that the reaction proceeds in both directions.

The substances on the right-hand side of the expression are the **hydrogen ion** (H^+) and the **hydroxide ion** (OH^-). A hydrogen ion is simply a proton. In reality, however, protons do not exist by themselves. In water, for example, protons associate with water molecules to form hydronium ions (H_3O^+). Thus, the dissociation of water is more accurately written as:

$$H_2O + H_2O \rightleftharpoons H_3O^+ + OH^-$$

One of the water molecules on the left-hand side of the expression has given up a proton, while the other water molecule has accepted a proton.

Substances that give up protons during chemical reactions and raise the hydronium ion concentration of water are called **acids;** molecules or ions that acquire protons during chemical reactions and lower the hydronium ion concentration of water are called **bases.** Most acids act only as acids, and most bases act only as bases; but water can act as both an acid and a base.

A chemical reaction that involves a transfer of protons is called an acid–base reaction. Every acid–base reaction requires a proton donor and a proton acceptor—an acid and a base, respectively.

Water is an extremely weak acid—very few water molecules dissociate to form hydronium ions and hydroxide ions. In contrast, strong acids like the hydrochloric acid (HCl) in your stomach readily give up a proton when they react with water.

$$HCl + H_2O \rightleftharpoons H_3O^+ + Cl^-$$

Strong bases readily acquire protons when they react with water. For example, sodium hydroxide (NaOH, commonly called lye) dissociates completely in water to form Na^+ and OH^-. The hydroxide ion produced by that reaction then accepts a proton from a hydronium ion in the water, forming two water molecules.

$$NaOH(aq) \longrightarrow Na^+ + OH^-$$
$$OH^- + H_3O^+ \rightleftharpoons 2 H_2O$$

(The "aq" in the first expression indicates that NaOH is in aqueous solution.)

To summarize, adding an acid to a solution increases the concentration of protons; adding a base to a solution lowers the concentration of protons. Water is both a weak acid and a weak base.

Determining the Concentration of Protons

In a solution, the tendency for acid–base reactions to occur is largely a function of the number of protons present. The problem is, there's no simple way to count the actual number of protons present in a sample. Researchers solve this problem using the mole concept.

A **mole** refers to the number 6.022×10^{23}—just as the unit called the dozen refers to the number 12 or the unit million refers to the number 1×10^6. The mole is a useful unit because the mass of one mole of any substance is the same as its molecular weight expressed in grams. **Molecular weight** is the sum of the atomic weights of all the atoms in a molecule.

For example, to get the molecular weight of H_2O, you sum the atomic weights of two atoms of hydrogen and one atom of oxygen. Since the atomic weights of hydrogen and oxygen are very close to their mass numbers (see Figure 2.2), the molecular weight of water would be $1 + 1 + 16$, or a total of 18. Thus, if you weighed a sample of 18 grams of water, it would contain around 6×10^{23} water molecules, or about 1 mole of water molecules.

When substances are dissolved in water, their concentration is expressed in terms of molarity (symbolized by "M"). **Molarity** is the number of moles of the substance present per liter of solution. A 1-molar solution of protons in water, for example, means that 1 mole of protons is contained in 1 liter of solution.

Chemists can measure the concentration of protons in a solution directly using molarity and an instrument called a pH meter. In a sample of pure water at 25°C, the concentration of H^+ is $1.0 \times 10^{-7}M$, or 1 ten-millionth molar.

The pH of a Solution Reveals Whether It Is Acidic or Basic

Because the concentration of protons in water is such a small number, exponential notation is cumbersome. So chemists and biologists prefer to express the concentration of protons in a solution, and thus whether it is acidic or basic, with a logarithmic notation called **pH**.[1]

By definition, the pH of a solution is the negative of the base-10 logarithm, or log, of the hydrogen ion concentration:

$$pH = -\log[H^+]$$

(To review logarithms, see **BioSkills 6** in Appendix B. The square brackets are a standard notation for indicating "concentration" of a substance in solution.)

Taking antilogs gives

$$[H^+] = \text{antilog}(-pH) = 10^{-pH}$$

Solutions that contain acids have a proton concentration larger than $1 \times 10^{-7}M$ and thus a pH < 7. This is because acidic

molecules tend to release protons into solution. In contrast, solutions that contain bases have a proton concentration less than $1 \times 10^{-7}M$ and thus a pH > 7. This is because basic molecules tend to accept protons from solution.

pH is a convenient way to indicate the concentration of protons in a solution, but take note of what the number represents. For example, if the concentration of H^+ in a sample of water is $1.0 \times 10^{-7}M$, then its pH is 7. If the pH changes to 5, then the sample contains 100 times more protons and has become 100 times more acidic. ✔ QUANTITATIVE If you understand how pH is related to $[H^+]$, you should be able to calculate the concentration of protons in a solution that has a pH of 8.5.

FIGURE 2.17 shows the pH scale and reports the pH of some selected solutions. Pure water is used as a standard, or point of reference, for pH 7 on the pH scale. The solution inside living cells is about pH 7, which is considered neutral—neither acidic nor basic. The normal function of a cell is dependent on maintaining this neutral internal environment. What is responsible for regulating pH?

FIGURE 2.17 The pH Scale. Because the pH scale is logarithmic, a change in one unit of pH represents a change in the concentration of hydrogen ions equal to a factor of 10. Coffee has a hundred times more H^+ than pure water has.

✔ QUESTION What happens to the concentration of protons in black coffee after you add milk?

---

[1] The term pH is derived from the French *puissance d'hydrogéne*, or "power of hydrogen."

Buffers Protect Against Damaging Changes in pH Life is sensitive to changes in pH. Changes in proton concentration affect the structure and function of polar or charged substances as well as the tendency of acid–base reactions to occur.

Compounds that minimize changes in pH are called **buffers** because they reduce the impact of adding acids or bases on the overall pH of a solution. Buffers are important in maintaining relatively constant conditions, or **homeostasis,** in cells and tissues. In cells, a wide array of naturally occurring molecules act as buffers.

Most buffers are weak acids, meaning that they are somewhat likely to give up a proton in solution, but once the proton concentration rises, the acid is regenerated. To see how buffers work, consider the disassociation of carbonic acid in water to form bicarbonate ions and protons:

$$CH_2O_3 \rightleftharpoons CHO_3^- + H^+$$
$$\text{carbonic acid} \qquad \text{bicarbonate}$$

When carbonic acid and bicarbonate are present in about equal concentrations in a solution, they function as a buffering system. If the concentration of protons increases slightly, the protons react with bicarbonate ions to form carbonic acid and pH does not change. If the concentration of protons decreases slightly, carbonic acid gives up protons and pH does not change.
✔ If you understand this concept, you should be able to predict what would happen to the concentration of bicarbonate ions if a base like sodium hydroxide (NaOH) were added to the solution of carbonic acid.

As chemical evolution began, then, water provided the physical environment for key reactions to take place. In some cases water also acted as an important reactant. Although acid–base reactions were not critical to the initial stages of chemical evolution, they became extremely important once the process was under way. Now let's consider what happened in solution, some 3.5 billion years ago.

2.3 Chemical Reactions, Energy, and Chemical Evolution

Proponents of the theory of chemical evolution contend that simple molecules present in the atmosphere and oceans of early Earth participated in chemical reactions that eventually produced larger, more complex organic (carbon-containing) molecules—such as the proteins, nucleic acids, sugars, and lipids introduced in the next four chapters. Currently, researchers are investigating two environments where these reactions may have occurred:

1. *The atmosphere*, which was probably dominated by gases ejected from volcanoes. Water vapor, carbon dioxide (CO_2), and nitrogen (N_2) are the dominant gases ejected by volcanoes today; a small amount of molecular hydrogen (H_2) and carbon monoxide (CO) may also be present.

2. *Deep-sea hydrothermal vents*, where extremely hot rocks contact deep cracks in the seafloor. In addition to gases such as CO_2 and H_2, certain deep-sea vents are rich in minerals containing reactive metals such as nickel and iron.

When gases like CO_2, N_2, H_2, and CO are put together and allowed to interact on their own, however, very little happens. They do not suddenly link together to create large, complex substances like those found in living cells. Instead, their bonds remain intact. To understand why the bonds of these molecules remain unchanged, you must first learn about how chemical reactions proceed.

How Do Chemical Reactions Happen?

Chemical reactions are written in a format similar to mathematical equations: The initial, or **reactant,** molecules are shown on the left and the resulting reaction **product(s)** shown on the right. For example, the most common reaction in the mix of gases and water vapor that emerges from volcanoes results in the production of carbonic acid, which can be precipitated with water as acid rain:

$$CO_2(g) + H_2O(l) \rightleftharpoons CH_2O_3(aq)$$
$$\text{carbonic acid}$$

The physical state of each reactant and product is indicated as gas (*g*), liquid (*l*), solid (*s*), or in aqueous solution (*aq*).

Note that the expression is balanced; that is, 1 carbon, 3 oxygen, and 2 hydrogen atoms are present on each side of the expression. This illustrates the conservation of mass in closed systems—mass cannot be created or destroyed, but it may be rearranged through chemical reactions.

Note also that the expression contains a double arrow, meaning that the reaction is reversible. When the forward and reverse reactions proceed at the same rate, the quantities of reactants and products remain constant, although not necessarily equal. A dynamic but stable state such as this is termed a **chemical equilibrium.**

Changing the concentration of reactants or products can disturb a chemical equilibrium. For example, adding more CO_2 to the mixture would drive the reaction to the right, creating more CH_2O_3 until the equilibrium proportions of reactants and products are reestablished. Removing CO_2 or adding more CH_2O_3 would drive the reaction to the left.

A chemical equilibrium can also be altered by changes in temperature. For example, the water molecules in the following set of interacting elements, or **system,** would be present as a combination of liquid water and water vapor:

$$H_2O(l) \rightleftharpoons H_2O(g)$$

If liquid water molecules absorb enough energy, like the heat released from a volcano, they transform to the gaseous state. (You may recall that water has a high heat of vaporization and requires a large amount of energy to change its state from liquid to gas.) As a result, this change is termed **endothermic** ("within heating") because heat is absorbed during the process. In contrast, the transformation of water vapor to liquid water releases heat and is **exothermic** ("outside heating"). Raising the temperature of this system drives the equilibrium to the right; cooling the system drives it to the left.

In relation to chemical evolution, though, these reactions and changes of physical state are not particularly interesting. Carbonic acid is not an important intermediate in the formation of more complex molecules. However, interesting things do begin to happen when energy is added to mixtures of volcanic gases.

What Is Energy?

Energy can be defined as the capacity to do work or to supply heat. This capacity exists in one of two ways—as a stored potential or as an active motion.

Stored energy is called **potential energy.** An object gains or loses its ability to store energy because of its position. An electron that resides in an outer electron shell will, if the opportunity arises, fall into a lower electron shell closer to the positive charges on the protons in the nucleus. Because of its position farther from the positive charges in the nucleus, an electron in an outer electron shell has more potential energy than does an electron in an inner shell (**FIGURE 2.18**). When stored in chemical bonds, this form of potential energy is called **chemical energy.**

Kinetic energy is energy of motion. Molecules have kinetic energy because they are constantly in motion.

- The kinetic energy of molecular motion is called **thermal energy.**

- The **temperature** of an object is a measure of how much thermal energy its molecules possess. If an object has a low temperature, its molecules are moving slowly. (We perceive this as "cold.") If an object has a high temperature, its molecules are moving rapidly. (We perceive this as "hot.")

- When two objects with different temperatures come into contact, thermal energy is transferred between them. This transferred energy is called **heat.**

There are many forms of potential energy and kinetic energy, and energy can change from one form into another. However, according to the **first law of thermodynamics,** energy is conserved—it cannot be created or destroyed, but only transferred and transformed. (A more thorough explanation of energy transformation is provided in Chapter 8 in the context of cellular metabolism.)

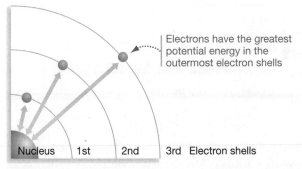

FIGURE 2.18 Potential Energy as a Function of Electron Shells.
Electrons in outer shells have more potential energy than do electrons in inner shells, because the negative charges of the electrons in outer shells are farther from the positive charges of the protons in the nucleus. Each shell represents a distinct level of potential energy.

Energy transformation is the heart of chemical evolution. According to the best data available, molecules that were part of the early Earth were exposed to massive inputs of energy. Kinetic energy, in the form of heat, was present in the gradually cooling molten mass that initially formed the planet. The atmosphere and surface of the early Earth were also bombarded with electricity from lightening and radiation from the Sun. Energy stored as potential energy in the chemical bonds of small molecules was also abundant.

Now that you understand that energy transformations are involved in chemical reactions, a big question remains: What determines if a reaction will take place?

What Makes a Chemical Reaction Spontaneous?

When chemists say that a reaction is spontaneous, they have a precise meaning in mind: Chemical reactions are spontaneous if they are able to proceed on their own, without any continuous external influence, such as added energy. Two factors determine if a reaction will proceed spontaneously:

1. Reactions tend to be spontaneous when the product molecules are less ordered than the reactant molecules. For example, nitroglycerin is a single, highly ordered molecule. But when nitroglycerin explodes, it breaks up into gases like carbon dioxide, nitrogen, oxygen, and water vapor. These molecules are much less ordered than the reactant nitroglycerin molecules. The heat that is given off from this explosion also contributes to increasing disorder in the environment. The amount of disorder in a system is called **entropy.** When the products of a chemical reaction are less ordered than the reactant molecules are, entropy increases and the reaction tends to be spontaneous. The **second law of thermodynamics,** in fact, states that entropy always increases in an isolated system.

2. Reactions tend to be spontaneous if the products have lower potential energy than the reactants. If the electrons in the reaction products are held more tightly than those in the reactants, then they have lower potential energy. Recall that highly electronegative atoms such as oxygen or nitrogen hold electrons in covalent bonds much more tightly than do atoms with a lower electronegativity, such as hydrogen or carbon. For example, when hydrogen and oxygen gases react, water is produced spontaneously:

$$2\,H_2(g) + O_2(g) \longrightarrow 2\,H_2O(g)$$

The electrons involved in the O—H bonds of water are held much more tightly by the more electronegative oxygen atom than when they were shared equally in the H—H and O=O bonds of hydrogen and oxygen (see **FIGURE 2.19a** on page 32). As a result, the products have much lower potential energy than the reactants. The difference in chemical energy between reactants and products is given off as heat, so the reaction is exothermic. And although the reaction between hydrogen and oxygen results in less entropy—three molecules of gas produce two molecules of water vapor—the reaction is still spontaneous due to the large drop in potential energy

(a) When hydrogen and oxygen gas react, the product has much lower potential energy than the reactants.

(b) The difference in potential energy is released as heat and light, which vaporizes the water produced.

Electrons are held "loosely" in bonds between atoms with equal electronegativities

Electrons are held tightly by highly electronegative atoms

H—H + O—O → Potential energy drops → H—O—H + Released energy

2 Hydrogens (H_2) 1 Oxygen (O_2) 2 Waters (H_2O)

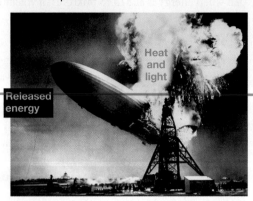

Heat and light

FIGURE 2.19 Potential Energy May Change during Chemical Reactions. In the Hindenburg disaster of 1937, the hydrogen gas from this lighter-than-air craft reacted with oxygen in the atmosphere, with devastating results.

✔**EXERCISE** Label which electrons have relatively low potential energy and which electrons have relatively high potential energy.

released as heat. Since heat increases disorder in the environment, the second law of thermodynamics remains intact. The Hindenburg disaster of 1937 illustrates the large and terrifying amount of heat energy that is given off from this relatively simple reaction (**FIGURE 2.19b**).

To summarize: In general, physical and chemical processes proceed in the direction that results in increased entropy and lower potential energy (**FIGURE 2.20**). These two factors—potential energy and entropy—are used to figure out whether a reaction is spontaneous (see Chapter 8 for more detail). Were the reactions that led to chemical evolution spontaneous? Section 2.4 explores how researchers address this question.

Reactants:
• high potential energy
• more order (lower entropy)

$C_6H_{12}O_6$ + 6 O_2
Glucose (a sugar)

This reaction occurs in your cells and when wood burns

Products:
• low potential energy
• less order (higher entropy)

+ Released heat

6 CO_2 + 6 H_2O

FIGURE 2.20 Spontaneous Processes Result in Lower Potential Energy, Increased Disorder, or Both.

check your understanding

If you understand that . . .

C Y U

• Chemical reactions result in the transformation of energy, either through the release of energy stored in chemical bonds or the uptake of energy from external sources.

• Chemical reactions tend to be spontaneous if they lead to lower potential energy and higher entropy (more disorder).

✔ **You should be able to . . .**

1. Determine if the reaction between methane (CH_4) and oxygen (O_2) shown here is spontaneous or not, addressing both potential energy and entropy:

$$CH_4 + 2\,O_2 \longrightarrow CO_2 + 2\,H_2O$$

2. Explain how the positions of the valence electrons in carbon and hydrogen change as methane is converted into carbon dioxide and water.

Answers are available in Appendix A.

2.4 Investigating Chemical Evolution: Approaches and Model Systems

To probe the kinds of reactions that may have set chemical evolution in motion, researchers have used two different approaches—one looking from the "top down" and the other from the "bottom up."

1. In the top-down approach, researchers examine modern cells to identify chemistry that is shared throughout the tree of life. Such ancient reactions are prime candidates for being involved in the chemical evolution that led up to **LUCA,** or last universal common ancestor (introduced in Chapter 1).

2. In the bottom-up approach, the primary focus is on the small molecules and environmental conditions that were present in

early Earth. Here, researchers attempt to identify reactions that could build the molecules found in life using only what was available at the time, without regard to reactions used by modern cells.

These approaches have been used to investigate two different model systems that attempt to explain the process component of the theory of chemical evolution:

1. The **prebiotic soup model** proposes that certain molecules were synthesized from gases in the atmosphere or arrived via meteorites. Afterward they would have condensed with rain and accumulated in oceans. This process would result in an "organic soup" that allowed for continued construction of larger, even more complex molecules.

2. The **surface metabolism model** suggests that dissolved gases came in contact with minerals lining the walls of deep-sea vents and formed more complex, organic molecules.

Since it is impossible to directly examine how and where chemical evolution occurred, the next best thing is to re-create the conditions in the lab and test predictions made by these models. In the following sections, you will learn about how biologists used the top-down and bottom-up approaches to identify reactions that support each of these models for chemical evolution.

Early Origin-of-Life Experiments

Chemical evolution was first taken seriously in 1953 when a graduate student named Stanley Miller performed a breakthrough experiment in the study of the prebiotic soup model.

Miller wanted to answer a simple question: Can complex organic compounds be synthesized from the simple molecules present in Earth's early atmosphere? In other words, is it possible to re-create the first steps in chemical evolution by simulating early-Earth conditions in the laboratory?

Miller's experimental setup (**FIGURE 2.21**) was designed to produce a microcosm of early Earth. The large glass flask represented the atmosphere and contained the gases methane (CH_4), ammonia (NH_3), and hydrogen (H_2), all of which have high potential energy. This large flask was connected to a smaller flask by glass tubing. The small flask held a tiny ocean—200 milliliters (mL) of liquid water.

To connect the mini-atmosphere with the mini-ocean, Miller boiled the water constantly. This added water vapor to the mix of gases in the large flask. As the vapor cooled and condensed, it flowed back into the smaller flask, where it boiled again. In this way, water vapor circulated continuously through the system. This was important: If the molecules in the simulated atmosphere reacted with one another, the "rain" would carry them into the mini-ocean, forming a simulated version of the prebiotic soup.

Had Miller stopped at merely boiling the molecules, little or nothing would have happened. Even at the boiling point of water (100°C), the starting molecules used in the experiment are stable and do not undergo spontaneous chemical reactions.

Something did start to happen in the apparatus, however, when Miller sent electrical discharges across the electrodes he'd inserted into the atmosphere. These miniature lightning bolts

RESEARCH

QUESTION: Can simple molecules and kinetic energy lead to chemical evolution?

HYPOTHESIS: If kinetic energy is added to a mix of simple molecules, reactions will occur that produce more complex molecules, perhaps including some with C–C bonds.

NULL HYPOTHESIS: Chemical evolution will not occur, even with an input of energy.

EXPERIMENTAL SETUP:

PREDICTION: Complex organic compounds will be found in the liquid water.

PREDICTION OF NULL HYPOTHESIS: Only the starting molecules will be found in the liquid water.

RESULTS

Samples taken from the liquid water contain formaldehyde, hydrogen cyanide, and several complex compounds with carbon–carbon bonds, including amino acids

CONCLUSION: Chemical evolution occurs readily if simple molecules with high free energy are exposed to a source of kinetic energy.

FIGURE 2.21 Miller's Spark-Discharge Experiment. The arrows in the "Experimental setup" diagram indicate the flow of water vapor or liquid. The condenser is a jacket with cold water flowing through it.

SOURCE: Miller, S. L. 1953. A production of amino acids under possible primitive Earth conditions. *Science* 117: 528–529.

✔**QUESTION** Which parts of the apparatus mimic the ocean, atmosphere, rain, and lightning?

added a crucial element to the reaction mix—pulses of intense electrical energy. After a day of continuous boiling and sparking, the solution in the boiling flask began to turn pink. After a week, it was deep red and cloudy.

When Miller analyzed samples from the mini-ocean, he found large quantities of hydrogen cyanide and formaldehyde. Even more exciting, the sparks and heating had led to the synthesis of additional, more complex organic compounds, including amino acids, which are the building blocks of proteins (see Chapter 3).

Recent Origin-of-Life Experiments

The production of more complex molecules from simple molecules in Miller's experiment supported his claim that the formation of a prebiotic soup was possible. The results came under fire, however, when other researchers pointed out that the early atmosphere was dominated by volcanic gases like CO, CO_2, and H_2, not the CH_4 and NH_3 used in Miller's experiment.

This controversy stimulated a series of follow-up experiments, which showed that the assembly of small molecules into more complex molecules can also occur under more realistic early Earth conditions.

Synthesis of Precursors Using Light Energy One such reaction that may have played a role in chemical evolution is the synthesis of formaldehyde (CH_2O) from carbon dioxide and hydrogen:

$$CO_2(g) + 2\,H_2(g) \longrightarrow \underset{\text{formaldehyde}}{CH_2O(g)} + H_2O(g)$$

This reaction has not been observed in cells—like Miller's experiment, it represents the bottom-up approach. But researchers have shown that when molecules of formaldehyde are heated, they react with one another to produce larger organic compounds, including energy-rich molecules like sugars (see Chapter 5). Note, however, that this reaction does not occur spontaneously—a large input of energy is required.

To explore the possibility of early formaldehyde synthesis, a research group constructed a computer model of the early atmosphere of Earth. The model consisted of a list of all possible chemical reactions that can occur among the molecules now thought to have dominated the early atmosphere: CO_2, H_2O, N_2, CO, and H_2. In this model, they included reactions that occur when these molecules are struck by sunlight. This was crucial because sunlight represents a source of energy.

The sunlight that strikes Earth is made up of packets of light energy called **photons.** Today, Earth is protected by a blanket of ozone (O_3) in the upper atmosphere that absorbs most of the higher-energy photons in sunlight. But since Earth's early atmosphere was filled with volcanic gases released as the molten planet cooled, and ozone is not among these gases, it is extremely unlikely that appreciable quantities of ozone existed. Based on this logic, researchers infer that when chemical evolution was occurring, large quantities of high-energy photons bombarded the planet.

To understand why this energy source was so important, recall that the atoms in hydrogen and carbon dioxide molecules have full valence shells through covalent bonding. This arrangement makes these molecules largely unreactive. However, energy from photons can break up molecules by knocking apart shared electrons. The fragments that result, called **free radicals,** have unpaired electrons in their outermost shells and are extremely reactive (**FIGURE 2.22**). To mimic the conditions on early Earth more accurately, the computer model included several reactions that produce highly reactive free radicals.

The result? The researchers calculated that, under conditions accepted as reasonable approximations of early Earth by most scientists, appreciable quantities of formaldehyde would have been produced. The energy in sunlight was converted to chemical energy in the form of new bonds in formaldehyde.

The complete reaction that results in the formation of formaldehyde is written as

$$CO_2(g) + 2\,H_2(g) + \text{sunlight} \longrightarrow CH_2O(g) + H_2O(g)$$

Notice that the reaction is balanced in terms of the atoms *and* the energy involved. The sunlight on the reactant side balances the higher energy required for the formation of formaldehyde and water. This result makes sense if you take a moment to think about it. Energy is the capacity to do work, and building larger, more complex molecules requires work to be done.

FIGURE 2.22 Free Radicals Are Extremely Reactive. When high-energy photons or pulses of intense electrical energy, such as lightning, strike molecules of hydrogen or carbon dioxide, free radicals can be created. Formation of free radicals is thought to be responsible for some key reactions in chemical evolution.

Using a similar model, other researchers have shown that hydrogen cyanide (HCN)—another important precursor of molecules required for life—could also have been produced in the early atmosphere. According to this research, large quantities of potential precursors for chemical evolution would have formed in the atmosphere and rained out into the early oceans. As a result, organic compounds with relatively high potential energy could have accumulated, and the groundwork would have been in place for the prebiotic soup model of chemical evolution to take off (**FIGURE 2.23a**).

Concentration and Catalysis in Hydrothermal Vents A major stumbling block in the prebiotic soup model is that precursor molecules would have become diluted when they entered the early oceans. Without some means of localized concentration, the formaldehyde and hydrogen cyanide mentioned in the previous section would have been unlikely to meet and react to form larger, more complex molecules. The surface metabolism model offers one possible solution to this dilution effect.

In the surface metabolism model, reactants are recruited to a defined space—a layer of reactive minerals deposited on the

(a) PROCESS: PREBIOTIC SOUP MODEL OF CHEMICAL EVOLUTION

1. Simple molecules were present in the atmosphere of ancient Earth.

2. Energy in sunlight drove reactions among the simple molecules.

3. Stimulated by heat, the products formed more complex molecules.

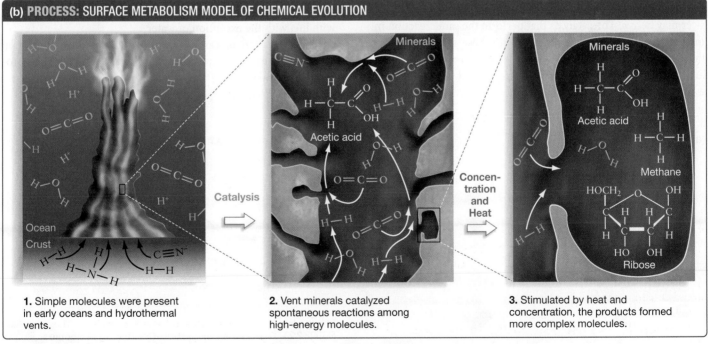

(b) PROCESS: SURFACE METABOLISM MODEL OF CHEMICAL EVOLUTION

1. Simple molecules were present in early oceans and hydrothermal vents.

2. Vent minerals catalyzed spontaneous reactions among high-energy molecules.

3. Stimulated by heat and concentration, the products formed more complex molecules.

FIGURE 2.23 The Start of Chemical Evolution—Two Models. The prebiotic soup and surface metabolism models illustrate how simple molecules containing C, H, O, and N reacted to form organic compounds that served as building blocks for more complex molecules.

walls of deep-sea vent chimneys. Dissolved gases would be attracted by the minerals and concentrated on vent-wall surfaces (**FIGURE 2.23b**).

Here's a key point of this model: Not only would vent-wall minerals bring reactants together, they would also be critical to the rate at which reaction products formed. Even if a potential reaction were spontaneous, it would probably not occur at a level useful for chemical evolution without the support of a **catalyst.** A catalyst provides the appropriate chemical environment for reactants to interact with one another effectively. (You will learn in Chapter 8 that a catalyst only influences the rate of a reaction—it does not provide energy or alter spontaneity.)

A reaction that provides an example of the role catalysts may have played during chemical evolution is the synthesis of acetic acid (CH_3COOH) from carbon dioxide and hydrogen:

$$2 CO_2(aq) + 4 H_2(aq) \longrightarrow CH_3COOH(aq) + 2 H_2O(l)$$
acetic acid

The reaction is driven by chemical energy stored in one of the reactants—H_2—and is spontaneous despite the apparent decrease in entropy. It is employed by certain groups of Bacteria and Archaea today as a step toward building even more complex organic molecules.

This reaction has grabbed wide attention among the chemical evolution research community, for two reasons in particular: (**1**) Acetic acid can be formed under conditions that simulate a hydrothermal vent environment (bottom-up approach). (**2**) It is a key intermediate in an ancient pathway that produces acetyl CoA, which is a molecule used by cells throughout the tree of life (top-down approach). (The role of acetyl CoA in modern cells is discussed in Chapter 9.)

Did vent minerals serve as catalysts in the synthesis of acetic acid in early Earth? Evidence from modern cells suggests the answer may be yes. The catalysts that perform the same reaction in modern cells contain minerals similar to those found in hydrothermal vents. These minerals may represent a form of molecular luggage taken from the deep-sea hydrothermal vents as LUCA evolved its independence.

Research is currently under way to establish laboratory systems to more closely mimic surface metabolism conditions in hydrothermal vents. Preliminary results show that in addition to the production of acetic acid, a variety of larger carbon-based molecules can be formed under early Earth conditions. Among these are precursors for the synthesis of nucleotides, the building blocks for the molecules of inheritance used by every living organism on Earth (see Chapter 4).

2.5 The Importance of Organic Molecules

Life has been called a carbon-based phenomenon, and with good reason. Except for water, almost all of the molecules found in organisms contain this atom. Molecules that contain carbon bonded to other elements, such as hydrogen, are called **organic** molecules. (Other types of molecules are referred to as *inorganic* compounds.)

Carbon has great importance in biology because it is the most versatile atom on Earth. Because of its four valence electrons, it will form four covalent bonds. This results in an almost limitless array of molecular shapes, made possible by different combinations of single and double bonds.

Linking Carbon Atoms Together

You have already examined the tetrahedral structure of methane and the linear shape of carbon dioxide. When molecules contain more than one carbon atom, these atoms can be bonded to one another in long chains, as in the component of gasoline called octane (C_8H_{18}; **FIGURE 2.24a**), or in a ring, as in the sugar glucose ($C_6H_{12}O_6$; **FIGURE 2.24b**). Carbon atoms provide the structural framework for virtually all the important compounds associated with life, with the exception of water.

The formation of carbon–carbon bonds was an important event in chemical evolution: It represented a crucial step toward the production of the types of molecules found in living organisms.

(a) Carbons linked in a chain

C_8H_{18} Octane

(b) Carbons linked in a ring

$C_6H_{12}O_6$ Glucose

FIGURE 2.24 The Shapes of Carbon-Containing Molecules.
(a) Octane is a hydrocarbon chain, and one of the primary ingredients in gasoline. **(b)** Glucose is a sugar that can form a ring-like structure.

Functional Groups

In general, the carbon atoms in an organic molecule furnish a skeleton that gives the molecule its overall shape. But the chemical behavior of the compound—meaning the types of reactions that it participates in—is dictated by groups of H, N, O, P, or S atoms that are bonded to one of the carbon atoms in a specific way.

The critically important H-, N-, O-, P-, and S-containing groups found in organic compounds are called **functional groups.** The composition and properties of six prominent functional groups that are commonly found in organic molecules and recognized by organic chemists are summarized in **TABLE 2.3**. To understand the role that organic compounds play in organisms, it is important to analyze how these functional groups behave.

SUMMARY TABLE 2.3 **Six Functional Groups Commonly Attached to Carbon Atoms**

Functional Group	Formula*	Family of Molecules	Properties of Functional Group	Example
Amino		Amines	Acts as a base—tends to attract a proton to form:	Glycine (an amino acid)
Carboxyl		Carboxylic acids	Acts as an acid—tends to lose a proton in solution to form:	Acetic acid
Carbonyl		Aldehydes	Aldehydes, especially, react with certain compounds to produce larger molecules to form:	Acetaldehyde
		Ketones		Acetone
Hydroxyl	R—OH	Alcohols	Highly polar, so makes compounds more soluble through hydrogen bonding with water; may also act as a weak acid and drop a proton	Ethanol
Phosphate		Organic phosphates	Molecules with more than one phosphate linked together store large amounts of chemical energy	3–Phosphoglyceric acid
Sulfhydryl	R—SH	Thiols	When present in proteins, can form disulfide (S–S) bonds that contribute to protein structure	Cysteine

*In these structural formulas, "R" stands for the rest of the molecule.

✔**EXERCISE** Based on the electronegativities of the atoms involved, predict whether each functional group is polar or nonpolar.

- *Amino and carboxyl functional groups* tend to attract or drop a proton, respectively, when in solution. Amino groups function as bases; carboxyl groups act as acids. During chemical evolution and in organisms today, the most important types of amino- and carboxyl-containing molecules are the amino acids (which Chapter 3 analyzes in detail). Amino acids contain both an amino group and a carboxyl group. (It's common for organic compounds to contain more than one functional group.) Amino acids can be linked together by covalent bonds that form between amino and carboxyl groups. In addition, both of these functional groups participate in hydrogen bonding.

- *Carbonyl groups* are found on aldehyde and ketone molecules such as formaldehyde, acetaldehyde, and acetone. This functional group is the site of reactions that link these molecules into larger, more complex organic compounds.

- *Hydroxyl groups* are important because they act as weak acids. In many cases, the protons involved in acid–base reactions that occur in cells come from hydroxyl groups on organic compounds. Because hydroxyl groups are polar, molecules containing hydroxyl groups will form hydrogen bonds and tend to be soluble in water.

- *Phosphate groups* carry two negative charges. When phosphate groups are transferred from one organic compound to another, the change in charge often dramatically affects the structure of the recipient molecule. In addition, phosphates that are bonded together store chemical energy that can be used in chemical reactions (some of these are discussed in Chapter 3).

- *Sulfhydryl groups* consist of a sulfur atom bonded to a hydrogen atom. They are important because sulfhydryl groups can link to one another via disulfide (S–S) bonds.

To summarize, functional groups make things happen. The number and types of functional groups attached to a framework of carbon atoms imply a great deal about how that molecule is going to behave.

When you encounter an organic compound that is new to you, it's important to do the following three things:

1. Examine the overall size and shape provided by the carbon framework.

2. Identify the types of covalent bonds present based on the electronegativities of the atoms. Use this information to estimate the polarity of the molecule and the amount of potential energy stored in its chemical bonds.

3. Locate any functional groups and note the properties these groups give to the molecule.

Understanding these three features will help you to predict the molecule's role in the chemistry of life.

Once carbon-containing molecules with functional groups had appeared early in Earth's history, what happened next? For chemical evolution to continue, small carbon-based molecules had to form still larger, more complex molecules like those found in living cells. How were the molecules of life—proteins, nucleic acids, carbohydrates, and lipids—formed, and how do they function in organisms today? The rest of this unit explores the next steps in chemical evolution, culminating in the formation of the first living cell.

CHAPTER 2 REVIEW

For media, go to MasteringBiology

If you understand . . .

2.1 Atoms, Ions, and Molecules: The Building Blocks of Chemical Evolution

- When atoms participate in chemical bonds to form molecules, the shared electrons give the atoms full valence shells and thus contribute to the atoms' stability.

- The electrons in a chemical bond may be shared equally or unequally, depending on the relative electronegativities of the two atoms involved.

- Nonpolar covalent bonds result from equal sharing; polar covalent bonds are due to unequal sharing. Ionic bonds form when an electron is completely transferred from one atom to another.

✔ You should be able to compare and contrast the types of bonds found in methane (CH_4), ammonia (NH_3), and sodium chloride (NaCl).

2.2 Properties of Water and the Early Oceans

- The chemical reactions required for life take place in water.

- Water is polar—meaning that it has partial positive and negative charges—because it is bent and has two polar covalent bonds.

- Polar molecules and charged substances, including ions, interact with water and stay in solution via hydrogen bonding and electrostatic attraction.

- Water's ability to participate in hydrogen bonding also gives it an extraordinarily high capacity to absorb heat and cohere to other water molecules.

- Water spontaneously dissociates into hydrogen ions (or protons, H^+) and hydroxide ions (OH^-). The concentration of protons in a solution determines the pH, which can be altered by acids and bases or stabilized by buffers.

✔ You should be able to predict what part of water molecules would interact with amino, carboxyl, and hydroxyl functional groups in solution and the types of bonds that would be involved.

2.3 Chemical Reactions, Energy, and Chemical Evolution

- The first step in chemical evolution was the formation of small organic compounds from molecules such as molecular hydrogen (H_2) and carbon dioxide (CO_2).

- Chemical reactions typically involve bonds being broken, atoms being rearranged, and new bonds being formed. This process involves energy, either from the reactants or external sources (e.g., heat).

- Energy comes in different forms. Although energy cannot be created or destroyed, one form of energy can be transformed into another.

✔ You should be able to explain how the energy in electricity can drive a reaction that is nonspontaneous.

2.4 Investigating Chemical Evolution: Approaches and Model Systems

- Experiments suggest that early in Earth's history, external sources of energy, such as sunlight or lightning, could have driven chemical reactions between simple molecules to form molecules with higher potential energy. In this way, energy in the form of radiation or electricity was transformed into chemical energy.

- The prebiotic soup and surface metabolism models for chemical evolution have been supported by the synthesis of organic molecules in laboratory simulations of the early Earth environment.

✔ You should be able to explain how the surface metabolism model is supported by both the top-down and bottom-up approaches used to investigate reactions involved in chemical evolution.

2.5 The Importance of Organic Molecules

- Carbon is the foundation of organic molecules based on its valence, which allows for the construction of molecules with complex shapes.

- Organic molecules are critical to life because they possess versatility of chemical behavior due to the presence of functional groups.

✔ You should be able to predict how adding hydroxyl groups to the octane molecule in Figure 2.24 would affect the properties of the molecule.

(MB) MasteringBiology

1. MasteringBiology Assignments

Tutorials and Activities Acids, Bases, and pH; Anatomy of Atoms; Atomic Number and Mass Number; BioSkill: Using Logarithms; Carbon Bonding and Functional Groups; Cohesion of Water; Covalent Bonds; Dissociation of Water Molecules; Diversity of Carbon-Based Molecules; Electron Arrangement; Energy Transformations; Functional Groups; Hydrogen Bonding and Water; Hydrogen Bonds; Ionic Bonds; Nonpolar and Polar Molecules; pH Scale; Polarity of Water; Properties of Water; Structure of the Atomic Nucleus

Questions Reading Quizzes, Blue-Thread Questions, Test Bank

2. eText Read your book online, search, take notes, highlight text, and more.

3. The Study Area Practice Test, Cumulative Test, BioFlix® 3-D Animations, Videos, Activities, Audio Glossary, Word Study Tools, Art

You should be able to . . .

✔ TEST YOUR KNOWLEDGE

Answers are available in Appendix A

1. Which of the following occurs when a covalent bond forms?
   a. The potential energy of electrons drops.
   b. Electrons in valence shells are shared between nuclei.
   c. Ions of opposite charge interact.
   d. Polar molecules interact.

2. If a reaction is exothermic, then which of the following statements is true?
   a. The products have lower potential energy than the reactants.
   b. Energy must be added for the reaction to proceed.
   c. The products have lower entropy (are more ordered) than the reactants.
   d. It occurs extremely quickly.

3. Which of the following is most likely to have been the energy source responsible for the formation of acetic acid in deep-sea hydrothermal vents?

   a. heat released from the vents
   b. solar radiation that passed through the ocean water
   c. chemical energy present in the reactants
   d. the increase in entropy in the products

4. What is thermal energy?
   a. a form of potential energy
   b. the temperature increase that occurs when any form of energy is added to a system
   c. mechanical energy
   d. the kinetic energy of molecular motion, measured as heat

5. What factors determine whether a chemical reaction is spontaneous or not?

6. What are the two models that have been proposed to explain the process component of chemical evolution?

7. Which of the following molecules would you predict to have the largest number of polar covalent bonds based on their molecular formulas?
   a. C_2H_6O (ethanol)
   b. C_2H_6 (ethane)
   c. $C_2H_4O_2$ (acetic acid)
   d. C_3H_8O (propanol)

8. Locate fluorine (F) on the partial periodic table provided in Figure 2.2. Predict its relative electronegativity compared to hydrogen, sodium, and oxygen. State the number and type of bond(s) you expect it would form if it reacted with sodium.

9. Oxygen is extremely electronegative, meaning that its nucleus pulls electrons shared in covalent bonds very strongly. Explain the changes in electron position that are illustrated in Figure 2.19 based on oxygen's electronegativity.

10. Draw the electron-sharing continuum and place molecular oxygen (O_2), magnesium chloride ($MgCl_2$), and carbon dioxide (CO_2) on it.

11. Consider the reaction between carbon dioxide and water, which forms carbonic acid:

$$CO_2(g) + H_2O(l) \rightleftharpoons CH_2O_3(aq)$$

In aqueous solution, carbonic acid immediately dissociates to form a proton and bicarbonate ion, as follows:

$$CH_2O_3(aq) \rightleftharpoons H^+(aq) + CHO_3^-(aq)$$

If an underwater volcano bubbled additional CO_2 into the ocean, would this sequence of reactions be driven to the left or the right? How would this affect the pH of the ocean?

12. What is the relationship between the carbon framework in an organic molecule (the "R" in Table 2.3) and its functional groups?

13. When H_2 and CO_2 react, acetic acid can be formed spontaneously while the production of formaldehyde requires an input of energy. Which of the following conclusions may be drawn from this observation?
    a. More heat is released when formaldehyde is produced compared to the production of acetic acid.
    b. Compared to the reactants from which it is formed, formaldehyde has more potential energy than does acetic acid.
    c. Entropy decreases when acetic acid is produced and increases when formaldehyde is produced.
    d. The mineral catalyst involved in acetic acid production provides energy to make the reaction spontaneous.

14. When chemistry texts introduce the concept of electron shells, they emphasize that shells represent distinct potential energy levels.

In introducing electron shells, this chapter also emphasizes that they represent distinct distances from the positive charges in the nucleus. Are these two points of view in conflict? Why or why not?

15. Draw a concept map relating water's structure to its properties. (For an introduction to concept mapping, see **BioSkills 15** in Appendix B.) Your concept map should include the following terms or phrases: polar covalent bonds, polarity (on the water molecule), hydrogen bonding, high heat of vaporization, high specific heat, less dense as a solid, effective solvent, unequal sharing of electrons, high energy input required to break bonds, high electronegativity of oxygen.

16. From what you have learned about water, why do coastal regions tend to have climates with lower annual variation in temperature than do inland areas at the same latitude?

3 Protein Structure and Function

In this chapter you will learn that

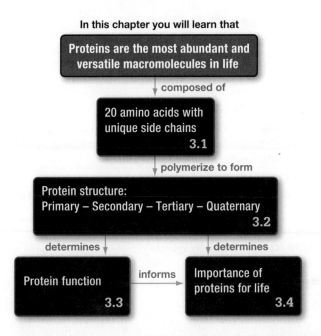

Proteins are the most abundant and versatile macromolecules in life

composed of

20 amino acids with unique side chains
3.1

polymerize to form

Protein structure:
Primary – Secondary – Tertiary – Quaternary
3.2

determines   *determines*

Protein function
3.3

informs

Importance of proteins for life
3.4

A space-filling model of hemoglobin—a protein that is carrying oxygen in your blood right now.

BIG PICTURE

This chapter is part of the Big Picture. See how on pages 104–105.

Chemical reactions in the atmosphere and ocean of ancient Earth are thought to have led to the formation of the first complex carbon-containing compounds. This idea, called chemical evolution, was first proposed by Alexander I. Oparin in 1924. The hypothesis was published again—independently and five years later—by J. B. S. Haldane.

Today, the Oparin–Haldane proposal is considered a formal scientific theory (see Chapter 1). Scientific theories are continuously refined as new information comes to light, and many of Oparin

✔ When you see this checkmark, stop and test yourself. Answers are available in Appendix A.

and Haldane's original ideas have been revised. In its current form, the theory can be broken into four steps.

Step 1 Chemical evolution began with the production of small organic compounds from reactants such as H_2, N_2, NH_3, and CO_2. (Chapter 2 focuses on this step.)

Step 2 These small, simple organic compounds reacted to form mid-sized molecules, such as amino acids, nucleotides, and sugars. (Amino acids are introduced in this chapter. Nucleotides and sugars are discussed in Chapters 4 and 5, respectively.)

Step 3 Mid-sized, building-block molecules linked to form the types of large molecules found in cells today, including proteins, nucleic acids, and complex carbohydrates. Each of these large molecules is composed of distinctive chemical subunits that join together: Proteins are composed of amino acids, nucleic acids are composed of nucleotides, and complex carbohydrates are composed of sugars.

Step 4 Life became possible when one of these large, complex molecules acquired the ability to replicate itself. By increasing in copy number, this molecule would then emerge from the pool of chemicals. At that point, life had begun—chemical evolution gave way to biological evolution.

What type of molecule was responsible for the origin of life? Answering this question is a recurring theme in this and the next three chapters.

To address this question, researchers first designed experiments to identify the types of molecules that could be produced in the waters of prebiotic Earth (Chapter 2). One series of results sparked particular excitement for origin-of-life researchers—the repeated discovery of amino acids among the products of early Earth simulations.

Amino acids have also been found in meteorites and produced in experiments that approximate the environment of interstellar space. Taken together, these observations have led researchers to conclude that amino acids were present and probably abundant during chemical evolution. Since amino acids are the building blocks of proteins, many researchers have therefore asked, Could a protein have been the initial spark of life?

For this question to be valid, proteins would need to possess three of the fundamental attributes of life, namely: information, replication, and evolution. To determine if they do, let's look at the molecules themselves. What are amino acids, and how are they linked to form proteins?

3.1 Amino Acids and Their Polymerization

Modern cells, such as those that make up your body, produce tens of thousands of distinct proteins. Most of these molecules are composed of just 20 different building blocks, called **amino acids.** All 20 of these building blocks have a common structure.

(a) Non-ionized form of amino acid

Amino group / Carboxyl group / Side chain

(b) Ionized form of amino acid

Amino group / Carboxyl group / Side chain

FIGURE 3.1 All Amino Acids Have the Same General Structure. The central α-carbon is shown in red.

The Structure of Amino Acids

To understand how amino acids are put together, recall that carbon atoms have a valence of four—they form four covalent bonds (Chapter 2). All 20 amino acids thus have a common core structure—with a central carbon atom (referred to as the α-carbon) bonded to the four different atoms or groups of atoms diagrammed in **FIGURE 3.1a**:

1. H—a hydrogen atom
2. NH_2—an amino functional group
3. COOH—a carboxyl functional group
4. a distinctive "R-group" (often referred to as a "side chain")

The combination of amino and carboxyl groups not only inspired the name amino acid, but is key to how these molecules behave. In water at pH 7, the concentration of protons causes the amino group to act as a base. It attracts a proton to form NH_3^+ (**FIGURE 3.1b**). The carboxyl group, in contrast, is acidic because its two oxygen atoms are highly electronegative. They pull electrons away from the hydrogen atom, which means that it is relatively easy for this group to lose a proton to form COO^-.

The charges on these functional groups are important for two reasons: **(1)** They help amino acids stay in solution, where they can interact with one another and with other solutes, and **(2)** they affect the amino acid's chemical reactivity.

The Nature of Side Chains

What about the R-group? The R-groups, or side chains, on amino acids vary from a single hydrogen atom to large structures containing carbon atoms linked into rings. While all amino acids share the same core structure, each of the 20 R-groups is unique. The properties of amino acids vary because their R-groups vary.

FIGURE 3.2 highlights the R-groups on the 20 most common amino acids found in cells.[1] As you examine these side chains,

---

[1]There are actually 22 amino acids found in proteins that occur in organisms, but two are very rare.

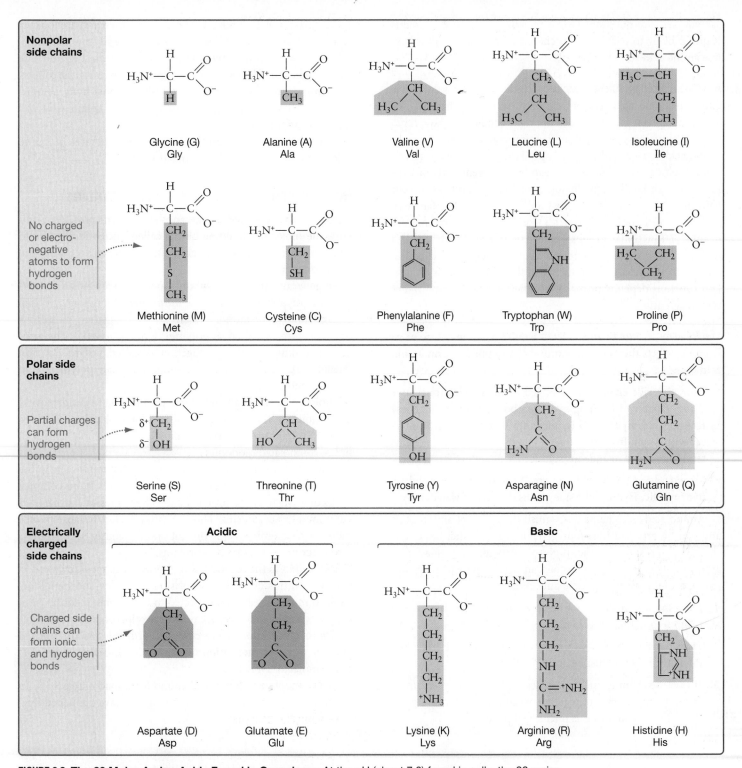

FIGURE 3.2 The 20 Major Amino Acids Found in Organisms. At the pH (about 7.0) found in cells, the 20 major amino acids found in organisms have the structural formulas shown here. The side chains are highlighted, and standard single-letter and three-letter abbreviations for each amino acid are given. For clarity, the carbon atoms in the ring structures of phenylalanine, tyrosine, tryptophan, and histidine are not shown; each bend in a ring is the site of a carbon atom. The hydrogen atoms in these structures are also not shown. A double line inside a ring indicates a double bond.

✔**EXERCISE** Explain why the green R-groups are nonpolar and why the pink R-groups are polar, based on the relative electronegativities of O, N, C, and H (see Chapter 2). Note that sulfur (S) has an electronegativity almost equal to that of carbon and slightly higher than that of hydrogen, making cysteine's side chain mildly hydrophobic.

ask yourself two questions (while referring to Table 2.3): Is this R-group likely to participate in chemical reactions? Will it help this amino acid stay in solution?

Functional Groups Affect Reactivity Several of the side chains found in amino acids contain carboxyl, sulfhydryl, hydroxyl, or amino functional groups. Under the right conditions, these functional groups can participate in chemical reactions. For example, amino acids with a sulfhydryl group (SH) in their side chains can form disulfide (S—S) bonds that help link different parts of large proteins. Such bonds naturally form between the proteins in your hair; curly hair contains many cross-links and straight hair far fewer.

In contrast, some amino acids contain side chains that are devoid of functional groups—consisting solely of carbon and hydrogen atoms. These R-groups rarely participate in chemical reactions. As a result, the influence of these amino acids on protein function depends primarily on their size and shape rather than reactivity.

The Polarity of Side Chains Affects Solubility The nature of its R-group affects the polarity, and thus the solubility, of an amino acid in water.

- Nonpolar side chains lack charged or highly electronegative atoms capable of forming hydrogen bonds with water. These R-groups are **hydrophobic,** meaning that they do not interact with water. Instead of dissolving, hydrophobic side chains tend to coalesce in aqueous solution.

- Polar or charged side chains interact readily with water and are **hydrophilic.** Hydrophilic side chains dissolve in water easily.

Amino acid side chains distinguish the different amino acids and can be grouped into four general types: acidic, basic, uncharged polar, and nonpolar. If given a structural formula for an amino acid, as in Figure 3.2, you can determine which type of amino acid it is by asking three questions:

1. Does the side chain have a negative charge? If so, it has lost a proton, so it must be acidic.

2. Does the side chain have a positive charge? If so, it has taken on a proton, so it must be basic.

3. If the side chain is uncharged, does it have an oxygen atom? If so, the highly electronegative oxygen will result in a polar covalent bond and thus is uncharged polar.

If the answers to all three questions are no, then you are looking at a nonpolar amino acid. ✔ If you understand how the interaction between amino acids and water is affected by the side chains, you should be able to use Figure 3.2 to order the following amino acids from most hydrophilic to most hydrophobic: valine, aspartate, asparagine, and tyrosine. Explain why you have chosen this order.

Now that you have seen the diversity of structures in amino acids, let's put them together to make a protein.

How Do Amino Acids Link to Form Proteins?

Amino acids are linked to one another to form proteins. Similarly, the molecular building blocks called nucleotides attach to one another to form nucleic acids, and simple sugars connect to form complex carbohydrates.

In general, a molecular subunit such as an amino acid, a nucleotide, or a sugar is called a **monomer** ("one-part"). When a large number of monomers are bonded together, the resulting structure is called a **polymer** ("many-parts"). The process of linking monomers together is called **polymerization** (**FIGURE 3.3**). Thus, amino acid monomers can polymerize to form proteins.

Biologists also use the word **macromolecule** to denote a very large molecule that is made up of smaller molecules joined together. Proteins are macromolecules—polymers—that consist of linked amino acid monomers.

The theory of chemical evolution states that monomers in the prebiotic soup polymerized to form larger and more complex molecules, such as the proteins and other types of macromolecules found in organisms. This is a difficult step, because monomers such as amino acids do not spontaneously self-assemble into macromolecules such as proteins.

According to the second law of thermodynamics (reviewed in Chapter 2), this fact is not surprising. Complex and highly organized molecules are not expected to form spontaneously from simpler constituents, because polymerization organizes the molecules involved into a more complex, ordered structure. Stated another way, polymerization decreases the disorder, or entropy, of the molecules involved.

For monomers to link together and form macromolecules, an input of energy is required. How could this have happened during chemical evolution?

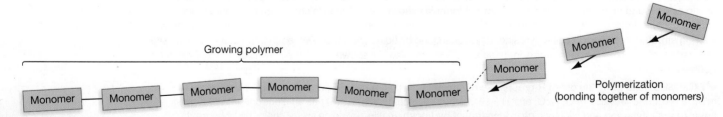

FIGURE 3.3 Monomers Are the Building Blocks of Polymers.

(a) Condensation reaction:
monomer in, water out

HO — Monomer — H

+ H — OH

(Water)

(b) Hydrolysis:
water in, monomer out

H — OH (Water)

Monomer — H

H + HO — Monomer — H

FIGURE 3.4 Polymers Can Be Extended or Broken Apart.

Could Polymerization Occur in the Energy-Rich Environment of Early Earth?

Monomers polymerize through **condensation reactions,** also known as **dehydration reactions.** These reactions are aptly named because the newly formed bond results in the loss of a water molecule (**FIGURE 3.4a**). The reverse reaction, called **hydrolysis,** breaks polymers apart by adding a water molecule (**FIGURE 3.4b**). The water molecule reacts with the bond linking the monomers, separating one monomer from the polymer chain.

In a solution such as the prebiotic soup, condensation and hydrolysis represent the forward and reverse reactions of a chemical equilibrium:

Monomer 1 + Monomer 2 ⇌ (condensation / hydrolysis) Monomer 1 — Monomer 2

Hydrolysis dominates because it increases entropy and is favorable energetically.

This means that, in the prebiotic soup, polymerization would occur only if there were a very high concentration of amino acids to push the reaction toward condensation. Since the equilibrium favors free monomers over polymers even under concentrated conditions, a polymer is unlikely to have grown much beyond a short chain.

According to recent experiments, though, there are several ways that amino acids could have polymerized early in chemical evolution.

- Researchers evaluating the surface metabolism model of chemical evolution have been able to generate stable polymers by mixing free amino acids with a source of chemical energy and tiny mineral particles. Apparently, growing macromolecules are protected from hydrolysis if they cling, or adsorb, to a mineral surface. One such experiment produced polymers that were 55 amino acids long.

- In conditions that simulate the hot, metal-rich environments of undersea volcanoes, researchers have observed not only amino acid formation but also their polymerization.

- Amino acids have also joined into polymers in experiments in cooler water if a carbon- and sulfur-containing gas—one that is commonly ejected from undersea volcanoes—is present.

The current consensus is that several mechanisms could have led to polymerization reactions between amino acids, early in chemical evolution. What kind of bond is responsible for linking these monomers?

The Peptide Bond As **FIGURE 3.5** shows, amino acids polymerize when a bond forms between the carboxyl group of one amino acid and the amino group of another. The C–N covalent bond that results from this condensation reaction is called a **peptide bond.** When a water molecule is removed in the condensation reaction, the carboxyl group is converted to a carbonyl functional group ($C=O$) in the resulting polymer, and the amino group is reduced to an N–H.

Peptide bonds are unusually stable compared to linkages in other types of macromolecules. This is because a pair of valence electrons on the nitrogen is partially shared in the C–N bond (see Figure 3.5). The degree of electron sharing is great enough that peptide bonds actually have some of the characteristics of a double bond. For example, the peptide bond is planar, limiting the movement of the atoms participating in the peptide bond.

When amino acids are linked by peptide bonds into a chain, the amino acids are referred to as residues to distinguish them from free monomers.

Carboxyl group   Amino group

$$H_3N^+ - \overset{\overset{H}{|}}{\underset{\underset{H}{|}}{C}} - \overset{O}{\underset{O^-}{C}} \quad + \quad H - \overset{\overset{H}{|}}{\underset{\underset{H}{|}}{N^+}} - \overset{\overset{H}{|}}{\underset{\underset{CH_3}{|}}{C}} - \overset{O}{\underset{O^-}{C}}$$

Peptide bond formation

Electron sharing here makes peptide bond like a double bond

$$H_3N^+ - \overset{\overset{H}{|}}{\underset{\underset{H}{|}}{C}} - \overset{O}{\underset{\text{Peptide bond}}{C}} - \overset{}{N} - \overset{\overset{H}{|}}{\underset{\underset{CH_3}{|}}{C}} - \overset{O}{\underset{O^-}{C}} \quad + \quad H_2O$$

FIGURE 3.5 Peptide Bonds Form When the Carboxyl Group of One Amino Acid Reacts with the Amino Group of a Second Amino Acid.

(a) Peptide chain

(b) Numbering system

FIGURE 3.6 **Amino Acids Polymerize to Form Chains.**

FIGURE 3.6a shows how the chain of peptide bonds in a short polymer gives the molecule a structural framework, or a "backbone." There are three key points to note about the peptide-bonded backbone:

1. **R-group orientation** The side chains present in each residue extend out from the backbone, making it possible for them to interact with each other and with water.

2. **Directionality** There is an amino group (NH_3^+) on one end of the backbone and a carboxyl group (COO^-) on the other. The end of the sequence that has the free amino group is called the N-terminus, or amino-terminus, and the end with the free carboxyl group is called the C-terminus, or carboxy-terminus. By convention, biologists always write amino acid sequences from the N-terminus to the C-terminus (**FIGURE 3.6b**), because the N-terminus is the start of the chain when proteins are synthesized in cells.

3. **Flexibility** Although the peptide bond itself cannot rotate because of its double-bond nature, the single bonds on either side of the peptide bond can rotate. As a result, the structure as a whole is flexible (**FIGURE 3.7**).

When fewer than 50 amino acids are linked together in this way, the resulting polymer is called an **oligopeptide** ("few peptides") or simply a **peptide.** Polymers that contain 50 or more amino acids are called **polypeptides** ("many peptides").

The term **protein** is often used to describe any chain of amino acid residues, but formally protein refers to the complete, often

FIGURE 3.7 **Peptide Chains Are Flexible.**

functional form of the molecule. In Section 3.2, you'll see that some proteins consist of a single polypeptide while others are functional only when multiple polypeptides are bonded to one another.

Proteins are the stuff of life. Let's take a look at how they are put together and then see what they do.

3.2 What Do Proteins Look Like?

The unparalleled diversity of proteins—in size, shape, and other aspects of structure—is important because function follows from structure. Proteins can serve diverse functions in cells because they are diverse in size and shape as well as in the chemical properties of their amino acid residues.

FIGURE 3.8 illustrates some of the variety in the sizes and shapes observed in proteins. In the case of the TATA box–binding protein in **FIGURE 3.8a** and the porin protein in **FIGURE 3.8b**, the shape of the molecule has a clear correlation with its function. The TATA box–binding protein has a groove where DNA molecules fit; porin has a hole that forms a pore. The groove in the TATA box–binding protein interacts with specific regions of a DNA molecule, while porin fits in cell membranes and allows certain hydrophilic molecules to pass through. Proteins that provide structural support for cells or tissues, such as the collagen triple helix in **FIGURE 3.8d**, often form long, cable-like fibers.

But many of the proteins found in cells do not have shapes that are noticeably correlated with their functions. For example, the trypsin protein in **FIGURE 3.8c** has an overall globular shape that tells little about its function, which is to bind and cleave peptide bonds of other proteins.

How can biologists make sense of this diversity of protein size and shape? Initially, the amount of variation seems overwhelming. Fortunately, it is not. No matter how large or complex a protein may be, its underlying structure can be broken down into just four basic levels of organization.

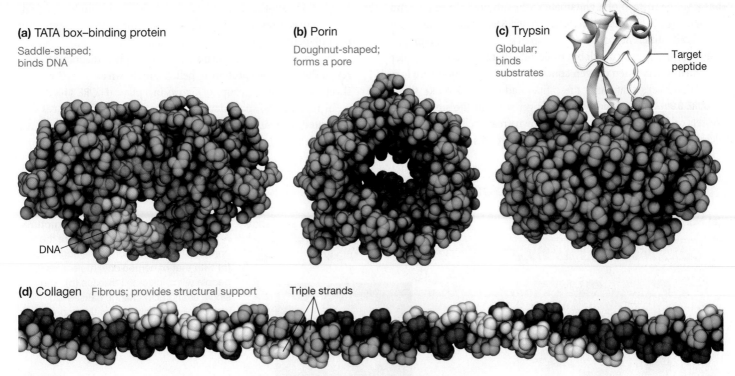

(a) TATA box–binding protein

Saddle-shaped; binds DNA

DNA

(b) Porin

Doughnut-shaped; forms a pore

(c) Trypsin

Globular; binds substrates

Target peptide

(d) Collagen   Fibrous; provides structural support   Triple strands

FIGURE 3.8 In Overall Shape, Proteins Are the Most Diverse Class of Molecules Known.

Primary Structure

Each protein has a unique sequence of amino acids. That simple conclusion was the culmination of 12 years of study by Frederick Sanger and co-workers during the 1940s and 1950s. Sanger's group worked out the first techniques for determining the amino acid sequence of insulin, a hormone that helps regulate sugar concentrations in the blood of humans and other mammals. When other proteins were analyzed, it rapidly became clear that each protein has a definite and distinct amino acid sequence.

Biochemists call the unique sequence of amino acids in a protein the **primary structure** of that protein. The sequence of amino acid residues in Figure 3.6, for example, defines the peptide's primary structure.

With 20 types of amino acids available and length ranging from two amino acid residues to tens of thousands, the number of primary structures that are possible is practically limitless. There may, in fact, be 20^n different combinations of amino acid residues for a polymer with a given length of n. For example, a peptide that is just 10 amino acids long has 20^{10} possible sequences. This is over 10,000 billion.

Why is the order and type of residues in the primary structure of a protein important? Recall that the R-groups present on each amino acid affect its chemical reactivity and solubility. It's therefore reasonable to predict that the R-groups present in a polypeptide will affect that molecule's properties and function.

This prediction is correct. In some cases, even a single change in the sequence of amino acids can cause radical changes in the way the protein as a whole behaves.

As an example, consider hemoglobin, an oxygen-binding protein in human red blood cells. In some individuals, hemoglobin has a valine instead of a glutamate at the 6th position of a strand containing 146 amino acid residues (**FIGURE 3.9a**). Valine's side chain is radically different from the R-group in glutamate. The change in R-group produces hemoglobin molecules that stick to one another and form fibers when oxygen concentrations in the blood are low. Red blood cells that carry these fibers adopt a sickle-like shape (**FIGURE 3.9b**). Sickled red blood cells get stuck in the small blood vessels called capillaries and starve downstream cells of oxygen. A debilitating illness called sickle-cell disease results.

A protein's primary structure is fundamental to its function. Primary structure is also fundamental to the higher levels of protein structure: secondary, tertiary, and quaternary.

Secondary Structure

Even though variation in the amino acid sequence of a protein is virtually limitless, it is only the tip of the iceberg in terms of generating structural diversity.

The next level of organization in proteins—**secondary structure**—is created in part by hydrogen bonding between components of the peptide-bonded backbone. Secondary structures are distinctively shaped sections of proteins that are stabilized largely by hydrogen bonding that occurs between the oxygen on the C=O group of one amino acid residue and the hydrogen on the N−H groups of another (**FIGURE 3.10a**). The oxygen atom in the C=O group has a partial negative charge due to its high electronegativity, while the hydrogen atom in the N−H group has a partial positive charge because it is bonded to nitrogen, which has high electronegativity.

Note a key point: Hydrogen bonding between sections of the same backbone is possible only when a polypeptide bends in a way that puts C=O and N−H groups close together. In most proteins, these polar groups are aligned and form hydrogen bonds with one another when the backbone bends to form one of two possible structures (**FIGURE 3.10b**):

1. an α-**helix** (alpha-helix), in which the polypeptide's backbone is coiled; or

2. a β-**pleated sheet** (beta-pleated sheet), in which segments of a peptide chain bend 180° and then fold in the same plane.

In both structures, the distance between residues that hydrogen-bond to one another is small. In an α-helix, for example, H-bonds form between residues that are just four linear positions apart in the polypeptide's primary sequence (Figure 3.10a).

When biologists use illustrations called ribbon diagrams to represent the shape of a protein, α-helices are shown as coils; β-pleated sheets are shown by groups of arrows in a plane (**FIGURE 3.10c**).

In most cases, secondary structure consists of α-helices and β-pleated sheets. Which one forms, if either, depends on the molecule's primary structure—specifically, the geometry and

(a) Normal amino acid sequence

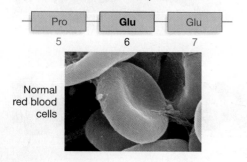

Normal red blood cells

(b) Single change in amino acid sequence

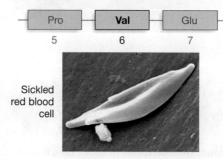

Sickled red blood cell

FIGURE 3.9 Changes in Primary Structure Affect Protein Function. Compare the primary structure of normal hemoglobin **(a)** with that of hemoglobin molecules in people with sickle-cell disease **(b)**. The single amino acid change causes red blood cells to change from their normal disc shape in (a) to a sickled shape in (b) when oxygen concentrations are low.

properties of the amino acids in the sequence. Certain amino acids are more likely to be involved in α-helices than in β-pleated sheets, and vice versa, due to the specific geometry of their side chains. Proline, for example, may be present in β-pleated sheets,

(a) Hydrogen bonds can form between nearby amino and carbonyl groups on the same polypeptide chain.

Hydrogen bond

Hydrogen bond

(b) Secondary structures of proteins result.

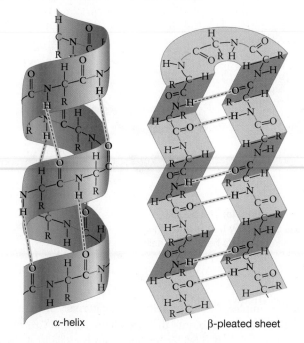

α-helix   β-pleated sheet

(c) Ribbon diagrams of secondary structure

α-helix   β-pleated sheet

Arrowheads point toward the carboxyl end of the primary structure

FIGURE 3.10 Secondary Structures of Proteins. A polypeptide chain can coil or fold in on itself when hydrogen bonds form between N–H and C=O groups on its peptide-bonded backbone.

but it will terminate α-helices due to its unusual side chain. The bond formed between proline's R-group and the nitrogen of the core amino group introduces kinks in the backbone that do not conform to the shape of the helix.

Although each of the hydrogen bonds in an α-helix or a β-pleated sheet is weak relative to a covalent bond, the large number of hydrogen bonds in these structures makes them highly stable. As a result, they increase the stability of the molecule as a whole and help define its shape. In terms of overall shape and stability, though, the tertiary structure of a protein is even more important.

Tertiary Structure

Most of the overall shape, or **tertiary structure,** of a polypeptide results from interactions between R-groups or between R-groups and the backbone. In contrast to the secondary structures, where hydrogen bonds link backbone components together, these side chains can be involved in a wide variety of bonds and interactions. In addition, the amino acid residues that interact with one another are often far apart in the linear sequence. Because each contact between R-groups causes the peptide-bonded backbone to bend and fold, each contributes to the distinctive three-dimensional shape of a polypeptide.

Five types of interactions involving side chains are particularly important:

1. *Hydrogen bonding*   Hydrogen bonds form between polar R-groups and opposite partial charges either in the peptide backbone or other R-groups.

2. *Hydrophobic interactions*   In an aqueous solution, water molecules interact with the hydrophilic polar side chains of a polypeptide and force the hydrophobic nonpolar side chains to coalesce into globular masses. When these nonpolar R-groups come together, the surrounding water molecules form more hydrogen bonds with each other, increasing the stability of their own interactions.

3. *van der Waals interactions*   Once hydrophobic side chains are close to one another, their association is further stabilized by electrical attractions known as **van der Waals interactions.** These weak attractions occur because the constant motion of electrons gives molecules a tiny asymmetry in charge that changes with time. If nonpolar molecules get extremely close to each other, the minute partial charge on one molecule induces an opposite partial charge in the nearby molecule and causes an attraction. Although the interaction is very weak relative to covalent bonds or even hydrogen bonds, a large number of van der Waals attractions can significantly increase the stability of the structure.

4. *Covalent bonding*   Covalent bonds can form between the side chains of two cysteines through a reaction between the sulfhydryl groups. These **disulfide ("two-sulfur") bonds** are frequently referred to as bridges, because they create strong links between distinct regions of the same polypeptide or two separate polypeptides.

(a) Interactions that determine the tertiary structure of proteins

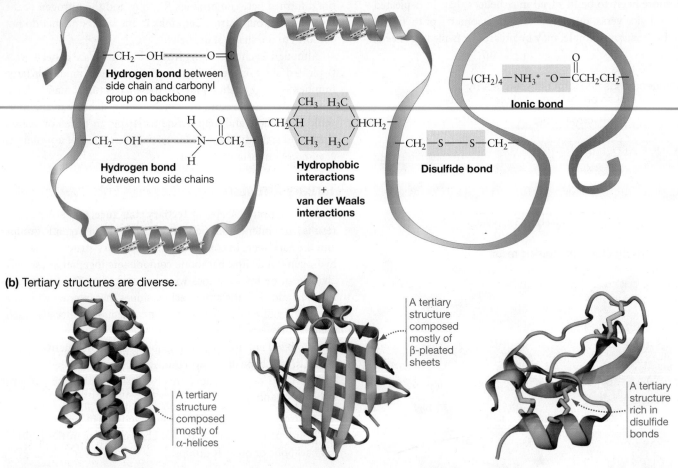

(b) Tertiary structures are diverse.

A tertiary structure composed mostly of α-helices

A tertiary structure composed mostly of β-pleated sheets

A tertiary structure rich in disulfide bonds

FIGURE 3.11 Tertiary Structure of Proteins Results from Interactions Involving R-Groups. (a) The overall shape of a single polypeptide is called its tertiary structure. This level of structure is created by bonds and other interactions that cause it to fold. **(b)** The tertiary structure of these proteins includes interactions between α-helices and β-pleated sheets.

5. *Ionic bonding* Ionic bonds form between groups that have full and opposing charges, such as the ionized acidic and basic side chains highlighted on the right in **FIGURE 3.11a**.

In addition, the overall shape of many proteins depends in part on the presence of secondary structures like α-helices and β-pleated sheets. Thus, tertiary structure depends on both primary and secondary structures.

With so many interactions possible between side chains and peptide-bonded backbones, it's not surprising that polypeptides vary in shape from rod-like filaments to ball-like masses. (See **FIGURE 3.11b**, and look again at Figure 3.8.)

Quaternary Structure

The first three levels of protein structure involve individual polypeptides. But some proteins contain multiple polypeptides that interact to form a single structure. The combination of polypeptides, referred to as subunits, gives a protein **quaternary structure.** The individual polypeptides are held together by the same types of bonds and interactions found in the tertiary level of structure.

In the simplest case, a protein with quaternary structure can consist of just two subunits that are identical. The Cro protein found in a virus called bacteriophage λ (pronounced *LAMB-da*) is an example (**FIGURE 3.12a**). Proteins with two polypeptide subunits are called dimers ("two-parts").

More than two polypeptides can be linked into a single protein, however, and the polypeptides involved may be distinct in primary, secondary, and tertiary structure. For example hemoglobin, an oxygen-binding protein, is a tetramer ("four-parts"). It consists of two copies of two different polypeptides (**FIGURE 3.12b**).

In addition, cells contain **macromolecular machines:** groups of multiple proteins that assemble to carry out a particular function. Some proteins are also found in complexes that include other types of macromolecules. The ribosome (introduced in Chapter 7) provides an example; it consists of several nucleic acid molecules and over 50 different proteins.

TABLE 3.1 summarizes the four levels of protein structure, using hemoglobin as an example. The key thing to note is that protein structure is hierarchical. Quaternary structure is based on tertiary structure, which is based in part on secondary

(a) Cro protein, a dimer

(b) Hemoglobin, a tetramer

FIGURE 3.12 Quaternary Structures of Proteins Are Created by Multiple Polypeptides. These diagrams represent primary sequences as ribbons. **(a)** The Cro protein is a dimer—it consists of two polypeptide subunits, colored light and dark green. The subunits are identical in this case. **(b)** Hemoglobin is a tetramer—it consists of four polypeptide subunits. The α subunits (light and dark green) are identical; so are the β subunits (light and dark blue).

structure. All three of the higher-level structures are based on primary structure.

The summary table and preceding discussion convey three important messages:

1. The combination of primary, secondary, tertiary, and quaternary levels of structure is responsible for the fantastic diversity of sizes and shapes observed in proteins.

2. Protein folding is directed by the sequence of amino acids present in the primary structure.

3. Most elements of protein structure are based on folding of polypeptide chains.

Does protein folding occur spontaneously? What happens to the function of a protein if normal folding is disrupted? Let's use these questions as a guide to dig deeper into how proteins fold.

SUMMARY TABLE 3.1 Protein Structure

Level	Description	Stabilized by	Example: Hemoglobin
Primary	The sequence of amino acids in a polypeptide	Peptide bonds	Gly — Ser — Asp — Cys
Secondary	Formation of α-helices and β-pleated sheets in a polypeptide	Hydrogen bonding between groups along the peptide-bonded backbone; thus, depends on primary structure	One α-helix
Tertiary	Overall three-dimensional shape of a polypeptide (includes contribution from secondary structures)	Bonds and other interactions between R-groups, or between R-groups and the peptide-bonded backbone; thus, depends on primary structure	One of hemoglobin's subunits
Quaternary	Shape produced by combinations of polypeptides (thus, combinations of tertiary structures)	Bonds and other interactions between R-groups, and between peptide backbones of different polypeptides; thus, depends on primary structure	Hemoglobin consists of four polypeptide subunits

If you understand that . . .

- Proteins have up to four levels of structure.
- Primary structure is the sequence of amino acids.
- Secondary structure results from hydrogen bonds between atoms in the peptide-bonded backbone of the same polypeptide. These bonds produce structures such as α-helices and β-pleated sheets.
- Tertiary structure is the overall shape of a polypeptide. Most tertiary structure is a consequence of bonds or other interactions between R-groups or between R-groups and the peptide-bonded backbone.
- Quaternary structure occurs when multiple polypeptides interact to form a single protein.

✔ **You should be able to . . .**

1. Explain how secondary, tertiary, and quaternary levels of structure depend on primary structure.

2. **QUANTITATIVE** Calculate the number of different primary sequences that could be generated by randomly assembling amino acids into peptides that are five residues long.

Answers are available in Appendix A.

3.3 Folding and Function

If you were able to synthesize one of the polypeptides in hemoglobin from individual amino acids, and you then placed the resulting chain in an aqueous solution, it would spontaneously fold into the shape of the tertiary structure shown in Table 3.1.

In terms of entropy, this result probably seems counterintuitive. Because an unfolded protein has many more ways to move about, it has much higher entropy than the folded version. Folding *does* tend to be spontaneous, however, because the chemical bonds, hydrophobic interactions, and van der Waals forces that occur release enough energy to overcome the decrease in entropy. In terms of energy, the folded molecule is more stable than the unfolded molecule.

Folding is crucial to the function of a completed protein. This relationship between protein structure and function was hammered home in a set of classic experiments by Christian Anfinsen and colleagues during the 1950s.

Normal Folding Is Crucial to Function

Anfinsen studied a protein called ribonuclease that is found in many organisms. Ribonuclease is an enzyme that cleaves ribonucleic acid polymers. Anfinsen found that ribonuclease could be unfolded, or **denatured**, by treating it with compounds that break hydrogen bonds and disulfide bonds. The denatured ribonuclease was unable to function normally—it could no longer break apart nucleic acids (**FIGURE 3.13**).

When the denaturing agents were removed, however, the molecule refolded and began to function normally again. These experiments confirmed that ribonuclease folds spontaneously and that folding is essential for normal function.

More recent work has shown that in cells, folding is often facilitated by specific proteins called **molecular chaperones.** Many molecular chaperones belong to a family of molecules called the heat-shock proteins. Heat-shock proteins are produced in large quantities after cells experience high temperatures or other

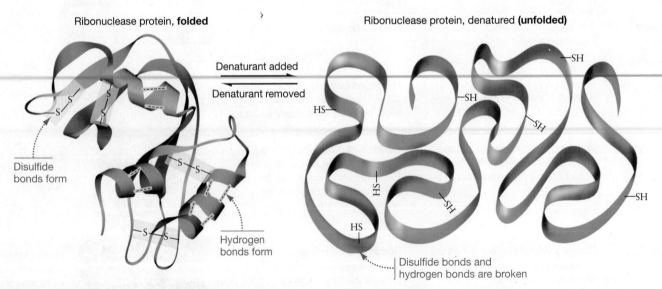

FIGURE 3.13 Protein Structure Determines Function. (left) Ribonuclease is functional when properly folded via hydrogen and disulfide bonds. **(right)** When the disulfide and various noncovalent bonds are broken, ribonuclease is no longer able to function. The double arrow indicates that this process is reversible.

treatments that make other proteins lose their tertiary structure. Heat-shock proteins recognize denatured proteins by binding to hydrophobic patches that would not normally be exposed in properly folded proteins. This interaction blocks inappropriate interactions with other molecules and allows the proteins to refold.

So what is the "normal shape" of a protein? Is only one shape possible for each protein, or could there be several different folded shapes with only one serving as the functional form?

Protein Shape Is Flexible

Although each protein has a characteristic folded shape that is necessary for its function, most proteins are flexible and dynamic, not rigid and static. As it turns out, many polypeptides are unable to fold into their active shape on their own. Over half of the proteins that have been analyzed to date have been found to contain disordered regions lacking any apparent structure. These proteins exist in an assortment of shapes. Only when they interact with particular ions or molecules, or are chemically modified, will they adopt the shape, or conformation, that allows them to perform their function in the cell.

Protein Folding Is Often Regulated   Since the function of a protein is dependent on its shape, controlling when or where it is folded will regulate the protein's activity.

For example, proteins involved in cell signaling are often regulated by controlling their shape. The inactive form of calmodulin—a protein that helps maintain normal blood pressure—has a disordered shape. When the concentration of calcium ions increases in the cell, calmodulin binds these ions, folds into an ordered, active conformation, and sends a signal to increase the diameter of blood vessels. **FIGURE 3.14** illustrates the major shape change that is induced in calmodulin when it binds to calcium.

Misfolding Can Be "Infectious"   In 1982, Stanley Prusiner published what may be the most surprising result to emerge from research on protein folding: Certain proteins can be folded into infectious, disease-causing agents. These proteins are called **prions** (pronounced *PREE-ons*), or proteinaceous infectious particles.

Infectious prions are alternate forms of normal proteins that are present in healthy individuals. The infectious and normal forms do not necessarily differ in amino acid sequence, but their *shapes* are radically different. The infectious form propagates by inducing conformational changes in normal proteins that cause them to adopt the alternate, infectious shape.

FIGURE 3.15 illustrates the differences in shape observed between the normal and infectious forms of the prion responsible for "mad cow disease" in cattle. The molecule in Figure 3.15a is called the prion protein (PrP) and is a normal component of mammalian cells. The improperly folded version of this protein, like the one in Figure 3.15b, represents the infectious form of the prion.

Prions cause a family of diseases known as the spongiform encephalopathies—literally, "sponge-brain-illnesses." Sheep, cows, goats, and humans afflicted with these diseases undergo massive degeneration of the brain. Although some spongiform encephalopathies

(a) Disordered, inactive calmodulin

Protein is not properly folded, and form is variable (one of many forms is shown)

(b) Ordered, active calmodulin

+ Ca²⁺

Calcium ion

After binding calcium, calmodulin folds to a single, functional form

FIGURE 3.14 Calmodulin Requires Calcium to Fold Properly. Many proteins, like calmodulin, do not complete their folding until after interacting with ions or other molecules. Once calmodulin binds to calcium, it assumes its functional shape.

can be inherited, in many cases the disease is transmitted when individuals eat tissues containing the infectious form of PrP. All the prion illnesses are fatal.

Prions are a particularly dramatic example of how a protein's function depends on its shape as well as how the final shape of a protein depends on folding.

(a) Normal prion protein

(b) Infectious prion protein

α-helixes in normal prion

β-pleated sheet in infectious prion

FIGURE 3.15 Prion Infectivity Is Linked to Structure. Ribbon model of **(a)** a normal, noninfectious prion protein; and **(b)** the infectious form that causes mad cow disease in cattle. Secondary structure is represented by coils (α-helices) and arrows (β-pleated sheets).

3.4 Proteins Are the Most Versatile Macromolecules in Cells

As a group, proteins perform more types of cell functions than any other type of molecule does. It makes sense to hypothesize that life began with proteins, simply because proteins are so vital to the life of today's cells.

Consider the red blood cells that are moving through your arteries right now. Each of these cells contains about 300 million copies of hemoglobin. Hemoglobin carries oxygen from your lungs to cells throughout the body. But every red blood cell also has thousands of copies of a protein called carbonic anhydrase, which is important for moving carbon dioxide from cells back to the lungs, where it can be breathed out. Other proteins form the cell's internal "skeleton" or reside on the cell's membrane to interact with neighboring cells.

Proteins are crucial to most tasks required for cells to exist:

- *Catalysis* Many proteins are specialized to **catalyze,** or speed up, chemical reactions. A protein that functions as a catalyst is called an **enzyme.** The carbonic anhydrase molecules in red blood cells are catalysts. So is the protein called salivary amylase, found in your mouth. Salivary amylase helps begin the digestion of starch and other complex carbohydrates into simple sugars. Most chemical reactions that make life possible depend on enzymes.

- *Defense* Proteins called antibodies and complement proteins attack and destroy viruses and bacteria that cause disease.

- *Movement* Motor proteins and contractile proteins are responsible for moving the cell itself, or for moving large molecules and other types of cargo inside the cell. As you turn this page, for example, specialized proteins called actin and myosin will slide past one another to flex or extend muscle cells in your fingers and arm.

- *Signaling* Proteins are involved in carrying and receiving signals from cell to cell inside the body. If sugar levels in your blood are low, a small protein called glucagon will bind to receptor proteins on your liver cells, triggering enzymes inside to release sugar into your bloodstream.

- *Structure* Structural proteins make up body components such as fingernails and hair, and define the shape of individual cells. Structural proteins keep red blood cells flexible and in their normal disc-like shape.

- *Transport* Proteins allow particular molecules to enter and exit cells or carry them throughout the body. Hemoglobin is a particularly well-studied transport protein, but virtually every cell is studded with membrane proteins that control the passage of specific molecules and ions.

Of all the functions that proteins perform in cells, catalysis may be the most important. The reason is speed. Life, at its most basic level, consists of chemical reactions. But most don't occur fast enough to support life unless a catalyst is present. Enzymes are the most effective catalysts on Earth. Why is this so?

Why Are Enzymes Good Catalysts?

Part of the reason enzymes are such effective catalysts is that they bring reactant molecules—called **substrates**—together in a precise orientation so the atoms involved in the reaction can interact.

The initial hypothesis for how enzymes work was proposed by Emil Fischer in 1894. According to Fischer's "lock-and-key" model, enzymes are analogous to a lock and the keys are substrates that fit into the lock and then react.

Several important ideas in this model have stood the test of time. For example, Fischer was correct in proposing that enzymes bring substrates together in a precise orientation that makes reactions more likely. His model also accurately explained why most enzymes catalyze one specific reaction effectively. Enzyme specificity is a product of the geometry and types of functional groups in the sites where substrates bind.

As researchers began to test Fischer's model, the location where substrates bind and react became known as the enzyme's **active site.** The active site is where catalysis actually occurs.

When techniques for solving the three-dimensional structure of enzymes became available, the active sites were identified as clefts or cavities within the globular shapes. The digestive enzyme trypsin, which is at work in your body now, is a good example. As **FIGURE 3.16** shows, the active site in trypsin is a small notch that contains three key amino acid residues with functional groups that catalyze the cleavage of peptide bonds in other proteins. No other class of macromolecule can match proteins for their catalytic potential. The variety of reactive functional groups present in amino acids is much better suited for this activity than those found in nucleotides or sugars.

The role of enzymes in catalyzing reactions is discussed in more detail in the next unit (see Chapter 8). There you will see that Fischer's model had to be modified as research on enzyme action progressed.

Key residues inside trypsin's active site cut protein like molecular scissors

Peptide substrate in active site

FIGURE 3.16 Substrates Bind to a Specific Location in an Enzyme Called the Active Site. The active site in trypsin, as in many enzymes, is a cleft that contains key amino acid residues that bind substrates and catalyze a reaction.

Was the First Living Entity a Protein Catalyst?

Several observations in the preceding sections could argue that a protein was the first molecule capable of replication. Experimental studies have shown that amino acids were likely abundant during chemical evolution, and that they could have polymerized to form small proteins. In addition, proteins are the most efficient catalysts known.

To date, however, attempts to simulate the origin of life with proteins have not been successful. The only experimental glimpse of a protein's potential to replicate involved an enzyme that could link two oligopeptides together to form a functional duplicate of itself. However, this result required a high concentration of preformed, specific oligopeptides that would not have been present during chemical evolution.

Although it is too early to arrive at definitive conclusions, most origin-of-life researchers are increasingly skeptical that life began with a protein. Their reasoning is that to make a copy of something, a mold or template is required. Proteins cannot furnish this information. Nucleic acids, in contrast, *can*. How they do so is the subject of the next chapter.

check your understanding

If you understand that . . .

- Proteins are the most versatile large molecules in cells, and each function is directly connected to structure.
- Enzymes speed up chemical reactions by binding substrates at their active site, where catalysis takes place.

✔ **You should be able to . . .**

Predict where amino acid changes would most likely occur in an enzyme that would result in it catalyzing a reaction with a new substrate, and explain how these changes could result in the new activity.

Answers are available in Appendix A.

CHAPTER 3   REVIEW

For media, go to MasteringBiology

If you understand . . .

3.1   Amino Acids and Their Polymerization

- Amino acids have a central carbon bonded to an amino group, a hydrogen atom, a carboxyl group, and an R-group.

- The structure of the R-group affects the chemical reactivity and solubility of the amino acid.

- In proteins, amino acids are joined by a peptide bond between the carboxyl group of one amino acid and the amino group of another amino acid.

✔ You should be able to explain how you could use the structural formula of an amino acid to determine if it is acidic, basic, uncharged polar, or nonpolar.

3.2   What Do Proteins Look Like?

- A protein's primary structure, or sequence of amino acids, is responsible for most of its chemical properties.

- Interactions that take place between C=O and N–H groups in the same peptide-bonded backbone create secondary structures, which are stabilized primarily by hydrogen bonding.

- Tertiary structure results from interactions between R-groups—or R-groups and the peptide-bonded backbone—that stabilize a folded protein into a characteristic overall shape.

- In many cases, a complete protein consists of several different polypeptides, bonded together. The combination of polypeptides represents the protein's quaternary structure.

✔ You should be able to predict where nonpolar amino acid residues would be found in a globular protein, such as the trypsin molecule shown in Figure 3.8c.

3.3   Folding and Function

- Protein folding is a spontaneous process.

- A protein's normal folded shape is essential to its function.

- Many proteins must first bind to other molecules or ions before they can adopt their active conformation.

- Improperly folded proteins can be detrimental to life, and certain proteins even cause deadly infectious diseases.

✔ You should be able to identify one way in which the process of folding in calmodulin and infectious prions is similar.

3.4   Proteins Are the Most Versatile Macromolecules in Cells

- In organisms, proteins function in catalysis, defense, movement, signaling, structural support, and transport of materials.

- Proteins can have diverse functions in cells because they have such diverse structures and chemical properties.

- Catalysis takes place at the enzyme's active site, which has unique chemical properties and a distinctive size and shape.

✔ You should be able to provide the characteristics of proteins that make them especially useful for the following cellular activities: catalysis, defense, and signaling.

You should be able to . . .

✓ TEST YOUR KNOWLEDGE

Answers are available in Appendix A

1. What two functional groups are present on every amino acid?
   a. a carbonyl (C=O) group and a carboxyl group
   b. an N–H group and a carbonyl group
   c. an amino group and a hydroxyl group
   d. an amino group and a carboxyl group

2. Twenty different amino acids are found in the proteins of cells. What distinguishes these molecules?

3. By convention, biologists write the sequence of amino acids in a polypeptide in which direction?
   a. carboxy- to amino-terminus
   b. amino- to carboxy-terminus
   c. polar residues to nonpolar residues
   d. charged residues to uncharged residues

4. In a polypeptide, what bonds are responsible for the secondary structure called an α-helix?
   a. peptide bonds
   b. hydrogen bonds that form between the core C=O and N–H groups on different residues
   c. hydrogen bonds and other interactions between side chains
   d. disulfide bonds that form between cysteine residues

5. Where is the information stored that directs different polypeptides to fold into different shapes?

6. What is an active site?
   a. the position in an enzyme where substrates bind
   b. the place where a molecule or ion binds to a protein to induce a shape change
   c. the portion of a motor protein that is involved in moving cargo in a cell
   d. the site on an antibody where it binds to bacterial cells or viruses

✓ TEST YOUR UNDERSTANDING

Answers are available in Appendix A

7. Explain how water participates in the development of the interactions that glue nonpolar amino acids together in the interior of globular proteins.

8. If amino acids were mixed together in a solution, resembling the prebiotic soup, would they spontaneously polymerize into polypeptides? Why or why not?

9. Provide an example of how a specific shape of a protein is correlated with its function.

10. A major theme in this chapter is that the structure of molecules correlates with their function. Use this theme to explain why proteins can perform so many different functions in organisms and why enzymes are such effective catalysts.

11. Why are proteins not considered to be a good candidate for the first living molecule?
    a. Their catalytic capability is insufficient.
    b. Their amino acid monomers were likely not present during chemical evolution.
    c. They cannot serve as a template for replication.
    d. They could not have polymerized on their own from amino acids during chemical evolution.

12. If proteins folded only into rigid, inflexible structures, how might this affect the cell's ability to regulate protein function?

✓ TEST YOUR PROBLEM-SOLVING SKILLS

Answers are available in Appendix A

13. Based on what you know of the peptide bonds that link together amino acid residues, why would proline's side chain reduce the flexibility of the backbone?

14. Make a concept map (see **BioSkills 15** in Appendix B) that relates the four levels of protein structure and shows how they can contribute to the formation of an active site. Your map should include the following boxed terms: Primary structure, Secondary structure, Tertiary structure, Quaternary structure, Active site, Amino acid sequence, R-groups, Helices and sheets, 3-D shape.

15. Proteins that interact with DNA often interact with the phosphates that are part of this molecule. Which of the following types of amino acids would you predict to be present in the DNA binding sites of these proteins?
    a. acidic amino acids
    b. basic amino acids
    c. uncharged polar amino acids
    d. nonpolar amino acids

16. Some prion-associated diseases are inherited, such as fatal familial insomnia. What is likely to be different between the infectious forms of these inherited prions compared to those that arise via transmission from one animal to another?

4 Nucleic Acids and the RNA World

In this chapter you will learn that

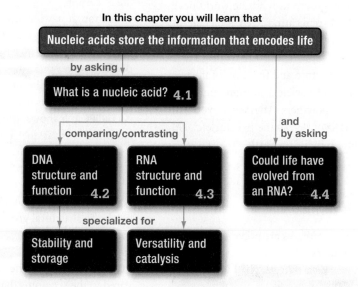

Nucleic acids store the information that encodes life

by asking ↓

What is a nucleic acid? **4.1**

comparing/contrasting

DNA structure and function **4.2**

RNA structure and function **4.3**

and by asking

Could life have evolved from an RNA? **4.4**

specialized for

Stability and storage

Versatility and catalysis

This is part of the sheet-metal-and-wire model that James Watson and Francis Crick used to figure out the secondary structure of DNA. The large "T" stands for the nitrogen-containing base thymine.

BIG PICTURE

This chapter is part of the Big Picture. See how on pages 104–105.

ife began when chemical evolution led to the production of a molecule that could promote its own replication. The nature of this first "living molecule," however, has been the subject of many investigations and heated debates. Even though proteins are the workhorse molecules of today's cells, relatively few researchers favor the hypothesis that life began as a protein molecule. Instead, the vast majority of biologists contend that life began as a polymer called a nucleic acid—specifically, a molecule of ribonucleic acid (RNA). This proposal is called the **RNA world hypothesis.**

The RNA world hypothesis contends that chemical evolution led to the existence of an RNA molecule that could replicate itself. Once this molecule existed, chance errors in the copying process

✔ When you see this checkmark, stop and test yourself. Answers are available in Appendix A.

created variations that would undergo natural selection—the evolutionary process by which individuals with certain attributes are selectively reproduced (see Chapter 1). At this point, chemical evolution was over and biological evolution was off and running.

To test this hypothesis, several groups around the world have been working to synthesize a self-replicating RNA molecule in the laboratory. If they ever succeed, they will have created a life-form in a test tube.

This chapter focuses on the structure and function of nucleic acids. Let's begin with an analysis of nucleic acid monomers and how they are linked together into polymers. Afterwards, you will learn about the experiments used to determine if a nucleic acid could have triggered the evolution of life on Earth.

4.1 What Is a Nucleic Acid?

Nucleic acids are polymers, just as proteins are polymers. But instead of being made up of monomers called amino acids, **nucleic acids** are made up of monomers called **nucleotides.**

FIGURE 4.1a diagrams the three components of a nucleotide: **(1)** a phosphate group, **(2)** a five-carbon sugar, and **(3)** a nitrogenous (nitrogen-containing) base. The phosphate is bonded to the sugar molecule, which in turn is bonded to the nitrogenous base.

The sugar component of a nucleotide is an organic compound bearing reactive hydroxyl (−OH) functional groups. Notice that the prime symbols (′) in Figure 4.1 indicate that the carbon being referred to is part of the sugar—not of the attached nitrogenous base. The phosphate group in a nucleotide is attached to the 5′ carbon.

Although a wide variety of nucleotides are found in living cells, origin-of-life researchers concentrate on two types: **ribonucleotides,** the monomers of **ribonucleic acid** (**RNA**), and **deoxyribonucleotides,** the monomers of **deoxyribonucleic acid** (**DNA**). In ribonucleotides, the sugar is ribose; in deoxyribonucleotides, it is deoxyribose (*deoxy* means "lacking oxygen"). As **FIGURE 4.1b** shows, these two sugars differ by a single oxygen atom. Ribose has an −OH group bonded to the 2′ carbon. Deoxyribose has an H instead at the same location. In both of these sugars, an −OH group is bonded to the 3′ carbon.

In addition to the type of sugar, nucleotides also differ in the type of nitrogenous base. These bases, diagrammed in **FIGURE 4.1c**, belong to structural groups called **purines** and **pyrimidines.** The purines are adenine (A) and guanine (G); the pyrimidines are cytosine (C), uracil (U), and thymine (T). Note that the two rings in adenine and guanine are linked together by nine atoms, compared to the six atoms that make a single ring in each pyrimidine. This makes remembering which bases are purines easy, since both adenine and guanine include "nine" in their names.

As Figure 4.1c shows, ribonucleotides and deoxyribonucleotides also differ in one of their pyrimidine bases. Ribonucleotides

FIGURE 4.1 The General Structure of a Nucleotide. Note that in the bases, the nitrogen that bonds to the sugar is colored blue.

use uracil (U) while deoxyribonucleotides use the closely related base thymine (T).

✔ You should be able to diagram a ribonucleotide and a deoxyribonucleotide. Use a ball for the phosphate group, a pentagon to represent the sugar subunit, and a hexagon to represent the nitrogenous base. Label the 2′, 3′, and 5′ carbons on the sugar molecule, and add the atoms or groups that are bonded to each.

To summarize: After the different sugars and bases are taken into account, eight different nucleotides are used to build nucleic acids—four ribonucleotides (A, G, C, and U) and four deoxyribonucleotides (A, G, C, and T). If nucleic acids played any role in the chemical evolution of life, then at least some of these nucleotides must have been present in the prebiotic oceans. Is there any evidence to suggest that this was possible?

Could Chemical Evolution Result in the Production of Nucleotides?

Based on data from Stanley Miller and researchers who followed (Chapter 2), most biologists contend that amino acids could have been synthesized early in Earth's history. The reactions behind the prebiotic synthesis of nucleotides, however, have been more difficult to identify.

Miller-like laboratory simulations have shown that nitrogenous bases and many different types of sugars can be synthesized readily under conditions that mimic the prebiotic soup. In these experiments, almost all the sugars that have five or six carbons—called pentoses and hexoses, respectively—are produced in approximately equal amounts. If nucleic acids were to form in the prebiotic soup, however, ribose would have had to predominate.

How ribose came to be the dominant sugar during chemical evolution (i.e., what selective process was at work) is still a mystery. Origin-of-life researchers refer to this issue as the "ribose problem." Recent work focusing on the conditions that exist in deep-sea hydrothermal vent systems (see Chapter 2) may point to a possible solution.

Here's the line of reasoning researchers are currently pursuing: Ribose molecules may have been selectively enriched from the mix of sugars in certain early Earth deep-sea vent systems. In one experiment, researchers simulated the conditions that exist in these vents. Then they tested whether minerals that are predicted to have existed in the vent chimneys are able to bind sugars. What they found was striking—the minerals preferentially bound to ribose over other pentoses and hexoses. Did this occur in the ancient vents? If so, the implications are exciting: A high concentration of ribose would be present in the same deep-sea vent environment where chemical evolution is thought to have taken place.

Despite the observed synthesis of nitrogenous bases and the recent discovery of ribose enrichment, the production of nucleotides remains a serious challenge for the theory of chemical evolution. At this time, experiments that attempt to simulate early

FIGURE 4.2 Nucleotides Polymerize via Phosphodiester Linkages. Ribonucleotides can polymerize via condensation reactions. The resulting phosphodiester linkage connects the 3′ carbon of one ribonucleotide and the 5′ carbon of another ribonucleotide.

Earth environments have yet to synthesize complete nucleotides. But research on this issue continues.

In the meantime, let's consider the next question: Once nucleotides formed, how would they polymerize to form RNA and DNA? This question has an answer.

How Do Nucleotides Polymerize to Form Nucleic Acids?

Nucleic acids form when nucleotides polymerize. As **FIGURE 4.2** shows, the polymerization reaction involves the formation of a bond between a hydroxyl on the sugar component of one nucleotide and the phosphate group of another nucleotide. The result of this condensation reaction is called a **phosphodiester linkage,** or a phosphodiester bond.

A phosphodiester linkage joins the 5′ carbon on the sugar of one nucleotide to the 3′ carbon on the sugar of another. When the nucleotides involved contain the sugar ribose, the polymer that is produced is RNA. If the nucleotides contain the sugar deoxyribose instead, then the resulting polymer is DNA.

DNA and RNA Strands Are Directional   FIGURE 4.3 (see page 60) shows how the chain of phosphodiester linkages in a nucleic acid acts as a backbone, analogous to the peptide-bonded backbone found in proteins.

Like the peptide-bonded backbone of a polypeptide, the sugar-phosphate backbone of a nucleic acid is directional. In a strand of RNA or DNA, one end has an unlinked 5′ phosphate while the other end has an unlinked 3′ hydroxyl—meaning the groups are not linked to another nucleotide. By convention, the

The sugar-phosphate backbone of RNA

5′ end of nucleic acid

5′

3′ and 5′ carbons joined by phosphodiester linkage

3′ end of nucleic acid: new nucleotides are added to the unlinked 3′ hydroxyl

3′

FIGURE 4.3 RNA Has a Sugar-Phosphate Backbone.

✔**EXERCISE** Identify the four bases in this RNA strand, using Figure 4.1c as a key. Then write down the base sequence, starting at the 5′ end.

6 billion of these letters to write the primary structure of the DNA in most of your cells.

Polymerization Requires an Energy Source In cells, the polymerization reactions that join nucleotides into nucleic acids are catalyzed by enzymes. Like other polymerization reactions, the process is not spontaneous. An input of energy is needed to tip the energy balance in favor of the process.

Polymerization can take place in cells because the potential energy of the nucleotide monomers is first raised by reactions that add two phosphate groups to the ribonucleotides or deoxyribonucleotides, creating nucleoside triphosphates.[1] In the case of nucleic acid polymerization, researchers refer to these nucleotides as "activated." **FIGURE 4.4a** shows an example of an activated nucleotide; this molecule is called **adenosine triphosphate,** or **ATP.**

Why do added phosphate groups raise the energy content of a molecule? Recall that phosphates are negatively charged and that like charges repel (Chapter 2). Linking two or more phosphates with covalent bonds generates strong repulsive forces. These bonds therefore carry a large amount of potential energy, which can be harvested to power other chemical reactions (**FIGURE 4.4b**). You will see in later chapters that the potential energy stored in ATP is used to drive other cellular activities, independent of nucleotide polymerization.

This is a key point, and one that you will encounter again and again in this text: The addition of one or more phosphate groups raises the potential energy of substrate molecules enough to make an otherwise nonspontaneous reaction possible. (Chapter 8 explains how this happens in more detail.)

Could Nucleic Acids Have Formed in the Absence of Cellular Enzymes? Accumulating data suggest that the answer is yes.

Activation of nucleotides has been observed when prebiotic conditions are simulated experimentally. In a suite of follow-up experiments, researchers have produced RNA molecules by incubating activated nucleotides with tiny mineral particles—in one case, molecules up to 50 nucleotides long were observed. These results support the hypothesis that polymerization of activated nucleotides in the prebiotic world may have been catalyzed by minerals. This model would be in line with the surface metabolism model for chemical evolution (introduced in Chapter 2).

More recent work has shown that under certain conditions, up to 100 nucleotides can be linked together, even without first being activated. To accomplish this, heat was introduced as a source of energy and small nonpolar molecules, called lipids, were added to help the monomers interact. This experiment is particularly interesting with respect to the setting for chemical evolution, because both of these factors—heat and lipids—are thought to have been present in prebiotic hydrothermal vents. (The chemical origins and properties of lipids are covered in Chapter 6.)

---

[1]A molecule consisting of a sugar and one of the bases in Figure 4.1c is called a nucleoside (a nucleotide is a sugar, a base, and one or more phosphate groups). Thus, a sugar attached to a base and three phosphate groups is called a nucleoside triphosphate.

sequence of bases found in an RNA or DNA strand is always written in the 5′→3′ direction. (This system is logical because in cells, RNA and DNA are always synthesized in this direction. Bases are added only at the 3′ end of the growing molecule.)

The order of the different nitrogenous bases in a nucleic acid forms the primary structure of the molecule. When biologists write the primary structure of a stretch of DNA or RNA, they simply list the sequence of nucleotides in the 5′→3′ direction, using their single-letter abbreviations. For example, a six-base-long DNA sequence might be ATTAGC. It would take roughly

(a) ATP is an example of an activated nucleotide.

The addition of phosphate groups raises the potential energy of the monomer

Adenine

Ribose

(b) Energy is released when phosphates are removed by hydrolysis.

ATP

Water

AMP

Inorganic pyrophosphate

Energy used to link nucleotide to RNA

10.9 kcal/mol ATP

FIGURE 4.4 Activated Monomers Drive Polymerization Reactions. Polymerization reactions are generally nonspontaneous, but those reactions involving nucleoside triphosphates, such as ATP, are spontaneous. The potential energy stored in activated nucleotides is released when the pyrophosphate (PP_i) is removed before the polymerizing condensation reaction shown in Figure 4.2.

Based on these results, there is a strong consensus that if ribonucleotides and deoxyribonucleotides were able to form during chemical evolution, they would be able to polymerize into DNA and RNA. Now, what do these nucleic acids look like, and what can they do?

check your understanding

If you understand that . . .

- Nucleotides are monomers that consist of a sugar, a phosphate group, and a nitrogen-containing base.
- Nucleotides polymerize to form nucleic acids through formation of phosphodiester linkages between the 3′ hydroxyl on one nucleotide and the 5′ phosphate on another.
- During polymerization, nucleotides are added only to the 3′ end of a nucleic acid strand.

✔ **You should be able to . . .**

Draw a simplified diagram of the phosphodiester linkage between two nucleotides, indicate the 5′→3′ polarity, and mark where the next nucleotide would be added to the growing chain.

Answers are available in Appendix A.

4.2 DNA Structure and Function

The primary structure of nucleic acids is somewhat similar to the primary structure of proteins. Proteins have a peptide-bonded backbone with a series of R-groups that extend from it. DNA and RNA molecules have a sugar-phosphate backbone, created by phosphodiester linkages, and a sequence of any of four nitrogenous bases that extend from it.

Like proteins, DNA and RNA also have secondary structure. While the α-helices and β-pleated sheets of proteins are formed by hydrogen bonding between groups in the backbone, the secondary structure of nucleic acids is formed by hydrogen bonding between the nitrogenous bases.

Let's analyze the secondary structure and function of DNA first, and then dig into the secondary structure and function of RNA.

What Is the Nature of DNA's Secondary Structure?

The solution to DNA's secondary structure, announced in 1953, ranks among the great scientific breakthroughs of the twentieth century. James Watson and Francis Crick presented a model for the secondary structure of DNA in a one-page paper published in the scientific journal *Nature*.

Early Data Provided Clues Watson and Crick's finding was a hypothesis based on a series of results from other laboratories. They were trying to propose a secondary structure that could explain several important observations about the DNA found in cells:

- Chemists had worked out the structure of nucleotides and knew that DNA polymerized through the formation of phosphodiester linkages. Thus, Watson and Crick knew that the molecule had a sugar-phosphate backbone.

- By analyzing the nitrogenous bases in DNA samples from different organisms, Erwin Chargaff had established two empirical rules: (**1**) The number of purines in a given DNA molecule is equal to the number of pyrimidines, and (**2**) the number of T's and A's in DNA are equal, and the number of C's and G's in DNA are equal.

- By bombarding DNA with X-rays and analyzing how it scattered the radiation, Rosalind Franklin and Maurice Wilkins had calculated the distances between groups of atoms in the molecule (see **BioSkills 11** in Appendix B for an introduction to this technique, called **X-ray crystallography**). The scattering patterns showed that three distances were repeated many times: 0.34 nanometer (nm), 2.0 nm, and 3.4 nm. Because the measurements repeated, the researchers inferred that DNA molecules had a regular and repeating structure. The pattern of X-ray scattering suggested that the molecule was helical, or spiral, in nature.

Based on this work, understanding DNA's structure boiled down to understanding the nature of the helix involved. What type of helix would have a sugar-phosphate backbone and explain both Chargaff's rules and the Franklin–Wilkins measurements?

DNA Strands Are Antiparallel Watson and Crick began by analyzing the size and geometry of deoxyribose, phosphate groups, and nitrogenous bases. The bond angles and measurements suggested that the distance of 2.0 nm probably represented the width of the helix and that 0.34 nm was likely to be the distance between bases stacked in a spiral.

How could they make sense of Chargaff's rules and the 3.4-nm distance, which appeared to be exactly 10 times the distance between a single pair of bases?

To solve this problem, Watson and Crick constructed a series of physical models like the one pictured in **FIGURE 4.5**. The models allowed them to tinker with different types of helical configurations. After many false starts, something clicked:

- They arranged two strands of DNA side by side and running in opposite directions—meaning that one strand ran in the 5′ → 3′ direction while the other strand was oriented 3′ → 5′. Strands with this orientation are said to be **antiparallel.**

- If the antiparallel strands are twisted together to form a **double helix,** the coiled sugar-phosphate backbones end up on the outside of the spiral and the nitrogenous bases on the inside.

FIGURE 4.5 Building a Physical Model of DNA Structure. Watson (left) and Crick (right) represented the arrangement of the four deoxyribonucleotides in a double helix, using metal plates and wires with precise lengths and geometries.

- For the bases from each backbone to fit in the interior of the 2.0-nm-wide structure, they have to form purine-pyrimidine pairs (see **FIGURE 4.6a**). This is a key point: The pairing allows hydrogen bonds to form between certain purines and pyrimidines. Adenine forms hydrogen bonds with thymine, and guanine forms hydrogen bonds with cytosine (**FIGURE 4.6b**).

- The A-T and G-C bases were said to be complementary. Two hydrogen bonds form when A and T pair, and three hydrogen bonds form when G and C pair. As a result, the G-C interaction is slightly stronger than the A-T bond. In contrast, A-C and G-T pairs allowed no or only one hydrogen bond.

(a) Only purine-pyrimidine pairs fit inside the double helix.

Purine-purine pair
NOT ENOUGH SPACE

Pyrimidine-pyrimidine pair
TOO MUCH SPACE

Purine-pyrimidine pair
JUST RIGHT

Space inside sugar-phosphate backbones

(b) Hydrogen bonds form between G-C pairs and A-T pairs.

FIGURE 4.6 Complementary Base Pairing Is Based on Hydrogen Bonding.

Watson and Crick had discovered **complementary base pairing.** In fact, the term **Watson–Crick pairing** is now used interchangeably with the phrase complementary base pairing. The physical restraints posed by these interactions resulted in a full helical twist every 10 bases, or 3.4 nm.

The Double Helix   **FIGURE 4.7a** shows how antiparallel strands of DNA form when complementary bases line up and form hydrogen bonds. As you study the figure, notice that DNA is put together like a ladder whose ends have been twisted in opposite directions. The sugar-phosphate backbones form the supports of the ladder; the base pairs represent the rungs of the ladder. The twisting allows the nitrogenous bases to line up in a way that makes hydrogen bonding between them possible.

The nitrogenous bases in the middle of the molecule are hydrophobic. This is a key point, because twisting into a double helix minimizes contact between the bases and surrounding water molecules. In addition to hydrogen bonding, van der Waals interactions between the tightly stacked bases in the interior further contribute to the stability of the helix. You see the same forces—hydrogen bonding, hydrophobicity, and van der Waals interactions—play similar roles in protein folding (Chapter 3). But DNA as a whole is hydrophilic and water soluble because the backbones, which face the exterior of the molecule, contain negatively charged phosphate groups that interact with water.

FIGURE 4.7b highlights additional features of DNA's secondary structure. It's important to note that the outside of the helical DNA molecule forms two types of grooves. The larger of the two is known as the major groove, and the smaller one is known as the minor groove. From this figure, you can identify how DNA's secondary structure explains the measurements observed by Franklin and Wilkins.

Since the model of the double helix was published, experimental tests have shown that the hypothesis is correct in almost every detail. To summarize:

- DNA's secondary structure consists of two antiparallel strands twisted into a double helix.

- The molecule is stabilized by hydrophobic interactions in its interior and by hydrogen bonding between the complementary base pairs A-T and G-C.

✔ You should be able to explain why complementary base pairing would not be possible if two DNA strands were aligned in a parallel fashion—instead of the antiparallel alignment shown in Figure 4.6b.

Now the question is, how does this secondary structure affect the molecule's function?

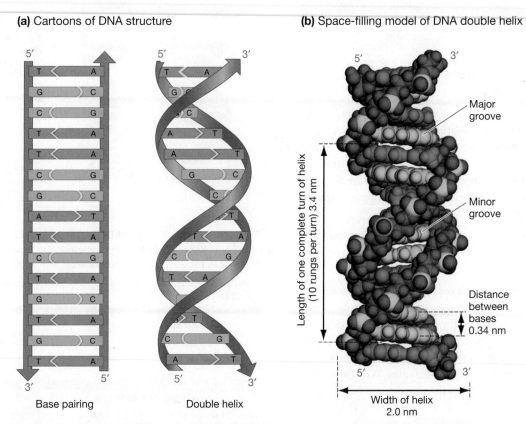

(a) Cartoons of DNA structure

Base pairing

Double helix

(b) Space-filling model of DNA double helix

Major groove

Minor groove

Distance between bases 0.34 nm

Length of one complete turn of helix (10 rungs per turn) 3.4 nm

Width of helix 2.0 nm

FIGURE 4.7 The Secondary Structure of DNA Is a Double Helix. (a) The cartoons illustrate complementary base pairing and how strands are twisted into a double helix. **(b)** The space-filling model shows tight packing of the bases inside the double helix. The double-helix structure explains the measurements inferred from X-ray analysis of DNA molecules.

DNA Functions as an Information-Containing Molecule

Watson and Crick's model created a sensation for a simple reason: It revealed how DNA could store and transmit biological information. In literature, information consists of letters on a page. In music, information is composed of the notes on a staff. But inside cells, information consists of a sequence of nucleotides in a nucleic acid. The four nitrogenous bases function like letters of the alphabet. A sequence of bases is like the sequence of letters in a word—it has meaning.

In all organisms that have been examined to date, from tiny bacteria to gigantic redwood trees, DNA carries the information required for the organism's growth and reproduction. Exploring how hereditary information is encoded and translated into action is the heart of several later chapters (Chapters 16 through 19).

Here, however, our focus is on how life began. The theory of chemical evolution holds that life began once a molecule emerged that could make a copy of itself. Does the information contained within DNA allow it to be replicated?

Watson and Crick ended their paper on the double helix with one of the classic understatements in the scientific literature: "It has not escaped our notice that the specific pairing we have postulated immediately suggests a possible copying mechanism." Here's the key insight: DNA's primary structure serves as a mold or template for the synthesis of a complementary strand. DNA contains the information required for a copy of itself to be made. **FIGURE 4.8** shows how a copy of DNA can be made by complementary base pairing.

Step 1 Heating or enzyme-catalyzed reactions can cause the double helix to separate.

Step 2 Free deoxyribonucleotides form hydrogen bonds with complementary bases on the original strand of DNA—also called a **template strand**. As they do, their sugar-phosphate groups form phosphodiester linkages to create a new strand—also called a **complementary strand**. Note that the $5' \rightarrow 3'$ directionality of the complementary strand is opposite that of the template strand.

Step 3 Complementary base pairing allows each strand of a DNA double helix to be copied exactly, producing two identical daughter molecules.

DNA copying is the basis for a second of the five characteristics of life (introduced in Chapter 1): replication. But can DNA catalyze the reactions needed to *self*-replicate? In today's cells and in laboratory experiments, the answer is no. Instead, the molecule is copied through a complicated series of energy-demanding reactions, catalyzed by a large suite of enzymes. Why can't DNA catalyze these reactions itself?

Is DNA a Catalytic Molecule?

The DNA double helix is highly structured. It is regular, symmetric, and held together by hydrogen bonding, hydrophobic

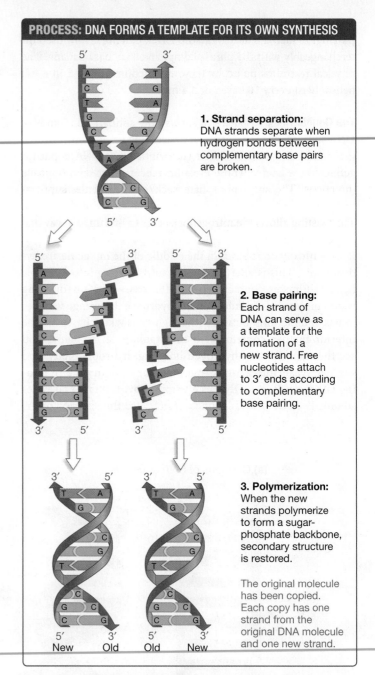

PROCESS: DNA FORMS A TEMPLATE FOR ITS OWN SYNTHESIS

1. Strand separation: DNA strands separate when hydrogen bonds between complementary base pairs are broken.

2. Base pairing: Each strand of DNA can serve as a template for the formation of a new strand. Free nucleotides attach to 3′ ends according to complementary base pairing.

3. Polymerization: When the new strands polymerize to form a sugar-phosphate backbone, secondary structure is restored.

The original molecule has been copied. Each copy has one strand from the original DNA molecule and one new strand.

New   Old   Old   New

FIGURE 4.8 Making a Copy of DNA. If new bases are added to each of the two strands of DNA via complementary base pairing, a copy of the DNA molecule can be produced.

✔**QUESTION** When double-stranded DNA is heated to 95°C, the bonds between complementary base pairs break and single-stranded DNA results. Considering this observation, is the reaction shown in step 1 spontaneous?

interactions, and phosphodiester linkages. In addition, the molecule has few functional groups exposed that can participate in chemical reactions. For example, the lack of a 2′ hydroxyl group on each deoxyribonucleotide makes the polymer much less reactive than RNA, and thus much more resistant to degradation.

Intact stretches of DNA have been recovered from fossils that are tens of thousands of years old. The molecules have the same sequence of bases as the organisms had when they were alive, despite death and exposure to a wide array of pH, temperature, and chemical conditions. DNA's stability is the key to its effectiveness as a reliable information-bearing molecule. DNA's structure is consistent with its function in cells.

The orderliness and stability that make DNA such a dependable information repository also make it extraordinarily inept at catalysis, however. Recall that enzyme function is based on a specific binding event between a substrate and a protein catalyst (Chapter 3). Thanks to variation in reactivity among R-groups in amino acids, and the enormous diversity of shapes found in proteins, a wide array of catalytic activities can be generated. In comparison, DNA's primary and secondary structures are simple. It is not surprising, then, that DNA has never been observed to catalyze any reaction in any organism. Although researchers have been able to construct single-stranded DNA molecules that can catalyze some reactions in the laboratory, the number and diversity of reactions involved is a minute fraction of the activity catalyzed by enzymes.

In short, DNA furnishes an extraordinarily stable template for copying itself and for storing information encoded in a sequence of bases. But owing to its inability to act as an effective catalyst, there is virtually no support for the hypothesis that the first life-form consisted of DNA. Instead, most biologists who are working on the origin of life support the hypothesis that life began with RNA. How does the structure of RNA differ from DNA?

check your understanding

If you understand that . . .

- DNA's primary structure consists of a sequence of deoxyribonucleotides.
- DNA's secondary structure consists of two DNA molecules that run in opposite orientations to each other. The two strands are twisted into a double helix, and they are held together by hydrogen bonds between A–T and G–C pairs and hydrophobic interactions that drive bases into the interior of the helix.
- The sequence of deoxyribonucleotides in DNA contains information. Owing to complementary base pairing, each DNA strand also contains the information required to form its complementary strand.

✔ **You should be able to . . .**

Make a sketch of a double-stranded DNA molecule in the form of a ladder with the sequence of A-G-C-T. Label the 5′ and 3′ ends, the sugar-phosphate backbones, the hydrogen bonds between complementary bases, and the location of hydrophobic interactions.

Answers are available in Appendix A.

The first living molecule would have needed to perform two key functions: carry information and catalyze reactions that promoted its own replication. At first glance, these two functions appear to conflict. Information storage requires regularity and stability; catalysis requires variation in chemical composition and flexibility in shape. How is it possible for a molecule to do both? The answer lies in structure.

Structurally, RNA Differs from DNA

Recall that proteins can have up to four levels of structure. Single-chain proteins possess a primary sequence of amino acids, secondary folds that are stabilized by hydrogen bonding between atoms in the peptide-bonded backbone, and tertiary folds that are stabilized by interactions involving R-groups. Quaternary structure is found in proteins consisting of multiple polypeptides.

DNA has only primary and secondary structure. But RNA, like single-chained proteins, can have up to three levels of structure.

Primary Structure Like DNA, RNA has a primary structure consisting of a sugar-phosphate backbone formed by phosphodiester linkages and, extending from that backbone, a sequence of four types of nitrogenous bases. But it's important to recall two significant differences between these nucleic acids:

1. The sugar in the sugar-phosphate backbone of RNA is ribose, not deoxyribose as in DNA.

2. The pyrimidine base thymine does not exist in RNA. Instead, RNA contains the closely related pyrimidine base uracil.

The first point is critical. Look back at Figure 4.1b and compare the functional groups attached to ribose and deoxyribose. Notice the hydroxyl (−OH) group on the 2′ carbon of ribose. This additional hydroxyl is much more reactive than the hydrogen atom on the 2′ carbon of deoxyribose. When RNA molecules fold in certain ways, the hydroxyl group can attack the phosphate linkage between nucleotides, which would generate a break in the sugar-phosphate backbone. While this −OH group makes RNA much less stable than DNA, it can also support catalytic activity by the molecule.

Secondary Structure Like DNA molecules, most RNA molecules have secondary structure that results from complementary base pairing between purine and pyrimidine bases. In RNA, adenine forms hydrogen bonds only with uracil, and guanine again forms hydrogen bonds with cytosine. (Other, non-Watson–Crick base pairs occur, although less frequently.) Three hydrogen bonds form between guanine and cytosine, but only two form between adenine and uracil.

This hydrogen bonding should seem familiar, since DNA bonds together in a similar manner—so how do the secondary structures of RNA and DNA differ? In the vast majority of cases, the purine and pyrimidine bases in RNA undergo hydrogen

FIGURE 4.9 Complementary Base Pairing and Secondary Structure in RNA: Stem-and-Loop Structures. This RNA molecule has secondary structure. The double-stranded "stem" and single-stranded "loop" form a hairpin. The bonded bases in the stem are antiparallel, meaning that they are oriented in opposite directions.

bonding with complementary bases on the *same strand*, rather than forming hydrogen bonds with complementary bases on a different strand, as in DNA.

FIGURE 4.9 shows how within-strand base pairing works. The key is that when bases on one part of an RNA strand fold over and align with ribonucleotides on another part of the same strand, the two sugar-phosphate strands are antiparallel. In this orientation, hydrogen bonding between complementary bases results in a stable double helix.

If the section where the fold occurs includes unpaired bases, then the stem-and-loop configuration shown in Figure 4.9 results. This type of secondary structure is called a **hairpin.** Several other types of RNA secondary structures are possible, each involving a different length and arrangement of base-paired segments.

Like the α-helices and β-pleated sheets observed in many proteins, RNA secondary structures can form spontaneously. They are directed by hydrophobic interactions and stabilized by hydrogen bonding between the bases. Even though hairpins and other types of secondary structure reduce the entropy of RNA molecules, the energy released in these interactions makes the overall process favorable.

Tertiary Structure RNA molecules can also have tertiary structure, which arises when secondary structures fold into more complex shapes. As a result, RNA molecules with different base sequences can have very different overall shapes and chemical properties. RNA molecules are much more diverse in size, shape, and reactivity than DNA molecules are. Structurally and chemically, RNA is intermediate between the complexity of proteins and the simplicity of DNA.

TABLE 4.1 summarizes the similarities and differences in the structures of RNA and DNA.

RNA's Structure Makes It an Extraordinarily Versatile Molecule

In terms of structure, you've seen that RNA is intermediate between DNA and proteins. RNA is intermediate in terms of function as well. RNA molecules cannot archive information nearly as efficiently as DNA molecules do, but they do perform key functions in information processing. Likewise, they cannot catalyze as many reactions as proteins do. But as it turns out, the reactions they do catalyze are particularly important.

In cells, RNA molecules function like a jackknife or a pocket tool with an array of attachments: They perform a wide variety of

SUMMARY TABLE 4.1   DNA and RNA Structure

Level of Structure	DNA		RNA	
Primary	Sequence of deoxyribonucleotides; bases are A, T, G, C	5′ A A T G T G C C G 3′	Sequence of ribonucleotides; bases are A, U, G, C	5′ U U A C A G G C 3′
Secondary	Two antiparallel strands twist into a double helix, stabilized by hydrogen bonding between complementary bases (A-T, G-C) and hydrophobic interactions		Most common are hairpins, formed when a single strand folds back on itself to form a double-helix "stem" and a single-stranded "loop"	
Tertiary	None*		Folds that form distinctive three-dimensional shapes	Example: tRNA

*In cells, DNA coils around proteins that bind to the double helix. In many cases the DNA-protein complex folds into highly organized, compact structures. But DNA does not form tertiary structure on its own.

tasks reasonably well. Some of the most surprising results in the last decade of biological science, in fact, involve new insights into the diversity of roles that RNAs play in cells. These molecules process information stored in DNA, synthesize proteins, and defend against attack by viruses, among other things.

Next let's focus on the roles that RNA could have played in the origin of life—as an information-containing entity and as a catalyst.

RNA Is an Information-Containing Molecule

Because RNA contains a sequence of bases analogous to the letters in a word, it can function as an information-containing molecule. And because hydrogen bonding occurs specifically between A-U pairs and G-C pairs in RNA, it is possible for RNA to furnish the information required to make a copy of itself.

FIGURE 4.10 illustrates how the information stored in an RNA molecule can be used to direct its own replication.

First, a complementary copy of the RNA is made when free ribonucleotides form hydrogen bonds with complementary bases on the original strand of RNA—the template strand. As they do, their sugar-phosphate groups form phosphodiester linkages to produce a double-stranded RNA molecule (steps 1 and 2).

To make a copy of the original single-stranded RNA, the hydrogen bonds between the double-stranded product must first be broken by heating or by a catalyzed reaction (step 3). The newly made complementary RNA molecule now exists independently of the original template strand. If steps 1–3 were repeated with the new strand serving as a template (steps 4–6), the resulting molecule would be a copy of the original. In this way, the primary sequence of an RNA serves as a mold.

RNA Can Function as a Catalytic Molecule

In terms of diversity in chemical reactivity and overall shape, RNA molecules are no match for proteins. The primary structure of RNA molecules is much more restricted because RNA has only four types of nucleotides versus the 20 types of amino acids found in proteins. Secondary through tertiary structure is more limited as a result, meaning that RNA cannot form the wide array of catalysts observed among proteins.

But because RNA has a degree of structural and chemical complexity, it is capable of catalyzing a number of chemical reactions. Sidney Altman and Thomas Cech shared the 1989 Nobel Prize in chemistry for showing that catalytic RNAs, or **ribozymes,** exist in organisms.

FIGURE 4.11 (on page 68) shows the structure of a ribozyme Cech isolated from a single-celled organism called *Tetrahymena*. This ribozyme catalyzes both the hydrolysis and the condensation of phosphodiester linkages in RNA. Researchers have since discovered a variety of ribozymes that catalyze an array of reactions in cells. For example, ribozymes catalyze the formation of peptide bonds when amino acids polymerize to form polypeptides. Ribozymes are at work in your cells right now.

The three-dimensional nature of ribozymes is vital to their catalytic activity. To catalyze a chemical reaction, substrates must

PROCESS: RNA FORMS A TEMPLATE FOR ITS OWN SYNTHESIS

1. Complementary bases pair.
2. Copied strand polymerizes.
3. Copy and template separate.
4. Copy serves as new template.
5. New copy polymerizes.
6. New copy is identical to original template.

FIGURE 4.10 RNA Molecules Contain Information That Allows Them to Be Replicated. For a single-stranded RNA to be copied, it must pass through double-stranded RNA intermediates.

Folding brings widely spaced nucleotides together at the active site of this catalytic RNA

FIGURE 4.11 Tertiary Structure of the *Tetrahymena* Ribozyme. The folded structure brings together bases from distant locations in the primary structure to form the active site.

be brought together in an environment that will promote the reaction. As with protein enzymes, the region of the ribozyme that is responsible for this activity is called the active site. When the *Tetrahymena* ribozyme was compared to protein enzymes that catalyze similar reactions, their active sites were found to be similar in structure. This observation about two very different molecules demonstrates the critical relationship between structure and function

The discovery of ribozymes was a watershed event in origin-of-life research. Before Altman and Cech published their results, most biologists thought that the only molecules capable of catalyzing reactions in cells were proteins. The fact that a ribozyme in *Tetrahymena* catalyzed a condensation reaction raised the possibility that an RNA molecule could make a copy of itself. Such a molecule could qualify as the first living entity. Is there any experimental evidence to support this hypothesis?

4.4 In Search of the First Life-Form

The theory of chemical evolution maintains that life began as a naked self-replicator—a molecule that existed by itself in solution, without being enclosed in a membrane. To make a copy of itself, that first living molecule had to (1) provide a template that could be copied, and (2) catalyze polymerization reactions that would link monomers into a copy of that template. Because RNA is capable of both processes, most origin-of-life researchers propose that the first life-form was made of RNA.

No self-replicating RNA molecules have been discovered in nature, however, so researchers test the hypothesis by trying to simulate the RNA world in the laboratory. The eventual goal is to create an RNA molecule that can catalyze its own replication.

How Biologists Study the RNA World

To understand how researchers go about testing the RNA world hypothesis, consider two recent experiments by researchers in David Bartel's laboratory. In one study, the team attempted to generate an RNA molecule that could catalyze the kind of template-directed polymerization needed for RNA replication—an RNA "replicase." Starting with a ribozyme capable of joining two ribonucleotides together, they generated billions of copies into which random mutations were introduced.

Next they incubated the mutants with free ribonucleotides and began selecting for replicase activity. Molecules that exhibited such activity were isolated and copied. After two weeks and 18 rounds of selection, the team succeeded in isolating a ribozyme that could add 14 nucleotides to an existing RNA strand.

Note that the team's experimental protocol was designed to mimic the process of natural selection introduced in Chapter 1. The population of RNAs from each round had variable characteristics that could be replicated and passed on to the next generation of ribozymes. In addition, the researchers were able to select the most efficient RNAs to be the "parents" of the next generation—and in the process introduce new mutations that potentially could make some of the "offspring" even better ribozymes.

This research created considerable excitement among biologists interested in the origin of life, because adding ribonucleotides to a growing strand is a key attribute of an RNA replicase. However, since the maximum product length generated was less than 10 percent of the ribozyme's own length, an RNA replicase capable of making a full-length copy of itself was far from being discovered. In fact, the difficulty in creating an effective RNA replicase has led many researchers to question the idea of a replicase being the first ribozyme to emerge in the RNA world.

In another study, Bartel's group asked a different question: Would it be possible to select for a ribozyme that could make ribonucleotides? This type of ribozyme is not known to exist in nature but would be a key component in the RNA world.

Recall that the direction of a chemical reaction and how much product it makes is influenced by the amount of reactants present (Chapter 2). Since the chemical evolution of nucleotides is thought to have been inefficient, nucleotides would have been a scarce resource on early Earth. Ribozymes that could catalyze the production of nucleotides would be more likely to be copied due to local accumulation of monomers.

Starting with a large pool of randomly generated RNA sequences, the researchers selected for RNAs that could catalyze the addition of a uracil base to a ribose sugar. By round 11, the group had recovered ribozymes that were 50,000 times better at catalyzing the reaction than those found in the fourth round and over 1 million times more efficient than the uncatalyzed reaction. In effect, molecular evolution had occurred in the reaction tubes.

Thanks to similar efforts at other laboratories around the world, biologists have produced an increasingly impressive set of catalytic activities from RNA molecules. The results from each of these studies help clarify our view of what occurred in the RNA world. If a living ribozyme ever existed, then each round of simulated molecular evolution brings us closer to resurrecting it.

The RNA World May Have Sparked the Evolution of Life

Although ribozymes like these lab-generated molecules may have been present in the RNA world, they have not been observed in nature. Of those that have been discovered in modern cells, most play key roles in the synthesis of proteins. This relationship suggests the order of events in chemical evolution—the RNA world preceded proteins.

The evolution of protein enzymes would have marked the end of the RNA world—providing the means for catalyzing reactions necessary for life to emerge in a cellular form. After this milestone, three of the five fundamental characteristics of life (see Chapter 1) were solidly in place:

1. *Information* Proteins and ribozymes were processing information stored in nucleic acids for the synthesis of more proteins.

2. *Replication* Enzymes, and possibly ribozymes, were replicating the nucleic acids that stored the hereditary information.

3. *Evolution* Random changes in the synthesis of proteins, and selective advantages resulting from some of these changes, allowed for the evolution of new proteins and protein families.

If these events occurred in a hydrothermal vent, the molecular assemblages of nucleic acids and proteins would have been constantly fed with thermal and chemical energy. To gain independence from their undersea hatchery, enzymes would have evolved to store this energy as something more portable—carbohydrates. The structure and function of carbohydrates will be the focus of the next chapter.

CHAPTER 4 REVIEW

For media, go to MasteringBiology

If you understand . . .

4.1 What Is a Nucleic Acid?

- Nucleic acids are polymers of nucleotide monomers, which consist of a sugar, a phosphate group, and a nitrogenous base. Ribonucleotide monomers polymerize to form RNA. Deoxyribonucleotide monomers polymerize to form DNA.

- Ribonucleotides have a hydroxyl (—OH) group on their 2′ carbon; deoxyribonucleotides do not.

- Nucleic acids polymerize when condensation reactions join nucleotides together via phosphodiester linkages.

- Nucleic acids are directional: they have a 5′ end and a 3′ end. During polymerization, new nucleotides are added only to the 3′ end.

✔ You should be able to state what cells do to activate nucleotides for incorporation into a polymer and explain why activation is required.

4.2 DNA Structure and Function

- DNA's primary structure consists of a sequence of linked nitrogenous bases. Its secondary structure consists of two DNA strands running in opposite directions that are twisted into a double helix.

- DNA is an extremely stable molecule that serves as a superb archive for information in the form of base sequences. It lacks a reactive 2′ hydroxyl group, and its secondary structure is stabilized by hydrophobic interactions and hydrogen bonds that form between complementary bases stacked on the inside of the helix.

- DNA is readily copied via complementary base pairing. Complementary base pairing occurs between A-T and G-C pairs in DNA.

- DNA's structural stability and regularity are advantageous for information storage, but they make DNA an ineffective catalyst.

✔ You should be able to explain why DNA molecules with a high percentage of guanine and cytosine are particularly stable.

4.3 RNA Structure and Function

- Like DNA, RNA's primary structure consists of a sequence of linked nitrogenous bases. RNA's secondary structure includes short regions of double helices and looped structures called hairpins.

- RNA molecules are usually single stranded. They have secondary structure because of complementary base pairing between A-U and G-C pairs on the same strand.

- Unlike DNA, the secondary structures of RNA can fold into more complex shapes, stabilized by hydrogen bonding, which give the molecule tertiary structure.

- RNA is versatile. The primary function of proteins is to catalyze chemical reactions, and the primary function of DNA is to carry information. But RNA is an "all-purpose" macromolecule that can do both.

✔ You should be able to explain why many RNA molecules exhibit tertiary structure, while most DNA molecules do not.

4.4 In Search of the First Life-Form

- To test the RNA world hypothesis, researchers are attempting to synthesize new ribozymes in the laboratory. Using artificial selection strategies, they have succeeded in identifying RNAs that catalyze several different reactions.

- Ribozymes that catalyze reactions necessary for the production of nucleotides may have preceded the evolution of RNA replicases.

✔ You should be able to provide two examples of activities in the RNA world you expect would benefit from catalysis and justify your choices.

MasteringBiology

1. MasteringBiology Assignments

Tutorials and Activities Double Helix; Heritable Information: DNA; Nucleic Acid Building Blocks; Nucleic Acid Structure; Structure of RNA and DNA

Questions Reading Quizzes, Blue-Thread Questions, Test Bank

2. eText Read your book online, search, take notes, highlight text, and more.

3. The Study Area Practice Test, Cumulative Test, BioFlix® 3-D Animations, Videos, Activities, Audio Glossary, Word Study Tools, Art

You should be able to . . .

✔ TEST YOUR KNOWLEDGE

Answers are available in Appendix A

1. What are the four nitrogenous bases found in RNA?
   a. uracil, guanine, cytosine, thymine (U, G, C, T)
   b. adenine, guanine, cytosine, thymine (A, G, C, T)
   c. adenine, uracil, guanine, cytosine (A, U, G, C)
   d. alanine, threonine, glycine, cysteine (A, T, G, C)

2. What determines the primary structure of a DNA molecule?
   a. the sugar-phosphate backbone
   b. complementary base pairing and the formation of hairpins
   c. the sequence of deoxyribonucleotides
   d. the sequence of ribonucleotides

3. DNA attains a secondary structure when hydrogen bonds form between the nitrogenous bases called purines and pyrimidines. What are the complementary base pairs that form in DNA?
   a. A-T and G-C
   b. A-U and G-C
   c. A-G and T-C
   d. A-T and G-U

4. Which of the following rules apply to the synthesis of nucleic acids?
   a. Nucleotides are added to the 5′ end of nucleic acids.
   b. The synthesis of nucleic acids cannot occur without the presence of an enzyme to catalyze the reaction.
   c. Strands are synthesized in a parallel direction such that one end of the double-stranded product has the 3′ ends and other has the 5′ ends.
   d. Complementary pairing between bases is required for copying nucleic acids.

5. Nucleic acids are directional, meaning that there are two different ends. What functional groups define the two different ends of a DNA strand?

6. What is responsible for the increased stability of DNA compared to RNA?

✔ TEST YOUR UNDERSTANDING

Answers are available in Appendix A

7. Explain how Chargaff's rules relate to the complementary base pairing seen in the secondary structure of DNA. Would you expect these rules to apply to RNA as well? Explain why or why not.

8. **QUANTITATIVE** If nucleotides from the DNA of a human were quantified and 30 percent of them consisted of adenine, what percentage of guanine nucleotides would be present?
   a. 20 percent
   b. 30 percent
   c. 40 percent
   d. 70 percent

9. What would be the sequence of the strand of DNA that is made from the following template: 5′-GATATCGAT-3′ (Your answer must be written 5′→3′.) How would this sequence be different if RNA were made from this DNA template?

10. A major theme in this chapter is that the structure of molecules correlates with their function. Explain how DNA's secondary structure limits its catalytic abilities compared with that of RNA. Why is it expected that RNA molecules can catalyze a modest but significant array of reactions?

11. To replicate a ribozyme, a complete complementary copy must be made. Would you expect the double-stranded intermediate to maintain its catalytic activity? Justify your answer with an explanation.

12. Suppose that Bartel's research group succeeded in producing a molecule that could make a copy of itself. Which of the five fundamental characteristics of life (provided in Chapter 1) would support the claim that this molecule is alive?

13. Make a concept map (see **BioSkills 15** in Appendix B) that relates DNA's primary structure to its secondary structure. Your diagram should include deoxyribonucleotides, hydrophobic interactions, purines, pyrimidines, phosphodiester linkages, DNA primary structure, DNA secondary structure, complementary base pairing, and antiparallel strands.

14. Viruses are particles that infect cells. In some viruses, the genetic material consists of two strands of RNA, bonded together via complementary base pairing. Would these antiparallel strands form a double helix? Explain why or why not.

15. Before Watson and Crick published their model of the DNA double helix, Linus Pauling offered a model based on a triple helix. If the three sugar-phosphate backbones were on the outside of such a molecule, would hydrogen bonding or hydrophobic interactions be more important in keeping such a secondary structure together?

16. How would you expect the structure of ribozymes in organisms that grow in very hot environments, such as hot springs or deep-sea vents, to differ from those in organisms that grow in cooler environments?
    a. These ribozymes would have more hairpin secondary structures.
    b. The hairpins would have more G's and C's in the primary structure.
    c. The hairpins would have more A's and U's in the primary structure.
    d. These ribozymes would exhibit no tertiary structure.

5 An Introduction to Carbohydrates

In this chapter you will learn that

The role carbohydrates play in life is based on how they are linked together

by examining

The structure of monosaccharides 5.1

and how they link to form

Polymers called polysaccharides 5.2

then asking

What major roles do carbohydrates play? 5.3

and looking at

Cell structure

Cell identity

Energy storage

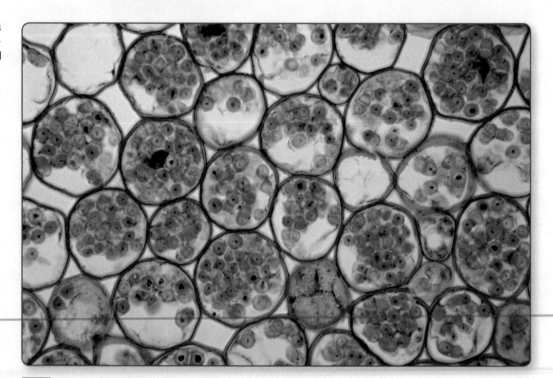

A cross section through a buttercup root. Cellulose-rich cell walls are stained green; starch-filled structures are stained purple. Cellulose is a structural carbohydrate; starch is an energy-storage carbohydrate.

This unit highlights the four types of macromolecules that were key to the evolution of the cell: proteins, nucleic acids, carbohydrates, and lipids. Understanding the structure and function of macromolecules is a basic requirement for exploring how life began and how organisms work. Recall that proteins and nucleic acids could satisfy only three of the five fundamental characteristics of life: information, replication, and evolution (Chapter 4). Carbohydrates, the subject of this chapter, play an important role in a fourth characteristic—energy.

The term **carbohydrate,** or **sugar,** encompasses the monomers called **monosaccharides** (literally, "one-sugar"), small polymers called **oligosaccharides** ("few-sugars"), and the large polymers called **polysaccharides** ("many-sugars"). The name carbohydrate is logical because the molecular formula of many of these molecules is $(CH_2O)_n$, where the n refers to the number of "carbohydrate" groups. The value of n can vary from 3, for the smallest sugar, to well over a thousand for some of the large polymers.

BIG PICTURE

This chapter is part of the Big Picture. See how on pages 104–105.

✔ When you see this checkmark, stop and test yourself. Answers are available in Appendix A.

An aldose
Carbonyl group at
end of carbon chain

A ketose
Carbonyl group in
middle of carbon chain

The name can also be misleading, though, because carbohydrates do not consist of carbon atoms bonded to water molecules. Instead, they are molecules with a carbonyl (C=O) and several hydroxyl (−OH) functional groups, along with several to many carbon–hydrogen (C−H) bonds. Consider formaldehyde, which was introduced as one of the molecules present in early Earth (Chapter 2). Even though formaldehyde has the same molecular formula as the one given above (CH_2O), it is not a carbohydrate since it does not contain a hydroxyl group.

Let's begin with monosaccharides, put them together into polysaccharides, and then explore how carbohydrates figured in the origin of life and what they do in cells today. As you study this material, be sure to ask yourself the central question of biological chemistry: How does this molecule's structure relate to its properties and function?

5.1 Sugars as Monomers

Sugars are fundamental to life. They provide chemical energy in cells and furnish some of the molecular building blocks required for the synthesis of larger, more complex compounds. Monosaccharides were important during chemical evolution, early in Earth's history. For example, as you've seen, the sugar called ribose is required for the formation of the nucleotides that make up nucleic acids (Chapter 4).

What Distinguishes One Monosaccharide from Another?

Monosaccharides, or simple sugars, are the monomers of carbohydrates. **FIGURE 5.1** illustrates two of the smallest monosaccharides. Although their molecular formulas are identical ($C_3H_6O_3$), their molecular structures are different. The carbonyl group that serves as one of monosaccharides' distinguishing features can be found either at the end of the molecule, forming an aldehyde sugar (an aldose), or within the carbon chain, forming a ketone sugar (a ketose). The presence of a carbonyl group along with multiple hydroxyl groups provides sugars with an array of reactive and hydrophilic functional groups. Based on this observation, it's not surprising that sugars are able to participate in a large number of chemical reactions.

The number of carbon atoms present also varies in monosaccharides. By convention, the carbons in a monosaccharide are numbered consecutively, starting with the end nearest the carbonyl group. Figure 5.1 features three-carbon sugars, or **trioses**. Ribose, which acts as a building block for nucleotides, has five carbons and is called a **pentose;** the glucose that is coursing through your bloodstream right now is a six-carbon sugar, or a **hexose.**

Besides varying in the location of the carbonyl group and the total number of carbon atoms present, monosaccharides can vary in the spatial arrangement of their atoms. There is, for example, a wide array of pentoses and hexoses. Each is distinguished by the configuration of its hydroxyl functional groups. **FIGURE 5.2**

Glucose

Galactose

FIGURE 5.2 Sugars May Vary in the Configuration of Their Hydroxyl Groups. The two six-carbon sugars shown here vary only in the spatial orientation of their hydroxyl groups on carbon number 4.

✔**EXERCISE** Mannose is a six-carbon sugar that is identical to glucose, except that the hydroxyl (–OH) group on carbon number 2 is switched in orientation. Circle carbon number 2 in glucose and galactose; then draw the structural formula of mannose.

illustrates glucose and galactose, which are six-carbon sugars. Notice that the two molecules have the same molecular formula ($C_6H_{12}O_6$) but not the same structure. Both are aldose sugars with six carbons, but they differ in the spatial arrangement of the hydroxyl group at the fourth carbon (highlighted in Figure 5.2).

This is a key point: Because the structures of glucose and galactose differ, their functions differ. In cells, glucose is used as a source of carbons for the construction of other molecules and chemical energy that sustains life. But for galactose to be used in these roles, it first has to be converted to glucose via an enzyme-catalyzed reaction. This example underscores a general theme: Even seemingly simple changes in structure—like the location of a single hydroxyl group—can have enormous consequences for function. This is because molecules interact in precise ways, based on their shape.

It's rare for sugars consisting of five or more carbons to exist in the form of the linear chains illustrated in Figure 5.2, however. In aqueous solution they tend to form ring structures. The bond responsible for ring formation occurs only between the carbon containing the carbonyl group and one of the carbons with a hydroxyl group. Glucose serves as the example in **FIGURE 5.3**. When the cyclic structure forms in glucose, the C-1 carbon (the carbon numbered 1 in the linear chain) forms a bond with the oxygen atom of the C-5 hydroxyl and transfers its hydrogen to the C-1 carbonyl, turning it into a hydroxyl group.

Transfer of hydrogen between the C-5 and C-1 functional groups preserves the number of atoms and hydroxyls found in the ring and linear forms. The newly formed C-1 hydroxyl group can be oriented in two distinct ways: above or below the plane of the ring. The different configurations produce the molecules α-glucose and β-glucose.

To summarize, many distinct monosaccharides exist because so many aspects of their structure are variable: aldose or ketose placement of the carbonyl group, variation in carbon number, different arrangements of hydroxyl groups in space, and alternative ring forms. Each monosaccharide has a unique structure and function.

Monosaccharides and Chemical Evolution

Laboratory simulations, like those you read about in Chapter 2, have shown that most monosaccharides are readily synthesized under conditions that mimic the conditions of early Earth. For example, when formaldehyde (CH_2O) molecules are heated in solution, they react with one another to form almost all the pentoses and hexoses.

In addition, researchers have discovered the three-carbon ketose illustrated in Figure 5.1, along with a wide array of compounds closely related to sugars, on a meteorite that struck Murchison, Australia, in 1969. Based on these observations, investigators suspect that sugars are synthesized on dust particles and other debris in interstellar space and could have rained down onto Earth as the planet was forming, as well as being synthesized in the hot water near ancient undersea volcanoes.

More recent evidence suggests that synthesis of sugars could have been catalyzed by minerals found in the walls of deep-sea hydrothermal vents. Most researchers interested in chemical evolution maintain that one or more of the above mechanisms led to the accumulation of monosaccharides in the early oceans.

Modern cells display a wide range of carbohydrates beyond monosaccharides. How do these monomers join together to form polymers? Is the process similar to how amino acids link together to form proteins and nucleotides join to form nucleic acids? Let's explore how the array of functional groups in monosaccharides influences the polymerization of carbohydrates.

(a) Linear form of glucose   **(b)** Ring forms of glucose

FIGURE 5.3 Sugars Exist in Linear and Ring Forms. (a) The linear form of glucose is rare. **(b)** In solution, almost all glucose molecules spontaneously react to form one of two ring structures, called the α and β forms of glucose. The two forms exist in equilibrium, but the β form is more common because it is slightly more stable than the α form.

check your understanding

If you understand that . . .

- Simple sugars differ from each other in three respects:
  1. the location of their carbonyl group,
  2. the number of carbon atoms present, and
  3. the spatial arrangement of their atoms— particularly the relative positions of hydroxyl (–OH) groups.

✔ **You should be able to . . .**

Draw the structural formula of a three-carbon monosaccharide ($C_3H_6O_3$) in linear form and then draw three other sugars that illustrate the three differences listed above.

Answers are available in Appendix A.

5.2 The Structure of Polysaccharides

Simple sugars can be covalently linked into chains of varying lengths, also known as complex carbohydrates. These chains range in size from small oligomers, or oligosaccharides, to the large polymers called polysaccharides. When only two sugars are linked together, they are known as **disaccharides.**

Similar to proteins and nucleic acids, the structure and function of larger carbohydrates depends on the types of monomers involved and how they are linked together. For example, maltose, also known as malt sugar, and lactose, an important sugar in milk, are two disaccharides that differ by just one monosaccharide. Maltose consists of two identical glucose molecules (**FIGURE 5.4a**), while lactose is made up of glucose and galactose (**FIGURE 5.4b**).

Monosaccharides polymerize when a condensation reaction occurs between two hydroxyl groups, resulting in a covalent interaction called a **glycosidic linkage.** The inverse reaction, hydrolysis, cleaves these linkages. (To review condensation and hydrolysis reactions, see Chapter 3.)

(a) Formation of α-glycosidic linkage

(b) Formation of β-glycosidic linkage

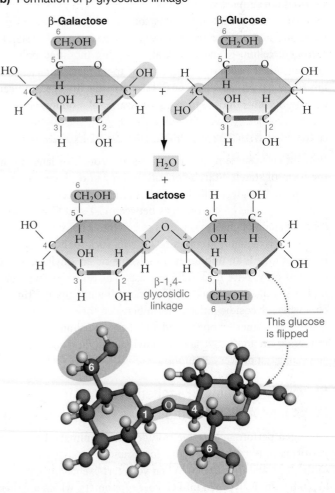

FIGURE 5.4 Monosaccharides Polymerize through Formation of Glycosidic Linkages. A glycosidic linkage occurs when hydroxyl groups on two monosaccharides undergo a condensation reaction. Maltose and lactose are disaccharides.

In that they hold monomers together, glycosidic linkages are analogous to the peptide bonds and phosphodiester linkages in proteins and nucleic acids. There is an important difference, however. Peptide bonds and phosphodiester linkages form between the same locations in their monomers, giving proteins and nucleic acids a standard backbone structure, but this is not the case for polysaccharides. Because glycosidic linkages form between hydroxyl groups, and because every monosaccharide contains at least two hydroxyls, the location and geometry of glycosidic linkages can vary widely among polysaccharides.

Maltose and lactose illustrate two of the most common glycosidic linkages, called the α-1,4-glycosidic linkage and the β-1,4-glycosidic linkage. The numbers refer to the carbons on either side of the linkage, indicating that both linkages are between the C-1 and C-4 carbons. Their geometry, however, is different: α and β refer to the contrasting orientations of the C-1 hydroxyls—on opposite sides of the plane of the glucose rings (i.e., "above" versus "below" the plane).

As Section 5.3 explains, the orientation of this hydroxyl in glycosidic linkages is particularly important in the structure, function, and durability of the molecules. In essence, the difference between polysaccharides used for storage and structural polysaccharides is a simple twist of a link.

To drive this point home, let's consider the structures of the most common polysaccharides found in organisms today: starch, glycogen, cellulose, and chitin, along with a modified polysaccharide called peptidoglycan. Each of these macromolecules is joined by particular α-1,4- or β-1,4-glycosidic linkages and can consist of a few hundred to many thousands of monomers.

Starch: A Storage Polysaccharide in Plants

In plant cells, some monosaccharides are stored for later use in the form of starch. **Starch** consists entirely of α-glucose monomers joined by glycosidic linkages. As the top panel in **TABLE 5.1** shows, the angle of the linkages between C-1 and C-4 carbons causes a chain of glucose subunits to coil into a helix.

Starch is actually a mixture of two such polysaccharides, however. One is an unbranched molecule called amylose, which contains only α-1,4-glycosidic linkages. The other is a branched molecule called amylopectin. The branching in amylopectin occurs when glycosidic linkages form between the C-1 carbon of a glucose monomer on one strand and the C-6 carbon of a glucose monomer on another strand. In amylopectin, branches occur at about one out of every 30 monomers.

Glycogen: A Highly Branched Storage Polysaccharide in Animals

Glycogen performs the same storage role in animals that starch performs in plants. In humans, for example, glycogen is stored in the liver and in muscles. When you start exercising, enzymes begin breaking glycogen into glucose monomers, which are then processed in muscle cells to supply energy. Glycogen is a polymer of α-glucose and is nearly identical to the branched form of starch. However, instead of an α-1,6-glycosidic linkage occurring

in about 1 out of every 30 monomers, a branch occurs in about 1 out of every 10 glucose subunits (see Table 5.1).

Cellulose: A Structural Polysaccharide in Plants

All cells are enclosed by a membrane (Chapter 1). In most organisms living today, the cell is also surrounded by a layer of material called a wall. A **cell wall** is a protective sheet that occurs outside the membrane. In plants, bacteria, fungi, and many other groups, the cell wall is composed primarily of one or more polysaccharides.

In plants, cellulose is the major component of the cell wall. **Cellulose** is a polymer of β-glucose monomers, joined by β-1,4-glycosidic linkages. As Table 5.1 shows, the geometry of the linkage is such that each glucose monomer in the chain is flipped in relation to the adjacent monomer. The flipped orientation is important because (**1**) it generates a linear molecule, rather than the helix seen in starch; and (**2**) it permits multiple hydrogen bonds to form between adjacent, parallel strands of cellulose. As a result, cellulose forms long, parallel strands that are joined by hydrogen bonds. The linked cellulose fibers are strong and give the cell structural support.

Chitin: A Structural Polysaccharide in Fungi and Animals

Chitin is a polysaccharide that stiffens the cell walls of fungi. It is also found in a few types of protists and in many animals. It is, for example, the most important component of the external skeletons of insects and crustaceans.

Chitin is similar to cellulose, but instead of consisting of glucose monomers, the monosaccharide involved is one called N-acetylglucosamine (abbreviated as NAG). These NAG monomers are joined by β-1,4-glycosidic linkages (see Table 5.1). As in cellulose, the geometry of these bonds results in every other residue being flipped in orientation.

Like the glucose monomers in cellulose, the NAG subunits in chitin form hydrogen bonds between adjacent strands. The result is a tough sheet that provides stiffness and protection.

Peptidoglycan: A Structural Polysaccharide in Bacteria

Most bacteria, like all plants, have cell walls. But unlike plants, in bacteria the ability to produce cellulose is extremely rare. Instead, a polysaccharide called **peptidoglycan** gives bacterial cell walls strength and firmness.

Peptidoglycan is the most complex of the polysaccharides discussed thus far. It has a long backbone formed by two types of monosaccharides that alternate with each other and are linked by β-1,4-glycosidic linkages (see Table 5.1). In addition, a short chain of amino acids is attached to one of the two sugar types. When molecules of peptidoglycan align, peptide bonds link the amino acid chains on adjacent strands. These links serve the same purpose as the hydrogen bonds between the parallel strands of cellulose and chitin in the cell walls of other organisms.

Polysaccharide	Chemical Structure	Three-dimensional Structure

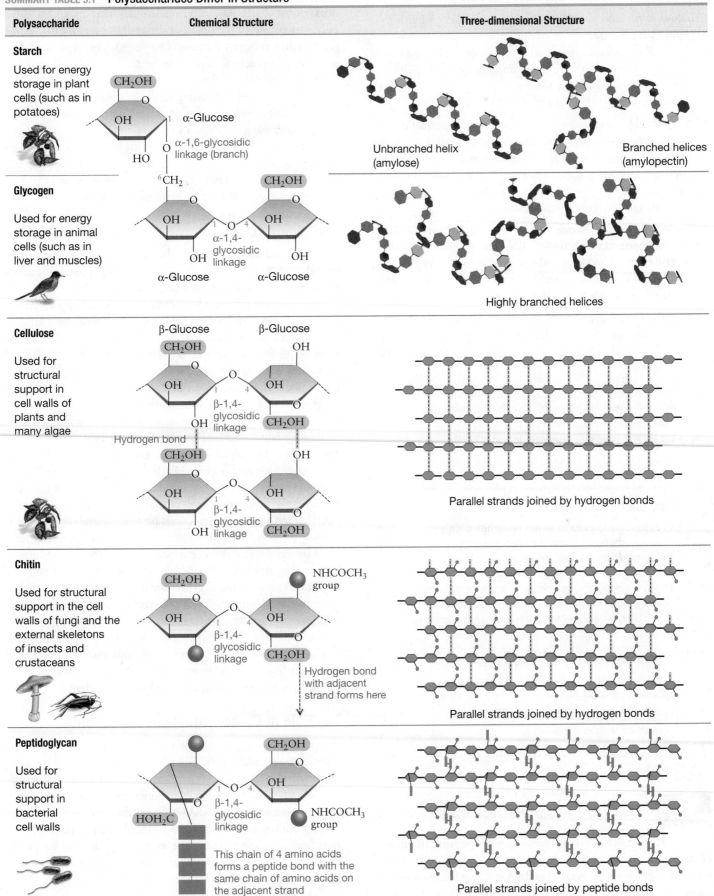

Starch

Used for energy storage in plant cells (such as in potatoes)

CH_2OH

OH   α-Glucose

HO   α-1,6-glycosidic linkage (branch)

Unbranched helix (amylose)

Branched helices (amylopectin)

Glycogen

Used for energy storage in animal cells (such as in liver and muscles)

$^6CH_2$

OH CH_2OH

OH   α-1,4-glycosidic linkage   OH

α-Glucose   α-Glucose

Highly branched helices

Cellulose

Used for structural support in cell walls of plants and many algae

β-Glucose   β-Glucose

CH_2OH OH

OH   OH

β-1,4-glycosidic linkage CH_2OH

Hydrogen bond

CH_2OH OH

OH   OH

β-1,4-glycosidic linkage CH_2OH

Parallel strands joined by hydrogen bonds

Chitin

Used for structural support in the cell walls of fungi and the external skeletons of insects and crustaceans

CH_2OH $NHCOCH_3$ group

OH   OH

β-1,4-glycosidic linkage CH_2OH

Hydrogen bond with adjacent strand forms here

Parallel strands joined by hydrogen bonds

Peptidoglycan

Used for structural support in bacterial cell walls

CH_2OH

OH

β-1,4-glycosidic linkage $NHCOCH_3$ group

HOH_2C

This chain of 4 amino acids forms a peptide bond with the same chain of amino acids on the adjacent strand

Parallel strands joined by peptide bonds

Polysaccharides and Chemical Evolution

Cellulose is the most abundant organic compound on Earth today, and chitin is probably the second most abundant by weight. Virtually all organisms depend on glycogen or starch as an energy source. But despite their current importance to life, polysaccharides probably played little to no role in the origin of life. This conclusion is supported by several observations:

- *No plausible mechanism exists for the polymerization of monosaccharides under conditions that prevailed early in Earth's history.* In cells and in laboratory experiments, the glycosidic linkages illustrated in Figure 5.4 and Table 5.1 form only with the aid of protein enzymes. No enzyme-like RNAs are known to catalyze these reactions.

- *To date, no polysaccharide has been discovered that can catalyze polymerization reactions.* Even though polysaccharides contain reactive hydroxyl and carbonyl groups, they lack the structural and chemical complexity that makes proteins, and to a lesser extent RNA, effective catalysts.

- *The monomers in polysaccharides are not capable of complementary base pairing.* Like proteins, but unlike nucleic acids, polysaccharides cannot act as templates for their own replication.

Even though polysaccharides probably did not play a significant role in the earliest forms of life, they became enormously important once cellular life evolved. In the next section, let's take a detailed look at how they function in today's cells.

check your understanding

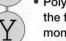

If you understand that . . .

- Polysaccharides form when enzymes catalyze the formation of glycosidic linkages between monosaccharides that are in the α or β form.
- Most polysaccharides are long chains of monosaccharides, but some branch extensively. Among linear forms, it is common for adjacent strands to be connected by hydrogen bonding or other types of linkages.

✓ **You should be able to . . .**

Provide four structural differences that could result in different oligosaccharides consisting of two glucose monomers and two galactose monomers.

Answers are available in Appendix A.

5.3 What Do Carbohydrates Do?

One of the basic functions that carbohydrates perform in organisms is to serve as a substrate for synthesizing more-complex molecules. For example, recall that RNA contains the five-carbon sugar ribose ($C_5H_{10}O_5$) and DNA contains the modified sugar deoxyribose ($C_5H_{10}O_4$). The nucleotides that make up these polymers consist of the ribose or deoxyribose sugar, a phosphate group, and a nitrogenous base (Chapter 4). The sugar itself acts as a subunit of each of these monomers.

In addition, sugars frequently furnish the raw "carbon skeletons" that are used as building blocks in the synthesis of important molecules. Your cells are producing amino acids right now, for example, using sugars as a starting point.

Carbohydrates have diverse functions in cells: In addition to serving as precursors to larger molecules, they (**1**) provide fibrous structural materials, (**2**) indicate cell identity, and (**3**) store chemical energy. Let's look at each function in turn.

Carbohydrates Can Provide Structural Support

Cellulose and chitin, along with the modified polysaccharide peptidoglycan, are key structural compounds. They form fibers that give cells and organisms strength and elasticity.

To appreciate why cellulose, chitin, and peptidoglycan are effective structural molecules, recall that they form long strands and that bonds can form between adjacent strands. In the cell walls of plants, for example, a collection of about 80 cellulose molecules are cross-linked by hydrogen bonding to create a tough fiber. These cellulose fibers, in turn, crisscross to form a tough sheet that is able to withstand pulling and pushing forces—what an engineer would call tension and compression.

Besides being stiff and strong, the structural carbohydrates are durable. Almost all organisms have the enzymes required to break the various α-glycosidic linkages that hold starch and glycogen molecules together, but only a few organisms have enzymes capable of hydrolyzing cellulose, chitin, and peptidoglycan. Due to the strong interactions between strands consisting of β-1,4-glycosidic linkages, water is excluded and the fibers tend to be insoluble. The absence of water within these fibers makes their hydrolysis more difficult. As a result, the structural polysaccharides are resistant to degradation and decay.

Ironically, the durability of cellulose supports digestion. The cellulose that you ingest when you eat plant cells—what biologists call dietary fiber—forms a porous mass that absorbs and retains water. This sponge-like mass adds moisture and bulk that helps fecal material move through the intestinal tract more quickly, preventing constipation and other problems.

The Role of Carbohydrates in Cell Identity

Structural polymers tend to be repetitive, with only one or two types of monosaccharides. The same is not true for all complex carbohydrates. Some types exhibit enormous structural diversity, because their component monomers—and the linkages between them—vary a lot. As a result, they are capable of displaying information to other cells through their structure. More specifically, polysaccharides act as an identification badge on the outer surface of the plasma membrane that surrounds a cell. (Chapter 6 describes plasma membranes in detail.)

FIGURE 5.5 Carbohydrates Are an Identification Badge for Cells.
Glycoproteins contain sugar groups that project outside the cell
from the surface of the plasma membrane enclosing the cell. These
sugar groups have distinctive structures that identify the type or
species of the cell.

FIGURE 5.5 shows how this information about cell identity is
displayed. Molecules called glycoproteins project outward from
the cell surface into the surrounding environment. A **glycoprotein** is a protein that has one or more carbohydrates covalently
bonded to it—usually relatively short oligosaccharides.

Glycoproteins are key molecules in what biologists call cell–
cell recognition and cell–cell signaling. Each cell in your body
has glycoproteins on its surface that identify it as part of your
body. Immune system cells use these glycoproteins to distinguish
your body's cells from foreign cells, such as bacteria. In addition,
each distinct type of cell in a multicellular organism—for example, the nerve cells and muscle cells in your body—displays a different set of glycoproteins on its surface.

The identification information displayed by glycoproteins
helps cells recognize and communicate with each other.

The key point here is to recognize that the variety in the types
of monosaccharides and how they can be linked together makes
it possible for an enormous number of unique oligosaccharides
to exist. As a result, each cell type and each species can display a
unique identity.

During the 1980s, Paul Wassarman and colleagues inves-
tigated the role of glycoproteins in one of the most important
cell–cell recognition events in the life of a plant or animal—the
attachment of sperm to eggs during fertilization. This step guar-
antees specificity—sperm recognize and bind only to eggs of
their own species.

In one experiment, the researchers mixed sperm with purified
egg-surface glycoproteins and discovered that most of the sperm
lost their ability to attach to eggs (**FIGURE 5.6**). Such loss of function is an example of what researchers call competitive inhibition. The glycoproteins had bound to—and thus blocked—the
same structure on the sperm that it uses to bind to eggs. This
result showed that sperm attach to eggs via egg glycoproteins.

RESEARCH

QUESTION: What part of surface glycoproteins
do sperm recognize when they attach to eggs?

HYPOTHESIS: Sperm attach to the carbohydrate component.

NULL HYPOTHESIS: Sperm attach to the protein component.

EXPERIMENTAL SETUP:

Glycoprotein

Carbohydrate   Protein

Sperm

Egg

1. **Isolate glyco-
proteins** from the
surface of a mouse
egg.

2. **Separate protein
and carbohydrate**
components.

3. **Treat sperm** with
components; allow
time for component
binding.

4. **Test attachment
inhibition:** Add mouse
eggs to treated sperm;
allow time for sperm
binding.

PREDICTION: The carbohydrate component of the glycoprotein
will bind to sperm and block their attachment to eggs.

PREDICTION OF NULL HYPOTHESIS: The protein component of
the glycoprotein will block sperm attachment to eggs.

RESULTS:

CONCLUSION: Sperm recognize and bind to the carbohydrates
of egg-surface glycoproteins when they attach to egg cells.

**FIGURE 5.6 Carbohydrates Are Required for Cellular Recognition
and Attachment.**

SOURCES: Florman, H. M., K. B. Bechtol, and P. M. Wassarman. 1984. Enzymatic dissection of
the functions of the mouse egg's receptor for sperm. *Developmental Biology* 106: 243–255. Also
Florman, H. M., and P. M. Wassarman. 1985. O-linked oligosaccharides of mouse egg ZP3 account
for its sperm receptor activity. *Cell* 41: 313–324.

✔**QUANTITATIVE** How would the bars change in the graph if sperm
attachment required only the protein portion of egg glycoproteins?

But which part of the egg glycoproteins is essential for recognition and attachment—the protein or the carbohydrate? In follow-up experiments, Wassarman's group used the same type of competitive-binding assay to answer this question. When sperm were mixed with purified carbohydrates alone, most were unable to attach to eggs. In contrast, most sperm treated with purified protein alone were not inhibited and still attached to eggs. Both results show that the carbohydrate component plays a fundamental role in the process of egg-cell recognition.

Carbohydrates and Energy Storage

Candy-bar wrappers promise a quick energy boost, and ads for sports drinks claim that their products provide the "carbs" needed for peak activity. If you were to ask friends or family members what carbohydrates do in your body, they would probably say something like "They give you energy." And after pointing out that carbohydrates are also used in cell identity, as a structural material, and as a source of carbon skeletons for the synthesis of other complex molecules, you'd have to agree.

Carbohydrates store and provide chemical energy in cells. What aspect of carbohydrate structure makes this function possible?

Carbohydrates Store Sunlight as Chemical Energy Recall that the essence of chemical evolution was energy transformations (Chapter 2). For example, it was proposed that the kinetic energy in sunlight may have been converted into chemical energy and stored in bonds of molecules such as formaldehyde (CH_2O).

This same type of transformation from light energy to chemical energy occurs in cells today, but instead of making formaldehyde, cells produce sugars. For example, plants harvest the kinetic energy in sunlight and store it in the bonds of carbohydrates by the process known as **photosynthesis.** (Photosynthesis is the focus of Chapter 10.)

Photosynthesis entails a complex set of reactions that can be summarized most simply as follows:

$$CO_2 + H_2O + sunlight \longrightarrow (CH_2O)_n + O_2$$

where $(CH_2O)_n$ represents a carbohydrate. The key to understanding the energy conversion that is taking place in this reaction is to compare the positions of the electrons in the reactants to those in the products.

1. The electrons in the C=O bonds of carbon dioxide and the C—O bonds of carbohydrates are held tightly because of oxygen's high electronegativity. Thus, they have relatively low potential energy.

2. The electrons involved in the C—H bonds of carbohydrates are shared equally because the electronegativity of carbon and hydrogen is about the same. Thus, these electrons have relatively high potential energy.

3. Electrons are also shared equally in the carbon–carbon C—C bonds of carbohydrates—meaning that they, too, have relatively high potential energy.

(a) Carbon dioxide

(b) A carbohydrate

(c) A fatty acid (a component of fat molecules)

FIGURE 5.7 In Organisms, Potential Energy Is Stored in C—H and C—C Bonds. (a) In carbon dioxide, the electrons involved in covalent bonds are held tightly by oxygen atoms. **(b)** In carbohydrates such as the sugar shown here, many of the covalently bonded electrons are held equally between C and H atoms. **(c)** The fatty acids found in fat molecules have more C—H bonds and fewer C—O bonds than carbohydrates do. ("R" stands for the rest of the molecule.)

✔**EXERCISE** Circle the bonds in this diagram that have high potential energy.

C—C and C—H bonds have much higher potential energy than C—O bonds have. As a result, carbohydrates have much more chemical energy than carbon dioxide has.

FIGURE 5.7 summarizes and extends these points. Start by comparing the structure of carbon dioxide in Figure 5.7a with the carbohydrate in Figure 5.7b. The main difference is the presence of C—C and C—H bonds in the carbohydrate. Now compare the carbohydrate in Figure 5.7b with the fatty acid—a subunit of a fat molecule—in Figure 5.7c. Compared with carbohydrates, fats contain many more C—C and C—H bonds and many fewer C—O bonds.

This point is important. C—C and C—H bonds have high potential energy because the electrons are shared equally by atoms with low electronegativities. C—O bonds, in contrast, have low potential energy because the highly electronegative oxygen atom holds the electrons so tightly. Both carbohydrates and fats are used as fuel in cells, but fats store twice as much energy per gram compared with carbohydrates. (Fats are discussed in more detail in Chapter 6.)

Enzymes Hydrolyze Polysaccharides to Release Glucose Starch and glycogen are efficient energy-storage molecules because they polymerize via α-glycosidic linkages instead of the β-glycosidic linkages observed in the structural polysaccharides. The α-linkages in storage polysaccharides are readily hydrolyzed to release glucose, while the structural polysaccharides resist enzymatic degradation.

The most important enzyme involved in catalyzing the hydrolysis of α-glycosidic linkages in glycogen molecules is a protein called **phosphorylase.** Many of your cells contain phosphorylase, so they can break down glycogen to provide glucose on demand.

The enzymes involved in breaking the α-glycosidic linkages in starch are called **amylases.** Your salivary glands and pancreas produce amylases that are secreted into your mouth and small intestine, respectively. These amylases are responsible for digesting the starch that you eat.

The glucose subunits that are hydrolyzed from glycogen and starch are processed in reactions that result in the production of chemical energy that can be used in the cell. Glycogen and starch are like a candy bar that has segments, so you can break off chunks whenever you need a boost.

Energy Stored in Glucose is Used to Make ATP When a cell needs energy, reactions lead to the breakdown of the glucose and capture of the released energy through synthesis of the nucleotide adenosine triphosphate (ATP) (introduced in Chapter 4).

More specifically, the energy that is released when sugars are processed is used to synthesize ATP from a precursor called adenosine diphosphate (ADP) plus a free inorganic phosphate (P_i) molecule. The overall reaction can be written as follows:

$$(CH_2O)_n + O_2 + ADP + P_i \longrightarrow CO_2 + H_2O + ATP$$

To put this in words, the chemical energy stored in the C—H and C—C bonds of carbohydrate is transferred to a new bond linking a third phosphate group to ADP to form ATP.

How much energy does it take to form ATP? Consider this example: A cell can use the 10 calories of energy stored in a LifeSavers candy to produce approximately 2×10^{23} molecules of ATP. Although this sounds like a lot of ATP, an average human's energy needs would burn through all of this ATP energy in a little over a minute! The energy in ATP drives reactions like polymerization and cellular processes like moving your muscles.

Carbohydrates are like the water that piles up behind a dam; ATP is like the electricity generated at a dam, which lights up your home. Carbohydrates store chemical energy; ATP makes chemical energy useful to the cell.

Later chapters analyze in detail how cells capture and store energy in sugars and how these sugars are then broken down to provide cells with usable chemical energy in the form of ATP (Chapters 8, 9, and 10). For both of these processes to occur, however, a selectively permeable membrane barrier is required. The following chapter introduces the lipids needed to build these membranes and the role they played in the evolution of the first cell.

check your understanding

C Y U

If you understand that . . .

- Carbohydrates provide building blocks for the synthesis of more complex compounds.
- Polysaccharides such as cellulose, chitin, and peptidoglycan form cell walls, which give cells structural strength.
- Glycoproteins project from the surface of cells. They provide a molecular badge that identifies the cell's type or species.
- Starch and glycogen store sugars for later use in reactions that produce ATP. Sugars contain large amounts of chemical energy because they contain carbon atoms that are bonded to hydrogen atoms or other carbon atoms. The C—H and C—C bonds have high potential energy because the electrons are shared equally by atoms with low electronegativity.

✔ **You should be able to . . .**

1. Identify two aspects of the structures of cellulose, chitin, and peptidoglycan that correlate with their function as structural molecules.

2. Describe how the carbohydrates you ate during breakfast today are functioning in your body right now.

Answers are available in Appendix A.

CHAPTER 5 **REVIEW**   *For media, go to MasteringBiology*

If you understand . . .

5.1 Sugars as Monomers

- Monosaccharides are organic compounds that have a carbonyl group and several hydroxyl groups. The molecular formula for a sugar is typically $(CH_2O)_n$, but the number of "carbon-hydrate" groups may vary between sugars, as indicated by the n.

- Although some monosaccharides may have the same molecular formula, the arrangement of functional groups can lead to differences in the molecular structure of the sugars.

- Individual monosaccharides may form ring structures that differ from one another in the orientation of a hydroxyl group.

✔ You should be able to explain how a relatively small difference in the location of a carbonyl or hydroxyl group can lead to dramatic changes in the properties and function of a monosaccharide.

5.2 The Structure of Polysaccharides

- Monosaccharides can be covalently bonded to one another via glycosidic linkages, which join hydroxyl groups on adjacent molecules.

- In contrast to proteins and nucleic acids, polysaccharides do not always form a single uniform backbone structure. The numerous hydroxyls found in each monosaccharide allow glycosidic linkages to form at different sites and new strands to branch from existing chains.

- The types of monomers involved and the geometries of the glycosidic linkages between monomers distinguish different polysaccharides from one another.

- The most common polysaccharides in organisms today are starch, glycogen, cellulose, and chitin; peptidoglycan is an abundant polysaccharide that has short chains of amino acids attached.

✔ You should be able to compare and contrast glycosidic linkages in polysaccharides with the linkages between monomers in proteins and nucleic acids.

5.3 What Do Carbohydrates Do?

- In carbohydrates, as in proteins and nucleic acids, structure correlates with function.

- Cellulose, chitin, and peptidoglycan are polysaccharides that function in support. They are made up of monosaccharide monomers joined by β-1,4-glycosidic linkages. When individual molecules of these polysaccharides align side by side, bonds form between them—resulting in strong, flexible fibers or sheets that resist hydrolysis.

- The oligosaccharides on cell-surface glycoproteins can function as specific signposts or identity tags because their constituent monosaccharides are so diverse in geometry and composition.

- Both starch and glycogen function as energy-storage molecules. They are made up of glucose molecules that are joined by α-glycosidic linkages. These linkages are readily hydrolyzed to release glucose for the production of ATP.

✔ You should be able to describe four key differences in the structure of polysaccharides that function in energy storage versus structural support.

(MB) MasteringBiology

1. MasteringBiology Assignments

Tutorials and Activities Carbohydrates; Carbohydrate Structure and Function; Types of Carbohydrates

Questions Reading Quizzes, Blue-Thread Questions, Test Bank

2. eText Read your book online, search, take notes, highlight text, and more.

3. The Study Area Practice Test, Cumulative Test, BioFlix® 3-D Animations, Videos, Activities, Audio Glossary, Word Study Tools, Art

You should be able to . . .

✔ TEST YOUR KNOWLEDGE

Answers are available in Appendix A

1. What is the difference between a monosaccharide, an oligosaccharide, and a polysaccharide?
   a. the number of carbon atoms in the molecule
   b. the type of glycosidic linkage between monomers
   c. the spatial arrangement of the various hydroxyl residues in the molecule
   d. the number of monomers in the molecule

2. What are three ways monosaccharides differ from one another?

3. What type of bond is formed between two sugars in a disaccharide?
   a. glycosidic linkage
   b. phosphodiester bond
   c. peptide bond
   d. hydrogen bond

4. What holds cellulose molecules together in bundles large enough to form fibers?
   a. the cell wall
   b. peptide bonds
   c. hydrogen bonds
   d. hydrophobic interactions between different residues in the cellulose helix

5. What are the primary functions of carbohydrates in cells?
   a. energy storage, cell identity, structure, and building blocks for synthesis
   b. catalysis, structure, and energy storage
   c. information storage and catalysis
   d. source of carbon, information storage, and energy storage

6. What is responsible for the difference in potential energy between carbohydrates and carbon dioxide?

7. Which of the differences listed here could be found in the same monosaccharide?

    a. different orientation of a hydroxyl in the linear form
    b. different number of carbons
    c. different orientation of a hydroxyl in the ring form
    d. different position of the carbonyl group in the linear form

8. What would most likely occur if the galactose in lactose were replaced with glucose?
    a. It would not be digested by human infants or adults.
    b. It would be digested by most adult humans.
    c. It would be digested by human infants, but not adults.
    d. It would be digested by human adults, but not infants.

9. Explain how the structure of carbohydrates supports their function in displaying the identity of a cell.

10. What is the difference between linking glucose molecules with α-1,4-glycosidic linkages versus β-1,4-glycosidic linkages? What are the consequences?

11. Give three reasons why researchers have concluded that polysaccharides were unlikely to play a large role in the origin of life.

12. Compare and contrast the structures and functions of starch and glycogen. How are these molecules similar? How are they different?

13. A weight-loss program for humans that emphasized minimal consumption of carbohydrates was popular in some countries

in the early 2000s. What was the logic behind this diet? (Note: This diet plan caused controversy and is not endorsed by some physicians and researchers).

14. Galactosemia is a potentially fatal disease that occurs in humans who lack the enzyme that converts galactose to glucose. To treat this disease, physicians exclude the monosaccharide galactose from the diet. Which of the following would you also predict to be excluded from the diet?
    a. maltose   **b.** starch   **c.** mannose   **d.** lactose

15. If you hold a salty cracker in your mouth long enough, it will begin to taste sweet. What is responsible for this change in taste?

16. Lysozyme, an enzyme found in human saliva, tears, and other secretions, catalyzes the hydrolysis of the β-1,4-glycosidic linkages in peptidoglycan. Predict the effect of this enzyme on bacteria, and explain the role its activity plays in human health.

6 Lipids, Membranes, and the First Cells

In this chapter you will learn how

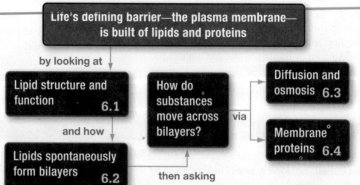

Life's defining barrier—the plasma membrane—
is built of lipids and proteins

by looking at

Lipid structure and
function   6.1

and how

Lipids spontaneously
form bilayers   6.2

How do
substances
move across
bilayers?

then asking

via

Diffusion and
osmosis 6.3

Membrane
proteins 6.4

A space-filling model of a phospholipid bilayer. In single-celled organisms, this cluster of molecules forms part of the boundary between life (inside the cell) and nonlife (outside the cell)—the cell membrane.

This chapter is part of the Big Picture. See how on pages 104–105.

Currently, most biologists support the hypothesis that biological evolution began with a catalytic RNA molecule that could replicate itself. As the offspring of this molecule multiplied, natural selection would have favored the most efficient versions. A second great milestone in the history of life occurred when descendants of these replicators became enclosed within a membrane.

Why is the presence of a membrane so important? The **plasma membrane,** or **cell membrane,** separates life from nonlife. It is a layer of molecules that surrounds the cell interior and separates it from the environment.

- The plasma membrane serves as a selective barrier: It keeps damaging compounds out of the cell and allows entry of compounds needed by the cell.

- Because the plasma membrane sequesters the appropriate chemicals in an enclosed area, reactants collide more frequently—the chemical reactions necessary for life occur much more efficiently.

✓ When you see this checkmark, stop and test yourself. Answers are available in Appendix A.

While researchers of chemical evolution are currently debating when membranes arose—whether early or late during the emergence of life—there is little argument about the importance of this event. After life secured a membrane, it continued to evolve into an efficient and dynamic reaction vessel—the cell.

How do membranes form? Which ions and molecules can pass through a membrane and which cannot, and why? These are some of the most fundamental questions in all of biological science. Let's delve into them, beginning with the membrane's foundation—lipids.

6.1 Lipid Structure and Function

Lipid is a catchall term for carbon-containing compounds that are found in organisms and are largely nonpolar and hydrophobic—meaning that they do not dissolve readily in water. (Recall from Chapter 2 that water is a polar solvent.) Lipids do dissolve, however, in liquids consisting of nonpolar organic compounds.

To understand why lipids are insoluble in water, examine the five-carbon compound called isoprene, illustrated in **FIGURE 6.1a**. Note that isoprene consists of carbon atoms bonded to hydrogen atoms. The figure also shows the structural formula of a chain of linked isoprenes, called an isoprenoid.

Molecules that contain only carbon and hydrogen are known as **hydrocarbons.** Hydrocarbons are nonpolar because electrons are shared equally in C—H bonds—owing to the approximately equal electronegativity of carbon and hydrogen. Since these bonds form no partial charges, hydrocarbons are hydrophobic. Thus lipids do not dissolve in water, because they have a significant hydrocarbon component.

Bond Saturation Is an Important Aspect of Hydrocarbon Structure

FIGURE 6.1b gives the structural formula of a **fatty acid,** a simple lipid consisting of a hydrocarbon chain bonded to a carboxyl (—COOH) functional group. Fatty acids and isoprenes are key building blocks of important lipids found in organisms. Just as subtle differences in the orientation of hydroxyls in sugars can lead to dramatic effects in their structure and function, the type of C—C bond used in hydrocarbon chains is a key factor in lipid structure.

When two carbon atoms form a double bond, the attached atoms are found in a plane instead of a three-dimensional tetrahedron. The carbon atoms involved are also locked into place. They cannot rotate freely, as they do in carbon–carbon single bonds. As a result, certain double bonds between carbon atoms produce a "kink" in an otherwise straight hydrocarbon chain (Figure 6.1b, left).

Hydrocarbon chains that consist of only single bonds between the carbons are called **saturated.** If one or more double bonds exist in the hydrocarbon chains, then they are **unsaturated.** The choice of terms is logical. If a hydrocarbon chain does not contain a double bond, it is saturated with the maximum number of hydrogen atoms that can attach to the carbon skeleton. If it is unsaturated, then a C—H bond is removed to form a C=C double bond, resulting in fewer than the maximum number of attached hydrogen atoms.

Foods that contain lipids with many double bonds are said to be polyunsaturated and are advertised as healthier than foods with saturated fats. Recent research suggests that polyunsaturated fats help protect the heart from disease. Exactly how this occurs is under investigation.

(a) Isoprenes can be linked into chains called isoprenoids.

(b) Fatty acids can be saturated or unsaturated.

FIGURE 6.1 Hydrocarbon Structure. (a) Isoprene subunits, like the one shown to the left, can be linked to each other, end to end, to form long hydrocarbon chains called isoprenoids. **(b)** Fatty acids typically contain a total of 14–20 carbon atoms, most found in their long hydrocarbon "tails." Unsaturated hydrocarbons contain carbon–carbon double bonds; saturated hydrocarbons do not.

(a) Saturated lipids

Butter

HO—C

(b) Saturated lipids with long hydrocarbon tails

Beeswax

(c) Unsaturated lipids

Safflower oil

HO—C

FIGURE 6.2 The Fluidity of Lipids Depends on the Length and Saturation of Their Hydrocarbon Chains. (a) Butter consists primarily of saturated lipids. **(b)** Waxes are lipids with extremely long saturated hydrocarbon chains. **(c)** Oils are dominated by "polyunsaturates"—lipids with hydrocarbon chains that contain multiple C=C double bonds.

Bond saturation also profoundly affects the physical state of lipids. Highly saturated fats, such as butter, are solid at room temperature (**FIGURE 6.2a**). Saturated lipids that have extremely long hydrocarbon tails, like **waxes** do, form particularly stiff solids at room temperature (**FIGURE 6.2b**). Highly unsaturated fats are liquid at room temperature (**FIGURE 6.2c**).

A Look at Three Types of Lipids Found in Cells

Unlike amino acids, nucleotides, and monosaccharides, lipids are characterized by a physical property—their insolubility in water—instead of a shared chemical structure. This insolubility is based on the high proportion of nonpolar C—C and C—H bonds relative to polar functional groups. As a result, the structure of lipids varies widely. For example, consider the most important types of lipids found in cells: fats, steroids, and phospholipids.

Fats   **Fats** are nonpolar molecules composed of three fatty acids that are linked to a three-carbon molecule called **glycerol**. Because of this structure, fats are also called triacylglycerols or triglycerides. When the fatty acids are polyunsaturated, they form liquid triacylglycerols called **oils**. In organisms, the primary role of fats is energy storage.

As **FIGURE 6.3a** shows, fats form when a dehydration reaction occurs between a hydroxyl group of glycerol and the carboxyl

(a) Fats form via dehydration reactions.

Glycerol

H_2O

Dehydration reaction

Fatty acid

(b) Fats consist of glycerol linked by ester linkages to three fatty acids.

Ester linkages

FIGURE 6.3 Fats Are One Type of Lipid Found in Cells. (a) When glycerol and a fatty acid react, a water molecule leaves. The covalent bond that results from this reaction is termed an ester linkage. **(b)** The structural formula and a space-filling model of tristearin, the most common type of fat in beef.

group of a fatty acid. The glycerol and fatty acid molecules become joined by an **ester linkage.** Fats are not polymers, however, and fatty acids are not monomers. As **FIGURE 6.3b** shows, fatty acids are not linked together to form a macromolecule in the way that amino acids, nucleotides, and monosaccharides are.

Steroids Steroids are a family of lipids distinguished by the bulky, four-ring structure shown in orange in **FIGURE 6.4a**. The various steroids differ from one another by the functional groups or side groups attached to different carbons in those hydrophobic rings. The steroid shown in the figure is cholesterol, which has a hydrophilic hydroxyl group attached to the top ring and an isoprenoid "tail" attached at the bottom. Cholesterol is an important component of plasma membranes in many organisms.

Phospholipids Phospholipids consist of a glycerol that is linked to a phosphate group and two hydrocarbon chains of either isoprenoids or fatty acids. The phosphate group is also bonded to a small organic molecule that is charged or polar (**FIGURE 6.4b**).

Phospholipids composed of fatty acids are found in the domains Bacteria and Eukarya; phospholipids with isoprenoid chains are found in the domain Archaea. (The domains of life were introduced in Chapter 1.) In all three domains, phospholipids are crucial components of the plasma membrane.

The lipids found in organisms have a wide array of structures and functions. In addition to storing chemical energy, lipids act as pigments that capture or respond to sunlight, serve as signals between cells, form waterproof coatings on leaves and skin, and act as vitamins used in many cellular processes. The most prominent function of lipids, however, is their role in cell membranes.

The Structures of Membrane Lipids

Not all lipids can form membranes. Membrane-forming lipids have a polar, hydrophilic region—in addition to the nonpolar, hydrophobic region found in all lipids.

To better understand this structure, take another look at the phospholipid illustrated in Figure 6.4b. Notice that the molecule has a "head" region containing highly polar covalent bonds as well as a negatively charged phosphate attached to a polar or charged group. The charges and polar bonds in the head region interact with water molecules when a phospholipid is placed in solution. In contrast, the long hydrocarbon tails of a phospholipid are nonpolar and hydrophobic. Water molecules cannot form hydrogen bonds with the hydrocarbon tail, so they do not interact extensively with this part of the molecule.

Compounds that contain both hydrophilic and hydrophobic elements are **amphipathic** (literally, "dual-sympathy"). Phospholipids are amphipathic. As Figure 6.4a shows, cholesterol is also amphipathic. Because it has a hydroxyl functional group attached to its rings, it has both hydrophilic and hydrophobic regions. ✔ If you understand these concepts, you should be able to look back at Figure 6.1b and explain why fatty acids are also amphipathic.

(a) A steroid

(b) A phospholipid

FIGURE 6.4 Some Lipids Contain Hydrophilic and Hydrophobic Regions. (a) All steroids have the distinctive four-ring structure shown in orange. Cholesterol has a polar hydroxyl group and an isoprenoid chain attached to these rings. **(b)** Most phospholipids consist of two fatty acid or isoprenoid chains that are linked to glycerol, which is linked to a phosphate group, which is linked to a small organic molecule that is polar or charged.

✔**QUESTION** If cholesterol and phospholipids were in solution, which part of the molecules would interact with water molecules?

The amphipathic nature of phospholipids is far and away their most important feature biologically. It is responsible for life's defining barrier—the plasma membrane. If the membrane defines life, then amphipathic lipids must have existed when life first originated during chemical evolution. Was that possible?

Were Lipids Present during Chemical Evolution?

Like amino acids, nucleic acids, and carbohydrates (Chapters 3–5), there is evidence that lipids were present during chemical evolution. Laboratory experiments have shown that simple lipids, such as fatty acids, can be synthesized from H_2 and CO_2 via reactions with mineral catalysts under conditions thought to be present in prebiotic hydrothermal vent systems (Chapter 2).

It is also possible that lipids literally fell from the sky early in Earth's history. Modern meteorites have been found to contain not only amino acids and carbohydrates but also lipids that exhibit amphipathic qualities. For example, lipids extracted from the meteorite that struck Murchison, Australia, in 1969 spontaneously formed lipid "bubbles" that resembled small cells. Why do amphipathic lipids do this?

check your understanding

C Y U

If you understand that . . .

- Fats, steroids, and phospholipids differ in structure and function.
- Fats and oils are nonpolar; fatty acids, phospholipids, and certain steroids, like cholesterol, are amphipathic because they have both polar and nonpolar regions.
- Fats store chemical energy; certain steroids and phospholipids are key components of plasma membranes.

✔ **You should be able to . . .**

1. Compare and contrast the structure of a fat, a steroid, and a phospholipid.
2. Based on their structure, explain what makes cholesterol and phospholipids amphipathic.

Answers are available in Appendix A.

6.2 Phospholipid Bilayers

Amphipathic lipids do not dissolve when they are placed in water. Their hydrophilic heads interact with water, but their hydrophobic tails do not. Instead of dissolving in water, then, amphipathic lipids assume one of two types of structures: micelles or lipid bilayers.

- Micelles (**FIGURE 6.5a**) are tiny droplets created when the hydrophilic heads of a set of lipids face the water and form hydrogen bonds, while the hydrophobic tails interact with each other in the interior, away from the water.

- A **lipid bilayer** is created when two sheets of lipid molecules align. As **FIGURE 6.5b** shows, the hydrophilic heads in each layer face the surrounding solution while the hydrophobic tails face one another inside the bilayer. In this way, the hydrophilic heads interact with water while the hydrophobic tails interact with one another.

(a) Lipid micelles

Hydrophilic heads interact with water

Hydrophobic tails interact with one another

Water

(b) Lipid bilayers

Hydrophilic heads interact with water

Hydrophobic tails interact with one another

FIGURE 6.5 Lipids Form Micelles and Bilayers in Solution. In **(a)** a micelle or **(b)** a lipid bilayer, the hydrophilic heads of lipids face out, toward water; the hydrophobic tails face in, away from water. Lipid bilayers are the foundation of plasma membranes.

Micelles tend to form from fatty acids or other simple amphipathic hydrocarbon chains. Bilayers tend to form from phospholipids that contain two hydrocarbon tails. For this reason, bilayers are often called phospholipid bilayers.

It's critical to recognize that micelles and phospholipid bilayers form spontaneously—no input of energy is required. This concept can be difficult to grasp because entropy clearly decreases when these structures form. The key is to recognize that micelles and lipid bilayers are much more stable energetically than are independent phospholipids in solution.

Independent lipids are unstable in water because their hydrophobic tails disrupt hydrogen bonds that could otherwise form between water molecules. As a result, the tails of amphipathic molecules are forced together and participate in hydrophobic interactions (introduced in Chapter 2). This point should also remind you of the aqueous behavior of hydrophobic side chains in proteins and bases in nucleic acids.

Artificial Membranes as an Experimental System

When phospholipids are added to an aqueous solution and agitated, lipid bilayers spontaneously form small spherical structures. The hydrophilic heads on both sides of the bilayer remain in contact with the aqueous solution—water is present both inside and outside the vesicle. Artificial membrane-bound vesicles like these are called liposomes (**FIGURE 6.6**).

To explore how membranes work, researchers began creating and experimenting with liposomes and planar bilayers—lipid bilayers constructed across a hole in a glass or plastic wall separating two aqueous solutions (**FIGURE 6.7a**). Some of the first questions they posed concerned the permeability of lipid bilayers. The **permeability** of a structure is its tendency to allow a given substance to pass through it.

Using liposomes and planar bilayers, researchers can study what happens when a known ion or molecule is added to one side of a lipid bilayer (**FIGURE 6.7b**). Does the substance cross the

25 nm

Water

Water

FIGURE 6.6 Liposomes Are Artificial Membrane-Bound Vesicles. Electron micrograph of liposomes in cross section (left) and a cross-sectional diagram of the lipid bilayer in a liposome (right).

(a) Planar bilayers: Artificial membranes

Water   Water

Lipid bilayer

(b) Artificial-membrane experiments

Solute (ion or molecule)

?

How rapidly can different solutes cross the membrane (if at all) when …

1. Different types of phospholipids are used to make the membrane?

2. Proteins or other molecules are added to the membrane?

FIGURE 6.7 Use of Planar Bilayers in Experiments. (a) The construction of a planar bilayer across a hole in a wall separating two water-filled compartments. **(b)** A wide variety of experiments are possible with planar bilayers; a few are suggested here.

membrane and show up on the other side? If so, how rapidly does the movement take place? What happens when a different type of phospholipid is used to make the artificial membrane? Does the membrane's permeability change when proteins or other types of molecules become part of it?

Biologists describe such an experimental system as elegant and powerful because it gives them precise control over which factor changes from one experimental treatment to the next.

Control, in turn, is why experiments are such an effective way to explore scientific questions. Recall that good experimental design allows researchers to alter one factor at a time and determine what effect, if any, each has on the process being studied (Chapter 1).

Selective Permeability of Lipid Bilayers

When researchers put molecules or ions on one side of a liposome or planar bilayer and measure the rate at which the molecules arrive on the other side, a clear pattern emerges: Lipid bilayers are highly selective.

Selective permeability means that some substances cross a membrane more easily than other substances do. Small nonpolar molecules move across bilayers quickly. In contrast, large molecules and charged substances cross the membrane slowly, if at all. This difference in membrane permeability is a critical issue because controlling what passes between the exterior and interior environments is a key characteristic of cells.

According to the data in **FIGURE 6.8**, small nonpolar molecules such as oxygen (O_2) move across selectively permeable membranes more than a billion times faster than do chloride ions (Cl^-). In essence, ions cannot cross membranes at all—unless they have "help" in the form of membrane proteins introduced later in the chapter. Very small and uncharged molecules such as water (H_2O) can cross membranes relatively rapidly, even if they are polar. Small polar molecules such as glycerol have intermediate permeability.

The leading hypothesis to explain this pattern is that charged compounds and large polar molecules are more stable dissolved in water than they are in the nonpolar interior of membranes. ✔ If you understand this hypothesis, you should be able to predict where amino acids and nucleotides would be placed in Figure 6.8 and explain your reasoning.

High permeability

Permeability scale (cm/sec)

10^0 O_2, CO_2, N_2 — Small, nonpolar molecules

10^{-2}

10^{-4} H_2O, glycerol — Small, uncharged polar molecules

10^{-6}

10^{-8} Glucose, sucrose — Large, uncharged polar molecules

10^{-10} Cl^-,

10^{-12} K^+, Na^+ — Ions

Low permeability   Phospholipid bilayer

FIGURE 6.8 Lipid Bilayers Show Selective Permeability. Only certain substances cross lipid bilayers readily. Size and polarity or charge affect the rate of diffusion across a membrane.

How Does Lipid Structure Affect Membrane Permeability?

The amphipathic nature of phospholipids allows them to spontaneously form membranes. But not all phospholipid bilayers are the same. The nature of the hydrocarbon tails, in addition to the presence of cholesterol molecules, profoundly influences how a membrane behaves.

Bond Saturation and Hydrocarbon Chain Length Affect Membrane Fluidity and Permeability The degree of saturation in a phospholipid—along with the length of its hydrocarbon tails—affects key aspects of a lipid's behavior in a membrane.

- When unsaturated hydrocarbon tails are packed into a lipid bilayer, kinks created by double bonds produce spaces among the tails. These spaces reduce the strength of the van der Waals interactions (see Chapter 3) that hold the hydrophobic tails together, weakening the barrier to solutes.

- Packed saturated hydrocarbon tails have fewer spaces and stronger van der Waals interactions. As the length of saturated hydrocarbon tails increases, the forces that hold them together also grow stronger, making the membrane even denser.

These observations have profound impacts on membrane fluidity and permeability—two closely related properties. As **FIGURE 6.9** shows, lipid bilayers are more permeable as well as more fluid when they contain short, kinked, unsaturated hydrocarbon tails. An unsaturated membrane allows more materials to pass because its interior is held together less tightly. Bilayers containing long, straight, saturated hydrocarbon tails are much less permeable and fluid. Experiments on liposomes have shown exactly these patterns.

Cholesterol Reduces Membrane Permeability Cholesterol molecules are present, to varying extents, in the membranes of every cell in your body. What effect does adding cholesterol have on a membrane? Researchers have found that adding cholesterol molecules to liposomes dramatically reduces the permeability of lipid bilayers. The data behind this conclusion are presented in **FIGURE 6.10**.

To read the graph in the "Results" section of Figure 6.10, put your finger on the *x*-axis at the point marked 20°C, and note that permeability to glycerol is much higher at this temperature in membranes that contain no cholesterol versus 20 percent or 50 percent cholesterol. Using this procedure at other temperature points should convince you that membranes lacking cholesterol are more permeable than the other two membranes at every temperature tested in the experiment.

What explains this result? Because the steroid rings in cholesterol are bulky, adding cholesterol fills gaps that would otherwise be present in the hydrophobic section of the membrane.

How Does Temperature Affect the Fluidity and Permeability of Membranes?

At about 25°C—or "room temperature"—the phospholipids in a plasma membrane have a consistency resembling olive oil. This fluid physical state allows individual lipid molecules to move laterally within each layer (**FIGURE 6.11**), a little like a person moving about in a dense crowd. By tagging individual phospholipids and following their movement, researchers have clocked average speeds of 2 micrometers (μm)/second at room temperature. At these speeds, a phospholipid could travel the length of a small bacterial cell in a second.

Recall that permeability is closely related to fluidity. As temperature drops, molecules in a bilayer move more slowly. As a result, the hydrophobic tails in the interior of membranes pack together more tightly. At very low temperatures, lipid bilayers even begin to solidify. As the graph in Figure 6.10 indicates, low temperatures can make membranes impervious to molecules that would normally cross them readily. Put your finger on the *x*-axis of that graph, just about the freezing point

Lipid bilayer with **short** and **unsaturated** hydrocarbon tails

Higher permeability and fluidity

Lipid bilayer with **long** and **saturated** hydrocarbon tails

Lower permeability and fluidity

FIGURE 6.9 Fatty Acid Structure Changes the Permeability of Membranes. Lipid bilayers consisting of phospholipids containing unsaturated fatty acids should have more gaps and be more permeable than those with saturated fatty acids.

Phospholipids are in constant lateral motion, but rarely flip to the other side of the bilayer

FIGURE 6.11 Phospholipids Move within Membranes. Membranes are dynamic—in part because phospholipid molecules randomly move laterally within each layer in the structure.

QUESTION: Does adding cholesterol to a membrane affect its permeability?

HYPOTHESIS: Cholesterol reduces permeability because it fills spaces in phospholipid bilayers.

NULL HYPOTHESIS: Cholesterol has no effect on permeability.

EXPERIMENTAL SETUP:

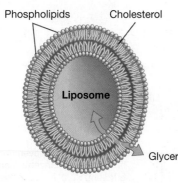

Phospholipids   Cholesterol

Liposome

Glycerol

1. Construct liposomes: Create with no cholesterol, 20% cholesterol, and 50% cholesterol.

2. Measure glycerol movement: Record how quickly glycerol moves across each type of membrane at different temperatures.

PREDICTION: Liposomes with higher cholesterol levels will have reduced permeability.

PREDICTION OF NULL HYPOTHESIS: All liposomes will have the same permeability.

RESULTS:

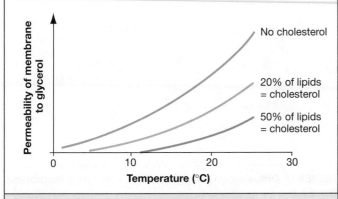

No cholesterol

20% of lipids = cholesterol

50% of lipids = cholesterol

Permeability of membrane to glycerol

Temperature (°C)

CONCLUSION: Adding cholesterol to membranes decreases their permeability to glycerol. The permeability of all membranes analyzed in this experiment increases with increasing temperature.

FIGURE 6.10 The Permeability of a Membrane Depends on Its Composition.

SOURCE: de Gier, J., et al. (1968). Lipid composition and permeability of liposomes. *Biochimica et Biophysica Acta* 150: 666–675.

✔**QUANTITATIVE** Suppose the investigators had instead created liposomes using phospholipids with fully saturated tails and compared them to two other sets of liposomes where either 20 percent or 50 percent of the phospholipids contained polyunsaturated tails. Label the three lines on the graph above with your prediction for the three different liposomes in this new experiment.

of water (0°C), and note that even membranes that lack cholesterol are almost completely impermeable to glycerol. Indeed, trace any of the three lines in Figure 6.10, and as you move to the right (increasing temperature), you also move up (increasing permeability).

These observations on glycerol and lipid movement demonstrate that membranes are dynamic. Phospholipid molecules whiz around each layer, while water and small nonpolar molecules shoot in and out of the membrane. How quickly molecules move within and across membranes is a function of temperature, the structure of hydrocarbon tails, and the number of cholesterol molecules in the bilayer.

6.3 How Molecules Move across Lipid Bilayers: Diffusion and Osmosis

Small uncharged molecules and hydrophobic compounds can cross membranes readily and spontaneously—without an input of energy. The question now is: How is this possible? What process is responsible for movement of molecules across lipid bilayers?

Diffusion

A thought experiment can help explain how substances can cross membranes spontaneously. Suppose you rack up a set of billiard balls in the middle of a pool table and then begin to vibrate the table.

1. Because of the vibration, the billiard balls will move about randomly. They will also bump into one another.

2. After these collisions, some balls will move outward—away from their original position.

3. As movement and collisions continue, the overall or net movement of balls will be outward. This occurs because the random motion of the balls disrupts their original, nonrandom position. As the balls move at random, they are more likely to move away from one another than to stay together.

4. Eventually, the balls will be distributed randomly across the table. The entropy of the billiard balls has increased. Recall that entropy is a measure of the randomness or disorder in

a system (Chapter 2). The second law of thermodynamics states that in an isolated system, entropy always increases.

This hypothetical example illustrates how vibrating billiard balls move at random. More to the point, it also explains how substances located on one side of a lipid bilayer can move to the other side spontaneously. All dissolved molecules and ions, or **solutes,** have thermal energy and are in constant, random motion. Movement of molecules and ions that results from their kinetic energy is known as **diffusion.**

A difference in solute concentrations creates what is called a **concentration gradient.** Solutes move randomly in all directions, but when a concentration gradient exists, there is a net movement from regions of high concentration to regions of low concentration. Diffusion down a concentration gradient, or away from the higher concentration, is a spontaneous process because it results in an increase in entropy.

Once the molecules or ions are randomly distributed throughout a solution, a chemical equilibrium is established. For example, consider two aqueous solutions separated by a lipid bilayer. **FIGURE 6.12** shows how molecules that can pass through the bilayer diffuse to the other side. At equilibrium, these molecules continue to move back and forth across the membrane, but at equal rates—simply because they are equally likely to move in any direction. This means that there is no longer a net movement of molecules across the membrane. ✔ If you understand diffusion, you should be able to predict how a difference in temperature across a membrane would affect the concentration of a solute at equilibrium.

Osmosis

What about water? As the data in Figure 6.8 show, water moves across lipid bilayers relatively quickly. The movement of water is a special case of diffusion that is given its own name: **osmosis.** Osmosis occurs only when solutions are separated by a membrane that permits water to cross, but holds back some or all of the solutes—that is, a selectively permeable membrane.

It's important to note that some of the water molecules in a solution are unavailable to diffuse across the membrane. Recall that solutes form ionic or hydrogen bonds with water molecules (Chapter 2). Water molecules that are bound to a solute that can't cross the membrane are themselves prevented from crossing.

Only unbound water molecules are able to diffuse across the membrane during osmosis. When these unbound water molecules move across a membrane, they flow from the solution with the lower solute concentration into the solution with the higher solute concentration.

To drive this point home, let's suppose the concentration of a particular solute is higher on one side of a selectively permeable membrane than it is on the other side (**FIGURE 6.13**, step 1). Further, suppose that this solute cannot diffuse through the membrane to establish equilibrium. What happens? Water will move from the side with a lower concentration of solute to the side with a higher concentration of solute (Figure 6.13, step 2). Osmosis dilutes the higher concentration and equalizes the concentrations

PROCESS: DIFFUSION ACROSS A LIPID BILAYER

1. Separation of solutes: Start with different solutes on opposite sides of a lipid bilayer. Both molecules diffuse freely across the bilayer.

2. Diffusion: Solutes diffuse across the membrane—each undergoes a net movement along its own concentration gradient.

3. Equilibrium: Equilibrium is established. Solutes continue to move back and forth across the membrane but at equal rates.

FIGURE 6.12 Diffusion across a Selectively Permeable Membrane Establishes an Equilibrium.

on both sides. The movement of water is spontaneous. It is driven by the increase in entropy achieved when solute concentrations are equal on both sides of the membrane.

Movement of water by osmosis is important because it can swell or shrink a membrane-bound vesicle. Consider the liposomes illustrated in **FIGURE 6.14**. (Remember that osmosis occurs only when a solute cannot pass through a separating membrane.)

- *Left* If the solution inside the membrane has a lower concentration of solutes than the exterior has, water moves out of the vesicle into the solution outside. The solution inside is said to be **hypotonic** ("lower-tone") relative to the outside of the vesicle. As water leaves, the vesicle shrinks and the membrane shrivels, resulting in lower vesicle firmness.

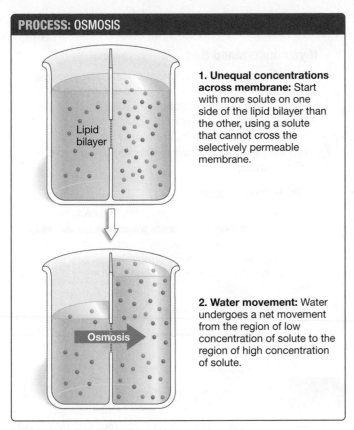

PROCESS: OSMOSIS

1. Unequal concentrations across membrane: Start with more solute on one side of the lipid bilayer than the other, using a solute that cannot cross the selectively permeable membrane.

Lipid bilayer

Osmosis

2. Water movement: Water undergoes a net movement from the region of low concentration of solute to the region of high concentration of solute.

FIGURE 6.13 Osmosis Is the Diffusion of Water.

✔ **QUESTION** Suppose you doubled the number of solute molecules on the left side of the membrane (at the start). At equilibrium, would the water level on the left side be higher or lower than what is shown in the second drawing?

- *Middle*   If the solution inside the membrane has a higher concentration of solutes than the exterior has, water moves into the vesicle via osmosis. The inside solution is said to be **hypertonic** ("excess-tone") relative to the outside of the vesicle. The incoming water causes the vesicle to swell and increase in firmness, or even burst.

- *Right*   If solute concentrations are equal on both sides of the membrane, the liposome maintains its size. When the inside solution does not affect the membrane's shape, that solution is called **isotonic** ("equal-tone").

Note that the terms hypertonic, hypotonic, and isotonic are relative—they can be used only to express the relationship between a given solution and another solution separated by a membrane. Biologists also commonly use these terms to describe the solution that is exterior to the cells or vesicles.

Membranes and Chemical Evolution

What do diffusion and osmosis have to do with the first membranes floating in the prebiotic soup? Both processes tend to *reduce* differences in chemical composition between the inside and outside of membrane-bound compartments.

If liposome-like structures first arose in the oceans of early Earth, their interiors probably didn't offer a radically different environment from the surrounding solution. In all likelihood, the primary importance of the first lipid bilayers was simply to provide a container for replicating RNA, the macromolecule most likely to have been the first "living" molecule (see Chapter 4). But ribonucleotide monomers would need to be available for these

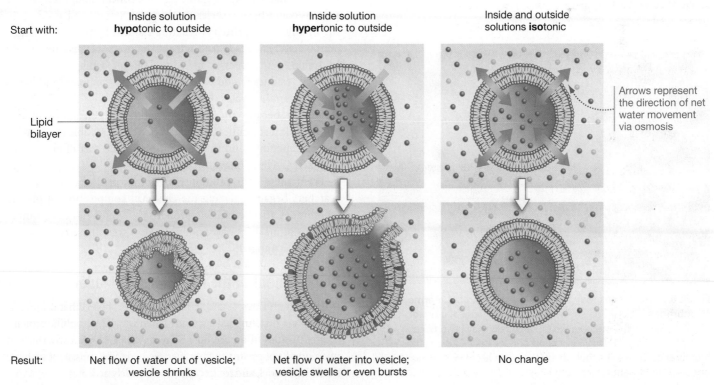

Start with:

Inside solution **hypo**tonic to outside

Inside solution **hyper**tonic to outside

Inside and outside solutions **iso**tonic

Lipid bilayer

Arrows represent the direction of net water movement via osmosis

Result: Net flow of water out of vesicle; vesicle shrinks

Net flow of water into vesicle; vesicle swells or even bursts

No change

FIGURE 6.14 Osmosis Can Shrink or Burst Membrane-Bound Vesicles.

RNAs to replicate. Can negatively charged ribonucleotides get across lipid bilayers and inside lipid-bounded vesicles?

The answer is yes. Jack Szostak and colleagues first set out to study the permeability of membranes consisting of fatty acids and other simple amphipathic lipids thought to be present in the early oceans. Like phospholipids, fatty acids will spontaneously assemble into lipid bilayers and water-filled vesicles. Their experiments showed that ions, and even ribonucleotides, can diffuse across the fatty acid vesicle membranes—meaning that monomers could have been available for RNA synthesis.

Lending support to this hypothesis, the same minerals found to catalyze the polymerization of RNA from activated nucleotides (see Chapter 4) will also promote the formation of fatty acid vesicles—and in the process, often incorporate themselves and RNA inside. Simple vesicle-like structures that harbor nucleic acids are referred to as **protocells** (**FIGURE 6.15**). Most origin-of-life researchers view protocells as possible intermediates in the evolution of the cell.

Laboratory simulations also showed that free lipids and micelles can become incorporated into fatty acid bilayers, causing protocells to grow. Shearing forces, as from bubbling, shaking, or wave action, cause protocells to divide. Based on these observations, it is reasonable to hypothesize that once replicating RNAs became surrounded by a lipid bilayer, this simple lifeform and its descendants would occupy cell-like structures that grew and divided.

Now let's investigate the next great innovation in the evolution of the cell: the ability to create and maintain a specialized internal environment that is conducive to life. What is necessary to construct an effective plasma membrane—one that imports ions and molecules needed for life while excluding ions and molecules that might damage it?

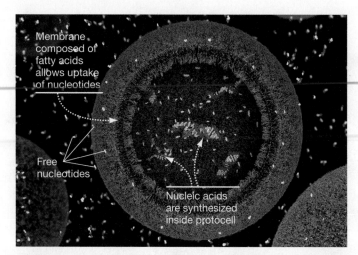

FIGURE 6.15 Protocells May Have Possessed Simple, Permeable Membranes. This image shows a computer model of a protocell. Like this model, the membranes of early cells may have been built of fatty acids. Passive transport of nucleotides across these membranes, as well as replication of nucleic acids inside, has been observed in the laboratory.

Labels within figure:
Membrane composed of fatty acids allows uptake of nucleotides

Free nucleotides

Nucleic acids are synthesized inside protocell

6.4 Membrane Proteins

What sort of molecule could become incorporated into a lipid bilayer and affect the bilayer's permeability? The title of this section gives the answer away. Proteins that are amphipathic can be inserted into lipid bilayers.

Proteins can be amphipathic because their monomers, amino acids, have side chains that range from highly nonpolar to highly polar or charged (see Figure 3.2). It's conceivable, then, that a protein could have a series of nonpolar amino acid residues in the middle of its primary structure flanked by polar or charged amino acid residues (**FIGURE 6.16a**). The nonpolar residues would be stable in the interior of a lipid bilayer, while the polar or charged residues would be stable alongside the polar lipid heads and surrounding water (**FIGURE 6.16b**).

Further, because the secondary and tertiary structures of proteins are almost limitless in their variety, it is possible for proteins to form openings and thus function as some sort of channel or pore across a lipid bilayer.

From these considerations, it's not surprising that when researchers began analyzing the chemical composition of plasma membranes, they found that proteins were often just as common, in terms of mass, as phospholipids. How were these two types of molecules arranged?

Development of the Fluid-Mosaic Model

In 1935 Hugh Davson and James Danielli proposed that cell membranes were structured like a sandwich in which hydrophilic proteins coat both sides of a pure lipid bilayer (**FIGURE 6.17a**). Early electron micrographs of plasma membranes seemed to be consistent with the sandwich model, and for decades it was widely accepted.

(a) Proteins can be amphipathic.

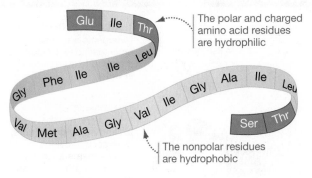

The polar and charged amino acid residues are hydrophilic

The nonpolar residues are hydrophobic

(b) Amphipathic proteins can integrate into lipid bilayers.

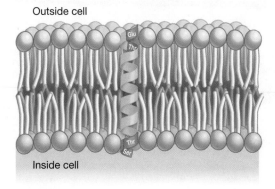

Outside cell

Inside cell

FIGURE 6.16 Amphipathic Proteins Are Anchored in Lipid Bilayers.

(a) Sandwich model

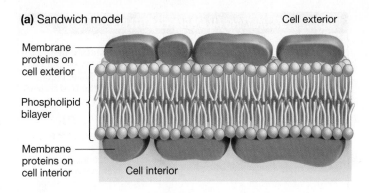

Cell exterior

Membrane proteins on cell exterior

Phospholipid bilayer

Membrane proteins on cell interior

Cell interior

(b) Fluid-mosaic model

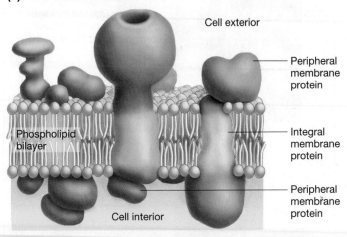

Cell exterior

Peripheral membrane protein

Phospholipid bilayer

Integral membrane protein

Peripheral membrane protein

Cell interior

FIGURE 6.17 Past and Current Models of Membrane Structure Differ in Where Membrane Proteins Reside. (a) The protein-lipid-lipid-protein sandwich model was the first hypothesis for the arrangement of lipids and proteins in cell membranes. **(b)** The fluid-mosaic model was a radical departure from the sandwich hypothesis.

The realization that membrane proteins could be amphipathic, however, led S. Jon Singer and Garth Nicolson to suggest an alternative hypothesis. In 1972, they proposed that at least some proteins span the membrane instead of being found only outside the lipid bilayer. Their hypothesis was called the **fluid-mosaic model** (**FIGURE 6.17b**). Singer and Nicolson suggested that membranes are a mosaic of phospholipids and different types of proteins. The overall structure was proposed to be dynamic and fluid.

The controversy over the nature of the cell membrane was resolved in the early 1970s with the development of an innovative technique for visualizing the surface of plasma membranes. The method is called freeze-fracture electron microscopy because the steps involve freezing and fracturing the membrane before examining it with a **scanning electron microscope (SEM),** which produces images of an object's surface (see **BioSkills 11** in Appendix B).

As **FIGURE 6.18** (see page 96) shows, the freeze-fracture technique allows researchers to split cell membranes and view the middle of the structure. The scanning electron micrographs that result show pits and mounds studding the inner surfaces of the lipid bilayer. Researchers interpreted these structures as the locations of membrane proteins. As step 4 in the figure shows, the mounds represent proteins that remained attached to one side of the split lipid bilayer and the pits are the holes they left behind.

These observations conflicted with the sandwich model but were consistent with the fluid-mosaic model. Based on these and subsequent observations, the fluid-mosaic model is now widely accepted.

Notice in Figure 6.17b that some proteins span the membrane and have segments facing both the interior and the exterior surfaces. Proteins like these are called **integral membrane proteins,** or **transmembrane proteins.** Proteins that bind to the membrane without passing through it are called **peripheral membrane proteins.**

Certain peripheral proteins are found only on the interior surface of a cellular membrane, while others are found only on the exterior surface. As a result, the interior and exterior surfaces of the plasma membrane are distinct—the peripheral proteins and the ends of transmembrane proteins differ. Peripheral membrane proteins are often attached to transmembrane proteins.

How do these proteins affect the permeability of membranes? The answer to this question starts with an investigation of the structure of proteins involved in the transport of molecules and ions across the plasma membrane.

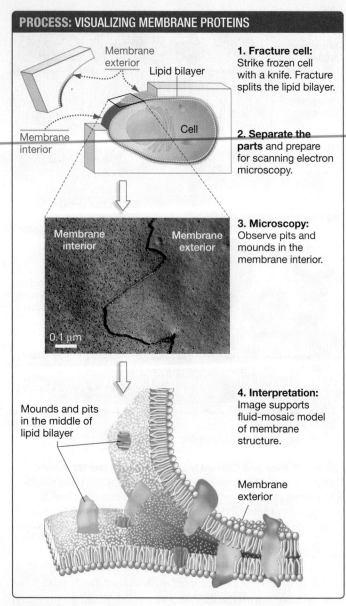

1. Fracture cell: Strike frozen cell with a knife. Fracture splits the lipid bilayer.

2. Separate the parts and prepare for scanning electron microscopy.

3. Microscopy: Observe pits and mounds in the membrane interior.

4. Interpretation: Image supports fluid-mosaic model of membrane structure.

FIGURE 6.18 Freeze-Fracture Preparations Allow Biologists to View Membrane Proteins.

✔**QUESTION** What would be an appropriate control to show that the pits and mounds were not simply irregularities in the lipid bilayer caused by the freeze-fracture process?

Systems for Studying Membrane Proteins

The discovery of transmembrane proteins was consistent with the hypothesis that proteins affect membrane permeability. To test this hypothesis, researchers needed some way to isolate and purify membrane proteins.

FIGURE 6.19 outlines one method that researchers developed to separate proteins from membranes. The key to the technique is the use of detergents. A **detergent** is a small amphipathic molecule. When detergents are added to the solution surrounding a lipid bilayer, the hydrophobic tails of the detergent molecule interact with the hydrophobic tails of the lipids and with the

1. Addition of detergents: Detergents are small, amphipathic molecules that tend to form micelles in water.

2. Binding by detergents: Detergents break up plasma membranes; they coat hydrophobic portions of membrane proteins and phospholipids.

Isolated protein

3. Isolation of proteins: Treating a plasma membrane with a detergent is an effective way to isolate membrane proteins so they can be purified and studied in detail.

FIGURE 6.19 Detergents Can Be Used to Isolate Proteins from Membranes.

hydrophobic portions of transmembrane proteins. These interactions displace the membrane phospholipids and end up forming water-soluble detergent–protein complexes that can be isolated.

Since intensive experimentation on membrane proteins began, researchers have identified three broad classes of proteins that affect membrane permeability: channels, carriers, and pumps. Let's consider each class in turn.

Facilitated Diffusion via Channel Proteins

As the data in Figure 6.8 show, ions almost never cross pure phospholipid bilayers on their own. But in cells, ions routinely cross membranes through specialized membrane proteins called **ion channels.**

Ion channels form pores, or openings, in a membrane. Ions move through these pores in a predictable direction: from regions of high concentration to regions of low concentration and from areas of like charge to areas of unlike charge.

In **FIGURE 6.20**, for example, a large concentration gradient favors the movement of sodium ions from the outside of a membrane to the inside. But in addition, the inside of this cell has a net negative charge while the outside has a net positive charge. As a result, the combination of these two factors influences the final concentration of sodium ions inside the cell once equilibrium has been established.

FIGURE 6.20 An Electrochemical Gradient Is a Combined Concentration and Electrical Gradient. Electrochemical gradients are established when ions build up on one side of a membrane.

Ions move in response to a combined concentration and electrical gradient, or what biologists call an **electrochemical gradient.** ✔ If you understand this concept, you should be able to add an arrow to Figure 6.20 indicating the electrochemical gradient for chloride ions.

Is an Ion Channel Involved in Cystic Fibrosis? To understand the types of experiments that biologists do to confirm that a membrane protein is an ion channel, consider work on the cause of cystic fibrosis.

Cystic fibrosis (CF) is the most common genetic disease in humans of Northern European descent. It affects cells that produce mucus, sweat, and digestive juices. Normally these secretions are thin and slippery and act as lubricants. In individuals with CF, however, the secretions become abnormally concentrated and sticky and clog passageways in organs like the lungs.

Experiments published in 1983 suggested that cystic fibrosis is caused by defects in a membrane protein that allow chloride ions (Cl^-) to move across plasma membranes. It was proposed that reduced chloride ion transport would account for the thick mucus.

How is the transport of chloride ions involved in mucus consistency? Water movement across cell membranes is largely determined by the presence of extracellular ions like chloride. If a defective channel prevents chloride ions from leaving cells, water isn't pulled from cells by osmosis to maintain the proper mucus consistency. In effect, the disease results from the mismanagement of osmosis.

Using molecular techniques introduced in Unit 3 (see Chapter 20), biologists were able to **(1)** find the gene that is defective in people suffering from CF and **(2)** use the gene to produce copies of the normal protein, which was called CFTR (short for cystic fibrosis transmembrane conductance regulator).

Is CFTR a chloride channel? To answer this question, researchers inserted purified CFTR into planar bilayers and

RESEARCH

QUESTION: Is CFTR a chloride channel?

HYPOTHESIS: CFTR increases the flow of chloride ions across a membrane.

NULL HYPOTHESIS: CFTR has no effect on membrane permeability.

EXPERIMENTAL SETUP:

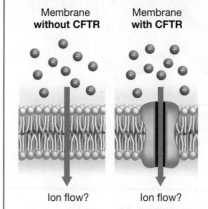

Membrane **without CFTR**

Membrane **with CFTR**

Ion flow?   Ion flow?

1. **Create planar bilayers** with and without CFTR.

2. **Add chloride ions** to one side of the planar bilayer to create an electrochemical gradient.

3. **Record electrical currents** to measure ion flow across the planar bilayers.

PREDICTION: Ion flow will be higher in membrane with CFTR.

PREDICTION OF NULL HYPOTHESIS: Ion flow will be the same in both membranes.

RESULTS:

CONCLUSION: CFTR facilitates diffusion of chloride ions along an electrochemical gradient. CFTR is a chloride channel.

FIGURE 6.21 Electric Current Measurements Indicate that Chloride Flows through CFTR.

SOURCE: Bear, C. A., et al. (1992). Purification and functional reconstitution of the cystic fibrosis transmembrane conductance regulator (CFTR). *Cell* 68: 809–818.

✔ **QUESTION** The researchers repeated the "with CFTR" treatment 45 times, but recorded a current in only 35 of the replicates. Does this observation negate the conclusion? Explain why or why not.

measured the flow of electric current across the membrane. Because ions carry a charge, ion movement across a membrane produces an electric current.

The graphs in **FIGURE 6.21**, which plot the amount of current flowing across the membrane over time, show the results from this experiment. Notice that when CFTR was absent, no electric current passed through the membrane. But when CFTR was

Outside cell

H₂O

Key residues allow water to pass but block ions and larger molecules

Inside cell

FIGURE 6.22 Membrane Channels Are Highly Selective. A cutaway view looking at the inside of a membrane channel, aquaporin. The key residues identified in the space-filling model selectively filter ions and other small molecules, allowing only water (red and white structures) to pass through.

inserted into the membrane, current began to flow. This was strong evidence that CFTR was indeed a chloride ion channel.

Protein Structure Determines Channel Selectivity

Subsequent research has shown that cells have many different types of pore-like **channel proteins** in their membranes, including ion channels like CFTR. Channel proteins are selective. Each channel protein has a structure that permits only a particular type of ion or small molecule to pass through it.

For example, Peter Agre and co-workers discovered channels called **aquaporins** ("water-pores") that allow water to cross the plasma membrane over 10 times faster than it does in the absence of aquaporins. Aquaporins admit water but not other small molecules or ions.

FIGURE 6.22 shows a cutaway view from the side of an aquaporin, indicating how it fits in a plasma membrane. Like other channels that have been studied in detail, aquaporins have a pore that is lined with polar functional groups—in this case, carbonyl groups that interact with water. A channel's pore is hydrophilic relative to the hydrophobic residues facing the phospholipid tails of the membrane.

But how can aquaporin be selective for water and not other polar molecules? The answer was found when researchers examined its structure. Key side chains in the interior of the pore function as a molecular filter. The distance between these groups across the channel allows only those substances capable of interacting with all of them to pass through to the other side.

Movement Through Many Membrane Channels Is Regulated

Recent research has shown that many aquaporins and ion channels are **gated channels**—meaning that they open or close in response to a signal, such as the binding of a particular molecule or a change in the electrical voltage across the membrane.

As an example of how voltage-gated channels work, **FIGURE 6.23** shows a potassium channel in closed and open configurations. The electrical charge on the membrane is normally negative on the inside relative to the outside, which causes the channel to adopt a closed shape that prevents potassium ions from passing through. When this charge asymmetry is reversed, the shape changes in a way that opens the channel and allows potassium ions to cross. The key point here is that in almost all cases, the flow of ions and small molecules through membrane channels is carefully controlled.

In all cases, however, the movement of substances through channels is passive—meaning it does not require an input of energy. **Passive transport** is powered by diffusion along an electrochemical gradient. Channel proteins simply enable ions or polar molecules to move across lipid bilayers efficiently, in response to

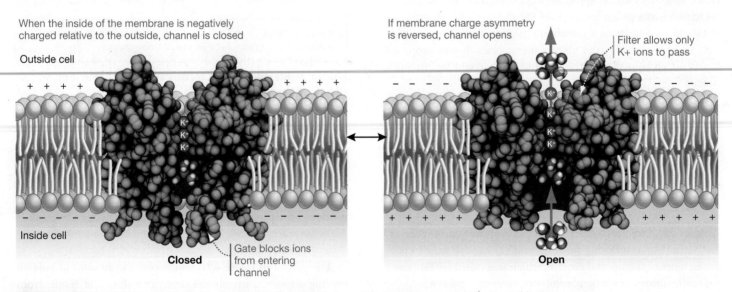

When the inside of the membrane is negatively charged relative to the outside, channel is closed

Outside cell

+ + + +   + + + +

K⁺
K⁺
K⁺

− − − −   − − − −

Inside cell

Closed   Gate blocks ions from entering channel

If membrane charge asymmetry is reversed, channel opens

Filter allows only K+ ions to pass

− − − −   − − − −

K⁺
K⁺
K⁺
K⁺

+ + + + +   + + + + +

Open

FIGURE 6.23 Some Membrane Channels Are Highly Regulated. A model of a voltage-gated K⁺ channel in the closed and open configurations. The channel filter displaces water molecules that normally surround the K⁺ ions in an aqueous solution.

an existing gradient. They are responsible for **facilitated diffusion:** the passive transport of substances that otherwise would not cross a membrane readily.

Facilitated Diffusion via Carrier Proteins

Facilitated diffusion can also occur through **carrier proteins**—specialized membrane proteins that change shape during the transport process. Perhaps the best-studied carrier protein is one that is involved in transporting glucose into cells.

The Search for a Glucose Carrier   Next to ribose, the six-carbon sugar glucose is the most prevalent sugar found in organisms. Virtually all cells alive today use glucose as a building block for important macromolecules and as a source of stored chemical energy (Chapter 5). But as Figure 6.8 shows, lipid bilayers are only moderately permeable to glucose. It is reasonable to expect, then, that plasma membranes have some mechanism for increasing their permeability to this sugar.

This prediction was supported in experiments on pure preparations of plasma membranes from human red blood cells. These plasma membranes turned out to be much more permeable to glucose than are pure lipid bilayers. Why?

After isolating and analyzing many proteins from red blood cell membranes, researchers found one protein that specifically increases membrane permeability to glucose. When they added this purified protein to liposomes, the artificial membrane transported glucose at the same rate as a membrane from a living cell. This experiment convinced biologists that the membrane protein—now called GLUT-1 (short for glucose transporter 1)—was indeed responsible for transporting glucose across plasma membranes.

How Does GLUT-1 Work?   Recall that proteins frequently change shape when they bind to other molecules and that such conformational changes are often a critical step in their function (Chapter 3).

FIGURE 6.24 illustrates the current hypothesis for how GLUT-1 works to facilitate the movement of glucose. The idea is that when glucose binds to GLUT-1, it changes the shape of the protein in a way that moves the sugar through the hydrophobic region of the membrane and releases it on the other side.

What powers the movement of molecules through carriers? The answer is diffusion. GLUT-1 facilitates diffusion by allowing glucose to enter the carrier from either side of the membrane. Glucose will pass through the carrier in the direction dictated by its concentration gradient. A large variety of molecules move across plasma membranes via specific carrier proteins.

Pumps Perform Active Transport

Diffusion—whether it is facilitated by proteins or not—is a passive process that will move substances in either direction across a membrane to make the cell interior and exterior more similar. But it is also possible for cells to move molecules or ions in a directed manner, often *against* their electrochemical gradient. Accomplishing this task requires an input of energy, because the cell must counteract the decrease in entropy that occurs when molecules or ions are concentrated. It makes sense, then, that transport against an electrochemical gradient is called **active transport.**

In cells, ATP (adenosine triphosphate) often provides the energy for active transport by transferring a phosphate group (HPO_4^{2-}) to an active transport protein called a **pump.** Recall that ATP contains three phosphate groups (Chapter 4), and that phosphate groups carry two negative charges (Chapter 2). When a phosphate group leaves ATP and binds to a pump, its negative charges interact with charged amino acid residues in the protein. As a result, the protein's potential energy increases and its shape changes.

The Sodium–Potassium Pump   A classic example of how structural change leads to active transport is provided in the **sodium–potassium pump,** or more formally, Na^+/K^+-ATPase. The Na^+/K^+ part of the name refers to the ions that are transported, ATP indicates that adenosine triphosphate is used, and *–ase* identifies the molecule as an enzyme.

PROCESS: A HYPOTHESIS FOR HOW GLUT-1 FACILITATES GLUCOSE DIFFUSION

1. Unbound protein: GLUT-1 is a transmembrane transport protein, shown with its binding site facing outside the cell.

2. Glucose binding: Glucose binds to GLUT-1 from outside the cell.

3. Conformational change: Glucose binding causes a conformational change, transporting glucose to the interior.

4. Release: Glucose moves inside the cell. Steps may repeat or reverse, depending on the concentration gradient.

FIGURE 6.24 Carrier Proteins Undergo Structural Changes to Move Substances. This model suggests that GLUT-1 binds a glucose molecule, undergoes a conformational change, and releases glucose on the other side of the membrane.

1. Unbound protein: Three binding sites within the protein have a high affinity for sodium ions.

2. Sodium binding: Three sodium ions from the inside of the cell bind to these sites.

3. Shape change: A phosphate group from ATP binds to the protein. In response, the protein changes shape.

4. Release: The sodium ions leave the protein and move to the exterior of the cell.

FIGURE 6.25 The Sodium–Potassium Pump Depends on an Input of Chemical Energy Stored in ATP.

As shown in **FIGURE 6.25**, sodium and potassium ions move in a multistep process:

Step 1 When Na⁺/K⁺-ATPase is in the conformation shown here, binding sites with a high affinity for sodium ions are available.

Step 2 Three sodium ions from the inside of the cell bind to these sites and activate the ATPase activity in the pump.

Step 3 A phosphate group from ATP is transferred to the pump. When the phosphate group attaches, the pump changes its shape in a way that opens the ion-binding pocket to the external environment and reduces its affinity for sodium ions.

Step 4 The sodium ions leave the protein and move to the exterior of the cell.

Step 5 In this conformation, the pump has binding sites with a high affinity for potassium ions facing the external environment.

Step 6 Two potassium ions from outside the cell bind to the pump.

Step 7 When the potassium is bound, the phosphate group is cleaved from the protein and its structure changes in response—back to the original shape with the ion-binding pocket facing the interior of the cell.

Step 8 In this conformation, the pump has low affinity for potassium ions. The potassium ions leave the protein and move to the interior of the cell. The cycle then repeats.

Other types of pumps move protons (H⁺), calcium ions (Ca²⁺), or other ions or molecules across membranes in a directed manner, regardless of the gradients. As a result, cells can import and concentrate valuable nutrients and ions inside the cell despite their relatively low external concentration. They can also expel molecules or ions, even when a concentration gradient favors diffusion of these substances into the cell.

Secondary Active Transport   Approximately 30 percent of all the ATP generated in your body is used to drive the Na⁺/K⁺-ATPase cycle. Each cycle exports three Na⁺ ions for every two K⁺ ions it imports. In this way, the sodium–potassium pump converts energy from ATP to an electrochemical gradient across the membrane. The outside of the membrane becomes positively charged relative to the inside. This gradient favors a flow of anions out of the cell and a flow of cations into the cell.

The electrochemical gradients established by the Na⁺/K⁺-ATPase represent a form of stored energy, much like the electrical energy stored in a battery. How do cells use this energy?

Gradients are crucial to the function of the cell, in part because they make it possible for cells to engage in **secondary active transport**—also known as cotransport. When cotransport occurs, a gradient set up by a pump provides the energy required to power the movement of a different molecule against its particular gradient.

Recall that GLUT-1 facilitates the movement of glucose into or out of cells in the direction of its gradient. Can glucose be moved against its gradient? The answer is yes—a cotransport protein in your gut cells uses the Na⁺ gradient created by Na⁺/K⁺-ATPases to import glucose against its chemical gradient. When Na⁺ ions bind to this cotransporter, its shape changes in a way that allows glucose to bind. Once glucose binds, another shape transports both the sodium and glucose to the inside of the cell. After dropping off sodium and glucose, the protein's original shape returns to repeat the cycle.

In this way, glucose present in the food you are digesting is actively transported into your body. The glucose molecules eventually diffuse into your bloodstream and are transported to your brain, where they provide the chemical energy you need to stay awake and learn some biology. (You will learn more about secondary active transport in Units 7 and 8.)

Plasma Membranes and the Intracellular Environment

Taken together, the selective permeability of the lipid bilayer and the specificity of the proteins involved in passive transport and

5. Unbound protein: In this conformation, the protein has binding sites with a high affinity for potassium ions.

6. Potassium binding: Two potassium ions bind to the pump.

7. Shape change: The phosphate group is cleaved from the protein, allowing the pump to return to its original shape.

8. Release: The potassium ions leave the protein and diffuse to the interior of the cell. *These 8 steps repeat.*

active transport enable cells to create an internal environment that is much different from the external one (**FIGURE 6.26**).

With the evolution of membrane proteins, the early cells acquired the ability to create an internal environment that was conducive to life—one that contained the substances required for manufacturing ATP and copying ribozymes. Cells with particularly efficient and selective membrane proteins would be favored by natural selection and would come to dominate the population. Cellular life had begun.

Some 3.5 billion years later, cells continue to evolve. What do today's cells look like, and how do they produce and store the chemical energy that makes life possible? Answering these and related questions is the focus of the following unit.

check your understanding

C Y U

If you understand that . . .

- Membrane proteins allow substances that ordinarily do not readily cross lipid bilayers to enter or exit cells.
- Substances may move across a membrane along an electrochemical gradient, via facilitated diffusion through channel or carrier proteins. Or, they may move against a gradient in response to work done by pumps.

✔ You should be able to . . .

Explain what is passive about passive transport, active about active transport, and "co" about cotransport.

Answers are available in Appendix A.

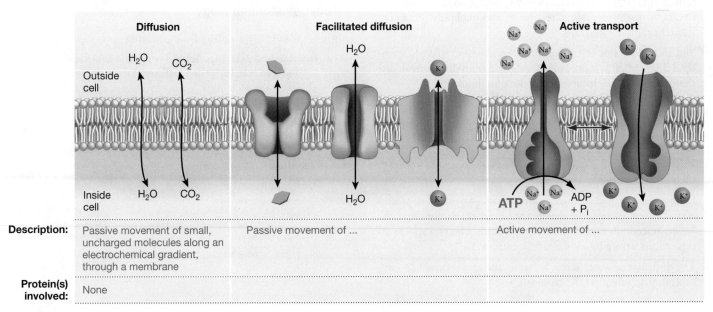

	Diffusion	**Facilitated diffusion**	**Active transport**
Description:	Passive movement of small, uncharged molecules along an electrochemical gradient, through a membrane	Passive movement of ...	Active movement of ...
Protein(s) involved:	None		

FIGURE 6.26 Summary of the Passive and Active Mechanisms of Membrane Transport.

✔**EXERCISE** Complete the chart.

If you understand . . .

6.1 Lipid Structure and Function

- Lipids are largely hydrophobic compounds due to their high number of nonpolar C—H bonds.

- The three main types of lipids found in cells are fats, steroids, and phospholipids. These molecules vary considerably in structure and function.

- In hydrocarbon chains, the length and degree of saturation have a profound effect on their physical properties.

- Amphipathic lipids possess a distinct hydrophilic region containing polar or charged groups. Phospholipids have a polar head and a nonpolar tail. The nonpolar tail usually consists of fatty acids or isoprenoids.

✔ You should be able to explain how adding hydrogen (H_2) to vegetable oil, a process called hydrogenation, results in a butter-like solid called margarine.

6.2 Phospholipid Bilayers

- In solution, phospholipids spontaneously assemble into bilayers that can serve as a physical barrier between an internal and external environment.

- Small nonpolar molecules tend to move across lipid bilayers readily; ions cross rarely, if at all.

- The permeability and fluidity of lipid bilayers depend on temperature, on the concentration of steroids, and on the chemical structure of the lipids present, such as the saturation status and length of the hydrocarbon chains. Phospholipids with longer or saturated tails form a dense and highly hydrophobic interior that lowers bilayer permeability, relative to phospholipids containing shorter or unsaturated tails.

✔ You should be able to explain how the structure of a phospholipid bilayer that is highly permeable and fluid differs from one that is highly impermeable and lacking in fluidity.

6.3 How Molecules Move across Lipid Bilayers: Diffusion and Osmosis

- Diffusion is the random movement of ions or molecules owing to their kinetic energy.

- Diffusion can result in the net directional movement of solutes across a membrane, if the membrane separates solutions that differ in concentration, charge, or temperature. This is a spontaneous process driven by an increase in entropy.

- The diffusion of water across a membrane in response to a concentration gradient is called osmosis.

✔ You should be able to imagine a beaker with solutions separated by a planar membrane and then predict what will happen after addition of a solute to one side if the solute (1) crosses the membrane readily or (2) is incapable of crossing the membrane.

6.4 Membrane Proteins

- The permeability of lipid bilayers can be altered significantly by membrane proteins.

- Channel proteins provide pores in the membrane and facilitate the diffusion of specific solutes into and out of the cell.

- Carriers undergo conformational changes that facilitate the diffusion of specific molecules into and out of the cell.

- Pumps use energy to actively move ions or molecules in a single direction, often against the electrical or chemical gradient.

- In combination, the selective permeability of phospholipid bilayers and the specificity of transport proteins make it possible to create an environment inside a cell that is radically different from the exterior.

✔ You should be able to draw and label the membrane of a cell that is placed in a solution containing calcium ions and lactose and show the activity of the following membrane proteins: (1) an H^+ pump that exports protons; (2) a calcium channel; and (3) a lactose carrier. Your drawing should include arrows and labels indicating the direction of solute movement and the direction of the appropriate electrochemical gradients.

MasteringBiology

1. MasteringBiology Assignments

Tutorials and Activities Active Transport; Diffusion, Diffusion and Osmosis; Facilitated Diffusion Lipids; Membrane Structure; Membrane Transport: Diffusion and Passive Transport; Membrane Transport: The Sodium–Potassium Pump; Membrane Transport Proteins; Osmosis; Membrane Transport: Cotransport; Osmosis and Water Balance in Cells; Selective Permeability of Membranes

Questions Reading Quizzes, Blue-Thread Questions, Test Bank

2. eText Read your book online, search, take notes, highlight text, and more.

3. The Study Area Practice Test, Cumulative Test, BioFlix® 3-D Animations, Videos, Activities, Audio Glossary, Word Study Tools, Art

You should be able to . . .

1. How is the structure of saturated fats different from that of unsaturated fats?
   a. All of the carbons in the hydrocarbon tails of saturated fats are bonded to one another with double bonds.
   b. Saturated fats have three hydrocarbon tails bonded to the glycerol molecule instead of just two.
   c. The hydrocarbon tails in a saturated fat have the maximum number of hydrogens possible.
   d. Saturated fats have no oxygens present.

2. What distinguishes amphipathic lipids from other lipids?
   a. Amphipathic lipids have polar and nonpolar regions.
   b. Amphipathic lipids have saturated and unsaturated regions.
   c. Amphipathic lipids are steroids.
   d. Amphipathic lipids dissolve in water.

3. If a solution surrounding a cell is hypertonic relative to the inside of the cell, how will water move?

   a. It will move into the cell via osmosis.
   b. It will move out of the cell via osmosis.
   c. It will not move, because equilibrium exists.
   d. It will evaporate from the cell surface more rapidly.

4. When does a concentration gradient exist?
   a. when membranes rupture
   b. when solute concentrations are high
   c. when solute concentrations are low
   d. when solute concentrations differ on the two sides of a membrane

5. What two conditions must be present for the effects of osmosis to occur?

6. In terms of structure, how do channel proteins differ from carrier proteins?

7. If a cell were placed in a solution with a high potassium concentration and no sodium, what would happen to the sodium–potassium pump's activity?
   a. It would stop moving ions across the membrane.
   b. It would continue using ATP to pump sodium out of the cell and potassium into the cell.
   c. It would move sodium and potassium ions across the membrane, but no ATP would be used.
   d. It would reverse the direction of sodium and potassium ions to move them against their gradients.

8. Cooking oil lipids consist of long, unsaturated hydrocarbon chains. Would you expect these molecules to form membranes spontaneously? Why or why not? Describe, on a molecular level, how you would expect these lipids to interact with water.

9. Explain why phospholipids form a bilayer in solution, and why the process is spontaneous.

10. Ethanol (C_2H_5OH) is the active ingredient in alcoholic beverages. Would you predict that this molecule crosses lipid bilayers quickly, slowly, or not at all? Explain your reasoning.

11. Integral membrane proteins are anchored in lipid bilayers. Of the following four groups of amino acids—nonpolar, polar, charged/acidic, charged/basic (see Figure 3.2)—which would likely be found in the portion that crosses the lipid bilayer? Explain your reasoning.

12. Examine the experimental chamber in Figure 6.7a. If the lipid bilayer were to contain the CFTR molecule, what would pass through the membrane if you added a 1-molar solution of sodium chloride on the left side and a 1-molar solution of potassium ions on the right? Assume that there is an equal amount of water on each side at the start of the experiment.

13. In an experiment, you create two groups of liposomes—one made from red blood cell membranes and the other from frog egg cell membranes. When placed in water, those made with red blood cell membranes burst more rapidly than those made from frog membranes. What is the best explanation for these results?
    a. The red blood cell liposomes are more hypertonic relative to water than the frog egg liposomes.
    b. The red blood cell liposomes are more hypotonic relative to water than the frog egg liposomes.
    c. The red blood cell liposomes contain aquaporins, which are not abundant in the frog egg liposomes.
    d. The frog egg liposomes contain ion channels, which are not present in the red blood cell liposomes.

14. When phospholipids are arranged in a bilayer, it is theoretically possible for individual molecules in the bilayer to flip-flop. That is, a phospholipid could turn 180° and become part of the membrane's other surface. From what you know about the

behavior of polar heads and nonpolar tails, predict whether flip-flops are frequent or rare. Then design an experiment, using a planar bilayer with one side made up of phospholipids that contain a dye molecule on their hydrophilic head, to test your prediction.

15. Unicellular organisms live in a wide range of habitats, from the hot springs in Yellowstone National Park to the freezing temperatures of the Antarctic. Make a prediction about the saturation status of membrane phospholipids in organisms that live in extremely cold environments versus those that live in extremely hot environments. Explain your reasoning.

16. When biomedical researchers design drugs, they sometimes add methyl (CH_3) groups or charged groups to the molecules. If these groups are not directly involved in the activity of the drug, predict the purpose of these modifications and explain why these strategies are necessary.

THE CHEMISTRY OF LIFE

The Big Picture

The first spark of life ignited when simple chemical reactions began to convert small molecules into larger, more complex molecules with novel 3-D structures and activities. According to the theory of chemical evolution, these reactions eventually led to the formation of the four types of macromolecules characteristic of life—proteins, nucleic acids, carbohydrates, and lipids.

As you look through this concept map, consider how the functions of the four types of macromolecules are determined by their structures, and how these structures stem from the chemical properties of the atoms and bonds used to build them.

Note that each box in the concept map indicates the chapters and sections where you can go for review. Also, be sure to do the blue exercises in the Check Your Understanding box below.

is based on the reactivity of

Atoms
- Carbon (C)
- Hydrogen (H)   96% of atoms in living matter
- Oxygen (O)
- Nitrogen (N)
- Phosphorus (P)
- Sulfur (S)
- Others   2.1, 2.4

differ in

Electronegativity

$O > N > C \cong H$

Number of unpaired electrons:
- Hydrogen: 1
- Oxygen: 2
- Nitrogen: 3
- Carbon: 4

2.1

combine to form

Molecules   2.1

notably

have specific

3-D structure
- (may possess primary, secondary, tertiary, quaternary structure)

2.1, 3.1–3, 4.1–3, 5.1–2, 6.1

determines function

determine type and number of

influence

Polarity   2.1

affects

Molecular function
2.2, 3.1, 4.1, 5.1

depends on

Chemical bonds
- Nonpolar covalent bonds
- Polar covalent bonds
- Ionic bonds
- Hydrogen bonds

δ^- δ^+

2.1

including

formed and broken in

Chemical reactions 2.3

as demonstrated by

WATER

is essential for life due to its

- Efficiency as a solvent
- Cohesion, adhesion and surface tension properties
- Higher density as a liquid than as a solid
- High capacity for energy absorption

2.2

Basic

14

H_2O 7

pH   2.2

has neutral

0

Acidic

check your understanding

C Y U

If you understand the big picture . . .

✔ **You should be able to . . .**

1. Explain how the relative electronegativities of atoms affect the bonding within and among water molecules.

2. Describe the attributes of RNA that make it a candidate for the origin of life. Why isn't DNA considered a candidate?

3. Circle the atoms in amino acids and nucleotides that engage in creating bonds with other monomers.

4. Draw a protein in the lipid bilayer. What role might it play?

Answers are available in Appendix A.

Biological macromolecules

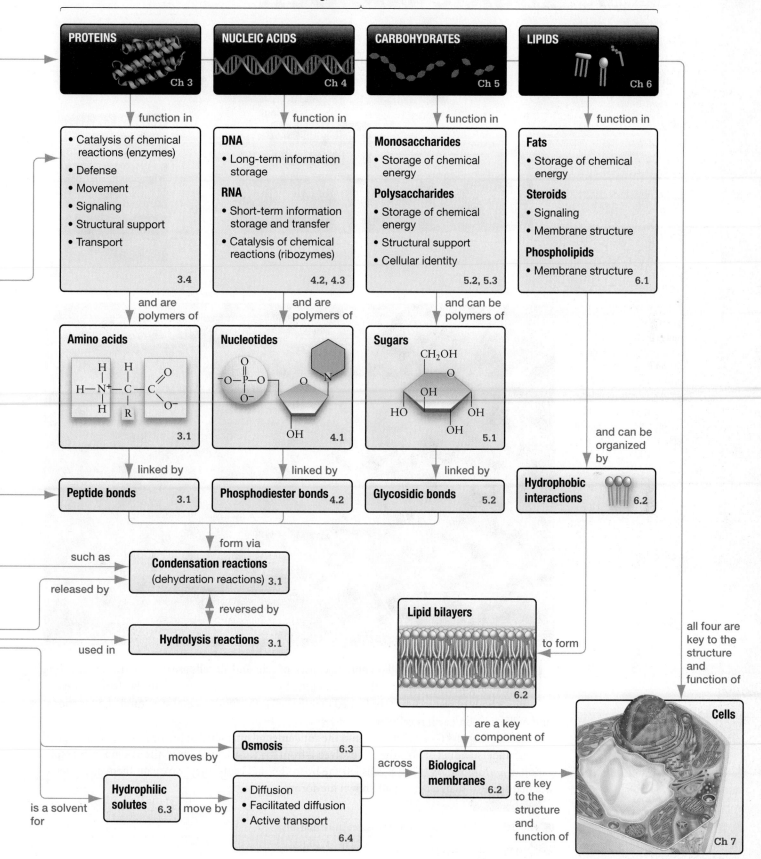

PROTEINS Ch 3

NUCLEIC ACIDS Ch 4

CARBOHYDRATES Ch 5

LIPIDS Ch 6

function in

function in

function in

function in

- Catalysis of chemical reactions (enzymes)
- Defense
- Movement
- Signaling
- Structural support
- Transport

3.4

DNA
- Long-term information storage

RNA
- Short-term information storage and transfer
- Catalysis of chemical reactions (ribozymes)

4.2, 4.3

Monosaccharides
- Storage of chemical energy

Polysaccharides
- Storage of chemical energy
- Structural support
- Cellular identity

5.2, 5.3

Fats
- Storage of chemical energy

Steroids
- Signaling
- Membrane structure

Phospholipids
- Membrane structure

6.1

and are polymers of

and are polymers of

and can be polymers of

Amino acids 3.1

Nucleotides 4.1

Sugars 5.1

and can be organized by

linked by

linked by

linked by

Peptide bonds 3.1

Phosphodiester bonds 4.2

Glycosidic bonds 5.2

Hydrophobic interactions 6.2

such as

released by

form via

Condensation reactions (dehydration reactions) 3.1

reversed by

Hydrolysis reactions 3.1

used in

Lipid bilayers 6.2

to form

all four are key to the structure and function of

are a key component of

moves by

Osmosis 6.3

across

Biological membranes 6.2

are key to the structure and function of

is a solvent for

Hydrophilic solutes 6.3

move by

- Diffusion
- Facilitated diffusion
- Active transport

6.4

Cells Ch 7

7 Inside the Cell

In this chapter you will learn that

Life's properties emerge from the collaboration of internal structures in a cell

by asking

What are the parts of the cell?

examining

Prokaryotic cell structures **7.1**

Eukaryotic cell structures **7.2**

How the parts fit into a whole **7.3**

examining

Nuclear transport **7.4**

looking closer at

Endomembrane system **7.5**

Dynamic cytoskeleton **7.6**

This cell has been treated with fluorescing molecules that bind to its fibrous cytoskeleton. Microtubules (large protein fibers) are yellow; actin filaments (smaller fibers) are blue. The cell's nucleus has been stained green.

The cell theory states that all organisms consist of cells and all cells are derived from preexisting cells (Chapter 1). Since this theory was initially developed and tested in the 1850s, an enormous body of research has confirmed that the cell is the fundamental structural and functional unit of life. Life on Earth is cellular.

Previous chapters (Unit 1) delved into the fundamental attributes of life by looking at biologists' current understanding of how the cell evolved—from the early chemistry to the assembly and replication of a protocell. As the first cells left the hydrothermal vents, they took with them characteristics that are now shared among all known life-forms.

All cells have

1. nucleic acids that store and transmit information;

2. proteins that perform most of the cell's functions;

✔ When you see this checkmark, stop and test yourself. Answers are available in Appendix A.

3. carbohydrates that provide chemical energy, carbon, support, and identity; and

4. a plasma membrane, which serves as a selectively permeable membrane barrier.

Thanks to the selective permeability of phospholipid bilayers and the activity of membrane transport proteins, the plasma membrane creates an internal environment that differs from conditions outside the cell. Our task now is to explore the structures inside the cell to understand how the properties of life emerged from the combination of these characteristics.

Let's begin by analyzing how the parts inside a cell function individually and then exploring how they work as a unit. This approach is analogous to studying individual organs in the body and then analyzing how they work together to form the nervous system or digestive system. As you study this material, keep asking yourself some key questions: How does the structure of this part or group of parts correlate with its function? What problem does it solve?

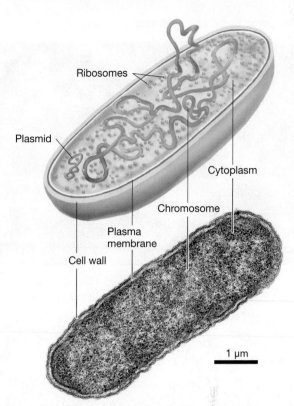

FIGURE 7.1 Overview of a Prokaryotic Cell. Prokaryotic cells are identified by a negative trait—the absence of a membrane-bound nucleus. Although there is wide variation in the size and shape of bacterial and archaeal cells, they all contain a plasma membrane, a chromosome, and protein-synthesizing ribosomes.

7.1 Bacterial and Archaeal Cell Structures and Their Functions

Cells are divided into two fundamental types called eukaryotes and prokaryotes (see Chapter 1). This division is mostly based on cell **morphology** ("form-science")—eukaryotic cells have a membrane-bound compartment called a nucleus, and prokaryotic cells do not.

But according to **phylogeny** ("tribe-source"), or evolutionary history, organisms are divided into three broad domains called **(1)** Bacteria, **(2)** Archaea, and **(3)** Eukarya. Members of the Bacteria and Archaea are prokaryotic; members of the Eukarya—including algae, fungi, plants, and animals—are eukaryotic.

A Revolutionary New View

For almost 200 years, biologists thought that prokaryotic cells were simple in terms of their morphology and that there was little structural diversity among species. This conclusion was valid at the time, given the resolution of the microscopes that were available and the number of species that had been studied.

Things have changed. Recent improvements in microscopy and other research tools have convinced biologists that prokaryotic cells, among which bacteria are the best understood, possess an array of distinctive structures and functions found among millions of species. This conclusion represents one of the most exciting discoveries in cell biology over the past 10 years.

To keep things simple at the start, though, **FIGURE 7.1** offers a low-magnification, stripped-down diagram of a bacterial cell.

Prokaryotic Cell Structures: A Parts List

The labels in Figure 7.1 highlight the components common to all or most bacteria studied to date. Let's explore these elements one by one, and also look at more specialized structures found in particular species, starting from the inside and working out.

The Chromosome Is Organized in a Nucleoid   The most prominent structure inside a bacterial cell is the **chromosome.** Most bacterial species have a single, circular chromosome that consists of a large DNA molecule associated with a small number of proteins. The DNA molecule contains information, and the proteins provide structural support for the DNA.

Recall that the information in DNA is encoded in its sequence of nitrogenous bases. Segments of DNA that contain information for building functional RNAs, some of which may be used to make polypeptides, are called **genes** (Chapter 4). Thus, chromosomes contain DNA, which contains genes.

In the well-studied bacterium *Escherichia coli*, the circular chromosome would be over 1 mm long if it were linear—500 times longer than the cell itself (**FIGURE 7.2a**; see page 108). This situation is typical in prokaryotes. To fit into the cell, the DNA double helix coils on itself with the aid of enzymes to form a compact, "supercoiled" structure. Supercoiled regions of DNA resemble a rubber band that has been held at either end and then twisted until it coils back upon itself.

The location and structural organization of the circular chromosome is called the **nucleoid** (pronounced *NEW-klee-oyd*). The genetic material in the nucleoid is often organized by clustering loops of DNA into distinct domains, but it is not separated from the rest of the cell interior by a membrane. The functional role of this organization of the bacterial chromosome and how it changes over time is currently the subject of intense research.

(a) Compared to the cell, chromosomal DNA is very long.

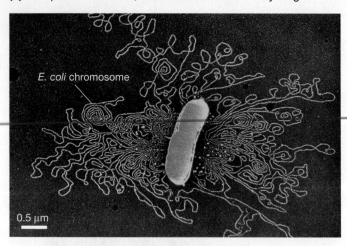

(b) DNA is packaged by supercoiling.

FIGURE 7.2 Bacterial DNA Is Supercoiled. (a) The chromosomes of bacteria and archaea are often over 1000 times the length of the cell, as shown in this micrograph of *E. coli* that has been treated to release its DNA. To fit inside cells, this DNA must be highly compacted by supercoiling. **(b)** A colorized electron micrograph showing the effect of supercoiling on the DNA of isolated plasmids.

In addition to one or more chromosomes, bacterial cells may contain from one to about a hundred small, usually circular, supercoiled DNA molecules called **plasmids** (**FIGURE 7.2b**). Plasmids contain genes but are physically independent of the cellular chromosome. In many cases the genes carried by plasmids are not required under normal conditions; instead, they help cells adapt to unusual circumstances, such as the sudden presence of a poison in the environment. As a result, plasmids can be considered auxiliary genetic elements.

Ribosomes Manufacture Proteins   **Ribosomes** are observed in all prokaryotic cells and are found throughout the cell interior. It is not unusual for a single cell to contain 10,000 ribosomes, each functioning as a protein-manufacturing center.

Ribosomes are complex structures composed of large and small subunits, each of which contains RNA and protein molecules. Biologists often refer to ribosomes, along with other multicomponent complexes that perform specialized tasks, as "macromolecular machines." (Chapter 17 analyzes the structure and function of ribosomes in detail.)

Photosynthetic Species Have Internal Membrane Complexes   In addition to the nucleoid and ribosomes found in all bacteria and archaea studied to date, it is common to observe extensive internal membranes in prokaryotes that perform photosynthesis. Photosynthesis is the suite of chemical reactions responsible for converting the energy in sunlight into chemical energy stored in sugars.

The photosynthetic membranes observed in prokaryotes contain the enzymes and pigment molecules required for these reactions to occur and develop as infoldings of the plasma membrane. In some cases, vesicles pinch off as the plasma membrane folds in. In other cases, flattened stacks of photosynthetic membrane remain connected to the plasma membrane, like those shown in **FIGURE 7.3**. The extensive surface area provided by these internal membranes makes it possible for more photosynthetic reactions to occur and thus increases the cell's ability to make food.

Organelles Perform Specialized Functions   Recent research indicates that several bacterial species have internal compartments that qualify as **organelles** ("little organs"). An organelle is a membrane-bound compartment inside the cell that contains enzymes or structures specialized for a particular function.

Bacterial organelles perform an array of tasks, including

- storing calcium ions or other key molecules;

- holding crystals of the mineral magnetite, which function like a compass needle to help cells sense a magnetic field and swim in a directed way;

- organizing enzymes responsible for synthesizing complex carbon compounds from carbon dioxide; and

- sequestering enzymes that generate chemical energy from ammonium ions.

The Cytoskeleton Structures the Cell Interior   Recent research has also shown that bacteria and archaea contain long, thin fibers that serve a variety of roles inside the cell. All bacterial species,

Photosynthetic membranes

FIGURE 7.3 Photosynthetic Membranes in Bacteria. The green stripes in this photosynthetic bacterium are membranes that contain the pigments and enzymes required for photosynthesis. This photo has been colorized to enhance the membranes.

for example, contain protein fibers that are essential for cell division to take place. Some species also have protein filaments that help maintain cell shape. Protein filaments such as these form the basis of the **cytoskeleton** ("cell skeleton").

The discovery of bacterial cytoskeletal elements is so new that much remains to be learned. Currently, researchers are working to understand how the different cytoskeletal elements enable cells to divide and if they play a role in organizing the cell interior into distinctive regions.

The Plasma Membrane Separates Life from Nonlife   The plasma membrane consists of a phospholipid bilayer and proteins that either span the bilayer or attach to one side. Inside the membrane, all the contents of a cell, excluding the nucleus, are collectively termed the **cytoplasm** ("cell-formed").

Because all archaea and virtually all bacteria are unicellular, the plasma membrane creates an internal environment that is distinct from the outside, nonliving environment. The combined effect of a lipid bilayer and membrane proteins prohibits the entry of many substances that would be dangerous to life while allowing the passage of molecules and ions required for life (see Chapter 6).

The Cell Wall Forms a Protective "Exoskeleton"   Because the cytoplasm contains a high concentration of solutes, in most habitats it is hypertonic relative to the surrounding environment. When this is the case, water enters the cell via osmosis and makes the cell's volume expand. In virtually all bacteria and archaea, this pressure is resisted by a stiff **cell wall.**

Bacterial and archaeal cell walls are a tough, fibrous layer that surrounds the plasma membrane. In prokaryotes, the pressure of the plasma membrane against the cell wall is about the same as the pressure in an automobile tire.

The cell wall protects the organism and gives it shape and rigidity, much like the exoskeleton (external skeleton) of a crab or insect. In addition, many bacteria have another protective layer outside the cell wall that consists of lipids with polysaccharides attached. Lipids that contain carbohydrate groups are termed **glycolipids.**

External Structures Enable Movement and Attachment   Besides having a cell wall to provide protection, as just described, many bacteria also interact with their environment via structures that grow from the plasma membrane. The flagella and fimbriae shown in **FIGURE 7.4** are examples that are commonly found on bacterial surfaces.

Bacterial **flagella** (singular: **flagellum**) are assembled from over 40 different proteins at the cell surface of certain species. The base of this structure is embedded in the plasma membrane, and its rotation spins a long, helical filament that propels cells through water. At top speed, flagellar movement can drive a bacterial cell through water at 60 cell lengths per second. In contrast, the fastest animal in the ocean—the sailfish—can swim at a mere 10 body lengths per second.

Fimbriae (singular: **fimbria**) are needlelike projections that extend from the plasma membrane of some bacteria and promote attachment to other cells or surfaces. These structures are

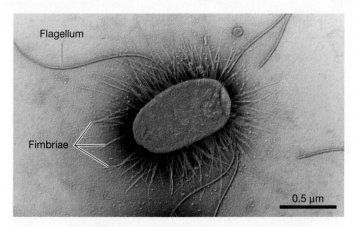

FIGURE 7.4 Extracellular Appendages Found on Bacteria. Some species of bacteria, such as the *E. coli* shown here, assemble large protein structures used for swimming through liquid (flagella) or adhering to surfaces (fimbriae).

more numerous than flagella and are often distributed over the entire surface of the cell. Fimbriae are crucial to the establishment of many infections based on their ability to glue bacteria to the surface of tissues.

The painting in **FIGURE 7.5** shows a cross section of a bacterial cell and provides a close-up view of the internal and external structures introduced in this section. One feature that prokaryotic and eukaryotic cells have in common: They are both packed with dynamic, highly integrated structures.

FIGURE 7.5 Close-up View of a Prokaryotic Cell. This painting is David Goodsell's representation of a cross section through part of a bacterial cell. It is based on electron micrographs of bacterial cells and is drawn to scale. Note that the cell is packed with proteins, DNA, ribosomes, and other molecular machinery.

check your understanding

If you understand that . . .

- Each structure in a prokaryotic cell performs a function vital to the cell.

✔ **You should be able to . . .**

Describe the structure and function of (1) the nucleoid, (2) photosynthetic membranes, (3) flagella, and (4) the cell wall.

Answers are available in Appendix A.

7.2 Eukaryotic Cell Structures and Their Functions

The Eukarya includes species that range from microscopic algae to 100-meter-tall redwood trees. Brown algae, red algae, fungi, amoebae, slime molds, green plants, and animals are all eukaryotic. Although multicellularity has evolved several times among eukaryotes (see Chapter 30), many species are unicellular.

The first thing that strikes biologists about eukaryotic cells is how much larger they are on average than bacteria and archaea. Most prokaryotic cells measure 1 to 10 μm in diameter, while most eukaryotic cells range from about 5 to 100 μm in diameter. A micrograph of an average eukaryotic cell, at the same scale as the bacterial cell in Figure 7.3, would fill this page. For many species of unicellular eukaryotes, this size difference allows them to make a living by ingesting bacteria and archaea whole.

Large size has a downside, however. As a cell increases in diameter, its volume increases more than its surface area. In other words, the relationship between them—the surface-area-to-volume ratio—changes. Since the surface is where the cell exchanges substances with its environment, the reduction in this ratio decreases the rate of exchange: Diffusion only allows for rapid movement across very small distances.

Prokaryotic cells tend to be small enough so that ions and small molecules arrive where they are needed via diffusion. The random movement of diffusion alone, however, is insufficient for this type of transport as the cell's diameter increases.

The Benefits of Organelles

How do eukaryotic cells solve the problems that size can engender? The answer lies in their numerous organelles. In effect, the huge volume inside a eukaryotic cell is compartmentalized into many small bins. Because eukaryotic cells are subdivided, the **cytosol**—the fluid portion between the plasma membrane and these organelles—is only a fraction of the total cell volume. This relatively small volume of cytosol reduces the effect of the total cell surface-area-to-volume ratio with respect to the exchange of nutrients and waste products.

Compartmentalization also offers two key advantages:

1. Incompatible chemical reactions can be separated. For example, new fatty acids can be synthesized in one organelle while

SUMMARY TABLE 7.1 How Do the Structures of Prokaryotic and Eukaryotic Cells Differ?

	Bacteria and Archaea	Eukaryotes
Location of DNA	In nucleoid (not membrane bound); plasmids also common	Inside nucleus (membrane bound); plasmids extremely rare
Internal Membranes and Organelles	Extensive internal membranes only in photosynthetic species; limited types and numbers of organelles	Large numbers of organelles; many types of organelles
Cytoskeleton	Limited in extent, relative to eukaryotes	Extensive—usually found throughout volume of cell
Overall Size	Usually small relative to eukaryotes	Most are larger than prokaryotes

excess or damaged fatty acids are degraded and recycled in a different organelle.

2. Chemical reactions become more efficient. First, the substrates required for particular reactions can be localized and maintained at high concentrations within organelles. Second, if substrates are used up in a particular part of the organelle, they can be replaced by substrates that have only a short distance to diffuse. Third, groups of enzymes that work together can be clustered within or on the membranes of organelles instead of floating free in the cytosol. When the product of one reaction is the substrate for a second reaction catalyzed by another enzyme, clustering the enzymes increases the speed and efficiency of both reaction sequences.

If bacteria and archaea can be compared to small, specialized machine shops, then eukaryotic cells resemble sprawling industrial complexes. The organelles and other structures found in eukaryotes are analogous to highly specialized buildings that act as administrative centers, factories, transportation corridors, waste and recycling facilities, warehouses, and power stations.

When typical prokaryotic and eukaryotic cells are compared, four key differences, identified in **TABLE 7.1**, stand out:

1. Eukaryotic chromosomes are found inside a membrane-bound compartment called the **nucleus.**

2. Eukaryotic cells are often much larger than prokaryotes.

3. Eukaryotic cells contain extensive amounts of internal membrane.

4. Eukaryotic cells feature a particularly diverse and dynamic cytoskeleton.

Eukaryotic Cell Structures: A Parts List

FIGURE 7.6 provides a simplified view of a typical animal cell and a plant cell. The artist has removed most of the cytoskeletal elements

(a) Generalized animal cell

Nuclear envelope ⎤
Nucleolus ⎬ Nucleus
Chromosomes ⎦

Rough endoplasmic reticulum

Ribosomes

Peroxisome

Smooth endoplasmic reticulum

Golgi apparatus

Centrioles

Structures that occur in animal cells but not plant cells

Lysosome

Mitochondrion

Cytoskeletal element

Plasma membrane

(b) Generalized plant cell

Nuclear envelope ⎤
Nucleolus ⎬ Nucleus
Chromosomes ⎦

Rough endoplasmic reticulum

Ribosomes

Smooth endoplasmic reticulum

Structures that occur in plant cells but not animal cells

Golgi apparatus

Vacuole

Cell wall

Chloroplast

Peroxisome

Mitochondrion

Plasma membrane

Cytoskeletal element

On average, prokaryotes are about 10 times smaller than eukaryotic cells in diameter and about 1000 times smaller than eukaryotic cells in volume.

FIGURE 7.6 Overview of Eukaryotic Cells. Generalized images of **(a)** animal and **(b)** plant cells that illustrate the cellular structures in the "typical" eukaryote. The structures have been color-coded for clarity. Compare with the prokaryotic cell, shown at true relative size at bottom left.

Nucleus

Loosely packed sections of chromosomes

Nucleolus

Densely packed sections of chromosomes

Nuclear envelope

2 μm

FIGURE 7.7 The Nucleus Stores and Transmits Information. The genetic, or hereditary, information is encoded in DNA, which is a component of the chromosomes inside the nucleus.

Ribosomes

100 nm

FIGURE 7.8 Ribosomes Are the Site of Protein Synthesis. Eukaryotic ribosomes are larger than bacterial and archaeal ribosomes, but similar in overall structure.

to make the organelles and other cellular parts easier to see. As you read about each cell component in the pages that follow, focus on identifying how its structure correlates with its function. Then use **TABLE 7.2** (see page 117) as a study guide. As with bacterial cells, let's start from the inside and move to the outside.

The Nucleus   The nucleus contains the chromosomes and functions as an administrative center for information storage and processing. Among the largest and most highly organized of all organelles (**FIGURE 7.7**), it is enclosed by a unique structure—a complex double membrane called the **nuclear envelope.** As Section 7.4 will detail, the nuclear envelope is studded with pore-like openings, and the inside surface is linked to fibrous proteins that form a lattice-like sheet called the **nuclear lamina.** The nuclear lamina stiffens the structure and maintains its shape.

Chromosomes do not float freely inside the nucleus—instead, each chromosome occupies a distinct area, which may vary in different cell types and over the course of cell replication. The nucleus also contains specific sites where gene products are processed and includes at least one distinctive region called the **nucleolus,** where the RNA molecules found in ribosomes are manufactured and the large and small ribosomal subunits are assembled.

Ribosomes   In eukaryotes, the cytoplasm consists of everything inside the plasma membrane excluding the nucleus. Scattered throughout this cytoplasm are millions of ribosomes (**FIGURE 7.8**).

Like bacterial ribosomes, eukaryotic ribosomes are complex macromolecular machines that manufacture proteins. They are not classified as organelles because they are not surrounded by membranes.

Endoplasmic Reticulum   The portions of the nuclear envelope extend into the cytoplasm to form an extensive membrane-enclosed factory called the **endoplasmic reticulum** (literally, "inside-formed-network"), or ER. As Figure 7.6 shows, the ER membrane is continuous with the nuclear envelope. Although the ER is a single structure, it has two regions that are distinct in structure and function. Let's consider each region in turn.

The **rough endoplasmic reticulum (RER),** or **rough ER,** is named for its appearance in transmission electron micrographs (see **FIGURE 7.9**, left). The knobby-looking structures in the rough ER are ribosomes that attach to the membrane.

The ribosomes associated with the rough ER synthesize proteins that will be inserted into the plasma membrane, secreted to the cell exterior, or shipped to an organelle. As they are being manufactured by ribosomes, these proteins move to the interior of the sac-like component of the rough ER. The interior of the rough ER, like the interior of any sac-like structure in a cell or body, is called the **lumen.** In the lumen of the rough ER, newly manufactured proteins undergo folding and other types of processing.

The proteins produced in the rough ER have a variety of functions. Some carry messages to other cells; some act as membrane

Rough endoplasmic reticulum

Lumen of rough ER

Ribosomes on outside of rough ER

Free ribosomes in cytoplasm

200 nm

Smooth endoplasmic reticulum

Lumen of smooth ER

200 nm

FIGURE 7.9 The Endoplasmic Reticulum Is a Site of Synthesis, Processing, and Storage. The ER is continuous with the nuclear envelope and possesses two distinct regions: on the left, the rough ER is a system of membrane-bound sacs and tubules with ribosomes attached; on the right, the smooth ER is a system of membrane-bound sacs and tubules that lacks ribosomes.

transport proteins or pumps; others are enzymes. The common theme is that many of the rough ER products are packaged into vesicles and transported to various distant destinations—often to the surface of the cell or beyond.

In electron micrographs, parts of the ER that are free of ribosomes appear smooth and even. Appropriately, these parts of the ER are called the **smooth endoplasmic reticulum (SER),** or **smooth ER** (see **FIGURE 7.9**, right).

The smooth ER contains enzymes that catalyze reactions involving lipids. Depending on the type of cell, these enzymes may synthesize lipids needed by the organism or break down lipids and other molecules that are poisonous. For example, the smooth ER is the manufacturing site for phospholipids used in plasma membranes. In addition, the smooth ER functions as a reservoir for calcium ions (Ca^{2+}) that act as a signal triggering a wide array of activities inside the cell.

The structure of the endoplasmic reticulum correlates closely with its function. The rough ER has ribosomes and functions primarily as a protein-manufacturing center; the smooth ER lacks ribosomes and functions primarily as a lipid-processing center.

Golgi Apparatus In many cases, the products of the rough ER pass through the Golgi apparatus before they reach their final destination. The **Golgi apparatus** consists of discrete flattened, membranous sacs called **cisternae** (singular: **cisterna**), which are stacked on top of one another (**FIGURE 7.10**). The organelle also

Golgi apparatus

cis

trans

Vesicle

cis side faces nucleus

Lumen of Golgi apparatus

Cisternae

trans side faces plasma membrane

Vesicles

100 nm

FIGURE 7.10 The Golgi Apparatus Is a Site of Protein Processing, Sorting, and Shipping. The Golgi apparatus is a collection of flattened vesicles called cisternae.

has a distinct polarity, or sidedness. The *cis* ("this side") surface is closest to the nucleus, and the *trans* ("across") surface is oriented toward the plasma membrane.

The *cis* side of a Golgi apparatus receives products from the rough ER, and the *trans* side ships them out to other organelles or the cell surface. In between, within the cisternae, the rough ER's products are processed and packaged for delivery. Micrographs often show "bubbles" on either side of the Golgi stack. These are membrane-bound vesicles that carry proteins or other products to and from the organelle. Section 7.5 analyzes the intracellular movement of molecules from the rough ER to the Golgi apparatus and beyond in more detail.

Lysosomes   Animal cells contain organelles called **lysosomes** that function as recycling centers (**FIGURE 7.11**). Lysosomes contain about 40 different enzymes, each specialized for hydrolyzing different types of macromolecules—proteins, nucleic acids, lipids, or carbohydrates. The amino acids, nucleotides, sugars, and other molecules that result from acid hydrolysis leave the lysosome via transport proteins in the organelle's membrane. Once in the cytosol, they can be used as sources of energy or building blocks for new molecules.

These digestive enzymes are collectively called acid hydrolases because under acidic conditions (pH of 5.0), they use water to break monomers from macromolecules. In the cytosol, where the pH is about 7.2, acid hydrolases are less active. Proton pumps in the lysosomal membrane maintain an acidic pH in the lumen of the lysosome by importing hydrogen ions.

Even though lysosomes are physically separated from the Golgi apparatus and the endoplasmic reticulum, these various

Vacuole

Vacuole

1 μm

FIGURE 7.12 Vacuoles Are Generally Storage Centers in Plant and Fungal Cells. Vacuoles vary in size and function. Some contain digestive enzymes and serve as recycling centers; most are large storage containers.

✔**QUESTION** Why are toxins like nicotine, cocaine, and caffeine stored in vacuoles instead of the cytosol?

organelles jointly form a key functional grouping referred to as the **endomembrane system.** The endomembrane ("inner-membrane") system is a center for producing, processing, and transporting proteins and lipids in eukaryotic cells. For example, acid hydrolases are synthesized in the ER, processed in the Golgi, and then shipped to the lysosome.

Vacuoles   The cells of plants, fungi, and certain other groups lack lysosomes. Instead, they contain a prominent organelle called a vacuole. Compared with the lysosomes of animal cells, the **vacuoles** of plant and fungal cells are large—sometimes taking up as much as 80 percent of a plant cell's volume (**FIGURE 7.12**).

Although some vacuoles contain enzymes that are specialized for digestion, most of the vacuoles observed in plant and fungal cells act as storage depots. In many cases, ions such as potassium (K^+) and chloride (Cl^-), among other solutes, are stored at such high concentrations they draw water in from the environment. As the vacuole expands in volume, the cytoplasm pushes the plasma membrane against the cell wall, which maintains the plant cell's shape. In other cells, vacuoles have more specialized storage functions:

- Inside seeds, cells may contain a large vacuole filled with proteins. When the embryonic plant inside the seed begins to grow, enzymes begin digesting these proteins to provide amino acids for the growing individual.

Lysosome

Material being digested within lysosomes

250 nm

FIGURE 7.11 Lysosomes Are Recycling Centers. Lysosomes are usually oval or globular and have a single membrane.

- In cells that make up flower petals or fruits, vacuoles are filled with colorful pigments.

- Elsewhere, vacuoles may be packed with noxious compounds that protect leaves and stems from being eaten by predators. The type of chemical involved varies by species, ranging from bitter-tasting tannins to toxins such as nicotine, morphine, caffeine, or cocaine.

Peroxisomes Virtually all eukaryotic cells contain globular organelles called **peroxisomes** (**FIGURE 7.13**). These organelles have a single membrane and originate as buds from the ER.

Although different types of cells from the same individual may have distinct types of peroxisomes, these organelles all share a common function: Peroxisomes are centers for reduction–oxidation (redox) reactions. (Chapter 8 explains in detail how redox reactions transfer electrons between atoms and molecules.) For example, the peroxisomes in your liver cells contain enzymes that remove electrons from, or oxidize, the ethanol in alcoholic beverages.

Different types of peroxisomes contain different suites of redox enzymes. In the leaves of plants, specialized peroxisomes called **glyoxysomes** are packed with enzymes that oxidize fats to form a compound that can be used to store energy for the cell. But plant seeds have a different type of peroxisome—one that is packed with enzymes responsible for releasing energy from stored fatty acids. The young plant uses this energy as it begins to grow.

In animals and plants, the products of these reactions often include hydrogen peroxide (H_2O_2), which is highly reactive.

If hydrogen peroxide escaped from the peroxisome, it would quickly react with and damage DNA, proteins, and cellular membranes. This event is rare, however, because inside the peroxisome, the enzyme catalase quickly "detoxifies" hydrogen peroxide by catalyzing its oxidation to form water and oxygen. The enzymes found inside the peroxisome make a specialized set of oxidation reactions possible and safe for the cell.

Mitochondria The energy required to build these organelles and do other types of work comes from adenosine triphosphate (ATP), most of which is produced in the cell's **mitochondria** (singular: **mitochondrion**).

As **FIGURE 7.14** shows, each mitochondrion has two membranes. The outer membrane defines the organelle's surface, while the inner membrane is connected to a series of sac-like **cristae.** The solution enclosed within the inner membrane is called the **mitochondrial matrix.** In eukaryotes, most of the enzymes and molecular machines responsible for synthesizing ATP are embedded in the membranes of the cristae or suspended in the matrix (see Chapter 9). Depending on the type of cell, from 50 to more than a million mitochondria may be present.

Each mitochondrion has many copies of a small, circular chromosome that is independent of the nuclear chromosomes. This mitochondrial DNA contains only around 37 genes in most eukaryotes—most of the genes responsible for the function of the organelle reside in the nuclear DNA.

Among the genes present in mitochondrial DNA are those that encode RNAs for mitochondrial ribosomes. These ribosomes are

FIGURE 7.13 Peroxisomes Are the Site of Oxidation Reactions. Peroxisomes are globular organelles that are defined by a single membrane.

Peroxisome

Peroxisome membrane

Enzyme core

Peroxisome lumen

100 nm

FIGURE 7.14 Mitochondria Are Power-Generating Stations. Mitochondria vary in size and shape, but all have a double membrane with sac-like cristae inside.

Mitochondrion

Outer and inner membranes

Matrix

Cristae

0.1 µm

Chloroplast

Stroma

Thylakoids

Granum

Outer and inner
membranes

1 μm

FIGURE 7.15 Chloroplasts Are Sugar-Manufacturing Centers in Plants and Algae. Many of the enzymes and other molecules required for photosynthesis are located in membranes inside the chloroplast. These membranes form thylakoids that consist of discs stacked into grana.

smaller than those found in the cytosol, yet they still function to produce some of the mitochondrial proteins. (Most of the proteins found in mitochondria are produced from ribosomes in the cytosol and imported into the organelle.)

Chloroplasts   Most algal and plant cells possess an organelle called the **chloroplast,** in which sunlight is converted to chemical energy during photosynthesis (**FIGURE 7.15**). The number of chloroplasts per cell varies from none to several dozen.

The chloroplast has a double membrane around its exterior, analogous to the structure of a mitochondrion. Instead of featuring sac-like cristae that connect to the inner membrane, though, the interior of the chloroplast is dominated by a network of hundreds of membrane-bound, flattened, sac-like structures called **thylakoids,** which are independent of the inner membrane.

Thylakoids have stacks, like pancakes, that are called **grana** (singular: **granum**). Many of the pigments, enzymes, and macromolecular machines responsible for converting light energy into chemical energy are embedded in the thylakoid membranes (see Chapter 10). The region outside the thylakoids, called the **stroma,** contains enzymes that use this chemical energy to produce sugars.

Like mitochondria, each chloroplast contains copies of a circular chromosome and small ribosomes that manufacture some, but not all, of the organelle's proteins. Both mitochondria

and chloroplasts also grow and divide independently of cell division through a process that resembles bacterial fission (see Chapter 12).

These attributes are odd compared with those of the other organelles and have led biologists to propose that mitochondria and chloroplasts were once free-living bacteria. According to the **endosymbiosis theory,** the ancestors of modern eukaryotes ingested these bacteria, but instead of destroying them, established a mutually beneficial relationship with them. (In Chapter 30, you will learn more about the origins of these eukaryotic organelles.)

Cytoskeleton   The final major structural feature that is common to all eukaryotic cells is the cytoskeleton, an extensive system of protein fibers. In addition to giving the cell its shape and structural stability, cytoskeletal proteins are involved in moving the cell itself and moving materials within the cell. In essence, the cytoskeleton organizes all the organelles and other cellular structures into a cohesive whole. Section 7.6 will analyze the structure and functions of the cytoskeleton in detail.

The Cell Wall   In fungi, algae, and plants, cells possess an outer cell wall in addition to their plasma membrane. The cell wall is located outside the plasma membrane and furnishes a durable, outer layer that provides structural support for the cell. The cells of animals, amoebae, and other groups lack a cell wall—their exterior surface consists of the plasma membrane only.

Although the composition of the cell wall varies among species and even among types of cells in the same individual, the general plan is similar: Rods or fibers composed of a carbohydrate run through a stiff matrix made of other polysaccharides and proteins (see Chapter 11 for details).

check your understanding

If you understand that . . .

- Each structure in a eukaryotic cell performs a function vital to the cell.
- In eukaryotes, many of the cellular functions are compartmentalized into organelles.

✓ **You should be able to . . .**

1. Explain how the structure of lysosomes and peroxisomes correlates with their function.

2. In Table 7.2, label each component with one of the following roles: administrative/information hub, power station, warehouse, large molecule manufacturing and shipping facility (with subtitles for lipid factory, protein finishing and shipping line, protein synthesis and folding center, waste processing and recycling center), support beams, perimeter fencing with secured gates, protein factory, food-manufacturing facility, and fatty-acid processing and detox center.

Answers are available in Appendix A.

Eukaryotic Cell Components

Icons Not to Scale		Structure		
		Membrane	Components	Function
	Nucleus	Double ("envelope"); openings called nuclear pores	Chromosomes	Information storage and transmission
			Nucleolus	Ribosome subunit assembly
			Nuclear lamina	Structural support
	Ribosomes	None	Complex of RNA and proteins	Protein synthesis
	Endomembrane system			
	Endoplasmic reticulum: rough	Single; contains receptors for entry of selected proteins	Network of branching sacs Ribosomes associated	Protein synthesis and processing
	Endoplasmic reticulum: smooth	Single; contains enzymes for synthesizing phospholipids	Network of branching sacs Enzymes for synthesizing or breaking down lipids	Lipid synthesis and processing
	Golgi apparatus	Single; contains receptors for products of rough ER	Stack of flattened, distinct cisternae	Protein, lipid, and carbohydrate processing
	Lysosomes	Single; contains proton pumps	Acid hydrolases (catalyze hydrolysis reactions)	Digestion and recycling
	Vacuoles	Single; contains transporters for selected molecules	Varies—pigments, oils, carbohydrates, water, or toxins	Varies—coloration, storage of oils, carbohydrates, water, or toxins
	Peroxisomes	Single; contains transporters for selected macromolecules	Enzymes that catalyze oxidation reactions Catalase (processes peroxide)	Oxidation of fatty acids, ethanol, or other compounds
	Mitochondria	Double; inner contains enzymes for ATP production	Enzymes that harvest energy from molecules to make ATP	ATP production
	Chloroplasts	Double; plus membrane-bound sacs in interior	Pigments Enzymes that use light energy to make sugars	Production of sugars via photosynthesis
	Cytoskeleton	None	Actin filaments Intermediate filaments Microtubules	Structural support; movement of materials; in some species, movement of whole cell
	Plasma membrane	Single; contains transport and receptor proteins	Phospholipid bilayer with transport and receptor proteins	Selective permeability—maintains intracellular environment
	Cell wall	None	Carbohydrate fibers running through carbohydrate or protein matrix	Protection, structural support

7.3 Putting the Parts into a Whole

Within a cell, the structure of each component correlates with its function. In the same way, the overall size, shape, and composition of a cell correlate with its function.

Cells might be analogous to machine shops or industrial complexes, but clothing manufacturing centers are very different in layout and composition from airplane production facilities. How does the physical and chemical makeup of a cell correlate with its function?

Structure and Function at the Whole-Cell Level

Inside an individual plant or animal, cells are specialized for certain tasks and have a structure that correlates with those tasks. For example, the muscle cells in your upper leg are extremely long, tube-shaped structures. They are filled with protein fibers that slide past one another as the entire muscle flexes or relaxes. It is this sliding motion that allows your muscles to contract or extend as you run. Muscle cells are also packed with mitochondria, which produce the ATP required for the sliding motion to occur.

In contrast, nearby fat cells are rounded, globular structures that store fatty acids. They consist of little more than a plasma membrane, a nucleus, and a fat droplet. Neither cell bears a close resemblance to the generalized animal cell pictured in Figure 7.6a.

To drive home the correlation between the overall structure and function of a cell, examine the transmission electron micrographs in **FIGURE 7.16**.

- The animal cell in Figure 7.16a, located in the pancreas, manufactures and exports digestive enzymes. It is packed with rough ER and Golgi, which make these functions possible.

- The animal cell in Figure 7.16b, from the testis, synthesizes and exports the steroid hormone testosterone—a lipid-soluble signal. This cell is dominated by smooth ER, where processing of steroids and other lipids takes place.

- The plant cell in Figure 7.16c, from the leaf of a potato, has hundreds of chloroplasts and is specialized for absorbing light and manufacturing sugar.

- The animal cells in Figure 7.16d come from brown fat. The cells have numerous mitochondria that have been altered so they convert energy stored in fat into heat instead of ATP.

In each case, the types of organelles in each cell and their size and number correlate with the cell's specialized function.

The Dynamic Cell

Biologists study the structure and function of organelles and cells with a combination of tools and approaches. For several decades, a technique called **differential centrifugation** was particularly important because it allowed researchers to isolate particular cell components and analyze their chemical composition. Differential centrifugation is based on breaking cells apart to create a complex mixture and separating components in a centrifuge (see **BioSkills 10** in Appendix B). The individual parts of the cell

(a) Animal pancreatic cell: Exports digestive enzymes.

(b) Animal testis cell: Exports lipid-soluble signals.

(c) Plant leaf cell: Manufactures ATP and sugar.

(d) Brown fat cells: Burn fat to generate heat in lieu of ATP.

FIGURE 7.16 Cell Structure Correlates with Function.

✔**EXERCISE** In part (a), label the rough ER and the dark, round secretory vesicles. In (b), label the smooth ER. In (c), label the chloroplasts, vacuole, and nucleus. In (d), label the mitochondria.

can then be purified and studied in detail, in isolation from other parts of the cell.

Historically and currently, however, the most important research in cell biology is based on imaging—simply looking at cells. Recent innovations allow biologists to put fluorescing tags or other types of markers on particular cell components and then look at them with increasingly sophisticated light microscopes and electron microscopes. Advances in microscopy provide increasingly high magnification and better resolution.

It's important to recognize, though, that some of these techniques have limitations. Differential centrifugation splits cells into parts that are analyzed independently, and electron microscopy gives a fixed "snapshot" of the cell or organisms being observed. Neither technique allows investigators to explore directly how things move from place to place in the cell or how parts interact. The information gleaned from these techniques can make cells seem static. In reality, cells are dynamic.

The amount of chemical activity and the speed of molecular movement inside cells are nothing short of fantastic. Bacterial ribosomes add up to 20 amino acids per second to a growing polypeptide, and eukaryotic ribosomes typically add 2 per second. Given that there are about 15,000 ribosomes in each bacterium and possibly a million in an average eukaryotic cell, hundreds or even thousands of new protein molecules can be produced each second in every cell. Here are some other remarkable cellular feats:

- In an average second, a typical cell in your body uses an average of 10 million ATP molecules and synthesizes just as many.

- It's not unusual for a cellular enzyme to catalyze 25,000 or more reactions per second; most cells contain hundreds or thousands of different enzymes.

- A minute is more than enough time for each membrane phospholipid in your body to travel the breadth of the organelle or cell where it resides.

- The hundreds of trillions of mitochondria inside you are completely replaced about every 10 days, for as long as you live.

Because humans are such large organisms, it's impossible for us to imagine what life is really like inside a cell. At the scale of a ribosome or an organelle or a cell, gravity is inconsequential. Instead, the dominant forces are the charge- or polarity-based electrostatic attractions between molecules and their energy of motion. At this level, events take nanoseconds, and speeds are measured in micrometers per second. This is the speed of life.

Contemporary methods for studying cells (including some of the imaging techniques featured in **BioSkills 11** in Appendix B) capture this dynamism by tracking how organelles and molecules move and interact over time. The ability to digitize video images of live cells, or take time-lapse photographs of living cells, is allowing researchers to see and study dynamic processes.

The rest of this chapter focuses on this theme of cellular dynamism and movement. Its goal is to put some of the individual pieces of a cell together and ask how they work as systems to accomplish key tasks.

To begin, let's first look at how molecules move into and out of the cell's control center—the nucleus—and then consider how proteins move from ribosomes into the lumen of the rough ER and then to the Golgi apparatus and beyond. The chapter closes by introducing the cytoskeletal elements and their associated motor proteins and how they are used to transport cargo inside the cell or move the cell itself.

7.4 Cell Systems I: Nuclear Transport

The nucleus is the information center of eukaryotic cells—a corporate headquarters, design center, and library all rolled into one. Appropriately enough, its interior is highly organized.

The organelle's overall shape and structure are defined by the mesh-like nuclear lamina. The nuclear lamina provides an attachment point for the chromosomes, each of which occupies a well-defined region in the nucleus.

In addition, specific centers exist where the genetic information in DNA is decoded and processed. At these locations, large suites of enzymes interact to produce RNA messages from specific genes at specific times. Meanwhile, the nucleolus functions as the site of ribosome assembly.

Structure and Function of the Nuclear Envelope

The nuclear envelope separates the nucleus from the rest of the cell. Starting in the 1950s, transmission electron micrographs of cross sections through the nuclear envelope showed that the structure is supported by the fibrous nuclear lamina and bounded by two membranes, each consisting of a lipid bilayer. How does this administrative center communicate with the rest of the cell across the double membrane barrier?

Micrographs like the one in **FIGURE 7.17** (see page 120) show that the nuclear envelope is broken with openings, approximately 60 nanometers (nm) in diameter, called **nuclear pores.** Because these pores extend through both the inner and outer nuclear membranes, they connect the inside of the nucleus with the cytosol. Follow-up research showed that each pore consists of over 50 different proteins. As the diagram on the right side of Figure 7.17 shows, these protein molecules form an elaborate structure called the **nuclear pore complex.**

What substances traverse nuclear pores? Chromosomal DNA clearly does not—it remains in the nucleus as long as the nuclear envelope remains intact. But DNA is used to synthesize RNA inside the nucleus, most of which is exported through nuclear pores to the cytoplasm.

Several types of RNA molecules are produced, each distinguished by size and function. For example, **ribosomal RNAs** are manufactured in the nucleolus, where they bind to proteins to form ribosomes. Molecules called **messenger RNAs (mRNA)** carry the information required to manufacture proteins. Both the newly assembled ribosomes and the mRNAs must be transported from the nucleus to the cytoplasm, where protein synthesis takes place.

FIGURE 7.17 Structure of the Nuclear Envelope and Nuclear Pore Complex.

Inbound traffic is also impressive. Nucleoside triphosphates that act as building blocks for DNA and RNA enter the nucleus, as do a variety of proteins responsible for copying DNA, synthesizing RNAs, extending the nuclear lamina, or assembling ribosomes.

To summarize, ribosomal subunits and various types of RNAs exit the nucleus; nucleotides and certain proteins enter it. In a typical cell, over 500 molecules pass through each of the 3000–4000 nuclear pores every second. The scale of traffic through the nuclear pores is mind-boggling. How is it regulated and directed?

Experiments in the early 1960s showed that size matters in the passage of molecules through nuclear pores. This conclusion was based on the results from injecting tiny gold particles that varied in diameter and tracking their movement across the pores. In electron micrographs, gold particles show up as defined black dots that can be easily distinguished from cellular structures. Immediately after injection, most of the gold particles were observed in the cytoplasm, and only a few were closely associated with nuclear pores. Ten minutes after injection, only the small particles (< 12.5 nm in diameter) appeared to be distributed throughout both the nucleus and the cytoplasm, and the larger particles were excluded from entering the nucleus.

The fact that the pore opening is almost 5 times larger than this 12.5-nm size limit supports the hypothesis that the nuclear pore complex serves as a gate to control passage through the envelope. If this is the case, then what is required to open these gates so that proteins larger than the size limit, like those responsible for replicating DNA, may pass?

How Do Large Molecules Enter the Nucleus?

It was clear to researchers that size was not the sole factor in selective transport across the nuclear envelope. Certain proteins were concentrated in the nucleus, while others were completely excluded—even if they were similar in size.

A series of experiments on a protein called nucleoplasmin helped researchers understand the nature of nuclear import. Nucleoplasmin is strictly found in the nucleus and plays an important role in the assembly of chromatin. When researchers labeled nucleoplasmin with a radioactive atom and injected it into the cytoplasm of living cells, they found that the radioactive protein was quickly concentrated into the nucleus. Is there a "send-to-nucleus" signal within the nucleoplasmin protein that is responsible for this directed transport?

As shown in **FIGURE 7.18**, the distinctive structure of nucleoplasmin was used to further investigate this process. First, researchers used enzymes called proteases to cleave the core sections of nucleoplasmin from the tails. After separating the tails from the core fragments, they labeled each component with radioactive atoms and injected them into the cytoplasm of different cells.

At various times after the injections, researchers examined the nuclei and cytoplasm of the cells to track down the radioactive label. The results were striking. They found that tail fragments were rapidly transported from the cytoplasm into the nucleus. Core fragments, in contrast, were not allowed to pass through the nuclear envelope and remained in the cytoplasm.

These data led to a key hypothesis: Nuclear proteins are synthesized by ribosomes in the cytosol and contain a "zip code"—a molecular address tag—that marks them for transport through the nuclear pore complex. This zip code allows the nuclear pore complex to open in some way that permits larger proteins and RNA molecules to pass through.

By analyzing different stretches of the tail, the biologists eventually found a 17-amino-acid-long section that had to be present to direct nucleoplasmin to the nucleus. Follow-up work confirmed that other proteins bound for the nucleus, even those expressed by some viruses, have similar amino acid sequences directing their transport. This common sequence came to be called the **nuclear localization signal (NLS)**. Proteins that leave the nucleus have a different signal, required for nuclear export.

QUESTION: Does the nucleoplasmin protein contain a "Send to nucleus" signal?

HYPOTHESIS: Nucleoplasmin contains a discrete "Send to nucleus" signal that resides in either the tail or core region.

NULL HYPOTHESIS: Nucleoplasmin does not require a signal to enter the nucleus, or the entire protein serves as the signal.

EXPERIMENTAL SETUP:

Nucleoplasmin protein

"Tails"

Core

1. Use protease to cleave tails off of nucleoplasmin protein core.

Labeled tails Labeled cores

2. Attach radioactive labels to protein tails and cores.

3. Inject labeled tails and cores into cytoplasm of different cells.

4. Wait, then locate labeled fragments ....

PREDICTION:

PREDICTION OF NULL HYPOTHESIS:

RESULTS:

Labeled tail fragments **located in nucleus**

Labeled core fragments still **located in cytoplasm**

CONCLUSION:

FIGURE 7.18 Does the Nucleoplasmin Protein Contain a "Send to Nucleus" Signal?

SOURCES: Mills, A. D., R. A. Laskey, P. Black, et al. 1980. *Journal of Molecular Biology* 139: 561–568; Dingwall, C., S. V. Sharnick, and R. A. Laskey. 1982. *Cell* 30: 449–458.

✔**EXERCISE** Without looking at the text, fill in the prediction(s) and conclusion(s) in this experiment.

More recent research has shown that the movement of proteins and other large molecules into and out of the nucleus is an energy-demanding process that involves special transport proteins. These nuclear transport proteins function like trucks that haul cargo into or out of the nucleus through the nuclear pore complex, depending on whether they have an import or export zip code. Currently, biologists are trying to unravel how all this traffic in and out of the nucleus is regulated to avoid backups and head-on collisions.

✔ If you understand the process of nuclear transport, you should be able to compare and contrast the movement of (1) nucleotides and (2) large proteins through the nuclear pore complex. Which would you expect to require the input of energy?

7.5 Cell Systems II: The Endomembrane System Manufactures, Ships, and Recycles Cargo

The nuclear membrane is not the only place in cells where cargo moves in a regulated and energy-demanding fashion. Most of the proteins found in peroxisomes, mitochondria, and chloroplasts are also actively imported from the cytosol. These proteins contain special signal sequences, like the nuclear localization signal, that target them to the appropriate organelles.

If you think about it for a moment, the need to sort proteins and ship them to specific destinations should be clear. Proteins are produced by ribosomes that are either free in the cytosol or on the surface of the ER. Many of these proteins must be transported to a compartment inside the eukaryotic cell. Acid hydrolases must be shipped to lysosomes and catalase to peroxisomes. To get to the right location, each protein has to have an address tag and a transport and delivery system.

To get a better understanding of protein sorting and transport in eukaryotic cells, let's consider perhaps the most intricate of all manufacturing and shipping complexes: the endomembrane system. In this system, proteins that are synthesized in the rough ER move to the Golgi apparatus for processing, and from there they travel to the cell surface or other destinations.

Studying the Pathway through the Endomembrane System

The idea that materials move through the endomembrane system in an orderly way was inspired by a simple observation. According to electron micrographs, cells that secrete digestive enzymes, hormones, or other products have particularly large amounts of rough ER and Golgi. This correlation led to the idea that these organelles may participate in a "secretory pathway" that starts in the rough ER and ends with products leaving the cell (**FIGURE 7.19**, see page 122). How does this hypothesized pathway work?

Tracking Protein Movement via Pulse–Chase Assay George Palade and colleagues did pioneering research on the secretory

FIGURE 7.19 The Secretory Pathway Hypothesis. The secretory pathway hypothesis proposes that proteins intended for secretion from the cell are synthesized and processed in a highly prescribed series of steps. Note that proteins are packaged into vesicles when they move from the RER to the Golgi and from the Golgi to the cell surface.

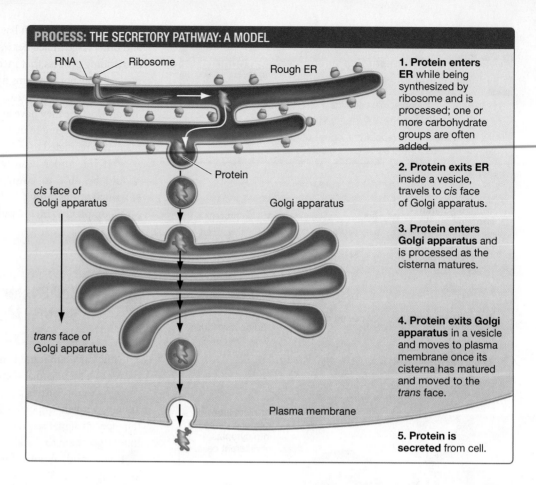

PROCESS: THE SECRETORY PATHWAY: A MODEL

RNA — Ribosome — Rough ER

Protein

cis face of Golgi apparatus

Golgi apparatus

trans face of Golgi apparatus

Plasma membrane

1. Protein enters ER while being synthesized by ribosome and is processed; one or more carbohydrate groups are often added.

2. Protein exits ER inside a vesicle, travels to *cis* face of Golgi apparatus.

3. Protein enters Golgi apparatus and is processed as the cisterna matures.

4. Protein exits Golgi apparatus in a vesicle and moves to plasma membrane once its cisterna has matured and moved to the *trans* face.

5. Protein is secreted from cell.

pathway using a **pulse–chase experiment** to track protein movement. This strategy is based on two steps:

1. *The "Pulse"* Expose experimental cells to a high concentration of a modified amino acid for a short time. For example, if a cell is briefly exposed to a large amount of radioactively labeled amino acid, virtually all the proteins synthesized during that interval will be radiolabeled.

2. *The "Chase"* The pulse ends by washing away the modified amino acid and replacing it with the normal version of the same molecule. The time following the end of the pulse is referred to as the chase. If the chase consists of unlabeled amino acid, then the proteins synthesized during the chase period will *not* be radiolabeled.

The idea is to mark a population of molecules at a particular interval and then follow their fate over time. This approach is analogous to adding a small amount of dye to a stream and then following the movement of the dye molecules.

To understand why the chase is necessary in these experiments, imagine what would happen if you added dye to a stream continuously. Soon the entire stream would be dyed—you could no longer tell where a specific population of dye molecules were moving.

In testing the secretory pathway hypothesis, Palade's team focused on pancreatic cells that were growing in **culture,** or

in vitro.[1] These cells are specialized for secreting digestive enzymes into the small intestine and are packed with rough ER and Golgi.

The basic experimental approach was to pulse the cell culture for 3 minutes with a radiolabeled version of the amino acid leucine, followed by a long chase with nonradioactive leucine (**FIGURE 7.20a**). The pulse produced a population of proteins that were related to one another by the timing of their synthesis. At different points during the chase, the researchers tracked the movement of these proteins by preparing samples of the cells for autoradiography and electron microscopy (see **BioSkills 10** and **11** in Appendix B). The drawings in **FIGURE 7.20b** illustrate what the researchers would have seen in micrographs taken at different times before and after the start of the chase.

Results of the Pulse–Chase Experiment The graph in Figure 7.20b was based on the electron microscopy results, which showed that proteins are trafficked through the secretory pathway in a highly organized and directed manner. Track the movement of proteins through the cell during the chase by covering the graph with a piece of paper and then slowly sliding it off from

[1]The term in vitro is Latin for "in glass." Experiments that are performed outside living organisms are done in vitro. The term in vivo, in contrast, is Latin for "in life." Experiments performed with living organisms are done in vivo.

(a) Setup for a pulse-chase experiment

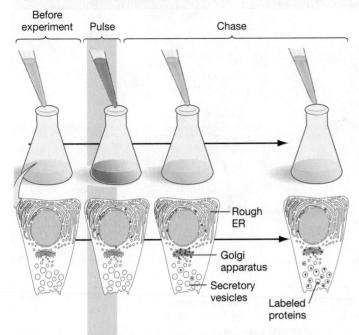

(b) Tracking pulse-labeled proteins during the chase

FIGURE 7.20 Tracking Protein Movement in a Pulse–Chase Experiment. Part **(a)** shows how investigators label newly synthesized proteins during the pulse with radioactive amino acids (red). At the start of the chase, this medium is replaced with non-radioactive amino acids (yellow) so only those proteins labeled in the pulse will be tracked. Part **(b)** provides the results of a pulse–chase experiment. The drawings represent micrographs taken that show the radiolabeled proteins (red dots) in the cells. The graph shows the relative abundance of radiolabeled proteins in three different organelles during the chase.

left to right. Notice what is happening to each line at the following three time points:

1. Immediately after the pulse, most of the newly synthesized proteins are inside this cell's rough ER.

2. At 37 minutes into the chase, the situation has changed. Most of the labeled proteins have left the rough ER and entered the

Golgi apparatus, and some of them have accumulated inside structures called secretory vesicles.

3. By the end of the chase, at 117 minutes, most of the labeled proteins have left the Golgi and are either in secretory vesicles or were secreted from the cells.

Over a period of two hours, the labeled population of proteins moved along a defined trail through the rough ER, Golgi apparatus, and secretory vesicles to reach the exterior of the cell. ✔ **QUANTITATIVE** If you understand how the pulse–chase experiment is used to track proteins, use the graph in Figure 7.20b to estimate the time it takes for proteins to pass through the Golgi apparatus.

The results support the hypotheses that a secretory pathway exists and that the rough ER and Golgi apparatus function together as an integrated endomembrane system. Next, let's break this secretory pathway down to examine four of the steps in more detail:

1. How do proteins enter the lumen of the ER?

2. How do the proteins move from the ER to the Golgi apparatus?

3. Once they're inside the Golgi, what happens to them?

4. And finally, how does the Golgi sort out the proteins so each will end up going to the appropriate place?

Entering the Endomembrane System: The Signal Hypothesis

The synthesis of proteins destined to be secreted or embedded in membranes begins in ribosomes free in the cytosol. Günter Blobel and colleagues proposed that at some point these ribosomes become attached to the outside of the ER. But what directs these ribosomes to the ER? The signal hypothesis predicts that proteins bound for the endomembrane system have a molecular zip code analogous to the nuclear localization signal. Blobel proposed that the first few amino acids in the growing polypeptide act as a signal that marks the ribosome for transport to the ER membrane.

This hypothesis received important support when researchers made a puzzling observation: When proteins that are normally synthesized in the rough ER are instead manufactured by isolated ribosomes in vitro—with *no* ER present—they are 20 amino acids longer, on average, than usual.

Blobel seized on these data. He claimed that the extra amino acids are the "send-to-ER" signal, and that the signal is removed inside the organelle. When the same protein is synthesized in vitro, the signal is not removed.

Blobel's group went on to produce convincing data that supported the hypothesis: They identified a sequence of amino acids that will move proteins into the ER lumen, called the **ER signal sequence.**

More recent work has documented the mechanisms responsible for receiving this send-to-ER signal and inserting the

PROCESS: THE SIGNAL HYPOTHESIS

RNA

Ribosome

Signal
sequence

SRP

Cytosol

Lumen of
rough ER

SRP receptor

Protein

1. Signal sequence is synthesized by ribosome.

2. Signal sequence binds to signal recognition particle (SRP) and halts synthesis.

3. SRP binds to receptor in ER membrane.

4. SRP is released. Protein synthesis continues. Protein enters ER.

5. Signal sequence is removed. Protein synthesis then proceeds to completion.

FIGURE 7.21 The Signal Hypothesis Explains How Proteins Destined for Secretion Enter the Endomembrane System. According to the signal hypothesis, proteins destined for secretion contain a short stretch of amino acids that interact with a signal recognition particle (SRP) in the cytoplasm. This interaction directs the synthesis of the remaining protein into the ER.

protein into the rough ER. **FIGURE 7.21** illustrates the key steps involved.

Step 1 Protein synthesis begins on a free ribosome in the cytosol. The ribosome synthesizes the ER signal sequence.

Step 2 The signal sequence binds to a **signal recognition particle (SRP)**—a complex of RNA and protein. The attached SRP causes protein synthesis to stop.

Step 3 The ribosome + signal sequence + SRP complex moves to the ER membrane, where it attaches to the SRP receptor. Think of the SRP as a key that is activated by an ER signal sequence. The SRP receptor in the ER membrane is the lock.

Step 4 Once the lock (the receptor) and key (the SRP) connect, the SRP is released and protein synthesis continues.

Step 5 The growing protein is fed into the ER lumen through a channel, and the signal sequence is removed.

If the protein will eventually be shipped to the inside of an organelle or secreted from the cell, it is completely transferred into the lumen of the rough ER. If it is an integral membrane protein, part of it remains in the cytosol and rough ER membrane while it is being processed.

Once proteins are inside the rough ER or inserted into its membrane, they fold into their three-dimensional shape with the help of chaperone proteins (see Chapter 3). In addition, proteins that enter the ER lumen interact with enzymes that catalyze the addition of carbohydrate side chains (Figure 7.19). Because carbohydrates are polymers of sugar monomers, the addition of one or more carbohydrate groups is called **glycosylation** ("sugar-together"). The resulting molecule is a **glycoprotein**

("sugar-protein"; see Chapter 5). The number and arrangement of these sugars changes as the protein matures, serving as an indicator for shipment to the next destination.

Moving from the ER to the Golgi

How do proteins travel from the ER to the Golgi apparatus? In Palade's pulse–chase experiment, labeled proteins found between the rough ER and the Golgi apparatus were inside membrane-bound structures. Based on these observations, Palade's group suggested that proteins are transported in vesicles that bud off from the ER, move away, fuse with the membrane on the *cis* face of the Golgi apparatus, and dump their cargo inside.

This hypothesis was supported when other researchers used differential centrifugation to isolate and characterize the vesicles that contained labeled proteins. They found that a distinctive type of vesicle carries proteins from the rough ER to the Golgi apparatus. Ensuring that only appropriate cargo is loaded into these vesicles and that the vesicles dock and fuse only with the *cis* face of the Golgi involves a complex series of events and is an area of active research.

What Happens Inside the Golgi Apparatus?

Section 7.2 indicated that the Golgi apparatus consists of a stack of flattened vesicles called cisternae, and that cargo enters one side of the organelle and exits the other. Recent research has shown that the composition of the Golgi apparatus is dynamic. New cisternae constantly form at the *cis* face of the Golgi, while old cisternae break apart at the *trans* face, to be replaced by the cisternae behind it. In this way a new cisterna follows those formed earlier, advancing toward the *trans* face of the Golgi. As it does, it changes in composition and activity through a process called **cisternal maturation.**

By separating individual cisternae and analyzing their contents, researchers have found that cisternae at various stages of maturation contain different suites of enzymes. Many of these enzymes catalyze glycosylation reactions that further modify the oligosaccharides that were attached to the protein in the ER. As the cisternae slowly move from *cis* to *trans*, these enzymes are replaced with those representing more mature cisternae. The result is that proteins are modified in a stepwise manner as they slowly move through the Golgi.

If the rough ER is like a foundry and stamping plant where rough parts are manufactured, then the Golgi can be considered a finishing area where products are polished, painted, and readied for shipping.

How Do Proteins Reach Their Destinations?

The rough ER and Golgi apparatus constitute an impressive assembly line. Certain proteins manufactured by this process remain in these organelles, replacing worn-out resident molecules. But those proteins that are simply passing through as cargo must be sorted and sent to their intended destination as the *trans* cisterna they are in breaks up into vesicles.

How are these finished products put into the right shipping containers, and how are the different containers addressed?

Studies on enzymes that are shipped to lysosomes have provided some answers to both questions. A key finding was that lysosome-bound proteins have a phosphate group attached to a specific sugar subunit on their surface, forming the compound mannose-6-phosphate. If mannose-6-phosphate is removed from these proteins, they are not transported to a lysosome.

This is strong evidence that the phosphorylated sugar serves as a zip code, analogous to the nuclear localization signal and ER signal sequence discussed earlier. Data indicate that mannose-6-phosphate binds to a receptor protein in the membrane of the *trans*-Golgi cisterna. Regions that are enriched with these receptor–cargo complexes will form vesicles that, in turn, have proteins on their cytosolic surfaces that direct their transport and fusion with pre-lysosomal compartments. In this way, the presence of mannose-6-phosphate targets proteins for vesicles that deliver their contents to organelles that eventually become lysosomes.

FIGURE 7.22 presents a simplified model of how cargo is sorted and loaded into specific vesicles that are shipped to different destinations. Each cargo protein has a molecular tag that directs it to particular vesicle budding sites by interacting with receptors in the *trans* cisterna. These receptors, along with other cytosolic proteins that are not shown, direct the transport vesicles to the correct destinations.

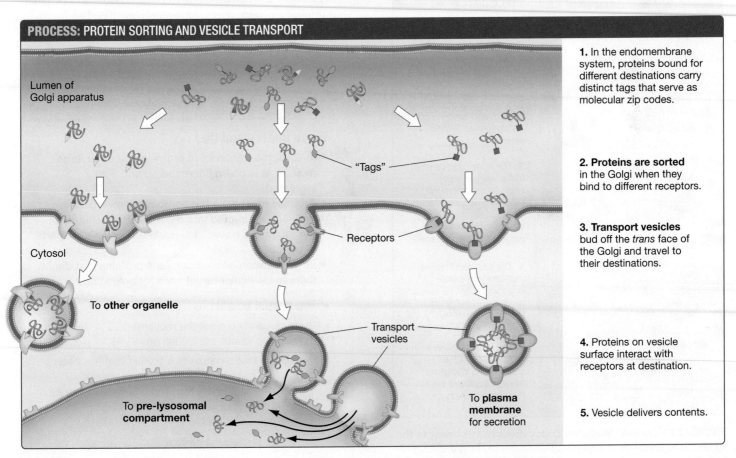

PROCESS: PROTEIN SORTING AND VESICLE TRANSPORT

Lumen of Golgi apparatus

"Tags"

Receptors

Cytosol

To **other organelle**

Transport vesicles

To **pre-lysosomal compartment**

To **plasma membrane** for secretion

1. In the endomembrane system, proteins bound for different destinations carry distinct tags that serve as molecular zip codes.

2. Proteins are sorted in the Golgi when they bind to different receptors.

3. Transport vesicles bud off the *trans* face of the Golgi and travel to their destinations.

4. Proteins on vesicle surface interact with receptors at destination.

5. Vesicle delivers contents.

FIGURE 7.22 In the Golgi Apparatus, Proteins Are Sorted into Vesicles That Are Targeted to a Destination.

In particular, notice that the transport vesicle shown on the right of Figure 7.22 is bound for the plasma membrane, where it will secrete its contents to the outside. This process is called **exocytosis** ("outside-cell-act"). When exocytosis occurs, the vesicle membrane and plasma membrane make contact. As the two membranes fuse, their lipid bilayers rearrange in a way that exposes the interior of the vesicle to the outside of the cell. The vesicle's contents then diffuse into the space outside the cell. This is how cells in your pancreas deliver digestive enzymes to the duct that leads to your small intestine—where food is digested.

Recycling Material in the Lysosome

Now that you have seen how cargo moves out of the cell, let's look at how cargo is brought into the cell. Previously, you learned about how cells import small molecules across lipid bilayers (see Chapter 6), but this is not possible for large molecules like proteins and complex carbohydrates. For these molecules to be recycled and used by the cell, they must first be digested in the lysosome—but how do they get there?

Endocytosis ("inside-cell-act") refers to any pinching off of the plasma membrane that results in the uptake of material from outside the cell. **Receptor-mediated endocytosis** is illustrated in **FIGURE 7.23**. As its name implies, the sequence of events begins when macromolecules outside the cell bind to receptors on the plasma membrane. More than 25 distinct receptors have now been characterized, each specialized for binding to different cargo.

Once receptor binding occurs, the plasma membrane folds in and pinches off to form an endocytic vesicle. These vesicles then drop off their cargo in a transient organelle called the **early endosome** ("inside-body"). The activity of proton pumps in the membrane of this organelle acidifies its lumen, which causes the cargo to be released from their receptors. Many of these emptied cargo receptors are then repackaged into vesicles and returned to the plasma membrane.

As proton pumps continue to lower the early endosome's pH, it undergoes a series of processing steps that cause it to mature into a **late endosome**. The late endosome is the pre-lysosomal compartment introduced earlier (Figure 7.22), where the acid hydrolases from the Golgi apparatus are dropped off. As before, the emptied cargo receptors transported from the Golgi are removed from the late endosome as it matures into a fully active lysosome.

In addition to receptor-mediated endocytosis, the lysosome is involved in recycling material via autophagy and phagocytosis (see **FIGURE 7.24**). During **autophagy** (literally, "same-eating"), damaged organelles are enclosed within an internal membrane and delivered to a lysosome. There the components are digested and recycled. In **phagocytosis** ("eat-cell-act"), the plasma membrane of a cell surrounds a smaller cell or food particle and engulfs it, forming a structure called a phagosome. This structure is delivered to a lysosome, where it is taken in and digested.

Regardless of whether the materials in lysosomes originate via autophagy, phagocytosis, or receptor-mediated endocytosis, the result is similar: Molecules are hydrolyzed. ✔ If you understand the interaction between the endomembrane system

PROCESS: RECEPTOR-MEDIATED ENDOCYTOSIS

Recycling of membrane proteins

Endocytic vesicle

H⁺

Early endosome

H⁺   H⁺

Vesicle from Golgi apparatus

Late endosome

Lysosome

1. Macromolecules outside the cell bind to membrane proteins that act as receptors.

2. The plasma membrane folds in and pinches off to form an endocytic vesicle.

3. The endocytic vesicle fuses with an early endosome, activating protons that lower its pH. Cargo is released and empty receptors are recycled to the surface.

4. The early endosome matures into a late endosome that receives digestive enzymes from the Golgi apparatus.

5. The late endosome matures into a functional lysosome and digests the endocytosed macromolecules.

FIGURE 7.23 Receptor-Mediated Endocytosis Is a Pathway to the Lysosome. Endosomes created by receptor-mediated endocytosis will mature into lysosomes.

check your understanding

If you understand that . . .

- In cells, the transport of proteins and other large molecules is energy demanding and tightly regulated.
- Proteins must have the appropriate molecular zip code to be directed into the nucleus, the lumen of the rough ER, or vesicles destined for different parts of the cell.
- Vesicles incorporate membrane proteins that direct them to particular target sites for unloading cargo.

✔ **You should be able to . . .**

1. Compare and contrast the movement of proteins into the nucleus versus the ER lumen.

2. Predict the final location of a protein that has been engineered to include an ER signal sequence, mannose-6-phosphate tag, and a nuclear localization signal. Justify your answer by addressing the impact of each signal on its transport.

Answers are available in Appendix A.

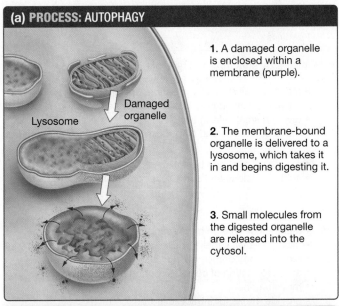

(a) PROCESS: AUTOPHAGY

Lysosome

Damaged organelle

1. A damaged organelle is enclosed within a membrane (purple).

2. The membrane-bound organelle is delivered to a lysosome, which takes it in and begins digesting it.

3. Small molecules from the digested organelle are released into the cytosol.

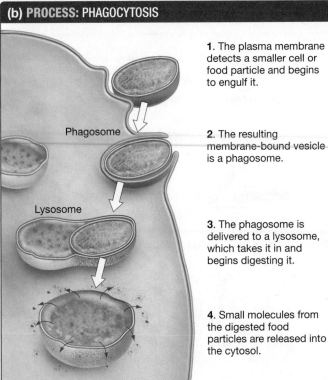

(b) PROCESS: PHAGOCYTOSIS

Phagosome

Lysosome

1. The plasma membrane detects a smaller cell or food particle and begins to engulf it.

2. The resulting membrane-bound vesicle is a phagosome.

3. The phagosome is delivered to a lysosome, which takes it in and begins digesting it.

4. Small molecules from the digested food particles are released into the cytosol.

FIGURE 7.24 Two More Ways to Deliver Materials to Lysosomes. Materials can be transported to lysosomes **(a)** via autophagy or **(b)** after phagocytosis.

and endocytosis, you should be able to predict how the loss of the mannose-6-phosphate receptor would affect receptor-mediated endocytosis.

It is important to note, however, that not all the materials that are surrounded by membrane and taken into a cell end up in lysosomes. In addition to receptor-mediated endocytosis and phagocytosis, small fluid-filled vesicles can be brought into a cell via **bulk-phase endocytosis.** There does not appear to be any cargo selection in bulk-phase endocytosis, and the vesicles are

not transported to lysosomes. These tiny vesicles are used elsewhere in the cell and are likely involved in recycling lipids deposited on the plasma membrane during exocytosis.

Throughout this section, vesicles have been key to the transport of cargo. If these transport steps depended on the random movement of diffusion alone, however, then the vesicles and their cargo might never reach their intended destinations. Are there instead defined tracks that direct the movement of these shipping containers? If so, what are these tracks, and what molecule or molecules function to transport the vesicles along them? Let's delve into these questions in the next section.

7.6 Cell Systems III: The Dynamic Cytoskeleton

The endomembrane system may be the best-studied example of how individual organelles work together in a dynamic, highly integrated way. This integration depends in part on the physical relationship of organelles, which is organized by the cytoskeletal system.

The cytoskeleton is a dense and complex network of fibers that helps maintain cell shape by providing structural support. However, the cytoskeleton is not a static structure like the scaffolding used at construction sites. Its fibrous proteins move and change to alter the cell's shape, shift its contents, and even move the cell itself. Like the rest of the cell, the cytoskeleton is dynamic.

As **TABLE 7.3** (see page 128) shows, there are three distinct cytoskeletal elements in eukaryotic cells: actin filaments, intermediate filaments, and microtubules. Recent research has shown structural and functional relationships between these three eukaryotic filaments and cytoskeletal elements in bacteria.

Each of the three cytoskeletal elements found in eukaryotes has a distinct size, structure, and function. Let's look at each one in turn.

Actin Filaments

Sometimes called **microfilaments** because they are the cytoskeletal element with the smallest diameter, **actin filaments** are fibrous structures made of the globular protein actin (Table 7.3). In animal cells, actin is often the most abundant of all proteins—typically it represents 5–10 percent of the total protein in the cell. Each of your liver cells contains about half a billion of these molecules.

Actin Filament Structure A completed actin filament resembles two long strands that coil around each other. Actin filaments form when individual actin protein subunits assemble, or polymerize, from head to tail through the formation of noncovalent bonds.

Because the actin proteins are not symmetrical, this head-to-tail arrangement of actin subunits results in filaments that have two different ends, or polarity. The two distinct ends of an actin filament are referred to as plus and minus ends. The structural difference between these two ends results in different rates of

Filament	Structure	Subunits	Functions
The three types of filaments that make up the cytoskeleton are distinguished by their size, structure, and type of protein subunit.			
Actin filaments (microfilaments)	Strands in double helix 7 nm − end + end	Actin	• maintain cell shape by resisting tension (pull) • move cells via muscle contraction or cell crawling • divide animal cells in two • move organelles and cytoplasm in plants, fungi, and animals
Intermediate filaments	Fibers wound into thicker cables 10 nm	Keratins, lamins, or others	• maintain cell shape by resisting tension (pull) • anchor nucleus and some other organelles
Microtubules	Hollow tube 25 nm − end + end	α- and β-tubulin dimers	• maintain cell shape by resisting compression (push) • move cells via flagella or cilia • move chromosomes during cell division • assist formation of cell plate during plant cell division • move organelles • provide tracks for intracellular transport

assembling new actin subunits: The plus end grows faster than the minus end.

Each filament is generally unstable and will grow or shrink depending on the concentration of free actin subunits. In addition to controlling the availability of free actin, cells regulate the length and longevity of microfilaments via actin-binding proteins that either stabilize or destabilize their structure.

In animal cells, actin filaments are particularly abundant just under the plasma membrane. They are organized into long, parallel bundles or dense, crisscrossing networks in which individual actin filaments are linked to one another by other proteins. The reinforced bundles and networks of actin filaments help stiffen the cell and define its shape.

Actin Filament Function In addition to providing structural support, actin filaments are involved in movement. In several cases, actin's role in movement depends on the protein myosin. Myosin is a **motor protein:** a protein that converts the potential energy in ATP into the kinetic energy of mechanical work, just as a car's motor converts the chemical energy in gasoline into spinning wheels.

The interaction between actin and myosin is frequently presented in the context of how it produces muscle contraction and movement (Chapter 48). For now, it's enough to recognize that when myosin binds and hydrolyzes ATP to ADP, it undergoes a series of shape changes that extends the "head" region, attaches it to actin, and then contracts to pull itself along the actin filament. The shape change of this protein causes the actin and myosin to slide past each other. After repeated rounds of this contraction cycle, the myosin progressively moves toward the plus end of the actin filament (**FIGURE 7.25a**). This type of movement is analogous to an inchworm contracting its body as it moves along a stick.

(a) Actin and myosin interact to cause movement.

When myosin "head" attaches to actin and moves, the actin filament slides

(b) Examples of movement caused by actin–myosin interactions

Cytokinesis in animals

Actin–myosin interactions pinch membrane in two

Cytoplasmic streaming in plants

Actin–myosin interactions move cytoplasm around cell

FIGURE 7.25 Many Cellular Movements Are Based on Actin–Myosin Interactions. (a) ATP hydrolysis in the "head" region of myosin causes the protein to attach to actin and change shape. The movement slides the myosin toward the plus end of actin. **(b)** Actin–myosin interactions can divide cells and move organelles and cytoplasm.

As **FIGURE 7.25b** shows, the ATP-powered interaction between actin and myosin is the basis for an array of cell movements:

- **Cytokinesis** ("cell-moving") is the process of cell division. In animals, this occurs by the use of actin filaments that are connected to the plasma membrane and arranged in a ring around the circumference of the cell. Myosin causes the filaments to slide past one another, drawing in the membrane and pinching the cell in two.

- **Cytoplasmic streaming** is the directed flow of cytosol and organelles within plant cells. The movement occurs along actin filaments and is powered by myosin. It is especially common in large cells, where the circulation of cytoplasm facilitates material transport.

In addition, the movement called **cell crawling** occurs when groups of actin filaments grow, creating bulges in the plasma membrane that extend and move the cell. Cell crawling occurs in a wide range of organisms and cell types, including amoebae, slime molds, and certain animal cells.

Intermediate Filaments

Many types of **intermediate filament** exist, each consisting of a different—though similar in size and structure—type of protein (Table 7.3). Humans, for example, have 70 genes that code for intermediate filament proteins. This is in stark contrast to actin filaments and microtubules, which are made from the same protein subunits in all eukaryotic cells.

Moreover, intermediate filaments are not polar; instead, each end of these filaments is identical. They are not involved in directed movement driven by myosin or other motor proteins, but instead serve a purely structural role in eukaryotic cells.

The intermediate filaments that you are most familiar with belong to a family of molecules called the keratins. The cells that make up your skin and line surfaces inside your body contain about 20 types of keratin. These intermediate filaments provide the mechanical strength required for these cells to resist pressure and abrasion. Certain cells in the skin can also produce secreted forms of keratin. Depending on the location of the cell and keratins involved, the secreted filaments form fingernails, toenails, or hair.

Nuclear lamins, which make up the nuclear lamina layer introduced in Section 7.4, also qualify as intermediate filaments. Nuclear lamins form a dense mesh under the nuclear envelope. Recall that in addition to giving the nucleus its shape, they anchor the chromosomes. They are also involved in the breakup and reassembly of the nuclear envelope when cells divide.

Some intermediate filaments project from the nucleus through the cytoplasm to the plasma membrane, where they are linked to intermediate filaments that run parallel to the cell surface. In this way, intermediate filaments form a flexible skeleton that helps shape the cell surface and hold the nucleus in place.

Microtubules

Microtubules are the largest cytoskeletal components in terms of diameter. As Table 7.3 shows, they are assembled from subunits consisting of two polypeptides, called α-tubulin and β-tubulin, that exist as stable protein **dimers** ("two-parts").

Tubulin dimers polymerize from head to tail to form filaments that interact with one another to create relatively large, hollow tubes. Because of this polarity, these microtubules have α-tubulin polypeptides at one end (the minus end) and β-tubulins at the other end (the plus end). Like actin filaments, microtubules are dynamic and grow faster at their plus ends compared with their minus ends.

Microtubules originate from a structure called the **microtubule organizing center (MTOC)**. Their plus ends grow outward, radiating throughout the cell. Although plant cells typically have hundreds of sites where microtubules start growing, most animal and fungal cells have just one site that is near the nucleus.

In animals, the microtubule organizing center has a distinctive structure and is called a **centrosome**. As **FIGURE 7.26** shows, animal centrosomes contain two bundles of microtubules called **centrioles.** Although additional microtubules emanate from these structures in animals, they do not grow directly from the centrioles.

(a) In animals, microtubules originate from centrosomes.

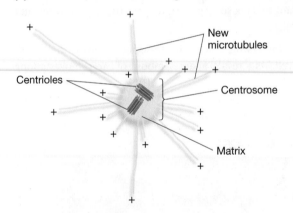

New microtubules

Centrioles

Centrosome

Matrix

(b) Centrioles consist of microtubules.

Centrosome

Centrioles

200 μm

Centrioles (oriented at 90° to each other)

Microtubule triplets

FIGURE 7.26 Centrosomes Are a Type of Microtubule-Organizing Center. (a) Microtubule-organizing centers, such as the centrosomes of animal cells, are the sites where new microtubules are made. Microtubules grow from the matrix surrounding the centrioles, and their positive ends point away from the centrosomes. **(b)** The two centrioles inside a centrosome consist of microtubules as triplets arranged in a circle.

In function, microtubules are similar to actin filaments: They provide stability and are involved in movement. Like steel girders in a skyscraper, the microtubules that radiate from an organizing center stiffen the cell by resisting compression forces. Microtubules also provide a structural framework for organelles. If microtubules are prevented from forming, the network-like configuration of the ER collapses and the Golgi apparatus disappears into vesicles.

Microtubules are best known for their role in separating chromosomes during mitosis and meiosis (see Chapters 12 and 13). But microtubules are involved in many other types of cellular movement as well. Let's first consider their role in moving materials inside cells and then explore how microtubules can help cells to swim.

Microtubules Serve as Tracks for Vesicle Transport Recall from Section 7.5 that vesicles are used to transport materials to a wide array of destinations inside cells. To study how this movement happens, Ronald Vale and colleagues focused on the giant axon, an extremely large nerve cell in squid that runs the length of the animal's body. If the squid is disturbed, the cell signals muscles to contract so it can jet away to safety. The researchers decided to study this particular cell for three reasons.

1. The giant axon is so large that it is relatively easy to see and manipulate.

2. Large numbers of vesicles are transported down the length of the cell. As a result, a large amount of cargo moves a long distance.

3. The researchers found that if they gently squeezed the cytoplasm out of the cell, vesicle transport still occurred in the extracellular cytoplasmic material. This allowed them to do experiments on vesicle transport without the plasma membrane being in the way.

In short, the squid giant axon provided a system that could be observed and manipulated efficiently in the lab. To watch vesicle transport in action, the researchers mounted a video camera to a microscope. As **FIGURE 7.27** shows, this technique allowed them to document that vesicle transport occurred along filamentous tracks.

To identify the filament involved, the biologists measured the diameter of the tracks and analyzed their chemical composition. Both types of data indicated that the tracks consist of microtubules. Microtubules also appear to be required for movement of materials elsewhere in the cell. For instance, if experimental cells are treated with a drug that disrupts microtubules, the movement of vesicles from the rough ER to the Golgi apparatus is impaired.

The general message of these experiments is that transport vesicles move through the cell along microtubules. How? Do the tracks themselves move, like a conveyer belt, or are vesicles carried along on some sort of molecular vehicle?

Motor Proteins Pull Vesicles Along the Tracks To study the way vesicles move along microtubules, Vale's group took the squid axon's transport system apart and then determined what components were required to put it back together. A simple experiment convinced the group that this movement is an energy-dependent process: If they depleted the amount of ATP in the cytoplasm, vesicle transport stopped.

To examine this process further, they mixed purified microtubules and vesicles with ATP, but no transport occurred. Something had been left out—but what? To find the missing element or elements, the researchers purified one subcellular part after another and added it to the microtubule + vesicle + ATP system.

Through trial and error, and further purification steps, the researchers finally succeeded in isolating a protein that generated vesicle movement. They named the molecule **kinesin,** from the Greek word *kinein* ("to move").

Like myosin, kinesin is a motor protein. Kinesin converts the chemical energy in ATP into mechanical energy in the form of movement. More specifically, when ATP is hydrolyzed by kinesin, the protein moves along microtubules in a directional manner: toward the plus end.

Biologists began to understand how kinesin works when X-ray diffraction studies showed that it has three major regions: a head section with two globular pieces, a tail associated with small polypeptides, and a stalk that connects the head and tail (**FIGURE 7.28a**).

Follow-up studies confirmed that the head region binds to the microtubule while the tail region binds to the transport vesicle. Recent work has shown that kinesin uses these domains to "walk" along the microtubule through a series of conformational changes as it hydrolyzes ATP (**FIGURE 7.28b**). Amazingly, these motors have been found to "walk" up to 375 steps per second.

Cells contain several different versions of the kinesin motor, each specialized for a different role in the cell. If kinesins move

(a) Electron micrograph

(b) Video image

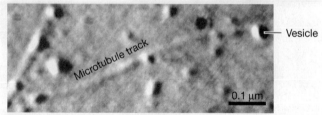

FIGURE 7.27 Transport Vesicles Move along Microtubule Track.
The images show extruded cytoplasm from a squid giant axon.
(a) An electron micrograph that allowed researchers to measure the diameter of the filaments and confirm that they are microtubules. In the upper part of this image, you can see a vesicle on a "track."
(b) A video microscope image using enhanced contrast that allowed researchers to watch vesicles move in real time.

(a) Structure of kinesin

Tail

Stalk

Head

5 nm

(b) Kinesin "walks" along a microtubule track.

Transport vesicle

Kinesin

Every step requires energy

ATP

ADP + P_i

Microtubule

− end

+ end

FIGURE 7.28 Motor Proteins Move Vesicles along Microtubules. (a) Kinesin has three distinct regions. **(b)** The current model depicting how kinesin "walks" along a microtubule track to transport vesicles. The two head segments act like feet that alternately attach, pivot, and release in response to the gain or loss of a phosphate group from ATP.

only toward the plus ends of microtubules, then what is responsible for moving the cargo in the opposite direction? By studying whole-cell locomotion, researchers discovered a motor that could move toward the minus end of microtubules.

Flagella and Cilia: Moving the Entire Cell

Flagella are long, whiplike projections from the cell surface that function in movement. While many bacteria and eukaryotes have flagella, the structure is completely different in the two groups.

- Bacterial flagella are helical rods made of a protein called flagellin; eukaryotic flagella consist of several microtubules constructed from tubulin dimers.

- Bacterial flagella move the cell by rotating the rod like a ship's propeller; eukaryotic flagella move the cell by undulating—they whip back and forth.

- Eukaryotic flagella are surrounded by the plasma membrane and are considered organelles; bacterial flagella are not.

Based on these observations, biologists conclude that the two structures evolved independently, even though their function is similar.

To understand how some cells move, let's focus on eukaryotic flagella. Eukaryotic flagella are closely related to structures called **cilia** (singular: **cilium**), which are short, hairlike projections that are also found in some eukaryotic cells (**FIGURE 7.29**). Flagella are generally much longer than cilia, and the two structures differ in

Cilia

50 µm

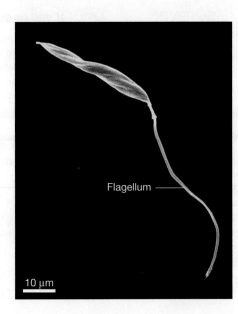

Flagellum

10 µm

FIGURE 7.29 Cilia and Flagella Differ in Length and Number. Cells typically only have 1–4 flagella but may have up to 14,000 cilia. The cells in these scanning electron micrographs have been colorized.

(a) Transmission electron micrograph of axoneme

Central
microtubules

Microtubule
doublet

75 nm

(b) Structure of axoneme

Spoke

Plasma
membrane

Link

Dynein
arms

Central
microtubules

Microtubule
doublet

– end

(c) Mechanism of axoneme bending

Microtubule doublet

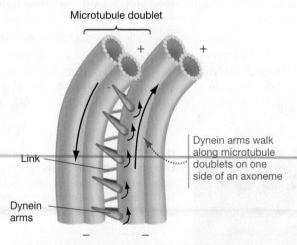

Link

Dynein
arms

Dynein arms walk
along microtubule
doublets on one
side of an axoneme

+ ATP: Causes dynein to walk toward minus end
and pull toward plus end

FIGURE 7.30 The Structure and Function of Cilia and Flagella.
(a) Transmission electron micrograph of a cross section through an
axoneme. **(b)** The microtubules in cilia and flagella are connected
by links and spokes, and the entire structure is surrounded by
the plasma membrane. **(c)** When dynein arms walk along the
microtubule doublets on one side of a flagellum, force is transmitted
to these links and spokes, causing the entire axoneme to bend.

✔ QUESTION If the links and spokes were removed from the
microtubule doublets, what would happen to the axoneme after
adding ATP?

their abundance and pattern of movement. But when researchers
examined the two structures with an electron microscope, they
found that their underlying organization is identical.

How are Cilia and Flagella Constructed? In the 1950s, anatomical studies established that most cilia and flagella have a characteristic "9 + 2" arrangement of microtubules. As **FIGURE 7.30a**
shows, nine microtubule pairs, or doublets, surround two central
microtubules. The doublets consist of one complete and one incomplete microtubule and are arranged around the periphery of
the structure.

The entire 9 + 2 structure is called the **axoneme** ("axle-
thread"). The nine doublets of the axoneme originate from a
structure called the **basal body.** The basal body is identical in
structure with a centriole and plays a central role in the growth
of the axoneme.

Through further study, biologists gained a more detailed view
of the axoneme's structure. Spoke-like proteins connect each doublet to the central pair of microtubules, and molecular links connect the nine doublets to one another (**FIGURE 7.30b**). Each doublet
also has a set of arms that project toward an adjacent doublet.

Axonemes are complex. How do their components interact to
generate motion?

What Provides the Force Required for Movement? In the 1960s
Ian Gibbons began studying the cilia of a common unicellular
eukaryote called *Tetrahymena*. Gibbons found that he could isolate axonemes by using a detergent to remove the plasma membrane that surrounds cilia and then subjecting the resulting
solution to differential centrifugation. These steps gave Gibbons
a cell-free system for studying how the axonemes in cilia and flagella work. He found that the isolated structures would beat only
if he supplied them with ATP, confirming that the beating of cilia
is an energy-demanding process.

check your understanding

C Y U

If you understand that . . .

- Each component of the cytoskeleton has a unique
  structure and set of functions. Actin filaments,
  intermediate filaments, and microtubules all play a
  role in structural support. In addition, actin filaments
  and microtubules work in conjunction with motor
  proteins to move cytoplasmic materials or the entire
  cell.
- Most elements of the cytoskeleton are dynamic—
  they grow and shrink depending on the needs of the
  cell.

✔ You should be able to . . .

Compare and contrast the structure and function
of actin filaments, intermediate filaments, and
microtubules.

Answers are available in Appendix A.

In another experiment, Gibbons treated the isolated axonemes with a molecule that disrupts interactions between proteins. The resulting axonemes could not beat even after being supplied with ATP. When Gibbons examined them in the electron microscope, he found that the treatment had removed the arms from the doublets. This result suggested that the arms are required for movement. Follow-up work showed that the arms are made of a large protein that Gibbons named **dynein** (from the Greek word *dyne*, meaning "force").

Like myosin and kinesin, dynein is a motor protein that uses ATP to undergo conformational changes. These shape changes move dynein along microtubules toward the minus end. Note that dynein moves in the opposite direction from the kinesin motor, which moves toward the plus end. In the cytoplasm, dynein motors are known to play various roles similar to the other motors, including the transport of vesicles. In the context of the axoneme, however, the outcome of dynein walking in the axoneme is very different.

So what is special about the axoneme? Remember that each of the nine doublets in the axoneme is connected to the central pair of microtubules by a spoke, and all the doublets are connected to each other by molecular links (Figure 7.30b). As a result, the sliding motion produced by dynein walking is constrained—if one doublet slides, it transmits force to the rest of the axoneme via the links and spokes (**FIGURE 7.30c**). If the dynein arms on just one side of the axoneme are activated, then the localized movement results in bending. The bending of cilia or flagella results in a swimming motion.

Scaled for size, flagella-powered swimming can be rapid. In terms of the number of body or cell lengths traveled per second, a sperm cell from a bull moves faster than a human world-record-holder does when swimming freestyle. At the cellular level, life is fast paced.

Taken together, the data reviewed in this chapter can be summed up in six words: Cells are dynamic, highly integrated structures. To maintain the level of organization that is required for life, chemical reactions must take place at mind-boggling speeds. How cells accomplish this feat is taken up elsewhere (see Chapter 8).

CHAPTER 7 REVIEW

For media, go to MasteringBiology

If you understand . . .

7.1 Bacterial and Archaeal Cell Structures and Their Functions

- There are two basic cellular designs: prokaryotic and eukaryotic. The single defining characteristic that differentiates prokaryotes from eukaryotes is the absence of a nucleus.

- Structures common to most, if not all, prokaryotes are ribosomes, a cell wall, a plasma membrane, an interior cytoskeleton, and a nucleoid.

- Many prokaryotes also possess flagella, fimbriae, and internal membrane structures, some of which are considered organelles.

✓ You should be able to predict what would happen to cells that are exposed to (1) a drug that prevents ribosomes from functioning, (2) an enzyme that degrades the cell wall, or (3) a drug that prevents the assembly of the cytoskeleton.

7.2 Eukaryotic Cell Structures and Their Functions

- Eukaryotic cells are usually much larger and more structurally complex than prokaryotic cells.

- Eukaryotic cells contain numerous specialized organelles, which allow eukaryotic cells to compartmentalize functions and grow to a large size. Organelles common to most, if not all, eukaryotes are as follows:

  1. The nucleus, which contains the cell's chromosomes and serves as its control center.

  2. The endomembrane system, which consists of a diverse group of interrelated organelles, including the endoplasmic reticulum, Golgi apparatus, lysosomes or vacuoles, and endosomes. These organelles work together to synthesize, process, sort, transport, and recycle material.

  3. Peroxisomes, which are organelles where key reactions take place that often result in the generation of toxic by-products. Specialized enzymes are included that safely disarm these by-products soon after they are generated.

  4. Mitochondria and chloroplasts, which have extensive internal membrane systems where the enzymes responsible for ATP generation and photosynthesis reside.

✓ You should be able to predict what would happen to a plant cell that is exposed to (1) a drug that poisons mitochondria, (2) a drug that inhibits catalase in the peroxisome, or (3) a drug that inhibits the formation of centrioles.

7.3 Putting the Parts into a Whole

- Cells have a tightly organized interior, where the presence and quantity of organelles often reflect the function of the cell.

- The activity in a cell illustrates the dynamic nature of life. Organelles and cytosolic proteins continually bustle about with a seemingly nonstop rush hour.

- Much of what is known about cellular activity has come from advances in cell imaging and techniques for isolating cellular components.

✓ You should be able to predict how a liver cell would differ compared with a salivary gland cell in terms of organelles.

7.4 Cell Systems I: Nuclear Transport

- Cells have sophisticated systems for making sure that proteins and other products end up in the right place.

- Traffic across the nuclear envelope occurs through nuclear pores, which contain a multiprotein nuclear pore complex that serves as gatekeeper.

- Small molecules can passively diffuse through the nuclear pore. Larger molecules enter the nucleus only if they contain a specific molecular signal that directs them through the pore via nuclear transport proteins.

✓ You should be able to propose a hypothesis that would address how certain cytoplasmic proteins can be induced to enter the nucleus by either the addition or the removal of phosphates.

7.5 Cell Systems II: The Endomembrane System Manufactures, Ships, and Recycles Cargo

- Molecules synthesized in the ER may be transported as cargo to the Golgi apparatus and then to a number of different sites, depending on the cargo.

- Before products leave the Golgi, they are sorted by their molecular "zip codes" that direct them to specific vesicles. The vesicles interact with receptor proteins at the target location so that the contents are delivered correctly.

- The lysosome is built from enzymes and membranes that are made and processed through the endomembrane system. These organelles are involved in recycling products via autophagy, phagocytosis, and receptor-mediated endocytosis.

✓ You should be able to justify why proteins (see Chapter 3)—and not RNA, DNA, carbohydrates, or lipids—are the molecules responsible for "reading" the array of molecular zip codes in cells.

7.6 Cell Systems III: The Dynamic Cytoskeleton

- The cytoskeleton is an extensive system of fibers that provides (1) structural support and a framework for arranging and organizing organelles and other cell components; (2) paths for moving

vesicles inside cells; and (3) machinery for moving the cell as a whole through the beating of flagella or cilia, or through cell crawling.

- Subunits are constantly being added to or removed from cytoskeletal filaments. Actin filaments and microtubules are polarized, meaning different ends of the filaments are designated as plus or minus ends. The plus ends have a higher growth rate than the minus ends.

- Movement often depends on motor proteins, which use chemical energy stored in ATP to change shape and position. Myosin motors move toward the plus ends of actin filaments. Kinesin and dynein motors move along microtubules toward the plus and minus ends, respectively.

- A specific type of dynein is found in the axonemes of eukaryotic cilia and flagella. These motors move microtubules to generate forces that bend the structures and enable cells to swim or generate water currents.

✓ You should be able to predict which of the three motors presented in this section would be responsible for transporting vesicles from the Golgi to the plasma membrane.

MasteringBiology

1. MasteringBiology Assignments

Tutorials and Activities Cilia and Flagella; Endomembrane System; Exocytosis and Endocytosis; Form Fits Function: Cells; Membrane Transport: Bulk Transport; Prokaryotic Cell Structure and Function; Pulse–Chase Experiment; Review: Animal Cell Structure and Function; Tour of a Plant Cell: Structures and Functions; Tour of an Animal Cell: Structures and Functions; Tour of an Animal Cell: The Endomembrane System; Transport into the Nucleus

Questions Reading Quizzes, Blue-Thread Questions, Test Bank

2. eText Read your book online, search, take notes, highlight text, and more.

3. The Study Area Practice Test, Cumulative Test, BioFlix® 3-D Animations, Videos, Activities, Audio Glossary, Word Study Tools, Art

You should be able to . . .

1. Which of the following accurately describes a difference between prokaryotic and eukaryotic cells?
   a. Prokaryotic cells have fimbriae that allow the cell to swim whereas eukaryotic cells have flagella.
   b. Eukaryotic cells are generally larger than prokaryotic cells.
   c. Eukaryotic cells have organelles.
   d. Prokaryotic cells have nuclei and eukaryotic cells have nucleoids.

2. What are three attributes of mitochondria and chloroplasts that suggest they were once free-living bacteria?

3. Which of the following is *not* true of secreted proteins?
   a. They are synthesized using ribosomes.
   b. They enter the ER lumen during translation.
   c. They contain a signal that directs them into the lysosome.
   d. They are transported between organelles in membrane-bound vesicles.

4. Which of the following results provided evidence of a nuclear localization signal in the nucleoplasmin protein?
   a. The protein was small and easily slipped through the nuclear pore complex.
   b. After cleavage of the protein, only the tail segments appeared in the nucleus.
   c. Removing the tail allowed the core segment to enter the nucleus.
   d. The SRP bound only to the tail, not the core segment.

5. Molecular zip codes direct molecules to particular destinations in the cell. How are these signals read?
   a. They bind to receptor proteins.
   b. They enter transport vesicles.
   c. They bind to motor proteins.
   d. They are glycosylated by enzymes in the Golgi apparatus.

6. How does the hydrolysis of ATP result in the movement of a motor protein along a cytoskeletal filament?

7. Compare and contrast the structure of a generalized plant cell, animal cell, and prokaryotic cell. Which features are common to all cells? Which are specific to just prokaryotes, or just plants, or just animals?

8. Cells that line your intestines are known to possess a large number of membrane proteins that transport small molecules and ions across the plasma membrane. Which of the following cell structures would you expect to be required for this function of the cells?
   a. the endoplasmic reticulum
   b. peroxisomes
   c. lysosomes
   d. the cell wall

9. Most of the proteins that reside in the nucleus possess a nuclear localization signal (NLS), even if they are small enough to pass through the pore complex unhindered. Why would a small protein have an NLS, when it naturally diffuses across the pore without one?

10. Make a flowchart that traces the movement of a secreted protein from its site of synthesis to the outside of a eukaryotic cell. Identify all the organelles that the protein passes through. Add notes indicating what happens to the protein at each step.

11. Although all three cytoskeletal fibers constantly replace their subunits, only actin filaments and microtubules demonstrate differences in the rate of growth between the two ends. What is responsible for this difference, and why is this not observed in intermediate filaments?

12. Describe how vesicles move in a directed manner between organelles of the endomembrane system. Explain why this movement requires ATP.

13. Which of the following cell structures would you expect to be most important in the growth of bacteria on the surface of your teeth?
    a. cell wall
    b. fimbriae
    c. flagella
    d. cilia

14. The enzymes found in peroxisomes are synthesized by cytosolic ribosomes. Suggest a hypothesis for how these proteins find their way to the peroxisomes.

15. Propose an experiment that would determine if the NLS in nucleoplasmin is limited to this protein only or if it could direct other structures into the nucleus.

16. George Palade's research group used the pulse–chase assay to dissect the secretory pathway in pancreatic cells. If they had instead performed this assay on muscle cells, which have high energy demands and primarily consist of actin and myosin filaments, where would you expect the labeled proteins to go during the chase?

8 Energy and Enzymes: An Introduction to Metabolic Pathways

In this chapter you will learn how

Enzymes use energy to drive the chemistry of life

looking at energy, asking

looking at enzymes, asking

What happens to energy in chemical reactions? **8.1**

How do enzymes help speed chemical reaction rates? **8.3**

Can chemical energy drive nonspontaneous reactions? **8.2**

What factors affect enzyme function? 8.4

How do enzymes work together in metabolic pathways? **8.5**

When table sugar is heated in the presence of oxygen, it undergoes the uncontrolled oxidation reaction known as burning. The heat energy in the flame is released as electrons are transferred from sugar to oxygen. Cells use the energy released from this type of reaction to drive the energy-demanding processes required for life.

This chapter is part of the Big Picture. See how on pages 198–199.

Cells are dynamic. Vesicles move cargo from the Golgi apparatus to the plasma membrane and other destinations, enzymes catalyze the synthesis of a complex array of macromolecules, and millions of proteins transport ions and molecules across cellular membranes. These activities change constantly in response to signals from other cells or the environment.

What drives all this action? The answer is twofold—energy and enzymes. Because staying alive takes work, there is no life without energy. Life, at its most basic level, consists of chemical reactions catalyzed by enzymes. By using enzymes to direct which reactions occur and which do not, life possesses the distinguishing feature of creating order from a naturally disordered environment.

✔ When you see this checkmark, stop and test yourself. Answers are available in Appendix A.

This chapter is about how enzymes work to help cells acquire and use energy. It is also your introduction to metabolic pathways—the ordered series of chemical reactions that build up or break down a particular molecule.

Let's begin by reviewing some fundamental concepts about energy and how it is used in cells.

8.1 What Happens to Energy in Chemical Reactions?

When biologists consider energy in chemical reactions, they often use the term **free energy** to describe the amount of energy that is available to do work. Recall that two types of energy exist: kinetic energy or potential energy (Chapter 2). **Kinetic energy** is energy of motion. There are several different forms of kinetic energy—at the molecular level, the energy of motion is called thermal energy. **Potential energy** is energy that is associated with position or configuration. In molecules, this is referred to as chemical energy and is stored in the position of electrons.

Chemical Reactions Involve Energy Transformations

The existence of two types of energy does not mean that energy is locked into either the kinetic or the potential type. Energy is often transformed from one type to the other. To drive this point home, consider a water molecule sitting at the top of a waterfall, as in **FIGURE 8.1**.

Step 1 The molecule has potential energy (E_p) because of its position.

Step 2 As the molecule passes over the waterfall, its potential energy is converted to the kinetic energy (E_k) of motion.

Step 3 When the molecule reaches the rocks below, it undergoes a change in potential energy because it has changed position. The difference in potential energy is transformed into an equal amount of kinetic energy that is manifested in a variety of forms: mechanical energy, which tends to break up the rocks; heat (thermal energy), which raises the temperature of the rocks and the water itself; and sound.

The amount of potential energy in an electron is based on its position relative to other electrons and the protons in the nuclei of nearby atoms (see **FIGURE 8.2a** on page 138). If an electron is close to negative charges on other electrons and far from the positive charges in nuclei, it has high potential energy. In general, the potential energy of a molecule is a function of the way its electrons are configured or positioned.

An electron in an outer electron shell is analogous to the water molecule at the top of a waterfall (**FIGURE 8.2b**). If the electron falls to a lower shell, its potential energy is converted to the kinetic energy of motion. After the electron occupies the lower electron shell, it undergoes a change in potential energy. As panel 3 in Figure 8.1b shows, the change in potential energy is transformed

into an equal amount of kinetic energy—usually thermal energy, but sometimes light.

These examples illustrate the **first law of thermodynamics,** which states that energy is conserved. Energy cannot be created or destroyed, but only transferred and transformed.

The total energy in a molecule is referred to as its **enthalpy** (represented by H). Enthalpy includes the potential energy of the molecule, often referred to as heat content, plus the effect of the molecule on its surroundings in terms of pressure and volume.

1. Potential energy
A water molecule sitting at the top of a waterfall has a defined amount of potential energy, E_p.

E_p (higher)

2. Kinetic energy
As the molecule falls, some of this stored energy is converted to kinetic energy (the energy of motion), E_k.

E_k

3. Other forms of kinetic energy
When the molecule strikes the rocks below, its energy of motion is converted to thermal, mechanical, and sound energy. The molecule's potential energy is now much lower. The change in potential energy has been transformed into an equal amount of other forms of kinetic energy.

Mechanical energy   Heat   Sound

E_p (lower)

Conclusion: Energy is neither created nor destroyed; it simply changes form.

FIGURE 8.1 Energy Transformations. During an energy transformation, the total amount of energy in the system remains constant.

(a) The potential energy of an electron is related to its position.

Electrons have the greatest potential energy in the outermost electron shells

Nucleus | 1st | 2nd | 3rd Electron shells

(b) PROCESS: ENERGY TRANSFORMATION IN AN ATOM

E_p (higher)

1. Potential energy
An electron in an outer shell has a defined amount of potential energy, E_p.

E_k

2. Kinetic energy
As the electron falls to a lower energy shell, its potential energy is converted to kinetic energy, E_k.

Heat or light

E_p (lower)

3. Other forms of kinetic energy
Once the electron arrives at a lower electron shell, the energy is converted to light or heat. The energy in the light or heat released is equal to the difference in potential energy between the outermost and inner shells.

Conclusion: Energy is neither created nor destroyed; it simply changes form.

FIGURE 8.2 Chemical energy transformations Potential energy energy stored in atoms or molecules may be transformed into kinetic energy by changes in electron position.

The contributions of heat, pressure, and volume to the enthalpy of a molecule are best understood by observing the changes in enthalpy in a chemical reaction. For example, let's examine the reaction responsible for the explosive bursts of scalding hot liquid a bombardier beetle can produce when provoked, as seen in **FIGURE 8.3**:

$$2\,H_2O_2(aq) \longrightarrow 2\,H_2O(l) + O_2(g)$$

In this reaction, hydrogen peroxide (H_2O_2) is broken down into water and O_2 gas, which expands to over 500 times the original volume of the H_2O_2. Heat given off from the reaction also increases the temperature of the liquid dramatically. These massive increases in temperature and volume generate the pressure that propels the boiling liquid out of an opening at the tip of the beetle's abdomen.

Changes in enthalpy in chemical reactions can be measured and are represented by ΔH. (The uppercase Greek letter delta, Δ, is often used in chemical and mathematical notation to represent change.) The value of ΔH is primarily based on the difference in heat content, since—apart from the reaction in the bombardier beetle—most biological reactions do not result in substantial changes in pressure and volume. When a reaction releases heat energy (products have less potential energy than the reactants), it is **exothermic** and the ΔH is negative. If heat energy is taken up during the reaction, generating products that have higher potential energy than the reactants, the reaction is **endothermic** and ΔH is positive.

Another factor that changes during a chemical reaction is the amount of disorder or **entropy** (symbolized by ΔS). When the products of a chemical reaction become less ordered than the reactant molecules were, entropy increases and ΔS is positive. The **second law of thermodynamics,** in fact, states that total entropy always increases in an isolated system. Keep in mind that the isolated system in this case is the universe, which includes the surroundings as well as the products of the reaction.

To determine whether a chemical reaction is spontaneous, it's necessary to assess the combined contributions of changes in heat and disorder. Chemists do this with a quantity called the **Gibbs free-energy change,** symbolized by ΔG.

$$\Delta G = \Delta H - T\Delta S$$

Here, T stands for temperature measured on the Kelvin scale (see **BioSkills 1**, in Appendix B). Water freezes at 273.15 K and boils at 373.15 K.

In words, the free-energy change in a reaction is equal to the change in enthalpy minus the change in entropy multiplied by the temperature. The $T\Delta S$ term simply means that entropy becomes more important in determining free-energy change as the temperature of the molecules increases. Thermal energy increases the amount of disorder in the system, so the faster molecules are moving, the more important entropy becomes in determining the overall free-energy change.

Chemical reactions are spontaneous when ΔG is less than zero. Such reactions are said to be **exergonic.** Reactions are

FIGURE 8.3 Reactions May Be Explosive due to Changes in Enthalpy. When provoked, the bombardier beetle mixes reactants with enzymes in a special chamber near the tip of its abdomen. The enzyme-catalyzed reaction releases heat energy and oxygen gas. The result is the projection of boiling hot liquid at a predator.

If you understand that . . .

- The ability of chemical reactions to proceed without an input of energy depends on the difference in enthalpy and entropy between the products and reactants.
- The combined effects of enthalpy and entropy changes are summarized in the equation for the Gibbs free-energy change.

✔ **You should be able to . . .**

1. Write out the Gibbs equation and define each of the components.
2. State when the overall free energy change in a reaction is most likely to be negative (meaning that the reaction is exergonic).

Answers are available in Appendix A.

[handwritten margin notes:] endergonic non-spontaneous $G > 0$ exergonic spontaneous $G < 0$

nonspontaneous when ΔG is greater than zero. Such reactions are termed **endergonic**. When ΔG is equal to zero, reactions are at equilibrium. ✔ If you understand these concepts, you should be able to explain (1) why the same reaction can be nonspontaneous at low temperature but spontaneous at high temperature, and (2) why some exothermic reactions are nonspontaneous.

Free energy changes when the potential energy and/or entropy of substances change. Spontaneous chemical reactions run in the direction that lowers the free energy of the system. Exergonic reactions are spontaneous and release energy; endergonic reactions are nonspontaneous and require an input of energy to proceed.

Temperature and Concentration Affect Reaction Rates

Even if a chemical reaction occurs spontaneously, it may not happen quickly. The reactions that convert iron to rust or sugar molecules to carbon dioxide and water are spontaneous, but at room temperature they occur very slowly, if at all.

For most reactions to proceed, one or more chemical bonds have to break and others have to form. For this to happen, the substances involved must collide in a specific orientation that brings the electrons involved near each other. (See Chapter 2 to review the forces involved in bond formation.)

The number of collisions occurring between the substances in a mixture depends on their temperature and concentration:

- When the concentration of reactants is high, more collisions should occur and reactions should proceed more quickly.

- When their temperature is high, reactants should move faster and collide more frequently.

Higher concentrations and higher temperatures should speed up chemical reactions. To test this hypothesis, students at Parkland College in Champaign, Illinois, performed the experiments shown in **FIGURE 8.4** (see page 140). Pay special attention to the two graphs in the "Results" section:

- **Temperature versus reaction rate** The graph on the left is based on experiments where the concentration of the reactants was the same, but the temperature varied. Each data point represents one experiment. Notice that the points represent a trend that rises from left to right—meaning, in this case, that the reaction rate speeded up when the temperature of the reaction mixture was higher.

- **Concentration versus reaction rate** The graph on the right is based on experiments where the temperature was constant, but the concentration of reactants varied. Each bar represents the average reaction rate over many replicates of each treatment, or set of concentrations. The thin lines at the top of each bar indicate the standard error of the mean—a measure of variability (see **BioSkills 4** in Appendix B). The take-home message of this graph is that reaction rates are higher when reactant concentrations are higher.

The reactions shown in Figure 8.4 were exergonic, meaning that the products had lower free energy than the reactants, so no input of energy was required. But, what drives nonspontaneous, endergonic reactions? Let's take a closer look.

8.2 Nonspontaneous Reactions May Be Driven Using Chemical Energy

By definition, endergonic reactions require an input of energy to proceed. Recall that radiation from the Sun and electricity from lightning could have driven nonspontaneous reactions during

QUESTION: Do chemical reaction rates increase with increased temperature and concentration?

RATE INCREASE HYPOTHESIS: Chemical reaction rates increase with increased temperature. They also increase with increased concentration of reactants.

NULL HYPOTHESIS: Chemical reaction rates are not affected by increases in temperature or concentration of reactants.

EXPERIMENTAL SETUP:

Experimental reaction: $3 HSO_3^-(aq) + IO_3^-(aq) \rightleftharpoons 3 HSO_4^-(aq) + I^-(aq)$

PREDICTION: Reaction rate, measured as 1/(time for reaction to go to completion), will increase with increased concentrations of reactants and increased temperature of reaction mix.

PREDICTION OF NULL HYPOTHESIS: There will be no difference in reaction rates among treatments in each setup.

RESULTS:

CONCLUSION: Chemical reaction rates increase with increased temperature or concentration.

FIGURE 8.4 Testing the Hypothesis that Reaction Rates Are Sensitive to Changes in Temperature and Concentration.

✔**QUESTION** Use **BioSkills 4** in Appendix B to explain why no error bars are used for the points shown on the graph on the left side of the "Results" section.

Exergonic reaction
(releases energy)

Energy

Higher energy reactants → Lower energy products

Energy

Lower energy reactants → Higher energy products

Energy

Endergonic reaction
(requires energy)

FIGURE 8.5 Energetic Coupling Allows Endergonic Reactions to Proceed Using the Energy Released from Exergonic Reactions.

chemical evolution (Chapter 2). What source of energy drives these reactions inside cells?

Exergonic reactions release free energy. **FIGURE 8.5** shows how **energetic coupling** between exergonic and endergonic reactions allows chemical energy released from one reaction to drive another. In cells, this process generally occurs in one of two ways, either through the transfer of high-energy electrons or the transfer of a phosphate group.

Redox Reactions Transfer Energy via Electrons

Chemical reactions that involve the loss or gain of one or more electrons are called **reduction–oxidation reactions,** or **redox reactions.** When an atom or molecule loses one or more electrons, it is oxidized. This makes sense if you notice that the term

oxidized sounds as if oxygen has done something to an atom or molecule. Recall that oxygen is highly electronegative and often pulls electrons from other atoms (Chapter 2). On the other hand, when an atom or molecule gains one or more electrons, it is reduced. To keep these terms straight, students often use the mnemonic "OIL RIG"—**Oxidation** *Is* *L*oss of electrons; **Reduction** *Is* Gain of electrons.

Oxidation events are always paired with a reduction; if one atom loses an electron, another has to gain it, and vice versa. Since electron position is related to energy levels, redox reactions represent the energetic coupling of two half-reactions, one exergonic and one endergonic. Oxidation is the exergonic half-reaction, and reduction is the endergonic half-reaction. Some of the energy that is lost by the oxidized molecule is used to increase potential energy of the reduced molecule. In cases where more free energy is released by the oxidation step than is necessary for the reduction step, the overall reaction is exergonic.

The gain or loss of an electron can be relative, however. During a redox reaction, an electron can be transferred completely from one atom to another, or an electron can simply shift its position in a covalent bond.

An Example of Redox in Action   To see how redox reactions work, consider the spontaneous reaction that occurs when reduced carbons in glucose ($C_6H_{12}O_6$) are oxidized as the sugar is burned in the presence of oxygen (O_2) (**FIGURE 8.6**). The orange dots in the illustration represent the positions of the electrons involved in covalent bonds.

Now compare the position of the electrons in the first reactant, glucose, with their position in the first product, carbon dioxide. Notice that many of the electrons have moved farther from the carbon nucleus in carbon dioxide. This means that carbon has been oxidized: it has "lost" electrons. The change occurred because the carbon and hydrogen atoms in glucose share electrons equally, while the carbon and oxygen atoms in CO_2 don't. In CO_2, the high electronegativity of the oxygen atoms pulled electrons away from the carbon atom.

FIGURE 8.6 Redox Reactions Involve the Gain or Loss of One or More Electrons. This diagram shows how the position of electrons changes when glucose reacts with oxygen. The carbons of glucose are oxidized while the oxygen atoms of O_2 are reduced.

Now compare the position of the electrons in the reactant O_2 molecules with their position in the product water molecules. In water, the electrons have moved closer to the oxygen nuclei than they were in the O_2 molecules, meaning that the oxygen atoms have been reduced. Oxygen has "gained" electrons. Thus, when glucose burns, carbon atoms are oxidized while oxygen atoms are reduced.

These shifts in electron position change the amount of chemical energy in the reactants and products. When glucose reacts with oxygen, electrons are held much tighter in the product molecules than in the reactant molecules. This means their potential energy has decreased. The entropy of the products is also much higher than that of the reactants, as indicated by the increase in the number of molecules. As a result, this reaction is exergonic. It releases energy in the form of heat and light.

Another Approach to Understanding Redox During the redox reactions that occur in cells, electrons (e^-) may also be transferred from an atom in one molecule, called the **electron donor,** to an atom in a different molecule, the **electron acceptor.** When this occurs, the electron may be accompanied by a proton (H^+), which would result in the addition of a neutral hydrogen (H) atom to the electron acceptor.

Molecules that obtain hydrogens via redox reactions tend to gain potential energy because the electrons in C–H bonds are equally shared and hence relatively far from the positive charges on the C and H nuclei. This observation should sound familiar,

from what you have learned about carbohydrates (see Chapter 5). Molecules that have a large number of C–H bonds, such as carbohydrates and fats, store a great deal of potential energy.

Conversely, molecules that are oxidized in cells often lose a proton along with an electron. Instead of having many C–H bonds, oxidized molecules in cells tend to have an increased number of C–O bonds (see Figure 8.6). Oxidized molecules tend to lose potential energy. To understand why, remember that oxygen atoms have extremely high electronegativity. Because oxygen atoms hold electrons so tightly, the electrons involved in bonds with oxygen atoms have low potential energy.

In many redox reactions in biology, understanding where oxidation and reduction have occurred is a matter of following hydrogen atoms—reduction often "adds Hs" and oxidation often "removes Hs." For example, **flavin adenine dinucleotide (FAD)** is a cellular electron acceptor that is reduced by two electrons accompanied by two protons to form $FADH_2$ (**FIGURE 8.7a**). $FADH_2$ readily donates these high-energy electrons to other molecules. As a result, it is called an **electron carrier** and is said to have "reducing power."

Another common electron carrier is **nicotinamide adenine dinucleotide (NAD$^+$),** which is reduced to form **NADH.** Like FAD, two electrons reduce NAD^+. These two carriers differ, however, in the number of hydrogen atoms transferred. NAD^+ acquires only one of the two hydrogens and releases the second into the environment as H^+ (**FIGURE 8.7b**).

(a) Flavin adenine dinucleotide

$$AH_2 + FAD \longrightarrow A + FADH_2$$

(b) Nicotinamide adenine dinucleotide

$$BH_2 + NAD^+ \longrightarrow B + NADH + H^+$$

FIGURE 8.7 Redox Reactions May Transfer Protons Along with Electrons. The potential energy of NAD^+ and FAD is increased by redox reactions that transfer high-energy electrons, which may or may not be accompanied by protons. The products $FADH_2$ and NADH are important electron carriers.

The two examples in Figure 8.6 illustrate an important point—all redox reactions involve the transfer of electrons, but they do not always involve the transfer of hydrogens. Redox reactions are central in biology—they transfer energy via electrons. The energy released from certain key redox reactions (see Chapters 9 and 10) is used to drive the endergonic formation of the nucleotide ATP from ADP and P_i. How is the energy stored in ATP used by the cell?

ATP Transfers Energy via Phosphate Groups

Adenosine triphosphate (ATP) (introduced in Chapter 4) makes things happen in cells because it has a great deal of potential energy. As **FIGURE 8.8a** shows, four negative charges are confined to a small area in the three phosphate groups in ATP. In part because these negative charges repel each other, the potential energy of the electrons in the phosphate groups is extraordinarily high.

ATP Hydrolysis Releases Free Energy When ATP reacts with water during a hydrolysis reaction, the bond between ATP's outermost phosphate group and its neighbor is broken, resulting in the formation of ADP and inorganic phosphate, P_i, which has the formula $H_2PO_4^-$ (**FIGURE 8.8b**). This reaction is highly exergonic. Under standard conditions of temperature and pressure in the laboratory, a total of 7.3 kilocalories of energy per mole of ATP (or 7.3 kcal/mol), is released during the reaction. A **kilocalorie (kcal)** of energy raises 1 kilogram (kg) of water 1°C.

ATP hydrolysis is exergonic because the entropy of the product molecules is higher than that of the reactants, and because there is a large drop in potential energy when ATP breaks down into ADP and P_i. The change in potential energy occurs in part because the electrons from ATP's phosphate groups are now spread across two molecules instead of being clustered on one molecule—meaning that there is now less electrical repulsion.

In addition, the destabilizing effect of the negative charges is reduced in ADP and P_i since these products interact with the partial positive charges on surrounding water molecules more efficiently than the clustered negative charges on ATP did.

How Does ATP Drive Endergonic Reactions? In the time it takes to read this sentence, millions of endergonic reactions have occurred in your cells. This chemical activity is possible, in part, because cells are able to use the energy released from the exergonic hydrolysis of ATP.

If the reaction diagrammed in Figure 8.8b occurred in a test tube, the energy released would be lost as heat. But cells don't lose that 7.3 kcal/mole as heat. Instead, they use it to make things happen. Specifically, the energy that is released when ATP is hydrolyzed may be used to transfer the cleaved phosphate to a target molecule, called a **substrate.**

The addition of a phosphate group to a substrate is called **phosphorylation.** When ATP is used as the phosphate donor, phosphorylation is exergonic because the electrons in ADP and the phosphate added to the substrate have much less potential energy than they did in ATP.

To see how this process works, consider an endergonic reaction between two reactant molecules—compound A and compound B—that results in a product AB needed by your cells. For this reaction to proceed, an input of energy is required.

When a phosphate group from ATP is added to one or both of the reactant molecules, the potential energy of the reactant is increased. This phosphorylated intermediate is referred to as an activated substrate. This is the critical point: Activated substrates have high enough potential energy that the reaction between compound A and, for example, the activated form of compound B is now exergonic. The two compounds then go on to react and form the product molecule AB.

(a) ATP stores a large amount of potential energy.

Phosphate groups

Clustered negative charges raise the potential energy of linked phosphate groups

Adenine

Ribose

FIGURE 8.8 Adenosine Triphosphate (ATP) Has High Potential Energy. (a) ATP's high potential energy results, in part, from the four negative charges clustered in its three phosphate groups. The negative charges repel each other, raising the potential energy of the electrons. **(b)** When ATP is hydrolyzed to ADP and inorganic phosphate, a large free-energy change occurs.

(b) Energy is released when ATP is hydrolyzed.

ATP + H_2O (Water) → ADP + P_i (Inorganic phosphate) + 7.3 kcal/mol ATP (Energy)

FIGURE 8.9 Exergonic Phosphorylation Reactions Are Coupled to Endergonic Reactions. In cells, many reactions only occur if one reactant is activated by phosphorylation. The phosphorylated reactant molecule has high enough free energy that the subsequent reaction is exergonic. In this graph, the free energy being tracked on the *y*-axis represents A, B, and the 7.3 kcal/mol that is released when ATP is hydrolyzed. For simplicity, the free energy in ADP and P_i is not shown. ΔG represents the change in free energy between the reactants and products for each indicated step.

✔**EXERCISE** Label the ΔG in the uncoupled reaction and the two steps of the coupled reaction to indicate if the change is representing a positive (> 0) or negative (< 0) value.

FIGURE 8.9 graphs how phosphorylation can couple exergonic and endergonic reactions. Note that the reaction between A and B to produce the product AB is endergonic—the ΔG is positive. But after the exergonic transfer of a phosphate group from ATP to B occurs, the free energy of the reactants A and BP is high enough to make the reaction that forms AB exergonic. When reactant molecules in an endergonic reaction are phosphorylated, the free energy released during phosphorylation is coupled to the endergonic reaction to make the combined overall reaction exergonic.

✔ If you understand the principles of energetic coupling, you should be able to compare and contrast how energy is transferred via redox reactions and ATP hydrolysis.

It is hard to overstate the importance of energetic coupling: Without it, life is impossible. If the cells in your body could no longer drive endergonic reactions by coupling them to exergonic reactions, you would die within minutes.

Now the question is, What role do enzymes play in these reactions?

8.3 How Enzymes Work

Regardless of whether reactions in cells are spontaneous or not, none would occur at the speed required for life without the support of enzymes. How do they do it?

Recall that the initial hypothesis for how enzymes speed up reactions—the "lock-and-key" model—was first proposed in 1894 by Emil Fischer (introduced in Chapter 3). In this model, the substrates would fit into enzymes and react in a manner analogous to a key being inserted into a lock. In other words, enzymes are **catalysts**—they bring substrates together in a precise orientation that makes reactions more likely. Fischer's model also explained why many enzymes are specific for a single reaction—specificity is a product of the geometry and chemical properties of the sites where substrates bind.

Enzymes Help Reactions Clear Two Hurdles

Recall that two hurdles must be cleared before reactions can take place: Reactants need to (**1**) collide in a precise orientation and

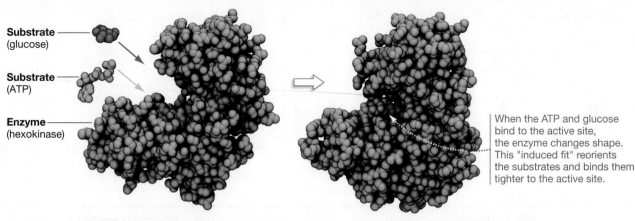

FIGURE 8.10 Reactant Molecules Bind to Specific Locations in an Enzyme. The reactant molecules, shown in red and yellow, fit into a precise location, called the active site, in the green enzyme. In this enzyme and in many others, the binding event causes the protein to change shape.

(2) have enough kinetic energy to overcome repulsion between electrons that come into contact as a bond forms (Chapter 2). To appreciate how enzymes work, let's consider each hurdle in turn.

Enzymes Bring Substrates Together Part of the reason enzymes are such effective catalysts is that they bring substrate molecules together in a substrate binding site known as the enzyme's **active site** (Chapter 3). In this way, enzymes help substrates collide in a precise orientation so that the electrons involved in the reaction can interact.

Enzymes generally are very large relative to substrates and roughly globular. The active site is in a cleft or cavity within the globular shape. A good example can be seen in the enzyme glucokinase, which catalyzes the phosphorylation of the sugar glucose. (Many enzymes have names that hint at the identity of the substrate and end with *-ase*.) As the left side of **FIGURE 8.10** shows, the active site in glucokinase is a small notch in an otherwise large, crescent-shaped enzyme.

In Fischer's original lock-and-key model, enzymes were conceived of as being rigid—almost literally as rigid as a lock. As research on enzyme action progressed, however, Fischer's model had to be modified. Perhaps the most important realization was that enzymes are not rigid and static, but flexible and dynamic. In fact, many enzymes undergo a significant change in shape, or conformation, when reactant molecules bind to the active site. You can see this conformational change, called an **induced fit,** in the glucokinase molecule on the right side of Figure 8.10. Once glucokinase binds its substrates—ATP and glucose—the enzyme rocks forward over the active site to bring the two substrates together.

In addition, recent research has clarified the nature of Fischer's key. When one or more substrate molecules enter the active site, they are held in place through hydrogen bonding or other weak interactions with amino acids in the active site. Once the substrate is bound, one or more R-groups in the active site come into play. The degree of interaction between the substrate and enzyme increases and reaches a maximum when a temporary,

unstable, intermediate condition called the **transition state** is formed. When Fischer's key is in its lock, it represents the transition state of the substrate.

There is more to achieving this transition state than simply an enzyme binding to its substrates, however. Even if the reaction is spontaneous, a certain amount of kinetic energy is required to strain the chemical bonds in substrates so they can achieve this transition state—called the **activation energy.** How do enzymes help clear the activation energy hurdle?

Enzymes Lower the Activation Energy Reactions happen when reactants have enough kinetic energy to reach the transition state. The kinetic energy of molecules, in turn, is a function of their temperature. (This is why reactions tend to proceed faster at higher temperatures.)

FIGURE 8.11 (see page 146) graphs the changes in free energy that take place during the course of a chemical reaction. As you read along the *x*-axis from left to right, note that a dramatic rise in free energy occurs when the reactants combine to form the transition state—followed by a dramatic drop in free energy when products form. The free energy of the transition state is high because the bonds that existed in the substrates are destabilized—it is the transition point between breaking old bonds and forming new ones.

The ΔG label on the graph indicates the overall change in free energy in the reaction—that is, the energy of the products minus the energy of the reactants. In this particular case, the products have lower free energy than the reactants, meaning that the reaction is exergonic. But because the activation energy for this reaction, symbolized by E_a, is high, the reaction would proceed slowly—even at high temperature.

This is an important point: The more unstable the transition state, the higher the activation energy and the less likely a reaction is to proceed quickly.

Reaction rates, then, depend on both the kinetic energy of the reactants and the activation energy of the particular reaction—meaning the free energy of the transition state. If the kinetic

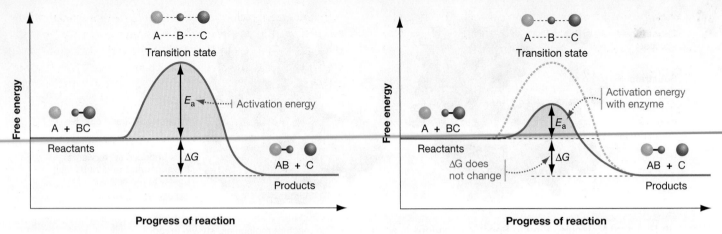

FIGURE 8.11 Changes in Free Energy during a Chemical Reaction. The energy profile shows changes in free energy that occur over the course of a hypothetical reaction between a molecule A and a molecule containing parts B and C. The overall reaction would be written as A + BC → AB + C. E_a is the activation energy of the reaction.

FIGURE 8.12 A Catalyst Changes the Activation Energy of a Reaction. The energy profile for the same reaction diagrammed in Figure 8.11, but now with a catalyst present. Even though the energy barrier to the reaction, E_a, is much lower, ΔG does not change.

✔ **QUESTION** Can a catalyst make a nonspontaneous reaction occur spontaneously? Explain why or why not.

energy of the participating molecules is high, such as at high temperatures, then molecular collisions are more likely to overcome the activation energy barrier. At this point, the transition state is formed and the reaction takes place.

Enzymes don't change the temperature of a solution, though. How do they fit in?

Interactions with amino acid R-groups at the enzyme active site stabilize the transition state and thus lower the activation energy required for the reaction to proceed. At the atomic level, R-groups that line the active site may form short-lived covalent bonds that assist with the transfer of atoms or groups of atoms from one reactant to another. More commonly, the presence of acidic or basic R-groups allows the reactants to lose or gain a proton more readily.

FIGURE 8.12 diagrams how enzymes lower the activation energy for a reaction by lowering the free energy of the transition state. Note that the presence of an enzyme does not affect the overall

energy change, ΔG, or change the energy of the reactants or the products. An enzyme changes only the free energy of the transition state.

Most enzymes are specific in their activity—they catalyze just a single reaction by lowering the activation energy that is required—and many are astonishingly efficient. Most of the important reactions in biology would not occur at all, or else proceed at imperceptible rates, without a catalyst. It's not unusual for enzymes to speed up reactions by a factor of a million; some enzymes make reactions go many *trillions* of times faster than they would without a catalyst.

It's also important to note that an enzyme is not consumed in a chemical reaction, even though it participates in the reaction. The composition of an enzyme is exactly the same after the reaction as it was before.

Enzyme catalysis can be analyzed as a three-step process. **FIGURE 8.13** summarizes this model:

PROCESS: A MODEL OF ENZYME ACTION

1. Initiation: Reactants bind to the active site in a specific orientation, forming an enzyme-substrate complex.

2. Transition state facilitation: Interactions between enzyme and substrate lower the activation energy required.

3. Termination: Products have lower affinity for active site and are released. Enzyme is unchanged after the reaction.

FIGURE 8.13 Enzyme Action Can Be Analyzed as a Three-Step Process.

1. *Initiation* Instead of reactants occasionally colliding in a random fashion, enzymes orient reactants precisely as they bind at specific locations within the active site.

2. *Transition state facilitation* Inside a catalyst's active site, reactant molecules are more likely to reach their transition state. In some cases the transition state is stabilized by a change in the enzyme's shape. Interactions between the substrate and R-groups in the enzyme's active site lower the activation energy required for the reaction. Thus, the catalyzed reaction proceeds much more rapidly than the uncatalyzed reaction.

3. *Termination* The reaction products have less affinity for the active site than the transition state does. Binding ends, the enzyme returns to its original conformation, and the products are released.

✔ If you understand the basic principles of enzyme catalysis, you should be able to complete the following sentences: (1) Enzymes speed reaction rates by _____ and lowering activation energy. (2) Activation energies drop because enzymes destabilize bonds in the substrates, forming the _____. (3) Enzyme specificity is a function of the active site's shape and the chemical properties of the _____ at the active site. (4) In enzymes, as in many molecules, function follows from _____.

What Limits the Rate of Catalysis?

For several decades after Fischer's model was published, most research on enzymes focused on rates of enzyme action, or what biologists call enzyme kinetics. Researchers observed that, when the amount of product produced per second—indicating the speed of the reaction—is plotted as a function of substrate concentration, a graph like that shown in **FIGURE 8.14** results.

FIGURE 8.14 Enzyme-Catalyzed Reactions Can Be Saturated. At high substrate concentration, enzyme-catalyzed reactions reach a maximum rate. Uncatalyzed reactions slowly increase as substrate concentration increases.

✔**EXERCISE** Label the parts of the *catalyzed reaction curve* that represent where (1) the reaction rate is most sensitive to changes in substrate concentration and (2) most or all of the active sites present are occupied.

In this graph, each data point represents an experiment where reaction rate was measured when substrates were at various concentrations. The two lines represent two series of experiments: one with the reactions catalyzed by an enzyme and the other uncatalyzed. As you read the curve for the catalyzed reaction from left to right, note that it has three basic sections:

1. When substrate concentrations are low, the speed of an enzyme-catalyzed reaction increases in a steep, linear fashion.

2. At intermediate substrate concentrations, the increase in speed begins to slow.

3. At high substrate concentration, the reaction rate plateaus at a maximum speed.

This pattern is in striking contrast to the situation for the uncatalyzed reactions, where the reaction speed is far slower, but tends to show a continuing linear increase with substrate concentration. The "saturation kinetics" of enzyme-catalyzed reactions were taken as strong evidence that the enzyme–substrate complex proposed by Fischer actually exists. The idea was that, at some point, active sites cannot accept substrates any faster, no matter how large the concentration of substrates gets. Stated another way, reaction rates level off because all available enzyme molecules are being used.

Do Enzymes Work Alone?

The answer to this question, in many cases, is no. Atoms or molecules that are not part of an enzyme's primary structure are often required for an enzyme to function normally. These enzyme "helpers" can be divided into three different types:

1. **Cofactors:** Inorganic ions, such as the metal ions Zn^{2+} (zinc), Mg^{2+} (magnesium), and Fe^{2+} (iron), which reversibly interact with enzymes. Cofactors that now participate in key reactions in virtually all living cells are thought to have been involved in catalysis early on in chemical evolution (see Chapter 2).

2. **Coenzymes:** Organic molecules that reversibly interact with enzymes, such as the electron carriers NADH or $FADH_2$.

3. **Prosthetic groups:** Non-amino acid atoms or molecules that are permanently attached to proteins, such as the molecule retinal. Retinal is involved in converting light energy into chemical energy (see Chapter 47).

In many cases, these enzyme helpers are part of the active site and play a key role in stabilizing the transition state. Their presence is therefore essential for the catalytic activity of many enzymes.

To appreciate why this is important, consider that many of the vitamins in your diet are required for the production of coenzymes. Vitamin deficiencies result in coenzyme deficiencies. Lack of coenzymes, in turn, disrupts normal enzyme function and causes disease. For example, thiamine (vitamin B_1) is required for the production of a coenzyme called thiamine pyrophosphate, which is required by three different enzymes. Lack of thiamine in the diet dramatically reduces the activity of these enzymes and causes an array of nervous system and heart disorders collectively known as beriberi.

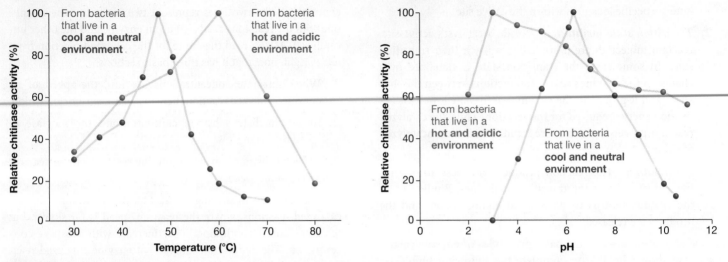

(a) Enzymes from different organisms may function best at different temperatures.

(b) Enzymes from different organisms may function best at different pHs.

FIGURE 8.15 Enzymes Have an Optimal Temperature and pH. The activity of enzymes is sensitive to changes in temperature **(a)** and pH **(b)**.

DATA: Nawani, N., B. P. Kapadnis, A. D. Das, et al. 2002. *Journal of Applied Microbiology* 93: 865–975. Also Nawani, N., and B. P. Kapadnis. 2001. *Journal of Applied Microbiology* 90: 803–808.

8.4 What Factors Affect Enzyme Function?

Given that an enzyme's structure is critical to its function, it's not surprising that an enzyme's activity is sensitive to conditions that alter protein shape. Recall that protein structure is dependent on the sequence of amino acids and a variety of chemical bonds and interactions that fold the polypeptide into its functional form (Chapter 3).

In particular, the activity of an enzyme often changes drastically as a function of temperature, pH, interactions with other molecules, and modifications of its primary structure. Let's take a look at how enzyme function is affected by, and sometimes even regulated by, each of these factors.

Enzymes Are Optimized for Particular Environments

Temperature affects the folding and movement of an enzyme as well as the kinetic energy of its substrates. The concentration of protons in a solution, as measured by pH, also affects enzyme structure and function. pH affects the charge on carboxyl and amino groups in residue side chains, and also the active site's ability to participate in reactions that involve the transfer of protons or electrons.

Do data support these assertions? **FIGURE 8.15a** shows how the activity of an enzyme, plotted on the *y*-axis, changes as a function of temperature, plotted on the *x*-axis. These data were collected for an enzyme called chitinase, which is used by bacteria to digest cell walls of fungi. In this graph, each data point represents the enzyme's relative activity—meaning the rate of the enzyme-catalyzed reaction, scaled relative to the highest rate observed—in

experiments conducted under conditions that differed only in temperature. Results are shown for two types of bacteria.

Note that, in both bacterial species, the enzyme has a distinct optimum or peak—a temperature at which it functions best. One of the bacterial species lives in the cool soil under palm trees, where the temperature is about 25°C, while the other lives in hot springs, where temperatures can be close to 100°C. The temperature optimum for the enzyme reflects these environments.

The two types of bacteria have different versions of the enzyme that differ in primary structure. Natural selection (introduced in Chapter 1) has favored a structure in each species that is best suited for its distinct environment. The two versions are adaptations that allow each species to thrive at different temperatures.

FIGURE 8.15b makes the same point for pH. The effect of pH on enzyme activity was tested on the same chitinases used in Figure 8.15a, but this time using conditions that varied only in pH. The soil-dwelling bacteria described earlier grow in a neutral pH environment, but the species that lives in hot springs is also exposed to acidic conditions.

Note that the organism that thrives in a hot, acidic environment has a version of the enzyme that performs best at high temperatures and low pH; the organism that lives in the cool soil has a version of the enzyme that functions best at cooler temperatures and nearly neutral pH. Each enzyme is sensitive to changes in temperature and pH, but each species' version of the enzyme has a structure that allows it to function best in its particular environment.

To summarize, the rate of an enzyme-catalyzed reaction depends not only on substrate concentration and the enzyme's intrinsic affinity for the substrate but also on temperature and pH (among other factors). Temperature affects the movement of the substrates and enzyme; pH affects the enzyme's shape and reactivity.

Substrates

(a) Competitive inhibition

(b) Allosteric regulation

Enzyme

or

or

or

Shape changes

Regulatory molecule

Shape changes

Regulatory molecule

Regulatory molecule

Enzyme in absence of regulation

Competitive inhibition
The substrates cannot bind when a regulatory molecule binds to the enzyme's active site.

Allosteric activation
The active site becomes available to the substrates when a regulatory molecule binds to a different site on the enzyme.

Allosteric inhibition
The active site becomes unavailable to the substrates when a regulatory molecule binds to a different site on the enzyme.

FIGURE 8.16 An Enzyme's Activity Is Precisely Regulated. Enzymes are turned on or off when specific regulatory molecules bind to them.

Most Enzymes Are Regulated

Controlling when and where enzymes will function is vital to the work of a cell. While temperature and pH affect the activity of enzymes, they are not often used as a means of regulating enzyme function. Instead, other molecules, in some cases other enzymes, regulate most of the cell's enzymatic activity. These regulatory molecules often change the enzyme's structure in some way, and their activity either activates or inactivates the enzyme.

Regulating Enzymes via Noncovalent Modifications   Many molecules that regulate enzyme activity bind non-covalently to the enzyme to either activate or inactivate it. Since the interaction does not alter the enzyme's primary structure, it is often referred to as a "reversible" modification.

Reversible modifications affect enzyme function in one of two ways:

1. The regulatory molecule is similar in size and shape to the enzyme's natural substrate and inhibits catalysis by binding to the enzyme's active site. This event is called **competitive inhibition** because the molecule involved competes with the substrate for access to the enzyme's active site (**FIGURE 8.16a**).

2. The regulatory molecule binds at a location other than the active site and changes the shape of the enzyme. This type of regulation is called **allosteric** ("different-structure") **regulation** because the binding event changes the shape of the enzyme in a way that makes the active site available or unavailable (**FIGURE 8.16b**).

Both strategies depend on the concentration of the regulatory molecule—the more regulatory molecule present, the more likely it will be to bind to the enzyme and affect its activity. The amount of regulatory molecule is often tightly controlled and, as you'll see in Section 8.5, the regulatory molecules themselves often manage the enzymes that produce them.

Regulating Enzymes via Covalent Modifications   In some cases, the function of an enzyme is altered by a chemical change in its primary structure. This change may be reversible or irreversible, depending on the type of modification.

Irreversible changes often result from the cleavage of peptide bonds that make up the primary structure of the enzyme. The enzyme trypsin, for example, is not functional until a small section of the protein is removed by a specific protease.

The most common modification of enzymes is the addition of one or more phosphate groups, similar to what was described for activated substrates in Section 8.2. In this case, however, the enzyme is phosphorylated instead of the substrate molecule. The transfer of a phosphate from ATP to the enzyme may be catalyzed by the enzyme itself or by a different enzyme.

When phosphorylation adds a negative charge to one or more amino acid residues in a protein, the electrons in that part of the protein change configuration. The enzyme's conformation

check your understanding

C Y U

If you understand that . . .

- An enzyme's function is dependent on its folded structure.
- Enzymes have been optimized to fold into functional structures at particular environmental conditions, such as temperature and pH.
- An enzyme's activity is often regulated by an alteration to its structure.

✔ **You should be able to . . .**

1. Explain why the relative activity appears to drop off in Figure 8.15b, when it has been shown that reaction rates tend to increase at higher temperatures (Figure 8.4).

2. Predict how the shape change that occurs when an enzyme is phosphorylated would affect its catalytic activity.

Answers are available in Appendix A.

Unphosphorylated form
(inactive)

Phosphorylated form
(active)

Sites of phosphorylation

Phosphate groups cause
green loop to move

FIGURE 8.17 Phosphorylation Changes the Shape and Activity of Proteins. When proteins are phosphorylated, they often change shape in a way that alters their activity. The figure shows the subtle structural change that occurs when mitogen-activated protein kinase (MAPK) is activated by adding two phosphate groups (yellow) to the enzyme.

usually changes as well, which may activate or inactivate its function. Note that although a substrate or an enzyme may be "activated" via phosphorylation, this activation does not represent the same effect. When a substrate is activated, its potential energy has increased, and this energy is used to convert an endergonic reaction to one that is exergonic. When an enzyme is activated, its catalytic function has been turned on—any change in the potential energy of the enzyme is not directly used in driving the reaction.

To see how phosphorylation affects the shape and activity of an enzyme, let's look at an enzyme called mitogen-activated protein kinase (MAPK), which is involved in cell signaling (see Chapter 11). As shown in **FIGURE 8.17**, phosphorylation of amino acid residues in a particular loop of the primary sequence causes a shape change, which functions like a switch to activate the enzyme.

Phosphorylation of an enzyme is a reversible modification to the protein's structure. Dephosphorylation—removal of phosphates—can quickly return the protein to its previous shape. The relative abundance of enzymes that catalyze phosphorylation and dephosphorylation, then, regulates the function of the protein.

8.5 Enzymes Can Work Together in Metabolic Pathways

The eukaryotic cell has been compared to an industrial complex, where distinct organelles are functionally integrated into a cooperative network with a common goal—life (see Chapter 7). Similarly, enzymes often work together in a manner resembling an assembly line in a factory. Each of the molecules of life presented in this book is built by a series of reactions, each catalyzed by a different enzyme. These multistep processes are referred to as **metabolic pathways.**

The following is an example of this type of teamwork, where an initial substrate A is sequentially modified by enzymes 1–3 to produce product D:

$$A \xrightarrow{enzyme\ 1} B \xrightarrow{enzyme\ 2} C \xrightarrow{enzyme\ 3} D$$

The B and C molecules are referred to as intermediates in the pathway—they serve as both a product and a reactant. For example, molecule B is the product of reaction 1 and the reactant for reaction 2.

Although these reactions have been written in a single direction, from left to right, the directionality is dependent on the relative concentrations of the reactants and products. At equilibrium, however, the concentration of the product for each reaction will be higher than the concentration of its respective reactant. Since D is the overall product for this pathway, it will have the highest concentration at equilibrium.

Metabolic Pathways Are Regulated

Since enzymes catalyze the reactions in metabolic pathways, the mechanisms that regulate enzyme function introduced in Section 8.4 also apply to the individual steps in a pathway. For example, to understand how blocking an individual reaction can affect an entire pathway, go back to the three-step model presented earlier and inactivate enzyme 2 by crossing it out. ✔ If you understand the assembly-line behavior of enzymes in a metabolic pathway, you should be able to predict how inactivating enzyme 2 would affect the concentration of molecules A, B, C, and D relative to what they would be if the pathway were fully functional.

When an enzyme in a pathway is inhibited by the product of the reaction sequence, **feedback inhibition** occurs. This is a convenient way for pathways to shut themselves down when their activity is no longer needed. As the concentration of the product molecule becomes abundant, it "feeds back" to stop the reaction sequence (**FIGURE 8.18**). By inhibiting a step early in the pathway, the amount of the initial substrate is not depleted unnecessarily, allowing it to be stored or used for other reactions.

Metabolic Pathways Evolve

While many enzymes are extraordinarily specific, some can catalyze a range of reactions and are able to interact with a family of related substrates. Research suggests that this flexibility allowed new enzymes to evolve and that enzymes specialized for catalyzing key reactions provided cells with a selective advantage. Could the same flexibility also help explain the evolution of the stepwise series of reactions seen in metabolic pathways?

In 1945, Norman Horowitz proposed a simple, stepwise process that could have directed pathway evolution. In Horowitz's model, enzymes first would have evolved to make the building blocks of life from readily available substrates, such as small organic compounds (see Chapter 2).

Low product (active)   **High product (inactive)**

FIGURE 8.18 Feedback Inhibition May Regulate Metabolic Pathways. Feedback inhibition occurs when the product of a metabolic pathway inhibits an enzyme that functions early in the pathway.

If an original substrate were depleted, natural selection would next favor the evolution of a new enzyme to make more of it from other existing molecules. By evolving a new reaction step to produce the original substrate—now serving as an intermediate in a two-step pathway—the original enzyme would have been able to continue its work. **FIGURE 8.19** illustrates this model—referred to as retro-evolution—in which repetition of this backward process produces a multistep metabolic pathway.

Researchers also speculate that as early pathways emerged, early enzymes may have been recruited to new pathways, where they evolved new catalytic activities that performed new tasks. This hypothesis is called patchwork evolution, since the new reaction series would consist of enzymes brought together from different pathways.

Evidence of patchwork evolution has been observed in modern organisms, where new metabolic activities have emerged in response to human-made chemicals. For example, a novel pathway has recently evolved in one species of bacterium to break down the pesticide pentachlorophenol, for use as a source of energy and carbon building blocks. Pentachlorophenol was first introduced into the environment in the 1930s as a timber preservative. The new pathway uses enzymes from two preexisting pathways, which had evolved the ability to work together. The metabolic activity of microbes is now being scrutinized and engineered to clean up a variety of human-made pollutants—giving rise to a new technology called **bioremediation** (see Chapter 29).

Regardless of how they evolved, metabolic pathways are now vital to the function of all cells. Those that break down molecules for sources of energy and carbon building blocks are called **catabolic pathways;** those that use energy and carbon building blocks to synthesize molecules are called **anabolic pathways.**

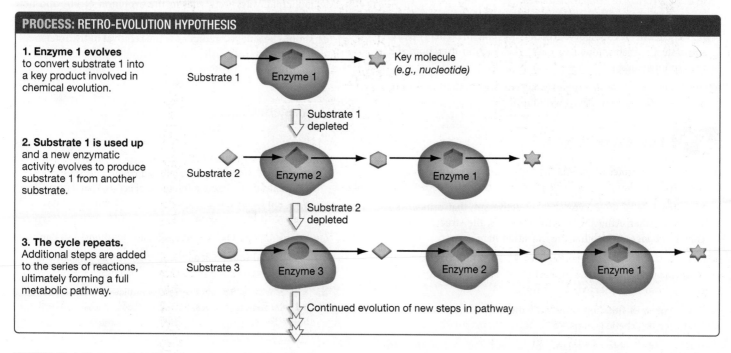

FIGURE 8.19 A Hypothetical Model for Metabolic Pathway Evolution.

You are being kept alive by key catabolic and anabolic pathways. The catabolic pathways of cellular respiration (introduced in Chapter 9) harvest high-energy electrons from reduced carbons (from foods such as starch and sugar) and pass them through redox reactions to generate ATP. These reduced carbons are produced by the anabolic pathways of photosynthesis that are driven by light energy (introduced in Chapter 10). The reactions involved in cellular respiration and photosynthesis perform the most important energy transformations to life on Earth.

If you understand . . .

8.1 What Happens to Energy in Chemical Reactions?

- Spontaneous reactions do not require an input of energy to occur.

- The Gibbs free energy change, ΔG, summarizes the combined effects of changes in enthalpy and entropy during a chemical reaction.

- Spontaneous reactions have a negative ΔG and are said to be exergonic; nonspontaneous reactions have a positive ΔG and are said to be endergonic.

✔ You should be able to explain why changes in enthalpy and entropy are used to determine whether a reaction is spontaneous.

8.2 Nonspontaneous Reactions May Be Driven Using Chemical Energy

- Redox reactions transfer energy by coupling exergonic oxidation reactions to endergonic reduction reactions.

- High-energy C–H bonds may be formed during the reduction step of a redox reaction when an H^+ is combined with a transferred electron.

- The hydrolysis of ATP is an exergonic reaction and may be used to drive a variety of cellular processes.

- When a phosphate group from ATP is added to a substrate that participates in an endergonic reaction, the potential energy of the substrate is raised enough to make the reaction exergonic and thus spontaneous.

✔ You should be able to explain what energetic coupling means, and why life would not exist without it.

8.3 How Enzymes Work

- Enzymes are protein catalysts. They speed reaction rates but do not affect the change in free energy of the reaction.

- The structure of an enzyme has an active site that brings substrates together. After binding to substrates, the structure of the enzyme changes to stabilize the transition state.

- Activation energy is the amount of kinetic energy required to reach the transition state of a reaction. Enzymes speed up a reaction by lowering the activation energy.

- Many enzymes function only with the help of cofactors, coenzymes, or prosthetic groups.

✔ You should be able to explain how an enzyme's active site can reduce the activation energy of a reaction.

8.4 What Factors Affect Enzyme Function?

- Enzymes are proteins, and thus their activity can be directly influenced by modifications or environmental factors, such as temperature and pH, that alter their three-dimensional structure.

- Most enzymes are regulated by molecules that either compete with substrates to occupy the active site, or alter enzyme shape.

- Protein cleavage and phosphorylation are examples of how enzymes may be regulated by modifying their primary structure.

✔ You should be able to compare and contrast the effect of allosteric regulation versus phosphorylation on enzyme function.

8.5 Enzymes Can Work Together in Metabolic Pathways

- In cells, enzymes often work together in metabolic pathways that sequentially modify a substrate to make a product.

- A pathway may be regulated by controlling the activity of one enzyme, often the first in the series of reactions. Feedback inhibition results from the accumulation of a product that binds to an enzyme in the pathway and inactivates it.

- Metabolic pathways were vital to the evolution of life, and new pathways continue to evolve in cells.

✔ You should be able to predict how the removal of the intermediate in a two-step metabolic pathway would affect the enzymatic rates of the first and last.

MB MasteringBiology

1. **MasteringBiology Assignments**

   Tutorials and Activities ATP and Energy; Chemical Reactions and ATP; Energy Transformations; Enzyme and Substrate Concentrations; Enzyme Inhibition; Factors That Affect Reaction Rate; How Enzymes Function; Regulating Enzyme Action; Redox Reactions

   Questions Reading Quizzes, Blue-Thread Questions, Test Bank

2. **eText** Read your book online, search, take notes, highlight text, and more.

3. **The Study Area** Practice Test, Cumulative Test, BioFlix® 3-D Animations, Videos, Activities, Audio Glossary, Word Study Tools, Art

You should be able to . . .

1. The first law of thermodynamics states which of the following?
   a. Energy exists in two forms: kinetic and potential.
   b. Reactions will take place only if energy is released.
   c. Energy is conserved: it cannot be created or destroyed.
   d. Disorder always increases in the universe.

2. If a reaction is exergonic, then which of these statements is true?
   a. The products have lower free energy than the reactants.
   b. Energy must be added for the reaction to proceed.
   c. The products have lower entropy (are more ordered) than the reactants.
   d. The reaction occurs extremely quickly.

3. What is a transition state?
   a. the complex formed as covalent bonds are being broken and re-formed during a reaction
   b. the place where an allosteric regulatory molecule binds to an enzyme
   c. an interaction between reactants with high kinetic energy, due to high temperature
   d. the shape adopted by an enzyme that has an inhibitory molecule bound at its active site

4. What often happens to an enzyme after it binds to its substrate? Is this a permanent change?

5. How does pH affect enzyme-catalyzed reactions?
   a. Protons serve as substrates for most reactions.
   b. Energy stored in protons is used to drive endergonic reactions.
   c. Proton concentration increases the kinetic energy of the reactants, allowing them to reach their transition state.
   d. The concentration of protons affects the folded structure of the enzyme.

6. When does feedback inhibition occur?

7. Explain the lock-and-key model of enzyme activity. What was incorrect about this model?

8. If you were to expose glucose to oxygen on your lab bench, why would you not expect to see it burn as shown in Figure 8.6?
   a. The reaction is endergonic and requires an input of energy.
   b. The reaction is not spontaneous unless an enzyme is added to the substrates.
   c. The sugar must first be phosphorylated to increase its potential energy.
   d. Energy is required for the sugar and oxygen to reach their transition state.

9. Explain why substrate phosphorylation using ATP is an exergonic reaction. How does the phosphorylation of reactants result in driving reactions that would normally be endergonic?

10. QUANTITATIVE In Figure 8.9, the energetic coupling of ATP hydrolysis and an endergonic reaction are shown. If the hydrolysis of ATP releases 7.3 kcal of free energy, use the graph in this figure to estimate what you would expect the ΔG values to be for the uncoupled reaction and the two steps in the coupled reaction.

11. Compare and contrast competitive inhibition and allosteric regulation.

12. Using what you have learned about changes in free energy, would you predict the ΔG value of catabolic reactions to be positive or negative? What about anabolic reactions? Justify your answers using the terms enthalpy and entropy.

13. Draw a redox reaction that occurs between compounds AH_2 and B^+ to form A, BH, and H^+. On the drawing, connect the reactant and product forms of each compound and state if it is the reduction or oxidation step and how many electrons are transferred. If this represents an exergonic reaction, identify which of the five substances would have the highest-energy electrons.

14. Researchers can analyze the atomic structure of enzymes during catalysis. In one recent study, investigators found that the transition state included the formation of a free radical (see Chapter 2) and that a coenzyme bound to the active site donated an electron to help stabilize the free radical. How would the reaction rate and the stability of the transition state change if the coenzyme were not available?

15. Recently, researchers were able to measure movement that occurred in a single amino acid in an enzyme as reactions were taking place in its active site. The amino acid that moved was located in the active site, and the rate of movement correlated closely with the rate at which the reaction was taking place. Discuss the significance of these findings, using the information in Figures 8.10 and 8.13.

16. You have discovered an enzyme that appears to function only when a particular sugar accumulates. Which of the following scenarios would you predict to be responsible for activating this enzyme?
   a. The sugar cleaves the enzyme so it is now in an active conformation.
   b. The sugar binds to the enzyme and changes the conformation of the active site.
   c. The sugar binds to the active site and competes with the normal substrate.
   d. The sugar phosphorylates the enzyme, triggering a conformational change.

9 Cellular Respiration and Fermentation

In this chapter you will learn how

Cells make ATP starting from sugars and other high potential energy compounds

by examining → How cells produce ATP when oxygen is present **9.1**

by examining → How cells produce ATP when oxygen is absent

looking closer at →

Glycolysis **9.2**

Pyruvate oxidation **9.3**

Citric acid cycle **9.4**

Electron transport and chemiosmosis **9.5**

focusing on →

Fermentation **9.6**

This hydroelectric dam on the Duero river between Spain and Portugal uses pumps to move water from the lower reservoir to the upper reservoir. During periods of high energy demand, the potential energy stored by this activity is used to generate electricity. A similar process is used by cells to produce ATP during cellular respiration.

This chapter is part of the Big Picture. See how on pages 198–199.

ife requires energy. From the very start, chemical evolution was driven by energy from chemicals, radiation, heat, or other sources (see Chapter 2). Harnessing energy and controlling its flow has been the single most important step in the evolution of life.

What fuels life in cells? The answer is the nucleotide adenosine triphosphate (ATP). ATP has high potential energy and allows cells to overcome life's energy barriers (see Chapter 8).

This chapter investigates how cells make ATP, starting with an introduction to the metabolic pathways that harvest energy from high-energy molecules like **glucose**—the most common source of chemical energy used by organisms. The four central pathways of cellular respiration will be

✔ When you see this checkmark, stop and test yourself. Answers are available in Appendix A.

presented with emphasis on how the oxidation of glucose leads to ATP production. Fermentation will also be introduced as an alternative pathway used to make ATP when key reactions in cellular respiration are either shut down or not available.

As cells process sugar, the energy that is released is used to transfer a phosphate group to adenosine diphosphate (ADP), generating ATP. (You can see the Big Picture of how the production of glucose in photosynthesis is related to its catabolism in cellular respiration on pages 198–199.)

9.1 An Overview of Cellular Respiration

In general, a cell contains only enough ATP to last from 30 seconds to a few minutes. Because it has such high potential energy, ATP is unstable and is not stored. Like many other cellular processes, the production and use of ATP is fast. Most cells are making ATP all the time.

Most of the glucose that is used to make ATP is produced by plants and other photosynthetic species. These organisms use the energy in sunlight to reduce carbon dioxide (CO_2) to glucose and other carbohydrates. While they are alive, photosynthetic species use the glucose that they produce to make ATP for themselves. When photosynthetic species decompose or are eaten, they provide glucose to animals, fungi, and many bacteria and archaea.

All organisms use glucose in the synthesis of complex carbohydrates, fats, and other energy-rich compounds. Storage carbohydrates, such as starch and glycogen, act like savings accounts for chemical energy. ATP, in contrast, is like cash. To withdraw chemical energy from the accounts to get cash, storage carbohydrates are first hydrolyzed into their glucose monomers. The glucose is then used to produce ATP through one of two general processes: cellular respiration or fermentation (**FIGURE 9.1**). The primary difference between these two processes lies in the degree to which glucose is oxidized.

What Happens When Glucose Is Oxidized?

When glucose undergoes the uncontrolled oxidation reaction called burning, some of the potential energy stored in its chemical bonds is converted to kinetic energy in the form of heat:

$$C_6H_{12}O_2 + 6\,O_2 \longrightarrow 6\,CO_2 + 6\,H_2O + Heat$$

glucose   oxygen   carbon dioxide   water

More specifically, a total of about 685 kilocalories (kcal) of heat is released when one mole of glucose is oxidized. To put this in perspective, if you burned this amount of glucose, it would give off enough heat to bring almost 2.5 gallons of room-temperature water to a boil.

Glucose does not burn in cells, however. Instead, the glucose in cells is oxidized through a long series of carefully controlled redox reactions. These reactions are occurring, millions of

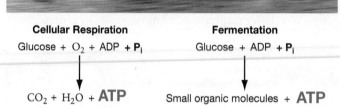

FIGURE 9.1 Glucose Is the Hub of Energy Processing in Cells. Glucose is a product of photosynthesis. Both plants and animals store glucose and oxidize it to provide chemical energy in the form of ATP.

times per minute, in your cells right now. Instead of releasing all of this energy as heat, much of it is being used to make the ATP you need to read, think, move, and stay alive. In cells, the change in free energy (Chapter 8) that occurs during the oxidation of glucose is used to synthesize ATP from ADP and P_i.

So how does fermentation differ from cellular respiration? Respiration, like burning, results in the complete oxidation of glucose into CO_2 and water. Fermentation, on the other hand, does not fully oxidize glucose. Instead, small, reduced organic molecules are produced as waste. As a result, cellular respiration releases more energy from glucose than fermentation.

The complete oxidation of glucose via cellular respiration can be thought of as a four-step process used to convert the chemical energy in glucose to chemical energy in ATP. Each of the four steps consists of a series of chemical reactions, and each step has a distinctive starting molecule and a characteristic set of products.

1. **Glycolysis** During **glycolysis,** one 6-carbon molecule of glucose is broken into two molecules of the three-carbon compound pyruvate. During this process, ATP is produced from ADP, and nicotinamide adenine dinucleotide (NAD^+) is reduced to form NADH.

2. **Pyruvate processing** Pyruvate is processed to release one molecule of CO_2, and the remaining two carbons are used to form the compound acetyl CoA. The oxidation of pyruvate results in more NAD^+ being reduced to NADH.

3. **Citric acid cycle** Acetyl CoA is oxidized to two molecules of CO_2. During this sequence of reactions, more ATP and NADH are produced, and flavin adenine dinucleotide (FAD) is reduced to form $FADH_2$.

4. **Electron transport and oxidative phosphorylation** Electrons from NADH and $FADH_2$ move through a series of proteins called an electron transport chain (ETC). The energy released in this chain of redox reactions is used to create a proton gradient across a membrane; the ensuing flow of protons back across the membrane is used to make ATP. Because this mode of ATP production links the phosphorylation of ADP with the oxidation of NADH and $FADH_2$, it is called **oxidative phosphorylation.**

FIGURE 9.2 summarizes the four steps in cellular respiration. Formally, **cellular respiration** is defined as any suite of reactions that uses electrons harvested from high-energy molecules to produce ATP via an electron transport chain.

The enzymes, products, and intermediates involved in cellular respiration and fermentation do not exist in isolation. Instead, they are part of a huge and dynamic inventory of chemicals inside the cell.

This complexity can be boiled down to a simple essence, however. Two of the most fundamental requirements of a cell are energy and carbon. They need a source of high-energy electrons for generating chemical energy in the form of ATP, and a source of carbon-containing molecules that can be used to synthesize DNA, RNA, proteins, fatty acids, and other molecules. Let's take a closer look at the central role cellular respiration plays in metabolic pathways as a whole.

Cellular Respiration Plays a Central Role in Metabolism

Recall that sets of reactions that break down molecules are called catabolic pathways (Chapter 8). These reactions often harvest stored chemical energy to produce ATP. On the other hand, sets of reactions that synthesize larger molecules from smaller components are called anabolic pathways. Anabolic reactions often use energy in the form of ATP.

How does the process of cellular respiration interact with other catabolic and anabolic pathways? Let's first consider how

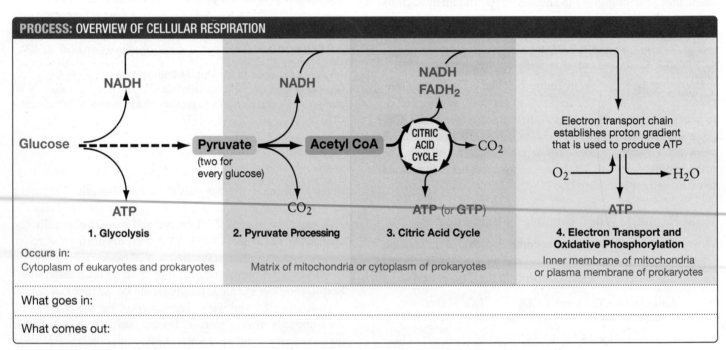

FIGURE 9.2 Cellular Respiration Oxidizes Glucose to Make ATP. Cells produce ATP from glucose via a series of processes: (1) glycolysis, (2) pyruvate processing, (3) the citric acid cycle, and (4) electron transport and oxidative phosphorylation. Each component produces high-energy molecules in the form of nucleotides (ATP or GTP) or electron carriers (NADH or $FADH_2$). Because the four components are connected, glucose oxidation is an integrated metabolic pathway. The first three steps oxidize glucose to produce NADH and $FADH_2$, which then feed the electron transport chain.

✔**EXERCISE** Fill in the chart along the bottom.

eukaryotes extract energy from molecules other than glucose and then examine how intermediates produced in glycolysis and the citric acid cycle are used as building blocks to synthesize cell components.

Catabolic Pathways Break Down a Variety of Molecules

Most organisms ingest, absorb, or synthesize many different carbohydrates. These molecules range from sucrose, maltose, and other simple sugars to large polymers such as glycogen and starch (see Chapter 5).

Recall that both glycogen and starch are polymers of glucose, but differ in the way their long chains of glucose branch. Using enzyme-catalyzed reactions, cells can produce glucose from glycogen, starch, and most simple sugars. Glucose and fructose can then be processed in glycolysis.

Carbohydrates are not the only important source of carbon compounds used in catabolic pathways, however. Fats are highly reduced macromolecules consisting of glycerol bonded to chains of fatty acids (see Chapter 6). In cells, enzymes routinely break down fats to release the glycerol and convert the fatty acids into acetyl CoA molecules. Glycerol can be further processed and enter glycolysis. Acetyl CoA enters the citric acid cycle.

Proteins can also be catabolized, meaning that they can be broken down and used to produce ATP. Once they are hydrolyzed to their constituent amino acids, enzyme-catalyzed reactions remove the amino ($-NH_2$) groups. The amino groups are excreted in urine as waste. The carbon compounds that remain are converted to pyruvate, acetyl CoA, and other intermediates in glycolysis and the citric acid cycle.

The top half of **FIGURE 9.3** summarizes the catabolic pathways of carbohydrates, fats, and proteins and shows how their breakdown products feed an array of steps in glucose oxidation and cellular respiration. When all three types of molecules are available in the cell to generate ATP, carbohydrates are used up first, then fats, and finally proteins.

Catabolic Intermediates Are Used in Anabolic Pathways

Where do cells get the precursor molecules required to synthesize amino acids, RNA, DNA, phospholipids, and other cell components? Not surprisingly, the answer often involves intermediates in carbohydrate metabolism. For example,

- In humans, about half the required amino acids can be synthesized from molecules siphoned from the citric acid cycle.

- Acetyl CoA is the starting point for anabolic pathways that result in the synthesis of fatty acids. Fatty acids can then be used to build phospholipid membranes or fats.

- Intermediates in glycolysis can be oxidized to start the synthesis of the sugars in ribonucleotides and deoxyribonucleotides. Nucleotides, in turn, are building blocks used in RNA and DNA synthesis.

- If ATP is abundant, pyruvate and lactate (from fermentation) can be used in the synthesis of glucose. Excess glucose may be converted to glycogen or starch and stored.

The bottom half of Figure 9.3 summarizes how intermediates in carbohydrate metabolism are drawn off to synthesize macromolecules. The take-home message is that the same molecule can serve many different functions in the cell. As a result, catabolic and anabolic pathways are closely intertwined.

Metabolism comprises thousands of different chemical reactions, yet the amounts and identities of molecules inside cells are relatively constant. By regulating key reactions involved in catabolic and anabolic pathways, the cell is able to maintain its internal environment even under different environmental conditions—a process referred to as **homeostasis.** Cellular respiration and

FIGURE 9.3 Cellular Respiration Interacts with Other Catabolic and Anabolic Pathways. A variety of high-energy compounds from carbohydrates, fats, or proteins can be broken down in catabolic reactions and used by cellular respiration for ATP production. Several of the intermediates in carbohydrate metabolism act as precursor molecules in anabolic reactions leading to the synthesis of glycogen or starch, RNA, DNA, fatty acids, and amino acids.

FIGURE 9.4 Pathways of Cellular Respiration Play a Central Role in the Metabolic Activity of Cells. A representation of a few of the thousands of chemical reactions that occur in cells. The dots represent molecules, and the lines represent enzyme-catalyzed reactions. At the center of all this, the first three steps of cellular respiration are emphasized by bold dots and thick lines. For reference, glucose, pyruvate, and acetyl CoA are represented by the distinctive colors used in Figure 9.3.

fermentation pathways may be crucial to the life of a cell, but they also have to be seen as central parts of a whole (**FIGURE 9.4**).

Once you've filled in the chart at the bottom of Figure 9.2, you'll be ready to analyze each of the four steps of cellular respiration in detail. As you delve in, keep asking yourself the same key questions: What goes in and what comes out? What happens to the potential energy that is released? Where does each step occur, and how is it regulated? Then take a look in the mirror. All these processes are occurring right now, in virtually all your cells.

9.2 Glycolysis: Processing Glucose to Pyruvate

Because the enzymes responsible for glycolysis have been observed in nearly every bacterium, archaean, and eukaryote, it is logical to infer that the ancestor of all organisms living today made ATP by glycolysis. It's ironic, then, that the process was discovered by accident.

In the 1890s Hans and Edward Buchner were working out techniques for breaking open baker's yeast cells and extracting the contents for commercial and medicinal use. (Yeast extracts are still added to some foods as a flavor enhancer or nutritional supplement.) In one set of experiments, the Buchners added sucrose to their extracts. At the time, sucrose was commonly used as a preservative—a substance used to preserve food from decay.

Instead of preserving the yeast extracts, though, the sucrose was quickly broken down and fermented, and alcohol appeared as a by-product. This was a key finding: It showed that metabolic pathways like fermentation could be studied in vitro—outside the organism. Until then, researchers thought that metabolism could take place only in intact organisms.

When researchers studied how the sugar was being processed, they found that the reactions could go on much longer than normal if inorganic phosphate were added to the mixture. This result implied that some of the compounds involved were being phosphorylated. Soon after, a molecule called fructose bisphosphate was isolated. (The prefix *bis*– means that the phosphate groups are attached to the fructose molecule at two different locations.) Subsequent work showed that all but the starting

FIGURE 9.5 Glycolysis Pathway. This sequence of 10 reactions oxidizes glucose to pyruvate. Each reaction is catalyzed by a different enzyme to produce two net ATP (4 ATP are produced, but 2 are invested), two molecules of NADH, and two molecules of pyruvate. In step 4, fructose-1,6-bisphosphate is divided into two products that both proceed through steps 6–10. The amounts for "What goes in" and "What goes out" are the combined totals for both molecules.

All 10 reactions of glycolysis occur in the cytosol

PROCESS: GLYCOLYSIS

What goes in: ···· **ATP** ·········· **ATP** ·········· Dihydroxyacetone phosphate

Glucose → Glucose-6-phosphate → Fructose-6-phosphate → Fructose-1,6-bisphosphate → Glyceraldehyde-3-phosphate

Glycolysis begins with an energy-investment phase: 2 ATP → 2 ADP

What comes out: ········· **ADP** ·········· **ADP**

and ending molecules in glycolysis—glucose and pyruvate—are phosphorylated.

In 1905 researchers found that the processing of sugar by yeast extracts stopped if they boiled the reaction mix. Because it was known that enzymes could be inactivated by heat, their discovery suggested that enzymes were involved in at least some of the processing steps. Years later, investigators realized that each step in glycolysis is catalyzed by a different enzyme. Eventually, each of the reactions and enzymes involved was worked out.

Glycolysis Is a Sequence of 10 Reactions

In both eukaryotes and prokaryotes, all 10 reactions of glycolysis occur in the cytosol (**FIGURE 9.5**). Note three key points about this reaction sequence:

1. Glycolysis starts by *using* ATP, not producing it. In the initial step, glucose is phosphorylated to form glucose-6-phosphate. After the second reaction rearranges the sugar to form fructose-6-phosphate, the third reaction adds a second phosphate group, forming the fructose-1,6-bisphosphate observed by early researchers. Thus, two ATP molecules are used up before any ATP is produced.

2. Once the energy-investment phase of glycolysis is complete, the subsequent reactions represent an energy-payoff phase. The sixth reaction in the sequence results in the reduction of two molecules of NAD^+; the seventh produces two molecules of ATP. This is where the energy "debt"—of two molecules of ATP invested early in glycolysis—is paid off. The final reaction in the sequence produces another two ATPs. For each molecule of glucose processed, the net yield is two molecules of NADH, two of ATP, and two of pyruvate.

3. In reactions 7 and 10 of Figure 9.5, an enzyme catalyzes the transfer of a phosphate group from a phosphorylated substrate to ADP, forming ATP. Enzyme-catalyzed reactions that result in

FIGURE 9.6 Substrate-Level Phosphorylation Involves an Enzyme and a Phosphorylated Substrate. Substrate-level phosphorylation occurs when an enzyme catalyzes the transfer of a phosphate group from a phosphorylated substrate to ADP, forming ATP.

ATP production are termed **substrate-level phosphorylation** (**FIGURE 9.6**). The key idea to note here is that the energy to produce the ATP comes from the phosphorylated substrate—not from a proton gradient, as it does when ATP is produced by oxidative phosphorylation.

The discovery and elucidation of the glycolytic pathway ranks as one of the great achievements in the history of biochemistry. For more detail concerning the enzymes that catalyze each step, see **TABLE 9.1** (on page 160). While the catabolism of glucose can occur via other pathways, this set of reactions is among the most ancient and fundamental of all life processes.

How Is Glycolysis Regulated?

An important advance in understanding how glycolysis is regulated occurred when biologists observed that high levels of ATP inhibit a key glycolytic enzyme called phosphofructokinase. **Phosphofructokinase** catalyzes reaction 3 in Figure 9.5—the synthesis of fructose-1,6-bisphosphate from fructose-6-phosphate. This is a crucial step in the sequence.

Step	Enzyme	Reaction
1	Hexokinase	Transfers a phosphate from **ATP** to glucose, increasing its potential energy.
2	Phosphoglucose isomerase	Converts glucose-6-phosphate to fructose-6-phosphate; referred to as an isomer of glucose-6-phosphate.
3	Phosphofructokinase	Transfers a phosphate from **ATP** to the opposite end of fructose-6-phosphate, increasing its potential energy.
4	Fructose-bis-phosphate aldolase	Cleaves fructose-1,6-bisphosphate into two different 3-carbon sugars.
5	Triose phosphate isomerase	Converts dihydroxyacetone phosphate (DAP) to glyceraldehyde-3-phosphate (G3P). Although the reaction is fully reversible, the DAP-to-G3P reaction is favored because G3P is immediately used as a substrate for step 6.
6	Glyceraldehyde-3-phosphate dehydrogenase	A two-step reaction that first oxidizes G3P using the **NAD$^+$** coenzyme to produce **NADH**. Energy from this reaction is used to attach a P_i to the oxidized product to form 1,3-bisphosphoglycerate.
7	Phosphoglycerate kinase	Transfers a phosphate from 1,3-bisphosphoglycerate to **ADP** to make 3-phosphoglycerate and **ATP**.
8	Phosphoglycerate mutase	Rearranges the phosphate in 3-phosphoglycerate to make 2-phosphoglycerate.
9	Enolase	Removes a water molecule from 2-phosphoglycerate to form a C=C double bond and produce phosphoenolpyruvate.
10	Pyruvate kinase	Transfers a phosphate from phosphoenolpyruvate to **ADP** to make pyruvate and **ATP**.

After reactions 1 and 2 occur, an array of enzymes can reverse the process and regenerate glucose for use in other pathways. Before step 3, then, the sequence is not committed to glycolysis. But once fructose-1,6-bisphosphate is synthesized, there is no point in stopping the process. Based on these observations, it makes sense that the pathway is regulated at step 3. How do cells do it?

As shown in Figure 9.5, ATP serves as a substrate for the addition of a phosphate to fructose-6-phosphate. In the vast majority of cases, increasing the concentration of a substrate would *speed* the rate of a chemical reaction, but in this case, it inhibits it. Why would ATP—a substrate that is required for the reaction—also serve as an inhibitor of the reaction? The answer lies in the fact that ATP is also the end product of the overall catabolic pathway.

Recall that when an enzyme in a pathway is inhibited by the product of the reaction sequence, feedback inhibition occurs (see Chapter 8). When the product molecule is abundant, it can inhibit its own production by interfering with the reaction sequence used to create it.

Feedback inhibition increases efficiency. Cells that are able to stop glycolytic reactions when ATP is abundant can conserve their stores of glucose for times when ATP is scarce. As a result, natural selection should favor individuals who have phosphofructokinase molecules that are inhibited by high concentrations of ATP.

How do high levels of the substrate inhibit the enzyme? As **FIGURE 9.7** shows, phosphofructokinase has two distinct binding sites for ATP. ATP can bind at the enzyme's active site, where it

is used to phosphorylate fructose-6-phosphate, or at a regulatory site, where it turns off the enzyme's activity.

The key to feedback inhibition lies in the ability of the two sites to bind to ATP. When concentrations are low, ATP binds

ATP at regulatory site

When ATP binds here, the reaction rate slows dramatically

Fructose-6-phosphate at active site

ATP at active site

FIGURE 9.7 Phosphofructokinase Has Two Binding Sites for ATP. A model of one of the four identical subunits of phosphofructokinase. In the active site, ATP is used as a substrate to transfer one of its phosphate groups to fructose-6-phosphate. In the regulatory site, ATP binding inhibits the reaction by changing the shape of the enzyme.

only to the active site, which has a greater affinity for ATP than does the regulatory site. As ATP concentrations increase, however, it also binds at the regulatory site on phosphofructokinase. When ATP binds at this second location, the enzyme's conformation changes in a way that dramatically lowers the reaction rate at the active site. In phosphofructokinase, ATP acts as an allosteric regulator (see Chapter 8). ✔ If you understand the principle behind the difference in affinity between the two ATP binding sites, you should be able to predict the consequences if the regulatory site had higher affinity for ATP than the active site did.

To summarize, glycolysis starts with one 6-carbon glucose molecule and ends with two 3-carbon pyruvate molecules. The reactions occur in the cytoplasm, and the energy that is released is used to produce a net total of two ATP and two NADH. Now the question is, what happens to the pyruvate?

9.3 Processing Pyruvate to Acetyl CoA

In eukaryotes, the pyruvate produced by glycolysis is transported from the cytosol to mitochondria. Mitochondria are organelles found in virtually all eukaryotes (see Chapter 7).

As shown in **FIGURE 9.8**, mitochondria have two membranes, called the inner membrane and outer membrane. The interior of the organelle is filled with layers of sac-like structures called **cristae.** Short tubes connect the cristae to the main part of the inner membrane. The region inside the inner membrane but outside the cristae is the **mitochondrial matrix.**

Pyruvate moves across the mitochondrion's outer membrane through small pores, but how it is transported across the inner membrane is still unclear. Current research suggests that either pyruvate is transported directly into the matrix using an unknown

transporter, or it is converted first into lactate, transported across the membrane, and then converted back into pyruvate.

Inside the mitochondrion, pyruvate reacts with a compound called **coenzyme A (CoA).** Coenzyme A is sometimes abbreviated as CoA-SH to call attention to its key sulfhydryl functional group. In this and many other reactions, CoA acts as a coenzyme by accepting and then transferring an acetyl group ($-COCH_3$) to a substrate (the A stands for acetylation). Pyruvate reacts with CoA, through a series of steps, to produce **acetyl CoA.**

The reaction sequence occurs inside an enormous and intricate enzyme complex called **pyruvate dehydrogenase.** In eukaryotes, pyruvate dehydrogenase is located in the mitochondrial matrix. In bacteria and archaea, pyruvate dehydrogenase is located in the cytosol.

As pyruvate is being processed, one of the carbons in the pyruvate is oxidized to CO_2 and NAD^+ is reduced to NADH. The remaining two-carbon acetyl unit is transferred to CoA (**FIGURE 9.9**).

FIGURE 9.9 Pyruvate Is Oxidized to Acetyl CoA. The reaction shown here is catalyzed by pyruvate dehydrogenase.

✔**EXERCISE** Above the reaction arrow, list three molecules whose presence speeds up the reaction. Label them "Positive control." Below the reaction arrow, list three molecules whose presence slows down the reaction. Label them "Negative control by feedback inhibition."

Cristae are sacs of inner membrane joined to the rest of the inner membrane by short tubes

Matrix

Cristae

Inner membrane

Intermembrane space

Outer membrane

100 nm

FIGURE 9.8 The Structure of the Mitochondrion. Mitochondria have outer and inner membranes that define the intermembrane space and matrix. Pyruvate processing occurs within the mitochondrial matrix. Recent research using cryo-electron tomography (the colorized micrograph on the right) shows the inner membrane is connected by short tubes to sac-like cristae.

Acetyl CoA is the final product of the pyruvate-processing step in glucose oxidation. Pyruvate, NAD$^+$, and CoA go in; CO_2, NADH, and acetyl CoA come out.

When supplies of ATP are abundant, however, the process shuts down. Pyruvate processing stops when the pyruvate dehydrogenase complex becomes phosphorylated and changes shape. The rate of phosphorylation increases when other products—specifically acetyl CoA and NADH—are at high concentration.

These regulatory changes are more examples of feedback inhibition. Reaction products feed back to stop or slow down the pathway.

On the contrary, high concentrations of NAD$^+$, CoA, or adenosine monophosphate (AMP)—which indicates low ATP supplies—*speed up* the reactions catalyzed by the pyruvate dehydrogenase complex.

Pyruvate processing is under both positive and negative control. Large supplies of products inhibit the enzyme complex; large supplies of reactants and low supplies of products stimulate it.

To summarize, pyruvate processing starts with the three-carbon pyruvate molecule and ends with one carbon released as CO_2 and the remaining two carbons in the form of acetyl CoA. The reactions occur in the mitochondrial matrix, and the potential energy that is released is used to produce one NADH for each pyruvate that is processed. Now the question is, what happens to the acetyl CoA?

9.4 The Citric Acid Cycle: Oxidizing Acetyl CoA to CO_2

While researchers were working out the sequence of reactions in glycolysis, biologists in other laboratories were focusing on redox reactions that oxidize small organic acids called **carboxylic acids.** Note that carboxylic acids all have carboxyl functional groups (R-COOH), hence the name.

A key finding emerged from their studies: Redox reactions that involve carboxylic acids such as citrate, malate, and succinate produce carbon dioxide. Recall from Section 9.1 that carbon dioxide is the endpoint of glucose oxidation via cellular respiration. Thus, it was logical for researchers to propose that the oxidation of small carboxylic acids could be an important component of glucose catabolism.

Early researchers identified eight small carboxylic acids that are rapidly oxidized in sequence, from least to most oxidized. What they found next was puzzling. When they added one of the eight carboxylic acids to cells, the rate of glucose oxidation increased, suggesting that the reactions are somehow connected to pathways involved in glucose catabolism. But, the added molecules did not appear to be used up. Instead, virtually all the carboxylic acids added were recovered later. How is this possible?

Hans Krebs solved the mystery when he proposed that the reaction sequence occurs in a cycle instead of a linear pathway. Krebs had another crucial insight when he suggested that the reaction sequence was directly tied to the processing of pyruvate—the endpoint of the glycolytic pathway.

To test these hypotheses, Krebs and a colleague set out to determine if adding pyruvate could link the two ends of the sequence of eight carboxylic acids. If pyruvate is the key link in forming a cycle, it would need to be involved in the conversion of oxaloacetate, the most oxidized of the eight carboxylic acids, to citrate, the most reduced carboxylic acid. When Krebs added pyruvate, the series of redox reactions occurred. The conclusion? The sequence of eight carboxylic acids is indeed arranged in a cycle (**FIGURE 9.10**).

Many biologists now refer to the cycle as the **citric acid cycle** because it starts with citrate, which is the salt of citric acid after the protons are released. The citric acid cycle is also known as the tricarboxylic acid (TCA) cycle, because citrate has three carboxyl groups, and also as the Krebs cycle, after its discoverer.

When radioactive isotopes of carbon became available in the early 1940s, researchers showed that carbon atoms cycle through the reactions just as Krebs had proposed. For more detail concerning the enzymes that catalyze each step, see **TABLE 9.2** (on page 164). In each cycle, the energy released by the oxidation of one molecule of acetyl CoA is used to produce three molecules of NADH, one of FADH$_2$, and one of **guanosine triphosphate (GTP),** or ATP, through substrate-level phosphorylation. Whether GTP or ATP is produced depends on the type of cell being considered.[1] For example, GTP appears to be produced in the liver cells of mammals, while ATP is produced in muscle cells.

In bacteria and archaea, the enzymes responsible for the citric acid cycle are located in the cytosol. In eukaryotes, most of the enzymes responsible for the citric acid cycle are located in the mitochondrial matrix. Because glycolysis produces two molecules of pyruvate, the cycle turns twice for each molecule of glucose processed in cellular respiration.

How Is the Citric Acid Cycle Regulated?

By now, it shouldn't surprise you to learn that the citric acid cycle is carefully regulated. The citric acid cycle can be turned off at multiple points, via several different mechanisms of feedback inhibition. Reaction rates are high when ATP is scarce; reaction rates are low when ATP is abundant.

FIGURE 9.11 highlights the major control points. Notice that in step 1, the enzyme that combines acetyl CoA and oxaloacetate to form citrate is shut down when ATP binds to it. This is another example of feedback inhibition, which also regulates enzymes at two additional points in the cycle. In step 3, NADH interferes with the reaction by binding to the enzyme's active site. This is an example of competitive inhibition (see Chapter 8). In step 4, ATP binds to the enzyme at an allosteric regulatory site.

To summarize, the citric acid cycle starts with the two-carbon acetyl molecule in the form of acetyl CoA and ends with the release of two CO_2. The reactions occur in the mitochondrial matrix, and the potential energy that is released is used to produce three NADH, one FADH$_2$, and one ATP or GTP for each acetyl oxidized. But a major question remains.

---

[1]Traditionally it was thought that the citric acid cycle produced GTP, which was later converted to ATP in the same cell. Recent work suggests that ATP is produced directly in some cell types, while GTP is produced in other cells. See C. O. Lambeth, Reconsideration of the significance of substrate-level phosphorylation in the citric acid cycle. *Biochemistry and Molecular Biology Education* 34 (2006): 21–29.

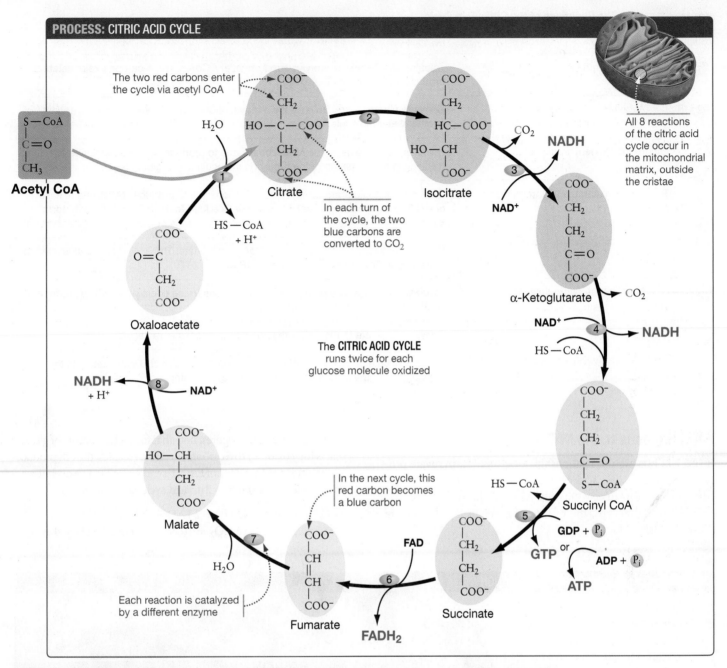

FIGURE 9.10 The Citric Acid Cycle Completes the Oxidation of Glucose. Acetyl CoA goes into the citric acid cycle, and carbon dioxide, NADH, FADH$_2$, and GTP or ATP come out. GTP or ATP is produced by substrate-level phosphorylation. If you follow individual carbon atoms around the cycle several times, you'll come to an important conclusion: each of the carbons in the cycle is eventually a "blue carbon" that is released as CO$_2$.

FIGURE 9.11 The Citric Acid Cycle Is Regulated by Feedback Inhibition. The citric acid cycle slows down when ATP and NADH are plentiful. ATP acts as an allosteric regulator, while NADH acts as a competitive inhibitor.

Step	Enzyme	Reaction
1	Citrate synthase	Transfers the 2-carbon acetyl group from acetyl CoA to the 4-carbon oxaloacetate to produce the 6-carbon citrate.
2	Aconitase	Converts citrate to isocitrate by the removal of one water molecule and the addition of another water molecule.
3	Isocitrate dehydrogenase	Oxidizes isocitrate using the **NAD$^+$** coenzyme to produce **NADH** and release one CO_2, resulting in the formation of the 5-carbon molecule α-ketoglutarate.
4	α-Ketoglutarate dehydrogenase	Oxidizes α-ketoglutarate using the **NAD$^+$** coenzyme to produce **NADH** and release one CO_2. The remaining 4-carbon molecule is added to coenzyme A (CoA) to form succinyl CoA.
5	Succinyl-CoA synthetase	CoA is removed, converting succinyl CoA to succinate. The energy released is used to transfer P_i to GDP to form **GTP**, or to ADP to form **ATP**.
6	Succinate dehydrogenase	Oxidizes succinate by transferring two hydrogens to the coenzyme **FAD** to produce **FADH$_2$**, resulting in the formation of fumarate.
7	Fumarase	Converts fumarate to malate by the addition of one water molecule.
8	Malate dehydrogenase	Oxidizes malate by using the **NAD$^+$** coenzyme to produce **NADH**, resulting in the regeneration of the oxaloacetate that will be used in step 1 of the cycle.

What Happens to the NADH and FADH$_2$?

FIGURE 9.12 reviews the relationships of glycolysis, pyruvate processing, and the citric acid cycle and identifies where each process takes place in eukaryotic cells. As the carbons in glucose are oxidized in these steps, the relative changes in free energy are shown in **FIGURE 9.13**.

As you study these figures, note that for each molecule of glucose that is fully oxidized to 6 carbon dioxide molecules, the cell produces 10 molecules of NADH, 2 of FADH$_2$, and 4 of ATP. The overall reaction for glycolysis and the citric acid cycle can be written as

$$C_6H_{12}O_2 + 10\ NAD^+ + 2\ FAD + 4\ ADP + 4\ P_i \longrightarrow$$
$$6\ CO_2 + 10\ NADH + 2\ FADH_2 + 4\ ATP$$

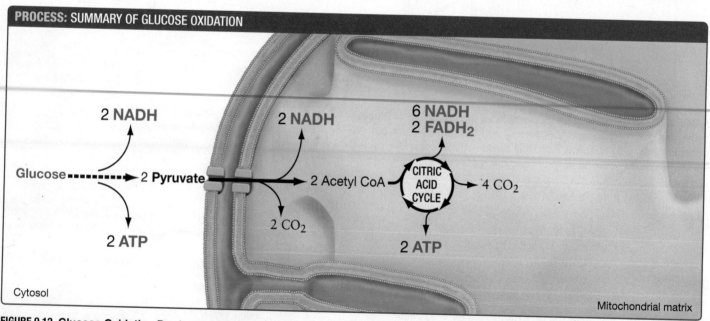

PROCESS: SUMMARY OF GLUCOSE OXIDATION

FIGURE 9.12 Glucose Oxidation Produces ATP, NADH, FADH$_2$, and CO$_2$. Glucose is completely oxidized to carbon dioxide via glycolysis, pyruvate processing, and the citric acid cycle. In eukaryotes, glycolysis occurs in the cytosol; pyruvate oxidation and the citric acid cycle take place in the mitochondrial matrix.

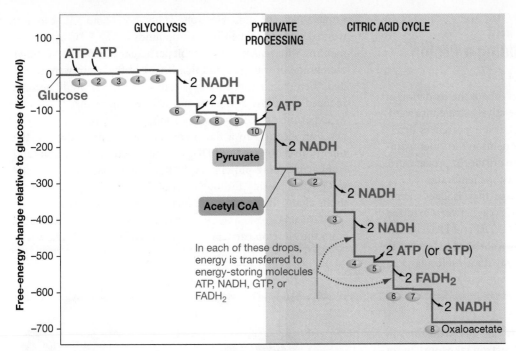

FIGURE 9.13 Free Energy Changes as Glucose Is Oxidized. If you read the vertical axis of this graph carefully, it should convince you that about 685 kcal/mol of free energy is released from the oxidation of glucose. Much of the energy is harnessed in the form of ATP, NADH, and $FADH_2$. The numbered green ovals identify the reaction steps in glycolysis and the citric acid cycle (see Tables 9.1 and 9.2).

DATA: Li, X., R. K. Dash, R. K. Pradhan, et al. 2010. *Journal of Physical Chemistry B.* 114: 16068–16082.

✔ **QUANTITATIVE** Based on the data in this graph, which of the three high-energy molecules produced during glucose oxidation would you expect to carry the highest amount of chemical energy? Justify your answer.

The ATP molecules are produced by substrate-level phosphorylation and can be used to drive endergonic reactions. The CO_2 molecules are a gas that is disposed of as waste—you exhale it; plants release it or use it as a reactant in photosynthesis.

What happens to the NADH and $FADH_2$ produced by glycolysis, pyruvate processing, and the citric acid cycle? Recall that the overall reaction for glucose oxidation is

$$C_6H_{12}O_6 + 6\,O_2 \longrightarrow 6\,CO_2 + 6\,H_2O + Energy$$

These three steps account for the glucose, the CO_2, and—because ATP is produced—some of the chemical energy that results from the overall reaction. But the O_2 and the H_2O are still unaccounted for. As it turns out, so is much of the chemical energy. The reaction that has yet to occur is

$$NADH + FADH_2 + O_2 + ADP + P_i \longrightarrow \\ NAD^+ + FAD + 2\,H_2O + ATP$$

In the above reaction, the electrons from NADH and $FADH_2$ are transferred to oxygen. NADH and $FADH_2$ are oxidized to NAD^+ and FAD, and oxygen is reduced to form water.

In effect, glycolysis, pyruvate processing, and the citric acid cycle transfer electrons from glucose to NAD^+ and FAD to form NADH and $FADH_2$. When oxygen accepts electrons from these reduced molecules, water is produced.

At this point, all the components of the overall reaction for glucose oxidation are accounted for, except for the energy. What happens to the energy that is released as electrons are transferred from NADH and $FADH_2$ to the highly electronegative oxygen atoms?

Specifically, how is the transfer of electrons linked to the production of ATP? In the 1960s—decades after the details of glycolysis and the citric acid cycle had been worked out—a startling answer to these questions emerged.

check your understanding

If you understand that . . .

- During glycolysis, glucose is oxidized to pyruvate, in the cytosol.
- During pyruvate processing, pyruvate is oxidized to acetyl CoA, in the mitochondrial matrix.
- In the citric acid cycle, the acetyl from acetyl CoA is oxidized to carbon dioxide (CO_2), in the mitochondrial matrix.
- Glycolysis, pyruvate processing, and the citric acid cycle are all regulated processes. The cell produces ATP only when ATP is needed.

✔ **You should be able to . . .**

Model the following components of cellular respiration by pretending that a large piece of paper is a cell. Draw a large mitochondrion inside it. Cut out small circles of paper and label them glucose, pyruvate, acetyl CoA, CO_2, ADP → ATP, NAD^+ → NADH, and FAD → $FADH_2$. Cut out small squares of paper and label them as glycolytic reactions, citric acid cycle reactions, and pyruvate dehydrogenase complex.

1. Put each of the squares in the appropriate location in the cell.

2. Add the circles and draw arrows to connect the appropriate molecules and reactions.

3. Using 12 paper triangles for pairs of electrons, show how electrons from glucose are transferred to NADH or $FADH_2$ (one pair should go to each NADH or $FADH_2$ formed) as glucose is oxidized to CO_2.

4. Label points where regulation occurs.

Answers are available in Appendix A.

9.5 Electron Transport and Chemiosmosis: Building a Proton Gradient to Produce ATP

The answer to one fundamental question about the oxidation of NADH and FADH$_2$ turned out to be relatively straightforward. By isolating different parts of mitochondria, researchers determined that NADH is oxidized by components in the inner membrane of the mitochondria, including the cristae. In prokaryotes, the oxidation of NADH occurs in the plasma membrane.

Biologists made a key discovery when they isolated membrane components—they were found to cycle between oxidized and reduced states after the addition of NADH and FADH$_2$. The membrane-associated molecules were hypothesized to be the key to processing NADH and FADH$_2$. What are these molecules, and how do they work?

The Electron Transport Chain

Collectively, the molecules responsible for the oxidation of NADH and FADH$_2$ are designated the **electron transport chain (ETC).** As electrons are passed from one molecule to another in the chain, the energy released by the redox reactions is used to move protons across the inner membrane of mitochondria.

Several points are fundamental to understanding how the ETC works:

- Most of the molecules are proteins that contain distinctive cofactors and prosthetic groups where the redox events take place (see Chapter 8). They include iron–sulfur complexes, ring-containing structures called flavins, or iron-containing heme groups called cytochromes. Each of these groups is readily reduced or oxidized.

- The inner membrane of the mitochondrion also contains a molecule called **ubiquinone,** which is not a protein. Ubiquinone got its name because it is nearly ubiquitous in organisms and belongs to a family of compounds called quinones. Also called **coenzyme Q** or simply Q, ubiquinone is lipid soluble and moves efficiently throughout the hydrophobic interior of the inner mitochondrial membrane.

- The molecules involved in processing NADH and FADH$_2$ differ in electronegativity, or their tendency to hold electrons. Some of the molecules pick up a proton with each electron, forming hydrogen atoms, while others obtain only electrons.

Because Q and the ETC proteins can cycle between a reduced state and an oxidized state, and because they differ in electronegativity, investigators realized that it should be possible to arrange them into a logical sequence. The idea was that electrons would pass from a molecule with lower electronegativity to one with higher electronegativity, via a redox reaction.

As electrons moved through the chain, they would be held more and more tightly. A small amount of energy would be released in each reaction, and the potential energy in each successive bond would lessen.

Organization of the Electron Transport Chain Researchers worked out the sequence of compounds in the ETC by experimenting with poisons that inhibit particular proteins in the inner membrane. It was expected that if part of the chain were inhibited, then the components upstream of the block would become reduced, but those downstream would remain oxidized.

Experiments with various poisons showed that NADH donates an electron to a flavin-containing protein (FMN) at the top of the chain, while FADH$_2$ donates electrons to an iron- and sulfur-containing protein (Fe·S) that then passes them directly to Q. After passing through each of the remaining components in the chain, the electrons are finally accepted by oxygen.

FIGURE 9.14 shows how electrons step down in potential energy from the electron carriers NADH and FADH$_2$ to O$_2$. The x-axis

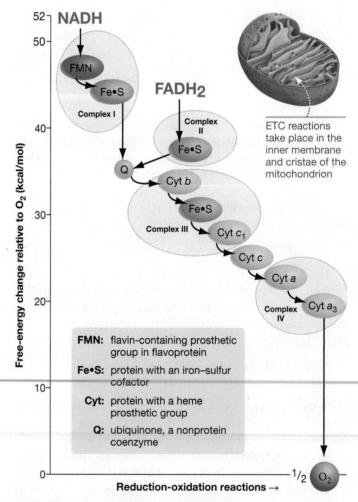

ETC reactions take place in the inner membrane and cristae of the mitochondrion

FMN: flavin-containing prosthetic group in flavoprotein

Fe•S: protein with an iron–sulfur cofactor

Cyt: protein with a heme prosthetic group

Q: ubiquinone, a nonprotein coenzyme

Reduction-oxidation reactions →

FIGURE 9.14 A Series of Reduction–Oxidation Reactions Occur in an Electron Transport Chain. Electrons step down in potential energy from the electron carriers NADH and FADH$_2$ through an electron transport chain to a final electron acceptor. When oxygen is the final electron acceptor, water is formed. The overall free-energy change of 52 kcal/mol (from NADH to oxygen) is broken into small steps.

DATA: Wilson D. F., M. Erecinska, and P. L. Dutton. 1974. *Annual Review of Biophysics and Bioengineering* 3: 203–230. Also Sled, V. D., N. I. Rudnitzky, Y. Hatefi, et al. 1994. *Biochemistry* 33: 10069–10075.

plots the sequence of redox reactions in the ETC; the *y*-axis plots the free-energy changes that occur.

The components of the electron transport chain are organized into four large complexes of proteins, often referred to as simply complexes I–IV. Q and the protein **cytochrome *c*** act as shuttles that transfer electrons between these complexes. Once the electrons at the bottom of the ETC are accepted by oxygen to form water, the oxidation of glucose is complete. Details on the names of the complexes and their role in the electron transport chain are provided in **TABLE 9.3** (on page 168).

Under controlled conditions in the laboratory, the total potential energy difference from NADH to oxygen is a whopping 53 kilocalories/mole (kcal/mol). Oxidation of the 10 molecules of NADH produced from each glucose accounts for almost 80 percent of the total energy released from the sugar. What does the ETC do with all this energy?

Role of the Electron Transport Chain Throughout the 1950s most biologists working on cellular respiration assumed that electron transport chains include enzymes that catalyze substrate-level phosphorylation. Recall that when substrate-level

phosphorylation occurs, a phosphate group is transferred from a phosphorylated substrate to ADP, forming ATP. Despite intense efforts, however, no one was able to find an enzyme among the components of the ETC that would catalyze the phosphorylation of ADP to produce ATP.

What researchers did find, however, is that the movement of electrons through the ETC actively transports protons from the matrix, across the inner membrane, and into the intermembrane space (see **FIGURE 9.15**). The exact route and mechanism used to pump protons is still being worked out. In some cases, it is not clear how the redox reactions taking place inside each complex result in the movement of protons.

The best-understood interaction between electron transport and proton transport takes place in complex III. Research has shown that when Q accepts electrons from complex I or complex II, it picks up protons from the matrix side of the inner membrane. The reduced form of Q then diffuses through the inner

FIGURE 9.15 How Does the Electron Transport Chain Work? The individual components of the electron transport chain diagrammed in Figure 9.14 are found in the inner membrane of mitochondria. Electrons are carried from one complex to another by Q and by cytochrome *c*; Q also shuttles protons across the membrane. The orange arrow indicates Q moving back and forth. Complexes I and IV use the potential energy released by the redox reactions to pump protons from the mitochondrial matrix to the intermembrane space.

✔**EXERCISE** Add an arrow across the membrane and label it "Proton gradient." In the boxes at the bottom, list "What goes in" and "What comes out" for each complex.

ETC Component	Descriptive Name	Reaction
Complex I	NADH dehydrogenase	Oxidizes **NADH** and transfers the two electrons through proteins containing FMN prosthetic groups and Fe·S cofactors to reduce an oxidized form of ubiquinone (Q). Four **H$^+$** are pumped out of the matrix to the intermembrane space.
Complex II	Succinate dehydrogenase	Oxidizes **FADH$_2$** and transfers the two electrons through proteins containing Fe·S cofactors to reduce an oxidized form of Q. This complex is also used in step 6 of the citric acid cycle.
Q	Ubiquinone	Reduced by complexes I and II and moves throughout the hydrophobic interior of the ETC membrane, where it is oxidized by complex III.
Complex III	Cytochrome *c* reductase	Oxidizes Q and transfers one electron at a time through proteins containing heme prosthetic groups and Fe·S cofactors to reduce an oxidized form of cytochrome *c* (cyt *c*). A total of four **H$^+$** for each pair of electrons is transported from the matrix to the intermembrane space.
Cyt *c*	Cytochrome *c*	Reduced by accepting a single electron from complex III and moves along the surface of ETC membrane, where it is oxidized by complex IV.
Complex IV	Cytochrome *c* oxidase	Oxidizes cyt *c* and transfers each electron through proteins containing heme prosthetic groups to reduce oxygen gas (O$_2$), which picks up two **H$^+$** from the matrix to produce water. Two additional **H$^+$** are pumped out of the matrix to the intermembrane space.

membrane, where its electrons are used to reduce a component of complex III near the intermembrane space. The protons held by Q are then released to the intermembrane space.

In this way, through redox reactions alone, Q shuttles electrons and protons from one side of the membrane to the other. The electrons proceed down the transport chain, and the protons contribute to an electrochemical gradient as they are released into the intermembrane space.

Once the nature of the electron transport chain became clear, biologists understood the fate of the electrons and the energy carried by NADH and FADH$_2$. Much of the chemical energy that was originally present in glucose is now accounted for in the proton electrochemical gradient. This is satisfying, except for one crucial question: If electron transport does not make ATP, what does?

The Discovery of ATP Synthase

In 1960 Efraim Racker made several key observations about how ATP is synthesized in mitochondria. When he used mitochondrial membranes to make vesicles, Racker noticed that some vesicles formed with their membrane inside out. Electron microscopy revealed that the inside-out membranes had many large proteins studded along their surfaces. Each protein appeared to have a base in the membrane, from which a lollipop-shaped stalk and a knob project (**FIGURE 9.16**). If the solution was vibrated or treated with a compound called urea, the stalks and knobs fell off.

Racker seized on this technique to isolate the stalks and knobs and do experiments with them. For example, he found that isolated stalks and knobs could hydrolyze ATP, forming ADP and

inorganic phosphate. The vesicles that contained just the base component, without the stalks and knobs, could not process ATP. The base components were, however, capable of transporting protons across the membrane.

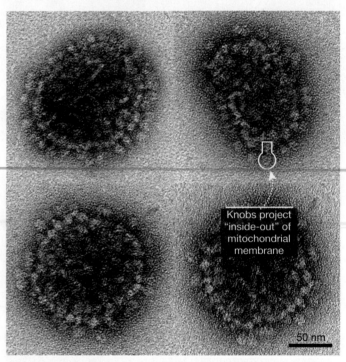

Knobs project "inside-out" of mitochondrial membrane

50 nm

FIGURE 9.16 The Discovery of ATP Synthase. When patches of mitochondrial membrane turn inside out and form vesicles, the lollipop-shaped stalk-and-knob structures of ATP synthase proteins face outward. Normally, the stalk and knob face inward, toward the mitochondrial matrix.

Based on these observations, Racker proposed that the stalk-and-knob component of the protein was an enzyme that both hydrolyzes and synthesizes ATP. To test his idea, Racker added the stalk-and-knob components back to vesicles that had been stripped of them and confirmed that the vesicles regained the ability to synthesize ATP. The entire complex is known as **ATP synthase.** Follow-up work also confirmed his hypothesis that the membrane-bound base component of ATP synthase is a proton channel. Is there a connection between proton transport and ATP synthesis?

The Chemiosmosis Hypothesis

In 1961 Peter Mitchell broke with the prevailing ideas that electron transport produces ATP via substrate phosphorylation. Instead, he proposed something completely new—an indirect connection between electron transport and ATP production. Mitchell's novel hypothesis? The real job of the electron transport chain is to pump protons across the inner membrane of mitochondria from the matrix to the intermembrane space. After a proton gradient is established, an enzyme in the inner membrane, like Racker's ATP synthase, would synthesize ATP from ADP and P_i.

Mitchell introduced the term **chemiosmosis** to describe the use of a proton gradient to drive energy-requiring processes, like the production of ATP. Here, osmosis refers to the force generated from the proton gradient rather than the transport of water. Although proponents of a direct link between electron transport and substrate-level phosphorylation objected vigorously to Mitchell's idea, several key experiments supported it.

FIGURE 9.17 illustrates how the existence of a key element in Mitchell's hypothesis was confirmed: A proton gradient alone can be used to synthesize ATP via ATP synthase. The researchers made vesicles from artificial membranes that contained Racker's ATP synthase isolated from mitochondria. Along with this enzyme, they inserted bacteriorhodopsin, a well-studied membrane protein that acts as a light-activated proton pump.

When light strikes bacteriorhodopsin, it absorbs some of the light energy and changes conformation in a way that pumps protons from the interior of a membrane to the exterior. As a result, the experimental vesicles established a strong electrochemical gradient favoring proton movement to the interior. When the vesicles were illuminated to initiate proton pumping, ATP began to be produced from ADP and P_i inside the vesicles.

Mitchell's prediction was correct: In this situation, ATP production depended solely on the existence of a **proton-motive force,** which is based on a proton electrochemical gradient. It could occur in the *absence* of an electron transport chain. This result, along with many others, has provided strong support for the hypothesis of chemiosmosis. Most of the ATP produced by cellular respiration is made by a flow of protons.

✔ If you understand chemiosmosis, you should be able to explain why ATP production during cellular respiration is characterized as indirect. More specifically, you should be able to explain the relationship between glucose oxidation, the proton gradient, and ATP synthase.

QUESTION: How are the electron transport chain and ATP production linked?

CHEMIOSMOTIC HYPOTHESIS: The linkage is indirect. The ETC creates a proton-motive force that drives ATP synthesis by the mitochondrial ATP synthase.

ALTERNATIVE HYPOTHESIS: The linkage is direct. The ETC is associated with enzymes that perform substrate-level phosphorylation.

EXPERIMENTAL SETUP:

Vesicle

ATP synthase

1. Produce vesicles from artificial membranes; add ATP synthase, an enzyme found in mitochondria.

Bacterio-rhodopsin

2. Add bacteriorhodopsin, a protein that acts as a light-activated proton pump.

Light
Light

3. Illuminate vesicle so that bacteriorhodopsin pumps protons out of vesicle, creating a proton gradient.

PREDICTION OF CHEMIOSMOTIC HYPOTHESIS: ATP will be produced within the vesicle.

PREDICTION OF ALTERNATIVE HYPOTHESIS: No ATP will be produced.

RESULTS:

ATP is produced within the vesicle, in the absence of the electron transport chain.

CONCLUSION: The linkage between electron transport and ATP synthesis is indirect; the movement of protons drives the synthesis of ATP.

FIGURE 9.17 Evidence for the Chemiosmotic Hypothesis.

Racker, E., and W. Stoeckenius. 1974. Reconstitution of purple membrane vesicles catalyzing light-driven proton uptake and adenosine triphosphate formation. *Journal of Biological Chemistry.* 249: 662–663.

✔**QUESTION** If bacteriorhodopsin were not available, what could the researchers have done with the ATP synthase vesicles to test their hypothesis?

Electron transport chains and ATP synthases are used by organisms throughout the tree of life. They are humming away in your cells now. Let's look in more detail at how they function.

The Proton-Motive Force Couples Electron Transport to ATP Synthesis As **FIGURE 9.18** shows, the structure of ATP synthase is now well understood. The ATP synthase "knob" component is called the F_1 unit; the membrane-bound, proton-transporting base component is the F_o unit. The F_1 and F_o units are connected by a shaft, as well as by a stator, which holds the two units in place.

The F_o unit serves as a rotor, whose turning is conveyed to the F_1 unit via the shaft. A flow of protons through the F_o unit causes the rotor and shaft to spin. By attaching long actin filaments to the shaft and examining them with a videomicroscope, researchers have been able to see the rotation, which can reach speeds of 350 revolutions per second. As the shaft spins within the F_1 unit, it is thought to change the conformation of the F_1 subunits in a way that catalyzes the phosphorylation of ADP to ATP.

Chemiosmosis is like the process of generating electricity in a hydroelectric dam (like the one pictured on page 154). The ETC is analogous to a series of gigantic pumps that force water up and behind the dam. The inner mitochondrial membrane functions as the dam, with ATP synthase spinning and generating electricity inside as water passes through—like a turbine. In a mitochondrion, protons are pumped instead of water. When protons move through ATP synthase, the protein spins and generates ATP.

It has been determined that the ETC transports enough protons to produce approximately three ATP for each NADH and two for each FADH$_2$, depending on the type of ATP synthase used. These yields, however, are not observed in cells, since the proton-motive force is also used to drive other processes, such as the import of phosphates into the mitochondrial matrix.

Unlike the turbines in a hydroelectric dam, however, ATP synthase can reverse its direction and hydrolyze ATP to build a proton gradient. If the proton gradient dissipates, the direction of the spin is reversed and ATP is hydrolyzed to pump protons from the matrix to the intermembrane space. Understanding how these reactions occur is currently the focus of intense research. ATP synthase makes most of the ATP that keeps you alive.

The Proton-Motive Force and Chemical Evolution How was energy first transformed into a usable form during the evolution of life? Since chemiosmosis is responsible for most of the ATP produced by cells throughout the tree of life, it is likely to have arisen early in evolution. But how could a complex electron transport chain evolve to produce the proton-motive force without a proton-motive force to supply the energy?

This apparent conundrum left many of the chemical evolution theorists perplexed until a key discovery was made deep in the ocean along the Mid-Atlantic Ridge—the Lost City hydrothermal vents (see Chapter 2). Researchers propose that the alkaline fluid (low proton concentration) released from these vents in the acidic oceans (high proton concentration) of early Earth may have provided such a gradient.

While there is still considerable debate concerning the role hydrothermal vents may have played in chemical evolution, their discovery has generated much excitement. By harnessing the natural electrochemical gradient deep in the early oceans, the proton-motive force of life may have evolved to mimic the environment of its origin.

Organisms Use a Diversity of Electron Acceptors

FIGURE 9.19 summarizes glucose oxidation and cellular respiration by tracing the fate of the carbon atoms and electrons in glucose. Notice that electrons from glucose are transferred to NADH and FADH$_2$, passed through the electron transport chain, and accepted by oxygen. Proton pumping during electron transport creates the proton-motive force that drives ATP synthesis.

The diagram also indicates the approximate yield of ATP from each component of the process. Recent research shows that about 29 ATP molecules are produced from each molecule of glucose.[2] Of these, 25 ATP molecules are produced by ATP synthase. The fundamental message here? The vast majority of the "payoff" from the oxidation of glucose occurs via oxidative phosphorylation.

Aerobic Versus Anaerobic Respiration During cellular respiration, oxygen is the electron acceptor used by all eukaryotes and

FIGURE 9.18 ATP Synthase Is a Motor. ATP synthase has two major components, designated F_o and F_1, connected by a shaft. The F_o unit spins as protons pass through. The shaft transmits the rotation to the F_1 unit, causing it to make ATP from ADP and P$_i$.

[2]Traditionally, biologists thought that 36 ATP would be synthesized for every molecule of glucose oxidized in eukaryotic cells. More recent work has shown that actual yield is only about 29 ATP [see M. Brand, Approximate yield of ATP from glucose, designed by Donald Nicholson. *Biochemistry and Molecular Biology Education* 31 (2003): 2–4]. Also, it's important to note that yield varies with conditions in the cell.

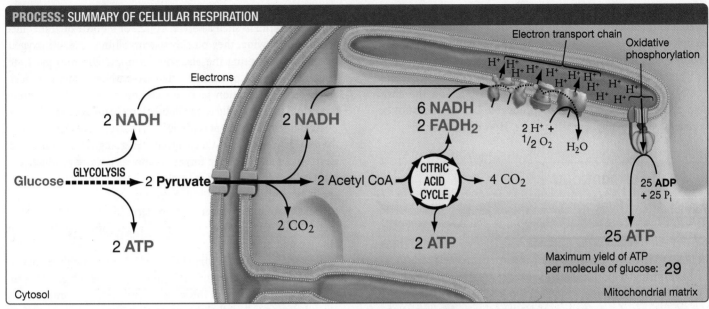

FIGURE 9.19 ATP Yield during Cellular Respiration. The actual yield of ATP per glucose (29 ATP) is lower than the theoretical calculation (38 ATP) because of energy required for the import of NADH from the cytoplasm and the use of the proton-motive force to actively transport P_i into the mitochondrial matrix.

a wide diversity of bacteria and archaea. Species that depend on oxygen as an electron acceptor for the ETC use **aerobic** respiration and are called aerobic organisms. (The Latin root *aero* means "air.")

It is important to recognize, though, that cellular respiration can occur without oxygen. Many thousands of bacterial and archaeal species rely on electron acceptors other than oxygen, and electron donors other than glucose. For example, nitrate (NO_3^-) and sulfate (SO_4^{2-}) are particularly common electron acceptors in species that live in oxygen-poor environments (see Chapter 29). In addition, many bacteria and archaea use H_2, H_2S, CH_4, or other inorganic compounds as electron donors—not glucose.

Cells that depend on electron acceptors other than oxygen are said to use **anaerobic** ("no air") respiration. Even though the starting and ending points of cellular respiration differ, aerobic and anaerobic cells still use electron transport chains to create a proton-motive force that drives the synthesis of ATP. In bacteria and archaea, the ETC and ATP synthase are located in the plasma membrane.

Aerobic Respiration Is Most Efficient Even though an array of compounds can serve as the final electron acceptor in cellular respiration, oxygen is the most efficient. Because oxygen holds electrons so tightly, the potential energy of electrons in a bond between an oxygen atom and a non-oxygenic atom, such as hydrogen, is low. As a result, there is a large difference between the potential energy of electrons in NADH and the potential energy of electrons bonded to an oxygen atom, such as found in water (see Figure 9.14). The large differential in potential energy means that the electron transport chain can generate a large proton-motive force.

Cells that do not use oxygen as an electron acceptor cannot generate such a large potential energy difference. As a result, they make less ATP from each glucose molecule than cells that use aerobic respiration. This finding is important: It means that anaerobic organisms tend to grow much more slowly than aerobic organisms. If cells that use anaerobic respiration compete with cells using aerobic respiration, those that use oxygen as an electron acceptor almost always grow faster and reproduce more.

What happens when oxygen or other electron acceptors get used up? When there is no terminal electron acceptor, the electrons in

check your understanding

If you understand that . . .

- As electrons from NADH and $FADH_2$ move through the electron transport chain, protons are pumped into the intermembrane space of mitochondria.
- The electrochemical gradient across the inner mitochondrial membrane drives protons through ATP synthase, resulting in the production of ATP from ADP.

✔ **You should be able to . . .**

Add paper squares labeled ETC and ATP synthase and a paper circle labeled $\frac{1}{2} O_2 \rightarrow H_2O$ to the model you made in Section 9.4. Explain the steps in electron transport and chemiosmosis using paper triangles to represent electron pairs and dimes to represent protons.

Answers are available in Appendix A.

each of the complexes of the electron transport chain have no place to go and the electron transport chain stops. Without an oxidized complex I, NADH remains reduced. The concentration of NAD^+ drops rapidly as cells continue to convert NAD^+ to NADH.

This situation is life threatening. When there is no longer any NAD^+ to drive glycolysis, pyruvate processing, and the citric acid cycle, then no ATP can be produced. If NAD^+ cannot be regenerated somehow, the cell will die. How do cells cope?

9.6 Fermentation

Fermentation is a metabolic pathway that regenerates NAD^+ by oxidizing stockpiles of NADH. The electrons removed from NADH are transferred to pyruvate, or a molecule derived from pyruvate, instead of an electron transport chain (**FIGURE 9.20**).

In respiring cells, fermentation serves as an emergency back-up that allows glycolysis to continue producing ATP even when the ETC is shut down. It allows the cell to survive and even grow in the absence of electron transport chains.

In many cases, the cell cannot use the molecule that is formed when pyruvate (or another electron acceptor) accepts electrons from NADH. This by-product may even be toxic and excreted from the cell as waste even though it has not been fully oxidized.

Many Different Fermentation Pathways Exist When you run up a long flight of stairs, your muscles begin metabolizing glucose so fast that the supply of oxygen is rapidly used up by their mitochondria. When oxygen is absent, the electron transport chains shut down and NADH cannot donate its electrons there. The pyruvate produced by glycolysis then begins to accept electrons from NADH, and fermentation takes place. This process, called **lactic acid fermentation,** regenerates NAD^+ by forming a product molecule called lactate: a deprotonated form of lactic acid (**FIGURE 9.21a**). Your body reacts by making you breathe faster and increasing your heart rate. By getting more oxygen to your muscle cells, the electron transport chain is revived.

FIGURE 9.21b illustrates a different fermentation pathway, **alcohol fermentation,** which occurs in the fungus *Saccharomyces*

cerevisiae—baker's and brewer's yeast. When yeast cells are placed in an environment such as bread dough or a bottle of grape juice and begin growing, they quickly use up all the available oxygen. Instead of depositing the electrons from NADH into pyruvate, yeast first convert pyruvate to the two-carbon compound acetaldehyde. This reaction gives off carbon dioxide, which causes bread to rise and produces the bubbles in champagne and beer.

Acetaldehyde then accepts electrons from NADH, forming the NAD^+ required to keep glycolysis going. The addition of electrons to acetaldehyde forms ethanol as a waste product. The yeast cells excrete ethanol as waste. In essence, the active ingredient in alcoholic beverages is like yeast urine.

Cells that employ other types of fermentation are used commercially in the production of soy sauce, tofu, yogurt, cheese, vinegar, and other products.

Bacteria and archaea that exist exclusively through fermentation are present in phenomenal numbers in the oxygen-free environment of your small intestine and in the first compartment of a cow's stomach, called the rumen. The rumen is a specialized digestive organ that contains over 10^{10} (10 billion) bacterial and archaeal cells per *milliliter* of fluid. The fermentations that occur in these cells produce an array of fatty acids. Cattle don't actually live off grass directly—they eat it to feed their bacteria and archaea and then use the fermentation by-products from these organisms as a source of energy.

Fermentation as an Alternative to Cellular Respiration Even though fermentation is a widespread type of metabolism, it is extremely inefficient compared with aerobic cellular respiration. Fermentation produces just 2 molecules of ATP for each molecule of glucose metabolized, while cellular respiration produces about 29—almost 15 times more energy per glucose molecule than fermentation. The reason for the disparity is that oxygen has much higher electronegativity than electron acceptors such as pyruvate and acetaldehyde. As a result, the potential energy drop between the start and end of fermentation is a tiny fraction of the potential energy change that occurs during cellular respiration.

Based on these observations, it is not surprising that organisms capable of both processes almost never use fermentation

FIGURE 9.20 Cellular Respiration and Fermentation Are Alternative Pathways for Producing Energy. When oxygen or another electron acceptor used by the ETC is present in a cell, the pyruvate produced by glycolysis enters the citric acid cycle and the electron transport system is active. But if no electron acceptor is available to keep the ETC running, the pyruvate undergoes reactions known as fermentation.

(a) Lactic acid fermentation occurs in humans.

No intermediate; pyruvate accepts electrons from NADH

2 Lactate

(b) Alcohol fermentation occurs in yeast.

2 Ethanol   2 Acetylaldehyde $2 \ CO_2$

FIGURE 9.21 Fermentation Regenerates NAD$^+$ So That Glycolysis Can Continue. These are just two examples of the many types of fermentation that occur among bacteria, archaea, and eukaryotes.

when an appropriate electron acceptor is available for cellular respiration. In organisms that usually use oxygen as an electron acceptor, fermentation is an alternative mode of ATP production when oxygen supplies temporarily run out.

Organisms that can switch between fermentation and cellular respiration that uses oxygen as an electron acceptor are called **facultative anaerobes.** The adjective facultative reflects the ability to use cellular respiration when oxygen is present and fermentation when it is absent (anaerobic). Many of your cells can function as facultative anaerobes to a certain extent; however, you cannot survive for long without oxygen. To make this point clear, try holding your breath—it should take only a minute for you to realize how important electron transport is to your cells.

check your understanding

If you understand that . . .

- Fermentation occurs in the absence of an electron acceptor at the end of an ETC. It consists of reactions that oxidize NADH to regenerate the NAD$^+$ required for glycolysis.

✔ **You should be able to . . .**

Explain why organisms that have an ETC as well as fermentation pathways seldom ferment pyruvate if an electron acceptor at the end of the ETC is readily available.

Answers are available in Appendix A.

CHAPTER 9 REVIEW

For media, go to MasteringBiology

If you understand . . .

9.1 An Overview of Cellular Respiration

- Cellular respiration is based on redox reactions that transfer electrons from a compound with high free energy, such as glucose, to a molecule with lower free energy, such as oxygen, through an electron transport chain.

- In eukaryotes, cellular respiration consists of four steps: glycolysis, pyruvate processing, the citric acid cycle, and electron transport coupled to oxidative phosphorylation.

- Glycolysis, pyruvate processing, and the citric acid cycle are central to the metabolism of most cells. Other catabolic pathways feed into them, and the intermediates of the central pathways are used in the synthesis of many key molecules.

✔ You should be able to explain why many different molecules—including lipids, amino acids, and CO_2—are radiolabeled when cells are fed glucose with ^{14}C radioactive carbons.

9.2 Glycolysis: Processing Glucose to Pyruvate

- The glycolytic pathway is a 10-step reaction sequence in which glucose is broken down into two molecules of pyruvate. It takes place in the cytosol, where ATP and NADH are produced.

- Glycolysis slows when ATP binds to phosphofructokinase.

✔ **QUANTITATIVE** You should be able to draw a graph predicting how the rate of ATP production in glycolysis changes as a function of ATP concentration. (Write "ATP concentration" on the x-axis and "ATP production" on the y-axis.)

9.3 Processing Pyruvate to Acetyl CoA

- During pyruvate processing, a series of reactions convert pyruvate to acetyl CoA. NADH is produced and CO_2 is released.

- The pyruvate dehydrogenase complex is inhibited when it is phosphorylated by ATP. It speeds up in the presence of substrates like NAD and ADP.

✔ You should be able to explain why it is not surprising that pyruvate dehydrogenase consists of a large, multi-enzyme complex.

9.4 The Citric Acid Cycle: Oxidizing Acetyl CoA to CO_2

- The citric acid cycle is an eight-step reaction cycle that begins with acetyl CoA. $FADH_2$, NADH, and GTP or ATP are produced; CO_2 is released. By the end of the citric acid cycle, glucose is completely oxidized to CO_2.

- Certain enzymes in the citric acid cycle are inhibited when NADH or ATP binds to them.

✔ You should be able to describe what would happen to NADH levels in a cell in the first few seconds after a drug has poisoned the enzyme that combines acetyl CoA and oxaloacetate to form citrate.

9.5 Electron Transport and Chemiosmosis: Building a Proton Gradient to Produce ATP

- NADH and $FADH_2$ donate electrons to an electron transport chain that resides in the inner membrane of mitochondria and the plasma membrane of many bacteria. The series of redox reactions in these chains gradually steps the electrons down in potential energy until they are transferred to a final electron acceptor (often O_2).

- The energy released from redox reactions in the electron transport chain is used to move protons across the inner mitochondrial membrane, creating an electrochemical gradient. ATP synthase uses the energy stored in this gradient to produce ATP via chemiosmosis—a process called oxidative phosphorylation.

✔ You should be able to predict the effect of a drug that inhibits ATP synthase on the pH in the mitochondrial matrix.

9.6 Fermentation

- In many eukaryotes and bacteria, fermentation occurs when cellular respiration slows down or stops due to an insufficient amount of the final electron acceptor. If the final electron acceptor is absent, then the electron transport chain would no longer oxidize NADH to NAD^+ and ATP could no longer be produced by glycolysis, the citric acid cycle, or oxidative phosphorylation.

- Fermentation pathways regenerate NAD^+, so glycolysis can continue to make ATP and keep the cell alive. This happens when an organic molecule such as pyruvate accepts electrons from NADH.

- Depending on the molecule that acts as an electron acceptor, fermentation pathways produce lactate, ethanol, or other reduced organic compounds as a by-product.

✔ You should be able to explain why you would expect organisms that produce ATP only via fermentation to grow much more slowly than organisms that produce ATP via cellular respiration.

(MB) MasteringBiology

1. MasteringBiology Assignments

Tutorials and Activities Build a Chemical Cycling System; Cellular Respiration (1 of 5): Inputs and Outputs; Cellular Respiration (2 of 5): Glycolysis; Cellular Respiration (3 of 5): Acetyl CoA Formation and the Citric Acid Cycle; Cellular Respiration (4 of 5): Oxidative Phosphorylation; Cellular Respiration (5 of 5): Summary; Citric Acid Cycle; Electron Transport; Fermentation; Glucose Metabolism; Glycolysis; Overview of Cellular Respiration; Pathways for Pyruvate

Questions Reading Quizzes, Blue-Thread Questions, Test Bank

2. eText Read your book online, search, take notes, highlight text, and more.

3. The Study Area Practice Test, Cumulative Test, BioFlix® 3-D animations, Videos, Activities, Audio Glossary, Word Study Tools, Art

You should be able to . . .

✔ TEST YOUR KNOWLEDGE

Answers are available in Appendix A

1. Make a flowchart indicating the relationships among the four steps of cellular respiration. Which steps are responsible for glucose oxidation? Which produce the most ATP?

2. Where does the citric acid cycle occur in eukaryotes?
   a. in the cytosol
   b. in the matrix of mitochondria
   c. in the inner membrane of mitochondria
   d. in the intermembrane space of mitochondria

3. What does the chemiosmotic hypothesis claim?
   a. Substrate-level phosphorylation occurs in the electron transport chain.
   b. Substrate-level phosphorylation occurs in glycolysis and the citric acid cycle.
   c. The electron transport chain is located in the inner membrane of mitochondria.
   d. Electron transport chains generate ATP indirectly, by the creation of a proton-motive force.

4. After glucose is fully oxidized by glycolysis, pyruvate processing, and the citric acid cycle, where is most of the energy stored?

5. What is the function of the reactions in a fermentation pathway?
   a. to generate NADH from NAD^+, so electrons can be donated to the electron transport chain
   b. to synthesize pyruvate from lactate
   c. to generate NAD^+ from NADH, so glycolysis can continue
   d. to synthesize electron acceptors, so that cellular respiration can continue

6. Which of the following would cause cells to switch from cellular respiration to fermentation?
   a. The final electron acceptor in the ETC is not available.
   b. The proton-motive force runs down.
   c. NADH and $FADH_2$ supplies are low.
   d. Pyruvate is not available.

TEST YOUR UNDERSTANDING · Answers are available in Appendix A

7. Describe the relationship between carbohydrate metabolism, the catabolism of proteins and fats, and anabolic pathways.

8. Compare and contrast substrate-level phosphorylation and oxidative phosphorylation.

9. Why does aerobic respiration produce much more ATP than anaerobic respiration?

10. If you were to expose cells that are undergoing cellular respiration to a radioactive oxygen isotope in the form of O_2, which of the following molecules would you expect to be radiolabeled?
    a. pyruvate
    b. water
    c. NADH
    d. CO_2

11. In step 3 of the citric acid cycle, the enzyme isocitrate dehydrogenase is regulated by NADH. Compare and contrast the regulation of this enzyme with what you have learned about phosphofructokinase in glycolysis.

12. Explain the relationship between electron transport and oxidative phosphorylation. What does ATP synthase look like, and how does it work?

TEST YOUR PROBLEM-SOLVING SKILLS · Answers are available in Appendix A

13. Cyanide ($C \equiv N^-$) blocks complex IV of the electron transport chain. Suggest a hypothesis for what happens to the ETC when complex IV stops working. Your hypothesis should explain why cyanide poisoning in humans is fatal.

14. The presence of many sac-like cristae results in a large amount of membrane inside mitochondria. Suppose that some mitochondria had few cristae. How would their output of ATP compare with that of mitochondria with many cristae? Justify your answer.

15. QUANTITATIVE Early estimates suggested that the oxidation of glucose via aerobic respiration would produce 38 ATP. Based on what you know of the theoretical yields of ATP from each step, show how this total was determined. Why do biologists now think this amount of ATP/glucose is not achieved in cells?

16. Suppose a drug were added to mitochondria that allowed protons to freely pass through the inner membrane. Which of the following mitochondrial activities would most likely be inhibited?
    a. the citric acid cycle
    b. oxidative phosphorylation
    c. substrate-level phosphorylation
    d. the electron transport chain

10 Photosynthesis

In this chapter you will learn how

Photosynthesis links life to the power of the Sun

by previewing

Conversion of light energy into chemical energy   **10.1**

by examining

How photosynthetic pigments capture light energy   **10.2**

then looking closer at

Energy flow and ATP production   **10.3**

→ Photosystem II

→ Photosystem I

and exploring

CO$_2$ fixation and reduction to form sugars   **10.4**

→ The Calvin cycle

A close-up of moss cells filled with chloroplasts, where photosynthesis converts the energy in sunlight to chemical energy in the bonds of sugar. The sugar produced by photosynthetic organisms fuels cellular respiration and growth. Photosynthetic organisms, in turn, are consumed by other organisms, including you. Directly or indirectly, most organisms on Earth get their energy from photosynthesis.

This chapter is part of the Big Picture. See how on pages 198–199.

Some 3 billion years ago, a novel combination of light-absorbing molecules and enzymes gave a bacterial cell the capacity to convert light energy into chemical energy in the C—C and C—H bonds of sugar. The origin of **photosynthesis**—the use of sunlight to manufacture carbohydrate—ranks as one of the great events in the history of life.

The vast majority of organisms alive today rely on photosynthesis, either directly or indirectly, to stay alive. Maples, mosses, and other photosynthetic organisms are termed **autotrophs** (literally, "self-feeders") because they make all their own food from ions and simple molecules. Humans, houseflies, and other non-photosynthetic organisms are called **heterotrophs** ("different-feeders") because they have to obtain the sugars and many of the other macromolecules they need from other organisms.

✔ When you see this checkmark, stop and test yourself. Answers are available in Appendix A.

Because there could be no heterotrophs without autotrophs, photosynthesis is fundamental to almost all life. Glycolysis may qualify as the most ancient set of energy-related chemical reactions from an evolutionary viewpoint, but ecologically—meaning, in terms of how organisms interact with one another—photosynthesis is easily the most important.

How does photosynthesis work? Let's begin with an overview and then delve into a step-by-step analysis of some of the most remarkable chemistry on Earth.

10.1 Photosynthesis Harnesses Sunlight to Make Carbohydrate

Research on photosynthesis began early in the history of biological science. In the 1770s, Joseph Priestley performed a series of experiments showing that the green parts of plants would "restore air" that had been consumed by animals or fire. This work led to the discovery of oxygen (O_2) and the finding that plants produce it in the presence of sunlight, carbon dioxide (CO_2), and water (H_2O).

By the 1840s, enough was known about the process for biologists to propose that photosynthesis allows plants to convert the electromagnetic energy of sunlight into chemical energy in the C–C and C–H bonds of carbohydrates. When glucose is the carbohydrate that is eventually produced, the overall reaction—the sum of many independent reactions—can be simplified and written as

$$6\,CO_2 + 6\,H_2O + \text{light energy} \longrightarrow C_6H_{12}O_6 + 6\,O_2$$
$$\text{glucose}$$

Now read the reaction again, and note the contrast with cellular respiration. Photosynthesis is an endergonic suite of redox reactions that produce sugars from carbon dioxide and light energy. Cellular respiration is an exergonic suite of redox reactions that produces carbon dioxide and ATP from sugars.

FIGURE 10.1 provides an incomplete electron-sharing diagram for the reaction shown above. ✔ If you understand the fundamental principles of reduction–oxidation (see Chapter 8), you should be able to complete Figure 10.1 (following the instructions in the caption exercise) and then use the data from the figure to explain why the reaction is endergonic.

So how does photosynthesis produce O_2 and glucose? Early investigators assumed that CO_2 and H_2O react directly to produce the CH_2O found in carbohydrates and release O_2 as a by-product. This idea, however, turned out to be incorrect. Instead, CO_2 and H_2O participate in entirely different reactions, and the oxygen atoms in O_2 come from water. How was this discovered?

Photosynthesis: Two Linked Sets of Reactions

Starting in the 1930s, two independent lines of research on photosynthesis converged, leading to a major advance in biologists' understanding of how oxygen gas and carbohydrates are produced.

The first research program, led by Cornelius van Niel, focused on photosynthesis in organisms called purple sulfur bacteria. Van Niel and his group found that these cells are autotrophs that manufacture their own carbohydrates from CO_2, sunlight, and hydrogen sulfide (H_2S).

Van Niel also showed that these cells did not produce oxygen as a by-product of photosynthesis. Instead, elemental sulfur (S) accumulated in their medium. In these organisms, the overall reaction for photosynthesis was

$$CO_2 + 2\,H_2S + \text{light energy} \longrightarrow (CH_2O)_n + H_2O + 2\,S$$

Van Niel's work was crucial for two reasons:

1. It showed that H_2S, the equivalent of H_2O in the plant reactions, and CO_2 do *not* combine directly during photosynthesis.

2. It showed that the oxygen atoms in CO_2 are *not* released as oxygen gas (O_2). The purple sulfur bacteria produced no oxygen, even though carbon dioxide participated in the reaction—just as it did in plants.

Based on these findings, biologists hypothesized that the oxygen atoms released during plant photosynthesis must come from H_2O. The hypothesis was confirmed when heavy isotopes

FIGURE 10.1 Electron Transfer during the Reduction of Glucose.

✔**EXERCISE** Fill in the electron positions for each bond in the reaction products, and complete the labels explaining which product is reduced and which is oxidized.

of oxygen—^{18}O in contrast to the normal isotope, ^{16}O—became available to researchers. They observed the ^{18}O in oxygen gas only when algae or plants were exposed to ^{18}O-labeled H_2O, not the ^{18}O-labeled CO_2.

In addition, the reactions responsible for producing oxygen gas occurred only in the presence of sunlight, but did not require the presence of CO_2. These data suggested that there were two distinct sets of reactions: one that uses light to produce O_2 from H_2O and one that converts CO_2 into sugars.

A second major line of research supported the idea of two sets of reactions. Between 1945 and 1955, a team led by Melvin Calvin began introducing radioactively labeled carbon dioxide ($^{14}CO_2$) to algae and identifying the molecules that subsequently became labeled with the radioisotope. These experiments allowed researchers to identify the sequence of reactions involved in reducing CO_2 to sugars.

Because Calvin played an important role in this research, the reactions that reduce carbon dioxide and produce sugar came to be known as the **Calvin cycle.** Later research showed that the Calvin cycle can function only if the light-capturing reactions are occurring.

To summarize: Early research showed that photosynthesis consists of two linked sets of reactions. One set is triggered by light; the other set—the Calvin cycle—requires the products of the light-capturing reactions. The light-capturing reactions produce oxygen from water; the Calvin cycle produces sugar from carbon dioxide.

The two reactions are linked by electrons that are released when water is split to form oxygen gas. During the light-capturing reactions, these electrons are promoted to a high-energy state by light and then transferred through a series of redox reactions to a phosphorylated version of NAD$^+$, called **NADP$^+$ (nicotinamide adenine dinucleotide phosphate).** This reaction forms **NADPH,** which functions as a reducing agent similar to the NADH produced in cellular respiration. Some of the energy released from these redox reactions is also used to produce ATP (**FIGURE 10.2**).

During the Calvin cycle, the electrons in NADPH and the potential energy in ATP are used to reduce CO_2 to carbohydrate. The resulting sugars are used in cellular respiration to produce ATP for the cell. Plants oxidize sugars in their mitochondria and consume O_2 in the process, just as animals and other eukaryotes do.

Where does all this activity take place?

Photosynthesis Occurs in Chloroplasts

Once experiments had established that photosynthesis takes place only in the green portions of plants, biologists focused on the bright green organelles called **chloroplasts** ("green-formed elements"). One leaf cell typically contains 40 to 50 chloroplasts, and a square millimeter of leaf averages about 500,000 (**FIGURE 10.3**).

When membranes derived from chloroplasts were found to release oxygen after exposure to sunlight, the hypothesis that chloroplasts are the site of photosynthesis became widely accepted.

FIGURE 10.2 Photosynthesis Has Two Linked Components. In the light-capturing reactions of photosynthesis, light energy is transformed to chemical energy in the form of ATP and NADPH. During the Calvin cycle, the ATP and NADPH produced in the light-capturing reactions are used to reduce carbon dioxide to carbohydrate.

As Figure 10.3 shows, a chloroplast is enclosed by an outer membrane and an inner membrane (see Chapter 7). The interior is dominated by flattened, sac-like structures called **thylakoids,** which often occur in interconnected stacks called **grana** (singular: **granum**). The space inside a thylakoid is its **lumen.** (Recall that lumen is a general term for the interior of any sac-like structure. Your stomach and intestines have a lumen.) The fluid-filled space between the thylakoids and the inner membrane is the **stroma.**

When researchers analyzed the chemical composition of thylakoid membranes, they found huge quantities of pigments. **Pigments** are molecules that absorb only certain wavelengths of light—other wavelengths are either reflected or transmitted (pass through). Pigments have colors because we see the wavelengths that they do *not* absorb.

The most abundant pigment in the thylakoid membranes turned out to be chlorophyll ("green-leaf"), which reflects or transmits green light. As a result, chlorophyll is responsible for the green color of plants, some algae, and many photosynthetic bacteria.

Before plunging into the details of how photosynthesis occurs, take a moment to consider just how astonishing the process is. Chemists have synthesized an amazing diversity of compounds from relatively simple starting materials, but their achievements pale in comparison to a cell that can synthesize sugar from just carbon dioxide, water, and sunlight. If photosynthesis is not *the* most sophisticated chemistry on Earth, it is certainly a contender.

In plants, cells that photosynthesize typically have 40–50 chloroplasts

10 µm

Chloroplast

— Outer membrane
— Inner membrane

0.5 µm

Thylakoids (flattened sacs)
Granum (stack of thylakoids)
Stroma (liquid matrix)

FIGURE 10.3 Photosynthesis Takes Place in Chloroplasts.

10.2 How Do Pigments Capture Light Energy?

The light-capturing reactions of photosynthesis begin with the simple act of sunlight striking chlorophyll. To understand the consequences of this event, it's helpful to review the nature of light.

Light is a type of electromagnetic radiation, a form of energy. Photosynthesis converts electromagnetic energy in the form of

Wavelengths (nm)

10^{-5}	10^{-3}	10^{-1}	10^1	10^3	10^5	10^7	10^9	10^{11}	10^{13}
Gamma rays	X-rays	Ultra-violet		Infrared		Micro-waves		Radio waves	

Shorter wavelength

Longer wavelength

Visible light

400   500   600   710 nm

Higher energy

Lower energy

FIGURE 10.4 The Electromagnetic Spectrum. Electromagnetic energy radiates through space in the form of waves. Humans can see radiation at wavelengths between about 400 nanometers (nm) to 710 nm. The shorter the wavelength of electromagnetic radiation, the higher its energy.

sunlight into chemical energy in the C—C and C—H bonds of sugar.

Physicists describe light's behavior as both wavelike and particle-like. Like water waves or airwaves, electromagnetic radiation is characterized by its **wavelength**—the distance between two successive wave crests (or wave troughs). The wavelength determines the type of electromagnetic radiation.

FIGURE 10.4 illustrates the range of wavelengths of electromagnetic radiation—the **electromagnetic spectrum.** The electromagnetic radiation that humans can see, the **visible light,** ranges in wavelength from about 400 to about 710 nanometers (nm, or 10^{-9} m). Shorter wavelengths of electromagnetic radiation contain more energy than longer wavelengths do. Thus, there is more energy in blue light than in red light.

To emphasize the particle-like nature of light, physicists point out that it exists in discrete packets called **photons.** Each photon of light has a characteristic wavelength and energy level. Pigment molecules absorb the energy of some of these photons. How?

Photosynthetic Pigments Absorb Light

When a photon strikes an object, the photon may be absorbed, transmitted, or reflected. A pigment molecule absorbs photons of particular wavelengths. Sunlight includes white light, which consists of all wavelengths in the visible portion of the electromagnetic spectrum at once.

If a pigment absorbs all the visible wavelengths, the pigment appears black because no visible wavelength of light is reflected back to your eye. If a pigment absorbs many or most of the wavelengths in the blue and green parts of the spectrum but transmits or reflects longer wavelengths, it appears red.

What wavelengths do various plant pigments absorb? In one approach to answering this question, researchers grind up leaves

Fluorescence is typical of isolated pigments, resonance energy transfer occurs in antenna complex pigments, and redox occurs in reaction center pigments.

Now the question is, what happens to the high-energy electrons that are transferred to the electron acceptor in the reaction center? Specifically, how are they used to manufacture sugar?

10.3 The Discovery of Photosystems I and II

During the 1950s, the fate of the high-energy electrons in photosystems was the central issue facing biologists interested in photosynthesis. A key breakthrough began with simple experiments by Robert Emerson on how green algae responded to various wavelengths of light. The algal cells being studied responded to wavelengths in the red and far-red regions of the visible spectrum.

Emerson found that if the algal cells were illuminated with either red or far-red wavelengths of light, the photosynthetic response was moderate. But if cells were exposed to a combination of both wavelengths, the rate of photosynthesis increased more than the sum of the rates produced by each wavelength independently. This phenomenon was called the enhancement effect, and is not limited to algal cells. In follow-up work by other researchers, it was also observed in isolated chloroplasts from plants (**FIGURE 10.11**). Why the enhancement effect occurred was a complete mystery at the time.

A solution to this puzzle was proposed by Robin Hill and Faye Bendall, who hypothesized that this enhancement effect resulted from two distinct types of reaction centers, each absorbing different wavelengths of light. According to the two-photosystem hypothesis, the enhancement effect occurs because photosynthesis is much more efficient when both photosystems operate together.

Subsequent work has shown that the two-photosystem hypothesis is correct for cyanobacteria ("blue-green bacteria") and the chloroplasts of eukaryotes, such as algae and plants. These two photosystems differ in structure and function, but work together in the light-capturing reactions.

To figure out how the two photosystems work, investigators focused on species of photosynthetic bacteria that possess one or the other of the two photosystems, but not both. Once each type of photosystem was understood in isolation, researchers explored how they work in combination. Let's do the same—first let's analyze **photosystem II,** then **photosystem I** (so named because it was discovered first), and then how the two interact.

How Does Photosystem II Work?

To study photosystem II, researchers focused on purple photosynthetic bacteria, including the purple sulfur bacteria that were studied by van Niel (see Section 10.1). These cells have a single photosystem that has many of the same components observed in photosystem II of cyanobacteria and the chloroplasts of algae and plants. (For simplicity, the eukaryotic chloroplast will serve as the model system for the remainder of the chapter.)

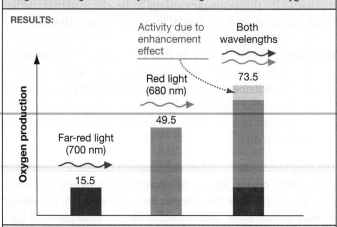

RESEARCH

QUESTION: Red and far-red light each stimulate a moderate rate of photosynthesis. How does a combination of both wavelengths affect the rate of photosynthesis?

HYPOTHESIS: When red and far-red light are combined, the rate of photosynthesis will be the sum of the single wavelength rates.

NULL HYPOTHESIS: When red and far-red light are combined, the rate of photosynthesis will be no more than the highest single wavelength rate.

EXPERIMENTAL SETUP:

Far-red light (713 nm) Red light (653 nm) Both wavelengths

O_2? O_2? O_2?

1. Expose algal cells to far-red light and then red light. Record oxygen produced as a measure of rate of photosynthesis.

2. Expose same cells to a combination of both lights.

PREDICTION: When the two wavelengths are combined, the amount of oxygen produced will be the sum of the single wavelength tests.

PREDICTION OF NULL HYPOTHESIS: When the two wavelengths are combined, the amount of oxygen produced will be no more than the single wavelength test that yielded the highest amount of oxygen.

RESULTS:

Activity due to enhancement effect

Both wavelengths

Red light (680 nm)

73.5

49.5

Far-red light (700 nm)

15.5

Oxygen production

CONCLUSION: Neither hypothesis is correct. The combination of both wavelengths yielded more oxygen than the sum of the single tests. A new hypothesis is required to explain this enhancement effect.

FIGURE 10.11 The "Enhancement Effect" of Two Different Wavelengths in Isolated Chloroplasts.

SOURCE: Govindjee, R., Govindjee, and G. Hoch. 1964. Emerson enhancement effect in chloroplast reactions. *Plant Physiology* 39: 10–14.

✔**QUESTION** Was it important for the researchers to keep the density of chloroplasts fairly constant in each treatment? Explain why or why not.

FIGURE 10.12 Photosystem II Feeds High-Energy Electrons to an Electron Transport Chain. When an excited electron leaves the chlorophyll in the reaction center of photosystem II, the electron is accepted by pheophytin, transferred to plastoquinone (PQ), and then stepped down in energy along an electron transport chain.

Converting Light Energy into Chemical Energy In photosystem II, the action begins when the antenna complex transmits resonance energy to the reaction center, where the electron acceptor pheophytin comes into play (**FIGURE 10.12**). Structurally,

pheophytin is identical to chlorophyll except that pheophytin lacks a magnesium atom in its head region. Functionally, the two molecules are extremely different.

Instead of acting as a pigment that energizes an electron when it absorbs a photon, pheophytin accepts high-energy electrons from the excited reaction center chlorophylls. The reduction of pheophytin (and the accompanying oxidation of the reaction center chlorophyll pigment) is a key step in the transformation of light energy into chemical energy.

Electrons that reduce pheophytin are passed through additional carriers to an electron transport chain (ETC) in the thylakoid membrane. In both structure and function, this ETC is similar to components in the mitochondrial ETC (see Chapter 9).

- Structurally, the ETC associated with photosystem II and the ETC in the mitochondrion both contain quinones and cytochromes.

- Functionally, the redox reactions that occur in both ETCs result in protons being actively transported from one side of an internal membrane to the other. The resulting proton electrochemical gradient forms a proton-motive force that drives ATP production via ATP synthase. Photosystem II triggers chemiosmosis and ATP synthesis in the chloroplast.

FIGURE 10.13 explains how the electron transport chain associated with photosystem II works in more detail. Start by focusing on the molecule called **plastoquinone (PQ)**—a quinone similar to ubiquinone in the ETC of cellular respiration. Recall that

FIGURE 10.13 Electron Transport between Photosystem II and the Cytochrome Complex. Plastoquinone (PQ) carries electrons from photosystem II along with protons from the stroma. The cytochrome complex oxidizes plastoquinone, releasing the protons in the thylakoid lumen that drive ATP synthesis.

quinones are small hydrophobic molecules that can transport electrons between molecules (see Chapter 9). Because plastoquinone is lipid soluble and not anchored to the thylakoid membrane, it is free to move from one side of the thylakoid membrane to the other.

When plastoquinone receives electrons from photosystem II, it carries them across the membrane to the lumen side of the thylakoid and delivers them to more electronegative molecules in the cytochrome complex. In this way, plastoquinone shuttles electrons from photosystem II to the cytochrome complex much like ubiquinone shuttled electrons between complexes I or II and complex III in mitochondria. The potential energy released by these reactions allows protons to be picked up from the stroma and dropped off in the lumen side of the thylakoid membrane.

The protons transported by plastoquinone result in a large concentration of protons in the thylakoid lumen. When photosystem II is active, the pH of the thylakoid interior reaches 5 while the pH of the stroma hovers around 8. Because the pH scale is logarithmic, the difference of 3 units means that the concentration of H^+ is $10 \times 10 \times 10 = 1000$ times higher in the lumen than in the stroma. In addition, the stroma becomes negatively charged relative to the thylakoid lumen.

The net effect of electron transport, then, is to set up a large proton electrochemical gradient, resulting in a proton-motive force that drives H^+ out of the thylakoid lumen and into the stroma. Based on what you know of cellular respiration, it should come as no surprise that this proton-motive force drives the production of ATP.

Specifically, proton flow down the electrochemical gradient is an exergonic process that drives the endergonic synthesis of ATP from ADP and P_i. The stream of protons through ATP synthase causes conformational changes that drive the phosphorylation of ADP. Since this process is initiated by the energy harvested from light, it is called **photophosphorylation.**

Photophosphorylation is similar to the oxidative phosphorylation that occurs in plant and animal mitochondria. Both depend on chemiosmosis.

The photosystem II story is not yet complete, however. The electrons from PQ are passed through the cytochrome complex, but what about the oxidized photosystem II reaction center? To continue this ETC, the electron removed from the reaction center needs to be replaced. Where do the electrons required by photosystem II come from?

Photosystem II Obtains Electrons by Oxidizing Water Think back to the overall reaction for photosynthesis:

$$6\,CO_2 + 6\,H_2O + \text{light energy} \longrightarrow C_6H_{12}O_6 + 6\,O_2$$

In the presence of sunlight, carbon dioxide and water are used to produce carbohydrate and oxygen gas.

Now recall that experiments with heavy isotopes of oxygen showed that the oxygen atoms in O_2 come from water, not from carbon dioxide. For this to happen, water must be oxidized. The oxygen-generating reaction can be written as

$$2\,H_2O \longrightarrow 4\,H^+ + 4\,e^- + O_2$$

This reaction is referred to as "splitting" water. It supplies electrons for photosystem II and is catalyzed by enzymes that are physically integrated into the photosystem II complex. Since oxygen is very electronegative, this reaction is highly endergonic. What supplies the energy necessary to oxidize water?

As it turns out, the light energy harvested by photosystem II is responsible for splitting water. When excited electrons leave photosystem II and enter the ETC, the photosystem becomes so electronegative that enzymes can remove electrons from water, leaving protons and oxygen.

Among all life-forms, photosystem II is the only protein complex that can catalyze the splitting of water molecules. The photosystem II of cyanobacteria and eukaryotic chloroplasts perform **oxygenic** ("oxygen-producing") photosynthesis, because they generate oxygen as a by-product of the process. Other organisms that have only a single photosystem do not oxidize water, and thus do not produce O_2 gas. Instead, these organisms use different electron donors, such as H_2S in the purple sulfur bacteria, to perform **anoxygenic** ("no oxygen-producing") photosynthesis.

✔ If you understand photosystem II, you should be able to make an energy flowchart that includes the antenna complex, ATP synthase, pheophytin, light, the proton gradient, an ETC, and a reaction center and then add notes explaining where the enzyme complex that splits water fits in.

What happens next in green algae and land plants? The answer lies in photosystem I. Let's take a closer look.

How Does Photosystem I Work?

Recall that researchers dissected photosystem II by studying similar, but simpler, photosystems in the purple photosynthetic bacteria. To understand the structure and function of photosystem I, they turned to heliobacteria ("sun-bacteria").

Like the purple bacteria, heliobacteria have only one photosystem that uses the energy in sunlight to promote electrons to a high-energy state. But instead of being passed to an electron transport chain that pumps protons across a membrane, the high-energy electrons in heliobacteria are used to reduce NAD^+. When NAD^+ gains two electrons and a proton, NADH is produced.

In the cyanobacteria and eukaryotic chloroplasts, a similar set of light-capturing reactions reduces a phosphorylated version of NAD^+, symbolized $NADP^+$, to yield NADPH. Both NADH and NADPH function as electron carriers.

FIGURE 10.14 explains how photosystem I works in chloroplasts—put your finger on the "2 photons" arrows and trace the steps that follow.

1. Pigments in the antenna complex absorb photons and pass the energy to the photosystem I reaction center.

2. Electrons are excited in reaction center chlorophyll molecules.

3. The reaction center pigments are oxidized, and the high-energy electrons are passed through a series of carriers inside the photosystem, then to a molecule called **ferredoxin,** and then to the enzyme called $NADP^+$ reductase.

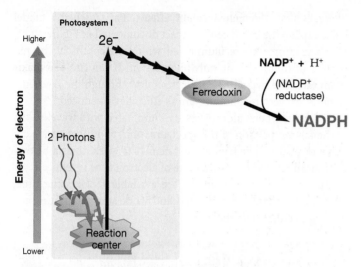

FIGURE 10.14 Photosystem I Produces NADPH. When excited electrons leave the chlorophyll molecule in the reaction center of photosystem I, they pass through a series of iron- and sulfur-containing proteins until they are accepted by ferredoxin. In an enzyme-catalyzed reaction, the reduced form of ferredoxin reacts with $NADP^+$ to produce NADPH.

4. $NADP^+$ reductase transfers two electrons and a proton to $NADP^+$. This reaction forms NADPH.

Photosystem I and $NADP^+$ reductase are anchored in the thylakoid membrane; ferredoxin is in the stroma, but it is closely associated with the thylakoid membrane.

To summarize: Electrons from photosystem I are used to produce NADPH, which is a reducing agent similar in function to the NADH and $FADH_2$ produced by the citric acid cycle. Electrons from photosystem II, in contrast, are used to produce a proton-motive force that drives the synthesis of ATP.

In combination, then, photosystems II and I produce chemical energy stored in ATP and NADPH. But there are still gaps in the flow of electrons through these two photosystems. Where do the electrons from photosystem II end up? How does the oxidized reaction center of photosystem I obtain electrons so NADPH will continue to be made?

The Z Scheme: Photosystems II and I Work Together

FIGURE 10.15 illustrates the **Z-scheme** model for how photosystems II and I interact. The name was inspired by the changes occurring in electron potential energy as plotted on a vertical axis, which takes on the shape of a Z that has fallen over.

To drive home how energy flows through the light-capturing reactions, trace the route of electrons through Figure 10.15 with your finger. Start on the lower left. The process starts when photons excite electrons in the chlorophyll molecules of photosystem II's antenna complex. When the energy in the excited electrons is transferred to the reaction center, a special pair of chlorophyll molecules, called P680, passes excited electrons to pheophytin. These are the same reaction center pigments described previously, and the name represents the wavelength of photons absorbed (680 nm).

When pheophytin is reduced, it transfers the high-energy electron to an electron transport chain. There the electron is gradually stepped down in potential energy through redox reactions among a series of quinones and cytochromes. Using the energy released by the redox reactions, plastoquinone (PQ) carries protons across the thylakoid membrane, from the stroma to the lumen. ATP synthase uses the resulting proton-motive force to phosphorylate ADP, creating ATP.

When electrons reach the end of the cytochrome complex, they are passed to a small diffusible protein called **plastocyanin**

FIGURE 10.15 The Z-Scheme Model Links Photosystems II and I. The Z scheme proposes that electrons from water are first energized by photosystem II to generate ATP and then energized again by photosystem I to reduce $NADP^+$ to NADPH.

(symbolized as PC in Figure 10.15). The reduced plastocyanin diffuses through the lumen of the thylakoid, and donates the electron to an oxidized reaction center pigment in photosystem I.

Stop tracing for a moment, and consider the following:

- Plastocyanin is critical—it forms a physical link between photosystem II and photosystem I.

- A single plastocyanin molecule can shuttle over 1000 electrons per second between the cytochrome complex and photosystem I.

- The flow of electrons between photosystems, by means of plastocyanin, is important because it replaces electrons that are carried away from the pair of pigments in the photosystem I reaction center. This pair of specialized chlorophyll molecules is called P700 (absorbs 700-nm photons).

Now keep going. The electrons that flow from photosystem II to P700, via plastocyanin, are eventually transferred to the protein ferredoxin, which passes electrons to an enzyme that catalyzes the reduction of NADP$^+$ to NADPH.

Finally, direct your attention back to the lower-left portion of the figure. Note that the electrons that initially left photosystem II are replaced by electrons that are stripped away from water, producing oxygen gas as a by-product.

✔ If you understand the Z-scheme model, you should be able to describe (1) the role of plastocyanin in linking the two photosystems, and (2) the point where the electrons that flow through the system have their highest potential energy.

Understanding the Enhancement Effect The Z-scheme model helps explain the enhancement effect documented in Figure 10.11. When chloroplasts are illuminated with wavelengths in the red portion of the spectrum, only photosystem II can run at a maximum rate. The overall rate of electron flow through the Z scheme is moderate because photosystem I's efficiency is reduced.

Similarly, when chloroplasts are illuminated with wavelengths in the far-red portion of the spectrum, only photosystem I is capable of peak efficiency; photosystem II is working at a below-maximum rate, so the overall rate of electron flow is reduced.

But when both wavelengths are available at the same time, both photosystems are activated and work at a maximum rate, leading to enhanced efficiency.

Noncyclic Electron Flow between Water and NADP$^+$ The complete path that electrons follow from photosystem II to photosystem I and how it is oriented in the thylakoid membrane is shown in **FIGURE 10.16**. Note that electrons pass from water to NADP$^+$ through a chain of redox reactions in a linear fashion, referred to as **noncyclic electron flow.**

Compare the movement of electrons and protons in Figure 10.16 with what you have learned about electron transport chains in mitochondria (see Figure 9.15). In both these organelles, the energy released from redox reactions is used to build a proton gradient for ATP production. At the end of the chains, electrons are donated to terminal electron acceptors.

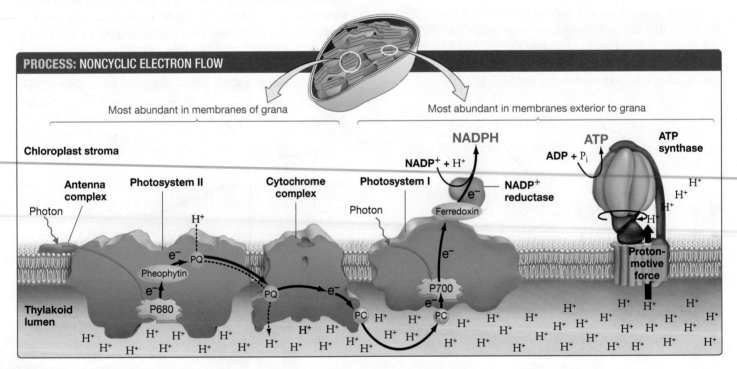

PROCESS: NONCYCLIC ELECTRON FLOW

Most abundant in membranes of grana

Most abundant in membranes exterior to grana

Chloroplast stroma

Antenna complex

Photosystem II

Cytochrome complex

Photosystem I

NADPH

NADP$^+$ + H$^+$

NADP$^+$ reductase

e$^-$

Ferredoxin

ATP

ADP + P$_i$

ATP synthase

Photon

Photon

H$^+$

H$^+$

H$^+$

H$^+$

e$^-$

PQ

Pheophytin

e$^-$

P680

PQ

e$^-$

P700

e$^-$

PC

PC

Proton-motive force

Thylakoid lumen

H$^+$ H$^+$ H$^+$ H$^+$ H$^+$ H$^+$ H$^+$ H$^+$ H$^+$ H$^+$ H$^+$ H$^+$ H$^+$ H$^+$ H$^+$ H$^+$ H$^+$ H$^+$ H$^+$ H$^+$ H$^+$ H$^+$ H$^+$ H$^+$ H$^+$ H$^+$ H$^+$ H$^+$ H$^+$

FIGURE 10.16 Electrons Are Passed from Water to NADP$^+$ in a Linear Pathway. In the thylakoid membrane, photosystem II uses light to energize electrons taken from water and pass them through an ETC including plastoquinone (PQ), the cytochrome complex, and plastocyanin (PC). The ETC produces a proton motive force that is used to make ATP. Photosystem I energizes electrons from PC and passes them on to ferredoxin to reduce NADP$^+$ to NADPH.

Chloroplasts and mitochondria differ, however, in how electron potential energy changes between the primary electron donor and the terminal electron acceptor. In the mitochondrial ETC, electron potential energy starts high and then steadily drops as the electrons are transferred to the terminal electron acceptor, which has the lowest potential energy. In chloroplasts, the reduced terminal electron acceptor (NADPH) has higher potential energy than the electron donor (H_2O) (see Figure 10.15).

Cyclic Electron Flow Recycles Electrons and Drives Photophosphorylation   Recent evidence indicates that an alternative electron path, called **cyclic electron flow,** also occurs in green algae and plants (**FIGURE 10.17**). In these organisms, ATP is produced via cyclic as well as noncyclic photophosphorylation.

During cyclic electron flow, photosystem I transfers electrons back to the electron transport chain associated with photosystem II, generating ATP through photophosphorylation instead of reducing $NADP^+$. This "extra" ATP is used for the chemical reactions that reduce carbon dioxide (CO_2) and produce sugars. Cyclic electron flow coexists with the noncyclic electron flow and produces additional ATP.

Where Are Photosystems II and I Located?   Although both photosystems reside in the thylakoid membrane, their distribution is far from random. Photosystem II and the cytochrome complex are much more abundant in the interior, stacked membranes of grana, while photosystem I and ATP synthase are much more common in the exterior, unstacked membranes.

This organization seems appropriate for ATP synthase. As shown in Figure 10.16, this enzyme complex is oriented with its bulky head toward the stroma, so avoiding the tightly stacked grana makes sense. In addition, ATP synthase and the $NADP^+$ reductase that is associated with photosystem I require substrates

that are found in the stroma, such as ADP, P_i, and $NADP^+$. These substrates would not be as readily available if the enzymes were buried in the membrane folds of the grana.

The benefit of physically separating the two photosystems is the focus of intense research and debate. Unlike ATP synthase, the functions of photosystems I and II are tightly integrated, requiring electrons to be transported between them in noncyclic electron flow. Compared to other electron transport chains, the distance between where plastocyanin is reduced—the cytochrome complex—and where it is oxidized—the photosystem I reaction center—is huge. This separation between the photosystems is currently thought to be involved in regulating the switch between noncyclic and cyclic electron flow.

Oxygenic Photosynthesis and the Evolution of Earth   Although oxygen is a by-product of photosynthesis, the impact of producing this molecule on the environment of early Earth cannot be overstated. Photosynthesis produces the oxygen that is keeping you alive right now. Biologists rank the evolution of Earth's oxygen-rich atmosphere as one of the most important events in the history of life. Why?

According to the geologic record, oxygen levels in the atmosphere and oceans began to rise only about 2 billion years ago, as organisms that performed oxygenic photosynthesis increased in abundance. O_2 was, in fact, almost nonexistent on Earth before enzymes evolved that could catalyze the oxidation of water. Since ozone is formed from O_2 gas, this protective layer would have arisen in our atmosphere only after the evolution of oxygenic photosynthesis. Without the ozone layer, Earth's surface would have been bombarded continually by the searing intensity of ultraviolet radiation—making the evolution of life on land nearly impossible.

As oxygen became more abundant, bacterial cells that evolved the ability to use it as an electron acceptor via cellular respiration began to dominate. O_2 is so electronegative that it creates a huge potential energy drop for the electron transport chains involved in cellular respiration. As a result, organisms that use O_2 as an electron acceptor in cellular respiration can produce much more ATP than can organisms that use other electron acceptors (see Chapter 9). In addition, this accumulation of oxygen was a disaster for anaerobic organisms because O_2 is such a powerful oxidant it is toxic to them.

Determining exactly how photosystem II splits water and generates oxygen may be the greatest challenge currently facing researchers interested in photosynthesis. This issue has important practical applications: If human chemists could replicate the reaction, it might be possible to produce huge volumes of O_2 and hydrogen gas (H_2) from water. The resulting H_2 could provide a clean, inexpensive fuel for vehicles.

Despite the importance of oxygen in the evolution and maintenance of life, in terms of photosynthesis, it is simply waste. The useful products of the light-capturing reactions are ATP and NADPH, which are required to reduce carbon dioxide to sugar. Your life, and the life of most organisms, also depends on this process. How does it happen?

FIGURE 10.17 Cyclic Electron Flow Leads to ATP Production.
Cyclic electron flow is an alternative to the Z scheme. Instead of being donated to $NADP^+$, electrons are returned to plastoquinone (PQ) and cycle between photosystem I and the ETC, resulting in the production of additional ATP via photophosphorylation.

If you understand that . . .

- Photosystem II contributes high-energy electrons to an electron transport chain that pumps protons, creating a proton-motive force that drives ATP synthase.
- Photosystem I uses high-energy electrons to make NADPH and can produce additional ATP by building a proton-motive force via cyclic electron flow.

✔ **You should be able to . . .**

Compare and contrast the flow of electrons in mitochondria and chloroplasts. What are the primary electron donors and terminal electron acceptors, and how do they differ in terms of energy?

Answers are available in Appendix A.

10.4 How Is Carbon Dioxide Reduced to Produce Sugars?

The reactions analyzed in Section 10.3 are triggered by light. This is logical, because their entire function is focused on transforming electromagnetic energy in the form of sunlight into chemical energy in the phosphate bonds of ATP and the electrons of NADPH. The reactions that produce sugar from carbon dioxide, in contrast, are not triggered directly by light. Instead, they depend on the ATP and NADPH produced by the light-capturing reactions of photosynthesis.

The Calvin Cycle Fixes Carbon

Carbon fixation is the addition of carbon dioxide to an organic compound. The word *fix* is appropriate because the process converts or fixes CO_2 gas to a biologically useful form. Once carbon atoms are fixed, they can be used as sources of energy and as building blocks to construct the molecules found in cells.

Carbon fixation is a redox reaction—the carbon atom in CO_2 is reduced. Research on how this happens in chloroplasts gained momentum just after World War II, when radioactive isotopes of carbon became available for research purposes.

Melvin Calvin's group made great strides early in this effort by tracking the incorporation of $^{14}CO_2$ into molecules during photosynthesis (**FIGURE 10.18**). After injecting $^{14}CO_2$ into a culture of algae that were undergoing photosynthesis, they stopped the reaction after different periods of time by killing the cells in hot alcohol. This treatment immediately denatured the enzymes involved in the reactions, effectively halting any further change in the radiolabeled intermediates.

The molecules labeled with the ^{14}C in this extract were separated by chromatography and detected using X-ray film. If radioactively labeled molecules were present in the chromatograph,

RESEARCH

QUESTION: What intermediates are produced as carbon dioxide is reduced to sugar?

HYPOTHESIS: No specific hypothesis.

EXPERIMENTAL SETUP:

$^{14}CO_2$

1. Add $^{14}CO_2$ to actively photosynthesizing algae.

2. Wait 5–60 seconds; then homogenize cells by immersing in hot alcohol.

3. Separate molecules via chromatography.

4. Lay X-ray film on chromatograph to locate radioactive label.

PREDICTION: No specific prediction.

RESULTS:

3-Phosphoglycerate

Compounds produced after 5 seconds

Compounds produced after 60 seconds

CONCLUSION: 3-Phosphoglycerate is the first intermediate product. Other intermediates appear later.

FIGURE 10.18 Experiments Revealed the Reaction Pathway Leading to Reduction of CO_2.

SOURCE: Benson, A. A., J. A. Bassham, M. Calvin, et al. 1950. The path of carbon in photosynthesis. V. Paper chromatography and radioautography of the products. *Journal of the American Chemistry Society* 72: 1710–1718.

✔ **QUESTION** Why wasn't this experiment based on a specific hypothesis and set of predictions?

the energy they emitted would expose the film and create a dark spot. The labeled compounds could then be isolated and identified.

By varying the amount of time the algae were exposed to labeled $^{14}CO_2$, Calvin and co-workers pieced together the sequence in which various intermediates formed. For example, when the team analyzed cells almost immediately after adding the $^{14}CO_2$, they found that the ^{14}C was predominantly in a three-carbon compound called 3-phosphoglycerate (3PGA). This result suggested that 3PGA was the initial product of carbon reduction. Stated another way, it appeared that carbon dioxide reacted with some unknown molecule to produce 3PGA.

This was an intriguing result, because 3-phosphoglycerate is also one of the 10 intermediates in glycolysis. The Calvin cycle manufactures carbohydrate; glycolysis breaks it down. Because the two processes are related in this way, it was logical that at least some intermediates in glycolysis and the Calvin cycle are the same.

RuBP Is the Initial Reactant with CO_2

Which compound reacts with CO_2 to produce 3-phosphoglycerate? This was the key, initial step. Calvin's group searched in vain for a two-carbon compound that might serve as the initial carbon dioxide acceptor and yield 3PGA.

Then, while Calvin was running errands one day, it occurred to him that the molecule reacting with carbon dioxide might contain five carbons, not two. Adding CO_2 to a five-carbon molecule would produce a six-carbon compound, which could then split in half to form 2 three-carbon molecules.

Experiments to test this hypothesis confirmed that the five-carbon compound **ribulose bisphosphate (RuBP)** is the initial reactant.

The Calvin Cycle Is a Three-Step Process

The complete Calvin cycle, as it came to be called, has three phases (**FIGURE 10.19**):

1. *Fixation phase* The Calvin cycle begins when CO_2 reacts with RuBP. This phase fixes carbon and produces two molecules of 3PGA.

2. *Reduction phase* The 3PGA is phosphorylated by ATP and then reduced by electrons from NADPH. The product is the phosphorylated three-carbon sugar **glyceraldehyde-3-phosphate (G3P).** Some of the G3P that is synthesized is drawn off to manufacture glucose and fructose.

3. *Regeneration phase* The rest of the G3P keeps the cycle going by serving as the substrate for the third phase in the cycle: reactions that use additional ATP in the regeneration of RuBP.

All three phases take place in the stroma of chloroplasts. One turn of the Calvin cycle fixes one molecule of CO_2. Three turns of the cycle fix three molecules of CO_2, yielding one molecule of G3P and fully regenerated RuBP (Figure 10.19).

The discovery of the Calvin cycle clarified how the ATP and NADPH produced by light-capturing reactions allow cells to reduce CO_2 gas to carbohydrate $(CH_2O)_n$. Because sugars store a great deal of potential energy, producing them takes a great deal of chemical energy. In the Calvin cycle, each mole of CO_2 requires the energy from 3 moles of ATP and 2 moles of

(a) The Calvin cycle has three phases.

All three phases of the Calvin cycle take place in the stroma of chloroplasts

1. Fixation

3 RuBP + 3 CO_2 → 6 3PGA

2. Reduction

6 3PGA + 6 ATP + 6 NADPH → 5 G3P (to step 3)
→ 1 G3P yield (to glucose/fructose)

3. Regeneration

5 G3P + 3 ATP → 3 RuBP

(b) The reaction occurs in a cycle.

3 CO_2

Carbons are symbolized as red balls (each CO_2 enters the cycle one at a time)

3 P — RuBP

1. Fixation of carbon dioxide

6 — 3PGA

3 ADP + 3 P_i

3 ATP

3. Regeneration of RuBP from G3P

2. Reduction of 3PGA to G3P

6 ATP

6 ADP + 6 P_i

6 NADPH

6 $NADP^+$ + 6 H^+

5 G3P

6 — G3P

1 G3P → Glucose, fructose

FIGURE 10.19 Carbon Dioxide Is Reduced in the Calvin Cycle. The number of reactants and products resulting from three turns of the cycle are shown. Of the six G3Ps that are generated during the reduction phase, one is used in the synthesis of glucose or fructose and the other five are used to regenerate RuBP. The 3 RuBPs that are regenerated participate in fixation reactions for additional turns of the cycle.

NADPH to fix it and reduce it to sugar. ✔ **QUANTITATIVE** If you understand the Calvin cycle, you should be able to provide the *minimum number* of RuBP, ATP, and NADPH molecules that would be required to run through six complete cycles. Explain why you would not need six RuBP molecules to fix and reduce six CO_2.

The conversion of CO_2 gas into carbohydrate is, without doubt, worthy of this energy investment. Plants use sugars to fuel cellular respiration and build leaves and other structures. Millions of non-photosynthetic organisms—from fungi to mammals—also depend on this reaction to provide the sugars they need for cellular respiration.

Ecologically, the addition of CO_2 to RuBP may be the most important chemical reaction on Earth. The enzyme that catalyzes it is fundamental to life. How does this protein work?

The Discovery of Rubisco

Most reactions involved in reducing CO_2 also occur during glycolysis or other metabolic pathways. The initial CO_2 fixation phase of the Calvin cycle, however, is one of only two reactions that are entirely unique to the Calvin cycle.

To find the enzyme that fixes CO_2 to RuBP, Arthur Weissbach and colleagues ground up spinach leaves, purified a large series of proteins from the resulting cell extracts, and tested each protein to see if it could catalyze this step. Eventually they isolated the catalyst, which happens to be the most abundant enzyme in leaf tissue. The researchers' data suggested that it constituted almost 50 percent of the total protein in spinach leaves.

The CO_2-fixing enzyme, ribulose-1,5-bisphosphate carboxylase/oxygenase (commonly referred to as **rubisco**), is found in all photosynthetic organisms that use the Calvin cycle to fix carbon and is thought to be the most abundant enzyme on Earth. As shown in **FIGURE 10.20a**, the rubisco enzyme is cube-shaped and consists of 16 polypeptides that form eight active sites where CO_2 is fixed.

Despite its large number of active sites, rubisco is a slow enzyme. Each active site catalyzes just three reactions per second; other enzymes typically catalyze thousands of reactions per second. Plants synthesize huge amounts of rubisco, possibly as an adaptation compensating for its lack of speed.

Besides being slow, rubisco is extremely inefficient because it will catalyze the addition of either O_2 or CO_2 to RuBP. This is a key point: Oxygen and carbon dioxide compete at the enzyme's active sites, which slows the rate of CO_2 reduction.

Why would an active site of rubisco accept both O_2 and CO_2? Given rubisco's importance in producing food for photosynthetic species, this trait would appear to be **maladaptive**—it reduces the fitness of individuals.

The reaction of O_2 with RuBP actually does more than just compete with the reaction of CO_2 at the same active site. One of the molecules produced from the addition of oxygen to RuBP is processed in reactions that consume ATP and release CO_2 in order to regenerate 3PGA. Part of this pathway occurs in chloroplasts, and part occurs in peroxisomes and mitochondria. The

(a) Rubisco has 16 subunits and a total of 8 active sites.

Active site

(b) Rubisco's active sites can interact with CO_2 or O_2.

Reaction with carbon dioxide during photosynthesis:

$$\text{RuBP} + CO_2 \xrightarrow{\text{Rubisco}} \text{2 3-phosphoglycerate}$$

used in Calvin cycle

Reaction with oxygen during photorespiration:

$$\text{RuBP} + O_2 \xrightarrow{\text{Rubisco}} \text{3-phosphoglycerate} + \text{2-phosphoglycolate}$$

used in Calvin cycle   when processed, CO_2 is released and **ATP** is used

FIGURE 10.20 Rubisco Is a Large Enzyme Complex That Can React with CO_2 or O_2. (a) The cube shape of rubisco consists of multiple polypeptides that form eight catalytic active sites. **(b)** In addition to fixing CO_2 in photosynthesis, rubisco catalyzes a competing reaction with O_2 with a very different outcome.

reaction sequence resembles respiration, because it consumes oxygen and produces carbon dioxide. As a result, it is called **photorespiration** (**FIGURE 10.20b**).

Because photorespiration consumes energy and releases fixed CO_2, it "undoes" photosynthesis. When photorespiration occurs, the overall rate of photosynthesis declines. This does not mean that there is no benefit to the plant, however. Some of the products from photorespiration are known to be involved in plant signaling and development. In addition, a protective role for photorespiration has been proposed when plants are under high light and low CO_2 conditions.

Oxygen and Carbon Dioxide Pass through Stomata

Atmospheric carbon dioxide is a key reactant in photosynthesizing cells. It would seem straightforward, then, for CO_2 to diffuse directly into plants along a concentration gradient. But the situation is not this simple, because plants are covered with a waxy coating called a cuticle. This lipid layer prevents water from evaporating out of tissues, but it also prevents the transport of gases like CO_2 and O_2.

How does CO_2 get into photosynthesizing tissues? The surface of a leaf is dotted with openings bordered by two distinctively shaped cells called **guard cells** (**FIGURE 10.21a**). The opening between these paired cells is called a pore, and the entire structure is a **stoma** (plural: **stomata**).

An open stoma allows CO_2 from the atmosphere to diffuse into air-filled spaces inside the leaf and excess O_2 to diffuse out (**FIGURE 10.21b**). Eventually the CO_2 diffuses along a concentration gradient into the chloroplasts of photosynthesizing cells. A strong concentration gradient favoring entry of CO_2 is maintained by the Calvin cycle, which constantly uses up the CO_2 in chloroplasts.

Stomata are normally open during the day, when photosynthesis is occurring, and closed at night. But if the daytime is extremely hot and dry, leaf cells may lose a great deal of water to evaporation through their stomata. When this occurs, they must either close the openings and halt photosynthesis or risk death from dehydration.

(a) Leaf surfaces contain stomata.

Guard cells + Pore = **Stoma**

(b) Carbon dioxide diffuses into leaves through stomata.

FIGURE 10.21 **Leaf Cells Obtain Carbon Dioxide through Stomata.**

When conditions are hot and dry, then, stomata must close and CO_2 and O_2 transport stops—meaning that photosynthesis slows and photorespiration increases. How do plants that live in hot, dry environments prevent dehydration while keeping CO_2 supplies high enough to avoid photorespiration?

Mechanisms for Increasing CO_2 Concentration

The oxygenation reaction that triggers photorespiration is favored when oxygen concentrations are high and CO_2 concentrations are low. But even with the stomata open, the atmosphere is 21 percent oxygen and only 0.03 percent carbon dioxide. How can photosynthesizing cells raise CO_2 concentrations to make photosynthesis more efficient? An answer emerged in a surprising experimental result.

The C_4 Pathway After the Calvin cycle had been worked out in algae, researchers in a variety of labs used the same radioactive carbon dioxide tracking approach to investigate how carbon fixation occurs in other species. Hugo Kortschak and colleagues and Y. S. Karpilov and associates exposed leaves of sugarcane and maize (corn) to $^{14}CO_2$ and sunlight; then they isolated and identified the intermediates.

Both research teams expected to find the first of the radioactive carbon atoms in 3-phosphoglycerate—the normal product of carbon fixation by rubisco. Instead, they found that in their species, the radioactive carbon atom ended up in four-carbon compounds such as malate and aspartate.

Instead of creating a three-carbon molecule as in the Calvin cycle, it appeared that these species were able to fix CO_2 to produce four-carbon molecules. This newly identified set of reactions became known as the C_4 **pathway** to distinguish it from Calvin's CO_2 fixation via what is now termed the C_3 **pathway** (**FIGURE 10.22**).

Researchers who followed up on the initial reports found that the C_4 pathway does not replace the Calvin cycle, but serves as an additional fixation step. C_4 plants can actually fix carbon dioxide using both pathways—to a three-carbon compound by an enzyme called **PEP carboxylase** (C_4) and to RuBP by rubisco (C_3). They also showed that the two pathways are found in distinct cell types within the same leaf. PEP carboxylase is common in **mesophyll cells** near the surface of leaves, while rubisco is found in **bundle-sheath cells** that surround the vascular tissue in the

FIGURE 10.22 **Initial Carbon Fixation in C_4 Plants Is Different from That in C_3 Plants.**

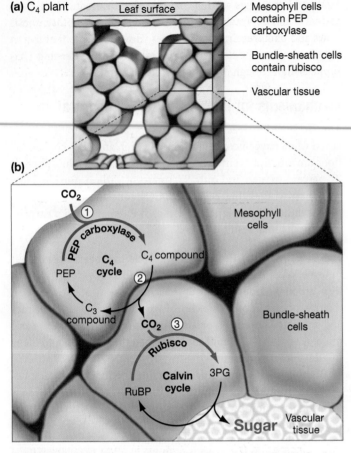

(a) C₄ plant

Leaf surface

Mesophyll cells contain PEP carboxylase

Bundle-sheath cells contain rubisco

Vascular tissue

(b)

CO_2

① PEP carboxylase

PEP

C₄ cycle

C₄ compound

②

C₃ compound

CO_2 ③

Rubisco

RuBP

Calvin cycle

3PG

Mesophyll cells

Bundle-sheath cells

Sugar

Vascular tissue

FIGURE 10.23 In C₄ Plants, Carbon Fixation and the Calvin Cycle Occur in Different Cell Types. (a) The carbon-fixing enzyme PEP carboxylase is located in mesophyll cells, while rubisco is in bundle-sheath cells. **(b)** CO_2 is fixed to the three-carbon compound PEP by PEP carboxylase, forming a four-carbon organic acid. A CO_2 molecule from the four-carbon sugar then feeds the Calvin cycle.

interior of the leaf (**FIGURE 10.23a**). Vascular tissue conducts water and nutrients in plants (see Chapter 38).

Based on the observations about C₄ plants, Hal Hatch and Roger Slack proposed a four-step model to explain how CO_2 that is fixed to a four-carbon sugar feeds the Calvin cycle (**FIGURE 10.23b**):

Step 1 PEP carboxylase fixes CO_2 to a three-carbon molecule (phosphoenolpyruvate, or PEP) in mesophyll cells.

Step 2 The four-carbon organic acids that result are transported to bundle-sheath cells via channels called plasmodesmata (see Chapter 11).

Step 3 The four-carbon organic acids release a CO_2 molecule that rubisco uses as a substrate to form 3PGA. This step initiates the Calvin cycle.

Step 4 The three-carbon compound remaining after CO_2 is released is returned to the mesophyll cell to regenerate PEP.

In effect, then, the C₄ pathway acts as a CO_2 concentrator. The reactions that take place in mesophyll cells require energy in the form of ATP, but they increase CO_2 concentrations in cells where rubisco is active. Because it increases the ratio of carbon dioxide to oxygen in photosynthesizing cells, less O_2 binds to rubisco's active sites. As a result, the C₄ pathway improves the efficiency of the Calvin cycle.

The C₄ pathway is an adaptation that keeps CO_2 concentrations in leaves high, but it comes at a cost. For each glucose molecule generated via photosynthesis, C₄ plants expend 30 ATP molecules compared to the 18 ATP molecules required by C₃ plants. This energy expenditure, however, is justified by the increased efficiency of photosynthesis in conditions where stomata are mostly closed to prevent dehydration. The affinity for CO_2 by PEP carboxylase is also much higher than that of rubisco, which means that stomata can be open for shorter periods in C₄ plants.

This strategy is not the only mechanism that plants use to continue growth in hot, dry climates, however. Some environments are so arid that even C₄ plants are unable to avoid dehydration. Nevertheless, certain plants use the C₄ pathway in a unique way that allows them to thrive in these deserts. How do they do it?

CAM Plants Researchers studying a group of flowering plants called the Crassulaceae discovered a second mechanism for limiting the effects of dehydration and photorespiration. This photosynthetic pathway, **crassulacean acid metabolism,** or **CAM,** resembles the C₄ pathway in a number of ways. It is a CO_2 concentrator that acts as an additional, preparatory step to the Calvin cycle. It also generates an organic acid with four carbons in its first CO_2 fixation step. But unlike the C₄ pathway, CAM occurs at a different time than the Calvin cycle does—not in a different place.

CAM occurs in cacti and other species that routinely keep their stomata closed on hot, dry days. At night, when conditions are cooler and more humid, CAM plants open their stomata and take in huge quantities of CO_2. The CO_2 is temporarily fixed to organic acids and stored in the central vacuoles of photosynthetic cells. During the day, when stomata are closed, these acids are processed in reactions that release the CO_2 and feed the Calvin cycle (**FIGURE 10.24**).

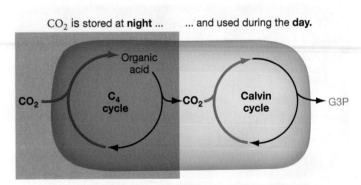

CO_2 is stored at **night** ... ... and used during the **day.**

Organic acid

CO_2

C₄ cycle

CO_2

Calvin cycle

G3P

FIGURE 10.24 In CAM Plants, Carbon Fixation Occurs at Night and the Calvin Cycle Occurs during the Day.

✔**QUESTION** At what part of the day would there be the highest concentration of four-carbon acids in the vacuoles of CAM plants?

The C$_4$ and CAM pathways function as CO$_2$ pumps. They minimize photorespiration when stomata are closed and CO$_2$ cannot diffuse in directly from the atmosphere. Both are found in species that live in hot, dry environments.

But while C$_4$ plants stockpile CO$_2$ by fixing and storing organic acids in cells *where* rubisco is not active, CAM plants store CO$_2$ *when* rubisco is inactive. In C$_4$ plants, the reactions catalyzed by PEP carboxylase and rubisco are separated in space; in CAM plants, the reactions are separated in time.

How Is Photosynthesis Regulated?

Like cellular respiration, photosynthesis is regulated. Although the mechanisms responsible for turning photosynthesis on or off are still under investigation, several patterns have emerged:

- The presence of light triggers the production of proteins required for photosynthesis.

- When sugar supplies are high, the production of proteins required for photosynthesis is inhibited, but the production of proteins required to process and store sugars is stimulated.

- Rubisco is activated by regulatory molecules that are produced when light is available, but inhibited in conditions of low CO$_2$ availability—when photorespiration is favored.

The central message here is that the rate of photosynthesis is finely tuned to use resources efficiently in response to changes in environmental conditions.

What Happens to the Sugar That Is Produced by Photosynthesis?

The products of the Calvin cycle enter one of several reaction pathways. The most important of these reaction sequences produces the monosaccharides glucose and fructose from G3P, a process called **gluconeogenesis.** This glucose is often combined with fructose to form the disaccharide ("two-sugar") **sucrose.**

When photosynthesis is taking place slowly, almost all the glucose that is produced is used to make sucrose. Sucrose is water soluble and readily transported to other parts of the plant. If sucrose is delivered to rapidly growing parts of the plant, it is broken down to fuel cellular respiration and growth.

An alternative pathway occurs when photosynthesis is proceeding rapidly and sucrose is abundant. Under these conditions, the glucose molecules are polymerized to form **starch** in the leaves and in storage cells in the roots. Starch production occurs inside the chloroplast; sucrose synthesis takes place in the cytosol.

In photosynthesizing cells, starch acts as a temporary sugar-storage product. At night, the starch that is stored in leaf cells is broken down and used to manufacture sucrose molecules. The sucrose is then broken down via cellular respiration or transported to other parts of the plant. In this way, chloroplasts provide sugars for cells throughout the plant by day and by night.

If a mouse eats the starch that is stored in the leaves or roots of a plant, however, the chemical energy in the reduced carbons of starch fuels the mouse's growth and reproduction. If an owl eats the mouse, the chemical energy in the mouse's tissues fuels the predator's growth and reproduction. (You can see the Big Picture of how energy is processed via photosynthesis and cellular respiration on pages 198–199.)

In this way, virtually all cell activity can be traced back to the sun's energy that was originally captured by photosynthesis. Photosynthesis is the staff of life.

check your understanding

If you understand that . . .

- The Calvin cycle is a three-phase process: CO$_2$ fixation (synthesis of 3PGA), carbon reduction (synthesis of G3P), and regeneration of RUBP.
- The C$_4$ and CAM pathways are mechanisms for increasing CO$_2$ concentrations in photosynthesizing cells. They limit the effect of photorespiration and allow photosynthesis to continue after stomata close.
- In photosynthesizing cells, G3P is used to make sucrose or stored as starch. Sucrose is transported to all plant cells and used to drive cellular respiration.

✔ You should be able to . . .

1. Describe how CO$_2$ is delivered to rubisco (a) via organic acids in mesophyll cells, (b) via organic acids stored in vacuoles, and (c) directly.

2. Predict the relative concentration of starch in leaves at the start of the day versus the end of the day.

Answers are available in Appendix A.

If you understand . . .

10.1 Photosynthesis Harnesses Sunlight to Make Carbohydrate

- The light-capturing reactions occur in internal membranes of the chloroplast that are organized into structures called thylakoids in stacks known as grana.

- The Calvin cycle takes place in a fluid portion of the chloroplast called the stroma.

- The CO_2-reduction reactions of photosynthesis depend on the products of the light-capturing reactions.

✓ You should be able to explain why it is not entirely accurate to adopt the common phrase "light-independent reactions" when referring to the Calvin cycle.

10.2 How Do Pigments Capture Light Energy?

- Pigment molecules capture light energy by exciting electrons after a photon is absorbed. Each pigment absorbs particular photons on the basis of their wavelengths.

- After a pigment molecule absorbs a photon, the energy may be released as fluorescence, resonance energy that excites a neighboring pigment, or the reduction of an electron acceptor.

- Pigments organized into antenna complexes will transfer the absorbed light energy via resonance to the reaction center, where an excited electron is transferred to an electron acceptor. The reduction of this electron acceptor completes the transformation of light energy into chemical energy.

✓ You should be able to explain why extracted chlorophyll molecules produce more fluorescence compared to the same number of chlorophyll molecules that remain in chloroplasts.

10.3 The Discovery of Photosystems I and II

- In photosystem II, high-energy electrons are accepted by pheophytin and passed along an electron transport chain, releasing energy that moves protons across the thylakoid membrane. The resulting proton-motive force drives the synthesis of ATP by ATP synthase. Photosystem II takes electrons from water, releasing oxygen and protons.

- In photosystem I, high-energy electrons are passed to ferredoxin. In an enzyme-catalyzed reaction, the reduced form of ferredoxin passes electrons to $NADP^+$, forming NADPH.

- The Z scheme connects photosystems II and I. Plastocyanin carries electrons from the end of photosystem II's ETC to photosystem I. They are promoted to a high-energy state in photosystem I's reaction center, and subsequently used to reduce $NADP^+$.

- Electrons from photosystem I may occasionally be passed back to photosystem II's ETC instead of being used to reduce $NADP^+$. A cyclic flow of electrons between the two photosystems boosts ATP supplies.

✓ You should be able to explain why measuring the rate of oxygen production in chloroplasts is appropriate for estimating the rate of photosynthesis.

10.4 How Is Carbon Dioxide Reduced to Produce Sugars?

- The Calvin cycle starts when CO_2 is attached to a five-carbon compound called ribulose bisphosphate (RuBP) in a reaction catalyzed by the enzyme rubisco.

- The six-carbon compound that results immediately splits in half to form two molecules of 3-phosphoglycerate (3PGA), which is then phosphorylated by ATP and reduced by NADPH to produce a sugar called glyceraldehyde-3-phosphate (G3P).

- Some G3P is used to synthesize glucose and fructose, which combine to form sucrose; the rest are phosphorylated by more ATP in a series of reactions that regenerate RuBP so the cycle can continue.

- Rubisco catalyzes the addition of oxygen as well as carbon dioxide to RuBP. The reaction with oxygen leads to a loss of fixed CO_2 and ATP and is called photorespiration.

- C_4 plants and CAM plants fix CO_2 to organic acids, before it is transferred to rubisco. As a result, they can increase CO_2 levels in their tissues, reducing the effect of photorespiration and allowing photosynthesis to continue when stomata close.

✓ QUANTITATIVE You should be able to connect the light-capturing reactions and Calvin cycle by estimating the number of photons required to produce one glucose molecule from CO_2. How would photorespiration affect the number of photons required per glucose?

(MB) MasteringBiology

1. **MasteringBiology Assignments**

   Tutorials and Activities Calvin Cycle; Chemiosmosis; Energy Flow in Plants; Experimental Inquiry: Which Wavelengths of Light Drive Photosynthesis?; Light Energy and Pigments; Light Reactions; Overview of Photosynthesis; Photosynthesis: Inputs, Outputs, and Chloroplast Structure; Photosynthesis: The Light Reactions; Photosynthesis in Dry Climates; Sites of Photosynthesis

   Questions Reading Quizzes, Blue-Thread Questions, Test Bank

2. **eText** Read your book online, search, take notes, highlight text, and more.

3. **The Study Area** Practice Test, Cumulative Test, BioFlix® 3-D Animations, Videos, Activities, Audio Glossary, Word Study Tools, Art

You should be able to . . .

1. In antenna complexes, how is energy transferred among the pigment molecules?
   a. photophosphorylation
   b. redox reactions
   c. fluorescence
   d. resonance

2. Why is chlorophyll green?
   a. It absorbs all wavelengths in the visible spectrum.
   b. It absorbs wavelengths only in the red portions of the spectrum (680 nm, 700 nm).
   c. It absorbs wavelengths in only the blue and red parts of the visible spectrum.
   d. It absorbs wavelengths only in the blue part of the visible spectrum.

3. What do the light-capturing reactions of photosynthesis produce?
   a. G3P
   b. RuBP
   c. ATP and NADPH
   d. sucrose or starch

4. Why do the absorption spectrum for chlorophyll and the action spectrum for photosynthesis coincide?
   a. Photosystems I and II are activated by different wavelengths of light.
   b. Wavelengths of light that are absorbed by chlorophyll trigger the light-capturing reactions.
   c. Energy from wavelengths absorbed by carotenoids is passed on to chlorophyll.
   d. The rate of photosynthesis depends on the amount of light received.

5. At what point in the light-capturing reactions is the electromagnetic energy of light converted into chemical energy? Where does this occur?

6. In noncyclic electron flow, photosystems I and II function as an integrated unit. What connects the two photosystems?

7. Explain how electrons from water can be used to produce both ATP and NADPH.

8. In addition to their protective function, carotenoids absorb certain wavelengths of light and pass the energy to other pigments via resonance. Based on this function, where would you expect carotenoids to be located in the chloroplast?
   a. the reaction centers of photosystems I and II
   b. the inner membrane of chloroplasts
   c. the antenna complex
   d. the stroma

9. Describe the three phases of the Calvin cycle and how the products of the light-capturing reactions participate in this process.

10. What conditions favor photorespiration? What are its consequences for the plant?

11. Compare and contrast how C_4 plants and CAM plants separate the acquisition of CO_2 from the production of sugar in the Calvin cycle.

12. Why do plants need both chloroplasts and mitochondria? How do their roles differ in the cell?

13. Predict how the following conditions would affect the production of O_2, ATP, and NADPH and state whether noncyclic or cyclic electron flow would occur in each: (1) Only blue photons hit a chloroplast; (2) blue and red photons hit a chloroplast, but no $NADP^+$ is available; (3) blue and red photons hit a chloroplast, but a proton channel has been introduced into the thylakoid membrane, so it is fully permeable to protons.

14. Some biologists claim that photorespiration is an evolutionary "holdover," because rubisco evolved over a billion years ago when O_2 levels were extremely low and CO_2 concentrations relatively high. Do you agree with this hypothesis? Why or why not?

15. An investigator exposes chloroplasts to 700-nm photons and observes low O_2 production, but high ATP production. Which of the following best explains this observation?

   a. The electrons from water are directly transferred to $NADP^+$, which is used to generate ATP.
   b. Photosystem II is not splitting water, and the ATP is being produced by cycling electrons via photosystem I.
   c. The O_2 is being converted to water as a terminal electron acceptor in the production of ATP.
   d. Electron transport has stopped and ATP is being produced by the Calvin cycle.

16. Consider plants that occupy the top, middle, or ground layer of a forest, and algae that live near the surface of the ocean or in deeper water. Would you expect the same photosynthetic pigments to be found in species that live in these different habitats? Why or why not? How would you test your hypothesis?

The Big Picture

It takes energy to stay alive. Use this concept map to study how the information on energy and energetics presented in this book fits together.

As you read the map, remember that chemical energy is potential energy. Potential energy is based on the position of matter in space, and chemical energy is all about the position of electrons in covalent bonds. When hydrogen gas reacts explosively with oxygen, all that's happening is that electrons are moving from high-energy positions to lower-energy positions.

In essence, organisms transform energy from the Sun into chemical energy in the C–C and C–H bonds of glucose, and then into chemical energy in the P–P bonds of ATP.

The potential energy in ATP allows cells to do work: pump ions, synthesize molecules, move cargo, and send and receive signals.

ENERGY FOR LIFE

begins as ↓

Electromagnetic energy in SUNLIGHT 10.2

Text section where you can find more information

drives ↓

PHOTOSYNTHESIS (in chloroplasts) 10.1

begins with ↓

Antenna complex
- Light excites electrons in pigment molecules 10.2

H_2O

donates energy from excited electrons to

donates energy from excited electrons to

enters ↓

Photosystem II
- "Splits" water to yield electrons
- Electron transport chain pumps H^+ 10.3

donates electrons to →

Photosystem I
- Electron transport ends with ferredoxin 10.3

Chemiosmosis
- H^+ gradient drives ATP synthase 9.5

releases ↓   yields ↓

O_2 **ATP** 9.1   **NADPH** 10.1

used in ↓

Calvin cycle
- Series of enzyme-catalyzed reactions 10.4

CO_2

fixed by rubisco to start →

yields substrate for synthesis of ↓

stored as ←

Glycogen, starch 5.2

broken down to yield →

GLUCOSE 5.1

check your understanding

C Y U

If you understand the big picture . . .
✔ You should be able to . . .

1. Explain how H_2O and O_2 are cycled between photosynthesis and cellular respiration.

2. Explain how CO_2 is cycled between photosynthesis and cellular respiration.

3. Describe what might happen to life on Earth if rubisco were suddenly unable to fix CO_2.

4. Fill in the blue ovals with appropriate linking verbs or phrases.

Answers are available in Appendix A.

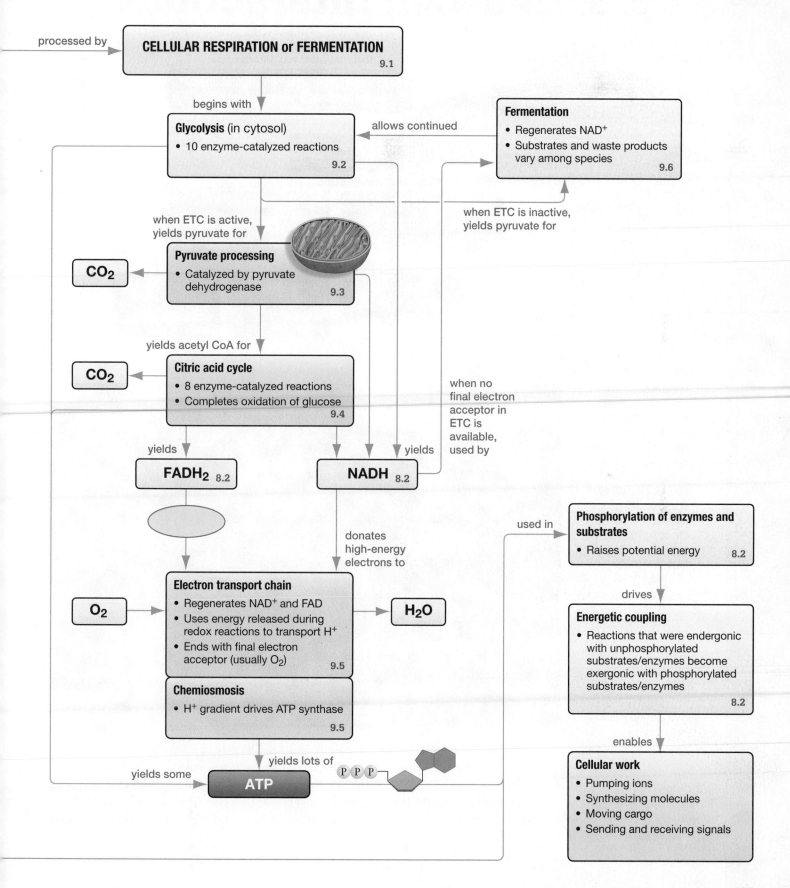

processed by →

CELLULAR RESPIRATION or FERMENTATION
9.1

begins with ↓

Glycolysis (in cytosol)
- 10 enzyme-catalyzed reactions
9.2

← allows continued

Fermentation
- Regenerates NAD^+
- Substrates and waste products vary among species
9.6

when ETC is active, yields pyruvate for

when ETC is inactive, yields pyruvate for

Pyruvate processing
- Catalyzed by pyruvate dehydrogenase
9.3

CO_2 ←

yields acetyl CoA for ↓

Citric acid cycle
- 8 enzyme-catalyzed reactions
- Completes oxidation of glucose
9.4

CO_2 ←

when no final electron acceptor in ETC is available, used by

yields ↓

FADH$_2$ 8.2

yields → **NADH** 8.2

donates high-energy electrons to

used in →

Phosphorylation of enzymes and substrates
- Raises potential energy
8.2

drives ↓

Electron transport chain
- Regenerates NAD^+ and FAD
- Uses energy released during redox reactions to transport H^+
- Ends with final electron acceptor (usually O_2)
9.5

O_2 →

→ H_2O

Energetic coupling
- Reactions that were endergonic with unphosphorylated substrates/enzymes become exergonic with phosphorylated substrates/enzymes
8.2

Chemiosmosis
- H^+ gradient drives ATP synthase
9.5

enables ↓

yields lots of ↓

yields some → **ATP**

P P P

Cellular work
- Pumping ions
- Synthesizing molecules
- Moving cargo
- Sending and receiving signals

11 Cell–Cell Interactions

In this chapter you will learn how

Cells modify their environment and interact with one another via social networks

by exploring

The cell surface
11.1

then asking

How do adjacent cells connect and communicate?
11.2

How do distant cells connect and communicate?

finally exploring

looking closer at

Cell-cell attachments and gaps

Signaling in multicellular organisms
11.3

Signaling between unicellular organisms
11.4

In this micrograph of cardiac tissue, muscle cells are stained red and their nuclei are stained blue. The green dye highlights a protein called dystrophin, which links the cytoskeleton of muscle cells to proteins that attach to the extracellular matrix. Deficiency in dystrophin leads to muscular dystrophy.

A diversity of events takes place at the cellular level. The plasma membrane surrounds a bustling enterprise consisting of organelles, molecular machines, and cytoskeletal elements (see Chapters 6 and 7). Molecular motors transport cargo throughout the cell at breathtaking speed. It would be a mistake, however, to think that cells are self-contained—that they are worlds in and of themselves. Instead, cells are dependent on interactions with other cells and the surrounding environment.

✔ When you see this checkmark, stop and test yourself. Answers are available in Appendix A.

For most unicellular species, the outside environment is teeming with other organisms. Inside your gut, for example, hundreds of billions of bacterial cells are jostling for space and resources. In addition to interacting with these individuals, every unicellular organism must contend with constant shifts in the physical environment, such as heat, light, ion concentrations, and food supplies. If unicellular organisms are unable to sense these conditions and respond appropriately, they die.

In multicellular species, the environment outside the cell is made up of other cells, both neighboring and distant. The cells that make up a redwood tree, a mushroom, or your body are intensely social. Although biologists often study cells in isolation, an individual tree, fungus, or person is actually an interdependent community of cells. If those cells do not communicate and cooperate, the whole will break into dysfunctional parts and die.

To understand the life of a cell, then, it is critical to analyze how the cell interacts with the world outside its membrane. How do cells obtain information about the world and respond to that information? In particular, how do cells interact with other cells? To answer these questions, let's begin with the cell surface—with the molecules that separate the cell from its environment.

Concrete (the "ground substance") resists compression

Steel rods (the "fibers") resist tension

FIGURE 11.1 Fiber Composites Resist Tension and Compression. Fiber composites, such as reinforced concrete, consist of ground substance that fills spaces between cross-linked rods.

11.1 The Cell Surface

The line between life and nonlife is drawn by the plasma membrane that surrounds every cell. Recall that the structure of this membrane consists of a phospholipid bilayer studded with membrane proteins that are integral, meaning that they are embedded in the bilayer, or peripheral, meaning that they are attached to one surface (see Chapter 6). These proteins participate in the primary function of the plasma membrane: to create an environment inside the cell that is different from conditions outside by regulating the transport of substances.

The plasma membrane does not exist in isolation, however. Cytoskeletal elements attach to the interior face of the bilayer (see Chapter 7), and a complex array of extracellular structures interacts with the membrane's exterior surface. Let's consider the nature of the material outside the cell and then analyze how the cell interacts with it and other cells.

The Structure and Function of an Extracellular Layer

It is actually extremely rare for cells to be bounded simply by a plasma membrane. Most cells secrete products that are assembled into a layer or wall just beyond the membrane. This extracellular material helps define the cell's shape and either attaches it to another cell or acts as a first line of defense against the outside world.

Virtually all types of extracellular structures—from the cell walls of bacteria, algae, fungi, and plants to the extracellular material that surrounds most animal cells—follow the same fundamental design principle. Like reinforced concrete, they are "fiber composites": They consist of a cross-linked network of long

filaments embedded in a stiff surrounding material, or ground substance (**FIGURE 11.1**). The molecules that make up the filaments and the encasing material vary from group to group, but the engineering principle is the same. Why?

- The rods or filaments in a fiber composite are extremely effective at withstanding stretching and straining forces, or tension. The fibers present in extracellular material of most cells are functionally similar to the steel rods in reinforced concrete—they resist being pulled or pushed lengthwise.

- The stiff surrounding substance is effective at withstanding the pressing forces called compression. Concrete performs this function in highways, and a gel-forming mixture of polysaccharides achieves the same end in extracellular material.

Thanks to the combination of tension- and compression-resisting elements, fiber composites are particularly rugged. And in many living cells, the fiber and composite elements are flexible as well as strong.

What molecules make up the rods and ground substance found on the surface of plant and animal cells? How are these extracellular layers synthesized, and what do they do?

The Cell Wall in Plants

Virtually all plant cells are surrounded by a cell wall—a fiber composite that is the basis of major industries. The paper in this book, the threads in your cotton clothing, and the wood in your neighborhood's houses are made up primarily of plant cell walls.

Before analyzing the structure of plant cell walls in detail, it's important to note that these structures are dynamic. If they are damaged by attacking insects, they may release signaling molecules that trigger the reinforcement of walls in nearby cells. Cell walls are also degraded in a controlled way as fruits ripen, making the fruits softer and more digestible for the animals that disperse the seeds inside.

Cellulose
microfibrils

Cross-links

Pectin

Primary
cell wall

Plasma
membrane

50 nm

1 μm

FIGURE 11.2 Primary Cell Walls of Plants Are Fiber Composites. In a plant's primary cell wall, cellulose microfibrils are cross-linked by polysaccharide chains. The spaces between the microfibrils are filled with pectin molecules, which form a gelatinous solid.

Primary Cell Walls   When plant cells first form, they secrete an initial fiber composite called a **primary cell wall.**

- The fibrous component of the primary cell wall consists of long strands of cellulose, which are bundled into stout, cable-like structures termed **microfibrils** and then cross-linked by other polysaccharide filaments. The microfibrils are synthesized by a complex of enzymes in the plasma membrane, forming a crisscrossed network (**FIGURE 11.2**).

- The space between microfibrils is filled with gelatinous polysaccharides such as **pectins**—the molecules that are used to thicken jams and jellies. Because the polysaccharides in pectin are hydrophilic, they attract and hold large amounts of water to keep the cell wall moist. The gelatinous components of the cell wall are synthesized in the rough endoplasmic reticulum and Golgi apparatus and secreted to the extracellular space.

The primary cell wall defines the shape of a plant cell. Under normal conditions, the nucleus and cytoplasm fill the entire volume of the cell and push the plasma membrane up against the wall. Because the concentration of solutes is higher inside the cell than outside, water tends to enter the cell via osmosis. The incoming water increases the cell's volume, exerting a force against the wall that is known as **turgor pressure.**

Although plant cells experience turgor pressure throughout their lives, it is particularly important in young cells that are actively growing. Young plant cells secrete proteins named expansins into their cell wall. **Expansins** disrupt hydrogen bonds that cross-link the microfibrils in the wall, allowing them to slide past one another. Turgor pressure then forces the wall to elongate and expand. The result is cell growth (see Chapter 40).

Secondary Cell Walls   As plant cells mature and stop growing, they may secrete an additional layer of material—a **secondary cell wall**—between the plasma membrane and the primary cell

wall. The structure of the secondary cell wall varies from cell to cell in the plant and correlates with that cell's function. Cells on the surface of a leaf have secondary cell walls that are impregnated with waxes that form a waterproof coating; the cells that support the plant's stem have secondary cell walls that contain a great deal of cellulose.

In cells that form wood, the secondary cell wall also includes **lignin,** a complex polymer that forms an exceptionally rigid network. Cells that have thick secondary cell walls of cellulose and lignin help plants withstand the forces of gravity and wind.

Although animal cells do not make a cell wall, they do form a fiber composite outside their plasma membrane. What is this substance, and what does it do?

The Extracellular Matrix in Animals

Most animal cells secrete a fiber composite called the **extracellular matrix (ECM).** Like the extracellular materials found in other organisms, structural support is one of the ECM's most important functions.

ECM design follows the same principles observed in the cell walls of bacteria, archaea, algae, fungi, and plants. There is a key difference, however: The animal ECM contains much more protein relative to carbohydrate than does a cell wall.

- The fibrous component of animal ECM is dominated by a cable-like protein termed **collagen** (**FIGURE 11.3a**).

- The matrix that surrounds collagen and other fibrous components contains gel-forming **proteoglycans** that consist of protein cores with many large polysaccharides attached to them. In some tissues, complexes of proteoglycans may also be produced (**FIGURE 11.3b**).

Most ECM components are synthesized in the rough endoplasmic reticulum (ER), processed in the Golgi apparatus, and

(a) Collagen proteins consist of three polypeptide chains that wind around one another to form the fibrous component of the animal ECM.

3 chains

1.5-nm collagen protein

(b) Complexes of gelatinous proteoglycans form the ground substance of the animal ECM.

Proteoglycan complex

Proteoglycan

0.5 μm

Proteoglycan

Polysaccharides   Core protein

FIGURE 11.3 The Extracellular Matrix Is a Fiber Composite.
(a) Although several types of fibrous proteins are found in the ECM of animal cells, the most abundant is collagen. Groups of collagen proteins coalesce to form collagen fibrils, and bundles of fibrils link to form collagen fibers. **(b)** The spaces between the collagen fibers are filled with complexes of gelatinous proteoglycans. The proteoglycan subunits consist of a protein core attached to many polysaccharides.

secreted from the cell via exocytosis. After secretion, however, proteins like collagen may then assemble into larger structures, such as the fibrils shown in **FIGURE 11.4**. In addition, the secreted proteoglycans may be attached to long polysaccharides synthesized by cellular enzymes in the extracellular space. These huge complexes, such as the one shown in Figure 11.3b, are responsible for the rubber-like consistency of cartilage.

Even in the same organism, the amount of ECM varies among different types of **tissues,** which consist of similar cells that function as a unit. Bone and cartilage, for example, have relatively few cells surrounded by a large amount of ECM. Skin cells, in contrast, are packed together with a minimal amount of ECM.

The composition of the ECM also varies among tissue types. For example, the ECM surrounding cells in lung tissue contains large amounts of a rubber-like protein called elastin, which allows the ECM to expand and contract during breathing. The structure of a cell's ECM correlates with the function of the tissue.

Although collagen and the other common ECM proteins are much more elastic and bendable than the stiff cell walls of plants, they support cell structure via attachments to the cell surface. As

Cytoskeleton

Plasma membrane

Extracellular matrix (ECM)

50 nm

Actin filaments

Integrin

Laminin

Proteoglycans

Collagen (structural fibril)

Collagen (anchoring fibrils)

FIGURE 11.4 Integrins Connect the Extracellular Matrix to the Cytoskeleton.

Figure 11.4 shows, membrane proteins called **integrins** bind to extracellular proteins, including laminins, which in turn bind to other components of the ECM. **Laminins** are ECM crosslinking proteins—not to be confused with lamins, which are intermediate filaments found in the nucleus (see Chapter 7).

The intracellular portions of the integrins also bind to proteins that are connected to the cytoskeleton, effectively forming a bridge between the two support systems. This linkage between the cytoskeleton and ECM is critical. Besides keeping individual cells in place, it helps adjacent cells adhere to each other via their common connection to the ECM.

Cells monitor this cytoskeleton–ECM linkage via signaling pathways that will be introduced in Section 11.3. When integrins bind to the ECM, they transmit signals that inform the cell it is in the right place and properly anchored. If these linkages break down, the signals are not transmitted and cells normally die as a result. For most of the cells in your body, anchorage to the ECM is a matter of life and death.

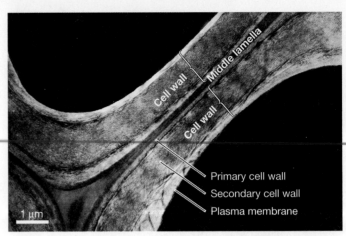

FIGURE 11.5 The Middle Lamella Connects Adjacent Plant Cells. The middle lamella contains gelatinous polysaccharides, called pectins, that help glue together the walls of adjacent cells.

check your understanding

If you understand that . . .

- Most cells secrete a layer of structural material that supports the cell and helps define its shape. The extracellular material is usually a fiber composite—a combination of cross-linked filaments surrounded by a ground substance.

✔ **You should be able to . . .**

Compare and contrast the molecular composition of a plant cell wall and the ECM of animal cells.

Answers are available in Appendix A.

11.2 How Do Adjacent Cells Connect and Communicate?

Intercellular connections are the basis of **multicellularity.** These physical connections between cells—either direct, or indirect via the ECM—are particularly important in the structure and function of tissues. The muscle tissue in your heart, for example, depends on these attachments to support the structure of the cells as they contract and relax with each beat (see the micrograph on the opening page of this chapter).

Let's look first at the structures that attach cells to each other and then examine how they allow adjacent cells to exchange materials and information.

Cell–Cell Attachments in Multicellular Eukaryotes

Materials and structures that bind cells together are particularly important in the **epithelium** (plural: **epithelia**)—a tissue that forms external and internal surfaces. These epithelial layers function as a barrier between the external and internal environments of plants and animals. In animals, epithelial cells also form layers that separate organs to prevent mixing of solutions from adjacent organs or structures.

The adhesive structures that hold cells together vary among multicellular organisms. To illustrate this diversity, consider the intercellular connections observed in the best-studied groups of organisms: plants and animals.

Indirect Intercellular Attachments   The extracellular space between adjacent plant cells comprises three layers (**FIGURE 11.5**). The primary cell walls of adjacent plant cells sandwich a central layer designated the middle lamella, which consists primarily of gelatinous pectins. Because this gel layer is continuous with the primary cell walls of the adjacent cells, it serves to glue them together. The two cell walls are like slices of bread; the middle lamella is like a layer of peanut butter. If enzymes degrade the middle lamella, as they do when flower petals and leaves detach and fall, the surrounding cells separate.

In many animal tissues, integrins connect the cytoskeleton of each cell to the extracellular matrix (see Section 11.1). A middle-lamella-like layer of gelatinous polysaccharides and proteoglycans runs between adjacent animal cells. Along with the cytoskeleton–ECM connections, the polysaccharide glue helps hold cells together in tissues. In addition, in certain animal tissues the polysaccharide glue is reinforced by collagen fibrils that span the ECM to connect adjacent cells.

In animals, where cell walls do not exist, a variety of membrane proteins allow for direct cell–cell attachments in epithelia and other tissues (**FIGURE 11.6**). Let's start by looking at the tight junctions and desmosomes that hold cells together and then examine the role of gap junctions in intercellular communication.

FIGURE 11.6 An Array of Structures Are Involved in Cell–Cell Adhesion and Communication between Animal Cells.

Tight junctions
seal cells
together.

Desmosomes
connect the
cytoskeletons
of cells.

Gap junctions
act as channels
between cells.

Space between cells

the proteins "stitch" the membranes of two cells together to form a watertight seal. In this way, tight junctions prevent solutions from flowing through the space between the two cells.

Because tight junctions form a watertight seal, this type of junction is commonly found in cells that form a barrier, such as the epithelial cells lining your stomach and intestines. There, they restrict the passive movement of substances between the contents of your gut and the rest of your body. Instead, only selected nutrients enter and leave the epithelia via specialized transport proteins and channels in the plasma membrane (Chapter 6).

Although tight junctions are indeed tight, they are variable. The tight junctions between the cells lining your bladder draw the cells closer together than those between the cells lining your small intestine, because they consist of different proteins. As a result, small ions can pass between the cells lining the surface of the small intestine more easily than between those lining the bladder—helping you absorb ions in your food and eliminate them in your waste.

Tight junctions are also dynamic. For example, they loosen to permit more transport between epithelial cells lining the small intestine after a meal and then "retighten" later. In this way, tight junctions can open and close in response to changes in environmental conditions.

Although tight junctions are very good at holding cells close together, they are weak adhesions that can be easily broken. Since epithelial cells often experience pulling and shearing forces, other intercellular adhesions are required to help hold cells together in a tissue. What are these other adhesions, and how do they resist being pulled apart?

Tight Junctions Form a Seal between Cells A **tight junction** is a cell–cell attachment composed of specialized proteins in the plasma membranes of adjacent animal cells (**FIGURE 11.7a**). As the drawing in **FIGURE 11.7b** indicates, these proteins line up and bind to one another. The resulting structure resembles quilting, where

(a) Electron micrograph of a tight junction in longitudinal section

A tight junction forms a
watertight seal between
epithelial cells

Tight junction

0.1 μm

(b) Three-dimensional view of a tight junction

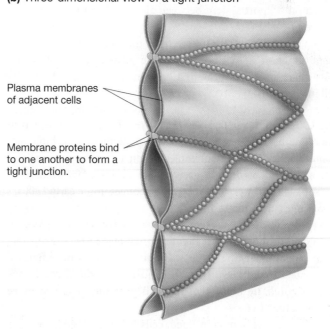

Plasma membranes
of adjacent cells

Membrane proteins bind
to one another to form a
tight junction.

FIGURE 11.7 In Animals, Tight Junctions Form a Seal between Adjacent Cells.

(a) Micrograph of desmosome in longitudinal section

Desmosome

0.1 μm

(b) Three-dimensional view of desmosome

Plasma membranes of adjacent cells

Anchoring proteins inside cell

Membrane proteins that link cells

Intermediate filaments

FIGURE 11.8 Adjacent Animal Cells Are Linked by Desmosomes, Which Bind Cytoskeletons Together.

Desmosomes Form Secure Adhesions   **FIGURE 11.8a** illustrates **desmosomes,** cell–cell attachments particularly common in animal epithelial cells and certain muscle cells. The structure and function of a desmosome are analogous to the rivets that hold pieces of sheet metal together.

As **FIGURE 11.8b** indicates, desmosomes are extremely sophisticated cell–cell connections. At their heart are integral membrane attachment proteins that form bridges between anchoring proteins inside adjacent cells. Intermediate filaments help reinforce these connections by attaching to the anchoring proteins in the cytoplasm. In this way, desmosomes help form a continuous structural support system between all the cells in the tissue (see Figure 11.6).

What are the membrane proteins that serve this cell attachment function in desmosomes? The answer to this question traces back to some of the first experiments conducted on cell–cell interactions.

Intercellular Adhesions Are Selective   Long before electron micrographs revealed the presence of desmosomes, biologists realized that some sort of molecule must bind animal cells to one another. This insight grew out of experiments from H. V. Wilson's lab that were conducted on sponges in the early 1900s.

Sponges are aquatic animals, and the sponge species used in this study consists of just two basic types of tissues. When Wilson treated adult sponges with chemicals that made the cells separate from one another, the result was a jumbled mass of individual and unconnected cells. But when normal chemical conditions were restored, he noted that the cells gradually began to move and stick together.

As the experiment continued, cells began to aggregate based on their origin—adhering to other cells of the same tissue type.

This phenomenon is now called **selective adhesion.** Eventually the experimental sponge cells re-formed functional adult sponges with two distinct tissues. How could this happen?

The Discovery of Cell–Cell Adhesion Proteins   What is the molecular basis of selective adhesion? The initial hypothesis, proposed in the 1970s, was that specialized membrane proteins were involved. The idea was that different types of cells have different types of adhesion proteins in their membranes, and only those with the same or complementary adhesion proteins are able to attach to one another.

This hypothesis was tested through experiments that relied on molecules called antibodies. An **antibody** is a protein produced by an immune response that binds specifically to a unique molecule type, often another protein (see Chapter 51). When an antibody binds to a protein, it can change the target protein's structure or interfere with its ability to interact with other molecules. This property of antibodies was crucial to these experiments.

FIGURE 11.9 shows how researchers tested the hypothesis that cell–cell adhesion takes place via interactions between membrane proteins:

Step 1 Isolate the membrane proteins from a certain cell type. Produce pure preparations of each protein.

Step 2 Inject one of the membrane proteins into a rabbit. The rabbit's immune system cells respond by creating antibodies to the membrane protein, which is recognized as being foreign. Purify those antibodies. Repeat this procedure for the other membrane proteins that were isolated. In this way, obtain a large collection of antibodies—each of which binds specifically to one (and only one) type of membrane protein.

QUESTION: Do animal cells have adhesion proteins on their surfaces?

HYPOTHESIS: Selective adhesion is due to specific membrane proteins.

NULL HYPOTHESIS: Selective adhesion is not due to specific membrane proteins.

EXPERIMENTAL SETUP:

Membrane proteins

1. Isolate the membrane proteins from a certain cell type that adheres to other cells of the same type. (There are many membrane proteins; only two are shown here.)

Antibodies

2. Produce antibodies that bind to specific membrane proteins. Purify the antibodies.

Experiment 1:   **Experiment 2:**   **3. Treat cells with an antibody**, one type at a time. Wait; then observe whether cells adhere normally.

PREDICTION:

PREDICTION OF NULL HYPOTHESIS:

RESULTS:

Experiment 1:
Cells adhere normally.

Experiment 2:
Cells do not adhere.

CONCLUSION: The protein that was blocked in experiment 2 (called a cadherin) is involved in cell–cell adhesion.

FIGURE 11.9 Evidence for Adhesion Proteins on Animal Cells.

SOURCES: Hatta, K., and M. Takeichi. 1986. Expression of N-cadherin adhesion molecules associated with early morphogenetic events in chick development. *Nature* 320: 447–449. Also Takeichi, M. 1988. The cadherins: Cell–cell adhesion molecules controlling animal morphogenesis. *Development* 102: 639–655.

✔ **EXERCISE** Fill in the prediction made by each hypothesis.

Step 3 Add one antibody type to a mixture of dissociated cells from a tissue and observe whether the cells reaggregate normally. Repeat this experiment with each of the other antibody types, one type at a time.

If treatment with a particular antibody prevents the cells from attaching to one another, the antibody is probably bound to an adhesion protein. The logic is that if the antibody "shakes hands" with the adhesion protein, the adhesion protein can't shake hands with other adhesion proteins and attach the cells to one another.

This approach allowed biologists to identify several major classes of cell adhesion proteins, including **cadherins**—the attachment molecules in desmosomes. There are various types of cadherins, and cells from different tissues have different forms of cadherin in their plasma membranes. Each cadherin can bind only to cadherins of the same type. In this way, cells of the same tissue type attach specifically to one another.

To summarize: Animal cells attach to one another in a selective manner because different types of cell adhesion proteins can bind and rivet certain cells together. Cadherins provide the physical basis for selective adhesion in many cells and are a critical component of the desmosomes that join mature cells.

✔ If you understand cell–cell attachments, you should be able to predict what would happen if you treated cells in a developing frog embryo with a molecule that blocked a cadherin present in muscle tissue.

In addition to providing structural support to tissues, intercellular connections can direct cell–cell communication. But how can cellular connections pass information between cells?

Cells Communicate via Cell–Cell Gaps

In both plants and animals, direct connections between cells in the same tissue help them to work in a coordinated fashion. One way of accomplishing this is to generate channels in the membranes of adjacent cells, allowing them to communicate via the diffusion of cytosolic ions and small molecules from cell to cell.

How cells respond to this exchange of information varies from signal to signal and from cell to cell, but the result falls into two general categories:

1. Signals may alter which proteins are produced and which are not, by regulating gene expression; or

2. Signals may activate or deactivate particular proteins that already exist in the cell—often those involved in metabolism, membrane transport, secretion, and the cytoskeleton.

Whatever the mechanism, the activity of the cell often changes dramatically after the signal arrives. Let's take a closer look at how these signals are able to travel between adjacent cells connected by gap junctions and plasmodesmata.

Gap Junctions Connect Cells via Protein Channels In most animal tissues, structures called **gap junctions** connect adjacent cells. The key feature of gap junctions is the specialized proteins that assemble in the membranes of adjacent cells, creating

(a) Gap junctions create gaps that connect animal cells.

Gap junctions

Membrane proteins from adjacent cells line up to form a channel.

20 μm

(b) Plasmodesmata create gaps that connect plant cells.

Plasmodesma with a tubule of endoplasmic reticulum passing through

Cell walls

0.1 μm

Smooth endoplasmic reticulum

Cell wall of cell 1 Cell wall of cell 2

Membrane of cell 1

Membrane of cell 2

FIGURE 11.10 Adjacent Animal Cells and Adjacent Plant Cells Communicate Directly.

interconnected pores between the cells (**FIGURE 11.10a**). These channels allow water, ions, and small molecules such as amino acids, sugars, and nucleotides to move between adjacent cells.

Gap junctions are communication portals. The flow of small molecules through gap junctions can help adjacent cells coordinate their activities by allowing the rapid passage of regulatory ions or molecules. In the muscle cells of your heart, for example, a flow of ions through gap junctions acts as a signal that coordinates contractions. Without this cell–cell communication, a normal heartbeat would be impossible.

In plants, direct interactions between membrane proteins are impossible due to the presence of cell walls. How do adjacent plant cells communicate?

Plasmodesmata Connect Cells via Membrane Channels   In plants, gaps in cell walls create direct connections between the cytoplasm

of adjacent cells. At these connections, named **plasmodesmata** (singular: **plasmodesma**), the plasma membrane and the cytoplasm of the two cells are continuous. Tubular extensions from the smooth endoplasmic reticulum (smooth ER) run through these membrane-lined portals (**FIGURE 11.10b**).

Like gap junctions, plasmodesmata are communication portals through the plasma membrane. In plants, the plasma membrane separates most tissues into two independent compartments: (**1**) the **symplast,** which is a continuous network of cytoplasm connected by plasmodesmata, and (**2**) the **apoplast,** which is the region outside the plasma membrane (**FIGURE 11.11**). The apoplast consists of cell walls, the middle lamella, and air spaces. Small molecules can move through plant tissues in either of these compartments without ever crossing a membrane (see Chapter 38).

Gap junctions and plasmodesmata allow for adjacent cells to transmit information, like a conversation between neighbors.

If you understand that . . .

- In plants and animals, adjacent cells are either directly physically connected or, via the ECM, indirectly connected.
- Adjacent cells may communicate with each other through openings in their plasma membranes.

✔ **You should be able to . . .**

1. Compare and contrast the structure and function of the middle lamella of plants and the tight junctions and desmosomes of animals.
2. Describe the structure and function of plasmodesmata and gap junctions.

Answers are available in Appendix A.

But how do multicellular organisms send messages between different tissues, where in most cases there is no direct contact? For example, suppose that the muscle cells in your arm are exercising so hard they run low on sugar or that leaf cells in a maple tree are attacked by caterpillars. How do these cells signal tissues or organs elsewhere in the organism to release materials that are needed to fend off exhaustion or caterpillars? Distant cell communication is the subject of Section 11.3.

11.3 How Do Distant Cells Communicate?

Cells that are not in physical contact communicate with one another. This is true for unicellular organisms, where hundreds or

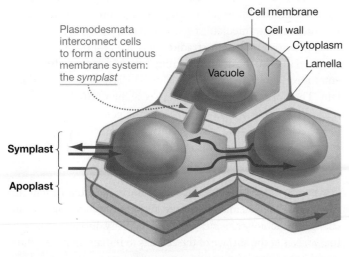

FIGURE 11.11 Most Plant Tissues Are Divided into Two Compartments—Symplast and Apoplast. Small molecules may be transported through plant tissues either within the shared cytoplasm (symplast) or through the extracellular space (apoplast).

thousands of cells may live in close proximity, as well as for multicellular organisms like humans and maple trees, which typically contain trillions of cells and dozens of tissue types.

Cell–cell communication qualifies as one of the most dynamic and important research areas in biology. Let's begin by analyzing how distant cells in humans and other multicellular eukaryotes exchange information, and then in Section 11.4 explore how unicellular organisms communicate.

Cell–Cell Signaling in Multicellular Organisms

Suppose that cells in your brain sense that you are becoming dehydrated. Brain cells can't do much about the water you lose during urination, but kidney cells can. In response to dehydration, certain brain cells release a signaling molecule that travels to the kidneys. The arrival of this molecule activates specialized membrane channels that prevent water from being lost in urine—an important aspect of fighting dehydration.

Thanks to cell–cell signals, the activities of cells in different parts of a multicellular body are coordinated.

Biologists have classified many different types of signaling molecules that keep distant tissues in touch. For example, neurotransmitters activate membrane channel receptors that open to allow a flow of ions into the cytosol of the cell, changing the electrical properties of the membrane. This type of signal is responsible for the transmission of information along neurons, allowing your brain to control the movements of the rest of your body (see Chapters 46 and 48). The best-studied means of distant signaling, however, may be via **hormones**—information-carrying molecules that can act on distant target cells because they are secreted by plant and animal cells into bodily fluids.

Hormones are usually small molecules and are typically present in minute concentrations. Even so, they have a large impact on the activity of target cells and the condition of the body as a whole. Hormones are like a scent or whispered phrase from someone you are attracted to—a tiny signal, but one that makes your cheeks flush and your heart pound.

As **TABLE 11.1** (see page 210) indicates, hormones have a wide array of chemical structures and effects. The important point about a signaling molecule, though, is not whether it is a gas or peptide or steroid, but whether it is lipid soluble or not. The ability of a signaling molecule to pass through lipid bilayers is crucial in determining how a target cell recognizes it. Where does this recognition occur—inside the cell or outside?

- Most lipid-soluble signaling molecules are able to diffuse across the hydrophobic region of the plasma membrane and enter the cytoplasm of their target cells.

- Large or hydrophilic signaling molecules are lipid insoluble and do not cross the plasma membrane. To affect a target cell, they have to be recognized at the cell surface.

How do cells receive and process signals from distant cells? The basic steps are common to all cell–cell signaling systems. Let's consider each step in turn.

TABLE 11.1 Hormones Have Diverse Structures and Functions

Hormone Name	Chemical Structure	Where Is Signal Received?	Function of Signal
Auxin	Small organic compound	At plasma membrane	Signals changes in long axis of plant body
Brassinosteroids	Steroid	At plasma membrane	Stimulate plant cell elongation
Estrogens	Steroid	Inside cell	Stimulate development of female characteristics in animals
Ethylene	C_2H_4 (a gas)	At plasma membrane	Stimulates fruit ripening; regulates aging
FSH	Glycoprotein	At plasma membrane	Stimulates egg maturation, sperm production in animals
Insulin	Protein, 51 amino acids	At plasma membrane	Stimulates glucose uptake in animal bloodstream
Prostaglandins	Modified fatty acid	At plasma membrane	Perform a variety of functions in animal cells
Systemin	Peptide, 111 amino acids	At plasma membrane	Stimulates plant defenses against herbivores
Thyroxine (T4)	Modified amino acid	Inside cell	Regulates metabolism in animals

Signal Reception

Hormones and other types of cell–cell signaling molecules deliver their message by binding to receptor molecules. Even though the molecule that carries the message "We're getting dehydrated—conserve water" is broadcast throughout the body, only certain kidney cells respond because only they have the appropriate receptor. The presence of an appropriate receptor dictates which cells will respond to a particular hormone. Bone and muscle cells don't respond to the "conserve water" message, because they don't have a receptor for it.

Cells in a wide array of tissues may respond to the same signaling molecule, though, if they have the appropriate receptor. If you are startled by a loud noise, cells in your adrenal glands secrete a hormone called adrenaline (also called epinephrine) that carries the message "Get ready to fight or run." In response, your heart rate increases, your breathing rate increases, and cells in your liver release sugars that your muscles can use to power rapid movement. This is the basis of an "adrenaline rush." Heart, lung, and liver cells respond to adrenaline because they each have the receptor that binds to it. Identical receptors in diverse cells and tissues allow long-distance signals to coordinate the activities of cells throughout a multicellular organism.

No matter where signal receptors are located, it's critical to note two important points about these proteins:

1. **Receptors are dynamic.** The number of receptors in a particular cell may decline if hormonal stimulation occurs at high levels over a long period of time. The ability of a receptor molecule to bind tightly to a signaling molecule may also decline in response to intensive stimulation. As a result, the sensitivity of a cell to a particular hormone may change over time.

2. **Receptors can be blocked.** The drugs called beta-blockers, for example, bind to certain receptors for the hormone adrenaline. When adrenaline binds to receptors in heart cells, it stimulates more rapid and forceful contractions. So if a physician wants to reduce a patient's heart cell contraction as a way to lower pressure, she is likely to prescribe a beta-blocker.

Most signal receptors are located in the plasma membrane, where they can bind to signaling molecules that cannot or do not cross the membrane. Other signal receptors exist inside the cell, where they respond to lipid-soluble signaling molecules that readily diffuse through the plasma membrane.

The most important general characteristic of signal receptors, though, is that their conformation—meaning, overall shape—changes when a hormone binds to them. A **signal receptor** is a protein that changes its shape and activity after binding to a signaling molecule.

This is a critical event in cell–cell signaling. The change in receptor structure means that the signal has been received. It's like throwing an "on" switch. What happens next?

Signal Processing

Once a cell receives a signal, something has to happen to initiate the cell's response. This signal processing step happens in one of two ways, depending on whether the signal is received inside the cell or at the membrane surface.

Processing Lipid-Soluble Signals When lipid-soluble signals enter a cell, the information they carry is processed directly—without any intermediate steps. For example, steroid hormones such as testosterone and estradiol (one of a group commonly referred to as estrogens) diffuse through the plasma membrane and enter the cytoplasm, where they bind to a cytosolic receptor protein. The hormone–receptor complex is then transported to the nucleus, where it triggers changes in the genes being expressed in the cell (**FIGURE 11.12**). By altering the expression of genes (see Chapter 17), the cell produces different proteins that will have a direct effect on the function or shape of the cell.

Processing Lipid-Insoluble Signals Hormones that *cannot* diffuse across the plasma membrane and enter the cytoplasm can't change the activity of genes or pumps directly. Instead, the signal that arrives at the surface of the cell has to initiate an intracellular signal—the signal processing step is indirect.

When a signaling molecule binds at the cell surface, it triggers **signal transduction**—the conversion of a signal from one form to another. A long and often complex series of events ensues, collectively called a signal transduction pathway.

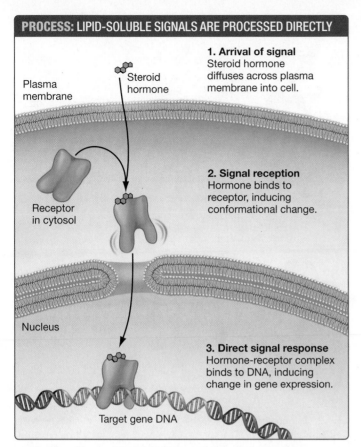

PROCESS: LIPID-SOLUBLE SIGNALS ARE PROCESSED DIRECTLY

Plasma membrane

Steroid hormone

1. Arrival of signal
Steroid hormone diffuses across plasma membrane into cell.

Receptor in cytosol

2. Signal reception
Hormone binds to receptor, inducing conformational change.

Nucleus

3. Direct signal response
Hormone-receptor complex binds to DNA, inducing change in gene expression.

Target gene DNA

FIGURE 11.12 Some Cell–Cell Signaling Molecules Enter the Cell and Bind to Receptors in the Cytoplasm. Because they are lipids, steroid hormones can diffuse across the plasma membrane and bind to signal receptors inside the cell. The hormone–receptor complex is transported to the nucleus and binds to genes, changing their activity.

✔**QUESTION** Based on what you have learned about nuclear transport (see Chapter 7), what type of signal would you expect to be exposed on the cytosolic receptor after the steroid hormone changes the receptor's conformation?

Signal transduction is a common occurrence in everyday life. For example, the e-mail messages you receive are transmitted from one computer to another over cables or wireless transmissions. These electronic signals can be transmitted efficiently over long distances but would be meaningless to you. Software in your computer has to transduce, or convert, the signals into a form that you can understand and respond to, such as words on the screen.

Signal transduction pathways work the same way (**FIGURE 11.13**). In a cell, signal transduction converts an extracellular signal to an intracellular signal. As in an e-mail transmission, a signal that is easy to transmit is converted to a signal that is easily understood and that triggers a response.

Intracellular Signals May Be Amplified Recall that hormones are present in minuscule concentrations but trigger a large response from cells. Signal amplification is one reason this is possible. When a hormone arrives at the cell surface, the message it transmits may be amplified as it changes form. The amplifier

in your portable music player performs an analogous function: Once it is amplified, a tiny sound signal can get a whole roomful of people dancing.

In cells, signal transduction begins at the plasma membrane; amplification occurs inside. Amplification may occur in a variety of ways. In general, the mechanism of amplification correlates with the mechanism of signal transduction. But the general observation is that the arrival of a single signaling molecule may result in a secondary signal that involves many ions or molecules.

For example, one major type of signal transduction system consists of membrane channel receptors that open to allow a flow of ions into the cytosol of the cell. In muscle cells, this type of amplification occurs when calcium ions flood into the cytosol, activating all the myosin filaments so the entire cell contracts as a whole (see Chapter 48).

Here let's focus on two other major types of signal transduction and amplification systems that are distinguished based on how they are initiated:

1. G-protein-coupled receptors initiate the production of intracellular or "second" messengers that then amplify the signal.

2. Enzyme-linked receptors amplify the signal by triggering the activation of a series of proteins inside the cell, through the addition of phosphate groups.

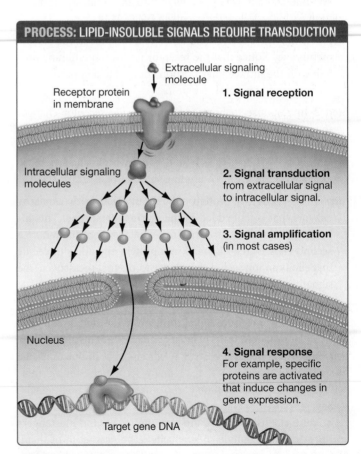

PROCESS: LIPID-INSOLUBLE SIGNALS REQUIRE TRANSDUCTION

Extracellular signaling molecule

Receptor protein in membrane

1. Signal reception

Intracellular signaling molecules

2. Signal transduction
from extracellular signal to intracellular signal.

3. Signal amplification
(in most cases)

Nucleus

4. Signal response
For example, specific proteins are activated that induce changes in gene expression.

Target gene DNA

FIGURE 11.13 Signal Transduction Converts an Extracellular Signal to an Intracellular Signal. Signal transduction is a multistep process.

Although there are many variations in the signaling pathways that fall within these two categories, the common features are emphasized here. Let's look at these two signal transduction systems in turn.

Signal Transduction via G-Protein-Coupled Receptors Many signal receptors span the plasma membrane and are closely associated with membrane-anchored proteins inside the cell called **G proteins.** When G proteins are activated by a signal receptor, they trigger a key step in signal transduction: the production of a messenger inside the cell. They link the receipt of an extracellular signal to the production of an intracellular signal.

G proteins got their name because the type of guanine nucleotide they are bound to regulates their activity: either guanosine triphosphate (GTP) or guanosine diphosphate (GDP). GTP is a nucleoside triphosphate that is similar in structure to adenosine triphosphate (ATP; introduced in Chapter 4). Recall that nucleoside triphosphates have high potential energy because their three phosphate groups have four negative charges close together.

When GTP binds to a G protein, the addition of the negative charges alters the protein's shape. Changes in shape produce changes in activity. G proteins are turned on or activated when they bind GTP; they are turned off or inactivated when a phosphate group, and thus a negative charge, is removed to form GDP. The G protein will remain in this off position until the GDP is removed and the protein binds to a new GTP.

To understand how G proteins fit into an overall signal transduction pathway, follow the events in **FIGURE 11.14**.

Step 1 A hormone arrives and binds to a receptor in the plasma membrane. Notice that the receptor is a transmembrane protein with the intracellular portion coupled to a G protein that is composed of multiple subunits.

Step 2 In response to hormone binding, the receptor changes shape and activates its G protein. Specifically, the receptor kicks out the GDP from the inactive G protein, allowing it to bind to a new GTP. When GTP is bound, the G protein changes shape radically: the active GTP-binding subunit splits off.

Step 3 The active G protein subunit interacts with a nearby enzyme that is embedded in the plasma membrane. This interaction stimulates the enzyme to catalyze production of a **second messenger**—a small, nonprotein signaling molecule that elicits an intracellular response to the first messenger (the signaling molecule that arrived at the cell surface).

Second messengers are effective because they are small and diffuse rapidly to spread the signal throughout the cell. In addition, they can be produced quickly in large quantities. This characteristic is important. Because the arrival of a single hormone molecule can stimulate the production of many second messenger molecules, the signal transduction event amplifies the original signal.

Several types of small molecules act as second messengers in cells. **TABLE 11.2** lists some of the best-studied second messengers and provides an example of how cells respond to each of them. Note that several second messengers activate **protein kinases**—enzymes that activate or inactivate other proteins by adding a phosphate group to them.

PROCESS: HOW DO G-PROTEIN-COUPLED RECEPTORS WORK?

1. G protein is "off" (bound to GDP). Signal arrives and binds to receptor.

2. Signal–receptor complex changes conformation. G protein binds GTP and splits into two parts.

3. Activated G protein binds to an enzyme and induces the production of second messenger, which triggers a response.

FIGURE 11.14 G-Protein-Coupled Receptors Trigger the Production of a Second Messenger.

It's also important to note two things:

1. Second messengers aren't restricted to a single role or single cell type—the same second messenger can initiate dramatically different events in different cell types; and

2. It is common for more than one second messenger to be involved in triggering a cell's response to the same extracellular signaling molecule.

To make sure that you understand how G proteins and second messengers work, imagine the following movie scene: A spy arrives at a castle gate. The guard receives a note from the spy, but

TABLE 11.2 Examples of Second Messengers

Name	Type of Response
Cyclic guanosine monophosphate (cGMP)	Opens ion channels; activates certain protein kinases
Diacylglycerol (DAG)	Activates certain protein kinases
Inositol trisphosphate (IP$_3$)	Opens calcium channels to transport stored calcium ions
Cyclic adenosine monophosphate (cAMP)	Activates certain protein kinases
Calcium ions (Ca^{2+})	Binds to a receptor called calmodulin; Ca^{2+}/calmodulin complex then activates proteins

he cannot read the coded message. Instead, the guard turns to the queen. She reads the note and summons the commander of the guard, who sends soldiers throughout the castle to warn everyone of approaching danger. ✔ You should be able to identify which characters in the scene correspond to the second messenger, G protein, hormone, receptor, and enzyme activated by the G protein.

It's difficult to overstate the importance of signal transduction by G-protein-coupled receptors. Biomedical researchers estimate that half of human drugs target signal receptors that are associated with G proteins.

Signal Transduction via Enzyme-Linked Receptors Enzyme-linked receptors transduce hormonal signals by directly catalyzing a reaction inside the cell. **FIGURE 11.15** focuses on the best-studied group of enzyme-linked receptors: the **receptor tyrosine kinases (RTKs).**

Step 1 A hormone binds to an RTK.

Step 2 The protein forms a dimer. In this conformation, the catalytic activity of the receptor is turned on, allowing it to phosphorylate itself using ATP inside the cell.

Step 3 Proteins inside the cell bind to the phosphorylated RTK to form a bridge between the receptor and a peripheral membrane protein called **Ras,** which is a G protein. The formation of the RTK bridge activates Ras by causing it to exchange its GDP for a GTP.

Step 4 When Ras is activated, it triggers the phosphorylation and activation of another protein.

Step 5 The phosphorylated protein is a protein kinase, which then catalyzes the phosphorylation and activation of other kinases, which phosphorylate yet another population of proteins.

This sequence of events is termed a **phosphorylation cascade,** and it culminates in a response by the cell. The enzymes involved are called **mitogen-activated protein kinases (MAPK).** They are so named because many of the signaling molecules that start

PROCESS: HOW DO ENZYME-LINKED RECEPTORS WORK?

Signaling molecule

GDP
Ras protein
Receptor tyrosine kinase (RTK)

1. Signaling molecule arrives and binds to receptor monomers, causing them to dimerize.

GDP   P   P   ATP   ADP

2. Activated receptor complex turns on the receptor's kinase activity, which phosphorylates itself at tyrosine residues.

GTP   GDP   P   P
Ras   Bridging proteins

3. Binding proteins bind to the phosphorylated receptor and one of them activates Ras by causing it to exchange GDP for GTP.

GTP   P   P
ATP   ADP
Inactive kinase 1   P   Active kinase 1
ATP   ADP
Inactive kinase 2   P   Active kinase 2
ATP   ADP
Inactive kinase 3
Triggers response

4. Activated Ras triggers phosphorylation and activation of a protein kinase.

5. A phosphorylation cascade results, where each kinase phosphorylates a different kinase until a response is triggered in the cell.

FIGURE 11.15 Enzyme-Linked Receptors Trigger a Phosphorylation Cascade.

these pathways are substances called mitogens, which activate cell division. (The *mito–* in mitogen stands for mitosis, a process involved in eukaryotic cell division.)

In some cases, each copy of an enzyme in the cascade catalyzes the phosphorylation of many copies of the next "downstream" protein, and so on. When this occurs, activated enzymes at a given stage in the cascade exist in greater numbers than the activated enzymes that preceded them, and the original signal is amplified many times over.

To make sure you understand how RTKs and phosphorylation cascades work, imagine that you have two red dominos, one black domino, and a large supply of green, blue, and yellow dominos. The red dominos represent the two subunits of an RTK dimer, and the black domino represents Ras. Each of the other colors represents a different protein kinase in a phosphorylation cascade. ✔ QUANTITATIVE You should be able to (1) explain how you would set up the dominos to simulate a phosphorylation cascade, and (2) state how many green, blue, and yellow dominos would be required to model the pathway if Ras and each protein kinase in the cascade were to activate 10 proteins.

In many cases, the proteins that take part in a phosphorylation cascade are held in close physical proximity by scaffolding proteins. Although this organization limits the amplification of the response, it increases the speed and efficiency of the reaction sequence.

In general, intracellular signals initiated by G-protein-coupled receptors result in the production of second messengers, while enzyme-linked receptors drive phosphorylation cascades. It's important to recognize, however, that these pathways overlap significantly. Some G-protein-coupled receptors trigger phosphorylation cascades, and some enzyme-linked receptors result in the production of second messengers.

To summarize: Many of the key signal transduction events observed in cells occur via G-protein-coupled receptors or enzyme-linked receptors. The signal transduction event has two results: **(1)** It converts an easily transmitted extracellular message into an intracellular message, and **(2)** in some cases it amplifies the original message many times over.

Signal Response

What is the ultimate response to the messages carried by hormones? Recall that when adjacent cells share information through cell–cell gaps, two general categories of response may occur (see Section 11.2). Likewise, second messengers or a cascade of protein phosphorylation events also may alter gene expression, or activate or deactivate existing proteins in the target cell.

For example, when plants experience drought, the tissues in the root system are the first to respond by secreting the hormone abscisic acid. This hormone travels huge distances to reach the leaves and eventually bind to its receptors in guard cells that control the stomatal pores that allow for gas exchange (see Chapter 10). When abscisic acid binds to these receptors, a signal transduction pathway ensues that increases the concentration of calcium inside the guard cells. In response, potassium ions (K^+) move out of the guard cells, creating an osmotic gradient that leads to the movement of water out of the guard cells. The guard cells deflate and close the pore, which prevents water loss from the plant.

At this point, you've analyzed the first three steps of cell–cell communication: signal reception, signal processing, and the response. Now the question is, how is the signal turned off? Consider the flush of testosterone and estrogens that you experienced during puberty, and the morphological changes these hormones induced. Abnormalities would result if these changes continued indefinitely. What limits the response to a cell–cell signal?

Signal Deactivation

Cells have built-in systems for turning off intracellular signals. For example, once activated G proteins turn on downstream enzymes, the GTP is hydrolyzed to GDP and P_i. When this reaction occurs, the G protein's conformation changes and it returns to an inactive state. Activation of its downstream target stops, and production of the second messenger ceases.

The presence of second messengers in the cytosol is also short lived. For example, pumps in the membrane of the smooth ER return calcium ions to storage, and enzymes called phosphodiesterases convert active cAMP and cGMP (see Table 11.2) to inactive AMP and GMP. When second messengers are cleared from the cytosol, the response stops.

To continue the response from G-protein-coupled receptors, the G proteins must be reactivated by the activated signal receptor to start the process again. Otherwise, the signal transduction system quickly shuts down.

Phosphorylation cascades are also sensitive to the presence of the external signal. Enzymes, called **phosphatases,** that remove phosphate groups from cascade proteins are always present in the cell. If hormone stimulation of a receptor tyrosine kinase ends, phosphatases will dephosphorylate enough components of the phosphorylation cascade that the response ceases.

Although an array of specific mechanisms are involved, here is the general observation: Signal transduction systems trigger a rapid response and can be shut down quickly. As a result, they are exquisitely sensitive to small changes in the concentration of hormones or in the number and activity of signal receptors.

It is critical, though, to appreciate what happens when a signal transduction system does not shut down properly. For example, recall that Ras is active when it binds GTP, but is deactivated once it has hydrolyzed GTP to GDP and P_i. If this hydrolysis activity were defective, however, Ras would stay in the GTP-bound "on" position and continue stimulating a phosphorylation cascade even when the external signal is absent.

Why is continuously active Ras a problem? Recall that the phosphorylation cascade that is activated by Ras involves mitogen-activated protein kinases, many of which induce cell replication. If cells express this type of defective Ras, they would receive a never-ending "divide now" signal that may lead to the development of cancer. An estimated 25–30 percent of all human cancers express this type of defective Ras.(To learn more about the family of diseases called cancer, see Chapters 12 and 19.)

Crosstalk: Synthesizing Input from Many Signals

Although the preceding discussion focused on how cells respond to individual signals, it's crucial to realize that every cell has an array of signal receptors on its plasma membrane and in its cytoplasm, and many cells receive an almost constant stream of different signals. Just as you receive information about your environment via text messages, e-mails, phone calls, and snail mail, cells get an array of chemical signals about changes in their environment.

The signal transduction pathways that are triggered by these signals and receptors intersect and connect. In reality, they are not strictly linear like the pathways illustrated in Figures 11.12 through 11.15. Signal transduction pathways form a network. This complexity is important: It allows cells to respond to many different signals in an integrated way.

The diverse signals that a cell receives are integrated by what biologists call **crosstalk**—meaning, the signals from different pathways interact to modify the cell response (**FIGURE 11.16**). This would be like getting advice from multiple people before making a decision. Here are three key things to note:

1. Elements or products from one pathway may inhibit steps in a different pathway—reducing the cell's response, even though the appropriate signal is present.

2. A response from one pathway may stimulate a greater response by a protein in a different pathway, increasing the cell's response to the other signal.

3. The presence of multiple steps in each signaling pathway provides a series of points where crosstalk can regulate the flow of information. These interactions are important, because they allow the cell to respond appropriately to many signals at the same time.

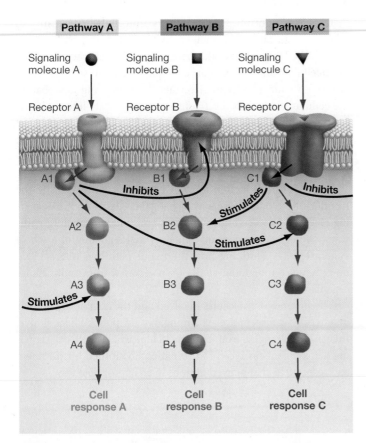

FIGURE 11.16 Signaling Pathways Interact via "Crosstalk."

✔ **EXERCISE** Predict which response would occur in cells exposed to the following signaling molecules: (1) A + B; (2) A + C; (3) C alone.

11.4 Signaling between Unicellular Organisms

Cell–cell signaling has been one of the hottest research areas in biological science over the past two decades. Surprisingly, much of what we know about signal transduction in multicellular organisms has come from the study of unicellular organisms. While single-cell microbes communicate with one another in a manner similar to what is observed in multicellular organisms, the topic of conversation often differs. Rather than asking for help, as when a dehydrated brain asks the kidney to conserve water, the conversations between unicellular microbes are often about sex and environmental change.

Responding to Sex Pheromones

While unicellular eukaryotes generally reproduce by cell division, some also are known to undergo sexual reproduction (see Chapters 12 and 13 to learn more about cellular reproduction). At its most basic level, sex involves the fusion of two cells such that genetic material of the two individuals is combined into one nucleus. What attracts individuals of the opposite sex to each other?

In *Saccharomyces cerevisiae*, or baker's yeast, there are two sexes, or mating types, referred to as "**a**" cells and "**α**" cells. By

(a) Yeast cells alter their growth in response to pheromones of the opposite mating type.

(b) Slime mold amoebae aggregate in response to sensing a quorum.

Red stain shows new cell growth

5 μm

100 μm

Cells migrate toward a central point

FIGURE 11.17 Unicellular Organisms Interact and Respond in a Variety of Ways. (a) Yeast cells respond to pheromone signals from the opposite mating type by growing toward the source of the signal during sexual reproduction. **(b)** Signals secreted by free-living slime mold protists stimulate them to aggregate under high population densities.

studying this eukaryotic model organism, biologists have figured out the cell signaling events that bring yeast of the opposite sex together (to learn more about model organisms, see **BioSkills 13** in Appendix B).

Although the yeast cells are visually indistinguishable, opposite mating types recognize one another via chemical signaling molecules called **pheromones.** The **a** mating type secretes the pheromone **a** factor, and the **α** type secretes the pheromone **α** factor. Receptors on the surface of these cells will bind only to the opposite mating type factor, so when an **a** cell detects the **α** factor, it recognizes that a suitable mate is nearby.

Since yeast cells are not motile, part of the cellular response to the pheromone is to direct new growth toward the signal (**FIGURE 11.17a**). The signaling pathway that is responsible for this morphological change uses both G-protein-coupled receptors and phosphorylation cascades that were presented in Section 11.3. One of the proteins affected by this signaling pathway is actin, which is used to construct new microfilaments at the site where the G proteins have been activated. These new filaments push out new growth toward the highest concentration of the pheromone to allow the cell to find, and eventually fuse with, its mate.

Responding to Population Density

Within a population of unicellular organisms, widespread communication can occur that closely monitors the environment—in particular, the density of the population. Signaling pathways that respond to population density in microbes are collectively referred to as **quorum sensing.** The name was inspired by the observation that cells of the same species may undergo dramatic

changes in activity when their numbers reach a threshold, or quorum.

Quorum sensing is based on signaling molecules that are secreted by cells and diffuse through the environment. The response to these molecules varies dramatically between species and ranges from bacterial bioluminescence—or light emission—in the light organs of squid to the secretion of molecules that help glue a community of microbes to a surface in biofilms (see Chapter 29).

In effect, quorum sensing allows unicellular organisms to communicate and coordinate activity. When it occurs, these single-celled organisms take on some of the characteristics of multicellular organisms. For example, quorum sensing via a G-protein-coupled receptor recruits the free-living amoebae of the cellular slime mold *Dictyostelium* to aggregate into multicellular mounds (**FIGURE 11.17b**). Amazingly, the slug-like bodies that are formed from these aggregates can crawl across a surface and eventually organize themselves into a fruiting body that releases spores into the air (see **BioSkills 13** in Appendix B).

This brief introduction to cell–cell signaling brings us to the frontier of biological research. It has taken decades of painstaking research to work out each step in individual signaling pathways. Biologists are now investigating cell signaling at a whole system level—examining how the major pathways interact and how they are integrated at the tissue level within multicellular organisms.

The number of molecules involved and the complexity of their interactions can seem overwhelming, but the punch line is simple: In organisms ranging from bacteria to blue whales, cell–cell signaling helps organisms receive information about their environment and respond to changing conditions in an appropriate way.

If you understand . . .

11.1 The Cell Surface

- The vast majority of cells secrete an extracellular layer.

- In bacteria, archaea, algae, and plants, the extracellular material is stiff and is called a cell wall. In animals, the secreted layer is flexible and is called the extracellular matrix (ECM).

- Extracellular layers are fiber composites. They consist of cross-linked filaments that provide tensile strength and a ground substance that fills space and resists compression.

- In plants, the extracellular filaments are cellulose microfibrils; in animals, the most abundant filaments are made of the protein collagen. In both plants and animals, the ground substance is composed of gel-forming polysaccharides.

✔ You should be able to predict what happens to animal cells when they are treated with an enzyme that (1) cuts integrin molecules, or (2) digests collagen fibrils.

11.2 How Do Adjacent Cells Connect and Communicate?

- In multicellular organisms, molecules in the extracellular layer and plasma membrane mediate interactions between adjacent cells.

- Adjacent cells may be physically bound to one another by glue-like middle lamellae in plants or by tight junctions and desmosomes in animals.

- The cytoplasm of adjacent cells is in direct communication through openings called plasmodesmata in plants and gap junctions in animals.

- Cells may respond to signals by activating certain enzymes, releasing or taking up specific ions or molecules, or changing the activity of target genes. As a result, cells and tissues throughout the body can alter their activity in response to changing conditions, and do so in a coordinated way.

✔ You should be able to predict the consequences of removing the gap junctions between the cells in the cardiac muscle in your heart.

11.3 How Do Distant Cells Communicate?

- Distant cells in multicellular organisms communicate through signaling molecules that bind to receptors found in or on specific target cells.

- Cell–cell signaling molecules that are not lipid soluble bind to receptors in the plasma membrane. The receptor then changes conformation and triggers production of a new type of intracellular signal—a second messenger or phosphorylation cascade.

- Because enzymes inside the cell quickly deactivate the signal and signaling pathways often interact, the cell's response is tightly regulated.

✔ You should be able to explain how the hormone adrenalin can stimulate cells in both the heart and the liver, yet trigger different responses (increasing heart rate versus releasing glucose).

11.4 Signaling between Unicellular Organisms

- Unicellular organisms use chemical signals to sense the pheromones of opposite mating types and population density. Quorum sensing allows closely related cells to coordinate changes in their activity when population density is high.

✔ You should be able to compare and contrast the role of intercellular signaling between unicellular organisms and the cells in a multicellular organism.

(MB) MasteringBiology

1. **MasteringBiology Assignments**

   Tutorials and Activities Build a Signaling Pathway; Cell Junctions; Cell Signaling: Reception; Cell Signaling: Transduction and Response; Cellular Responses; Overview of Cell Signaling; Reception; Signal Transduction Pathways

   Questions Reading Quizzes, Blue-Thread Questions, Test Bank

2. **eText** Read your book online, search, take notes, highlight text, and more.

3. **The Study Area** Practice Test, Cumulative Test, BioFlix® 3-D Animations, Videos, Activities, Audio Glossary, Word Study Tools, Art

You should be able to . . .

1. What is a fiber composite? How do cellular fiber composites resemble reinforced concrete?

2. In animals, where are most components of the extracellular material synthesized?
   a. smooth ER
   b. the rough ER
   c. in the extracellular layer itself
   d. in the plasma membrane

3. Treating dissociated cells with certain antibodies makes the cells unable to reaggregate. Why?
   a. The antibodies bind to cell adhesion proteins.
   b. The antibodies bind to the fiber component of the extracellular matrix.
   c. The antibodies bind to the cell surface and inhibit motility.
   d. The antibodies act as enzymes that break down desmosomes.

4. What does it mean to say that a signal is transduced?
   a. The signaling molecule enters the cell directly and binds to a receptor inside.
   b. The physical form of the signal changes between the outside of the cell and the inside.
   c. The signal is amplified, such that even a single molecule evokes a large response.
   d. The signal triggers a sequence of phosphorylation events inside the cell.

5. What characteristics do tight junctions bestow on tissues that use these adhesions to connect adjacent cells?
   a. They allow communication between adjacent cells.
   b. They provide strong connections to resist pulling forces.
   c. They use the extracellular matrix to indirectly connect adjacent cells.
   d. They form a watertight barrier between the cells.

6. What are the two general categories of cellular responses to an intercellular signal?

7. Which of the following statements represents a fundamental difference between the fibers found in the extracellular layers of plants and those of animals?
   a. Plant fibers resist compression forces; animal fibers resist pulling forces.
   b. Animal fibers consist of proteins; plant fibers consist of polysaccharides instead.
   c. Plant extracellular fibers never move; animal fibers can slide past one another.
   d. Cellulose microfibrils run parallel to one another; collagen filaments crisscross.

8. Explain how it is possible for a phosphorylation cascade to amplify an intercellular signal.

9. Compare and contrast the structure and function of tight junctions, desmosomes, and gap junctions.

10. Animal cells adhere to each other selectively. Summarize experimental evidence that supports this statement. Explain the molecular basis of selective adhesion.

11. Make a flowchart summarizing the reception, processing, response, and deactivation steps for a signaling molecule that binds to an intracellular receptor.

12. What is the significance of the observation that many signal transduction pathways create a network, where they intersect or overlap?

13. What would be the impact on the structure of a plant tissue if the cells lacked the ability to modify the extracellular environment?
    a. Cells would swell and burst if placed in a hypotonic environment.
    b. Cells would not be able to adhere to one another.
    c. No defined tissues, consisting of similar cells with coordinated activities, would be possible.
    d. All of the above.

14. Suppose that a particular cell–cell signaling molecule induces a cellular response without requiring signal transduction (i.e., no second messengers or phosphorylation cascades). Compared to the signal transduction pathways you learned about in this chapter, how would an event like this affect (a) the types of responses that are possible, (b) amplification, and (c) regulation?

15. In most species of fungi, chitin is a major polysaccharide found in cell walls. Review the structure of chitin (see Chapter 5), and then describe what would have to take place for the directional growth that occurs when yeast, a type of fungi, respond to sex pheromones.

16. Suppose you created an antibody that bound to the receptor tyrosine kinase illustrated in Figure 11.15. You expected this antibody to inhibit the cell response, but instead it resulted in activating the response, even when no signal was present. Explain this result.

12 The Cell Cycle

In this chapter you will learn how

The life cycle of a cell culminates in division

starting with

The four phases of the cell cycle 12.1

asking

How does cell division take place? 12.2

by examining

Control of the cell cycle 12.3

via and

Mitosis Cytokinesis

and applying

Cancer: out-of-control cell division 12.4

This cell, from a hyacinth plant, is undergoing a type of nuclear division called mitosis. Understanding how mitosis occurs is a major focus of this chapter.

This chapter is part of the Big Picture. See how on pages 366–367.

The cell theory maintains that all organisms are made of cells and all cells arise from preexisting cells (Chapter 1). Although the cell theory was widely accepted among biologists by the 1860s, most believed that new cells arose within preexisting cells by a process that resembled the growth of mineral crystals. But Rudolf Virchow proposed that new cells arise by splitting preexisting cells—that is, by **cell division.**

In the late 1800s, microscopic observations of newly developing organisms, or **embryos,** confirmed Virchow's hypothesis. Multicellular eukaryotes start life as single-celled embryos and grow through a series of cell divisions.

Early studies revealed two fundamentally different ways that nuclei divide before cell division: meiosis and mitosis. In animals, **meiosis** leads to the production of sperm and eggs, which are the male and female reproductive cells termed **gametes. Mitosis** leads to the production of all other cell types, referred to as **somatic** (literally, "body-belonging") **cells.** (You can see the Big Picture of

✔ When you see this checkmark, stop and test yourself. Answers are available in Appendix A.

how these two nuclear divisions are related to each other and the transmission of genetic information on pages 366–367.)

Mitosis and meiosis are usually accompanied by **cytokinesis** ("cell movement")—the division of the cytoplasm into two distinct cells. When cytokinesis is complete, a so-called parent cell has given rise to two daughter cells.

Mitotic and meiotic cell divisions are responsible for one of the five fundamental attributes of life: reproduction (see Chapter 1). But even though mitosis and meiosis share many characteristics, they are fundamentally different. During mitosis, the genetic material is copied and then divided equally between two cells. This is referred to as cellular *replication*, since these daughter cells are genetically identical with the original parent cell. In contrast, meiosis results in daughter cells that are genetically different from each other and that have half the amount of hereditary material as the parent cell.

This chapter focuses on mitotic cell division; meiotic cell division is the subject of the next chapter (Chapter 13). Let's begin with a look at the key events in a cell's life cycle, continue with an in-depth analysis of mitosis and the regulation of the cell cycle, and end by examining how uncontrolled cell division can lead to cancer.

12.1 How Do Cells Replicate?

For life on Earth to exist, cells must replicate. The general requirements for cellular replication are to (1) copy the DNA (deoxyribonucleic acid), (2) separate the copies, and (3) divide the cytoplasm to create two complete cells.

This chapter focuses on eukaryotic cell replication, which is responsible for three key events:

1. *Growth*   The trillions of genetically identical cells that make up your body are the product of mitotic divisions that started in a single fertilized egg.

2. *Wound repair*   When you suffer a scrape, cellular replication generates the cells that repair your skin.

3. *Reproduction*   When yeast cells grow in bread dough or in a vat of beer, they are reproducing by cellular replication. In yeasts and other single-cell eukaryotes, mitotic division is the basis of asexual reproduction. **Asexual reproduction** produces offspring that are genetically identical with the parent.

These events are so basic to life that cell replication has been studied for well over a century. Like much work in biology, the research on how cells divide began by simply observing the process.

What Is a Chromosome?

As studies of cell division in eukaryotes began, biologists found that certain chemical dyes made threadlike structures visible within nuclei. In 1879, Walther Flemming used a dye made from a coal tar to observe these threadlike structures and how they changed in the dividing cells of salamander embryos. The threads

first appeared in pairs just before division and then split to produce single, unpaired threads in the daughter cells. Flemming introduced the term mitosis, from the Greek *mitos* ("thread"), to describe this process.

Others studied the roundworm *Ascaris* and noted that the total number of threads in a cell was the same before and after mitosis. All the cells in a roundworm had the same number of threads.

In 1888 Wilhelm Waldeyer coined the term **chromosome** ("colored-body") to refer to these threadlike structures (visible in the chapter-opening photo). A chromosome consists of a single, long DNA double helix that is wrapped around proteins, called **histones,** in a highly organized manner. DNA encodes the cell's hereditary information, or genetic material. A gene is a length of DNA that codes for a particular protein or ribonucleic acid (RNA) found in the cell.

Before mitosis, each chromosome is replicated. As mitosis starts, the chromosomes condense into compact structures that can be moved around the cell efficiently. Then one of the chromosome copies is distributed to each of two daughter cells.

FIGURE 12.1 illustrates unreplicated chromosomes, replicated chromosomes before they have condensed prior to mitosis, and replicated chromosomes that have condensed at the start of mitosis. Each of the DNA copies in a replicated chromosome is called a **chromatid.** Before mitosis, the two chromatids are joined along their entire length by proteins called cohesins. Once mitosis begins, however, many of these connections are removed except for those at a specialized region of the chromosome called the **centromere.** Chromatid copies that remain attached at their centromere are referred to as **sister chromatids.** Even though a replicated chromosome consists of two chromatids, it is still considered a single chromosome.

Cells Alternate between M Phase and Interphase

The division of eukaryotic cells is like a well-choreographed stage performance. The most visually stimulating part of the show occurs when cells are in their dividing phase, called the **M** (*m*itotic or *m*eiotic) **phase.** With a light microscope, chromosomes can be stained and observed as discrete units only during M phase, when they condense into compact structures.

The rest of the time, the cell is in **interphase** ("between-phase"). No dramatic changes in the nucleus are visible by light microscopy during interphase. The chromosomes uncoil into the extremely long, thin structures shown in Figure 12.1 and are no longer stained as individual threads. However, this does not mean that the cell is idle. Interphase is an active time: The cell is either growing and preparing to divide or fulfilling its specialized function in a multicellular individual. Cells actually spend most of their time in interphase.

The Discovery of S Phase

Once M phase and interphase were identified by microscopy, researchers could start assigning roles to these distinct phases. They could see that the separation of chromosomes and cytokinesis

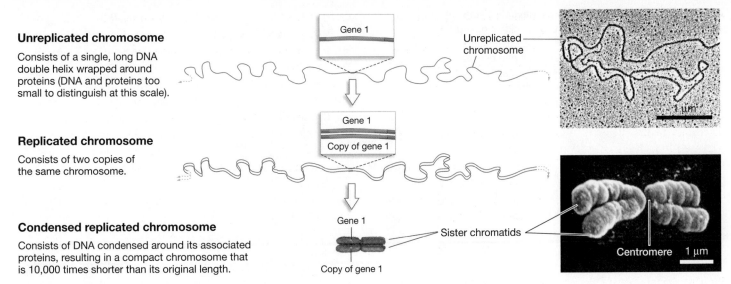

Unreplicated chromosome

Consists of a single, long DNA double helix wrapped around proteins (DNA and proteins too small to distinguish at this scale).

Gene 1

Unreplicated chromosome

1 µm

Replicated chromosome

Consists of two copies of the same chromosome.

Gene 1

Copy of gene 1

Condensed replicated chromosome

Consists of DNA condensed around its associated proteins, resulting in a compact chromosome that is 10,000 times shorter than its original length.

Gene 1

Copy of gene 1

Sister chromatids

Centromere   1 µm

FIGURE 12.1 Changes in Chromosome Morphology. After chromosomes replicate, the two identical copies are attached to each other along their entire length. Early in mitosis, replicated chromosomes condense and sister chromatids remain attached at a region called the centromere.

took place during the M phase, but when are the chromosomes copied?

To answer this question, researchers needed to distinguish cells that were replicating their DNA from those that were not. They were able to do this by adding radioactive phosphorus, in the form of phosphates, to cells. Those cells that were synthesizing DNA would incorporate the radioactive isotope into nucleotides (see Chapter 4 to review where phosphates are in DNA).

The idea was to:

1. label DNA as chromosomes were being copied;

2. wash away any radioactive isotope that hadn't been incorporated and remove RNA, which would also incorporate phosphorus; and then

3. visualize the labeled, newly synthesized DNA by exposing the treated cells to X-ray film. Emissions from radioactive phosphorus create a black dot in the film. This is the technique called autoradiography (see **BioSkills 9** in Appendix B).

In 1951, Alma Howard and Stephen Pelc performed this experiment and looked for cells with black dots—indicating active DNA synthesis—immediately after the exposure to a radioactive isotope ended. They found black dots in some of the interphase cells, but none in M-phase cells. Several years later, these results were verified using radioactive thymidine, which is incorporated into DNA but not RNA. These results were strong evidence that DNA replication occurs during interphase.

Thus, biologists had identified a new stage in the life of a cell. They called it **synthesis** (or **S**) **phase.** S phase is part of interphase. Replication of the genetic material is separated, in time, from the partitioning of chromosome copies during M phase.

Howard and Pelc coined the term **cell cycle** to describe the orderly sequence of events that leads a eukaryotic cell through the duplication of its chromosomes to the time it divides.

The Discovery of the Gap Phases

In addition to discovering the S phase, Howard and Pelc made another key observation—not all the interphase cells were labeled. This meant that there was at least one "gap" in interphase when DNA was not being copied.

Howard and Pelc, along with researchers in other labs, followed up on these early results by asking where S phase was positioned in interphase. There were three possible scenarios:

1. The S phase is immediately before M phase, with a single gap between the end of M and start of S phase;

2. the S phase is immediately after M phase, with a gap between the end of S and the start of M phase; or

3. two gaps exist, one before and one after the S phase.

To address which of these models, if any, is correct, many experiments were done using cells in culture. Cultured cells are powerful experimental tools because they can be manipulated much more easily than cells in an intact organism (see **BioSkills 12** in Appendix B). In most of these studies, researchers used cultures that were asynchronous, meaning that the cells were randomly distributed in the cycle.

To understand the value of these asynchronous cultures, imagine the cell cycle were a clock. Every complete rotation around the clock would represent one cell division, and each tick would represent a different point in the cycle. At any given time, an asynchronous culture would have at least one cell present at each of the ticks on the clock. As time passes, these cells would move around this cell-cycle clock at the same rate and in the same direction.

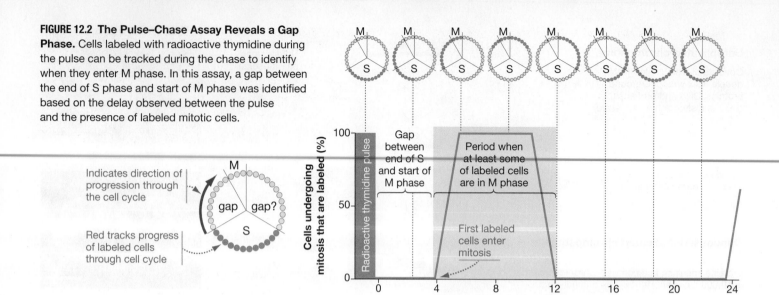

FIGURE 12.2 The Pulse–Chase Assay Reveals a Gap Phase. Cells labeled with radioactive thymidine during the pulse can be tracked during the chase to identify when they enter M phase. In this assay, a gap between the end of S phase and start of M phase was identified based on the delay observed between the pulse and the presence of labeled mitotic cells.

Indicates direction of progression through the cell cycle

Red tracks progress of labeled cells through cell cycle

Radioactive thymidine pulse

Gap between end of S and start of M phase

Period when at least some of labeled cells are in M phase

First labeled cells enter mitosis

Cells undergoing mitosis that are labeled (%)

Time since end of thymidine pulse (hours)

In one experiment, researchers marked the S-phase cells in a human cell culture by exposing it to radioactively labeled thymidine. A short time later, they stopped the labeling by flooding the solution surrounding the cultured cells with nonradioactive thymidine. This pulse–chase approach (introduced in Chapter 7) labeled only those cells that were in S phase during the radioactive pulse. Imagine these marked cells moving like a hand on the clock that could be tracked as they progressed through the cell cycle.

Once the pulse ended, the researchers analyzed samples of the culture at different times during the chase. For each batch of cells, they recorded how many labeled cells were undergoing mitosis, meaning that the cells that were in S phase during the pulse had entered M phase.

One striking result emerged early on: None of the labeled cells started mitosis immediately. Because the cultures were asynchronous, at least some of the cells must have been at the very end of their S phase when exposed to the pulse. If the S phase had been immediately followed by the M phase, some of these labeled cells would have entered M just as the chase began. Instead, it took several hours before any of the labeled cells began mitosis.

The time lag between the end of the pulse and the appearance of the first labeled mitotic nuclei corresponds to a period between the end of S phase and the beginning of M phase. This gap represents the time when chromosome replication is complete, but mitosis has not yet begun. **FIGURE 12.2** shows how cells labeled with radioactive thymidine can be tracked as they progress through the M phase.

After this result, the possibilities for the organization of the cell cycle were narrowed to either one gap between the end of S and start of M phase, or two gaps that flank the S phase. Which of the models best represents the eukaryotic cell cycle? Once researchers determined the lengths of the S and M phases, they found that the combined time, including the gap between these phases, was still short compared with the length of the cell cycle. This discrepancy represents an additional gap phase that is between the end of M and the start of S phase.

The cell cycle was thus finally mapped out. The gap between the end of M and start of S phase is called the **G₁ phase.** The second gap, between the end of S and start of M phase, is called the **G₂ phase.**

The Cell Cycle

FIGURE 12.3 pulls these results together into a comprehensive view of the cell cycle. The cell cycle involves four phases: M phase and an interphase consisting of the G_1, S, and G_2 phases. In the cycle diagrammed here, the G_1 phase is about twice as long as G_2, but the timing of these phases varies depending on the cell type and growth conditions.

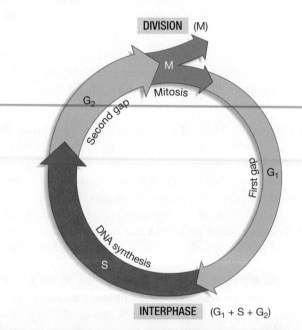

FIGURE 12.3 The Cell Cycle Has Four Phases. A representative cell cycle. The time required for the G_1 and G_2 phases varies dramatically among cells and organisms.

Why do the gap phases exist? Besides needing to copy their chromosomes during interphase, cells also must prepare for division by replicating organelles and increasing in size. Before mitosis can take place, the parent cell must grow large enough to divide into two cells that will be normal in size and function. The two gap phases provide the time required to accomplish these tasks. They allow the cell to complete all the requirements for cell division other than chromosome replication.

Now let's turn to the M phase. Once the genetic material has been copied, how do cells divide it between daughter cells?

12.2 What Happens during M Phase?

The M phase typically consists of two distinct events: the division of the nucleus and the division of the cytoplasm. During cell replication, mitosis divides the replicated chromosomes to form two daughter nuclei with identical chromosomes and genes. Mitosis is usually accompanied by cytokinesis—cytoplasmic division that results in two daughter cells.

FIGURE 12.4 provides an overview of how chromosomes change before, during, and after mitosis and cytokinesis, beginning with a hypothetical plant cell or animal cell in G_1 phase. The first drawing shows a total of four chromosomes in the cell, but chromosome number varies widely among species—chimpanzees and potato plants have a total of 48 chromosomes in each cell; a maize (corn) plant has 20, dogs have 78, and fruit flies have 8.

Eukaryotic chromosomes consist of DNA wrapped around the globular histone proteins. In eukaryotes this DNA–protein material is called **chromatin.** During interphase, the chromatin of each chromosome is in a "relaxed" or uncondensed state, forming long, thin strands (see Figure 12.1, top).

The second drawing in Figure 12.4 shows chromosomes that have been copied before mitosis. Each chromosome now consists of two sister chromatids. Each chromatid contains one long DNA double helix, and sister chromatids represent exact copies of the same genetic information.

At the start of mitosis, then, each chromosome consists of two sister chromatids that are attached to each other at the centromere.

✔ You should be able to explain the relationship between chromosomes and (1) genes, (2) chromatin, and (3) sister chromatids.

Events in Mitosis

As the third drawing in Figure 12.4 indicates, mitosis begins when chromatin condenses to form a much more compact structure. Replicated, condensed chromosomes correspond to the paired threads observed by early biologists.

During mitosis, the two sister chromatids separate to form independent daughter chromosomes. One copy of each chromosome goes to each of the two daughter cells. (See the final drawing in Figure 12.4.) As a result, each cell receives an identical copy of the genetic information that was contained in the parent cell.

Biologists have identified five subphases within M phase based on distinctive events that occur. Interphase is followed by the mitotic subphases of prophase, prometaphase, metaphase, anaphase, and telophase.

Recall that before mitosis begins, chromosomes have already replicated during the S phase of interphase. Now let's look at how cells separate the chromatids in these replicated chromosomes by investigating each subphase of mitosis in turn (**FIGURE 12.5**, on page 224).

INTERPHASE · M PHASE · Daughter cells

G₁ PHASE · S PHASE · G₂ PHASE

Parent cell · Parent cell · Parent cell

Sister chromatids

4 unreplicated chromosomes (chromosomes are shown partially condensed to make them visible)

4 replicated chromosomes, each consisting of two sister chromatids

At start of mitosis, replicated chromosomes condense.

During mitosis, sister chromatids separate. Two daughter cells are formed by cytokinesis.

FIGURE 12.4 An Overview of the Cell Cycle. Chromosomes are replicated during S phase to produce the G_2 cell. During M phase, the replicated chromosomes are partitioned to the two daughter cells. Each daughter cell contains the same complement of chromosomes as the parent cell.

How Does the G₁ Checkpoint Work? In G_0 cells, the arrival of growth factors stimulates the production of a key regulatory protein called E2F. When E2F is activated, it triggers the expression of genes required for S phase.

When E2F is first produced, however, its activity is blocked by a tumor suppressor protein called Rb. **Rb protein** is one of the key molecules that enforces the G_1 checkpoint. It is called Rb because a nonfunctional version was first discovered in children with retinoblastoma, a cancer in the light-sensing tissue, or retina, of the eye.

When E2F is bound to Rb, it is in the "off" position—it can't activate the genes required for S phase. As long as Rb stays bound to E2F, the cell remains in G_0. But as **FIGURE 12.15** shows, the situation changes dramatically if growth factors continue to arrive. To understand how growth factors affect E2F activity, think back to how cells progress from G_2 to M phase. As in passage from G_2 to M phase, phosphorylation of other proteins catalyzed by an activated cyclin–Cdk dimer permits passage from G_1 to S.

Step 1 Growth factors arrive from other cells.

Step 2 The growth factors stimulate the production of E2F and of G_1 cyclins, which are different from those used in MPF.

Step 3 Rb binds to E2F, inactivating it. The G_1 cyclins begin forming cyclin–Cdk dimers. Initially, the Cdk component is phosphorylated and inactive.

Step 4 When dephosphorylation turns on the G_1 cyclin–Cdk complexes, they catalyze the phosphorylation of Rb.

Step 5 The phosphorylated Rb changes shape and releases E2F.

Step 6 The unbound E2F is free to activate its target genes. Production of S-phase proteins gets S phase under way.

In this way, growth factors function as a social signal that says, "It's OK to override Rb. Go ahead and pass the G_1 checkpoint and divide."

How Do Social Controls and Cell-Cycle Checkpoints Fail? Cells can become cancerous when social controls fail—meaning, when cells begin dividing in the absence of the go-ahead signal from growth factors. One of two things can go wrong: The G_1 cyclin is overproduced, or Rb is defective.

When cyclins are overproduced and stay at high concentrations, the Cdk that binds to cyclin phosphorylates Rb continuously. This activates E2F and sends the cell into S phase.

Cyclin overproduction results from (1) excessive amounts of growth factors or (2) cyclin production in the absence of growth signals. Cyclins are produced continuously when a signaling pathway is defective. Because this pathway includes the Ras protein (highlighted in Chapter 11), it is common to find overactive Ras proteins in cancerous cells.

What happens if Rb is defective? When Rb is missing or does not bind normally to E2F, any E2F that is present pushes the cell through the G_1 checkpoint and into S phase, leading to uncontrolled cell division.

Because cancer is actually a family of diseases with a complex and highly variable molecular basis, there will be no "magic bullet," or single therapy, that cures all forms of the illness. Still, recent progress in understanding the cell cycle and the molecular basis of cancer has been dramatic, and cancer prevention and early detection programs are increasingly effective. The prognosis for many cancer patients is remarkably better now than it was even a few years ago. Thanks to research, almost all of us know someone who is a cancer survivor.

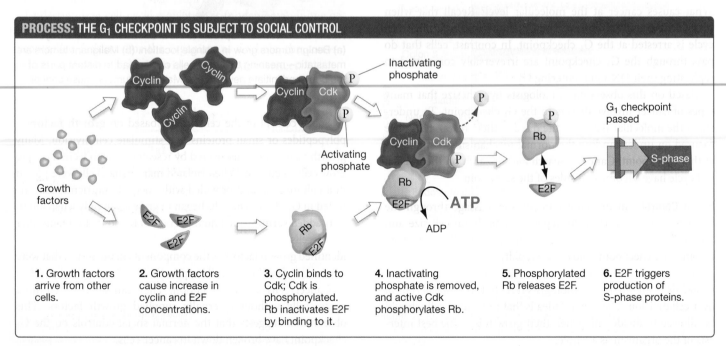

PROCESS: THE G₁ CHECKPOINT IS SUBJECT TO SOCIAL CONTROL

1. Growth factors arrive from other cells.

2. Growth factors cause increase in cyclin and E2F concentrations.

3. Cyclin binds to Cdk; Cdk is phosphorylated. Rb inactivates E2F by binding to it.

4. Inactivating phosphate is removed, and active Cdk phosphorylates Rb.

5. Phosphorylated Rb releases E2F.

6. E2F triggers production of S-phase proteins.

FIGURE 12.15 Growth Factors Move Cells through the G₁ Checkpoint.

If you understand . . .

12.1 How Do Cells Replicate?

- When a cell divides, it must copy its chromosomes, separate the copies, and divide the cytoplasm to generate daughter cells such that each carries the same chromosomal complement as the parent.
- Eukaryotic cells divide by alternating between interphase and M phase.
- Interphase consists of S phase, when chromosomes are replicated, and gap phases called G_1 and G_2, when cells grow and prepare for division.
- Eukaryotic cells divide by cycling through four phases: G_1, S, G_2, and M. Mature cells arrest at G_1 and enter a nonreplicating phase called G_0.

✔ You should be able to explain the roles of each of the four stages of the cell cycle.

12.2 What Happens during M Phase?

- Mitosis and cytokinesis are responsible for the partitioning of chromosomes and division of the parent cell into two daughter cells.
- Mitosis can be described as a sequence of five phases:

  1. *Prophase*   Chromosomes condense. The spindle apparatus begins to form, and polar microtubules overlap each other.
  2. *Prometaphase*   In cells of many organisms, the nuclear envelope disintegrates. Microtubules attach to the kinetochores of chromosomes and begin moving them to the middle of the spindle.
  3. *Metaphase*   All the chromosomes are positioned in the middle of the spindle. The spindle is anchored to the cell membrane by astral microtubules.
  4. *Anaphase*   Sister chromatids are pulled apart by the disassembly of kinetochore microtubules at the kinetochore. The separated chromatids are now daughter chromosomes. The spindle poles are moved farther apart to fully separate the replicated chromosomes.
  5. *Telophase*   Daughter chromosomes are fully separated and are clustered at opposite poles of the spindle. A nuclear envelope forms around each set and the chromosomes de-condense.

- In most cells, mitosis is followed by cytokinesis—division of all cell contents.

✔ You should be able to predict how mitosis would be different in cells where the nuclear envelope remains intact (e.g., yeast).

12.3 Control of the Cell Cycle

- The onset of S and M phases is primarily determined by the activity of protein dimers consisting of cyclin and cyclin-dependent kinases (Cdks).

- Cyclin concentrations oscillate during the cell cycle, regulating the formation of the dimer. The activity of the Cdk is further regulated by addition of a phosphate in its activating site and removal of one from its inhibitory site.
- Progression through the cell cycle is controlled by checkpoints in three phases.

  1. The G_1 checkpoint regulates progress based on nutrient availability, cell size, DNA damage, and social signals.
  2. The G_2 checkpoint delays progress until chromosome replication is complete and any damaged DNA that is present is repaired.
  3. The two M-phase checkpoints will (1) delay anaphase until all chromosomes are correctly attached to the spindle apparatus and (2) delay the onset of cytokinesis and G_1 until all the chromosomes have been properly partitioned.

✔ You should be able to predict what would happen if the kinase that adds the inhibitory phosphates to Cdk were defective.

12.4 Cancer: Out-of-Control Cell Division

- Cancer is characterized by (1) loss of control at the G_1 checkpoint, resulting in cells that divide in an uncontrolled fashion; and (2) metastasis, or the ability of tumor cells to spread throughout the body.
- The G_1 checkpoint depends in part on Rb, which prevents progression to S phase, and G_1 cyclin–Cdk complexes that trigger progression to S phase. Defects in Rb and G_1 cyclin are common in human cancer cells.

✔ You should be able to compare and contrast the effect of removing growth factors from asynchronous cultures of human cells that are normal versus those that are cancerous.

MB MasteringBiology

1. MasteringBiology Assignments

Tutorials and Activities Causes of Cancer; Cell Culture Methods; Four Phases of the Cell Cycle; Mitosis (1 of 3): Mitosis and the Cell Cycle; Mitosis (2 of 3): Mechanism of Mitosis; Mitosis (3 of 3): Comparing Cell Division in Animals, Plants, and Bacteria; Mitosis and Cytokinesis Animation; Roles of Cell Division; The Cell Cycle; The Phases of Mitosis

Questions Reading Quizzes, Blue-Thread Questions, Test Bank

2. eText Read your book online, search, take notes, highlight text, and more.

3. The Study Area Practice Test, Cumulative Test, BioFlix® 3-D Animations, Videos, Activities, Audio Glossary, Word Study Tools, Art

You should be able to . . .

1. Which statement about the daughter cells following mitosis and cytokinesis is correct?
   a. They are genetically different from each other and from the parent cell.
   b. They are genetically identical with each other and with the parent cell.
   c. They are genetically identical with each other but different from the parent cell.
   d. Only one of the two daughter cells is genetically identical with the parent cell.

2. Progression through the cell cycle is regulated by oscillations in the concentration of which type of molecule?
   a. p53, Rb, and other tumor suppressors
   b. receptor tyrosine kinases
   c. cyclin-dependent kinases
   d. cyclins

3. After the S phase, what comprises a single chromosome?
   a. two daughter chromosomes
   b. a double-stranded DNA molecule
   c. two single-stranded molecules of DNA
   d. two sister chromatids

4. What major events occur during anaphase of mitosis?
   a. Chromosomes replicate, so each chromosome consists of two identical sister chromatids.
   b. Chromosomes condense and the nuclear envelope disappears.
   c. Sister chromatids separate, and the spindle poles are pushed farther apart.
   d. The chromosomes end up at opposite ends of the cell, and two nuclear envelopes form around them.

5. What evidence suggests that during anaphase, kinetochore microtubules shorten at the kinetochore?

6. Under normal conditions, what happens to the cell cycle if the chromosomes fail to separate properly at anaphase?

7. Identify at least two events in the cell cycle that must be completed successfully for daughter cells to share an identical complement of chromosomes.

8. Make a concept map illustrating normal events at the G_1 checkpoint. Your diagram should include p53, DNA damage, Rb, E2F, social signals, G_1 Cdk, G_1 cyclin, S-phase proteins, phosphorylated (inactivated) cyclin–Cdk, dephosphorylated (activated) cyclin–Cdk, phosphorylated (inactivated) Rb.

9. Explain how microinjection experiments supported the hypothesis that specific molecules in the cytoplasm are involved in the transition from interphase to M phase. What was the control for this experiment?

10. Why are most protein kinases considered regulatory proteins?

11. Why are cyclins called cyclins? Explain their relationship to MPF activity.

12. In multicellular organisms, nondividing cells stay in G_0 phase. For the cell, why is it better to be held in G_1 rather than S, G_2, or M phase?
    a. G_1 cells are larger and more likely to perform the normal functions of the cell.
    b. G_1 cells have not replicated their DNA in preparation for division.
    c. G_1 cells are the only ones that do not have their chromatin in a highly condensed state.
    d. MPF is required to enter S phase, so the cell is committed to entering M phase if the cycle moves beyond G_1.

13. **QUANTITATIVE** A particular cell spends 4 hours in G_1 phase, 2 hours in S phase, 2 hours in G_2 phase, and 30 minutes in M phase. If a pulse–chase assay were performed with radioactive thymidine on an asynchronous culture, what percentage of mitotic cells would be radiolabeled after 9 hours?
    a. 0%
    b. 50%
    c. 75%
    d. 100%

14. When fruit fly embryos first begin to develop, a large cell is generated that contains over 8000 nuclei that are genetically identical with one another. What is most likely responsible for this result?

15. What is most likely responsible for the reduction in death rates over the past several years in cancers of the breast and prostate? How is this related to the development of cancer?

16. Cancer is primarily a disease of older people. Further, a group of individuals may share a genetic predisposition to developing certain types of cancer, yet vary a great deal in time of onset—or not get the disease at all. What conclusion could be drawn based on these observations? How does this relate to the requirements for a cell to become cancerous?

13 Meiosis

In this chapter you will learn how

Meiosis promotes genetic diversity and allows the benefits of sex

starting with → How does meiosis occur? **13.1**

and comparing to → Mitosis **Ch. 12**

by examining → How meiosis produces genetic variation

looking at → Independent assortment, crossing over, and fertilization **13.2**

then asking → What happens when things go wrong? **13.3**

by asking → Why does meiosis exist? **13.4**

→ Purifying-selection hypothesis

→ Changing-environment hypothesis

Scanning electron micrograph (with color added) showing human sperm attempting to enter a human egg. This chapter introduces the type of nuclear division called meiosis, which in animals occurs before the formation of sperm and eggs.

This chapter is part of the Big Picture. See how on pages 366–367.

Why sex?

Simple questions—such as why sexual reproduction exists—are sometimes the best. This chapter asks what sexual reproduction is and why some organisms employ it. The focus here is on how organisms reproduce, or replicate—one of the five fundamental attributes of life introduced in Chapter 1.

For centuries people have known that during sexual reproduction, a male reproductive cell—a **sperm**—and a female reproductive cell—an **egg**—unite to form a new individual. The process of uniting sperm and egg is called **fertilization.** The first biologists to observe fertilization studied the large, translucent eggs of sea urchins. Owing to the semitransparency of the sea urchin egg cell, researchers were able to see the nuclei of a sperm and an egg fuse.

✔ When you see this checkmark, stop and test yourself. Answers are available in Appendix A.

When these observations were published in 1876, they raised an important question, because biologists had already established that the number of chromosomes is constant from cell to cell within a multicellular organism. The question is, How can the chromosomes from a sperm cell and an egg cell combine, but form an offspring that has the same chromosome number as its mother and its father?

A hint at the answer came in 1883, with the observation that cells in the body of roundworms of the genus *Ascaris* have four chromosomes, while their sperm and egg nuclei have only two chromosomes apiece.

Four years later, August Weismann formally proposed a hypothesis to explain the riddle: During the formation of **gametes**—reproductive cells such as sperm and eggs—there must be a distinctive type of cell division that leads to a reduction in chromosome number. Specifically, if the sperm and egg contribute an equal number of chromosomes to the fertilized egg, Weismann reasoned, they must each contain half of the usual number of chromosomes. Then, when sperm and egg combine, the resulting cell has the same chromosome number as its mother's cells and its father's cells have.

In the decades that followed, biologists confirmed this hypothesis by observing gamete formation in a wide variety of plant and animal species. Eventually this form of cell division came to be called meiosis (literally, "lessening-act").

Meiosis is nuclear division that leads to a halving of chromosome number and ultimately to the production of sperm and egg. (Meiosis is an important part of The Big Picture of Genetic Information on pages 366-367). To a biologist, asking "Why sex?" is equivalent to asking "Why meiosis?" Let's delve in by first looking at how meiosis happens.

13.1 How Does Meiosis Occur?

To understand meiosis, it is critical to grasp some key ideas about chromosomes. For example, when cell biologists began to study the cell divisions that lead to gamete formation, they made an important observation: Each organism has a characteristic number of chromosomes.

Consider the drawings in **FIGURE 13.1**, based on a paper published by Nettie Maria Stevens in 1908. They show the chromosomes of the fruit fly *Drosophila melanogaster*, or *Drosophila* for short. This model organism has been a focus of biological research for more than 100 years (see **BioSkills 13** in Appendix B). Stevens was studying the cell divisions leading up to the formation of egg and sperm. In total, she found eight chromosomes in *Drosophila* cells. Your cells have 46 chromosomes, and some ferns have over 1000.

Chromosomes Come in Distinct Sizes and Shapes

Stevens found that each *Drosophila* cell has eight chromosomes, but just five distinct types, distinguished by their size and shape. Three of these chromosomes always occurred in pairs and are labeled chromosomes 2–4 in Figure 13.1. In males, Stevens

Drosophila chromosomes drawn by Nettie Stevens in 1908 ...

... and photographed through a modern microscope.

2 3 4 X Y

Autosomes (2–4) Sex chromosomes

FIGURE 13.1 Cells Contain Different Types of Chromosomes, and in Diploid Organisms, Chromosomes Come in Pairs. Numbers and letters designate the types of *Drosophila* chromosomes. These chromosomes are from the cell of a male, so there is an X and a Y chromosome plus three homologous pairs of autosomes.

observed an unpaired set of chromosomes, which came to be known as the X and the Y chromosomes. Stevens found that females lack a Y chromosome but contain a pair of X chromosomes. This is the same situation in some other insects and in mammals. The X and Y chromosomes are called **sex chromosomes** and are associated with an individual's sex. Non-sex chromosomes, such as chromosomes 2–4 in *Drosophila*, are **autosomes.**

Chromosomes that are the same size and shape are called **homologous** ("same-proportion") **chromosomes,** or **homologs,** and the pair is called a homologous pair. Later work showed that homologous chromosomes are similar in content as well as in size and shape. Homologous chromosomes carry the same genes. A **gene** is a section of DNA that influences some hereditary trait in an individual. For example, each copy of chromosome 2 found in *Drosophila* carries genes that influence eye color, wing size and shape, and bristle size.

The versions of a gene found on homologous chromosomes may differ, however. Biologists use the term **allele** to denote different versions of the same gene. For example, the allele for an eye-color gene on one homolog of chromosome 2 may be associated with red eyes, the normal color in *Drosophila*, whereas the allele of the same eye-color gene on the other homolog may be associated with purple eyes (**FIGURE 13.2**); the particular alleles of the bristle-size gene will influence whether the fly's bristles are long or short, and so on.

Homologous chromosomes carry the same genes, but each homolog may contain different alleles.

The Concept of Ploidy

At this point in her study, Stevens had determined the *Drosophila* **karyotype**—meaning the number and types of chromosomes present. As karyotyping studies became more common, cell biologists realized that, like *Drosophila*, the vast majority of plants and animals have more than one of each type of chromosome.

Homologous
chromosomes
(replicated)

Gene for eye color —— —— **Gene** for eye color
(allele for red eyes)   (allele for purple eyes)

Drosophila autosome 2

FIGURE 13.2 Homologous Chromosomes May Contain Different Alleles of the Same Gene. The homologs of *Drosophila* chromosome 2 are shown; the location of only one of many genes is indicated.

Insects, humans, oak trees, and other organisms that have two versions of each type of chromosome are called **diploid** ("double-form"). Diploid organisms have two alleles of each gene. One allele is carried on each of the homologous pairs of chromosomes. Although a diploid individual can carry only two different alleles of a gene, there can be many different alleles in a population.

Organisms whose cells contain just one of each type of chromosome—for example, bacteria, archaea, and many algae and fungi—are called **haploid** ("single-form"). Haploid organisms have only one copy of each chromosome and just one allele of each gene.

Biologists use a compact notation to indicate the number of chromosomes and chromosome sets in a particular organism or type of cell:

- By convention, the letter n stands for the number of distinct types of chromosomes in a given cell and is called the **haploid number.** If sex chromosomes are present, they are counted as a single type in the haploid number. In humans, n is 23.

- To indicate the number of complete chromosome sets observed, a number is placed before the n. Thus, a cell can be n, or $2n$, or $3n$, and so on.

The combination of the number of sets and n is termed the cell's **ploidy.** Diploid cells or species are designated $2n$, because two chromosomes of each type are present—one from each parent. A **maternal chromosome** comes from the mother, and a **paternal chromosome** comes from the father.

Humans are diploid; $2n$ is 46. Haploid cells or species are labeled simply n, because they have just one set of chromosomes—no homologs are present. In haploid cells, the number 1 in front of n is implied and is not written out.

To summarize, the haploid number n indicates the number of distinct types of chromosomes present. In contrast, a cell's ploidy (n, $2n$, $3n$, etc.) indicates the number of each type of chromosome present. Stating a cell's ploidy is the same as stating the number of haploid chromosome sets present. ✔ If you understand how these terms relate, you should be able to state the haploid number, ploidy, and total number of chromosomes present in a male *Drosophila*.

Later work revealed that it is common for species in some lineages—particularly certain land plants, such as ferns—to contain more than two of each type of chromosome. Instead of having two homologous chromosomes per cell, **polyploid** ("many-form") species have three or more of each type of chromosome in each cell.

Depending on the number of homologs present, polyploid species are called triploid ($3n$), tetraploid ($4n$), hexaploid ($6n$), octoploid ($8n$), and so on.

Stevens and other early cell biologists did more than just describe the karyotypes observed in their study organisms. Through careful examination, they were able to track how chromosome numbers change during meiosis. These studies confirmed Weismann's hypothesis that a special type of cell division occurs during gamete formation.

An Overview of Meiosis

Cells replicate each of their chromosomes before undergoing meiosis. At the start of meiosis, chromosomes are in the same state they are in before mitosis.

Recall that an unreplicated eukaryotic chromosome consists of a single, long DNA double helix organized around proteins called histones (see Chapter 12). When chromosome replication is complete, each chromosome will consist of two identical **sister chromatids.** Sister chromatids contain identical copies of the DNA double helix present in the unreplicated chromosome and therefore the same genetic information. They remain physically joined along their entire length during much of meiosis (**FIGURE 13.3**).

To understand meiosis, it is critical to understand the relationship between chromosomes and sister chromatids. The trick is to recognize that unreplicated and replicated chromosomes are both considered *single* chromosomes—even though the replicated chromosome contains *two* sister chromatids. This makes sense if you consider that a chromosome carries a particular set of genetic information in its DNA and that the amount of *unique* information is the same whether there is one copy of it present or two.

Maternal chromosome

Paternal chromosome

Nuclear envelope

Replication

Replicated chromosome

Centromere

Sister chromatids

Homologous pair of replicated chromosomes

FIGURE 13.3 Each Chromosome Replicates before Undergoing Meiosis.

Term	Definition	Example or Comment
Chromosome	Structure made up of DNA and proteins; carries the cell's hereditary information (genes)	Eukaryotes have linear chromosomes; most bacteria and archaea have just one, circular, chromosome
• **Sex chromosome**	Chromosome associated with an individual's sex	X and Y chromosomes of humans (males are XY, females XX); Z and W chromosomes of birds and butterflies (males are ZZ, females ZW)
• **Autosome**	A non-sex chromosome	Chromosomes 1–22 in humans
Unreplicated chromosome	A chromosome that consists of one double-helical molecule of DNA packaged with proteins	
Replicated chromosome	A chromosome that has been copied; consists of two identical chromatids, each containing one double-helical DNA molecule	Centromere
Sister chromatids	The two identical chromatid copies in a replicated chromosome	Sister chromatids
Homologous chromosomes (homologs)	In a diploid organism, chromosomes that are similar in size, shape, and gene content	You have a chromosome 22 from each parent Homologous chromosomes
Non-sister chromatids	Chromatids belonging to homologous chromosomes	Non-sister chromatids
Bivalent (or tetrad)	Homologous replicated chromosomes that are joined together during prophase I and metaphase I of meiosis	Bivalent
Haploid number	The number of different types of chromosomes in a cell; symbolized n	Humans have 23 different types of chromosomes ($n = 23$)
Diploid number	The number of chromosomes present in a diploid cell (see below); symbolized $2n$	In humans all cells except gametes are diploid and contain 46 chromosomes ($2n = 46$)
Ploidy	The number of each type of chromosome present	The number of haploid chromosome sets present
• **Haploid**	Having one of each type of chromosome (n)	Bacteria and archaea are haploid, as are many algae; plant and animal gametes are haploid
• **Diploid**	Having two of each type of chromosome ($2n$)	Most familiar plants and animals are diploid
• **Polyploid**	Having more than two of each type of chromosome; cells may be triploid ($3n$), tetraploid ($4n$), hexaploid ($6n$), and so on	Seedless bananas are triploid; many ferns are tetraploid; bread wheat is hexaploid

Note that an unreplicated chromosome is never called a chromatid; you can refer to chromatids only as the structures in a replicated chromosome.

TABLE 13.1 summarizes the vocabulary that biologists use to describe chromosomes and illustrates the relationship between chromosomes and chromatids. ✔ If you understand this relationship, you should be able to draw the same chromosome in the unreplicated and replicated state, label the sister chromatids, indicate the number of DNA molecules present in each structure, and explain why both structures represent a single chromosome.

Meiosis Comprises Two Cell Divisions Meiosis consists of two cell divisions, called **meiosis I** and **meiosis II**. As **FIGURE 13.4** shows, the two divisions occur consecutively but differ sharply.

During meiosis I, the homologs in each chromosome pair separate from each other. One homolog goes to one daughter cell; the other homolog goes to the other daughter cell. It is a matter of chance which daughter cell receives which homolog.

At the end of meiosis I, each of the two daughter cells has one of each type of chromosome instead of two, and thus half as many chromosomes as the parent cell had. Put another way: During meiosis I, the diploid ($2n$) parent cell produces two haploid (n)

MEIOSIS I

Parent cell is diploid (2n) and contains a homologous pair of replicated chromosomes

Homologs separate

Sister chromatids separate

MEIOSIS II

Daughter cells are haploid (n) and contain just one homolog

Four daughter cells contain one chromosome each (n). In animals, these cells become gametes.

FIGURE 13.4 The Major Events in Meiosis. Before undergoing meiosis, chromosomes are replicated so there are two chromatids per chromosome. Meiosis reduces chromosome number by half. In diploid organisms, the products of meiosis are haploid. Maternal chromosomes are shown in red; paternal chromosomes, blue. Note that in this cell, $2n = 2$.

daughter cells. Notice, however, that each chromosome still consists of two sister chromatids—meaning that chromosomes are still replicated at the end of meiosis I.

During meiosis II, sister chromatids from each chromosome separate. One sister chromatid (now called a daughter chromosome) goes to one daughter cell; the other sister chromatid goes to the other daughter cell as a daughter chromosome. Remember that the cell that started meiosis II had one of each type of chromosome, but each chromosome was still replicated (meaning still consisting of two sister chromatids). The cells produced by meiosis II also have one of each type of chromosome, but now the daughter chromosomes are no longer replicated.

To reiterate, sister chromatids separate into daughter chromosomes during meiosis II. This is just what happens during mitosis. Meiosis II is actually equivalent to mitosis occurring in a haploid cell. In meiosis I, on the other hand, sister chromatids stay together. This sets meiosis I apart from both mitosis and meiosis II.

As in mitosis, chromosome movements during meiosis I and II are coordinated by microtubules of the **spindle apparatus** that attach to **kinetochores** located at the **centromere** of each chromosome. Recall that the centromere is a region on the chromosome; kinetochores are structures in that region (see Chapter 12). Chromosome movement is driven by fraying of the ends of microtubules at each kinetochore, just as it is in mitosis (see Figure 12.7).

Meiosis I is a Reduction Division A host of early cell biologists worked out this sequence of events through careful observation of cells with the light microscope. Based on these studies, they came to a key realization: The outcome of meiosis I is a reduction

in chromosome number. For this reason, meiosis I is known as a reduction division. Reduction is another important way in which meiosis I differs from meiosis II and mitosis.

In most plants and animals, the original cell entering meiosis is diploid and the four final daughter cells are haploid. In animals, the haploid daughter cells, each containing one of each homologous chromosome, eventually go on to form egg cells or sperm cells via a process called **gametogenesis** ("gamete-origin").

When two haploid gametes fuse during fertilization, a full complement of chromosomes is restored (**FIGURE 13.5**). The cell that results from fertilization is diploid and is called a **zygote**. In this way, each diploid individual receives a haploid chromosome set from its mother and a haploid set from its father.

FIGURE 13.6 puts these events into the context of an animal's **life cycle**—the sequence of events that occurs over the life span of an individual, from fertilization to the production of offspring. As you study the figure, note how ploidy changes as the result of

Female gamete (egg) (n)

Male gamete (sperm) (n)

Fertilization

Diploid offspring contains homologous pair of chromosomes

Zygote (2n)

FIGURE 13.5 Fertilization Restores a Full Complement of Chromosomes.

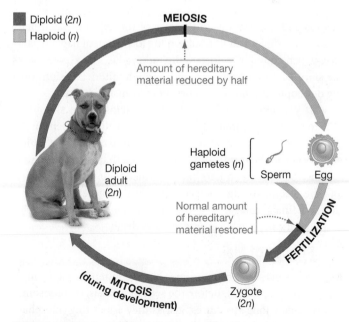

■ Diploid (2n)
■ Haploid (n)

MEIOSIS

Amount of hereditary material reduced by half

Diploid adult (2n)

Haploid gametes (n)

Sperm   Egg

Normal amount of hereditary material restored

FERTILIZATION

MITOSIS (during development)

Zygote (2n)

FIGURE 13.6 Ploidy Changes during the Life Cycle of a Dog. Most of the dog life cycle involves diploid cells.

PROCESS: MEIOSIS I   Homologous chromosomes separate

Nuclear envelope   Replicated chromosomes   Non-sister chromatids   Bivalent (4 chromatids from 2 homologous chromosomes)   Spindle apparatus   Chiasma

2n
In this example, n = 3 so 2n = 6

● Maternal chromosomes
● Paternal chromosomes

1. Interphase: Uncondensed chromosomes replicate in parent cell.

2. Early prophase I: Chromosomes condense, spindle apparatus forms, nuclear envelope begins to break down. Synapsis (pairing) of homologous chromosomes.

3. Late prophase I: Chiasmata (crossover points) visible, nuclear envelope broken down. Often multiple chiasmata between non-sister chromatids.

4. Metaphase I: Migration of bivalents to metaphase plate is complete.

5. Anaphase I: Homologs separate and begin moving to opposite poles of the spindle apparatus.

FIGURE 13.7 The Phases of Meiosis. The micrographs of each phase are from a species of salamander.

meiosis and fertilization. In the case of the dog illustrated here, meiosis in a diploid adult results in the formation of haploid gametes, which combine to form a diploid zygote. The zygote marks the start of a new generation, and through mitotic divisions during development, the zygote goes on to form an adult dog.

✔ If you understand the events of meiosis, you should be able to predict how many DNA molecules will be present in the gametes of the fruit fly *Drosophila*, a diploid organism that has eight replicated chromosomes in each cell that enters meiosis.

Once Stevens and others had published their work on meiosis and the accompanying changes in ploidy, the mystery of fertilization was finally solved. To appreciate the consequences of meiosis fully, let's analyze the events in more detail.

The Phases of Meiosis I

Meiosis begins after chromosomes have been replicated during S phase (see Chapter 12). Before the start of meiosis, chromosomes are extremely long structures, just as they are during interphase of the normal cell cycle. The major steps that occur once meiosis begins are shown in **FIGURE 13.7**.

Early Prophase I During early prophase I, the nuclear envelope begins to break down, chromosomes condense and the spindle apparatus begins to form. Then a crucial event occurs: Homologous chromosome pairs come together. The end result of this process is called **synapsis** and is illustrated in step 2 of Figure 13.7. Synapsis is possible because regions of homologous chromosomes that are similar at the molecular level come together. In most organisms, synapsis requires breaking and then connecting together DNA of the two homologs at one or more spots along their length.

The structure that results from synapsis is called a **bivalent** (*bi* means "two" in Latin) or **tetrad** (*tetra* means "four" in Greek). A bivalent consists of paired homologous chromosomes, with each homolog consisting of two sister chromatids. Chromatids from different homologs are referred to as **non-sister chromatids.** In the figure, the red-colored chromatids are non-sister chromatids with respect to the blue-colored chromatids.

Late Prophase I During late prophase I, the nuclear envelope breaks down and microtubules of the spindle apparatus attach to

242   UNIT 3   Gene Structure and Expression

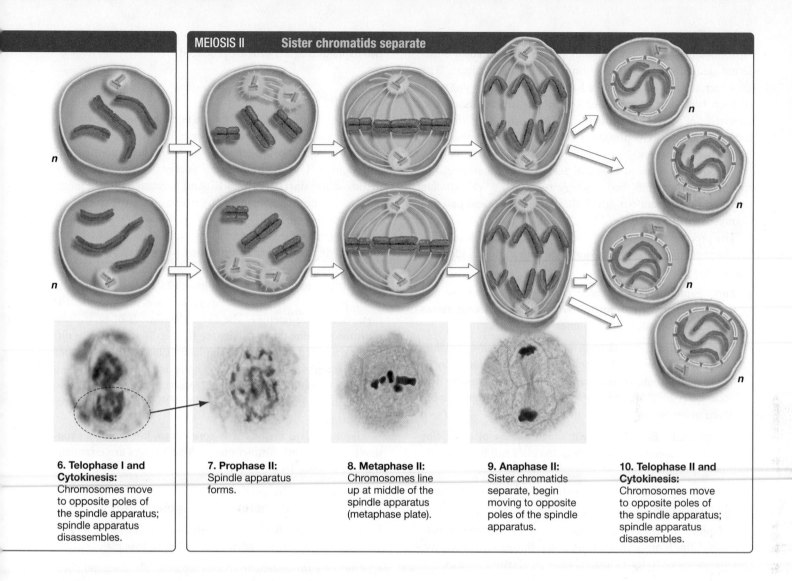

n

n

n

n

n

n

6. Telophase I and Cytokinesis: Chromosomes move to opposite poles of the spindle apparatus; spindle apparatus disassembles.

7. Prophase II: Spindle apparatus forms.

8. Metaphase II: Chromosomes line up at middle of the spindle apparatus (metaphase plate).

9. Anaphase II: Sister chromatids separate, begin moving to opposite poles of the spindle apparatus.

10. Telophase II and Cytokinesis: Chromosomes move to opposite poles of the spindle apparatus; spindle apparatus disassembles.

kinetochores. Non-sister chromatids begin to separate at many points along their length. They stay joined at certain locations, however, each of which forms an X-shaped structure called a **chiasma** (plural: **chiasmata**). (In the Greek alphabet, the letter X is called "chi.") Normally, at least one chiasma forms in every pair of homologous chromosomes; often there are several chiasmata. The chiasmata mark sites where DNA was broken and rejoined between homologs early in prophase I.

As step 3 of Figure 13.7 shows, the chromatids that meet to form a chiasma are non-sister chromatids. At each chiasma there is an exchange of parts of chromosomes between paternal and maternal homologs. These reciprocal exchanges between different homologs create non-sister chromatids that have both paternal and maternal segments. This process of chromosome exchange is called **crossing over.**

In step 4 of Figure 13.7, the result of crossing over is illustrated by chromosomes with a combination of red and blue segments. When crossing over occurs, the chromosomes that result have a mixture of maternal and paternal alleles. Crossing over is a major way that meiosis creates genetic diversity.

Metaphase I   The next major stage in meiosis I is metaphase I. This is when kinetochore microtubules move the pairs of homologous chromosomes (bivalents) to a region called the **metaphase plate** in the middle of the spindle apparatus (step 4). The metaphase plate is not a physical structure but an imaginary plane dividing the spindle apparatus.

Here are two key points about chromosome movement: Each bivalent moves to the metaphase plate independently of the other bivalents, and the alignment on one side or the other of the metaphase plate is random for maternal and paternal homologs from each chromosome. This movement explains the most basic principles of genetics (see Chapter 14).

Anaphase and Telophase I   Sister chromatids of each chromosome remain together. During anaphase I, the homologous chromosomes in each bivalent separate and begin moving to opposite poles of the spindle apparatus (step 5). Meiosis I concludes with telophase I, when the homologs finish moving to opposite sides of the spindle (step 6). When meiosis I is complete, **cytokinesis** (division of cytoplasm) occurs and two haploid daughter cells form.

Meiosis I: A Recap The end result of meiosis I is that one chromosome of each homologous pair is distributed to a different daughter cell. A reduction division has occurred: The daughter cells of meiosis I are haploid, having only one copy of each type of chromosome. The sister chromatids remain attached in each chromosome, however, meaning that the haploid daughter cells produced by meiosis I still contain replicated chromosomes.

The chromosomes in each cell are a random assortment of maternal and paternal chromosomes as a result of (1) crossing over and (2) the random distribution of maternal and paternal homologs during metaphase.

The preceding discussion shows that although meiosis I is a continuous process, biologists summarize the events by identifying distinct phases:

- *Early Prophase I* Replicated chromosomes condense and the spindle apparatus forms. Synapsis of homologs forms pairs of homologous chromosomes, or bivalents.

- *Late Prophase I* Breakdown of the nuclear envelope. Microtubules of the spindle apparatus attach to kinetochores. Chiasmata become visible, marking sites where crossing over occurs. Crossing over results in an exchange of segments between maternal and paternal chromosomes.

- *Metaphase I* Bivalents move to the metaphase plate and line up. One homolog is on one side of the plate and the other homolog is on the other.

- *Anaphase I* Homologs separate and begin moving to opposite spindle poles.

- *Telophase I* Homologs finish moving to opposite spindle poles. Spindle apparatus disassembles. In some species, a nuclear envelope re-forms around each set of chromosomes.

Throughout, chromosome movement takes place as microtubules that are attached to the kinetochore dynamically assemble

and disassemble. When meiosis I is complete, the cell divides and two haploid daughter cells are produced.

The Phases of Meiosis II

Recall that chromosome replication occurred before meiosis I. Throughout meiosis I, sister chromatids remained attached. Because no chromosome replication occurs between meiosis I and meiosis II, each chromosome consists of two sister chromatids at the start of meiosis II. And because only one member of each homologous pair of chromosomes is present, the cell is haploid.

During prophase II, a spindle apparatus forms in both daughter cells. Microtubules attach to kinetochores on each side of every chromosome and begin moving the chromosomes toward the middle of each cell (step 7 of Figure 13.7).

In metaphase II, the chromosomes are lined up at the metaphase plate (step 8). The sister chromatids of each chromosome separate during anaphase II (step 9) and move to different daughter cells during telophase II (step 10). Once they are separated, each chromatid is considered an independent daughter chromosome. Meiosis II results in four haploid cells, each with one daughter chromosome of each type in the chromosome set.

Like meiosis I, meiosis II is continuous, but biologists routinely divide it into distinct phases. These stages are essentially those of mitosis. To summarize,

- *Prophase II* The spindle apparatus forms. If a nuclear envelope formed at the end of meiosis I, it breaks apart.

- *Metaphase II* Replicated chromosomes, consisting of two sister chromatids, are lined up at the metaphase plate.

- *Anaphase II* Sister chromatids separate. The daughter chromosomes that result begin moving to opposite poles of the spindle apparatus.

SUMMARY TABLE 13.2 **Key Differences between Mitosis and Meiosis**

Feature	Mitosis	Meiosis
Number of cell divisions	One	Two
Number of chromosomes in daughter cells compared with parent cell	Same	Half
Synapsis of homologs	No	Yes
Number of crossing-over events	None	One or more per pair of homologous chromosomes
Makeup of chromosomes in daughter cells	Identical	Different—various combinations of maternal and paternal chromosomes, paternal and maternal segments mixed within chromosomes
Role in organism life cycle	Asexual reproduction in some eukaryotes; cell division for growth	Halving of chromosome number in cells that will produce gametes

- *Telophase II*   Chromosomes finish moving to opposite poles of the spindle apparatus. A nuclear envelope forms around each haploid set of chromosomes.

When meiosis II is complete, each cell divides to form two daughter cells. Because meiosis II occurs in both daughter cells of meiosis I, the process results in a total of four daughter cells from each original, parent cell. To describe meiosis in a nutshell, one diploid cell with replicated chromosomes gives rise to four haploid cells with unreplicated chromosomes.

TABLE 13.2 and **FIGURE 13.8** provide a comparison of mitosis and meiosis. A key difference between the two processes is that homologous chromosomes pair early in meiosis but do not pair at all during mitosis. Because homologs pair through synapsis in prophase of meiosis I, they can migrate to the metaphase plate together and then separate during anaphase of meiosis I, resulting in a reduction division. ✔ If you understand this key distinction between meiosis and mitosis, you should be able to describe the consequences for meiosis if homologs do not pair.

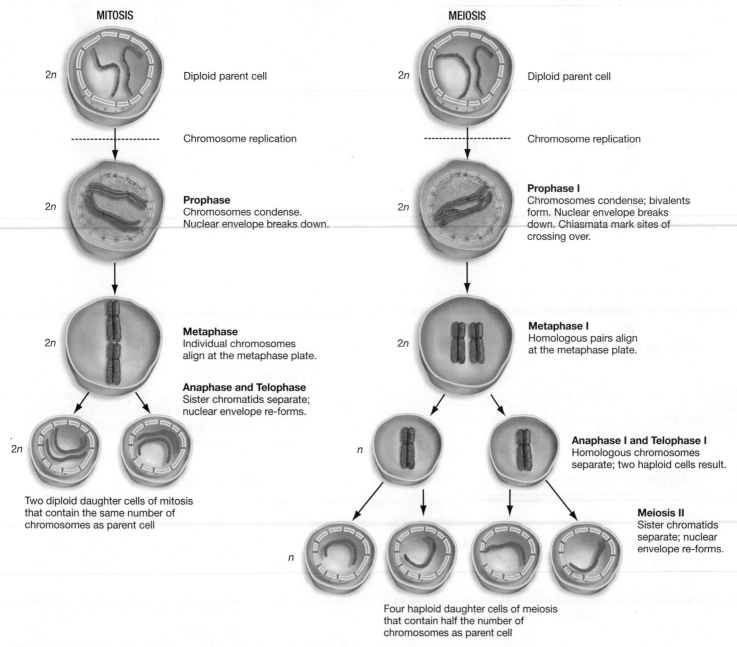

FIGURE 13.8 A Comparison of Mitosis and Meiosis. Mitosis produces two daughter cells with chromosomal complements identical to the parent cell. Meiosis produces four haploid cells with chromosomal complements unlike one another and unlike the diploid parent cell.

1. Condensation

2. Pairing

3. Synapsis (bivalent formation)

4. Partial separation of homologs

FIGURE 13.9 A Closer Look at Key Events in Prophase of Meiosis I.

A Closer Look at Synapsis and Crossing Over

The pairing of homologs and crossing over in prophase I are important events unique to meiosis. **FIGURE 13.9** takes a closer look at how chromosomes come together and exchange parts during meiosis I.

Step 1 Sister chromatids are held together along their full length by proteins known as cohesins. At the entry to prophase I, chromosomes begin to condense.

Step 2 Homologs pair. In many organisms, pairing begins when a break is made in the DNA of one chromatid. This break initiates a crossover between non-sister chromatids.

Step 3 A network of proteins forms the **synaptonemal complex,** which holds the two homologs tightly together.

Step 4 The synaptonemal complex disassembles in late prophase I. The two homologs partially separate and are held together only at chiasmata. Attachments at chiasmata are eventually broken to restore individual, unconnected chromosomes.

At a chiasma the non-sister chromatids from each homolog have been physically broken at the same point and *attached to each other.* As a result, corresponding segments of maternal and paternal chromosomes are exchanged.

Crossing over can occur at many locations along the length of paired homologs, and it routinely occurs at least once between each pair of non-sister chromatids. In humans, each chromosome undergoes an average of 1½ crossovers during meiosis.

Why does meiosis exist at all? What are its consequences?

check your understanding

If you understand that . . .

 C
 Y
U

• Meiosis is called a reduction division because the total number of chromosomes present is cut in half.

• During meiosis, a single diploid parent cell with replicated chromosomes gives rise to four haploid daughter cells, with unreplicated chromosomes.

✔ **You should be able to . . .**

1. Demonstrate the phases of meiosis I illustrated in Figure 13.7 by using pipe cleaners or pieces of cooked spaghetti.

2. Identify the event that makes meiosis a reduction division, unlike mitosis, and explain why it is responsible for reduction division.

3. Explain how meiosis generates cells with one of every kind of chromosome rather than random mixtures of different chromosomes.

Answers are available in Appendix A.

13.2 Meiosis Promotes Genetic Variation

The cell biologists who worked out the details of meiosis in the late 1800s and early 1900s realized that the process solved the riddle of fertilization. Weismann's hypothesis—that a reduction division precedes gamete formation in animals—was confirmed.

By now, having come to appreciate that meiosis is an intricate, tightly regulated process, you shouldn't be surprised to learn that it involves dozens, if not hundreds, of different proteins. Given this complexity, it is logical to hypothesize that meiosis accomplishes something extremely important: Thanks to the independent shuffling of maternal and paternal chromosomes and crossing over during meiosis I, the chromosomes in one gamete are different from the chromosomes in another gamete and different from the chromosomes in parental cells. Subsequently, fertilization brings haploid sets of chromosomes from a mother and father together to form a diploid offspring. The chromosome complement of this offspring is unlike that of either parent. It is a random combination of genetic material from each parent.

This change in chromosomal complement is crucial. The critical factor here is that changes in chromosome sets occur only during sexual reproduction—*not* during asexual reproduction.

- **Asexual reproduction** is any mechanism of producing offspring that does not involve the production and fusion of gametes. Asexual reproduction in eukaryotes is based on mitosis. The chromosomes in cells produced by mitosis are identical to the chromosomes in the parental cell.

- **Sexual reproduction** is the production of offspring through the production and fusion of gametes. Sexual reproduction results in offspring that have chromosome complements unlike those of their siblings or their parents.

Why is this difference important?

Chromosomes and Heredity

The changes in chromosomes produced by meiosis and fertilization are significant because chromosomes contain the cell's hereditary material. Stated another way, chromosomes contain the instructions for specifying particular traits. These inherited traits range from eye color and height in humans to the number or shape of the bristles on a fruit fly's leg to the color or shape of the seeds found in pea plants.

In the early 1900s, biologists began using the term gene to refer to the inherited instructions for a particular trait. Recall that the term allele refers to a particular version of a gene and that homologous chromosomes may carry different alleles.

Chromosomes are the repositories of genes, and identical copies of chromosomes are distributed to daughter cells during mitosis. Thus, cells that are produced by mitosis are genetically identical to the parent cell, and offspring produced during asexual reproduction are genetically identical to one another as well as to their parent. Offspring produced by asexual reproduction

are **clones**—or exact copies—of their parent. A familiar example of asexual reproduction is growing a new plant from a cutting.

In contrast, the offspring produced by sexual reproduction are genetically different from one another and unlike either their mother or their father.

Let's begin by analyzing two aspects of meiosis that create variation among chromosomes: (1) separation and distribution of homologous chromosomes and (2) crossing over. We'll then look at how these processes interact with fertilization to produce genetically variable offspring.

The Role of Independent Assortment

Each somatic cell in your body contains 23 homologous pairs of chromosomes and 46 chromosomes in total. Half of these chromosomes came from your mother, and half came from your father. Each chromosome contains genes, and genes influence particular traits. For example, one gene that affects your eye color might be located on one chromosome, while one of the genes that affects your hair color might be located on a different chromosome (**FIGURE 13.10a**).

(a) Example: Individual with different alleles of two genes

(b) During meiosis I, bivalents can line up two different ways before the homologs separate.

Brown eyes Black hair   Green eyes Red hair   Brown eyes Red hair   Green eyes Black hair

FIGURE 13.10 Independent Assortment of Homologous Chromosomes Results in Varied Combinations of Genes.

Suppose that the chromosomes you inherited from your mother contain alleles associated with brown eyes and black hair, but the chromosomes you inherited from your father include alleles associated with green eyes and red hair. (This is a simplification for the purpose of explanation. In reality, several genes with various alleles interact in complex ways to produce human eye color and hair color.)

Will any particular gamete you produce contain the genetic instructions inherited from your mother or the instructions inherited from your father? To answer this question, study the diagram of meiosis in **FIGURE 13.10b**. It shows that when pairs of homologous chromosomes line up during meiosis I and the homologs separate, a variety of combinations of maternal and paternal chromosomes can result. Each daughter cell gets a random assortment of maternal and paternal chromosomes.

This phenomenon is known as the principle of independent assortment. In the example given here, meiosis results in gametes with alleles for brown eyes and black hair, the traits from your mother, and green eyes and red hair, the traits from your father. But two additional combinations also occur: brown eyes and red hair, or green eyes and black hair. The appearance of new combinations of alleles is called **genetic recombination.** Four different combinations of paternal and maternal chromosomes are possible when two chromosomes are distributed to daughter cells during meiosis I.

✔ If you understand how independent assortment produces genetic variation in the daughter cells of meiosis, you should be able to explain how genetic variation would be affected if maternal chromosomes always lined up together on one side of the metaphase plate during meiosis I and paternal chromosomes always lined up on the other side.

How many different combinations of maternal and paternal homologs are possible when more chromosomes are involved? With each additional pair of chromosomes, the number of combinations doubles. In general, a diploid organism can produce 2^n combinations of maternal and paternal chromosomes, where n is the haploid chromosome number. This means that you ($n = 23$) can produce 2^{23}, or about 8.4 million, gametes that differ in their combination of maternal and paternal chromosome sets. The random assortment of whole chromosomes generates an impressive amount of genetic variation among gametes.

The Role of Crossing Over

Recall from Section 13.1 that segments of paternal and maternal chromatids exchange when crossing over occurs during meiosis I. Thus, crossing over produces new combinations of alleles within a chromosome—combinations that did not exist in either parent. This phenomenon is known as recombination. Crossing over is an important source of genetic recombination.

Genetic recombination is important because it dramatically increases the genetic variability of gametes produced by meiosis. The independent assortment of homologous chromosomes during meiosis generates varied combinations of chromosomes in gametes; genetic recombination due to crossing over varies the combinations of alleles along each chromosome that is involved

in a crossover. With crossing over, the number of genetically different gametes that you can produce is much more than the 8.4 million—it is virtually limitless.

How Does Fertilization Affect Genetic Variation?

Crossing over and the independent assortment of maternal and paternal chromosomes ensure that each gamete is genetically unique. Even if two gametes produced by the same individual fuse to form a diploid offspring—in which case **self-fertilization,** or "selfing," is taking place—the offspring are very likely to be genetically different from the parent (**FIGURE 13.11**). Selfing is common in many plant species, and it also occurs in animal species in which single individuals—called hermaphrodites—contain both male and female sex organs.

Self-fertilization, however, is rare or nonexistent in many sexually reproducing species. Instead, gametes from different individuals combine to form offspring. This process is called **outcrossing.** Outcrossing increases the genetic diversity of offspring even further because it combines chromosomes from different individuals. These chromosomes are likely to contain different alleles.

How many genetically distinct offspring can be produced when outcrossing occurs? Let's answer this question using humans as an example. Recall that a single human can produce about 8.4 million different gametes by independent assortment alone. When a sperm and egg come together at fertilization, the number of possible genetic combinations that can result is equal to the product of the numbers of different gametes produced by each parent. (To understand this logic, see **BioSkills 5** in Appendix B.) In humans this means that two parents can potentially produce 8.4 million $\times$ 8.4 million $= 70.6 \times 10^{12}$ genetically distinct offspring, even without crossing over. This number is far greater than the total number of people who have ever lived.

check your understanding

If you understand that . . .

- The daughter cells produced by meiosis are genetically different from the parent cell because (1) maternal and paternal homologs align randomly at metaphase of meiosis I and (2) crossing over leads to recombination within chromosomes.

✔ **You should be able to . . .**

1. Draw a diploid parent cell with $2n = 6$ (three types of chromosomes) and then sketch the genetically distinct types of daughter cells that may result by independent assortment of these chromosomes at meiosis.

2. Discuss how crossing over would influence the genetic diversity of these gametes.

3. Compare and contrast the degree of genetic variation that results from asexual reproduction, selfing, and outcrossing.

Answers are available in Appendix A.

The red and blue chromosomes can line up in different ways during metaphase

OR

$2n = 4$

1. Parent cell with four chromosomes.

2. Crossing over during meiosis I.

3. Independent assortment of homologous chromosomes during meiosis I.

4. Gametes produced by meiosis II.

5. Fertilization of random pairs of gametes (only some possibilities shown).

FIGURE 13.11 Crossing Over, Independent Assortment, and the Random Pairing of Gametes during Fertilization Increase Genetic Variation, Even in Offspring Produced by Self-Fertilization.

✔**EXERCISE** In step 5, only a few of the many types of offspring that could be produced are shown. Sketch two additional types that are different from those shown.

In any complicated process such as meiosis, things can and do go wrong. What happens if there is a mistake, and the chromosomes are not properly distributed?

13.3 What Happens When Things Go Wrong in Meiosis?

Errors in meiosis are surprisingly common. If this were like a spelling mistake, it might be only an annoyance. But in humans, a conservative estimate is that 25 percent of conceptions are spontaneously terminated because of problems in meiosis. What are the consequences for offspring if gametes contain an abnormal set of chromosomes?

In 1866 Langdon Down described a distinctive set of co-occurring conditions observed in some people. The syndrome was characterized by mental retardation, a high risk for heart problems and leukemia, and a degenerative brain disorder similar to Alzheimer's disease. **Down syndrome,** as the disorder came to be called, is observed in about 0.15 percent of live births (1 infant in every 666).

For over 80 years the cause of the syndrome was unknown. Then, in the late 1950s, a study of the chromosome sets of nine Down syndrome children suggested that the condition is associated with the presence of an extra copy of chromosome 21. This situation is called **trisomy** ("three-bodies")—in this case, trisomy-21—because each cell has three copies of the chromosome. The explanation proposed for the trisomy was that a mistake had occurred during meiosis in either the mother or the father.

How Do Mistakes Occur?

For a gamete to get one complete set of chromosomes, two steps in meiosis must be perfectly executed.

1. The chromosomes in each homologous pair must separate from each other during the first meiotic division, so that only one homolog ends up in each daughter cell.

2. Sister chromatids must separate from each other and move to opposite poles of the dividing cell during meiosis II.

If both homologs in meiosis I or both sister chromatids in meiosis II move to the same pole of the parent cell, the products of meiosis will be abnormal. This sort of meiotic error is referred to as **nondisjunction,** because the homologs or sister chromatids fail to separate, or disjoin.

FIGURE 13.12 shows what happens when homologs do not separate correctly during meiosis I. Notice that at the end of meiosis, two daughter cells have two copies of the same chromosome—the smaller one in Figure 13.12—while the other two lack that chromosome entirely. Gametes that contain an extra chromosome are symbolized as $n + 1$; gametes that lack one chromosome are symbolized as $n - 1$.

If an $n + 1$ gamete is fertilized by a normal n gamete, the resulting zygote will be $2n + 1$. This situation is trisomy. If the $n - 1$ gamete is fertilized by a normal n gamete, the resulting zygote will be $2n - 1$. This situation is called **monosomy.** Cells that have too many or too few chromosomes of a particular type are said to be **aneuploid** ("without-form").

Meiotic mistakes occur often. Researchers estimate that 25 percent of all human conceptions produce a zygote that is aneuploid. Most of these errors result from the failure of a homologous pair to separate in anaphase of meiosis I; less often, sister chromatids stay together during anaphase of meiosis II.

The consequences of meiotic mistakes are almost always severe. Trisomy-21 is unusual in allowing development to proceed when there are three copies of chromosome 21. Even for this chromosome, live births are not seen when there is only one copy.

In one study of human pregnancies that ended in early embryonic or fetal death, 38 percent involved abnormal chromosome sets that resulted from mistakes in meiosis. Trisomy accounted for about one-third of the abnormal karyotypes. Three copies of every chromosome ($3n$), a condition called triploidy, was also common. Monosomy and abnormally sized or shaped chromosomes were also seen. Mistakes in meiosis are common and are the leading cause of spontaneous abortion (miscarriage) in humans.

Why Do Mistakes Occur?

Trisomy and other meiotic mistakes are random errors that occur during meiosis. Recent research indicates that problems are especially common in attaching microtubules to kinetochores early in meiosis I and in separating chromosomes that have a single chiasma near their ends or near centromeres.

Researchers see strong patterns in occurrence of human trisomies at birth. Here are some of the patterns that emerge:

- Trisomy is much more common with the smaller chromosomes (numbers 13–22) than it is with the larger chromosomes (numbers 1–12), and trisomy-21 is far and away the most common type of trisomy observed. Chromosome 21 is the smallest human autosome.

- With the exception of trisomy-21, most trisomies and monosomies observed in humans involve the sex chromosomes.

- Errors in meiosis leading to eggs are more common than errors in meiosis leading to sperm.

- Maternal age is an important factor in the occurrence of trisomy. For example, in the case of Down syndrome, as **FIGURE 13.13** shows, the incidence increases dramatically in babies born to mothers over 35 years old.

Why do these patterns occur? One important point to remember is that these observations are made at birth, not fertilization. The frequency of nondisjunction is about equal among chromosomes, but aneuploidy tends to be lethal to embryos if it involves chromosomes that contain a large number of genes. Trisomy-21 is common most likely because it involves a small chromosome with a correspondingly small number of genes.

PROCESS: NONDISJUNCTION

$2n = 4$
$n = 2$

1. Meiosis I starts normally. Bivalents line up in middle of cell.

2. Nondisjunction occurs with one set of homologs.

3. Meiosis II occurs normally.

4. Aneuploidy results. All gametes have too many or too few chromosomes.

$n + 1$

$n + 1$

$n - 1$

$n - 1$

FIGURE 13.12 Nondisjunction Leads to Gametes with Abnormal Chromosome Numbers. If homologous chromosomes fail to separate during meiosis I, the gametes that result will have an extra chromosome or will lack a chromosome. Nondisjunction can also occur during meiosis II if sister chromatids fail to separate.

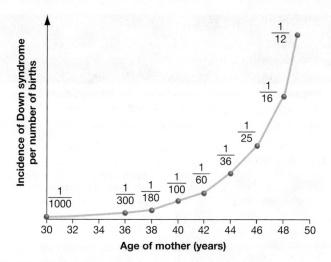

FIGURE 13.13 The Frequency of Down Syndrome Increases as a Function of a Mother's Age.

DATA: National Down Syndrome Society. 2012. www.ndss.org/index.php?option=com_content&view=article&id=61&Itemid=78

There are more questions than answers. Why are there more errors in women than men? Why is there such a strong correlation between maternal age and frequency of trisomy-21?

In part, the answers lie with an unusual feature of human egg development (oogenesis). Diploid precursors to eggs enter meiosis I before birth and arrest in prophase I until ovulation. For some eggs, this is 40 years or more later. Spindle apparatus function and the ability to separate chromosomes properly appear to decline after this long wait. Much remains to be discovered, but one thing is clear: Successful meiosis is critical to the health of offspring.

Why sex and meiosis? Although it seems obvious that sex and meiosis are needed universally for reproduction, that's not the case. Meiosis and sexual reproduction occur in only a small fraction of the lineages on the tree of life. Bacteria and archaea normally undergo only asexual reproduction; most algae, all fungi, and some animals and land plants reproduce asexually as well as sexually. Asexual reproduction is even observed in some vertebrates. Several species of guppy in the genus *Poeciliopsis*, for example, reproduce exclusively by mitosis.

Although sexual reproduction plays a central role in the life of most familiar organisms—including us—until recently, scientists had no clear idea of why it occurs. In fact, on the basis of theory, biologists had good reason to think that sexual reproduction should *not* exist.

The Paradox of Sex

In 1978 John Maynard Smith pointed out that the existence of sexual reproduction presents a paradox. Maynard Smith developed a mathematical model showing that because asexually reproducing individuals do not have to produce male offspring, their progeny on average can produce twice as many offspring as individuals that reproduce sexually. **FIGURE 13.14** diagrams this result by showing the number of females (♀), males (♂), and asexually reproducing organisms (**O**) produced over several generations by asexual versus sexual reproduction.

In this example, each asexually reproducing individual and each sexually reproducing couple produces four offspring over the course of their lifetimes. Note that in the sexual population, it takes two individuals—one male and one female—to produce

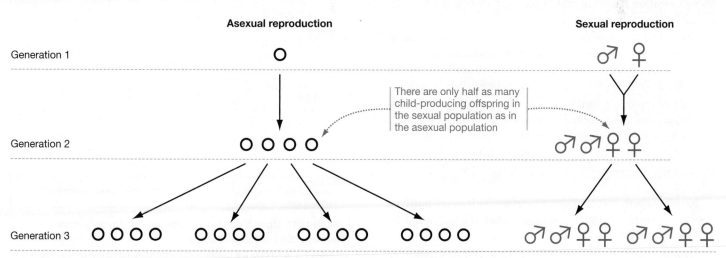

FIGURE 13.14 Asexual Reproduction Produces More Offspring. Each female (♀), male (♂), and asexual (**O**) symbol represents an individual. This hypothetical example assumes that (1) every asexual individual or sexually reproducing couple produces four offspring over the course of a lifetime, (2) sexually reproducing individuals produce half males and half females, and (3) all offspring survive to breed.

✔**QUANTITATIVE** How many asexually produced offspring would be present in generation 4? How many sexually produced offspring?

four offspring. Two out of every four children that each female produces sexually are males who cannot have children of their own. As a result, after one generation (generation 2 in Figure 13.14) the sexual population has just half as many child-producing individuals as the asexual population. Maynard Smith referred to this as the "two-fold cost of males." Asexual reproduction is much more efficient than sexual reproduction because no males are produced.

Based on this analysis, what will happen when asexual and sexual individuals exist in the same population and compete with one another? If all other things are equal, individuals that reproduce asexually should increase in frequency in the population while individuals that reproduce sexually should decline in frequency. In fact, Maynard Smith's model predicts that sexual reproduction is so inefficient that it should be completely eliminated.

To resolve this paradox, biologists began examining the assumption "If all other things are equal." Stated another way, biologists began looking for ways that meiosis and outcrossing could lead to the production of offspring that reproduce more than asexually produced individuals do. After decades of debate and analysis, two solid hypotheses to explain the paradox of sex are beginning to emerge.

The Purifying Selection Hypothesis

The first clue to unraveling the paradox of sex is a simple observation: If a gene is damaged or altered in a way that causes it to function poorly, it will be inherited by *all* of that individual's offspring when asexual reproduction occurs. Suppose the altered gene arose in generation 1 of Figure 13.14. If this gene is important, its alteration might cause the four asexual females present in generation 2 to produce fewer than four offspring apiece—perhaps because the members of generation 2 die young. If so, then generation 3 will not have twice as many individuals in the asexual lineage compared with the sexual lineage.

An allele that functions poorly and lowers the fitness of an individual is said to be deleterious. Asexual individuals are doomed to transmitting all their deleterious alleles to all of their offspring.

Suppose, however, that the same deleterious allele arose in the sexually reproducing female in generation 1 of Figure 13.14. If the female also has a normal copy of the gene, and if she mates with a male that has normal copies of the gene, then on average half her offspring will lack the deleterious allele. Sexual individuals are likely to have some offspring that lack the deleterious alleles that are present in a parent.

Natural selection against deleterious alleles is called purifying selection. Purifying selection should reduce the numerical advantage of asexual reproduction.

To test this hypothesis, researchers recently compared the same genes in two closely related species of *Daphnia*, a tiny crustacean that is a common inhabitant of ponds and lakes (see Chapter 50). One of these species reproduces asexually and the other reproduces sexually. As predicted, the researchers found that individuals in the asexual species contained many more deleterious alleles than individuals in the sexual species. Results like

QUESTION: Does exposure to evolving pathogens favor outcrossing?

HYPOTHESIS: In environments where evolving pathogens are present, sexual reproduction by outcrossing will be favored.

NULL HYPOTHESIS: The presence of evolving pathogens will not favor outcrossing.

EXPERIMENTAL SETUP:

1. **Start with a pathogen free population of roundworms** with a 20% rate of outcrossing.

Grow without pathogen   Grow with pathogen

2. **Divide the population;** grow one subgroup in the absence of a pathogen and another subgroup in the presence of an evolving pathogen.

3. **Assess the rate of outcrossing** over many generations.

PREDICTION: The rate of outcrossing will increase in response to exposure by a pathogen.

PREDICTION OF NULL HYPOTHESIS: The rate of outcrossing will not be influenced by a pathogen.

RESULTS:

CONCLUSION: Exposure to evolving pathogens favors outcrossing.

FIGURE 13.15 Does Exposure to Pathogens Favor Sexual Reproduction through Outcrossing? Each point in the graph shows the average percentage of reproduction by outcrossing for five populations. The bars indicate the degree of variation in the data (see **BioSkills 4** in Appendix B for a description of error bars).

Morran, L. T., et al. 2011. Running with the red queen: Host-parasite coevolution selects for biparental sex. *Science* 333: 216–218.

✔ **QUESTION** What would you predict if a non-evolving pathogen were used?

these have convinced biologists that purifying selection is an important factor promoting the success of sexual reproduction.

The Changing-Environment Hypothesis

The second hypothesis to explain sexual reproduction also focuses on the benefits of producing genetically diverse offspring. Here's the key idea: Offspring that are genetic clones of their parents are unlikely to thrive if the environment changes.

What type of environmental change might favor genetically diverse offspring? The leading hypothesis points to pressure put on hosts by rapidly changing pathogens and parasites—bacteria, viruses, fungi, and other entities that cause disease. In your own lifetime, for example, several new disease-causing agents have emerged that afflict humans. These include the SARS virus, HIV, and new strains of the tuberculosis bacterium. Hundreds of genes help defend you against these types of invaders. Certain alleles help you fight off particular strains of disease-causing bacteria, eukaryotes, or viruses. In this evolutionary arms race, pathogens and parasites constantly evolve new ways to infect the most common types of host.

What happens if all the offspring produced by an individual are genetically identical? If a new strain of disease-causing agent evolves, then all the asexually produced offspring are likely to be susceptible to that new strain. But if the offspring are genetically varied, then it is likely that at least some offspring will have combinations of alleles that enable them to fight off the new strain of pathogen or parasite and produce offspring of their own.

Recall from Section 13.2 that over multiple generations, outcrossing—mating between two different individuals—increases the amount of genetic diversity relative to self-fertilization. A logical question to ask is: Does exposure to evolving pathogens favor outcrossing in sexually reproducing organisms? To address this question, Levi Morran, Curtis Lively, and colleagues carried out a pivotal study on a tiny (only about 1 mm long) roundworm named *Caenorhabditis elegans*.

C. elegans is an important model organism (see **BioSkills 13** in Appendix B) that was chosen for this study because it leads an unusual sex life. There are no females, only males and hermaphrodites. Because hermaphrodites have both male and female sex organs, this means that *C. elegans* can reproduce either by self-fertilization or by outcrossing with males. The proportion of roundworms that reproduce by self-fertilization versus outcrossing can vary in different strains or in different environments over time. So, the research team was able to test whether the rate of outcrossing increased in response to intense selection by a pathogen.

The setup of Morran, Lively, and colleagues' experiment is shown in **FIGURE 13.15**. The team began with a population of roundworms that had not been exposed to the pathogen and that reproduced predominantly by self-fertilization. The researchers then split the starting population into different groups. Half the groups were grown in the presence of a pathogen—a deadly bacterium—and the other half were grown without it. Once ingested by a roundworm, the bacterial pathogen could kill a susceptible individual within 24 hours.

At each generation, bacteria were collected from the carcasses of roundworms killed in the previous generation. Companion experiments showed that the pathogen evolved to become even more infectious over the course of the study.

The results are shown at the bottom of Figure 13.15. The rate of outcrossing stayed low over 32 generations in populations that did not encounter the pathogen. In contrast, populations that were exposed to the evolving pathogen showed a rapid increase in the rate of outcrossing.

At the end of the experiment, the roundworms in the pathogen-exposed population were significantly more resistant to the evolved pathogen than their ancestors. In other words, the roundworms in the predominantly outcrossing population had evolved along with the pathogen. In striking contrast, when a parallel experiment was done with a strain of roundworms that could reproduce only by self-fertilization, those populations were unable to evolve resistance to the pathogen. In fact, they became extinct.

These results and many others support the changing-environment hypothesis. Although the paradox of sex remains an active area of research, more biologists are becoming convinced that sexual reproduction is helpful for two reasons: (1) Offspring are not doomed to inherit harmful alleles, and (2) the production of genetically varied offspring means that at least some may be able to resist rapidly evolving pathogens and parasites.

| **CHAPTER 13** | **REVIEW** | *For media, go to MasteringBiology* |

If you understand . . .

13.1 How Does Meiosis Occur?

- Meiosis is a nuclear division resulting in cells that have only one of each type of chromosome and half as many chromosomes as the parent cell. In animals it leads to the formation of eggs and sperm.

- In diploid (2n) organisms, individuals have two versions of each type of chromosome. The two versions are called homologs. One homolog is inherited from the mother and one from the father. Haploid organisms (n) have just one of each type of chromosome.

- Each chromosome is replicated before meiosis begins. At the start of meiosis I, each chromosome consists of a pair of sister chromatids.

- Homologous pairs of chromosomes synapse early in meiosis I, forming a bivalent—two closely paired homologous chromosomes. Non-sister chromatids undergo crossing over.

- When crossing over is complete, the pair of homologous chromosomes is moved to the metaphase plate.

- At the end of meiosis I, the homologous chromosomes are separated and distributed to two daughter cells. The daughter cells are haploid, because each receives one of each type of chromosome.

- During meiosis II, sister chromatids separate and are distributed to two daughter cells.

- From one diploid cell with replicated chromosomes, meiosis produces four haploid daughter cells with unreplicated chromosomes.

✔You should be able to diagram a diploid cell with a homologous pair of chromosomes and show when in meiosis this cell produces haploid daughter cells. Be sure to show all chromatids.

13.2 Meiosis Promotes Genetic Variation

- Each cell produced by meiosis receives a different combination of chromosomes. Because genes are located on chromosomes, each cell produced by meiosis receives a different complement of genes. The resulting offspring are genetically distinct from one another and from their parents.

- When meiosis and outcrossing occur, the chromosome complements of offspring differ from one another and from their parents for three reasons:

  1. Gametes receive a random assortment of maternal and paternal chromosomes when homologs separate in meiosis I. This is independent assortment.
  2. Because of crossing over, each chromosome contains a random assortment of paternal and maternal alleles.
  3. Outcrossing results in a combination of chromosome sets from different individuals.

✔You should be able to draw a diploid cell with four chromosomes entering meiosis and illustrate (a) how different combinations of chromosomes can result from independent assortment; and (b) focusing on a single homologous pair, show how many recombinant chromosomes (chromosomes with mixtures of maternal and paternal segments) are produced when crossing over occurs once along this homologous pair.

13.3 What Happens When Things Go Wrong in Meiosis?

- If mistakes occur during meiosis, the resulting egg and sperm cells may contain the wrong number of chromosomes. It is rare for offspring with an incorrect number of chromosomes to develop normally.

- Mistakes during meiosis lead to gametes and offspring with an abnormal number of chromosomes. Children with Down syndrome, for example, have an extra copy of chromosome 21.

- The leading hypothesis to explain meiotic mistakes is that they are accidental failures of homologous chromosomes or sister chromatids to separate properly during meiosis.

✔Using pipe cleaners or spaghetti to model a homologous pair of chromosomes, you should be able to demonstrate (a) what happens if one pair of homologous chromosomes fails to separate at anaphase of meiosis I and (b) what happens if sister chromatids of one chromosome fail to separate at anaphase of meiosis II but later separate in a daughter cell.

13.4 Why Does Meiosis Exist?

- Asexual reproduction is much more efficient than sexual reproduction because all individuals produced asexually are capable of bearing progeny. With sexual reproduction, half the offspring (the males) are unable to bear progeny.

- The leading hypotheses to explain the existence of meiosis and sexual reproduction are that:

  1. parents can produce offspring that lack harmful alleles; and
  2. genetically diverse offspring are likely to include some that are better able to resist evolving pathogens and parasites.

✔You should be able to predict whether, in species that alternate between asexual and sexual reproduction, sexual reproduction occurs during times when environmental conditions are stable or times when conditions change rapidly.

(MB) MasteringBiology

1. MasteringBiology Assignments

Tutorials and Activities Asexual and Sexual Life Cycles; Genetic Variation from Sexual Recombination; Meiosis; Meiosis (1 of 3): Genes, Chromosomes, and Sexual Reproduction; Meiosis (2 of 3): The Mechanism; Meiosis (3 of 3): Determinants of Heredity and Genetic Variation; Meiosis Animation; Mistakes in Meiosis; Origins of Genetic Variation

Questions Reading Quizzes, Blue-Thread Questions, Test Bank

2. eText Read your book online, search, take notes, highlight text, and more.

3. The Study Area Practice Test, Cumulative Test, BioFlix® 3-D Animations, Videos, Activities, Audio Glossary, Word Study Tools, Art

You should be able to . . .

1. In the roundworm *Ascaris*, eggs and sperm have two chromosomes, but all other cells have four. Observations such as this inspired which important hypothesis?
   a. Before gamete formation, a special type of cell division leads to a quartering of chromosome number.
   b. Before gamete formation, a special type of cell division leads to a halving of chromosome number.
   c. After gamete formation, half the chromosomes are destroyed.
   d. After gamete formation, either the maternal or the paternal set of chromosomes disintegrates.

2. What are homologous chromosomes?
   a. chromosomes that are similar in their size, shape, and gene content
   b. similar chromosomes that are found in different individuals of the same species
   c. the two "threads" in a replicated chromosome (they are identical copies)
   d. the products of crossing over, which contain a combination of segments from maternal chromosomes and segments from paternal chromosomes

3. What is a bivalent?
   a. the X that forms when chromatids from homologous chromosomes cross over
   b. a group of four chromatids produced when homologs synapse
   c. the four points where homologous chromosomes touch as they synapse
   d. the group of four genetically identical daughter cells produced by mitosis

4. What is an outcome of genetic recombination?
   a. the synapsing of homologs during prophase of meiosis I
   b. the new combination of maternal and paternal chromosome segments that results when homologs cross over
   c. the new combinations of chromosome segments that result when self-fertilization occurs
   d. the combination of a haploid phase *and* a diploid phase in a life cycle

5. What proportion of chromosomes in a human skin cell are paternal chromosomes?

6. Meiosis II is similar to _____.

7. Explain the relationship between homologous chromosomes and the relationship between sister chromatids.

8. Lay four pens and four pencils on a tabletop, and imagine that they represent replicated chromosomes in a diploid cell where $n = 2$. Explain the phases of meiosis II by moving the pens and pencils around. (If you don't have enough pens and pencils, use strips of paper or fabric.)

9. Meiosis is called a reduction division, but all the reduction occurs during meiosis I—no reduction occurs during meiosis II. Explain why meiosis I is a reduction division but meiosis II is not.

10. Dogs have 78 chromosomes in their diploid cells. If a diploid dog cell enters meiosis, how many chromosomes and chromatids will be present in each daughter cell at the end of meiosis I?

   a. 39 chromosomes and 39 chromatids
   b. 39 chromosomes and 78 chromatids
   c. 78 chromosomes and 78 chromatids
   d. 78 chromosomes and 156 chromatids

11. Triploid ($3n$) watermelons are produced by crossing a tetraploid ($4n$) strain with a diploid ($2n$) plant. Briefly explain why this mating produces a triploid individual. Why can mitosis proceed normally in triploid cells, but meiosis cannot?

12. Some plant breeders are concerned about the susceptibility of asexually cultivated plants, such as seedless bananas, to new strains of disease-causing bacteria, viruses, or fungi. Briefly explain their concern by discussing the differences in the genetic "outcomes" of asexual and sexual reproduction.

13. The gibbon has 44 chromosomes per diploid set, and the siamang has 50 chromosomes per diploid set. In the 1970s a chance mating between a male gibbon and a female siamang produced an offspring. Predict how many chromosomes were observed in the somatic cells of the offspring. Do you predict that this individual would be able to form viable gametes? Why or why not?

14. Meiosis results in a reassortment of maternal and paternal chromosomes. If $n = 3$ for a given organism, there are eight different combinations of paternal and maternal chromosomes. If no crossing over occurs, what is the probability that a gamete will receive *only* paternal chromosomes?
   a. 0; **b.** 1/16; **c.** 1/8; **d.** 1/3

15. Some researchers hypothesize that older women are less responsive to triggers of spontaneous abortion than younger women. How could the data shown in Figure 13.13, which graphs a mother's age versus the incidence of Down syndrome, be used to support this hypothesis?

16. A species of rotifer, a small freshwater invertebrate, abandoned sexual reproduction millions of years ago. A remarkable feature of the rotifer's life cycle is its ability to withstand extreme drying. When the rotifer's watery environment dries out, so does the rotifer, and it can be blown in the wind to a new environment. Once blown to water, the rotifer rehydrates and resumes an active life. A major pathogen of these rotifers is a species of fungus. Some scientists hypothesize that fungus-infected rotifers rid themselves of the pathogen when they dry.
   a. Design an experimental study to test this hypothesis.
   b. Provide an explanation for how these asexually reproducing rotifers are able to evade pathogens even though they are genetically identical.

14 Mendel and the Gene

In this chapter you will learn how

Mendel's principles can predict patterns of inheritance

starting with ↓

Mendel's experimental system 14.1

then examining ↓

Experiments with a single trait 14.2

explained by ↓

The principle of segregation

Experiments with two traits 14.3

explained by ↓

The principle of independent assortment

expanding to ↓

The chromosome theory of inheritance 14.4

exploring ↓

Extensions to Mendel's principles 14.5

applied to ↓

Human inheritance 14.6

Experiments on garden peas and sweet peas (shown here) helped launch the science of genetics.

This chapter is part of the Big Picture. See how on pages 366–367.

The science of biology is built on a series of great ideas. Two of these—the cell theory and the theory of evolution—were introduced in Chapter 1. The cell theory describes the basic structure of organisms; the theory of evolution by natural selection clarifies why species change through time. Life is cellular; populations evolve. These are two of the five fundamental attributes of life.

This chapter introduces a third great idea in biology: the chromosome theory of inheritance. The chromosome theory explained how genetic information is transmitted from one generation to the next. It shed light on a third fundamental attribute of life: Organisms process information.

✔ When you see this checkmark, stop and test yourself. Answers are available in Appendix A.

An Austrian monk named Gregor Mendel laid the groundwork for the theory in 1865, when he announced that he had worked out the rules of inheritance through a series of experiments on garden peas. Another key insight emerged during the final decades of the nineteenth century, when biologists described the details of meiosis (see Chapter 13).

The chromosome theory of inheritance, formulated in 1902 by Walter Sutton and Theodor Boveri, linked these two insights. This theory contends that meiosis causes the patterns of inheritance that Mendel observed. It also asserts that the hereditary factors called genes are located on chromosomes. (Genes determine inherited traits and are center stage in the Big Picture on pages 366–367.)

The chromosome theory launched the study of **genetics,** the branch of biology that focuses on the inheritance of traits. Let's start at the beginning: What are the rules of inheritance that Mendel discovered?

14.1 Mendel's Experimental System

When biological science began to emerge as an important discipline, questions about **heredity**—meaning inheritance, or the transmission of traits from parents to offspring—were primarily the concern of animal breeders and horticulturists. A **trait** is any characteristic of an individual, ranging from height to the primary structure of a particular membrane protein.

In the city where Gregor Mendel lived, there was particular interest in how selective breeding could result in hardier and more productive varieties of sheep, fruit trees, and vines; and an agricultural society had been formed there to promote research into making selective breeding more efficient. Mendel was an active member of this society, and the monastery he belonged to was also devoted to scientific teaching and research.

What Questions Was Mendel Trying to Answer?

Mendel set out to address the most fundamental of all issues concerning heredity: What are the basic patterns in the transmission of traits from parents to offspring?

At the time, two hypotheses had been formulated to answer this question:

1. *Blending inheritance* claimed that the traits observed in a mother and father blend together to form the traits observed in their offspring. As a result, an offspring's traits are intermediate between the mother's and father's traits.

2. *Inheritance of acquired characters* claimed that traits present in parents are modified, through use, and passed on to their offspring in the modified form.

Each of these hypotheses made predictions. Blending inheritance contended that when black sheep and white sheep mate, their hereditary determinants will blend to form a new hereditary determinant for gray wool. Therefore, their offspring should be gray. Inheritance of acquired characters predicts that if giraffes extend their necks by straining to reach leaves high in the tops of trees, they subsequently produce longer-necked offspring.

These hypotheses were being promoted by the greatest scientists of Mendel's time. Are they correct?

The Garden Pea Served as the First Model Organism in Genetics

After investigating and discarding several candidate species to study, Mendel chose the garden pea, *Pisum sativum*. His reasons were practical: Peas are inexpensive and easy to grow from seed, have a relatively short generation time, and produce reasonably large numbers of seeds. These features made it possible for him to continue experiments over several generations and collect data from large numbers of individuals.

Peas served as a **model organism:** a species that is used for research because it is practical and because conclusions drawn from studying it turn out to apply to many other species as well. **BioSkills 13** in Appendix B introduces some of the important model organisms used in biological science today.

Two additional features of the pea made it possible for Mendel to design his experiments: Individuals were available that differed in easily recognizable traits, such as flower color or seed shape, and he could control which parents were involved in a mating.

How Did Mendel Control Matings? **FIGURE 14.1a** shows the male and female reproductive organs of a garden pea flower. Sperm cells are produced in pollen grains, which are small sacs that

(a) Self-pollination

Female organ (receives pollen)

Male organs (produce pollen grains, which produce sperm cells)

Eggs

(b) Cross-pollination

Collect pollen from one individual and transfer it ...

... to the female organs of an individual whose male organs have been removed.

FIGURE 14.1 Peas Can Be Self-Pollinated or Cross-Pollinated.
(a) Under normal conditions, garden peas pollinate themselves.
(b) Mendel developed a method of controlling the matings of his model organism.

mature in the male reproductive structure of the plant. Eggs are produced in the female reproductive structure.

Under normal conditions, garden peas **self-fertilize:** a flower's pollen falls on the female reproductive organ of that same flower. As **FIGURE 14.1b** shows, however, Mendel could prevent self-fertilization by removing the male reproductive organs from a flower before any pollen formed. Later he could transfer pollen from another pea plant to the target flower's female reproductive organ with a brush. This type of mating is referred to as a cross-fertilization, or simply a **cross.** Using this technique, Mendel could control the matings of his model organism.

What Traits Did Mendel Study? Mendel conducted his experiments on varieties of peas that differed in seven traits: seed shape, seed color, pod shape, pod color, flower color, flower and pod position, and stem length. Biologists refer to the observable traits of an individual, such as the shape of a pea seed or the eye color of a person, as its **phenotype** (literally, "show-type"). Phenotype is just one term in the rich vocabulary of genetics. You can review many of these terms in **TABLE 14.1**. In the first pea populations that Mendel studied, two distinct phenotypes existed for each of the seven traits.

Mendel began his work by obtaining individuals from what breeders called pure lines or true-breeding lines. A **pure line** consists of individuals that produce offspring identical to themselves when they are self-pollinated or crossed to another member of the pure-line population. For example, earlier breeders had developed pure lines for wrinkled seeds and round seeds. During two years of trial experiments, Mendel confirmed that individuals that germinated from his wrinkled seeds produced only wrinkled-seeded offspring when they were mated to themselves or to another pure-line individual that germinated from a wrinkled seed, and he confirmed that the same was true for round seeds.

Why is this important? Remember that Mendel wanted to find out how traits are transmitted from parents to offspring. Once he had confirmed that he was working with pure lines, he could compare the results of crosses within a pure line with crosses between individuals from different pure lines.

SUMMARY TABLE 14.1 **Vocabulary Used in Mendelian Genetics**

Term	Definition	Example or Comment
Gene	A hereditary factor that influences a particular trait.	This definition will become more precise in later chapters.
Allele	A particular form of a gene.	The two alleles in a diploid may be the same or different.
Genotype	A listing of the alleles in an individual.	In diploids, the genotype lists two alleles of each gene; in haploids, the genotype lists one allele of each gene.
Phenotype	An individual's observable traits.	Can be observed at levels from molecules to the whole organism; influenced, not dictated, by the genotype.
Homozygous	Having two of the same allele.	Refers to a particular gene.
Heterozygous	Having two different alleles.	Refers to a particular gene.
Dominant allele	An allele that produces its phenotype in heterozygous and homozygous form.	Dominance does not imply high frequency or high fitness.
Recessive allele	An allele that produces its phenotype only in homozygous form.	Phenotype "recedes" or disappears in heterozygous individuals.
Pure line	Individuals of the same phenotype that, when crossed, always produce offspring with the same phenotype.	Pure-line individuals are homozygous for the gene in question.
Hybrid	Offspring from crosses between homozygous parents with different genotypes.	Hybrids are heterozygous.
Reciprocal cross	A cross in which the phenotypes of the male and female are reversed compared with a prior cross.	If reciprocal crosses give identical results, the sex of the parent does not influence transmission of the trait.
Testcross	A cross between a homozygous recessive individual and an individual with the dominant phenotype but an unknown genotype.	Usually used to determine whether a parent with a dominant phenotype is homozygous or heterozygous.
X-linked	Referring to a gene located on the X chromosome.	X-linked genes and traits show different patterns of inheritance in males and females.
Y-linked	Referring to a gene located on the Y chromosome.	In humans, Y-linked genes determine male-specific development.
Autosomal	Referring to a gene located on any non-sex chromosome (an autosome) or a trait determined by an autosomal gene.	Mendel studied only autosomal genes and traits.

Suppose that Mendel arranged matings between a pure-line individual with round seeds and a pure-line individual with wrinkled seeds. He knew that one parent carried a hereditary determinant for round seeds, while the other carried a hereditary determinant for wrinkled seeds. But each offspring from this mating would contain both types of hereditary determinants. They would be **hybrids**—offspring from matings between true-breeding parents that differ in one or more traits.

Would these hybrid offspring have wrinkled seeds, round seeds, or a blended combination of wrinkled and round? What would the seed shape in subsequent generations be when hybrid individuals self-pollinated or were crossed with members of the pure lines?

14.2 Mendel's Experiments with a Single Trait

Mendel's first set of experiments consisted of crossing pure lines that differed in just one trait. This is an important research strategy in biological science: Start with a simple situation. Once you understand what's going on, you can consider more complex questions, such as, What happens in crosses between individuals that differ in two traits?

Mendel began his single-trait crosses by crossing individuals from round-seeded and wrinkled-seeded pure lines. The adults used in an initial experimental cross are the **parental generation**. Their progeny (that is, offspring) are the **F_1 generation**. F_1 stands for "first filial"; the Latin roots *filius* and *filia* mean "son" and "daughter," respectively. Subsequent generations are called the F_2 generation, F_3 generation, and so on.

The Monohybrid Cross

In his first set of crosses, Mendel took pollen from round-seeded plants and placed it on the female reproductive organs of plants from the wrinkled-seeded line. As **FIGURE 14.2a** shows, all the seeds produced by progeny from this cross were round.

This was a remarkable result, for two reasons:

1. The traits did not blend together to form an intermediate phenotype. Instead, the round-seeded form appeared intact. This result was in stark contrast to the predictions of the blending-inheritance hypothesis shown in **FIGURE 14.2b**.

2. The genetic determinant for wrinkled seeds seemed to have disappeared. Where did it go?

Dominant and Recessive Traits To figure out what was going on, Mendel did something that turned out to be brilliant: He planted the F_1 seeds and allowed the individuals to self-pollinate when they matured.

Remember that he knew that each of these individuals had inherited a genetic determinant for round seeds and a genetic determinant for wrinkled seeds. A mating like this—between parents that each carry two different genetic determinants for the same trait—is called a **monohybrid cross.**

When he collected the seeds that were produced by many plants in the resulting F_2 generation, he observed that 5474 were round and 1850 were wrinkled (see Figure 14.2a). This observation was astonishing. The wrinkled seed shape had reappeared in the F_2 generation after disappearing completely in the F_1 generation. No one had observed the phenomenon before, simply because it had been customary for biologists to stop their breeding experiments with F_1 offspring.

Mendel invented some important terms to describe this result.

- He designated wrinkled shape as a **recessive** trait relative to the round-seed trait. This was an appropriate term because none of the F_1 individuals had wrinkled seeds—meaning the wrinkled-seed phenotype appeared to recede or temporarily become latent or hidden.

- He referred to round seeds as **dominant** to the wrinkled-seed trait. This term was apt because the round-seed phenotype appeared to dominate over the wrinkled-seed determinant when both were present.

It's important to note, though, that in genetics the term dominant has nothing to do with its everyday English usage as powerful

(a) Results of Mendel's single-trait (monohybrid) cross

(b) Prediction of blending-inheritance hypothesis

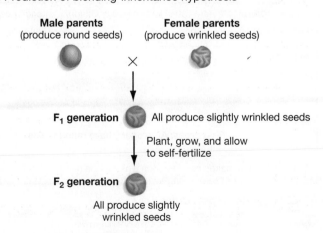

FIGURE 14.2 A Monohybrid Cross. The results of Mendel's crosses involving a single trait **(a)** contrasted strongly with the predictions of the blending-inheritance hypothesis **(b)**.

or superior. Subsequent research has shown that individuals with the dominant phenotype do not necessarily have higher fitness than individuals with the recessive phenotype. Nor are genetic determinants associated with a dominant phenotype necessarily more common than recessive ones. For example, a fatal illness—a type of brain degeneration called Huntington's disease—is caused by a rare, dominant genetic determinant. In genetics, the terms dominant and recessive identify *only* which phenotype is observed in individuals carrying two different genetic determinants for a given trait.

Mendel also noticed that the round and wrinkled seeds of the F_2 generation were present in a ratio of 2.96 : 1, or essentially 3 : 1. The 3 : 1 ratio means that for every four individuals, on average three had the dominant phenotype and one had the recessive phenotype. The results can also be stated in terms of frequencies or proportions: In this case, about ¾ of the F_2 seeds were round and ¼ were wrinkled.

A Reciprocal Cross Mendel wanted to test the hypothesis that it mattered which parent and gamete type had a particular genetic determinant—that gender influenced the inheritance of

seed shape. To do this, he performed a second set of crosses between two pure-breeding lines—this time with pollen taken from an individual from a pure line of wrinkled-seeded peas (**FIGURE 14.3**).

These experiments completed a **reciprocal cross**—a set of matings where the mother's phenotype in the initial cross is the father's phenotype in a subsequent cross, and the father's phenotype in the initial cross is the mother's phenotype in a subsequent cross.

In this case the results of the reciprocal crosses were identical: All the F_1 progeny had round seeds, just as in the initial cross. The reciprocal cross established that it does not matter whether the genetic determinants for seed shape are located in the male or female parent.

Do Mendel's Results Hold for Other Traits? Before he tried to interpret this pattern, it was important for Mendel to establish that the results were not restricted to inheritance of seed shape. So he repeated the experiments with each of the six other traits listed earlier. As **TABLE 14.2** shows, in each case, he obtained similar results:

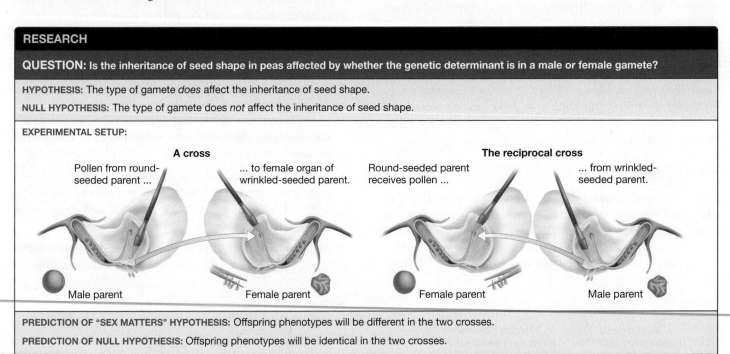

RESEARCH

QUESTION: Is the inheritance of seed shape in peas affected by whether the genetic determinant is in a male or female gamete?

HYPOTHESIS: The type of gamete *does* affect the inheritance of seed shape.

NULL HYPOTHESIS: The type of gamete does *not* affect the inheritance of seed shape.

EXPERIMENTAL SETUP:

A cross

Pollen from round-seeded parent ...

... to female organ of wrinkled-seeded parent.

Male parent   Female parent

The reciprocal cross

Round-seeded parent receives pollen ...

... from wrinkled-seeded parent.

Female parent   Male parent

PREDICTION OF "SEX MATTERS" HYPOTHESIS: Offspring phenotypes will be different in the two crosses.

PREDICTION OF NULL HYPOTHESIS: Offspring phenotypes will be identical in the two crosses.

RESULTS:

Results are identical

First cross: All progeny have round seeds.   Reciprocal cross: All progeny have round seeds.

CONCLUSION: It makes no difference whether the genetic determinant for seed shape comes from the male gamete or from the female gamete.

FIGURE 14.3 A Reciprocal Cross.

SOURCE: Mendel, G. 1866. Versuche über Pflanzen-hybriden. *Verhandlungen des naturforschenden Vereines in Brünn.* 4: 3–47. English translation available from ESP: Electronic Scholarly Publishing (www.esp.org).

✔**QUESTION** Some people think that experiments are failures if the hypothesis being tested is not supported. What does it mean to say that an experiment failed? Was this experiment a failure?

TABLE 14.2 **F₂ Results from Mendel's Monohybrid Reciprocal Cross Experiments***

Trait	Dominant Phenotype	Recessive Phenotype	Ratio
Seed shape	5474 round	1850 wrinkled	2.96 : 1
Seed color	6022 yellow	2001 green	3.01 : 1
Pod shape	882 inflated	299 constricted	2.95 : 1
Pod color	428 green	152 yellow	2.82 : 1
Flower color	705 purple	224 white	3.15 : 1
Flower and pod position	651 axial	207 terminal	____ : __
Stem length	787 tall	____ dwarf	2.96 : 1

*Mendel pooled the results from the reciprocal crosses for each trait because the results were the same whether the dominant trait originated from the male parent or the female parent.
DATA: Mendel, G. 1866. *Verhandlungen des naturforschenden Vereines in Brünn.* 4: 3–47.

✔**EXERCISE** Two entries in the last rows are left blank. Fill in these entries with the correct values calculated from the available data.

- The F₁ progeny showed only the dominant trait and did not exhibit an intermediate phenotype.
- Reciprocal crosses produced the same results.
- The ratio of F₂ generation individuals with dominant and recessive phenotypes was about 3 to 1.

How could these patterns be explained? Mendel answered this question with a series of propositions about the nature and behavior of the hereditary determinants. These hypotheses are considered some of the most brilliant insights in the history of biological science.

Particulate Inheritance

Mendel's results were clearly inconsistent with either the hypothesis of blending inheritance or the hypothesis of acquired characters. To explain the patterns that he observed, Mendel proposed a competing hypothesis called **particulate inheritance.** He maintained that the hereditary determinants for traits do not blend together or become modified through use. In fact, hereditary determinants maintain their integrity from generation to generation. Instead of blending together, they act as discrete entities or particles.

Mendel's hypothesis was the only way to explain the observation that phenotypes disappeared in one generation and reappeared intact in the next. It also represented a fundamental break with ideas that had prevailed for hundreds of years.

Genes, Alleles, and Genotypes Today, geneticists use the word **gene** to indicate the hereditary determinant for a trait. For example, the hereditary factor that determines whether the seeds of garden peas are round or wrinkled is referred to as the gene for seed shape.

Mendel also proposed that each individual can have two versions of any gene. Today different versions of the same gene are called **alleles.** Different alleles are responsible for the variation in the traits that Mendel studied. In the case of the gene for seed shape, one allele of this gene is responsible for the round form of the seed while another allele is responsible for the wrinkled form.

The alleles found in a particular individual are called the **genotype.** An individual's genotype has a profound effect on the phenotype—the observable physical traits.

The hypothesis that pea plants have two copies of each gene—either two of the same allele or two different alleles—was important because it gave Mendel a framework for explaining dominance and recessiveness. He proposed that some alleles are dominant and others are recessive. Recall that dominance and recessiveness determine which phenotype appears in an individual when two different alleles are present. In garden peas, the allele for round seeds is dominant; the allele for wrinkled seeds is recessive. Therefore, so long as one allele for round seeds is present, seeds are round. When both alleles present are for wrinkled seeds (thus no allele for round seeds is present), seeds are wrinkled.

These hypotheses explain why the wrinkled-seed phenotype disappeared in the F₁ generation. But why did wrinkled seeds reappear in the F₂, and why was there a 3:1 ratio of round and wrinkled seeds in the F₂ generation?

The Principle of Segregation To explain the reappearance of the recessive phenotype and the characteristic 3:1 ratio of phenotypes in F₂ individuals, Mendel reasoned that the two members of each gene pair must segregate—that is, separate—into different gamete cells during the formation of eggs and sperm. As a result, each gamete contains one allele of each gene. This idea is called the **principle of segregation.**

To show how this principle works, Mendel used a letter to indicate the gene for a particular trait. For example, he used uppercase *R* to symbolize a dominant allele for seed shape and lowercase *r* to symbolize a recessive allele for seed shape. (Notice that the symbols for genes are always italicized.)

Using this notation, Mendel described the genotype of the individuals in the round-seed pure line as *RR* (having two of the dominant allele). The genotype of the wrinkled-seed pure line is *rr* (two of the recessive allele). Because *RR* and *rr* individuals have two copies of the same allele, they are said to be **homozygous** for the seed-shape gene (*homo* is the Greek root for "same," while *zygo* means "yoked"). Crosses of individuals from the same pure line always produce offspring with the same phenotype because they are homozygous—no other allele is present.

FIGURE 14.4a uses a diagram called a Punnett square to show what happened to these alleles when Mendel crossed the *RR* and *rr* pure lines. R. C. Punnett invented this straightforward technique for predicting the genotypes and phenotypes of different crosses years after Mendel published his work. According to Mendel's hypothesis, *RR* parents produce eggs and sperm that all carry the *R* allele, while *rr* parents produce gametes with the *r* allele only. When two gametes—one from each parent—come together at fertilization, they create offspring with the *Rr* genotype. Such individuals, with two different alleles for the same gene, are said to be **heterozygous** (*hetero* is the Greek root for "different"). Heterozygous individuals, or heterozygotes, show that the *R* allele is dominant because only the round phenotype is expressed even though the wrinkled allele is present.

Why do the two phenotypes appear in a 3:1 ratio in the F_2 generation? Mendel proposed that during gamete formation in the F_1 heterozygotes, the paired *Rr* alleles separate into different gamete cells. As a result, and as shown in the Punnett square of **FIGURE 14.4b**, half the gametes should carry the *R* allele and half should carry the *r* allele. A given sperm has an equal chance of fertilizing either an *R*-bearing egg or an *r*-bearing egg.

Predicting Offspring Genotypes and Phenotypes with a Punnett Square

The box you've just studied in Figure 14.4b is an example of a simple Punnett square. To produce a Punnett square, follow these steps:

1. Write each of the *unique* gamete genotypes produced by one parent in a horizontal row along the top of the diagram.

2. Write each of the *unique* gamete genotypes produced by the other parent in a vertical column down the left side of the diagram.

3. Create a table under the horizontal row of gametes and to the right of the vertical column of gametes.

4. Fill in the table with the entries for the parental gamete genotypes that are written at the top and at the left side. This step represents fertilization and produces the offspring genotypes.

(a) A cross between two **homozygotes**

Offspring genotypes: All *Rr* (heterozygous)

Offspring phenotypes: All round seeds

(b) A cross between two **heterozygotes**

Offspring genotypes: ¼ *RR* : ½ *Rr* : ¼ *rr*

Offspring phenotypes: ¾ round : ¼ wrinkled

FIGURE 14.4 Mendel Analyzed the F_1 and F_2 Offspring of a Cross between Pure Lines. Notice that when you construct a Punnett square, you only need to list each unique type of gamete once at the head of a row or column. For example, even though the *RR* alleles segregate in the male parent of part (a), you have to list just one *R* gamete to represent the male's contribution, not two.

✔**QUESTION** In constructing a Punnett square, does it matter whether the male or female gametes go on the left or across the top? Why or why not?

5. Tally the proportions or ratios of each offspring genotype and phenotype.

✔ If you understand these concepts, you should be able to state how filling in the top and side of the Punnett square is related to the principle of segregation and predict the phenotype and genotype ratios for a cross between *Rr* and *rr* peas.

Mendel's Model to Explain the Results of a Cross between Pure Lines*

Mendel's Claims	Comments
1. Peas have two copies of each gene and thus may have two different alleles of the gene.	This also turns out to be true for many other organisms.
2. Genes are particles of inheritance that do not blend together.	Genes maintain their integrity from generation to generation.
3. Each gamete contains one copy of each gene (one allele).	This is because of the principle of segregation—the members of each gene pair segregate during the formation of gametes.
4. Males and females contribute equally to the genotype of their offspring.	When gametes fuse, offspring acquire a total of two of each gene—one from each parent.
5. Some alleles are dominant to other alleles.	When a dominant and a recessive allele for the same gene are found in the same individual (a heterozygote), that individual has the dominant phenotype.

*Mendel did not use these modern terms. He expressed these ideas in different words.

check your understanding

If you understand that . . .

- Mendel discovered that individuals have two alleles of each gene and that these alleles separate (segregate) into gametes. This is the principle of segregation.
- The alleles that Mendel analyzed were either dominant or recessive, meaning heterozygous individuals had the dominant phenotype.

✔ You should be able to . . .

Use the genetic problems at the end of this chapter to practice the following skills:

1. Starting with parents of known genotypes, create and analyze Punnett squares of crosses involving a single trait to predict the genotypes and phenotypes that will occur in their F₁ and F₂ offspring; then use the Punnett square to determine the expected frequency of each genotype and phenotype. (Do Problem 13 in Test Your Problem-Solving Skills.)

2. Given the outcome of a cross, infer the genotypes and phenotypes of the parents. (Do Problem 15 in Test Your Problem-Solving Skills.)

Answers are available in Appendix A.

As an example of the concluding step in analyzing a cross, the Punnett square in Figure 14.4b predicts that ¼ of the F₂ offspring will be *RR*, ½ will be *Rr*, and ¼ will be *rr*. Because the *R* allele is dominant to the *r* allele, ¾ of the offspring should be round-seeded (the sum of the *RR* and the *Rr* offspring) and ¼ should be wrinkled-seeded (the *rr* offspring). These results are what Mendel found in his experiments with peas. In the simplest and most elegant fashion possible, Mendel's interpretation explains the 3 : 1 ratio of round to wrinkled seeds observed in the F₂ offspring and the mysterious reappearance of the wrinkled seeds.

The term genetic model refers to a set of hypotheses that explains how a particular trait is inherited. **TABLE 14.3** summarizes Mendel's model for explaining the basic patterns in the transmission of traits from parents to offspring; these hypotheses are sometimes referred to as Mendel's rules. They represent a radical break from the ideas of blending inheritance and the inheritance of acquired characters that had dominated scientific thinking about heredity.

14.3 Mendel's Experiments with Two Traits

Working with one trait at a time allowed Mendel to establish that blending inheritance does not occur. It also allowed him to infer that each pea plant had two copies of each gene and to recognize the principle of segregation.

Mendel's next step extended these results. The important question now was whether the principle of segregation holds true if individuals differ with respect to two traits, instead of just one. Do different genes segregate together, or independently?

The Dihybrid Cross

Mendel crossed a pure-line parent that produced round, yellow seeds with a pure-line parent that produced wrinkled, green seeds. According to his model, the F₁ offspring of this cross should be heterozygous for both genes. A mating between two such individuals—both heterozygous for two traits—is called a **dihybrid cross.**

Mendel's earlier experiments had established that the allele for yellow seeds was dominant to the allele for green seeds; these alleles were designated *Y* for yellow and *y* for green. As **FIGURE 14.5** (see page 264) indicates, two distinct possibilities existed for how the alleles of these two different genes—the gene for seed shape and the gene for seed color—would be transmitted to offspring.

- The first possibility was that the allele for seed shape and the allele for seed color originally present in each parent would separate from each other and be transmitted independently. This hypothesis is called independent assortment because the two alleles would be sorted into gametes independently of each other (Figure 14.5a).

- The second possibility was that the allele for seed shape and the allele for seed color originally present in each parent

(a) Hypothesis of independent assortment:
Alleles of different genes don't stay together when gametes form.

r = Recessive allele for seed shape (**wrinkled**)

y = Recessive allele for seed color (**green**)

Female parent

rryy

Female gametes

F₁   *ry*

Male parent

RY

RrYy

RRYY

F₁ offspring all *RrYy*

R = Dominant allele for seed shape (**round**)

Y = Dominant allele for seed color (**yellow**)

F₂ female parent

RrYy

Female gametes

F₂   ¼ *RY*   ¼ *Ry*   ¼ *rY*   ¼ *ry*

F₂ male parent

RrYy

Male gametes

¼ *RY*	*RRYY*	*RRYy*	*RrYY*	*RrYy*
¼ *Ry*	*RRYy*	*RRyy*	*RrYy*	*Rryy*
¼ *rY*	*RrYY*	*RrYy*	*rrYY*	*rrYy*
¼ *ry*	*RrYy*	*Rryy*	*rrYy*	*rryy*

F₂ offspring genotypes: ⁹⁄₁₆ *R–Y–* :   ³⁄₁₆ *R–yy* :   ³⁄₁₆ *rrY–* :   ¹⁄₁₆ *rryy*
F₂ offspring phenotypes: ⁹⁄₁₆ ● :   ³⁄₁₆ ● :   ³⁄₁₆ ◉ :   ¹⁄₁₆ ◉

Dashes in a genotype mean that either allele can be present

(b) Hypothesis of dependent assortment:
Alleles of different genes stay together when gametes form.

Female parent

rryy

Female gametes

F₁   *ry*

Male parent

RY

RrYy

RRYY

F₁ offspring all *RrYy*

F₂ female parent

RrYy

Female gametes

F₂   ½ *RY*   ½ *ry*

F₂ male parent

RrYy

Male gametes

| ½ *RY* | *RRYY* | *RrYy* |
| ½ *ry* | *RrYy* | *rryy* |

F₂ offspring genotypes: ¼ *RRYY* :   ½ *RrYy* :   ¼ *rryy*
F₂ offspring phenotypes: ¾ ● :   ¼ ◉

(c) Mendel's results

F₂ phenotypes	●	●	◉	◉	556 total
Number	315	108	101	32	
Fraction of offspring	⁹⁄₁₆	³⁄₁₆	³⁄₁₆	¹⁄₁₆	

Data are consistent with the predictions of independent assortment.

FIGURE 14.5 Mendel Analyzed the F₁ and F₂ Offspring of a Cross between Pure Lines for Two Traits. Each of two hypotheses predicted a different pattern for the outcome when alleles of different genes are transmitted to offspring: **(a)** The alleles could be sorted into gametes independently of each other, or **(b)** particular alleles could always be transmitted together. **(c)** Mendel's results supported independent assortment.

would be transmitted to gametes together. This hypothesis can be called dependent assortment because the transmission of one allele would depend on the transmission of another (Figure 14.5b).

As Figure 14.5 shows, the F_1 offspring of Mendel's mating are expected to have the dominant round and yellow phenotypes whether the different genes are transmitted together or independently. When Mendel did the cross and observed the F_1 individuals, this is exactly what he found. All the F_1 offspring had round, yellow seeds.

The two hypotheses make radically different predictions, however, about what will be observed when the F_1 individuals are allowed to self-fertilize and produce an F_2 generation. If the alleles of different genes assort independently to form gametes, then each heterozygous parent should produce four different gamete genotypes, as shown in Figure 14.5a. A 4-row-by-4-column Punnett square results, and it predicts that there should be 9 different offspring genotypes and 4 phenotypes. Further, the yellow-round, green-round, yellow-wrinkled, and green-wrinkled phenotypes should be present in frequencies of $\frac{9}{16}$, $\frac{3}{16}$, $\frac{3}{16}$, and $\frac{1}{16}$, respectively. This is a ratio of 9 : 3 : 3 : 1.

On the other hand, if the alleles from each parent stay together, then the prediction is for only three possible offspring genotypes and a 3 : 1 ratio of two phenotypes—yellow-round or green-wrinkled—in the F_2, as Figure 14.5b shows.

When Mendel examined the phenotypes of the F_2 offspring, he found that they conformed to the predictions of the hypothesis of independent assortment. Four phenotypes were present in proportions that closely approximated the predicted ratio of 9 : 3 : 3 : 1 (Figure 14.5c). On the basis of these data, Mendel accepted the hypothesis that alleles of different genes are transmitted independently of one another. This result became known as the **principle of independent assortment.**

✔ If you understand the principle of independent assortment, it should make sense to you that an individual with the genotype *AaBb* produces gametes with the genotypes *AB, Ab, aB,* and *ab.* You should be able to predict the genotypes of the gametes produced by individuals with the genotypes *AABb, PpRr,* and *AaPpRr.*

Using a Testcross to Confirm Predictions

Mendel did experiments with combinations of traits other than seed shape and color and obtained results similar to those in Figure 14.5c. Each paired set of traits produced a 9 : 3 : 3 : 1 ratio of progeny phenotypes in the F_2 generation. He even did a limited set of crosses examining three traits at a time. Although all these data were consistent with the principle of independent assortment, his most powerful support for the hypothesis came from a different type of experiment.

In designing this study, Mendel's goal was to test the prediction that an *RrYy* plant produces four different types of gametes in equal proportions. To accomplish this, Mendel invented a technique called a testcross. A **testcross** uses a parent that contributes only recessive alleles to its offspring and helps to determine the unknown genotype of the second parent.

Testcrosses are useful because the genetic contribution of the homozygous recessive parent is known. As a result, a testcross allows experimenters to test the genetic contribution of the other parent. If the other parent has the dominant phenotype but an unknown genotype, the results of the testcross allow researchers to infer whether that parent is homozygous or heterozygous for the dominant allele.

In this case, Mendel performed a cross between a parent that was homozygous for the recessive green and wrinkled phenotypes (*rryy*) and a parent that had an unknown genotype but was known from its yellow- and round-seeded phenotype to possess the dominant *R* and *Y* alleles. Two (of four) possible genotypes for this yellow- and round-seeded parent are *RrYy* and *RRYY.* The types and proportions of offspring that could result from a testcross involving *RrYy* or *RRYY* pea plants can be predicted with the Punnett square shown in **FIGURE 14.6** (see page 266). If the principle of independent assortment is valid, the testcross should produce four types of offspring in equal proportions if the tested parent is *RrYy,* and only one type of offspring if the tested parent is *RRYY.*

What were the actual proportions observed? Mendel did a testcross using the F_1 offspring of a pure line yellow-, round-seeded parent and a green-, wrinkled-seeded parent. These F_1 offspring were expected to have an *RrYy* genotype as shown in Figure 14.5a. When he examined the seeds produced by the testcross, Mendel found that among the 110 progeny, 31 were round

check your understanding

If you understand that . . .

- Mendel found that alleles of different genes are transmitted to gametes independently of one another. This is the principle of independent assortment.
- The genotype of a strain with a dominant phenotype can be revealed in testcrosses between the dominant strain and a homozygous recessive strain.

✔ You should be able to . . .

Use the genetics problems at the end of this chapter to practice the following skills:

1. Starting with parents of known genotypes for two different traits, create and analyze Punnett squares to predict the genotypes and phenotypes that will occur in their F_1 and F_2 offspring and then calculate the expected frequency of each genotype and phenotype. (Do Problem 14 in Test Your Problem-Solving Skills.)

2. Given the outcome of a cross, infer the genotypes and phenotypes of the parents. (Do Problem 16 in Test Your Problem-Solving Skills.)

Answers are available in Appendix A.

Homozygous recessive parent

Parent of known phenotype but unknown genotype

rryy

R?Y?

Could be → *RrYy*

Could be → *RRYY*

¼ *RY*   ¼ *Ry*   ¼ *rY*   ¼ *ry*

All *RY*

All *ry*

| ¼ *RrYy* | ¼ *Rryy* | ¼ *rrYy* | ¼ *rryy* |

All *RrYy*

Offspring predicted if unknown parent is **heterozygous** at both genes

Offspring predicted if unknown parent is **homozygous dominant** at both genes

FIGURE 14.6 The Predictions Made by the Principle of Independent Assortment Can Be Evaluated in a Testcross. If the principle of independent assortment is correct, and *RrYy* parents produce four types of gametes in equal proportions, then a mating between *RrYy* and *rryy* parents should produce four types of offspring in equal proportions, as shown on the left. Test crosses can also reveal the genotype of a parent with dominant phenotypes, as seen in the example of different results obtained from *RrYy* and *RRYY* parental genotypes.

and yellow, 26 were round and green, 27 were wrinkled and yellow, and 26 were wrinkled and green. This is almost exactly ¼ of each type, which matched the predicted proportions for offspring of an *RrYy* parent. The testcross had confirmed the principle of independent assortment.

Mendel's work provided a powerful conceptual framework for thinking about transmission genetics—the patterns that occur as alleles pass from one generation to the next. This framework was based on (1) the segregation of discrete, paired genes into separate gametes, and (2) the independent assortment of genes that affect different traits.

The experiments you've just reviewed were brilliant in design, execution, and interpretation. Unfortunately, they were ignored for 34 years.

14.4 The Chromosome Theory of Inheritance

Historians of science debate why Mendel's work was overlooked for so long. It is probably true that his use of ratios and proportions were difficult for biologists of that time to understand and absorb. It may also be true that the theory of blending inheritance was so well entrenched that his results were dismissed as peculiar or unbelievable.

Whatever the reason, Mendel's work was not appreciated until 1900, when three biologists, working with a variety of plants and animals, independently "discovered" Mendel's work and reached the same main conclusions.

The rediscovery of Mendel's work more than three decades after its publication ignited the young field of genetics. Mendel's experiments had established the basic patterns of inheritance, but what process is responsible for these patterns? Two biologists, working independently, came up with the answer. Walter Sutton and Theodor Boveri each realized that meiosis could account for

Mendel's rules. When this hypothesis was published in 1902, research in genetics exploded.

Meiosis Explains Mendel's Principles

What Sutton and Boveri grasped is that meiosis explains the principle of segregation and the principle of independent assortment. The cell at the top of **FIGURE 14.7** illustrates Sutton and Boveri's central insight: Mendel's hereditary determinants, or genes, are located on chromosomes. In this example, the gene for seed shape is shown at a particular position along a certain chromosome. This location is known as a **locus** ("place"; plural, **loci**).

The paternal and maternal chromosomes shown in Figure 14.7 happen to possess different alleles at the seed shape gene locus: One allele specifies round seeds (*R*) and the other specifies wrinkled seeds (*r*).

The subsequent steps in Figure 14.7 show how these alleles segregate into different daughter cells during meiosis I, when homologous chromosomes separate. The physical separation of alleles during anaphase of meiosis I is responsible for Mendel's principle of segregation.

FIGURE 14.8 follows the segregation of two different gene pairs—in this case, for seed shape and seed color—as meiosis proceeds. If the alleles for different genes are located on different chromosomes, they assort independently of one another at meiosis I. This is the physical basis of Mendel's principle of independent assortment. Over many meiotic divisions, four types of gametes will be produced in equal proportions.

Sutton and Boveri formalized these observations in the **chromosome theory of inheritance.** Like other theories in biology, the chromosome theory describes a predictable pattern—a set of observations about the natural world—and a process that explains the pattern. The chromosome theory states that Mendel's rules can be explained by the independent alignment and separation of homologous chromosomes at meiosis I.

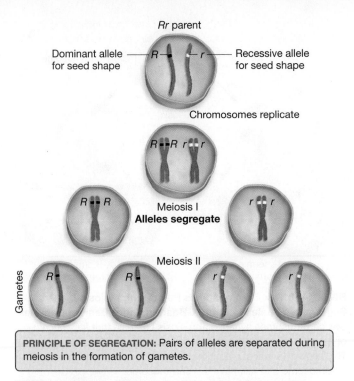

FIGURE 14.7 Meiosis Explains the Principle of Segregation.
The two members of a parent's gene pair segregate into different
gametes because homologous chromosomes separate during
meiosis I.

PRINCIPLE OF SEGREGATION: Pairs of alleles are separated during
meiosis in the formation of gametes.

Gametes

Rr parent

Dominant allele
for seed shape — R — r — Recessive allele
for seed shape

Chromosomes replicate

Meiosis I
Alleles segregate

Meiosis II

When Sutton and Boveri published their ideas, however,
the hypothesis that genes are located on chromosomes was un-
tested. What experiments confirmed that chromosomes contain genes?

Testing the Chromosome Theory

During the first decade of the twentieth century, an unassuming
insect rose to prominence as a model organism for testing the
chromosome theory of inheritance. This organism—the fruit fly
Drosophila melanogaster—has been at the center of genetic studies ever since (see **BioSkills 13** in Appendix B).

Drosophila melanogaster has all the attributes of a useful
model organism for studies in genetics: small size, ease of rearing
in the lab, a short generation time (about 10 days), and abundant offspring (up to a few hundred per mating). The elaborate
external anatomy of this insect also makes it possible to identify
interesting phenotypic variation among individuals (**FIGURE 14.9a**;
see page 268).

Drosophila was adopted as a model organism by Thomas
Hunt Morgan and his students. But because *Drosophila* is not a
domesticated species like the garden pea, common phenotypic
variants such as Mendel's round and wrinkled seeds were not
available to Morgan. Instead, he had access only to flies with the
most common phenotype for each trait, phenotypes referred to
as **wild type**. Consequently, an early goal of Morgan's research

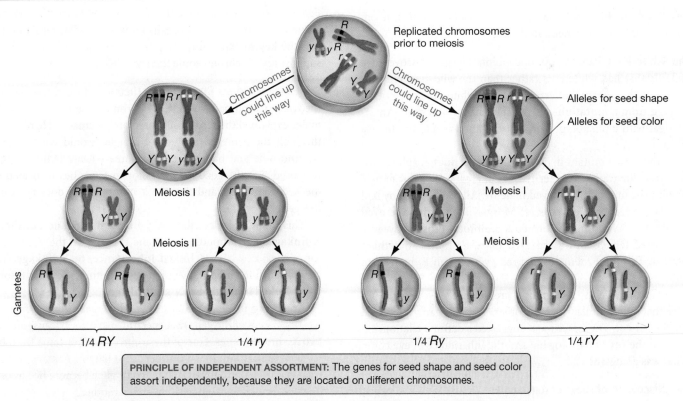

Replicated chromosomes
prior to meiosis

Chromosomes
could line up
this way

Chromosomes
could line up
this way

Alleles for seed shape

Alleles for seed color

Meiosis I

Meiosis I

Meiosis II

Meiosis II

Gametes

1/4 *RY*   1/4 *ry*   1/4 *Ry*   1/4 *rY*

PRINCIPLE OF INDEPENDENT ASSORTMENT: The genes for seed shape and seed color
assort independently, because they are located on different chromosomes.

FIGURE 14.8 Meiosis Is Responsible for the Principle of Independent Assortment. The genes for different traits
assort independently because nonhomologous chromosomes assort independently during meiosis.

(a) The fruit fly *Drosophila melanogaster*

1 mm

(b) Eye color is a variable trait.

Wild type

Mutant

FIGURE 14.9 The Fruit Fly *Drosophila melanogaster* Is an Important Model Organism in Genetics.

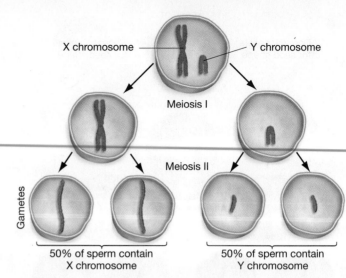

X chromosome — Y chromosome

Meiosis I

Meiosis II

Gametes

50% of sperm contain X chromosome

50% of sperm contain Y chromosome

FIGURE 14.10 Sex Chromosomes Pair during Meiosis I, Then Segregate in Males to Form X-Bearing and Y-Bearing Gametes. Sex chromosomes synapse at meiosis I in male fruit flies because of a small, gene-free region shared by the X and Y chromosomes. This allows normal segregation, so half the sperm cells bear an X chromosome and half have a Y chromosome.

was simply to find and characterize individuals with different phenotypes to use in genetic studies.

The White-Eyed Mutant At one point, Morgan discovered a male fly that had white eyes rather than the wild-type red eyes (**FIGURE 14.9b**). Morgan inferred that the white-eyed phenotype resulted from a **mutation**—a heritable change in a gene. An individual with a phenotype due to a mutation is referred to as a **mutant.**

With his first mutant in hand, Morgan set out to explore how the eye-color trait was inherited. He mated a red-eyed female fly with the mutant white-eyed male fly. All the F₁ progeny had red eyes. By continued crosses, Morgan obtained white-eyed female flies. When he performed a reciprocal cross between a white-eyed female and a red-eyed male, he found something puzzling: All the F₁ females had red eyes, but all F₁ males had white eyes.

Recall that Mendel's reciprocal crosses had always given results that were similar to each other. But Morgan's reciprocal crosses did not. The experiment suggested a definite relationship between the sex of the progeny and the inheritance of eye color. What was going on?

The Discovery of Sex Chromosomes Nettie Stevens began studying the karyotypes of insects about the time that Morgan began his work with *Drosophila*. First, in the beetle *Tenebrio*

molitor, and later in other insects, including *Drosophila*, she noticed a striking difference in the chromosome complements of males and females.

Recall that Stevens and others discovered that there were sex chromosomes (the X and the Y) and autosomes (see Chapter 13). Female flies have a pair of X chromosomes and male flies have an X and a Y chromosome. Morgan's knowledge of Stevens' findings was the key to explaining his puzzling results and in providing support for the chromosome theory.

Sex Linkage and the Chromosome Theory Morgan realized that the transmission pattern of the X chromosome in males and females explained the results of his reciprocal crosses. He reasoned that half the gametes produced by males would contain an X chromosome and half a Y chromosome (**FIGURE 14.10**). Morgan proposed that the gene for eye color in fruit flies is located on the X chromosome and that the Y chromosome does not carry this gene.

This situation is described as **X-linked inheritance,** or simply **X-linkage.** Correspondingly, a gene residing on the Y chromosome is said to have **Y-linked inheritance,** or **Y-linkage.** The general term for inheritance of genes on either sex chromosome is **sex-linked inheritance,** or **sex-linkage.**

According to the hypothesis of X-linkage, a female fruit fly has two copies of the gene that specifies eye color because she has two X chromosomes. One of these chromosomes came from her female parent, the other from her male parent. A male, in contrast, has only one copy of the eye-color gene because he has only one X chromosome, inherited from his mother.

The Punnett squares in **FIGURE 14.11** show that Morgan's experimental results can be explained if the gene for eye color is located

(a) One half of reciprocal cross

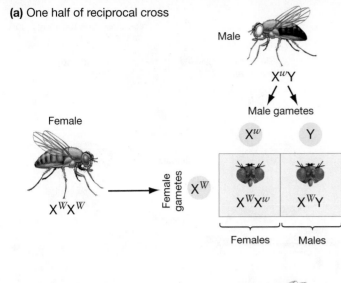

(b) Other half of reciprocal cross

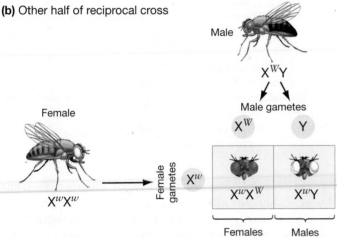

FIGURE 14.11 Reciprocal Crosses Confirm that Eye Color in Drosophila Is an X-Linked Trait. When Morgan crossed red-eyed females with white-eyed males **(a)** and then crossed white-eyed females with red-eyed males **(b)**, he observed strikingly different results.

on the X chromosome, and if the allele for red color is dominant to the allele for white color. In this figure, the allele for red eyes is denoted X^W while the allele for white eyes is denoted X^w. The Y chromosome present in males is simply designated by Y. Using this notation,[1] the genotypes used in the experiment are written as $X^W X^W$ for red-eyed females; $X^w Y$ for white-eyed males; $X^w X^w$ for white-eyed females; and $X^W Y$ for red-eyed males.

Notice how the symbol for *Drosophila* alleles represents the mutant phenotype. Instead of showing alleles for the eye color gene with an *R* for wild-type red eyes and an *r* for mutant white eyes, the alleles are shown with a *W* (red eyes) or a *w* (white eyes).

---

[1]Scientific papers on fruit fly genetics use a different notation. The wild-type allele is designated with a superscript +, and no X is used for X-linked traits. The red-eye allele, for example, is denoted w^+; the white-eye allele *w*. The notation used here is simplified and conforms to conventions used in human genetics.

By applying the principles of segregation and random fertilization, you should see that the results predicted by the hypothesis of X-linkage match the observed results.

When reciprocal crosses give different results, such as those illustrated in Figure 14.11, it is likely that the gene in question is located on a sex chromosome—it is sex-linked. Recall that non-sex chromosomes are called autosomes (see Chapter 13). Genes on non-sex chromosomes are said to show **autosomal inheritance.**

Morgan's discovery of X-linked inheritance carried an even more fundamental message. In *Drosophila*, the gene for white eye color is clearly correlated with inheritance of the X chromosome. This correlation was important evidence in support of the hypothesis that chromosomes contain genes. The discovery of X-linked inheritance convinced most biologists that the chromosome theory of inheritance was correct.

check your understanding

C
Y
U

If you understand that . . .

- Meiosis is the process responsible for Mendel's principle of segregation. It occurs because alleles on homologous chromosomes separate at anaphase of meiosis I.
- Meiosis is the process responsible for Mendel's principle of independent assortment. Alleles of different genes go to gametes independently because pairs of homologous chromosomes line up randomly at metaphase of meiosis I.

✔ You should be able to . . .

1. Draw the chromosomes involved in a cross between *Pp* and *Pp* peas, and use your diagram to explain the segregation of alleles.

2. Draw the chromosomes involved in a cross between *YyRr* and *YyRr* peas, and use your diagram to explain the independent assortment of alleles.

Answers are available in Appendix A.

14.5 Extending Mendel's Rules

Biologists point out that Mendel analyzed the simplest possible genetic system. The traits that he was studying were not sex-linked. Moreover, they were influenced by just two alleles of each gene, and each allele was completely dominant or recessive.

With this well-chosen model system, Mendel was able to discover the most fundamental rules of inheritance. Mendel probably would have failed, as so many others had done before him, had he been trying to analyze more complex patterns of inheritance.

Once Mendel's work was rediscovered, researchers began to analyze traits and alleles whose inheritance was more complicated.

If experimental crosses produced F_2 progeny that did not conform to the expected 3:1 or 9:3:3:1 ratios, researchers had a strong hint that something interesting was going on. The discovery of sex-linkage is a prominent example. How can other traits that don't appear to follow Mendel's rules contribute to a more complete understanding of heredity?

Linkage: What Happens When Genes Are Located on the Same Chromosome?

Once the chromosome theory had been tested and supported, biologists began to reevaluate Mendel's principle of independent assortment. It seemed unlikely that genes on the same chromosome would assort independently.

Linkage is the tendency of particular alleles of different genes to be inherited together. Linkage is seen when genes are on the same chromosome. Notice that the terms linkage and sex-linkage have different meanings. If genes are linked, it means that they are located on the same chromosome. If a gene is sex-linked, it means that it is located on a sex chromosome but says nothing about its location relative to other genes.

The first examples of linked genes were those on the X chromosome of fruit flies. After Morgan established that the *white-eye* gene was located on *Drosophila*'s X chromosome, he and colleagues established that one of the several genes that affects body color is also located on the X. Red eyes and gray body are the wild-type phenotypes; white eyes and a yellow body occur as rare mutant phenotypes. The alleles for red eyes (X^W) and gray body (X^Y) are dominant to the alleles for white eyes (X^w) and yellow body (X^y). (Be sure not to confuse the notation for the Y chromosome in males, Y, with the gray body allele, X^Y.)

Linked Genes Do Not Assort Independently   Because linked genes are located on the same chromosome, it is logical to predict that they should always be transmitted together during gamete formation. Stated another way, linked genes should violate the principle of independent assortment.

Recall from Section 14.4 that genes on different chromosomes show independent assortment because different chromosomes assort independently during meiosis I. How will the X-linked body-color and eye-color genes be inherited? **FIGURE 14.12** shows a cell of a female fruit fly with one X chromosome carrying the white eye and gray body alleles, written X^{wY}, and the homologous X chromosome carrying the red eye and yellow body alleles, written X^{Wy}. This female would be expected to generate just two classes of gametes in equal numbers during meiosis, instead of the four classes that are predicted under the principle of independent assortment. Is this what actually occurs?

The Role of Crossing Over   To determine whether linked traits always stay linked, Morgan performed crosses like the one described in the "Experimental Setup" section of **FIGURE 14.13**. In this case, $X^{wY}X^{Wy}$ females mated with $X^{wY}Y$ males.

The "Results" table in Figure 14.13 summarizes the phenotypes and genotypes observed in this experimental cross.

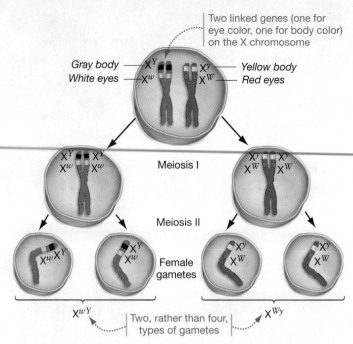

Gray body — X^Y | X^y — Yellow body
White eyes — X^w | X^W — Red eyes

Two linked genes (one for eye color, one for body color) on the X chromosome

Meiosis I

Meiosis II

Female gametes

X^{wY} Two, rather than four, types of gametes X^{Wy}

FIGURE 14.12 Linked Genes Are Often Inherited Together. If the eye-color and body-color genes were found on different chromosomes, then this female would generate four different types of gametes instead of just two types as shown here.

✔**EXERCISE** List the four genotypes that would be generated if the white-eye and yellow-body genes were not linked.

Morgan recorded results only from male offspring. By doing this he could figure out which X-linked alleles were present on the chromosomes produced during meiosis in the mother. Since there is a single X chromosome in the male offspring, the phenotype associated with any X-linked allele, dominant or recessive, is expressed:

- Most of these males carried an X chromosome with one of the two combinations of alleles found in the chromosomes of their mothers: X^{wY} or X^{Wy}.

- A small percentage of males had novel combinations of phenotypes and genotypes: X^{wy} and X^{WY}. Morgan referred to these individuals as **recombinant** because the combination of alleles on their X chromosome was different from the combinations of alleles present in their mother.

Morgan concluded that alleles on the same chromosome don't always stay together. To explain the recombinant phenotypes, Morgan proposed that gametes with new combinations of alleles were generated when crossing over occurred during prophase of meiosis I in the females.

Recall that crossing over involves a physical exchange of segments of non-sister chromatids between homologous chromosomes (see Chapter 13). Crossing over typically occurs at least once in every synapsed pair of homologous chromosomes, and usually multiple times. (Male fruit flies are an exception to this rule. For unknown reasons, no crossing over occurs in male fruit flies.)

QUESTION: Will genes undergo independent assortment if they are on the same chromosome?

LINKAGE HYPOTHESIS: Linked genes will violate the principle of independent assortment.

NULL HYPOTHESIS: Linked genes will adhere to the principle of independent assortment.

EXPERIMENTAL SETUP:

Red-eyed gray-bodied female $X^{wY}X^{Wy}$

White-eyed gray-bodied male $X^{wY}Y$

PREDICTION: Because these two genes are X-linked, male offspring will have only one copy of each gene, from their mother; the two possible male offspring genotypes are $X^{wY}Y$ and $X^{Wy}Y$.

PREDICTION OF NULL HYPOTHESIS: Four male genotypes are possible ($X^{wY}Y : X^{Wy}Y : X^{wy}Y : X^{WY}Y$) and will occur with equal frequency.

RESULTS:

Male offspring

	Phenotype	Genotype	Number	
		$X^{wY}Y$	4292	Four male genotypes were observed (rather than two), but not the equal frequencies predicted by independent assortment
		$X^{Wy}Y$	4605	
Recombinant genotypes		$X^{wy}Y$	86	
		$X^{WY}Y$	44	

CONCLUSION: Neither hypothesis is fully supported. Independent assortment does not apply to linked genes—linked genes segregate together except when crossing over and genetic recombination have occurred.

FIGURE 14.13 Linked Genes Are Inherited Together Unless Recombination Occurs.

SOURCE: Morgan, T. H. 1911. An attempt to analyze the constitution of the chromosomes on the basis of sex-limited inheritance in *Drosophila*. *Journal of Experimental Zoology* 11: 365–414.

✔**QUESTION** Why didn't Morgan observe equal numbers of white-eyed, yellow-bodied males and red-eyed, gray-bodied males?

As **FIGURE 14.14** shows, a crossover between the eye-color and body-color genes in the $X^{wY}X^{Wy}$ females can explain the recombinant gametes. Male progeny produced from fertilization with these gametes are predicted to have either yellow bodies and white eyes or gray bodies and red eyes. This is what Morgan observed.

Notice, however, that the results of Figure 14.14 don't fit either the model of independent assortment or complete linkage. Independent assortment predicts a 1:1:1:1 ratio of all four

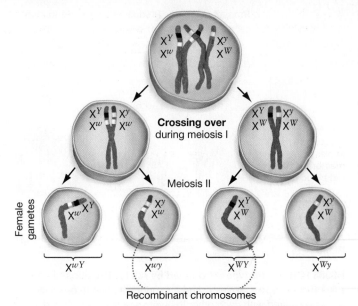

Crossing over during meiosis I

Meiosis II

Female gametes

X^{wY} X^{wy} X^{WY} X^{Wy}

Recombinant chromosomes

FIGURE 14.14 Genetic Recombination Results from Crossing Over. To explain the results in Figure 14.13, Morgan proposed that crossing over occurred between the body color (*y*) and eye color (*w*) genes in a small percentage of meiotic divisions in the female parent. The recombinant chromosomes that resulted would produce the recombinant phenotypes observed in the male offspring.

combinations of phenotypes, while complete linkage would give only the two phenotypes associated with the nonrecombinant or parental chromosomes. Instead, most flies have parental phenotypes and a smaller number have recombinant phenotypes.

As **Quantitative Methods 14.1** (see page 274) explains, the percentage of recombinant offspring that occur in crosses like the one diagrammed in Figure 14.14 can be used to estimate the relative distance between genes. The reasoning is that the farther genes are apart on the same chromosome, the more likely it is that a crossover will occur someplace between these genes. Data on the frequency of crossing over between many genes on the same chromosome can be used to create a **genetic map**—a diagram showing the relative positions of genes along a particular chromosome.

Knowing a gene's locus relative to others can be very useful. An important example is genes involved in human genetic diseases. Most of these genes have been identified on the basis of mapping their location relative to other known genes.

The take-home message of Morgan's experiments is simple: Linked genes are inherited together unless crossing over occurs between them. When crossing over takes place, genetic recombination occurs. Linkage is an important exception to Mendel's rules.

How Many Alleles Can a Gene Have?

Mendel worked with genes that each had two alleles. In most populations, however, it's not unusual to find dozens of alleles of a single gene. The existence of more than two alleles of the same gene is known as **multiple allelism.** When you consider that

Product of I^A allele　Product of I^B allele　Product of i allele

Glycoproteins in plasma membrane of red blood cells

Genotype I^AI^A　I^AI^B　I^BI^B　I^Ai　I^Bi　ii

Blood type A　AB　B　A　B　O

FIGURE 14.15 Phenotypes Produced by Alleles Responsible for ABO Blood Types. Alleles I^A and I^B produce a phenotype called AB when paired with each other in heterozygotes. Both I^A and I^B produce a dominant phenotype when paired with allele i. The different colored hexagons on the products of the I^A and I^B alleles represent related but distinct sugars.

genes are made of DNA sequences that can change over time, the idea of multiple alleles isn't surprising.

The ABO blood group types in humans are coded for by a gene with three common alleles. The gene is known as I, and it has I^A, I^B, and i alleles. As the number of alleles for a gene increases, the number of possible genotypes rises sharply. When more than two distinct phenotypes are present in a population owing to multiple allelism, the trait is **polymorphic** ("many-formed"). ABO blood type is a polymorphic trait.

Each I gene allele controls the production of a different polysaccharide attached to a glycoprotein (see Chapter 5) found in the plasma membrane of red blood cells. The I^A and I^B alleles code for different forms of an enzyme that adds a different sugar to the end of a core polysaccharide. The i allele codes for a nonfunctional form of this enzyme, so no sugar is added to the core polysaccharide. ABO blood group types are important in blood transfusions. Some mismatches of blood type between a donor and recipient are tolerated, but others can cause fatal reactions.

The type of polysaccharide associated with each allele is shown in **FIGURE 14.15** along with all possible genotypes and phenotypes. As you can see, there are six possible genotypes for the three alleles of the I gene. The I^A allele codes for the type A polysaccharide, and the I^B allele codes for the type B polysaccharide. What happens in a person with both alleles?

Are Alleles Always Dominant or Recessive?

The terms dominant and recessive describe which phenotype is observed when two different alleles of a gene occur in a heterozygous individual. In all traits that Mendel studied, only the phenotype associated with one allele—the "dominant" one—appeared in heterozygotes. However, not all combinations of alleles work this way.

Codominance Many alleles show a relationship that is called **codominance.** The type AB blood group shown in Figure 14.15 is an example of codominance. In this case, an AB heterozygote expresses both the A and the B polysaccharides together on the surface of red blood cells. This is the essence of codominance—the simultaneous expression of the phenotype associated with each allele in a heterozygote. An AB individual expresses both the A and the B phenotypes.

The three alleles of the ABO blood group system illustrate another interesting point—alleles of one gene can show more than one form of dominance. Notice in Figure 14.15 that the I^A and I^B alleles are both completely dominant to the i allele while the I^A and I^B alleles are codominant with each other.

Incomplete Dominance Complete dominance and codominance are not the end of the story. Consider the flowers called four-o'clocks, pictured in **FIGURE 14.16a**. Plant breeders have developed a pure line that has red flowers and a pure line that has white flowers. When individuals from these strains are mated, all their offspring are pink (**FIGURE 14.16b**). In Mendel's peas, crosses between dominant and recessive parents produced only offspring with the dominant phenotype. Why the difference?

Biologists answered this question by allowing the pink flowered F_1 plants to self-fertilize and examining the phenotypes of F_2 four-o'clocks. Of the F_2 plants, ¼ have red flowers, ½ have pink flowers, and ¼ have white flowers. This 1:2:1 ratio of phenotypes exactly matches the 1:2:1 ratio of genotypes that is produced when flower color is controlled by one gene with two alleles.

To convince yourself that this explanation is sound, study the genetic model shown in Figure 14.16b. According to the diagram, the pattern of inheritance of flower color genotypes in four-o'clocks and peas is identical, but the four-o'clock alleles show a different form of dominance known as **incomplete dominance.**

When incomplete dominance occurs, heterozygotes have a phenotype that is between the two different homozygous parents. In the case of four-o'clocks, neither red nor white alleles dominate. Instead, the heterozygous F_1 progeny show a phenotype in between the two parental strains.

By this point, it should be clear that the answer to the question of whether alleles are always dominant or recessive is a resounding no. Instead, there are three possible dominance relationships between different alleles: complete, incomplete, and codominance.

Does Each Gene Affect Just One Trait?

Mendel's results led him to hypothesize that one gene influences one trait. The gene for seed color in garden peas, for example, did not appear to affect other aspects of the individual's phenotype. In reality, however, many genes influence more than one trait.

A gene that influences many traits is said to be **pleiotropic** ("more-turning"). For example, mutations in one gene, *FBN1*, cause Marfan syndrome in humans. In this case the mutant

(a) Flower color is variable in four-o'clocks.

(b) Incomplete dominance in flower color

FIGURE 14.16 When Incomplete Dominance Occurs, Heterozygotes Have Intermediate Phenotypes. The cross in part (b) is explained by hypothesizing that a single gene influences flower color and that alleles *R* and *r* exhibit incomplete dominance.

FBN1 allele is dominant to the wild-type allele. What is the phenotype of an individual with a mutant *FBN1* allele?

People with a mutant *FBN1* allele are tall, have disproportionately long limbs and fingers and an abnormally shaped chest, and often have severe heart problems. Therefore, the gene associated with Marfan syndrome influences many traits and is pleiotropic. On the basis of this set of phenotypes, many medical scientists believe Abraham Lincoln had Marfan syndrome.

Is There More to Phenotype than Genotype?

After analyzing the results of Mendel's experiments, it would be tempting to conclude that *R* alleles dictate that seeds are round and *T* alleles dictate that individual plants are tall—that there is a strict correspondence between alleles and phenotypes.

It's important to recognize, though, that when Mendel analyzed height in his experiments, he ensured that each plant received a similar amount of sunlight and grew in similar soil. This was critical because even individuals with alleles for tallness will be stunted if they are deprived of nutrients, sunlight, or water—so much so that they look similar to individuals with alleles for dwarfing. Mendel also worked with pure lines that had been inbred for many generations. This breeding method reduces the genetic variation in each line.

For Mendel to analyze the hereditary determinants of height, he had to control the environmental determinants of height. Let's consider how the environment and alleles at other gene loci affect phenotype.

The Environment Affects Phenotypes The phenotypes produced by most genes are strongly affected by the individual's environment. Consequently, an individual's phenotype is often as much a product of the environment as it is a product of the genotype. Environmental influences include temperature, sunlight, nutrient availability, competition, and even a mother's hormone levels during development of an embryo. To capture this point, biologists refer to the combined effect of genes and environment as gene-by-environment interaction.

Gene-by-environment interactions have a profound effect on how physicians treat people with the genetic disease phenylketonuria (PKU). These individuals are homozygous for a recessive allele of an enzyme-coding gene. The enzyme helps convert the amino acid phenylalanine to the amino acid tyrosine. In PKU, this enzyme is absent and, as a result, phenylalanine and a related molecule, phenylpyruvic acid, accumulate. These compounds interfere with the development of the nervous system and produce severe mental retardation.

But are people without the ability to metabolize phenylalanine genetically fated to mental retardation?

In many countries, newborns are routinely tested for the defect. If identified at birth, individuals with PKU can be placed on a low-phenylalanine diet. The change in environment—reduced phenylalanine in the diet—has a dramatic influence on phenotype. Treated individuals develop normally. PKU is a genetic disease, but by controlling the environment, it is neither inevitable nor invariant.

Interactions Between Genes Affect Phenotypes In Mendel's pea plants, there was a one-to-one correspondence between genes and traits. The pea seeds he analyzed were round or wrinkled regardless of the types of alleles present at other loci. Only one gene influenced each trait.

In many cases, however, different genes work together to control a single trait. Consider a classic experiment published in 1905 on comb shape in chickens. William Bateson and R. C. Punnett crossed parents from pure lines with comb shapes called rose and pea and found that the F₁ offspring had a new phenotype, called walnut combs, a new phenotype not seen in either parent. When these individuals bred, their offspring had walnut, rose,

FIGURE 14.17 The Physical Distance between Genes Determines the Frequency of Crossing Over. The arrows show that crossing over is possible at any point between the genes, but the chance of a crossover between a pair of genes increases when the distance between the genes is large.

Gene 1
Gene 2
Crossing over is rare between genes that are close together

Gene 2
Crossing over occurs frequently between genes that are far apart
Gene 3

In experiments like the one diagrammed in Figure 14.13, researchers calculate the recombination frequency as the number of offspring with recombinant phenotypes divided by the total number of offspring. With crosses involving the X-linked traits of white eyes and yellow bodies, 1.4 percent of offspring have recombinant phenotypes and genotypes.

But in crosses with different pairs of X-linked traits, the fraction of recombinant offspring varies, and this variation provides information on the relative distance separating linked genes. For example, in crosses of fruit flies with X-linked genes for white eyes and another mutant phenotype called singed bristles, recombinant phenotypes are seen 19.6 percent of the time.

To explain these observations, Alfred Sturtevant, an undergraduate student working with Morgan, proposed that the physical distance between genes determines how frequently crossing over occurs between them. His idea was that crossing over occurs at random and can take place at any location along a chromosome. The shorter the distance between a pair of genes, the lower the probability that crossing over will take place between them (**FIGURE 14.17**). Using this reasoning, he set out to create a genetic map.

To define the unit of distance on his genetic map, Sturtevant used the percentage of offspring that have recombinant phenotypes with respect to two genes. One map unit (later called 1 centiMorgan [cM]) is the physical distance that produces 1 percent recombinant offspring.

The eye-color and bristle-shape genes of fruit flies are 19.6 cM apart on the X chromosome, because recombination between these genes results in 19.6 percent recombinant offspring, on average. The genes for yellow body and white eye color, in contrast, are just 1.4 cM apart.

Where is the *yellow-body* gene relative to the *singed-bristles* gene? Twenty-one percent of the offspring were recombinant for these traits, meaning that the *yellow-body* and *singed-bristles* genes are 21.0 cM apart. Sturtevant inferred that the gene for white eyes must be located *between* the genes for yellow body and singed bristles, as shown in **FIGURE 14.18a**.

Mapping genes relative to one another is like fitting pieces into a puzzle: Placing *white* between *yellow* and *singed bristles* is the only way to make the distances between each pair sum correctly. The key observation is that 21.0 cM—the distance between *yellow* and *singed bristles*—is equal to 1.4 cM + 19.6 cM, or the sum of the distances between *yellow* and *white* and *white* and *singed bristles*.

FIGURE 14.18b provides a partial genetic map of the X chromosome in *Drosophila melanogaster*, along with the data used to establish the map positions. Using this logic and similar data, Sturtevant assembled the first genetic map.

(a) Mapping genetic distance

Yellow body
White eyes

1.4

19.6

21

Singed bristles

Frequency of recombinant offspring correlates directly with the distance between two genes; 19.6% recombinant offspring, for example, translates to 19.6 map units (centiMorgans, cM).

(b) Constructing a genetic map

% Frequency of crossing over between some genes on the X chromosome of fruit flies		
	Miniature Wings	Ruby Eyes
Yellow body	36.1	7.5
White eyes	34.7	6.1
Singed bristles	15.1	13.5
Miniature wings	—	28.6

These distances are in cM

Yellow body
White eyes

1.4
6.1
7.5

13.5

Singed bristles

36.1

15.1

FIGURE 14.18 The Locations of Genes Can Be Mapped by Analyzing the Frequency of Recombination. (a) The *yellow body* gene is on the end of the fruit fly X chromosome. To explain the recombination frequencies observed in experimental crosses, the *yellow body, white eyes,* and *singed bristles* genes must be in the locations shown here. **(b)** A partial genetic map of the X chromosome in fruit flies.

✔**EXERCISE** In part (b), label the orange and blue genes. (Which is *ruby* and which is *miniature wings*?)

pea, and a fourth phenotype called single combs in a 9:3:3:1 ratio (**FIGURE 14.19a**).

The genetic model in **FIGURE 14.19b** shows how the interaction between two different genes that control one trait (comb shape) can account for the results. If comb morphology results from interactions between two genes (symbolized *R* and *P*), if a dominant and a recessive allele exist for each gene, and if the four comb phenotypes are associated with the genotypes indicated at the bottom of the figure, then a cross between *RRpp* and *rrPP* parents would give the results that Bateson and Punnett observed.

When gene-by-gene interactions occur, one trait is influenced by the alleles of two or more different genes. If a chicken has an *R* allele, its phenotype depends on the allele present at the *P* gene.

Gene-by-gene interaction is very common and has important implications in human genetics. Imagine that two people have the same genotype at one locus that increases risk for a heart disease. If there is gene-by-gene interaction, then the risk of developing heart disease also depends on the genotype at other loci. Even if they experience identical environments, these two people may have very different overall risks from genetic factors alone.

Can Mendel's Principles Explain Traits That Don't Fall into Distinct Categories?

Mendel worked with **discrete traits**—traits that are clearly different from each other. In garden peas, seed color is either yellow or green—no intermediate phenotypes exist. But many traits in peas and other organisms don't fall into discrete categories. In humans, for example, height, weight, and skin color vary continuously. People are not limited to being either 160 cm tall or 180 cm tall—countless other heights are possible.

For height and many other characteristics, individuals differ by degree. These types of continuously varying traits that don't fall into discrete categories are called **quantitative traits.** Like discrete traits, quantitative traits are greatly influenced by the environment. The effects of nutrition on human height, intelligence, and disease resistance, for example, have been well documented.

Many quantitative traits share a common characteristic: When the frequencies of different trait values observed in a population are plotted on a histogram, or frequency distribution (see **BioSkills 3** in Appendix B), they often form a bell-shaped curve, or normal distribution (**FIGURE 14.20**).

In 1909 Herman Nilsson-Ehle had an important insight: If many genes each contribute a small amount to the value of a quantitative trait, then a normal distribution results for the population as a whole.

(a) Crosses between chickens with different comb phenotypes give odd results.

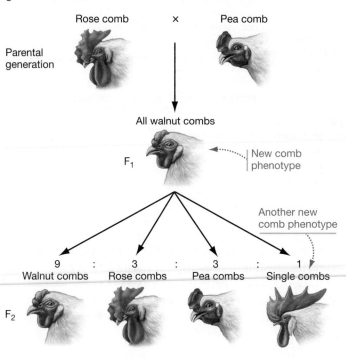

(b) A genetic model based on gene-by-gene interactions can explain the results.

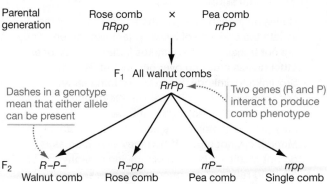

FIGURE 14.19 Genes at Different Loci Can Interact to Influence a Trait. (a) This cross is notable because new phenotypes show up in the F$_1$ and the F$_2$ generation. **(b)** To explain the results, researchers hypothesized that comb shape depends on two genes that interact. The phenotype associated with any one allele of one gene depends on the alleles at a second gene.

✔**EXERCISE** What is different about the 9:3:3:1 ratio in the F$_2$ of this cross compared with the ratio observed in a standard dihybrid cross?

FIGURE 14.20 Quantitative Traits Have a Normal Distribution. A histogram showing the heights of first-grade schoolchildren in Guatemala in 2001.

DATA: Pan American Health Organization/WHO. 2004. *Epidemiological Bulletin* 25: 9-13, Graph 1.

(a) Wheat kernel color is a quantitative trait.

(b) Hypothesis to explain inheritance of kernel color

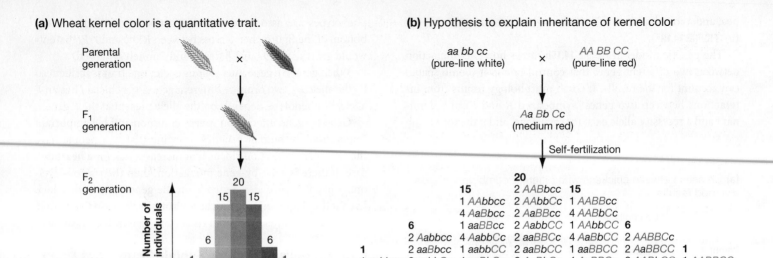

FIGURE 14.21 **Quantitative Traits Result from the Action of Many Genes. (a)** When wheat plants with white kernels were crossed with wheat plants with red kernels, the F_2 offspring showed a range of kernel colors. The frequency of these phenotypes approximates a normal distribution. **(b)** This model can explain the results of part (a). The bold numbers above each genotype match the numbers shown in part (a) and indicate the relative number of plants with each phenotype.

✔ **QUESTION** Why are there fewer very dark or very light wheat kernels compared with kernels of intermediate coloration?

Nilsson-Ehle established this finding using strains of wheat that differed in kernel color. **FIGURE 14.21a** includes a histogram showing the distribution of F_2 phenotypes from a cross he performed between pure lines of white wheat and dark-red wheat. Notice that the frequency of colors in F_2 progeny approximates a bell-shaped curve. To explain these results, Nilsson-Ehle proposed the hypotheses illustrated in **FIGURE 14.21b**:

- The parental strains differ with respect to three genes that control kernel color: *AABBCC* produces dark-red kernels, and *aabbcc* produces white kernels.

- The three genes assort independently.

- The *a, b,* and *c* alleles do not contribute to pigment production, but the *A, B,* and *C* alleles contribute to pigment production in an equal and additive way. This is a form of incomplete dominance. As a result, the degree of red pigmentation is determined by the number of *A, B,* or *C* alleles present. Each uppercase (dominant) allele that is present makes a wheat kernel slightly darker red.

Later work supported Nilsson-Ehle's model. He did not have to propose any new genetic principles to explain the inheritance of a quantitative trait. All that was needed was extension of Mendel's hypotheses about segregation and independent assortment.

Quantitative traits are produced by the independent actions of many genes, although it is now clear that some genes have much greater effects on the trait in question than other genes do. As a result, the transmission of quantitative traits is said to result from polygenic ("many-genes") inheritance. In **polygenic inheritance,** each gene adds a small amount to the value of the phenotype.

In the decades immediately after the rediscovery of Mendel's work, the question of why offspring resemble their parents could be answered in more satisfying ways. **TABLE 14.4** summarizes some of the key exceptions and extensions to Mendel's rules and gives you a chance to compare and contrast their effects on patterns of inheritance.

check your understanding

C Y U

If you understand that . . .

- Genes near each other on the same chromosome violate the principle of independent assortment. They are not transmitted to gametes independently of each other unless crossing over occurs between them.

- Sex linkage, linkage, incomplete dominance, codominance, multiple allelism, pleiotropy, environmental effects, gene interactions, and polygenic inheritance are aspects of inheritance that Mendel did not study. When they occur, crosses do not result in classical Mendelian monohybrid or dihybrid ratios of offspring phenotypes.

✔ **You should be able to . . .**

Explain why the following crosses don't produce a 3:1 phenotype ratio in F_2 offspring:

1. Rose-comb × pea-comb chickens
2. Red-kernel × white-kernel wheat plants

Answers are available in Appendix A.

Type of Inheritance	Definition	Consequences or Comments
Sex linkage	Genes located on sex chromosomes.	Patterns of inheritance in males and females differ.
Linkage	Two genes found on same chromosome.	Linked genes violate principle of independent assortment.
Incomplete dominance	Heterozygotes have intermediate phenotype.	Polymorphism—heterozygotes have unique phenotype.
Codominance	Heterozygotes have phenotype of both alleles.	Polymorphism is possible—heterozygotes have unique phenotype.
Multiple allelism	In a population, more than two alleles present at a locus.	Polymorphism is possible.
Polymorphism	In a population, more than two phenotypes associated with a single gene are present.	Can result from actions of multiple alleles, incomplete dominance, and codominance.
Pleiotropy	A single gene affects many traits.	This is common.
Gene-by-gene interaction	In discrete traits, the phenotype associated with an allele depends on which alleles are present at another gene.	One allele can be associated with different phenotypes.
Gene-by-environment interaction	Phenotype influenced by environment experienced by individual.	Same genotypes can be associated with different phenotypes.
Polygenic inheritance of quantitative traits	Many genes are involved in specifying traits that exhibit continuous variation.	Unlike alleles that determine discrete traits, each allele adds a small amount to phenotype.

14.6 Applying Mendel's Rules to Human Inheritance

When researchers set out to study how a particular gene is transmitted in wheat or fruit flies or garden peas, they begin by making a series of controlled experimental crosses. For obvious reasons, this strategy is not possible with humans. But suppose you are concerned about an illness that runs in your family and go to a genetic counselor to find out how likely your children are to have the disease. To advise you, the counselor needs to know how the trait is transmitted, including whether the gene involved is autosomal or sex-linked and what type of dominance is associated with the disease allele.

To understand the transmission of human traits, investigators have to analyze human genotypes and phenotypes that already exist. A **mode of transmission** describes a trait as autosomal or sex-linked and gives the type of dominance of the allele. To learn the mode of transmission, scientists construct a **pedigree,** or family tree, of affected and unaffected individuals. By analyzing pedigrees, biomedical researchers have been able to discover how more than 2000 human genetic diseases are inherited.

A pedigree records the genetic relationships between the individuals in a family along with each person's sex and phenotype with respect to the trait in question. If the trait is governed by a single gene, then analyzing the pedigree may reveal whether a given phenotype is due to a dominant or recessive allele and whether the gene responsible is located on a sex chromosome or

on an autosome. Let's look at a series of specific case histories to see how this work is done.

Identifying Human Alleles as Recessive or Dominant

To analyze the inheritance of a discrete trait, biologists begin by assuming that a single autosomal gene is responsible and that the alleles present in the population have a simple dominant–recessive relationship.

This is the simplest possible situation. If the pattern of inheritance fits this model, then the assumptions—of inheritance by a single autosomal gene and simple dominance—are supported. Let's first analyze the pattern of inheritance that is typical of autosomal recessive traits and then examine patterns that emerge in pedigrees for autosomal dominant traits.

Patterns of Inheritance: Autosomal Recessive Traits   If a phenotype is due to an autosomal recessive allele, then

- Individuals with the trait must be homozygous.

- If the parents of an affected individual do not have the trait, then the parents are heterozygous for the trait.

Heterozygous individuals who carry a recessive allele for an inherited disease are referred to as **carriers** of the disease. These individuals carry the allele and may transmit it even though they do not exhibit signs of the disease. When two carriers mate,

(a) Autosomal recessive trait (e.g., sickle-cell disease)

Each row represents a generation

I — Carrier male ⬛ ⬛ Carrier female

II — ⬜ Affected male ⬛ ⬜ ⬛

III — Affected female ⬛ ⬜ ⬜

(b) Autosomal dominant trait (e.g., Huntington's disease)

I — Affected female ⬤ ⬜ Unaffected male

II — ⬜ ⬛ ⬤ ⬜

III — ⬤ ⬜ ⬛ ⬛ ⬜

CHARACTERISTICS:
- Males and females are equally likely to be affected
- Affected offspring often have unaffected parents
- Unaffected parents of affected offspring are heterozygous (carriers)
- Affected offspring are homozygous
- If both parents are heterozygous, about ¼ of the offspring will be affected
- Trait often skips generations

CHARACTERISTICS:
- Males and females are equally likely to be affected
- Affected offspring have at least one affected parent
- Affected offspring are heterozygous if only one parent is affected
- Unaffected offspring are homozygous recessive
- If one parent is heterozygous, about ½ of the offspring will be affected
- Trait does not skip generations

FIGURE 14.22 Pedigrees of Families with Autosomal Recessive and Autosomal Dominant Traits. Pedigrees use standard symbols: squares = males, circles = females; unfilled symbols = unaffected individuals (those without the trait), filled symbols = affected individuals, half-filled symbols = heterozygotes for a recessive trait (carriers); horizontal lines connect parents, vertical lines connect parents to children.

about ¼ of their offspring are expected to express the recessive phenotype.

FIGURE 14.22a is a pedigree from a family in which an autosomal recessive trait, such as sickle-cell disease, occurs. The key feature to notice in this pedigree is that both boys and girls can exhibit the trait even though their parents do not. This is the pattern you would expect when the parents of an individual with the trait are heterozygous. It is also logical to observe that when an affected (homozygous) individual has children, those children do not necessarily have the trait. This pattern is predicted if affected people marry individuals who are homozygous for the dominant allele. This is likely to occur if the recessive allele is rare in the population.

In general, a recessive phenotype should show up in offspring only when both parents have that recessive allele and pass it on to their offspring. By definition, a recessive allele produces a given phenotype only when the individual is homozygous for that allele.

Patterns of Inheritance: Autosomal Dominant Traits When a trait is autosomal dominant, individuals who are homozygous or heterozygous for it must have the dominant phenotype. Even if one parent is heterozygous and the other is homozygous recessive, on average half their children should show the dominant phenotype. And unless a new mutation has occurred in a gamete, any child with the trait must have a parent with this trait. The latter observation is in strong contrast to the pattern seen in autosomal recessive traits.

FIGURE 14.22b shows the inheritance of the degenerative brain disorder called Huntington's disease (see Section 14.2). This pedigree has two features that indicate Huntington's disease is passed to the next generation through an autosomal dominant allele.

First, if a child shows the trait, then one of its parents shows the trait as well. Second, if families have a large number of children, the trait usually shows up in every generation—owing to the high probability of heterozygous parents having affected children.

Identifying Human Traits as Autosomal or Sex-Linked

When it is not possible to arrange reciprocal crosses, can data in a pedigree indicate whether a trait is autosomal or sex-linked? The answer is based on a simple premise. If a trait appears about equally often in males and females, then it is likely to be autosomal. But if males express the trait in question more often than females, then the allele responsible is likely to be recessive and found on the X chromosome.

X-linked Recessive Traits X-linked recessive traits are relatively common. They include the form of red–green color blindness that affects about 10 percent of males and the devastating blood clotting disorder hemophilia. A pedigree for red–green color blindness is shown in **FIGURE 14.23a**.

A key characteristic of X-linked recessive traits is that males express the trait more often than females. This is because males have only one copy of the X chromosome. Therefore, any X-linked allele will be expressed in a male.

Because human females have two X chromosomes, they have two copies of each X-linked gene and a recessive allele will not be expressed in heterozygotes. In the pedigree of Figure 14.23a the fact that only males are affected gives an immediate clue that this is an X-linked recessive trait.

Further, the appearance of an X-linked recessive trait usually skips a generation in the pedigree. Notice how on the right-hand

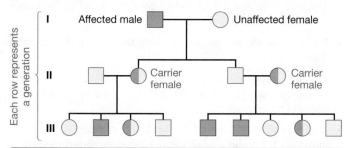

(a) X-linked recessive trait (e.g., red–green color blindness)

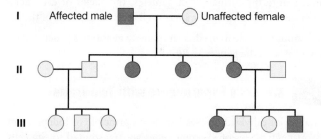

(b) X-linked dominant trait (e.g., hypophosphatemia)

CHARACTERISTICS:
- Males are affected more frequently than females
- Trait is never passed from father to son
- Affected sons are usually born to carrier mothers
- About ½ of the sons of a carrier mother will be affected
- All daughters of affected males and unaffected non-carrier females are carriers
- Trait often skips generations

CHARACTERISTICS:
- Males and females are equally likely to be affected
- All daughters of an affected father are affected, but no sons
- Affected sons always have affected mothers
- About ½ of the offspring of an affected mother will be affected
- Affected daughters can have an affected mother or father
- Trait does not skip generations

FIGURE 14.23 Pedigrees of Families with X-linked Recessive (a) and X-linked Dominant (b) Traits.

✔**QUESTION** What genotype in a mother and a father would be predicted to produce a 1:1 ratio of normal : color-blind offspring? What would the ratio of color-blind male : female be from this mating?

side of the pedigree the trait appears in generations I and III but not in generation II. This pattern occurs because an affected male passes his only X chromosome on to his daughters. But because his daughters almost always receive a wild-type allele from their mother, the daughters don't show the trait. They will pass the defective allele on to about half their sons, however.

Most X-linked traits are recessive. There are only a few rare examples of X-linked dominant traits.

X-linked Dominant Traits One example of an X-linked dominant trait is a bone disorder known as hypophosphatemia, or vitamin D–resistant rickets. What can be predicted about the inheritance of an X-linked dominant trait? As the pedigree in **FIGURE 14.23b** shows, the most telling feature is that an affected male will pass the trait to all his daughters and none of his sons.

This is because daughters receive their father's only X chromosome. In contrast, a heterozygous female will pass the trait to half her daughters and half her sons. This occurs because there is an equal chance that a heterozygous mother will pass on an X chromosome with either the dominant or the recessive trait.

What about Y-linked traits? Although the patterns for Y-linked inheritance can easily be predicted, the reality is that very few genes occur on the Y chromosome. These genes are involved with male-specific sexual development. Except for maleness, there are no known human Y-linked traits.

Within a few decades of the rediscovery of Mendel's work, the burning question in genetics was no longer the nature of inheritance but the nature of the gene. What are genes made of, and how are they copied so that parents pass their alleles on to their offspring? These are the questions turned to in Chapter 15.

CHAPTER 14 REVIEW

For media, go to MasteringBiology

If you understand . . .

14.1 Mendel's Experimental System

- When Mendel began his work, there were two leading hypotheses of inheritance: blending inheritance and the inheritance of acquired characteristics.
- Mendel chose pea plants as a model organism and began his work with pure lines that he crossed to produce hybrids.
- Mendel sought to discover rules of heredity that explained how phenotypes were transmitted from parent to offspring.

✔ You should be able explain why pure lines are homozygous.

14.2 Mendel's Experiments with a Single Trait

- From analysis of the trait expressed in hybrids, Mendel concluded that one trait was dominant and the other recessive.
- From the results of monohybrid crosses, Mendel concluded that inheritance is particulate—genes do not blend together.
- The traits that Mendel studied are specified by paired hereditary determinants (genes) that separate from each other during gamete formation.

- Analysis of monohybrid crosses led to the principle of segregation: Before the formation of gametes, the alleles of each gene separate so that each egg or sperm cell receives only one of them.

✔ You should be able to predict the gamete genotypes generated by a parent with the genotype *Bb*.

14.3 Mendel's Experiments with Two Traits

- Analysis of dihybrid crosses led to the principle of independent assortment: alleles of different genes are transmitted to egg cells and sperm cells independently of each other.

- Testcrosses allow an investigator to determine whether an individual of dominant phenotype is homozygous or heterozygous.

✔ You should be able to predict the gamete genotypes generated by a parent with the genotype *BbRr*.

14.4 The Chromosome Theory of Inheritance

- The chromosome theory states that chromosomes contain genes and that Mendel's rules can be explained by the segregation and independent assortment of homologous chromosomes at meiosis I.

- The chromosome theory was supported by the discovery of sex linkage. Crosses with X-linked traits supported the theory's contention that genes are found on chromosomes.

- X-linked traits give different results in reciprocal crosses.

✔ You should be able to use a Punnett square to show why reciprocal crosses involving an X-linked recessive gene give different results.

14.5 Extending Mendel's Rules

- There are important exceptions and extensions to the basic patterns of inheritance that Mendel discovered.

- All genes follow the principle of segregation, but genes close together on the same chromosome do not follow the principle of independent assortment.

- Crossing over between homologous chromosomes creates new combinations of alleles along each homolog.

- The frequency of crossing over can be used to create genetic maps that show the position and spacing of genes along chromosomes.

- Many genes have more than two alleles; this is multiple allelism. When more than two phenotypes are due to multiple allelism, the trait is polymorphic.

- Not all heterozygotes show a dominant phenotype. In addition to complete dominance, incomplete dominance and codominance are also common.

- Many genes are pleiotropic, meaning that they influence more than one trait.

- The phenotype associated with an allele is influenced by the environment that the individual experiences, and by the actions of alleles of other genes.

- Traits are often influenced by the action of many genes. These polygenic traits show quantitative instead of discrete variation. The frequency of different phenotypes is distributed normally (a plot produces a "bell curve" or normal distribution).

✔ You should be able to explain why crossing over is said to break up linkage between alleles.

14.6 Applying Mendel's Rules to Human Inheritance

- Pedigrees map out the transmission of human traits.

- Applying Mendel's rules to pedigrees can reveal the mode of transmission—whether a trait is dominant or recessive, autosomal or X-linked.

✔ You should be able to draw a pedigree that shows two sons and two daughters produced by a red–green color-blind father and a homozygous mother with normal color vision and explain why all the daughters are expected to be carriers of color blindness and none of the sons are expected to be color-blind.

(MB) MasteringBiology

1. **MasteringBiology Assignments**

   Tutorials and Activities Determining Genotype: Pea Pod Color; What Is the Inheritance Pattern of Sex-Linked Traits? Gregor's Garden; Incomplete Dominance; Inheritance of Fur Color in Mice; Linked Genes and Crossing Over; Linked Genes and Linkage Mapping; Mendel's Experiments; Mendel's Law of Independent Assortment; Mendel's Law of Segregation; Pedigree Analysis: Dominant and Recessive Autosomal Conditions; Pedigree Analysis: Galactosemia, Sex Linkage, Sex-Linked Genes

   Questions Reading Quizzes, Blue-Thread Questions, Test Bank

2. **eText** Read your book online, search, take notes, highlight text, and more.

3. **The Study Area** Practice Test, Cumulative Test, BioFlix® 3-D Animations, Videos, Activities, Audio Glossary, Word Study Tools, Art

You should be able to . . .

✔ TEST YOUR KNOWLEDGE

Answers are available in Appendix A

1. In studies of how traits are inherited, what makes certain species candidates for model organisms?
   a. They are the first organisms to be used in a particular type of experiment, so they are a historical "model" of what researchers expect to find.
   b. They are easy to study because a great deal is already known about them.
   c. They are the best or most fit of their type.
   d. They are easy to maintain, have a short life cycle, produce many offspring, and yield data that are relevant to many other organisms.

2. Why is the allele for wrinkled seed shape in garden peas considered recessive?
   a. It "recedes" in the F_2 generation when homozygous parents are crossed.
   b. The trait associated with the allele is not expressed in heterozygotes.
   c. Individuals with the allele have lower fitness than that of individuals with the dominant allele.
   d. The allele is less common than the dominant allele. (The wrinkled allele is a rare mutant.)

3. The alleles found in haploid organisms cannot be dominant or recessive. Why?
   a. Dominance and recessiveness describe which allele is expressed in phenotype when different alleles occur in the same individual.
   b. Because only one allele is present, alleles in haploid organisms are always dominant.
   c. Alleles in haploid individuals are transmitted like mitochondrial DNA or chloroplast DNA.
   d. Most haploid individuals are bacteria, and bacterial genetics is completely different from eukaryotic genetics.

4. Why can you infer that individuals that are "pure line" are homozygous for the gene in question?
   a. Because they are highly inbred.
   b. Because only two alleles are present at each gene in the populations to which these individuals belong.
   c. Because in a pure line, phenotypes are not affected by environmental conditions or gene interactions.
   d. Because no other phenotype is ever observed in a pure-line population, this implies that only one allele is present.

5. The genes for the traits that Mendel worked with are either located on different chromosomes or so far apart on the same chromosome that crossing over almost always occurs between them. How did this circumstance help Mendel recognize the principle of independent assortment?
   a. Otherwise, his dihybrid crosses would not have produced a 9:3:3:1 ratio of F_2 phenotypes.
   b. The occurrence of individuals with unexpected phenotypes led him to the discovery of recombination.
   c. It led him to the realization that the behavior of chromosomes during meiosis explained his results.
   d. It meant that the alleles involved were either dominant or recessive, which gave 3:1 ratios in the F_1 generation.

6. What is meant by the claim that Mendel worked with the simplest possible genetic system?
   a. Discrete traits, two alleles, simple dominance and recessiveness, no sex chromosomes, and unlinked genes are the simplest situation known.
   b. The ability to self-fertilize or cross-pollinate made it simple for Mendel to set up controlled crosses.
   c. Mendel was aware of meiosis and the chromosome theory of inheritance, so it was easy to reach the conclusions he did.
   d. Mendel's experimental designs and his rules of inheritance are actually neither complex nor sophisticated.

7. Mendel's rules do not correctly predict patterns of inheritance for tightly linked genes or the inheritance of alleles that show incomplete dominance. Does this mean that his hypotheses are incorrect?
   a. Yes, because they are relevant to only a small number of organisms and traits.

b. Yes, because not all data support his hypotheses.
   c. No, because he was not aware of meiosis or the chromosome theory of inheritance.
   d. No, it just means that his hypotheses are limited to certain conditions.

8. The artificial sweetener NutraSweet consists of a phenylalanine molecule linked to aspartic acid. The labels of diet sodas that contain NutraSweet include a warning to people with PKU. Why?
   a. NutraSweet stimulates the same taste receptors that natural sugars do.
   b. People with PKU have to avoid phenylalanine in their diet.
   c. In people with PKU, phenylalanine reacts with aspartic acid to form a toxic compound.
   d. People with PKU cannot lead normal lives, even if their environment is carefully controlled.

9. When Sutton and Boveri published the chromosome theory of inheritance, research on meiosis had not yet established that paternal and maternal homologs of different chromosomes assort independently. Then, in 1913, Elinor Carothers published a paper about a grasshopper species with an unusual karyotype: One chromosome had no homolog (meaning no pairing partner at meiosis I); another chromosome had homologs that could be distinguished under the light microscope. If chromosomes assort independently, how often should Carothers have observed each of the four products of meiosis shown in the following figure?

Four types of gametes possible
(each meiotic division can produce only two of the four)

   a. Only the gametes with one of each type of chromosome would occur.
   b. The four types of gametes should be observed to occur at equal frequencies.
   c. The chromosome with no pairing partner would disintegrate, so only gametes with one copy of the other chromosome would be observed.
   d. Gametes with one of each type of chromosome would occur twice as often as gametes with just one chromosome.

10. Which of the following is the strongest evidence that a trait might be influenced by polygenic inheritance?
   a. F_1 offspring of parents with different phenotypes have an intermediate phenotype.
   b. F_1 offspring of parents with different phenotypes have the dominant phenotype.
   c. The trait shows qualitative (discrete) variation.
   d. The trait shows quantitative variation.

The best way to test and extend your knowledge of transmission genetics is to work problems. Most genetics problems are set up as follows: You are given some information about the genotypes or phenotypes of one or both parents, along with data on the phenotypes of F_1 or F_2 offspring. Your task is to generate a set of hypotheses—a genetic model—to explain the results. Your hypotheses should address each of the following questions:

- Is the trait under study discrete or quantitative?
- Is the phenotype a product of one gene or many genes?
- For each gene involved, how many alleles are present—one, two, or many?
- Do the alleles involved show complete dominance, incomplete dominance, or codominance?
- Are the genes involved sex-linked or autosomal?
- If more than one gene is involved, are they linked or unlinked? If they are linked, does crossing over occur frequently?

It's also helpful to ask yourself whether gene interactions or pleiotropy might be occurring and whether it is safe to assume that the experimental design carefully controlled for effects of variation in other genes and in the environment.

In working the problem, be sure to start with the simplest possible explanation. For example, if you are dealing with a discrete trait, you might hypothesize that the cross involves a single autosomal gene with two alleles that show complete dominance. Your next step is to infer what the parental genotypes would be (according to your hypothesis), if they are not already given, and then do a Punnett square to predict what the offspring phenotypes and their frequencies should be based on your hypothesis. Next, check whether these predictions match the observed results given in the problem. If the

answer is yes, you have a valid solution. But if the answer is no, you need to go back and change one of your hypotheses, redo the Punnett square, and check to see if the predictions and observations match. Keep repeating these steps until you have a model that fits the data.

11. **Example Problem** *Plectritis congesta* plants produce fruits that either have or do not have prominent structures called wings. The alleles involved are W^+ = winged fruit; W^- = wingless fruit. Researchers collected an array of individuals from the field and performed a series of crosses. The results are given in the following table. Complete the table by writing down the genotype of the parent or parents involved in each cross.

Parental Phenotype(s)	Number of Offspring with Winged Fruits	Number of Offspring with Wingless Fruits	Parental Genotype(s)
Wingless (self-fertilized)	0	80	
Winged (self-fertilized)	90	30	
Winged × wingless	46	0	
Winged × winged	44	0	

A worked solution is available in Appendix A.

12. **Example Problem** Two black female mice are crossed with a brown male. In several litters, female I produced 9 blacks and 7 browns; female II produced 57 blacks. What deductions can you make concerning the inheritance of black and brown coat color in mice? What are the genotypes of the parents in this case?

A worked solution is available in Appendix A.

13. In peas, purple flowers are dominant to white. If a purple-flowered, heterozygous plant were crossed with a white-flowered plant, what is the expected ratio of genotypes and phenotypes among the F_1 offspring? If two of the purple-flowered F_1 offspring were randomly selected and crossed, what is the expected ratio of genotypes and phenotypes among the F_2 offspring?

14. In garden peas, yellow seeds (Y) are dominant to green seeds (y), and inflated pods (I) are dominant to constricted pods (i). Suppose you have crossed $YYII$ parents with $yyii$ parents.
   - Draw the F_1 Punnett square and predict the expected F_1 phenotype(s).
   - List the genotype(s) of gametes produced by F_1 individuals.
   - Draw the F_2 Punnett square. Based on this Punnett square, predict the expected phenotype(s) in the F_2 generation and the expected frequency of each phenotype.

15. The smooth feathers on the back of the neck in pigeons can be reversed by a mutation to produce a "crested" appearance in which feathers form a distinctive spike at the back of the head. A pigeon breeder examined offspring produced by a single pair of non-crested birds and recorded the following: 22 non-crested and 7 crested. She then made a series of crosses using offspring from the first cross. When she crossed two of the crested birds, all 20 of the offspring were crested. When she crossed a non-crested bird with a crested bird, 7 offspring were non-crested and 6 were crested.

   - For these three crosses, provide genotypes for parents and offspring that are consistent with these results.
   - Which allele is dominant?

16. A plant with orange, spotted flowers was grown in the greenhouse from a seed collected in the wild. The plant was self-pollinated and gave rise to the following progeny: 88 orange with spots, 34 yellow with spots, 32 orange with no spots, and 8 yellow with no spots. What can you conclude about the dominance relationships of the alleles responsible for the spotted and unspotted phenotypes? For the orange and yellow phenotypes? What can you conclude about the genotype of the original plant that had orange, spotted flowers?

17. As a genetic counselor, you routinely advise couples about the possibility of genetic disease in their offspring based on their family histories. This morning you met with an engaged couple, both of whom are phenotypically normal. The man, however, has a brother who died of Duchenne-type muscular dystrophy, an X-linked condition that results in death before the age of 20. The allele responsible for this disease is recessive. His prospective bride, whose family has no history of the disease, is worried that the couple's sons or daughters might be afflicted.
   - How would you advise this couple?
   - The sister of this man is planning to marry his fiancée's brother. How would you advise this second couple?

18. Suppose you are heterozygous for two genes that are located on different chromosomes. You carry alleles *A* and *a* for one gene and alleles *B* and *b* for the other. Draw a diagram illustrating what happens to these genes and alleles when meiosis occurs in your reproductive tissues. Label the stages of meiosis, the homologous chromosomes, sister chromatids, nonhomologous chromosomes, genes, and alleles. Be sure to list all the genetically different gametes that could form and indicate how frequently each type should be observed. On the diagram, identify the events responsible for the principle of segregation and the principle of independent assortment.

19. Review the text's description of ABO blood types. Suppose a woman with blood type O married a man with blood type AB. What phenotypes and genotypes would you expect to observe in their offspring, and in what proportions? Answer the same question for a heterozygous mother with blood type A and a heterozygous father with blood type B.

20. An alien friend named Tukan has two sets of eyes, one set forward-looking and one set backward-looking, and smooth skin. His mate, Valco, lacks eyes but has skin covered with tiny hooks that attract all sorts of debris. Tukan and Valco have thrived on earth and have had four children, all with no eyes and smooth skin. Typical of their ways, the children interbred and produced 32 children of their own.
    • Under the models of inheritance proposed by Mendel, identify which alleles are dominant and which are recessive.
    • Provide gene symbols that would reflect the dominant–recessive allelic relationships.
    • Of the 32 children, how many would you expect to have two sets of eyes and smooth skin?

21. Phenylketonuria (PKU) is a genetic disease caused by homozygosity for a recessive mutation in the enzyme that converts the amino acid phenylalanine to tyrosine. In the absence of this enzyme, phenylalanine and some of its derivatives accumulate in the body and cause mental retardation. If individuals are identified soon enough after birth, they can be treated by a low-phenylalanine diet for the early years of their lives. As adults, though, homozygous recessive individuals are allowed to adopt a diet with normal amounts of phenylalanine. Not long after such treatments were initiated, a troubling phenomenon was observed. A high number of children born to treated mothers were mentally retarded even though the children were heterozygous for the PKU gene. Children born of treated PKU males suffered no ill effects.
    • Can you offer an explanation as to why genetically heterozygous children of treated PKU mothers might be prone to mental retardation?
    • Propose a solution to reduce the likelihood of mental retardation in children of treated PKU mothers.

22. The blending-inheritance hypothesis proposed that the genetic material from parents is unavoidably and irreversibly mixed in the offspring. As a result, offspring and later descendants should always appear intermediate in phenotype to their forebears. Mendel, in contrast, proposed that genes are discrete and that their integrity is maintained in the offspring and in subsequent generations. Suppose the year is 1890. You are a horse breeder and have just read Mendel's paper. You don't believe his results,

however, because you often work with cremello (very light-colored) and chestnut (reddish-brown) horses. You know that if you cross a cremello individual from a pure-breeding line with a chestnut individual from a pure-breeding line, the offspring will be palomino—meaning they have an intermediate (golden-yellow) body color. What additional crosses would you do to test whether Mendel's model is valid in the case of genes for horse color? List the crosses and the offspring genotypes and phenotypes you'd expect to obtain. Explain why these experimental crosses would provide a test of Mendel's model.

23. Two mothers give birth to sons at the same time in a busy hospital. The son of couple 1 is afflicted with hemophilia A, which is a recessive X-linked disease. Neither parent has the disease. Couple 2 has a normal son even though the father has hemophilia A. The two couples sue the hospital in court, claiming that a careless staff member swapped their babies at birth. You appear in court as an expert witness. What do you tell the jury? Make a diagram that you can submit to the jury.

24. You have crossed two *Drosophila melanogaster* individuals that have long wings and red eyes—the wild-type phenotype. In the progeny, the mutant phenotypes called curved wings and lozenge eyes appear as follows:

Females	Males
600 long wings, red eyes	300 long wings, red eyes
200 curved wings, red eyes	100 curved wings, red eyes
	300 long wings, lozenge eyes
	100 curved wings, lozenge eyes

   • According to these data, is the curved-wing allele autosomal recessive, autosomal dominant, sex-linked recessive, or sex-linked dominant?
   • Is the lozenge-eyed allele autosomal recessive, autosomal dominant, sex-linked recessive, or sex-linked dominant?
   • What is the genotype of the female parent?
   • What is the genotype of the male parent?

25. In parakeets, two autosomal genes that are located on different chromosomes control the production of feather pigment. Gene *B* codes for an enzyme that is required for the synthesis of a blue pigment, and gene *Y* codes for an enzyme required for the synthesis of a yellow pigment. Recessive mutations that result in no production of the affected pigment are known for both genes. Suppose that a bird breeder has two green parakeets and mates them. The offspring are green, blue, yellow, and albino (unpigmented).
   • Based on this observation, what are the genotypes of the green parents? What is the genotype of each type of offspring? What fraction of the total progeny should exhibit each type of color?
   • Suppose that the parents were the progeny of a cross between two true-breeding strains. What two types of crosses between true-breeding strains could have produced the green parents? Indicate the genotypes and phenotypes for each cross.

26. Recall that hemophilia is an X-linked recessive disease. If a woman with hemophilia had children with a man without hemophilia, what is the chance that their first child will have the disease? What is the chance that their first child will be a carrier?

deoxyribonucleotide should be added next. A **primer**—a strand a few nucleotides long that is bonded to the template strand—provides DNA polymerase with a free 3′ hydroxyl (−OH) group that can combine with an incoming deoxyribonucleotide to form a phosphodiester bond. As shown in the figure below and in Figure 15.8, step 2, primers used during cellular DNA synthesis are short RNA strands, not DNA strands.

But what adds the primer? Before DNA synthesis can get under way, an enzyme called **primase** synthesizes a short stretch of RNA that acts as a primer for DNA polymerase. The primer is about 10 nucleotides long in *E. coli*. Primase is a type of **RNA polymerase**—an enzyme that catalyzes the polymerization of ribonucleotides into RNA (see Chapter 4 to review RNA's structure). Unlike DNA polymerases, primase and other RNA polymerases do not require a primer to begin synthesis.

Once a primer is present on a single-stranded template, DNA polymerase begins working in the 5′ → 3′ direction and adds deoxyribonucleotides to complete the complementary strand. As Figure 15.8, step 2, shows, DNA polymerase has a shape that grips the DNA strand during synthesis, similar to your hand clasping a rope. Deoxyribonucleotide addition is catalyzed at an active site in a groove between the enzyme's "thumb" and "fingers." As DNA polymerase moves along the DNA molecule, a doughnut-shaped structure behind it, called the sliding clamp, holds the enzyme in place on the template strand.

The enzyme's product is called the **leading strand,** or **continuous strand,** because it leads into the replication fork and is synthesized continuously. ✔ If you understand leading-strand synthesis, you should be able to list the enzymes involved and predict the consequences if any of them are defective.

How Is the Lagging Strand Synthesized?

Synthesis of the leading strand is straightforward. After an RNA primer is in place, DNA polymerase moves along, adding deoxyribonucleotides to the 3′ end of that strand. The enzyme moves into the replication fork, which "unzips" ahead of it. By comparison, events on the opposite strand are more involved.

Recall that the two strands of the DNA double helix are antiparallel—meaning they lie parallel to one another but oriented in opposite directions. The fact that DNA polymerases can synthesize DNA only in the 5′ → 3′ direction creates a paradox. Only one strand of DNA at the replication fork—the leading strand—can be synthesized in a direction that follows the moving replication fork.

The other strand must be synthesized in a direction that runs *away* from the moving replication fork, as illustrated in **FIGURE 15.9**. The strand of DNA that extends in the direction away from the

FIGURE 15.9 The Lagging Strand Is Synthesized in a Direction Moving Away from the Replication Fork. This occurs because the DNA strands are antiparallel and DNA polymerase can work only in the 5′ → 3′ direction.

replication fork is called the **lagging strand,** or **discontinuous strand,** because it lags behind the synthesis occurring at the fork. As the replication fork moves, it exposes gaps of single-stranded template DNA (Time 2 in Figure 15.9). How are the growing gaps filled in?

The Discontinuous Replication Hypothesis The puzzle posed by lagging-strand synthesis was resolved when Reiji Okazaki and colleagues tested a hypothesis called discontinuous replication. This hypothesis held that primase synthesizes new RNA primers for lagging strands as the moving replication fork opens single-stranded regions of DNA, and that DNA polymerase uses these primers to synthesize short lagging-strand DNA fragments that are linked together into a continuous strand. These ideas are illustrated in **FIGURE 15.10**.

Note that Figure 15.10 shows details of how lagging-strand synthesis occurs in *E. coli*. The overall process, however, applies to all groups of organisms—bacteria, archaea, and eukaryotes. The basic reactions of lagging-strand synthesis are universal. The differences lie in the names or specific properties of the key proteins and enzymes.

To explore the discontinuous replication hypothesis, Okazaki's group set out to test a key prediction: Could they find short DNA fragments produced during replication? Their critical experiment was based on the pulse–chase strategy (see Chapter 7). They added a brief "pulse" of radioactive deoxyribonucleotides to *E. coli* cells, followed by a "chase" of nonradioactive deoxyribonucleotides. According to the discontinuous replication model, some of these radioactive deoxyribonucleotides should first appear in short, fragments of DNA.

The Discovery of Okazaki Fragments As predicted, the researchers succeeded in finding short DNA fragments when they purified DNA from the experimental cells, separated the two strands of DNA, and analyzed the size of the molecules

1. Primer added. Primase synthesizes RNA primer.

2. First fragment synthesized. DNA polymerase III works in 5′→ 3′ direction, synthesizing first Okazaki fragment of lagging strand.

3. Second fragment synthesized. Primase and DNA polymerase III synthesize another Okazaki fragment.

4. Primer replaced. DNA polymerase I removes ribonucleotides of primer, replaces them with deoxyribonucleotides in 5′→ 3′ direction.

5. Gap closed. DNA ligase closes gap in sugar–phosphate backbone.

FIGURE 15.10 Lagging-Strand Synthesis.

by centrifugation. A small number of labeled DNA fragments about 1000 base pairs long were present immediately after the pulse. These short DNAs came to be known as **Okazaki fragments** and are shown in steps 2 and 3 of Figure 15.10. These small DNAs became larger during the chase as they were linked together into longer pieces. Subsequent work showed that Okazaki fragments in eukaryotes are even smaller—just 100 to 200 base pairs long.

How are Okazaki fragments connected? First, as step 4 of Figure 15.10 shows, in *E. coli* a specialized DNA polymerase called DNA polymerase I attaches to the 3′ end of an Okazaki fragment. As DNA polymerase I moves along in the 5′ → 3′ direction, it

removes that RNA primer ahead of it and replaces the ribonucleotides with the appropriate deoxyribonucleotides.

Once the RNA primer is removed and replaced by DNA, an enzyme called **DNA ligase** catalyzes the formation of a phosphodiester bond between the adjacent fragments (Figure 15.10, step 5). ✔ If you understand lagging-strand synthesis, you should be able to draw what the two newly synthesized molecules of DNA at a single replication fork would look like if DNA ligase were defective.

In eukaryotes, the mechanism for primer removal is different, but the mechanism of synthesizing short Okazaki fragments that are later joined into an unbroken chain of DNA is the same.

Working together, the enzymes that open the replication fork and manage the synthesis of the leading and lagging strands (**TABLE 15.1**) produce faithful copies of DNA before cell division. Although separate enzymes are drawn at different locations around the replication fork in Figures 15.8 and 15.10, in reality, all these enzymes are joined into the **replisome,** a large macromolecular machine. In *E. coli*, the replisome contains two copies of DNA polymerase III that are actively engaged in DNA synthesis. As shown in **FIGURE 15.11**, the lagging strand loops out and

SUMMARY TABLE 15.1 **Proteins Required for DNA Synthesis in Bacteria**

Name	Structure	Function
Opening the helix		
Helicase		Catalyzes the breaking of hydrogen bonds between base pairs to open the double helix
Single-strand DNA-binding proteins (SSBPs)		Stabilizes single-stranded DNA
Topoisomerase		Breaks and rejoins the DNA double helix to relieve twisting forces caused by the opening of the helix
Leading strand synthesis		
Primase		Catalyzes the synthesis of the RNA primer
DNA polymerase III		Extends the leading strand
Sliding clamp		Holds DNA polymerase in place during strand extension
Lagging strand synthesis		
Primase		Catalyzes the synthesis of the RNA primer on an Okazaki fragment
DNA polymerase III		Extends an Okazaki fragment
Sliding clamp		Holds DNA polymerase in place during strand extension
DNA polymerase I		Removes the RNA primer and replaces it with DNA
DNA ligase		Catalyzes the joining of Okazaki fragments into a continuous strand

FIGURE 15.11 **The Replisome.** The enzymes required for DNA synthesis are organized into a macromolecular machine. Note how the lagging strand loops out as the leading strand is being synthesized.

check your understanding

C Y U **If you understand that . . .**

- DNA synthesis begins at specific origins of replication on the chromosome and then proceeds in both directions.
- Synthesis at the replication fork occurs in three steps: (1) Helicase opens the double helix, SSBPs stabilize the exposed single strands, and topoisomerase prevents twists downstream of the fork; (2) DNA polymerase synthesizes the leading strand after primase has added an RNA primer; and (3) A series of enzymes synthesize the lagging strand.
- Lagging-strand synthesis cannot be continuous, because it moves away from the replication fork. In bacteria, enzymes called primase, DNA polymerase III, DNA polymerase I, and ligase work in sequence to synthesize Okazaki fragments and link them into a continuous whole.

✔ **You should be able to . . .**

1. Explain the function of primase.
2. Explain why DNA polymerase I is used predominantly on the lagging strand.

Answers are available in Appendix A.

around the complex, allowing the replisome to move as a single unit as it follows the replication fork. After the DNA polymerase on the lagging strand completes synthesis of an Okazaki fragment, it is released from the DNA and reassembles on the most recently synthesized primer.

15.4 Replicating the Ends of Linear Chromosomes

The circular DNA molecules in bacteria and archaea can be synthesized by the enzymes introduced in Section 15.3, and so can most of the linear DNA molecules found in eukaryotes. But replication at the very ends of linear eukaryotic chromosomes is another story altogether. Replication of chromosome ends requires a specialized DNA replication enzyme that has been the subject of intense research.

The End Replication Problem

The region at the end of a eukaryotic chromosome is called a **telomere** (literally, "end-part"). **FIGURE 15.12** illustrates the problem that arises during the replication of telomeres.

- When the replication fork reaches the end of a linear chromosome, a eukaryotic DNA polymerase synthesizes the leading strand all the way to the end of the parent DNA template (step 1 and step 2, top strand). As a result, leading-strand synthesis results in a double-stranded copy of the DNA molecule.

- On the lagging strand, primase adds an RNA primer close to the tip of the chromosome (see step 2, bottom strand).

- DNA polymerase synthesizes the final Okazaki fragment on the lagging strand (step 3). An enzyme that degrades ribonucleotides removes the primer.

PROCESS: PROBLEMS WITH COPYING THE ENDS OF LINEAR CHROMOSOMES

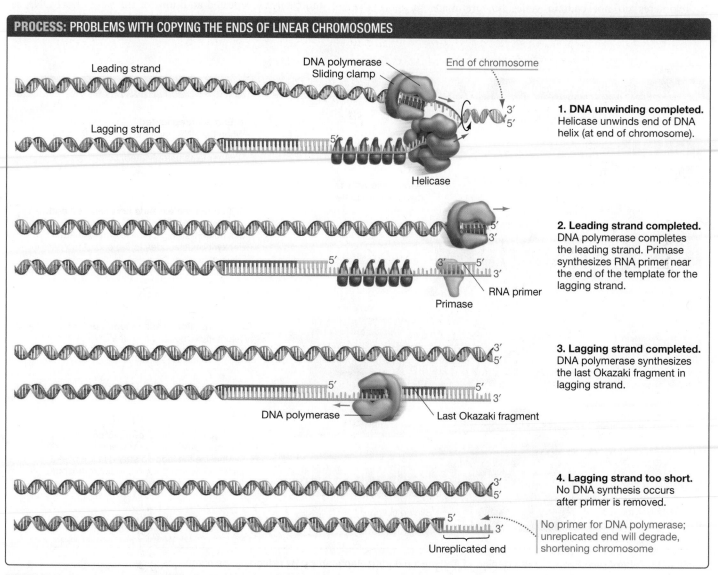

FIGURE 15.12 Chromosomes Shorten during Normal DNA Replication. An RNA primer is added to the lagging strand near the end of the chromosome. Once the primer is removed, it cannot be replaced with DNA. As a result, the chromosome shortens.

- DNA polymerase is unable to add DNA near the tip of the chromosome, because it cannot synthesize DNA without a primer (step 4). As a result, the single-stranded DNA that is left stays single stranded.

The single-stranded DNA at the end of the lagging strand is eventually degraded, which results in the shortening of the chromosome. If this process were to continue unabated, every chromosome would shorten by about 50 to 100 deoxyribonucleotides each time DNA replication occurred. Over time, linear chromosomes would vanish.

Telomerase Solves the End Replication Problem

How do eukaryotes maintain their chromosomes? One answer emerged after Elizabeth Blackburn, Carol Greider, and Jack Szostak reported two striking discoveries:

1. Telomeres do not contain genes but are made of short stretches of bases that are repeated over and over. In human telomeres, for example, the base sequence TTAGGG is repeated thousands of times.

2. A remarkable enzyme called telomerase that carries its own template is involved in replicating telomeres.

Telomerase is extraordinary because it catalyzes the synthesis of DNA from an RNA template that it contains. Telomerase adds DNA onto the end of a chromosome to prevent it from getting shorter. **FIGURE 15.13** shows one model for how telomerase works to maintain the ends of eukaryotic chromosomes.

Step 1 The unreplicated segment of the telomere at the 3′ end of the template for the lagging strand forms a single-stranded "overhang".

Step 2 Telomerase binds to the overhanging single-stranded DNA and begins DNA synthesis. The template for this reaction is a portion of the RNA held within telomerase.

Step 3 Telomerase synthesizes DNA in the 5′ → 3′ direction and catalyzes repeated additions of the same short DNA sequence to the end of the growing single strand.

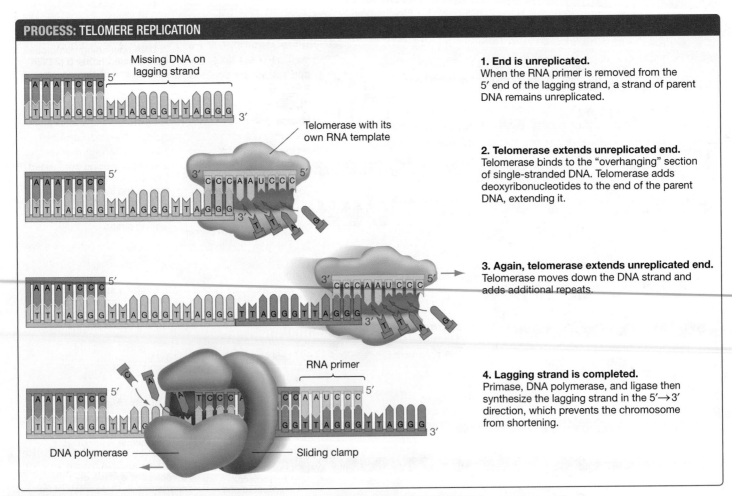

PROCESS: TELOMERE REPLICATION

Missing DNA on lagging strand

Telomerase with its own RNA template

RNA primer

DNA polymerase

Sliding clamp

1. End is unreplicated.
When the RNA primer is removed from the 5′ end of the lagging strand, a strand of parent DNA remains unreplicated.

2. Telomerase extends unreplicated end.
Telomerase binds to the "overhanging" section of single-stranded DNA. Telomerase adds deoxyribonucleotides to the end of the parent DNA, extending it.

3. Again, telomerase extends unreplicated end.
Telomerase moves down the DNA strand and adds additional repeats.

4. Lagging strand is completed.
Primase, DNA polymerase, and ligase then synthesize the lagging strand in the 5′→3′ direction, which prevents the chromosome from shortening.

FIGURE 15.13 Telomerase Prevents Shortening of Telomeres during Replication. By extending the number of repeated sequences in the 5′ → 3′ direction, telomerase provides room for enzymes to add an RNA primer to the lagging-strand template. Normal DNA replication enzymes can then fill in the missing section of the lagging strand.

✔**QUESTION** Would this telomerase work as well if its RNA template had a different sequence?

Step 4 Once the single-stranded overhang on the parent strand is lengthened, the normal enzymes of DNA synthesis use this strand as a template to synthesize a complementary strand. The result is that the lagging strand becomes slightly longer than it was originally.

Telomerase Regulation

The way telomerase is regulated is just as remarkable as the enzyme itself. Telomerase is active in only a limited number of cell types. In humans, for example, active telomerase is found primarily in the cells that produce gametes. Most **somatic cells,** meaning cells that are not involved in gamete formation, lack telomerase activity. As predicted, the chromosomes of somatic cells gradually shorten with each mitotic division, becoming progressively smaller as an individual ages.

These observations led to the hypothesis that the number of cell divisions possible for a somatic cell would be limited by the initial length of its telomeres. Carol Greider and colleagues tested this hypothesis by obtaining cells with a variety of telomere lengths from donors aged newborn to 90 years old and growing these cells in culture. (For an introduction to cell culture, see **BioSkills 12** in Appendix B.) Results of their study are shown in **FIGURE 15.14**. As predicted, there was a positive relationship between initial telomere length and the number of cell divisions before cells stop dividing—longer initial telomere length allowed a greater number of cell divisions, regardless of the donor's age.

You've probably noticed that the data points in Figure 15.14 do not fall perfectly on a line. This scatter or noise is typical in many studies. In interpreting results like these, researchers must consider what might account for the scatter and use statistical tests (see **BioSkills 4** in Appendix B) to determine how reliable the results are likely to be.

If telomere shortening controls the number of divisions possible for a cell, then a related prediction is that by restoring telomerase in somatic cells, these cells should be freed from growth limitations. As predicted, when researchers added telomerase to human cells growing in culture, the cells continued dividing long past the age when otherwise identical cells stop growing. Most biologists are convinced that telomere shortening has a role in limiting the number of cell divisions for somatic cells.

There is a dark side of telomerase activity, however. Unlike the somatic cells they derive from, most cancer cells have active telomerase. Many cancer biologists have proposed that telomerase activity allows the unlimited divisions of cancer cells. A simple prediction is that by inhibiting telomerase, the progression of cancer can be slowed or stopped. When combined with other approaches, could drugs that knock out telomerase be an effective way to fight cancer? Unfortunately, the complexity of cancer often thwarts such simple predictions. So far, answers to this question are unclear. Research continues.

check your understanding

If you understand that . . .

- Linear chromosomes shorten during replication because the end of the lagging strand lacks a primer and cannot be synthesized.
- Shortening is prevented in certain cells—particularly those that produce sperm and egg—because telomerase adds short, repeated DNA sequences to the template strand. Primase can then add an RNA primer to the lagging strand, and DNA polymerase can fill in the missing sections.

✔ You should be able to . . .

1. Explain why telomerase is not needed by bacterial cells.
2. Explain why telomerase has to have a built-in template.

Answers are available in Appendix A.

15.5 Repairing Mistakes and DNA Damage

DNA polymerases work fast. In *E. coli*, for example, each replication fork advances about 500 nucleotides per second. But the replication process is also astonishingly accurate. In organisms ranging from *E. coli* to animals, the error rate during DNA replication averages about one mistake per *billion* deoxyribonucleotides.

This level of accuracy is critical. Humans, for example, develop from a fertilized egg that has roughly 12 billion deoxyribonucleotides in its DNA. This DNA is replicated over and over to create the trillions of cells that eventually make up the adult body. If more than one or two mutations occurred during each

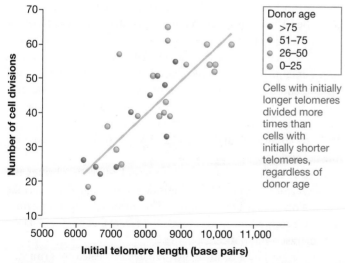

FIGURE 15.14 Telomere Length Predicts the Number of Divisions before Cells Stop Dividing.

DATA: Allsopp, R. C., et al. 1992. *Proceedings of the National Academy of Sciences*, 82: 10114–10118.

Cells with initially longer telomeres divided more times than cells with initially shorter telomeres, regardless of donor age cell division cycle as a person developed, genes would be riddled with errors by the time the individual reached maturity. Genes that contain errors are often defective.

Based on these observations, it is no exaggeration to claim that the accurate replication of DNA is a matter of life and death. Natural selection favors individuals with enzymes that copy DNA quickly and accurately.

These observations raise a key question. How can the enzymes of DNA replication be as precise as they are?

Correcting Mistakes in DNA Synthesis

As DNA polymerase marches along a DNA template, hydrogen bonding occurs between incoming deoxyribonucleotides and the deoxyribonucleotides on the template strand. DNA polymerases are selective about the bases they add to a growing strand because (1) the correct base pairings (A-T and G-C) are energetically the most favorable, and (2) these correct pairings have a distinct shape. As a result, DNA polymerase inserts an incorrect deoxyribonucleotide (**FIGURE 15.15a**) only about once in every 100,000 bases added.

An error rate of one in 100,000 seems low, but it is much higher than the rate of one in a billion listed at the start of this section. What happens when DNA polymerase makes a mistake?

DNA Polymerase Proofreads Biologists learned more about how DNA synthesis could be so accurate when they found mutant cells in which DNA synthesis was *in*accurate.

Specifically, researchers found *E. coli* mutants with error rates that were 100 times greater than normal. Recall that a mutant is an individual with a novel trait caused by a mutation (see Chapter 14). In the case of *E. coli* mutants with high error rates

(a) DNA polymerase adds a mismatched base...

(b) ...but detects the mistake and corrects it.

FIGURE 15.15 DNA Polymerase Can Proofread. If a mismatch such as the pairing of A with C occurs **(a)**, DNA polymerase can act as a 3' → 5' exonuclease, meaning that it can remove bases in that direction **(b)**. The DNA polymerase then adds the correct base.

in DNA replication, biologists found a defect in a portion of the DNA polymerase III enzyme called the ε (epsilon) subunit. Further analyses showed that the ε subunit acts as an exonuclease—meaning an enzyme that removes deoxyribonucleotides from the ends of DNA strands (**FIGURE 15.15b**).

If a newly added deoxyribonucleotide is not correctly paired with a base on the complementary strand, the positioning of the incorrect deoxyribonucleotide provides a poor substrate for DNA polymerase to extend. This is because the geometry of incorrect base pairs differs from that of the correct A-T and G-C pairs. DNA polymerase's active site can detect these shapes and will add a new deoxyribonucleotide only when the previous base pair is correct. In wild-type *E. coli*, the polymerase pauses when it detects the wrong shape, and the exonuclease activity of the ε subunit removes the mismatched deoxyribonucleotide.

These findings led to the conclusion that DNA polymerase III can **proofread.** If the wrong base is added during DNA synthesis, the enzyme pauses, removes the mismatched deoxyribonucleotide that was just added, and then proceeds again with synthesis.

Eukaryotic DNA polymerases have the same type of proofreading ability. Typically, proofreading reduces the overall error rate of DNA synthesis to about one mistake in 10 million bases added. Is this accurate enough? The answer remains no.

Mismatch Repair If—despite its proofreading ability—DNA polymerase leaves a mismatched base behind in the newly synthesized strand, a battery of enzymes springs into action to correct the problem. **Mismatch repair** occurs when mismatched bases are corrected after DNA synthesis is complete.

The proteins responsible for mismatch repair were discovered in the same way proofreading was—by analyzing *E. coli* mutants. In this case, the mutants had normal DNA polymerase III but abnormally high mutation rates.

The first mutation that caused a deficiency in mismatch repair was identified in the late 1960s and was called *mutS*. (The *mut* is short for "mutator.") Twenty years later, researchers had identified 10 proteins involved in the identification and repair of base-pair mismatches in *E. coli*.

These proteins recognize the mismatched base, remove a section containing the incorrect base from the newly synthesized strand, and fill in the correct bases using the older strand as a template. In *E. coli*, chemical marks on the older strand allow the enzymes to distinguish the original strand from the newly synthesized strand. Eukaryotes use a different scheme to recognize the old and new strands of DNA.

This final layer of error detection and correction brings the overall error rate of DNA synthesis down to roughly one mistake per billion deoxyribonucleotides. The mismatch-repair enzymes are like a copy editor who corrects the errors that a writer—DNA polymerase—did not catch.

The importance of mismatch repair is revealed by grim discoveries: Mutations in components of the mismatch repair system are observed in many common human cancers, where they play an important role in cancer development and progression.

Repairing Damaged DNA

Even after DNA is synthesized and proofread and mismatches repaired, the job of ensuring accuracy doesn't end. Genes are under constant assault. DNA is damaged by sunlight, X-rays, and many chemicals like the hydroxyl (OH) radicals produced during aerobic metabolism, aflatoxin B1 found in moldy peanuts and corn, and benzo[α]pyrene in cigarette smoke. If this damage were ignored, mutations would quickly accumulate to lethal levels. To fix problems caused by chemical attack, radiation, or other events, organisms have evolved a wide array of DNA damage-repair systems. As an example, consider the **nucleotide excision repair** system that works on DNA damage caused by ultraviolet light and many different chemicals.

Ultraviolet (UV) light in sunlight—and tanning booths—can cause a covalent bond to form between adjacent pyrimidine bases within the same DNA strand. The thymine-thymine pair illustrated in **FIGURE 15.16** is a common example. This defect, called a thymine dimer, creates a kink in the structure of DNA. The kink stalls standard DNA polymerases, blocking DNA replication. If the damage is not repaired, the cell may die.

Nucleotide excision repair fixes thymine dimers and many other types of damage that distort the DNA helix. In the first step of excision repair, an enzyme recognizes the kink in the DNA helix (step 1 in **FIGURE 15.17**). Once a damaged region is recognized, another enzyme removes a segment of single-stranded DNA containing the defective sequence (step 2). The intact DNA strand provides a template for synthesis of a corrected strand, and the 3′ hydroxyl of the DNA strand next to the gap serves as a primer (step 3). DNA ligase links the newly synthesized DNA to

the original undamaged DNA (step 4). As with mismatch repair, multiple enzymes work together and DNA synthesis plays a central role in repair.

What happens when a human DNA repair system is defective?

Xeroderma Pigmentosum: A Case Study

Xeroderma pigmentosum (XP) is a rare autosomal recessive disease in humans. Individuals with this condition are extremely sensitive to ultraviolet (UV) light. Their skin develops lesions including rough, scaly patches and irregular dark spots after even slight exposure to sunlight.

In 1968 James Cleaver proposed a connection between XP and DNA nucleotide excision repair. He knew that mutants of *E. coli* had defects in nucleotide excision repair that caused an increased sensitivity to radiation. Cleaver's hypothesis was that people with XP have similar mutations. He proposed that they are extremely sensitive to sunlight because they are unable to repair damage induced by UV light.

Cleaver and other researchers made extensive use of cell cultures (**BioSkills 12**, see Appendix B) to study the hypothesized connection between DNA damage, faulty nucleotide excision repair, and XP. They collected skin cells from people with XP and from people with normal UV light sensitivity. When these cells were grown in culture, the biologists exposed them to increasing amounts of UV radiation and recorded how many survived.

FIGURE 15.17 In Nucleotide Excision Repair, Defective Bases Are Removed and Replaced.

PROCESS: NUCLEOTIDE EXCISION REPAIR

1. **Error detection.** Enzymes detect an irregularity in DNA structure and cut the damaged strand.

2. **Nucleotide excision.** An enzyme excises a stretch of nucleotides that includes the damage.

3. **Nucleotide replacement.** DNA polymerase fills in the gap in the 5′→3′ direction.

4. **Nucleotide linkage.** DNA ligase links the new and old nucleotides into a continuous strand.

FIGURE 15.16 UV Light Damages DNA. When UV light strikes a section of DNA that has adjacent thymines, the energy can break bonds within each base and lead to the formation of bonds *between* them. The thymine dimer that is produced causes a kink in the DNA.

✔**QUESTION** Why are infrared wavelengths much less likely than UV to damage DNA? (Hint: See Figure 10.4.)

(a) Vulnerability of cells to UV light damage

(b) Ability of cells to repair UV light damage

FIGURE 15.18 DNA Damage from UV Light Is Not Repaired Properly in Individuals with XP. (a) When cell cultures from unaffected individuals and XP patients are irradiated with various doses of UV light (expressed here as ergs/mm^2), the percentage of cells that survive is strikingly different. **(b)** When cell cultures from unaffected individuals and XP patients are irradiated with various doses of UV light and then provided with a radioactive deoxyribonucleotide, only cells from unaffected individuals incorporate the labeled deoxyribonucleotide into their DNA.

DATA: (a) Cleaver, J. E. 1970. *Int. J. Rad. Biol.* 18: 577–565, Fig 3. (b) Cleaver, J. E. 1972. *J. Invest. Dermatol.* 58: 124–128, Fig 1.

✔ **QUESTION** Why are people who cultivate a sun tan increasing their risk of developing cancer? (Hint: Tanning is a response to UV light.)

FIGURE 15.18a shows the results of one such study by Cleaver. Note that the intensity of the radiation is graphed on the *x*-axis, and the percentage of cells surviving is graphed on the *y*-axis. Note, too, that the *y*-axis is logarithmic. (For help with reading graphs, see **BioSkills 3** and for help with logarithms, see **BioSkills 6**, both in Appendix B.) Cell survival declined with increasing radiation dose in both types of cells, but XP cells died off much more rapidly.

The connection to nucleotide excision repair systems was confirmed in a separate study when Cleaver exposed cells from unaffected and XP individuals to various amounts of UV light and then incubated the cells with a radioactive deoxyribonucleotide to label DNA synthesized during DNA repair. If repair is defective in XP individuals, then their cells should incorporate little radioactive deoxyribonucleotide into their DNA. Cells from unaffected individuals, in contrast, should incorporate large amounts of labeled deoxyribonucleotide into their DNA as it is repaired.

As **FIGURE 15.18b** shows, this is exactly what happens. Here the amount of radioactive deoxyribonucleotides incorporated into DNA is graphed against radiation dose. Increasingly large amounts of radioactivity are found in the DNA of healthy cells as UV dose increases, but almost no such increase occurs in XP cells. These data are consistent with the hypothesis that nucleotide excision repair is virtually nonexistent in XP individuals.

Genetic analyses of XP patients have shown that the condition can result from mutations in any of eight genes. This discovery is not surprising in light of the large number of enzymes involved in repairing damaged DNA.

As you saw for mismatch repair, defects in DNA repair genes are frequently associated with cancer. Individuals with

xeroderma pigmentosum, for example, are 1000 to 2000 times more likely to get skin cancer than are individuals with intact excision repair systems. To explain this pattern, biologists suggest that if DNA damage in the genes involved in the cell cycle goes unrepaired, mutations will result that may allow the cell to grow in an uncontrolled manner. Tumor formation could result. Recall

check your understanding

If you understand that . . .

- DNA polymerases occasionally add the wrong base during DNA synthesis.
- Proofreading by DNA polymerase and mismatch repair of misincorporated bases sharply reduces the number of errors.
- DNA is damaged frequently, and most of this damage can be fixed by DNA repair systems such as nucleotide excision repair.

✔ **You should be able to . . .**

1. Predict how the mutation rate would be affected if there were no differences in stability and shape between all possible base pairs.
2. Predict the effect on mutation rate of a failure in the system for distinguishing old and newly synthesized DNA.
3. State which nucleotide excision repair enzymes are specific for DNA repair and which work in both normal DNA replication and in DNA repair.

Answers are available in Appendix A.

that most cancers develop only after several genes have been damaged (see Chapter 12). If the overall mutation rate in a cell is elevated because of defects in DNA repair, then mutations that trigger cancer become more likely.

At this point, it's clear that genes are made of DNA and that DNA is accurately copied and passed on to offspring. How can information be stored in DNA, and how can this information be used? (These are the topics of the next two chapters.)

If you understand . . .

15.1 What Are Genes Made Of?

- Experiments on viruses that had labeled proteins or DNA showed that DNA is the hereditary material.

- DNA's primary structure consists of a sugar–phosphate backbone and a sequence of nitrogen-containing bases.

- DNA's secondary structure consists of two strands in an antiparallel orientation. The strands twist into a helix and are held together by complementary pairing between bases.

- ✓ You should be able to interpret an imaginary experiment like the one done by Hershey and Chase that shows that ^{32}P is found only in the pellet and that ^{35}S is found in both the pellet and the solution.

15.2 Testing Early Hypotheses about DNA Synthesis

- By labeling DNA with ^{15}N or ^{14}N, researchers were able to validate the hypothesis that DNA replication is semiconservative.

- In semiconservative replication, each strand of a parent DNA molecule provides a template for the synthesis of a daughter strand, resulting in two complete DNA double helices.

- ✓ You should be able to write a sequence of double-stranded DNA that is 10 base pairs long, separate the strands, and, without comparing them, write in the bases that are added during DNA replication.

15.3 A Model for DNA Synthesis

- DNA synthesis requires many different enzymes, and it occurs in one direction only.

- DNA synthesis requires both a template and a primer sequence. It takes place at the replication fork where the double helix is opened.

- Synthesis of the leading strand in the $5' \rightarrow 3'$ direction is continuous, but synthesis of the lagging strand is discontinuous because on that strand, the DNA polymerase moves away from the replication fork.

- On the lagging strand, short DNA fragments called Okazaki fragments form and are joined together. Okazaki fragments are primed by a short strand of RNA.

- ✓ You should be able to draw and label a diagram of a replication bubble that shows (1) the $5' \rightarrow 3'$ polarity of the two parental DNA strands and (2) the leading and lagging daughter strands at each replication fork.

15.4 Replicating the Ends of Linear Chromosomes

- At the ends of linear chromosomes in eukaryotes, the enzyme telomerase adds short, repeated sections of DNA so that the lagging strand can be synthesized without shortening the chromosome.

- Telomerase is active in reproductive cells that eventually undergo meiosis. As a result, gametes contain chromosomes of normal length.

- Chromosomes in cells without telomerase shorten with continued cell division until their telomeres reach a critical length at which cell division no longer occurs.

- ✓ You should be able to explain the significance of telomerase reactivation in cancer cells.

15.5 Repairing Mistakes and DNA Damage

- DNA replication is remarkably accurate because (1) DNA polymerase selectively adds a deoxyribonucleotide that correctly pairs with the template strand; (2) DNA proofreads each added deoxyribonucleotide; and (3) mismatch repair enzymes remove incorrect bases once synthesis is complete and replace them with the correct base.

- DNA repair occurs after DNA has been damaged by chemicals or radiation.

- Nucleotide excision repair cuts out damaged portions of DNA and replaces them with correct sequences.

- If DNA repair enzymes are defective, mutation rate increases. Because of this, several types of human cancers are associated with defects in the genes responsible for DNA repair.

- ✓ You should be able to explain the logical connections between failure of repair systems, increases in mutation rate, and high likelihood of cancer developing.

1. **MasteringBiology Assignments**

   Tutorials and Activities DNA and RNA Structure; DNA Double Helix; DNA Replication; DNA Replication: A Closer Look; DNA Replication: A Review; DNA Replication: An Overview; DNA Synthesis; Experimental Inquiry: Does DNA Replication Follow the Conservative, Semiconservative, or Dispersive Model; Hershey–Chase Experiment

   Questions Reading Quizzes, Blue-Thread Questions, Test Bank

2. **eText** Read your book online, search, take notes, highlight text, and more.

3. **The Study Area** Practice Test, Cumulative Test, BioFlix® 3-D Animations, Videos, Activities, Audio Glossary, Word Study Tools, Art

You should be able to . . .

✓ TEST YOUR KNOWLEDGE
Answers are available in Appendix A

1. What does it mean to say that strands in a double helix are antiparallel?
   a. Their primary sequences consist of a sequence of *complementary* bases.
   b. They each have a sugar–phosphate backbone.
   c. They each have a $5' \rightarrow 3'$ directionality.
   d. They have opposite directionality, or polarity.

2. Which of the following is *not* a property of DNA polymerase?
   a. It adds dNTPs only in the $5' \rightarrow 3'$ direction.
   b. It requires a primer to work.
   c. It is associated with a sliding clamp only on the leading strand.
   d. Its exonuclease activity is involved in proofreading.

3. The enzyme that removes twists in DNA ahead of the replication fork is _____.

4. What is the function of primase?
   a. synthesis of the short section of double-stranded DNA required by DNA polymerase
   b. synthesis of a short RNA, complementary to single-stranded DNA
   c. closing the gap at the $3'$ end of DNA after excision repair
   d. removing primers and synthesizing a short section of DNA to replace them

5. How are Okazaki fragments synthesized?
   a. using the leading strand template, and synthesizing $5' \rightarrow 3'$
   b. using the leading strand template, and synthesizing $3' \rightarrow 5'$
   c. using the lagging strand template, and synthesizing $5' \rightarrow 3'$
   d. using the lagging strand template, and synthesizing $3' \rightarrow 5'$

6. An enzyme that uses an internal RNA template to synthesize DNA is _____.

✓ TEST YOUR UNDERSTANDING
Answers are available in Appendix A

7. Researchers design experiments so that only one thing is different between the treatments that are being compared. In the Hershey–Chase experiment, what was this single difference?

8. What is the relationship between defective DNA repair and cancer?

9. Why is the synthesis of the lagging strand of DNA discontinuous? How is it possible for the synthesis of the leading strand to be continuous?

10. Explain how telomerase prevents linear chromosomes from shortening during replication.

11. Predict what would occur in a bacterial mutant that lost the ability to chemically mark the template strand of DNA.
    a. The mutation rate would increase.

b. The ability of DNA polymerase to discriminate between correct and incorrect base pairs would decrease.
    c. The energy differences between correct and incorrect base pairs would decrease.
    d. The energy differences between correct and incorrect base pairs would increase.

12. What aspect of DNA structure makes it possible for the enzymes of nucleotide excision repair to recognize many different types of DNA damage?
    a. the polarity of each DNA strand
    b. the antiparallel orientation of strands in the double helix
    c. the energy differences between correct and incorrect base pairs
    d. the regularity of DNA's overall structure

13. If you could engineer an activity into DNA polymerase to allow both strands to follow the replication fork, what would this additional activity be?
   a. the ability to begin DNA synthesis without a primer
   b. the ability to proofread in the 5′ → 3′ direction
   c. the ability to synthesize DNA in the 3′ → 5′ direction
   d. the ability to synthesize DNA without using a template

14. In the late 1950s, Herbert Taylor grew bean root-tip cells in a solution of radioactive thymidine and allowed them to undergo one round of DNA replication. He then transferred the cells to a solution without the radioactive deoxyribonucleotide, allowed them to replicate again, and examined their chromosomes for the presence of radioactivity. His results are shown in the following figure, where red indicates a radioactive chromatid.

Nonradioactive chromosome   Radioactive chromatids

1. DNA replication in radioactive solution   **2.** Mitosis   **3.** DNA replication in nonradioactive solution

   a. Draw diagrams explaining the pattern of radioactivity observed in the sister chromatids after the first and second rounds of replication.
   b. What would the results of Taylor's experiment be if eukaryotes used a conservative mode of DNA replication?

15. The graph that follows shows the survival of four different *E. coli* strains after exposure to increasing doses of ultraviolet light. The wild-type strain is normal, but the other strains have a mutation in either a gene called *uvrA*, a gene called *recA*, or both.

DATA: Howard-Flanders, P., and R. P. Boyce. 1966. *Radiation Research Supplement* 6: 156–184, Fig. 8.

   a. Which strains are most sensitive to UV light? Which strains are least sensitive?
   b. What are the relative contributions of these genes to the repair of UV damage?

16. **QUANTITATIVE** Assuming that each replication fork moves at a rate of 500 base pairs per second, how long would it take to replicate the *E. coli* chromosome (with 4.6 million base pairs) from a single origin of replication?

16 How Genes Work

In this chapter you will learn how

Genetic information flows from DNA → RNA → proteins

by asking ↓ | then examining ↓ | and analyzing ↓

What do genes do? 16.1

The central dogma of molecular biology 16.2

The genetic code, with its 3-letter "words" 16.3

which together explain ↓

How mutations can modify genes and genomes 16.4

This image shows a normal human male spectral karyotype—a micrograph of metaphase chromosomes stained to show different homologous chromosome pairs. This chapter explores how DNA sequences in chromosomes are related to phenotypes.

DNA has been called the blueprint of life. If an organism's DNA is like a set of blueprints, then its cells are like construction sites, and the enzymes inside a cell are like construction workers. But how does the DNA inside each cell assemble this team of skilled laborers and specify the materials needed to build and maintain the cell, and remodel it when conditions change?

Mendel provided insights that made the study of these questions possible. He discovered that particular alleles are associated with certain phenotypes and that alleles do not change when transmitted from parent to offspring. Later, the chromosome theory of inheritance established that genes are found in chromosomes, whose movement during meiosis explains Mendel's results.

The science of molecular biology began with the discovery that DNA is the hereditary material and that DNA is a double-helical structure containing sequences of four bases. From these early advances, it was clear that genes are made of DNA and that genes carry the instructions for making and maintaining an individual.

This chapter is part of the Big Picture. See how on pages 366–367.

✔ When you see this checkmark, stop and test yourself. Answers are available in Appendix A.

But biologists still didn't know how the information in DNA is translated into action. How does **gene expression**—the process of converting archived information into molecules that actually do things in the cell—occur?

This chapter introduces some of the most pivotal ideas in all of biology—ideas that connect genotypes to phenotypes by revealing how genes work at the molecular level. They also speak to the heart of a key attribute of life: processing genetic information to produce a living organism. (You can see how these concepts fit into the Big Picture of Genetic Information on pages 366–367.)

Understanding how genes work triggered a major transition in biological science. Instead of thinking about genes solely in relation to their effects on eye color in fruit flies or on seed shape in garden peas, biologists could begin analyzing the molecular composition of genes and their products. The molecular revolution in biology took flight.

16.1 What Do Genes Do?

Although biologists of the early twentieth century made tremendous progress in understanding how genes are inherited, an explicit hypothesis explaining what genes do did not appear until 1941. That year George Beadle and Edward Tatum published a series of breakthrough experiments on a bread mold called *Neurospora crassa*.

Beadle and Tatum's research was inspired by an idea that was brilliant in its simplicity. As Beadle said: "One ought to be able to discover what genes do by making them defective." The idea was to knock out a gene by damaging it and then infer what the gene does by observing the phenotype of the mutant individual.

Today, alleles that do not function at all are called **knock-out, null,** or **loss-of-function alleles.** Creating knock-out mutant alleles and analyzing their effects is still one of the most common research strategies in studies of gene function. But Beadle and Tatum were the pioneers.

The One-Gene, One-Enzyme Hypothesis

To start their work, Beadle and Tatum exposed a large number of *N. crassa* cells to radiation. As described earlier (Chapter 15), radiation can damage the double-helical structure of DNA—often in a way that makes the affected gene nonfunctional.

Their next step was to examine the mutant cells. Eventually they succeeded in finding *N. crassa* mutants that could not make specific compounds. For example, one of the mutants could not make pyridoxine, also called vitamin B_6, even though normal individuals can. Further, Beadle and Tatum showed that the inability to synthesize pyridoxine was due to a defect in a single gene, and that the inability to synthesize other molecules was due to defects in other genes.

These results inspired their **one-gene, one-enzyme hypothesis.** Beadle and Tatum proposed that the mutant *N. crassa* could not make pyridoxine because it lacked an enzyme required to synthesize the compound. They further proposed that the lack of the enzyme was due to a genetic defect. Based on analyses of knock-out mutants, the one-gene, one-enzyme hypothesis claimed that each gene contains the information needed to make an enzyme.

An Experimental Test of the Hypothesis

Three years later, Adrian Srb and Norman Horowitz published a rigorous test of the one-gene, one-enzyme hypothesis. These biologists focused on the ability of *N. crassa* to synthesize the amino acid arginine. In the lab, normal cells of this bread mold grow well on a laboratory culture medium that lacks arginine. This is possible because *N. crassa* cells are able to synthesize their own arginine.

Previous work had shown that organisms synthesize arginine in a series of steps called a **metabolic pathway.** As **FIGURE 16.1** shows, compounds called ornithine and citrulline are intermediate products in the metabolic pathway leading to arginine. Specific enzymes are required to synthesize ornithine, convert ornithine to citrulline, and change citrulline to arginine. Srb and Horowitz hypothesized that specific *N. crassa* genes are responsible for producing each of the three enzymes involved.

To test this idea, Srb and Horowitz used radiation to create a large number of mutant cells. However, radiation is equally likely to damage DNA and mutate genes in any part of the organism's genome, and most organisms have thousands or tens of thousands of genes. Of the many mutants the biologists created, how could they find the handful that specifically knocked out a step in the pathway for arginine synthesis?

To find the mutants they were looking for, the researchers performed what is now known as a genetic screen. A **genetic screen** is any technique for picking certain types of mutants out of many randomly generated mutants.

Srb and Horowitz began their screen by raising colonies of irradiated cells on a medium that included arginine. Then they transferred a sample of each colony to a medium that *lacked* arginine. If an individual could grow in the presence of arginine but failed to grow without arginine, they concluded that it couldn't make its own arginine.

Metabolic pathway for arginine synthesis: Precursor → (Enzyme 1) → Ornithine → (Enzyme 2) → Citrulline → (Enzyme 3) → Arginine

FIGURE 16.1 Different Enzymes Catalyze Each Step in the Metabolic Pathway for Arginine.

✔**QUESTION** If a cell lacked enzyme 2 but was placed in growth medium with ornithine, could it grow? Could it grow if it received citrulline instead?

The biologists followed up by confirming that the offspring of these cells also had this defect. Based on these data, they were confident that they had isolated individuals with mutations in one or more of the genes for the enzymes shown in Figure 16.1.

To test the one-gene, one-enzyme hypothesis, the biologists grew each mutant under four different conditions: on normal media without added arginine, and on normal medium supplemented with ornithine, citrulline, or arginine.

As **FIGURE 16.2** shows, the results from these growth experiments were dramatic. Some of the mutant cells were able to grow on some of these media but not on others. More specifically, the mutants fell into three distinct classes, which the researchers called *arg1*, *arg2*, and *arg3*.

As the "Interpretation" section of the figure shows, the data make sense if each type of mutant lacked a different, specific step in a metabolic pathway because of a defect in a particular gene. In short, Srb and Horowitz had documented a correlation between a specific genetic defect and a defect at a specific point in a metabolic pathway. This experiment convinced most investigators that the one-gene, one-enzyme hypothesis was correct.

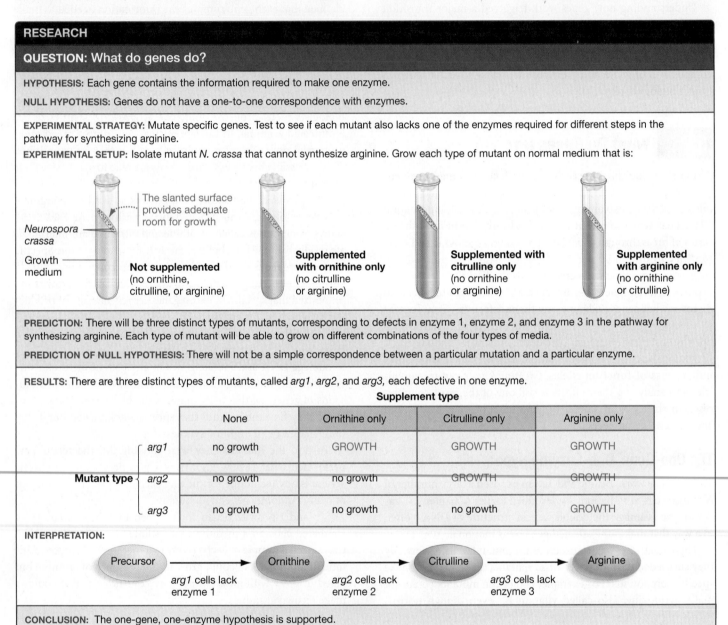

RESEARCH

QUESTION: What do genes do?

HYPOTHESIS: Each gene contains the information required to make one enzyme.

NULL HYPOTHESIS: Genes do not have a one-to-one correspondence with enzymes.

EXPERIMENTAL STRATEGY: Mutate specific genes. Test to see if each mutant also lacks one of the enzymes required for different steps in the pathway for synthesizing arginine.

EXPERIMENTAL SETUP: Isolate mutant *N. crassa* that cannot synthesize arginine. Grow each type of mutant on normal medium that is:

The slanted surface provides adequate room for growth

Neurospora crassa

Growth medium

Not supplemented (no ornithine, citrulline, or arginine)

Supplemented with ornithine only (no citrulline or arginine)

Supplemented with citrulline only (no ornithine or arginine)

Supplemented with arginine only (no ornithine or citrulline)

PREDICTION: There will be three distinct types of mutants, corresponding to defects in enzyme 1, enzyme 2, and enzyme 3 in the pathway for synthesizing arginine. Each type of mutant will be able to grow on different combinations of the four types of media.

PREDICTION OF NULL HYPOTHESIS: There will not be a simple correspondence between a particular mutation and a particular enzyme.

RESULTS: There are three distinct types of mutants, called *arg1*, *arg2*, and *arg3*, each defective in one enzyme.

	Supplement type			
Mutant type	None	Ornithine only	Citrulline only	Arginine only
arg1	no growth	GROWTH	GROWTH	GROWTH
arg2	no growth	no growth	GROWTH	GROWTH
arg3	no growth	no growth	no growth	GROWTH

INTERPRETATION:

Precursor → Ornithine → Citrulline → Arginine

arg1 cells lack enzyme 1

arg2 cells lack enzyme 2

arg3 cells lack enzyme 3

CONCLUSION: The one-gene, one-enzyme hypothesis is supported.

FIGURE 16.2 Experimental Support for the One-Gene, One-Enzyme Hypothesis. The association between specific genetic defects in *N. crassa* and specific defects in the metabolic pathway for arginine synthesis provided evidence that supported the one-gene, one-enzyme hypothesis.

SOURCE: Srb, A. M., and N. H. Horowitz. 1944. The ornithine cycle in *Neurospora* and its genetic control. *Journal of Biological Chemistry* 154: 129–139.

✓**QUESTION** Experimental designs must be repeatable so that other investigators can try the experiment themselves to check the results. Name three things that these researchers would need to describe so that others could repeat this experiment.

Follow-up work showed that genes contain the information for all the proteins produced by an organism—not just enzymes. Biologists finally understood what most genes do: They contain the instructions for making proteins.

In many cases, though, a protein is made up of several different polypeptides, each of which is a product of a different gene. Consequently, for greater accuracy, the one-gene, one-enzyme hypothesis is best called the one-gene, one-polypeptide hypothesis.

16.2 The Central Dogma of Molecular Biology

How does a gene specify the production of a protein? As soon as Beadle and Tatum's hypothesis had been supported in *N. crassa* and a variety of other organisms, this question became a central one.

Part of the answer lay in the molecular structure of the gene. Biochemists knew that the primary components of DNA were four nitrogen-containing bases: the pyrimidines thymine (abbreviated T) and cytosine (C), and the purines adenine (A) and guanine (G). They also knew that these bases were connected in a linear sequence by a sugar–phosphate backbone. Watson and Crick's model for the secondary structure of the DNA molecule (see Chapters 4 and 15) revealed that two strands of DNA are wound into a double helix, held together by hydrogen bonds between the complementary base pairs A-T and G-C.

Given DNA's structure, it appeared extremely unlikely that DNA directly catalyzed the reactions that produce proteins. Its shape was too regular to suggest that it could bind a wide variety of substrate molecules and lower the activation energy for chemical reactions. So how, then, did information translate into action?

The Genetic Code Hypothesis

Crick proposed that the sequence of bases in DNA might act as a code. His idea was that DNA was *only* an information-storage molecule. The instructions it contained would have to be read and then translated into proteins.

Crick offered Morse code as an analogy. Morse code is a message-transmission system using dots and dashes to represent the letters of the alphabet, and in that way it can convey all the complex information of human language. Crick proposed that different combinations of bases could specify the 20 amino acids, just as different combinations of dots and dashes specify the 26 letters of the alphabet. A particular stretch of DNA, then, could contain the information needed to produce the amino acid sequence of a particular polypeptide.

In code form, the tremendous quantity of information required to build and operate a cell could be stored compactly. This information could also be copied through complementary base pairing and transmitted efficiently from one generation to the next.

It soon became apparent, however, that the information encoded in the base sequence of DNA is not translated into the amino acid sequence of proteins directly. Instead, the link between DNA as information repository and proteins as cellular machines is indirect.

RNA as the Intermediary between Genes and Proteins

The first clue that the biological information in DNA must go through an intermediary in order to produce proteins came from knowledge of cell structure. In eukaryotic cells, DNA is enclosed within a membrane-bound organelle called the nucleus (see Chapter 7). But the cells' ribosomes, where protein synthesis takes place, are outside the nucleus, in the cytoplasm.

To make sense of this observation, François Jacob and Jacques Monod suggested that RNA molecules act as a link between genes and the protein-manufacturing centers. Jacob and Monod's hypothesis is illustrated in **FIGURE 16.3**. They predicted that short-lived molecules of RNA, which they called **messenger RNA,** or **mRNA** for short, carry information out of the nucleus from DNA to the site of protein synthesis. Messenger RNA is one of several distinct types of RNA in cells.

Follow-up research confirmed that the messenger RNA hypothesis is correct. One particularly important piece of evidence was the discovery of an enzyme that catalyzes the synthesis of RNA. This protein is called **RNA polymerase** because it polymerizes ribonucleotides into strands of RNA.

RNA polymerase synthesizes RNA molecules according to the information provided by the sequence of bases in a particular stretch of DNA. Unlike DNA polymerase, RNA polymerase

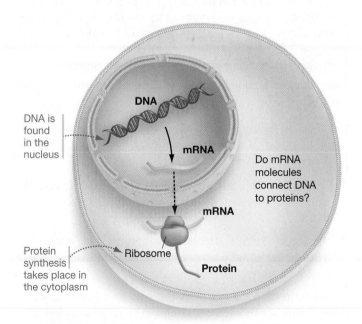

FIGURE 16.3 The Messenger RNA Hypothesis. In cells of eukaryotes such as plants, animals, and fungi, most DNA is found in the nucleus, but proteins are manufactured using ribosomes in the cytoplasm outside the nucleus. Biologists proposed that the information coded in DNA is carried from inside the nucleus out to the ribosomes by messenger RNA (mRNA).

does not require a primer to begin connecting ribonucleotides together to produce a strand of RNA.

To test the mRNA hypothesis, researchers created a reaction mix containing three critical elements: (1) the enzyme RNA polymerase; (2) ribonucleotides containing the bases adenine (A), uracil (U), guanine (G), and cytosine (C); and (3) strands of synthetic DNA that contained deoxyribonucleotides in which the only base was thymine (T).

After allowing the polymerization reaction to proceed, the biologists isolated RNA molecules that contained only the base adenine.

This result supported the hypothesis that RNA polymerase synthesizes RNA according to the rules of complementary base pairing (introduced in Chapter 4), because thymine pairs with adenine. Similar experiments showed that synthetic DNAs containing only cytosine result in the production of RNA molecules containing only guanine.

Dissecting the Central Dogma

Once the mRNA hypothesis was accepted, Francis Crick articulated what became known as the central dogma of molecular biology. The **central dogma** summarizes the flow of information in cells. It simply states that DNA codes for RNA, which codes for proteins:

$$\text{DNA} \longrightarrow \text{RNA} \longrightarrow \text{proteins}$$

Crick's simple statement encapsulates much of the research reviewed in this chapter and the preceding one. DNA is the hereditary material. Genes consist of specific stretches of DNA that code for products used in the cell. The sequence of bases in DNA specifies the sequence of bases in an RNA molecule, which specifies the sequence of amino acids in a protein. In this way, genes ultimately code for proteins.

Proteins are the workers of cells, functioning not only as enzymes but also as motors, structural elements, transporters, and molecular signals.

The Roles of Transcription and Translation

Biologists use specialized vocabulary to summarize the sequence of events captured in the central dogma.

1. DNA is transcribed to RNA by RNA polymerase. **Transcription** is the process of copying hereditary information in DNA to RNA.

2. Messenger RNA is translated to proteins in ribosomes. **Translation** is the process of using the information in nucleic acids to synthesize proteins.

The term transcription is appropriate. In everyday English, transcription simply means making a copy of information. The scientific use is similar because it conveys the idea that DNA acts as a permanent record—an information archive or blueprint. This permanent record is copied, during transcription, to produce the short-lived form called mRNA.

Translation is also an appropriate term. In everyday English, translation refers to converting information from one language to another. In biology, translation is the transfer of information from one type of molecule to another—from the "language" of nucleic acids to the "language" of proteins. Translation is also referred to simply as protein synthesis.

The following equation summarizes the relationship between transcription and translation as well as the relationships between DNA, RNA, and proteins:

Gene expression occurs via transcription and translation.

Linking Genotypes and Phenotypes

An organism's genotype is determined by the sequence of bases in its DNA, while its phenotype is a product of the proteins it produces.

To appreciate this point, consider that the proteins encoded by genes are what make the "stuff" of the cell and dictate which chemical reactions occur inside. For example, in populations of the oldfield mouse native to southeastern North America, individuals have a gene for a protein called the melanocortin receptor. Melanocortin is a hormone—an important type of molecular signal (discussed in Chapter 11)—that works through the melanocortin receptor to influence how much dark pigment is deposited in fur. An important aspect of a mouse's phenotype—its coat color—is determined in part by the DNA sequence at the gene for this receptor (**FIGURE 16.4a**).

Later work revealed that alleles of a gene differ in their DNA sequence. As a result, the proteins produced by different alleles of the gene may differ in their amino acid sequence. If the primary structures of proteins vary, their functions are likely to vary as well.

To drive this point home, look at the DNA sequence in the portion of the melanocortin receptor gene shown in **FIGURE 16.4b**, and compare it with the sequence in Figure 16.4a. The sequences differ—meaning that they are different alleles. Now look at the protein products of each allele, and note that one of the amino acids in the protein's primary structure differs—one allele specifies an arginine residue; the other specifies a cysteine residue.

At the protein level, the phenotypes associated with these alleles differ. The consequences for the mouse are striking: Melanocortin receptors that have arginine in this location deposit a large amount of pigment, but receptors that have cysteine in this location deposit small amounts of pigment. Whether a mouse is dark or light depends, largely, on a single base change in its DNA sequence. In this case, a tiny difference in genotype produces a large change in phenotype. The central dogma links genotypes to phenotypes.

Exceptions to the Central Dogma

The central dogma provided an important conceptual framework for the burgeoning field of molecular genetics and inspired a series of fundamental questions about how genes and cells work. But important modifications to the central dogma have occurred in the decades since Frances Crick first proposed it:

(a) Genetic information flows from DNA to RNA to proteins.

Mice with this DNA sequence have **dark** coats.

(b) Differences in genotype may cause differences in phenotype.

GENOTYPE

PHENOTYPE
Physical traits that are a product of the proteins produced.

Mice with this DNA sequence have **light** coats.

FIGURE 16.4 The Relationship between Genotype and Phenotype. The central dogma revealed the flow of information within the cell. The DNA sequences given in parts **(a)** and **(b)** are from different alleles (genotypes) that influence coat color (phenotypes) in oldfield mice. Forest-dwelling mice are dark, which camouflages them in their forested habitats. Beach-dwelling mice are light, which camouflages them in their sandy habitat.

• Many genes code for RNA molecules that do not function as mRNAs—they are not translated into proteins.

• In some cases, information flows from RNA back to DNA.

The discovery of a wide array of different RNA types ranks among the most profound advances in the past decade of biological science. Some RNAs form major parts of the ribosome, others help to form mRNA from a much longer precursor RNA (Chapter 17), and yet others regulate which genes are expressed (see Chapter 19). New types of RNA are still being discovered. For the genes coding for these types of RNA, information flow would be diagrammed as simply DNA → RNA.

In the early 1970s, the discovery of "reverse" information flow created the kind of excitement now being generated by the discovery of so many kinds of RNA. Some viruses, for example, have genes consisting of RNA. When some RNA viruses infect a cell, a specialized viral polymerase called **reverse transcriptase** synthesizes a DNA version of the RNA genes. In these viruses, information flows from RNA to DNA.

The human immunodeficiency virus (HIV), which causes AIDS, is an RNA virus that uses reverse transcriptase. Several of the most commonly prescribed drugs for AIDS patients fight the infection by poisoning the HIV reverse transcriptase. The drugs prevent viruses from replicating efficiently by disrupting reverse information flow.

The punch line? Crick's hypothesis is a central concept in biology, but cells, viruses, and researchers aren't dogmatic about it.

check your understanding

If you understand that . . .

• Genes code for proteins, but they do so indirectly.
• The sequence of bases in DNA is used to produce RNA, including messenger RNA (mRNA), via transcription. The sequence of bases in an RNA molecule is complementary to one of the DNA strands of a gene.
• Messenger RNAs are translated into proteins.
• Differences in DNA sequence can lead to differences in the amino acid sequence of proteins.

✔ **You should be able to . . .**

List the steps that link a change in the base sequence of a gene to a change in the phenotype of an organism.

Answers are available in Appendix A.

16.3 The Genetic Code

Once biologists understood the general pattern of information flow in the cell, the next challenge was to understand the final link between DNA and proteins. Exactly how does the sequence of bases in a strand of mRNA code for the sequence of amino acids in a protein?

If this question could be answered, biologists would have cracked the **genetic code**—the rules that specify the relationship between a sequence of nucleotides in DNA or RNA and the sequence of amino acids in a protein. Researchers from all over the world took up the challenge. A race was on.

How Long Is a Word in the Genetic Code?

The first step in cracking the genetic code was to determine how many bases make up a "word." In a sequence of mRNA, how long is a message that specifies one amino acid?

Based on some simple logic, George Gamow suggested that each code word contains three bases. His reasoning derived from the observation that 20 amino acids are commonly used in cells and from the hypothesis that each amino acid must be specified by a particular sequence of mRNA. **FIGURE 16.5** illustrates Gamow's reasoning:

- There are only four different bases in ribonucleotides (A, U, G, and C), so a one-base code could specify only four different amino acids.

- A two-base code could represent just 4 × 4, or 16, different amino acids.

- A three-base code could specify 4 × 4 × 4, or 64, different amino acids.

A three-base code provides more than enough words to code for all 20 amino acids. A three-base code is known as a **triplet code**.

Gamow's hypothesis suggested that the genetic code could be redundant. That is, more than one triplet of bases might specify the same amino acid. As a result, different three-base sequences in an mRNA—say, AAA and AAG—might code for the same amino acid—say, lysine.

The group of three bases that specifies a particular amino acid is called a **codon.** According to the triplet code hypothesis, many of the 64 codons that are possible might specify the same amino acids.

Work by Francis Crick and Sydney Brenner confirmed that codons are three bases long. Their experiments used chemicals that caused an occasional addition or deletion of a base in DNA. As predicted for a triplet code, a one-base addition or deletion in the base sequence led to a loss of function in the gene being studied. This is because a single addition or deletion mutation throws the sequence of codons, or the **reading frame,** out of register. To understand how a reading frame works, consider the sentence

"The fat cat ate the rat."

The reading frame of this sentence is a three-letter word and a space. If the fourth letter in this sentence—the *f* in *fat*—were deleted, the reading frame would transform the sentence into

"The atc ata tet her at."

This is gibberish.

When the reading frame in a DNA sequence is thrown out of register by the addition or deletion of a base, the composition of each codon changes just like the letters in each word of the example sentence above. The protein produced from the altered DNA sequence has a completely different sequence of amino acids. In terms of its normal function, this protein is gibberish.

Crick and Brenner were also able to produce DNA sequences that had deletions or additions of two base pairs or three base

FIGURE 16.5 How Many Bases Form a "Word" in the Genetic Code?

pairs. The only time functional proteins were produced was when three bases were added or removed. In the sentence

"The fat cat ate the rat."

the combination of removing one letter from each of the first three words might result in

"Tha tca ate the rat."

Just as the altered sentence still conveys some meaning, genes with three deletion mutations were able to produce a functional protein.

The researchers interpreted these results as strong evidence in favor of the triplet code hypothesis. Most other biologists agreed.

The confirmation of the triplet code launched an effort to determine which amino acid is specified by each of the 64 codons. Ultimately, it was successful.

How Did Researchers Crack the Code?

The initial advance in deciphering the genetic code came in 1961, when Marshall Nirenberg and Heinrich Matthaei developed a method for synthesizing RNAs of known sequence. They began by creating a long polymer of uracil-containing ribonucleotides. These synthetic RNAs were added to an in vitro system for synthesizing proteins. The researchers analyzed the resulting amino acid chain and determined that it was polyphenylalanine—a polymer consisting of the amino acid phenylalanine.

This result provided evidence that the RNA triplet UUU codes for the amino acid phenylalanine. By complementary base pairing, it was clear that the corresponding DNA sequence would be AAA. This initial work was followed by experiments using RNAs consisting of only A or C. RNAs with only AAAAA . . . produced

polypeptides composed of only lysine; poly-C RNAs (RNAs consisting of only CCCCC . . .) produced polypeptides composed entirely of proline.

Nirenberg and Philip Leder later devised a system for synthesizing specific codons. With these they performed a series of experiments in which they added each codon to a cell extract containing the 20 different amino acids, ribosomes, and other molecules required for protein synthesis. Recall that ribosomes are macromolecular machines that synthesize proteins (Chapter 7). Then the researchers determined which amino acid became bound to the ribosomes when a particular codon was present. For example, when the codon CAC was in the reaction mix, the amino acid histidine would bind to the ribosomes. This result indicated that CAC codes for histidine.

These ribosome-binding experiments allowed Nirenberg and Leder to determine which of the 64 codons coded for each of the 20 amino acids.

Researchers also discovered that certain codons are punctuation marks signaling "start of message" or "end of message." These codons indicate that protein synthesis should start at a given codon or that the protein chain is complete.

- There is one **start codon** (AUG), which signals that protein synthesis should begin at that point on the mRNA molecule. The start codon specifies the amino acid methionine.

- There are three **stop codons,** also called termination codons (UAA, UAG, and UGA). The stop codons signal that the protein is complete, they do not code for any amino acid, and they end translation.

The complete genetic code is given in **FIGURE 16.6.** Deciphering it was a tremendous achievement, requiring more than five years of work by several teams of researchers.

SECOND BASE

	U	C	A	G	
U	UUU UUC — Phenylalanine (Phe) UUA UUG — Leucine (Leu)	UCU UCC UCA UCG — Serine (Ser)	UAU UAC — Tyrosine (Tyr) UAA — Stop codon UAG — Stop codon	UGU UGC — Cysteine (Cys) UGA — Stop codon UGG — Tryptophan (Trp)	U C A G
C	CUU CUC CUA CUG — Leucine (Leu)	CCU CCC CCA CCG — Proline (Pro)	CAU CAC — Histidine (His) CAA CAG — Glutamine (Gln)	CGU CGC CGA CGG — Arginine (Arg)	U C A G
A	AUU AUC AUA — Isoleucine (Ile) AUG — Methionine (Met) Start codon	ACU ACC ACA ACG — Threonine (Thr)	AAU AAC — Asparagine (Asn) AAA AAG — Lysine (Lys)	AGU AGC — Serine (Ser) AGA AGG — Arginine (Arg)	U C A G
G	GUU GUC GUA GUG — Valine (Val)	GCU GCC GCA GCG — Alanine (Ala)	GAU GAC — Aspartic acid (Asp) GAA GAG — Glutamic acid (Glu)	GGU GGC GGA GGG — Glycine (Gly)	U C A G

FIRST BASE — THIRD BASE

FIGURE 16.6 The Genetic Code. To read a codon in mRNA, locate its first base in the red band on the left; then move rightward to the box under the codon's second base in the blue band along the top. Finally, locate the codon's third base in the green band on the right side to learn the amino acid. By convention, codons are always written in the $5' \rightarrow 3'$ direction.

(a) Using the genetic code to predict an amino acid sequence

The bottom strand of the DNA sequence...

5′ | A T G | G C C | A A T | G A C | T T T | C A A | T A A | 3′

3′ | T A C | C G G | T T A | C T G | A A A | G T T | A T T | 5′

...would be transcribed as

5′ | A U G | G C C | A A U | G A C | U U U | C A A | U A A | 3′

...and translated as

| Met (start) | Ala | Asn | Asp | Phe | Gln | (stop) |

(b) Your turn—a chance to practice using the genetic code

The bottom strand of the DNA sequence...

5′ | A T G | C T G | G A G | G G G | G T T | A G A | C A T | 3′

3′ | T A C | G A C | C T C | C C C | C A A | T C T | G T A | 5′

...would be transcribed as

5′ | | | | | | | | 3′

...and translated as

| | | | | | | |

Remember that RNA contains U (uracil) instead of T (thymine), and that U forms a complementary base pair with A (adenine)

FIGURE 16.7 Using the Genetic Code.

✔**EXERCISE** Fill in the mRNA and amino acid sequences in part (b).

Analyzing the Code Once biologists had cracked the genetic code, they realized that it has a series of important properties.

- *The code is redundant.* All amino acids except methionine and tryptophan are coded by more than one codon.

- *The code is unambiguous.* A single codon never codes for more than one amino acid.

- *The code is non-overlapping.* Once the ribosome locks onto the first codon, it then reads each separate codon one after another.

- *The code is nearly universal.* With a few minor exceptions, all codons specify the same amino acids in all organisms.

- *The code is conservative.* When several codons specify the same amino acid, the first two bases in those codons are almost always identical.

The last point is subtle, but important. Here's the key: If a mutation in DNA or an error in transcription or translation affects the third position in a codon, it is less likely to change the amino acid in the final protein. This feature makes individuals less vulnerable to small, random changes or errors in their DNA sequences. Compared with randomly generated codes, the existing genetic code minimizes the phenotypic effects of small changes in DNA sequence and errors during translation. Stated another way, the genetic code does not represent a random assemblage of bases, like letters drawn from a hat. It has been honed by natural selection and is remarkably efficient.

Using the Code Using the genetic code and the central dogma, biologists can:

1. Predict the codons and amino acid sequence encoded by a particular DNA sequence (see **FIGURE 16.7a**).

2. Determine the set of mRNA and DNA sequences that would code for a particular sequence of amino acids.

Why is a *set* of mRNA or DNA sequences predicted from a given amino acid sequence? The answer lies in the code's redundancy. If a polypeptide contains phenylalanine, you don't know if the codon responsible is UUU or UUC.

✔If you understand how to read the genetic code, you should be able to do the following tasks: (1) Identify the codons in Figure 16.4 and decide whether they are translated correctly. (2) Complete the exercise for **FIGURE 16.7b**. (3) Write an mRNA that codes for the amino acid sequence Ala-Asn-Asp-Phe-Gln yet is different from the one given in Figure 16.7a. Indicate the mRNA's 5′ → 3′ polarity. Then write the double-stranded DNA that corresponds to this mRNA. Indicate the 5′ → 3′ polarity of both DNA strands.

Once they understood the central dogma and genetic code, biologists were able to explore and eventually understand the molecular basis of mutation. How do novel traits—such as dwarfing in garden peas and white eye color in fruit flies—come to be?

check your understanding

If you understand that . . .

- The sequence of bases in mRNA constitutes a code. Particular combinations of three bases specify specific amino acids in the protein encoded by the gene.

- The genetic code is redundant. It consists of 64 combinations of bases, but only 20 amino acids plus start and stop "punctuation marks" need to be specified.

✔ **You should be able to . . .**

Consider the consequences of a mutation in the DNA template sequence ATA to one of the following sequences: GTA, TTA, or GCA.

1. In each case, specify the resulting change in the mRNA codon.

2. In each case, describe the effect on the resulting protein.

Answers are available in Appendix A.

16.4 How Can Mutation Modify Genes and Chromosomes?

This chapter has explored how the information archived in DNA is put into action in the form of working RNAs and proteins. Now the questions are, what happens if the information in DNA changes? In what ways can this information be changed? What are the consequences for the cell and organism?

A **mutation** is any permanent change in an organism's DNA. It is a modification in a cell's information archive—a change in its genotype. Mutations create new alleles.

Mutations can alter DNA sequences that range in size from a single base pair in DNA to whole sets of chromosomes. Let's look at these different types of mutation and their consequences.

Point Mutation

FIGURE 16.8 shows how a common type of mutation occurs. If a mistake is made during DNA synthesis or DNA repair, a change in the sequence of bases in DNA results. A single-base change such as this is called a **point mutation.**

What happens when point mutations occur in regions of DNA that code for proteins? To answer this question, look back at Figure 16.4 and recall that a change in a single base in DNA is associated with a difference in coat color in populations of oldfield mice. The DNA sequence in Figure 16.4a is found in dark-colored mice that live in forest habitats; the sequence in Figure 16.4b is found in light-colored mice that live in beach habitats.

Because beach-dwelling populations are evolutionarily younger than the nearby forest-dwelling populations, researchers hypothesize the following sequence of events:

1. Forest mice colonized beach habitats.

2. Either before or after the colonization event, a random point mutation occurred in a mouse that altered the melanocortin receptor gene and resulted in some offspring with light coats.

3. Light-colored mice are camouflaged in beach habitats; in sandy environments, they suffer lower predation than dark-colored mice.

4. Over time, the allele created by the point mutation increased in frequency in beach-dwelling populations.

Point mutations that cause these types of changes in the amino acid sequence of proteins are called **missense mutations.** But note that if the same G-to-A change had occurred in the third position of the same DNA codon, instead of the first position, there would have been no change in the protein produced. The mRNA codons CGC and CGU both code for arginine. A point mutation that does not change the amino acid sequence of the gene product is called a **silent mutation.**

Some point mutations disrupt major portions of a protein. Recall that a single addition or deletion mutation throws the sequence of codons out of register and alters the meaning of all subsequent codons. Such mutations are called **frameshift mutations.** Another type of point mutation with a large effect is a **nonsense mutation.** Nonsense mutations occur when a codon that specifies an amino acid is changed by mutation to one that specifies a stop codon. This causes early termination of the polypeptide chain and often results in a non-functional protein.

In terms of the impact on organisms, biologists divide mutations into three categories:

1. **Beneficial** Some mutations increase the fitness of the organism—meaning, its ability to survive and reproduce—in certain environments. The G-to-A mutation is beneficial in beach habitats because it camouflages mice.

2. **Neutral** If a mutation has no effect on fitness, it is termed neutral. Silent mutations are usually neutral.

3. **Deleterious** Because organisms tend to be well adapted to their current habitat, and because mutations are random changes in the genotype, many mutations lower fitness. These mutations are termed harmful or deleterious. The G-to-A mutation would be deleterious in the forest habitat.

Recent studies indicate that the majority of point mutations are slightly deleterious or neutral. **TABLE 16.1** (see page 314) summarizes the types of point mutations that occur in protein-coding sequences of a gene and reviews their consequences for the amino acid sequences of proteins and for fitness.

Point mutations can and do occur in DNA sequences that do not code for proteins. These mutations, however, are not referred to as missense, silent, frameshift, or nonsense mutations

FIGURE 16.8 Unrepaired Mistakes in DNA Synthesis Lead to Point Mutations.

Name	Definition	Example	Consequence
		Original DNA sequence — TAT TGG CTA GTA CAT	
		Original mRNA transcript — UAU UGG CUA GUA CAU	
		Tyr—Trp—Leu—Val—His — Original polypeptide	
Silent	Change in nucleotide sequence that does not change the amino acid specified by a codon	TAC TGG CTA GTA CAT UAC UGG CUA GUA CAU Tyr—Trp—Leu—Val—His	No change in phenotype; neutral with respect to fitness
Missense	Change in nucleotide sequence that changes the amino acid specified by codon	TAT TGT CTA GTA CAT UAU UGU CUA GUA CAU Tyr—Cys—Leu—Val—His	Change in primary structure of protein; may be beneficial, neutral, or deleterious
Nonsense	Change in nucleotide sequence that results in an early stop codon	TAT TGA CTA GTA CAT UAU UGA CUA GUA CAU Tyr—STOP	Leads to mRNA breakdown or a shortened polypeptide; usually deleterious
Frameshift	Addition or deletion of a nucleotide	TAT TCG GCT AGT ACA T UAU UCG GCU AGU ACA U Tyr—Ser—Ala—Ser—Thr	Reading frame is shifted, altering the meaning of all subsequent codons; almost always deleterious

because these terms apply only to mutations that can change the protein-coding potential of a gene. If point mutations alter DNA sequences that are important for *gene expression*, they can have important effects on phenotype even though they do not change the amino acid sequence of a protein.

Chromosome Mutations

Besides documenting various types of point mutations, biologists study larger-scale mutations that change chromosomes. You might recall, for example, that polyploidy is an increase in the number of each type of chromosome, while aneuploidy is the addition or deletion of individual chromosomes (Chapter 13).

Polyploidy, aneuploidy, and other changes in chromosome number result from chance mistakes in moving chromosomes into daughter cells during meiosis or mitosis. Polyploidy and aneuploidy are forms of mutation that don't change DNA sequences, but alter the number of chromosome copies.

In addition to changes in overall chromosome number, the composition of individual chromosomes can change in important ways. For example, chromosome segments can become detached when accidental breaks in chromosomes occur. The segments may be flipped and rejoined—a phenomenon known as a chromosome **inversion**—or become attached to a different chromosome, an event called chromosome **translocation.** When a segment of chromosome is lost, this is a **deletion,** and when additional copies of a segment are present, this is a **duplication.**

Like point mutations, chromosome mutations can be beneficial, neutral, or deleterious. For example, more than 200 different inverted sections of chromosomes were found in comparisons

of the DNA from eight phenotypically normal people. These mutations appear to be neutral. Not all chromosome mutations are so harmless, however. Chromosomes of cancer cells exhibit deleterious chromosome mutations that include aneuploidy, inversions, translocations, deletions, and duplications. **FIGURE 16.9**

FIGURE 16.9 Chromosome-Level Mutations. A spectral karyotype of a breast cancer cell from a female that shows chromosome rearrangements and aneuploidy typical of cancer. In a normal spectral karyotype, each chromosome is stained a single, solid color, which varies for each chromosome pair.

✔**EXERCISE** Compare this karyotype to the one shown in the chapter-opening image. Remember that females normally have two X chromosomes and males normally have one X chromosome. Which chromosomes show evidence of aneuploidy? Which chromosomes show evidence of rearrangements?

drives this point home by showing the **karyotype**—the complete set of chromosomes in a cell—of a cancerous human cell.

To summarize, point mutations and chromosome mutations are random changes in DNA that can produce new genes, new alleles, and new traits. At the level of individuals, mutations can cause disease or death or lead to increases in fitness. At the level of populations, mutations furnish the heritable variation that Mendel and Morgan analyzed and that makes evolution possible. The central role of mutation in evolution is explored in depth in Unit 5.

CHAPTER 16 REVIEW

For media, go to MasteringBiology

If you understand . . .

16.1 What Do Genes Do?

- Experiments with mutants of the bread mold *N. crassa* led to the one-gene, one-enzyme hypothesis.

- The original one-gene, one-enzyme hypothesis has been broadened to account for genes that code for proteins other than enzymes and for genes that have RNA as a final product.

- ✓ You should be able to use Figure 16.1 to explain what compounds could be added to the medium to allow the growth of a mutant unable to synthesize citrulline because of a mutation in the gene for enzyme 1.

16.2 The Central Dogma of Molecular Biology

- DNA is transcribed to messenger RNA (mRNA) by RNA polymerase, and then mRNA is translated to proteins by ribosomes. In this way, genetic information is converted from DNA to RNA to protein.

- The flow of information from DNA to RNA to protein is called the central dogma of molecular biology.

- Many RNAs do not code for proteins. Instead, these RNAs perform other important functions in the cell.

- Reverse transcriptase reverses information flow by copying RNA into DNA. Some viruses with an RNA genome use this enzyme during their replication.

- ✓ You should be able to explain how a compound that blocks RNA synthesis will affect protein synthesis.

16.3 The Genetic Code

- Each amino acid in a protein is specified by a codon—a group of three bases in mRNA.

- By synthesizing RNAs of known base composition and then observing the results of translation, researchers were able to decipher the genetic code.

- The genetic code is redundant—meaning that most of the 20 amino acids are specified by more than one codon.

- Certain codons signal where translation starts and stops.

- ✓ Using the genetic code shown in Figure 16.6, you should be able to write all possible mRNA sequences that would produce the following sequence of amino acids: Met-Trp-Lys-Gln.

16.4 How Can Mutation Modify Genes and Chromosomes?

- Mutations are random, heritable changes in DNA that range from changes in a single base to changes in the structure and number of chromosomes.

- Point mutations in protein-coding regions may have no effect on the protein (silent mutation), may change a single amino acid (missense mutation), may shorten the protein (nonsense mutation), or may shift the reading frame and cause many amino acids to be wrong (frameshift mutation).

- Mutations can have beneficial, neutral, or harmful effects on organisms.

- ✓ You should be able to explain how redundancy in the genetic code allows for silent mutations and whether a silent mutation is likely to be beneficial, neutral, or harmful.

MB **MasteringBiology**

1. MasteringBiology Assignments

Tutorials and Activities Genetic Code; One-Gene One-Enzyme Hypothesis; Overview of Protein Synthesis; Role of the Nucleus and Ribosomes in Protein Synthesis; Triplet Nature of the Genetic Code

Questions Reading Quizzes, Blue-Thread Questions, Test Bank

2. eText Read your book online, search, take notes, highlight text, and more.

3. The Study Area Practice Test, Cumulative Test, BioFlix® 3-D Animations, Videos, Activities, Audio Glossary, Word Study Tools, Art

You should be able to . . .

✔ TEST YOUR KNOWLEDGE

Answers are available in Appendix A

1. What does the one-gene, one-enzyme hypothesis state?
   a. Genes are composed of stretches of DNA.
   b. Genes are made of protein.
   c. Genes code for ribozymes.
   d. A single gene codes for a single protein.

2. Which of the following is an important exception to the central dogma of molecular biology?
   a. Many genes code for RNAs that function directly in the cell.
   b. DNA is the repository of genetic information in all cells.
   c. Messenger RNA is a short-lived "information carrier."
   d. Proteins are responsible for most aspects of the phenotype.

3. DNA's primary structure is made up of just four different bases, and its secondary structure is regular and highly stable. How can a molecule with these characteristics hold all the information required to build and maintain a cell?
   a. The information is first transcribed, then translated.
   b. The messenger RNA produced from DNA has much more complex secondary structures, allowing mRNA to hold much more information.
   c. A protein coded for in DNA has much more complex primary and secondary structures, allowing it to hold much more information.
   d. The information in DNA is in a code form that is based on the sequence of bases.

4. Why did researchers suspect that DNA does not code for proteins directly?

5. Which of the following describes an important experimental strategy in deciphering the genetic code?
   a. comparing the amino acid sequences of proteins with the base sequence of their genes
   b. analyzing the sequence of RNAs produced from known DNA sequences
   c. analyzing mutants that changed the code
   d. examining the polypeptides produced when RNAs of known sequence were translated

6. What is a stop codon?

✔ TEST YOUR UNDERSTANDING

Answers are available in Appendix A

7. Explain why Morse code is an appropriate analogy for the genetic code.

8. Draw a hypothetical metabolic pathway in *Neurospora crassa* composed of five substrates, five enzymes, and a product called Biological Sciazine. Number the substrates 1–5, and label the enzymes A–E, in order. (For instance, enzyme A catalyzes the reaction between substrates 1 and 2.)
   • Suppose a mutation made the gene for enzyme C nonfunctional. What molecule would accumulate in the affected cells?
   • Suppose a mutant strain can survive if substrate 5 is added to the growth medium but it cannot grow if substrates 1, 2, 3, or 4 are added. Which enzyme in the pathway is affected in this mutant?

9. How did experiments with *Neurospora crassa* mutants support the one-gene, one-enzyme hypothesis?

10. Why does a single-base deletion mutation within a protein-coding sequence usually have a more severe effect than a deletion of three adjacent bases?
    a. because single-base deletions prevent the ribosome from binding to mRNA
    b. because single-base deletions stabilize mRNA
    c. because single-base deletions change the reading frame
    d. because single-base deletions alter the meaning of individual codons

11. When researchers discovered that a combination of three deletion mutations or three addition mutations would restore the function of a gene, most biologists were convinced that the genetic code was read in triplets. Explain the logic behind this conclusion.

12. Explain why all point mutations change the genotype, but why only some point mutations change the phenotype.

✔ TEST YOUR PROBLEM-SOLVING SKILLS

Answers are available in Appendix A

13. Recall that DNA and RNA are synthesized only in the 5′ → 3′ direction and that DNA and RNA sequences are written in the 5′ → 3′ direction, unless otherwise noted. Consider the following DNA sequence:

    5′ TTGAAATGCCCGTTTGGAGATCGGGTTACAGCTAGTCAAAG 3′

    3′ AACTTTACGGGCAAACCTCTAGCCCAATGTCGATCAGTTTC 5′

    • Identify bases in the bottom strand that can be transcribed into start and stop codons.
    • Write the mRNA sequence that would be transcribed between start and stop codons if the bottom strand served as the template for RNA polymerase.
    • Write the amino acid sequence that would be translated from the mRNA sequence you just wrote.

14. What problems would arise if the genetic code contained only 22 codons—one for each amino acid, a start signal, and a stop signal?

15. Scientists say that a phenomenon is a "black box" if they can describe it and study its effects but don't know the underlying mechanism that causes it. In what sense was genetics—meaning the transmission of heritable traits—a black box before the central dogma of molecular biology was understood?

16. **QUANTITATIVE** One of the possibilities that researchers interested in the genetic code considered was that the code was overlapping, meaning that a single base could be part of up to three codons. How many amino acids would be encoded in the sequence 5′ AUGUUACGGAAU 3′ by a non-overlapping and maximally overlapping code?
    a. 4 (non-overlapping) and 16 (overlapping)
    b. 4 and 12
    c. 4 and 10
    d. 12 and 4

17 Transcription, RNA Processing, and Translation

In this chapter you will learn how

Information in genes directs the synthesis of RNAs and proteins

by looking at

How DNA is transcribed into RNA **17.1**

and

How eukaryotes process RNA **17.2**

then asking

How is messenger RNA translated into proteins? **17.3**

and looking closer at

The structure and function of transfer RNA **17.4**

The structure and function of ribosomes **17.5**

Extensive transcription is occurring along this gene within a frog cell. The horizontal strand in the middle of this micrograph is DNA; the strands that have been colored yellow and red, and that are coming off on either side, are RNA molecules.

This chapter is part of the Big Picture. See how on pages 366–367.

Proteins are the stuff of life. They give shape to our cells, control the chemical reactions that go on inside them, and regulate how materials move into, out of, and through them. Some of these proteins may not be produced at all in some types of cells; others may be present in quantities ranging from millions of copies to fewer than a dozen.

A cell builds the proteins it needs from instructions encoded in its DNA. The central dogma of molecular biology states that the flow of information in cells is from DNA to mRNA to protein (Chapter 16). Once this pattern of information flow had been established, biologists puzzled over how cells actually accomplish the two major steps of the central dogma: transcription and

✔ When you see this checkmark, stop and test yourself. Answers are available in Appendix A.

translation. Specifically, how does RNA polymerase know where to start transcribing a gene, and where to end? And once an RNA message is produced, how is the linear sequence of ribonucleotides translated into the linear sequence of amino acids in a protein?

This chapter delves into the molecular mechanisms of gene expression—the blood and guts of the central dogma. It starts with the monomers that build RNA and ends with a protein.

17.1 An Overview of Transcription

The first step in converting genetic information into proteins is to synthesize an RNA version of the instructions archived in DNA. Enzymes called **RNA polymerases** are responsible for synthesizing mRNA (see Chapter 15).

FIGURE 17.1 shows how the polymerization reaction occurs. Note the incoming monomer—a ribonucleoside triphosphate, or NTP—at the far right of the diagram. NTPs are like dNTPs (introduced in Chapter 15), except that they have a hydroxyl (—OH) group on the 2′ carbon. This makes the sugar in an NTP a ribose instead of the deoxyribose sugar of DNA.

Once an NTP that matches a base on the DNA template is in place, RNA polymerase cleaves off two phosphates and catalyzes the formation of a phosphodiester linkage between the 3′ end of the growing mRNA chain and the new ribonucleoside monophosphate. As this 5′ → 3′ matching-and-catalysis process continues, an RNA that is complementary to the gene is synthesized. This is transcription.

Notice that only one of the two DNA strands is used as a template and transcribed, or "read," by RNA polymerase.

- The strand that is read by the enzyme is the **template strand.**
- The other strand is called the **non-template strand** or **coding strand.** Coding strand is an appropriate name, because, with one exception, its sequence matches the sequence of the RNA that is transcribed from the template strand and codes for a polypeptide.

The coding strand and the RNA don't match exactly, because RNA has uracil (U) rather than the thymine (T) found in the coding strand. Likewise, an adenine (A) in the DNA template strand specifies a U in the complementary RNA strand.

Like DNA polymerases (see Chapter 15), an RNA polymerase performs a template-directed synthesis in the 5′ → 3′ direction. But unlike DNA polymerases, RNA polymerases do not require a primer to begin transcription.

Bacteria have a single RNA polymerase. In contrast, eukaryotes, have at least three distinct types. Let's first take a look at general principles of transcription using bacteria as an example and then examine things that differ in eukaryotes.

Initiation: How Does Transcription Begin in Bacteria?

How does RNA polymerase know where and in which direction to start transcription on the DNA template? The answer to this question defined what biologists call the **initiation** phase of transcription.

FIGURE 17.1 Transcription Is the Synthesis of RNA from a DNA Template. The reaction catalyzed by RNA polymerase (not shown) results in the formation of a phosphodiester linkage between ribonucleotides. RNA polymerase produces an RNA strand whose sequence is complementary to the bases in the DNA template.

✔**QUESTION** In which direction is RNA synthesized, 5′ → 3′ or 3′ → 5′? In which direction is the DNA template "read"?

Soon after the discovery of bacterial RNA polymerase, researchers found that the enzyme cannot initiate transcription on its own. Instead, a detachable protein subunit called **sigma** must bind to the polymerase before transcription can begin.

Bacterial RNA polymerase and sigma form a **holoenzyme** (literally, "whole enzyme"; **FIGURE 17.2a**). A holoenzyme consists of a **core enzyme** (RNA polymerase, in this case), which contains the active site for catalysis, and other required proteins (such as sigma).

What does sigma do? When researchers mixed the polymerase and DNA together, they found that the core enzyme could bind to any sequence of DNA. When sigma was added to this mixture, the holoenzyme formed and bound only to specific sections of DNA. These binding sites were named **promoters,** because they are sections of DNA that promote the start of transcription.

Most bacteria have alternative sigma proteins that bind to promoters with slightly different DNA base sequences, and may activate a group of genes in response to environmental change. For example, one type of sigma initiates the transcription of genes that help the cell cope with high temperatures. Controlling which sigma proteins are used is one of the ways that bacterial cells regulate which groups of genes are expressed.

The discovery of promoters suggested that sigma was responsible for guiding RNA polymerase to specific locations where transcription should begin. What is the nature of these specific locations? What do promoters look like, and what do they do?

Bacterial Promoters David Pribnow offered an initial answer to these questions in the mid-1970s. When Pribnow analyzed the base sequence of promoters from various bacteria and from viruses that infect bacteria, he found that the promoters were 40–50 base pairs long and had a particular section in common: a series of bases identical or similar to TATAAT. This six-base-pair sequence is now known as the −10 box, because it is centered about 10 bases from the point where bacterial RNA polymerase starts transcription (**FIGURE 17.2b**).

DNA that is located in the direction RNA polymerase moves during transcription is said to be **downstream** from the point of reference; DNA located in the opposite direction is said to be **upstream**. Thus, the −10 box is centered about 10 bases upstream from the transcription start site. The place where transcription begins is called the +1 site.

Soon after the discovery of the −10 box, researchers recognized that the sequence TTGACA also occurred in promoters and was about 35 bases upstream from the +1 site. This second key sequence is called the −35 box. Although all bacterial promoters have a −10 box and a −35 box, the sequences within the promoter but outside these boxes vary.

Events inside the Holoenzyme In bacteria, transcription begins when sigma, as part of the holoenzyme complex, binds to the −35 and −10 boxes. Sigma, and not RNA polymerase, makes the initial contact with DNA of the promoter. Sigma's binding to a promoter determines where and in which direction RNA polymerase will start synthesizing RNA.

Once the holoenzyme is bound to a promoter for a bacterial gene, the DNA helix is opened by RNA polymerase, creating two

(a) RNA polymerase and sigma form a holoenzyme.

(b) Sigma recognizes and binds to the promoter.

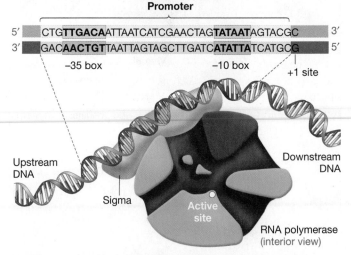

FIGURE 17.2 Sigma Is the Promoter-Recognizing Subunit of Bacterial RNA Polymerase Holoenzyme. (a) A space-filling model of bacterial RNA polymerase holoenzyme. **(b)** A cartoon of bacterial RNA polymerase, showing that sigma binds to the −35 box and −10 box of the promoter.

separated strands of DNA as shown in **FIGURE 17.3** (see page 320), steps 1 and 2. As step 2 shows, the template strand is threaded through a channel that leads to the active site inside RNA polymerase. Ribonucleoside triphosphates (NTPs)—the RNA building blocks—enter a channel in the enzyme and diffuse to the active site.

When an incoming NTP pairs with a complementary base on the template strand of DNA, RNA polymerization begins. The reaction catalyzed by RNA polymerase is exergonic and spontaneous because NTPs have significant potential energy, owing to their three phosphate groups. As step 3 of Figure 17.3 shows, the initiation phase of transcription is complete as RNA polymerase extends the mRNA from the +1 site.

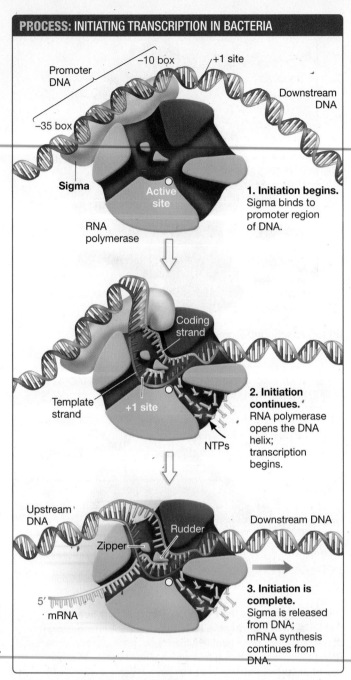

Promoter DNA

−10 box

+1 site

Downstream DNA

−35 box

Sigma

Active site

RNA polymerase

1. Initiation begins. Sigma binds to promoter region of DNA.

Coding strand

Template strand

+1 site

NTPs

2. Initiation continues. RNA polymerase opens the DNA helix; transcription begins.

Upstream DNA

Zipper

Rudder

Downstream DNA

5′

mRNA

3. Initiation is complete. Sigma is released from DNA; mRNA synthesis continues from DNA.

FIGURE 17.3 Sigma Orients the DNA Template inside RNA Polymerase. Sigma binds to the promoter, and RNA polymerase opens the DNA helix and threads the template strand through the active site.

Elongation and Termination

Once RNA polymerase begins moving along the DNA template synthesizing RNA, the **elongation** phase of transcription is under way. RNA polymerase is a macromolecular machine with different parts. In the interior of the enzyme, a group of amino acids forms a rudder to help steer the template and non-template strands through channels inside the enzyme (see Figure 17.3, step 3). Meanwhile, the enzyme's active site catalyzes the addition

of nucleotides to the 3′ end of the growing RNA molecule at the rate of about 50 nucleotides per second. A group of projecting amino acids forms a region called the zipper to help separate the newly synthesized RNA from the DNA template.

During the elongation phase of transcription, all the prominent channels and grooves in the enzyme are filled (Figure 17.3, step 3). Double-stranded DNA goes into and out of one groove, ribonucleoside triphosphates enter another, and the growing RNA strand exits to the rear. The enzyme's structure is critical for its function.

Termination ends transcription. In bacteria, transcription stops when RNA polymerase transcribes a DNA sequence that functions as a transcription-termination signal.

The bases that make up the termination signal in bacteria are transcribed into a stretch of RNA with an important property: As soon as it is synthesized, this portion of the RNA folds back on itself and forms a short double helix that is held together by complementary base pairing (**FIGURE 17.4**). Recall that this type of RNA secondary structure is called a hairpin (Chapter 4). The hairpin structure disrupts the interaction between RNA polymerase and the RNA transcript, resulting in the physical separation of the enzyme and its product.

Transcription in Eukaryotes

Fundamental features of transcription are the same in bacteria and eukaryotes. In fact, these similarities provide compelling evidence for a common ancestor of all cells. There are, however, some differences that are worth noting:

- Eukaryotes have three polymerases—RNA polymerase I, II, and III—that are often referred to as pol I, pol II, and pol III. Each polymerase transcribes only certain types of RNA in eukaryotes. RNA pol II is the only polymerase that transcribes protein-coding genes.

- Promoters in eukaryotic DNA are more diverse than bacterial promoters. Most eukaryotic promoters include a sequence called the **TATA box,** centered about 30 base pairs upstream of the transcription start site, and other important sequences that vary more widely.

- Instead of using a sigma protein, eukaryotic RNA polymerases recognize promoters using a group of proteins called **basal transcription factors.** Basal transcription factors assemble at the promoter, and RNA polymerase follows. (This idea, as well as the extensive use of other types of transcription factors to regulate transcription, will be covered in Chapter 19.)

- Termination of eukaryotic protein-coding genes involves a short sequence called the polyadenylation signal or **poly(A) signal.** Soon after the signal is transcribed, the RNA is cut by an enzyme downstream of the poly(A) signal as the polymerase continues to transcribe the DNA template. Eventually RNA polymerase falls off the DNA template and terminates transcription. Bacteria end transcription at a distinct site for each gene, but in eukaryotes, transcription ends variable distances from the poly(A) signal.

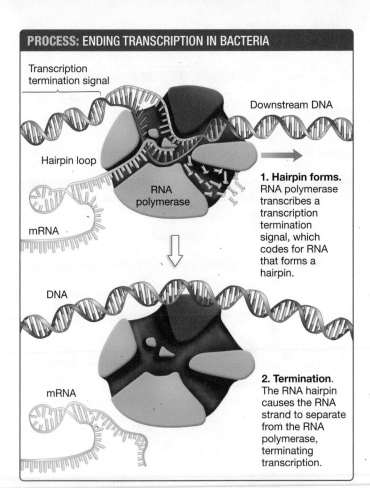

PROCESS: ENDING TRANSCRIPTION IN BACTERIA

Transcription termination signal

Downstream DNA

Hairpin loop

RNA polymerase

mRNA

1. Hairpin forms. RNA polymerase transcribes a transcription termination signal, which codes for RNA that forms a hairpin.

DNA

mRNA

2. Termination. The RNA hairpin causes the RNA strand to separate from the RNA polymerase, terminating transcription.

FIGURE 17.4 Transcription Terminates When an RNA Hairpin Forms.

check your understanding

If you understand that . . .

- Transcription initiation depends on interactions between proteins associated with RNA polymerase and a promoter sequence in DNA.
- In bacteria, sigma protein binds to RNA polymerase and contacts the promoter. In eukaryotes, basal transcription factors bind to the promoter and recruit RNA polymerase.
- During transcription elongation, ribonucleoside triphosphates are the substrate for a polymerization reaction catalyzed by RNA polymerase. The enzyme adds ribonucleotides that are complementary to the template strand in DNA.
- Transcription ends in bacteria when a termination signal at the end of the gene is transcribed, leading to the disassociation of RNA polymerase and the DNA template.

✔ **You should be able to . . .**

1. Predict how a mutation in a bacterial cell that deletes three nucleotides 10 nucleotides upstream from the transcription start site is likely to affect transcription.
2. Explain why ribonucleoside triphosphates, rather than ribonucleoside monophosphates, are the monomers required for RNA synthesis.

Answers are available in Appendix A.

17.2 RNA Processing in Eukaryotes

The molecular machinery required for transcription is much more complex in eukaryotes than in bacteria. But these differences are minor when compared with what happens to the eukaryotic RNA after transcription. In bacteria, when transcription terminates, the result is a mature mRNA that's ready to be translated into a protein. In fact, translation often begins while the mRNA is still being transcribed.

The fate of the transcript in eukaryotes is more complicated. When eukaryotic genes of any type are transcribed, the initial product is termed a **primary transcript.** This RNA must undergo multistep processing before it is functional. For protein-coding genes, the primary transcript is called a **pre-mRNA.**

The processing of primary transcripts has important consequences for gene expression in eukaryotes. Let's delve in to see how and why.

The Startling Discovery of Split Eukaryotic Genes

Eukaryotic genes do not consist of one continuous DNA sequence that codes for a product, as do bacterial genes. Instead, the regions in a eukaryotic gene that code for proteins are intermittently interrupted by stretches of hundreds or even thousands of intervening bases.

Although these intervening bases are part of the gene, they do not code for a product. To make a functional RNA, eukaryotic cells must dispose of certain sequences inside the primary transcript and then combine the separated sections into an integrated whole.

What sort of data would provoke such a startling claim? The first evidence came from work that Phillip Sharp and colleagues carried out in the late 1970s to determine the location of genes within the DNA of a virus that infects mammalian cells. Viruses are often used as tools to provide insights into fundamental processes of the cells they infect.

They began one of their experiments by heating the virus' DNA sufficiently to break the hydrogen bonds between complementary bases. This treatment separated the two strands. The single-stranded DNA was then incubated with the mRNA encoded by the virus. The team's intention was to promote base pairing between the mRNA and the single-stranded DNA.

The researchers expected that the mRNA would form base pairs with the DNA sequence that acted as the template for its synthesis—that the mRNA and DNA would match up exactly.

(a) Micrograph of DNA-RNA hybrid   **(b)** Interpretation of micrograph

Single-stranded DNA only

Single-stranded DNA base paired with mRNA

FIGURE 17.5 The Discovery of Introns. The loops in the micrograph and drawing represent regions of DNA that are transcribed but are not found in the final mRNA. These regions are introns.

✔ **QUESTION** If the noncoding regions of the gene did not exist, what would the micrograph in part (a) look like?

But when the team examined the DNA–RNA hybrid molecules using an electron microscope, they observed the structure shown in **FIGURE 17.5a**. Instead of matching up exactly, parts of the DNA formed loops.

What was going on? As **FIGURE 17.5b** shows, Sharp's group interpreted these loops as stretches of DNA that are present in the template strand but are *not* in the corresponding mRNA.

Sharp's group and a team headed by Richard Roberts then carried out similar studies on eukaryotic genes. The results were the same as for the viral genes. They went on to propose that there is not a one-to-one correspondence between the nucleotide sequence of a eukaryotic gene and its mRNA. As an analogy, it could be said that eukaryotic genes do not carry messages such as "Biology is my favorite course of all time." Instead, they carry messages that read something like:

BIOLτηεπροτεινχοδινγρεγιονσοφγενεσOGY IS MY
FAVORαρειντερρθπτεθβυνονψοθινγθITE COURSE
OF ανθηαωετοβεσπλιχεθτογετηερ ALL TIME

Here the sections of noncoding sequence are represented with Greek letters. They must be removed from the mRNA before it can carry an intelligible message to the translation machinery.

When it became clear that the genes-in-pieces hypothesis was correct, Walter Gilbert suggested that regions of eukaryotic genes that are part of the final mRNA be referred to as **exons** (because they are *ex*pressed) and the sections of primary transcript not in mRNA be referred to as **introns** (because they are *int*ervening). Introns are sections of genes that are not represented in the final RNA product. Because of introns, eukaryotic genes are much larger than their corresponding mature RNAs. Introns were first discovered in genes that produce mRNA, but researchers later found that genes for other types of RNA also could be split.

RNA Splicing

The transcription of eukaryotic genes by RNA polymerase generates a primary transcript (**FIGURE 17.6a**) that contains both exons and introns. As transcription proceeds, the introns are removed

(a) Introns must be removed from eukaryotic RNA transcripts.

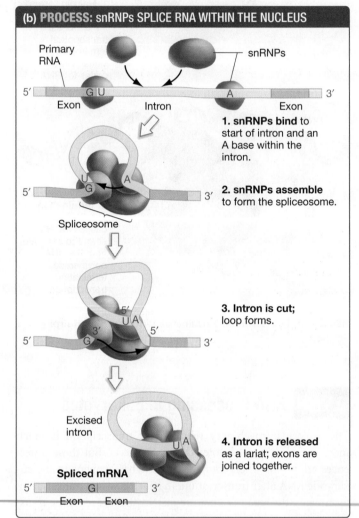

(b) PROCESS: snRNPs SPLICE RNA WITHIN THE NUCLEUS

Primary RNA   snRNPs

5′   G U   A   3′
Exon   Intron   Exon

1. snRNPs bind to start of intron and an A base within the intron.

Spliceosome

2. snRNPs assemble to form the spliceosome.

3. Intron is cut; loop forms.

Excised intron

4. Intron is released as a lariat; exons are joined together.

Spliced mRNA
5′   G   3′
Exon   Exon

FIGURE 17.6 Introns Are Spliced Out of the Primary Transcript.

from the growing RNA strand by a process known as **splicing.** In this phase of information processing, pieces of the primary transcript are removed and the remaining segments are joined together. Splicing occurs within the nucleus while transcription is still under way and results in an RNA that contains an uninterrupted genetic message.

FIGURE 17.6b provides more detail about how introns are removed from primary transcripts to form mRNA. Splicing of primary transcripts is catalyzed by RNAs called small nuclear RNAs (snRNAs) working with a complex of proteins. These protein-plus-RNA macromolecular machines are known as **small nuclear ribonucleoproteins,** or **snRNPs** (pronounced "snurps").

5′ cap

Poly(A) signal   Poly(A) tail

5′ m⁷G–P–P–P– ... AAAAAAAAAAAAAAAAAAA 3′

5′ untranslated region   **Coding region**   3′ untranslated region

FIGURE 17.7 In Eukaryotes, a Cap and a Tail Are Added to mRNAs. As part of eukaryotic pre-mRNA processing, a cap consisting of a modified guanine (G) nucleotide (symbolized as m⁷G) bonded to three phosphate groups is added to the 5′ end, and a tail made up of a long series of adenine (A) residues is added to the 3′ end after cleavage of the primary transcript.

The snRNAs of the snRNPs recognize RNA sequences critical for splicing. Splicing can be broken into four steps:

1. The process begins when snRNPs bind to the 5′ exon–intron boundary, which is marked by the bases GU, and to a key adenine ribonucleotide (A) near the end the intron.

2. Once the initial snRNPs are in place, other snRNPs arrive to form a multipart complex called a **spliceosome.** The spliceosomes found in human cells contain about 145 different proteins and RNAs, making them the most complex macromolecular machines known.

3. The intron forms a loop plus a single-stranded stem (a lariat) with the adenine at its connecting point.

4. The lariat is cut out, and a phosphodiester linkage links the exons on either side, producing a continuous coding sequence—the mRNA.

Splicing is now complete. In most cases, the excised intron is degraded to ribonucleoside monophosphates.

As you'll see later (Chapter 19), for many genes, the RNA can be spliced in more than one way. This allows the production of different, related mRNAs and proteins from one gene.

Current data suggest that both the cutting and rejoining reactions that occur during splicing are catalyzed by the snRNA molecules in the spliceosome—meaning that the reactions are catalyzed by a ribozyme. Section 17.5 will demonstrate that ribozymes also play a key role in translation. As the RNA world hypothesis (Chapter 4) predicts, proteins are not the only important catalysts in cells.

What is the origin of introns? One hypothesis is that introns in eukaryotes arose from an ancient type of DNA sequence that is present in many bacteria and archaea. These ancient sequences are related to viruses (Chapter 36) and, like viruses, can infect cells and insert into their genomes. Remarkably, when this DNA sequence inserts into a gene and is transcribed, the RNA catalyzes its own splicing out of the primary transcript. This is possible in part because these virus-like elements have sequences similar to snRNAs.

The bacterium that was the source of mitochondria (see Chapter 30) likely carried some of these virus-like DNA sequences. When this bacterium was taken up by an ancestral eukaryote, the virus-like sequences are hypothesized to have spread rapidly. Later in evolution, the portion of the sequence that was a precursor to today's snRNA may have separated from the portion of the sequence that was spliced. This spliced sequence is hypothesized to be the ancestor of the modern eukaryotic intron.

Adding Caps and Tails to Transcripts

For pre-mRNAs, intron splicing is accompanied by other important processing steps.

- As soon as the 5′ end of a eukaryotic pre-mRNA emerges from RNA polymerase, enzymes add a structure called the **5′ cap** (**FIGURE 17.7**). The cap consists of a modified guanine (7-methylguanylate) nucleotide with three phosphate groups.

- An enzyme cleaves the 3′ end of the pre-mRNA downstream of the poly(A) signal (introduced in Section 17.1). Another enzyme adds a long row of 100–250 adenine nucleotides that are not encoded on the DNA template strand. This string of adenines is known as the **poly(A) tail.**

With the addition of the cap and tail and completion of splicing, processing of the pre-mRNA is complete. The product is a mature mRNA.

Figure 17.7 also shows that in the mature RNA molecule, the coding sequence for the polypeptide is flanked by sequences that are not destined to be translated. These 5′ and 3′ untranslated regions (or UTRs) help stabilize the mature RNA and regulate its translation. The mRNAs in bacteria also possess 5′ and 3′ UTRs.

Not long after the caps and tails on eukaryotic mRNAs were discovered, evidence began to accumulate that they protect mRNAs from degradation by ribonucleases—enzymes that

check your understanding

If you understand that . . .

- Eukaryotic genes consist of exons, which are parts of the primary transcript that remain in mature RNA, and introns, which are regions of the primary transcript that are removed in forming mature RNA.
- Macromolecular machines, called spliceosomes, splice introns out of pre-mRNAs.
- Enzymes add a 5′ cap and a poly(A) tail to spliced transcripts, producing a mature mRNA that is ready to be translated.

✔ **You should be able to . . .**

1. Explain why ribonucleoprotein is an appropriate name for the subunits of the spliceosome.

2. Explain the function of the 5′ cap and the poly(A) tail.

Answers are available in Appendix A.

degrade RNA—and enhance the efficiency of translation. For example:

- Experimental mRNAs that have a cap and a tail last longer when they are introduced into cells than do experimental mRNAs that lack a cap or a tail.

- Experimental mRNAs with caps and tails produce more proteins than do experimental mRNAs without caps and tails.

Follow-up work has shown that the 5′ cap and the poly(A) tail are bound by proteins that prevent ribonucleases in the cytoplasm from recognizing and destroying the mRNA. The 5′ cap and the poly(A) tail also are important for initiating translation.

RNA processing is the general term for any of the modifications, such as splicing or poly(A) tail addition, needed to convert a primary transcript into a mature RNA. It is summarized in **TABLE 17.1** along with other important differences in how RNAs are produced in eukaryotes as compared with bacteria.

17.3 An Introduction to Translation

To synthesize a protein, the sequence of bases in a messenger RNA molecule is translated into a sequence of amino acids in a polypeptide. The genetic code specifies the correspondence between each triplet codon in mRNA and the amino acid it codes for (see Chapter 16). But how are the amino acids assembled into a polypeptide according to the information in messenger RNA?

(a) Bacterial ribosomes during translation

(b) In bacteria, transcription and translation are tightly coupled.

Studies of translation in cell-free systems proved extremely effective in answering this question. Once in vitro translation systems had been developed from human cells, *E. coli*, and a variety of other organisms, biologists could see that the sequence of events is similar in bacteria, archaea, and eukaryotes. As with similarities in transcription across the domains of life, the shared mechanisms of translation argue for a common ancestor of all cells living today.

Ribosomes Are the Site of Protein Synthesis

The first question that biologists answered about translation concerned where it occurs. The answer grew from a simple observation: There is a strong correlation between the number of **ribosomes** in a given type of cell and the rate at which that cell synthesizes proteins. Based on this observation, investigators proposed that ribosomes are the site of protein synthesis.

To test this hypothesis, Roy Britten and collaborators did a pulse–chase experiment similar in design to experiments introduced earlier (Chapter 7). Recall that a pulse–chase experiment labels a population of molecules as they are being produced. The location of the tagged molecules is then followed over time.

In this case, the tagging was done by supplying a pulse of radioactive sulfur atoms that would be incorporated into the amino acids methionine and cysteine, followed by a chase of unlabeled sulfur atoms. If the ribosome hypothesis were correct, the radioactive signal should be associated with ribosomes for a short period of time—when the amino acids are being polymerized into proteins. Later, when translation was complete, all the radioactivity should be found in proteins that are not associated with ribosomes.

This is exactly what the researchers found. Based on these data, biologists concluded that proteins are synthesized at ribosomes and then released.

Translation in Bacteria and Eukaryotes

About a decade after the ribosome hypothesis was confirmed, electron micrographs showed bacterial ribosomes in action (**FIGURE 17.8a**). The images showed that in bacteria, ribosomes

FIGURE 17.8 Transcription and Translation Occur Simultaneously in Bacteria. In bacteria, ribosomes attach to mRNA transcripts and begin translation while RNA polymerase is still transcribing the DNA template strand.

TABLE 17.1   Transcription and RNA Processing in Bacteria and Eukaryotes

Point of Comparison	Bacteria	Eukaryotes
RNA polymerase(s)	One	Three; each produces a different class of RNA
Promoter structure	Typically contains a −35 box and a −10 box	More variable; often includes a TATA box about −30 from the transcription start site
Proteins involved in recognizing promoter	Sigma; different versions of sigma bind to different promoters	Many basal transcription factors
RNA processing	None	Extensive; several processing steps occur in the nucleus before RNA is exported to the cytoplasm: **(1)** Enzyme-catalyzed addition of 5′ cap on mRNAs, **(2)** Splicing (intron removal); by spliceosome to produce mRNA, **(3)** Enzyme-catalyzed addition of 3′ poly(A) tail on mRNAs

attach to mRNAs and begin synthesizing proteins even before transcription is complete. In fact, multiple ribosomes attach to each mRNA, forming a **polyribosome** (**FIGURE 17.8b**). In this way, many copies of a protein can be produced from a single mRNA.

Transcription and translation can occur concurrently in bacteria because there is no nuclear envelope to separate the two processes.

The situation is different in eukaryotes. In these organisms, primary transcripts are processed in the nucleus to produce a mature mRNA, which is then exported to the cytoplasm (**FIGURE 17.9**). This means that in eukaryotes, transcription and translation are separated in time and space. Once mRNAs are outside the nucleus, ribosomes can attach to them and begin translation. As in bacteria, polyribosomes form.

How Does an mRNA Triplet Specify an Amino Acid?

When an mRNA interacts with a ribosome, instructions encoded in nucleic acids are translated into a different chemical language—the amino acid sequences found in proteins. The discovery of the genetic code revealed that triplet codons in mRNA specify particular amino acids in a protein. How does this conversion occur?

One early hypothesis was that mRNA codons and amino acids interact directly. This hypothesis proposed that the bases in a particular codon were complementary in shape or charge to the side group of a particular amino acid (**FIGURE 17.10a**, see page 326). But Francis Crick pointed out that the idea didn't make chemical sense. For example, how could the nucleic acid bases interact with a hydrophobic amino acid side group, which does not form hydrogen bonds?

Crick proposed an alternative hypothesis. As **FIGURE 17.10b** shows, he suggested that some sort of adapter molecule holds amino acids in place while interacting directly and specifically with a codon in mRNA by hydrogen bonding. In essence, Crick predicted the existence of a chemical go-between that produced a physical connection between the two types of molecules. As it turns out, Crick was right.

(a) mRNAs are exported to the cytoplasm.

(b) Polypeptides grow from ribosomes translating mRNA.

FIGURE 17.9 Transcription and Translation Are Separated in Space and Time in Eukaryotes.

(a) Hypothesis 1: Amino acids interact directly with mRNA codons.

(b) Hypothesis 2: Adapter molecules hold amino acids and interact with mRNA codons.

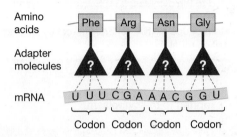

FIGURE 17.10 How Do mRNA Codons Interact with Amino Acids?

17.4 The Structure and Function of Transfer RNA

Crick's adapter molecule was discovered by accident. Biologists were trying to work out an in vitro protein-synthesis system and reasoned that ribosomes, mRNA, amino acids, ATP, and a molecule called guanosine triphosphate, or GTP, would be needed. (GTP is similar to ATP but contains guanine instead of adenine.)

These results were logical: Ribosomes provide the catalytic machinery, mRNAs contribute the message to be translated, amino acids are the building blocks of proteins, and ATP and GTP supply potential energy to drive the endergonic polymerization reactions responsible for forming proteins.

But, in addition, a cellular fraction that contained a previously unknown type of RNA turned out to be indispensable. If this type of RNA is missing, protein synthesis does not occur. What is this mysterious RNA, and why is it essential to translation?

The novel class of RNAs eventually became known as **transfer RNA (tRNA)**. The role of tRNA in translation was a mystery until some researchers happened to add a radioactive amino acid—leucine—to an in vitro protein-synthesis system. The treatment was actually done as a control for an unrelated experiment. To the researchers' amazement, some of the radioactive leucine attached to tRNA molecules.

What happens to the amino acids bound to tRNAs? To answer this question, Paul Zamecnik and colleagues tracked the fate of radioactive leucine molecules that were attached to tRNAs. They found that the amino acids are transferred from tRNAs to proteins.

The data supporting this conclusion are shown in the "Results" section of **FIGURE 17.11**. The graph shows that radioactive amino acids are lost from tRNAs and incorporated into polypeptides synthesized by ribosomes. To understand this conclusion:

FIGURE 17.11 Evidence that Amino Acids Are Transferred from tRNAs to Proteins.

SOURCE: Hoagland, M. B., M. L. Stephenson, J. F. Scott, et al. 1958. A soluble ribonucleic acid intermediate in protein synthesis. *Journal of Biological Chemistry* 231: 241–257.

✔ **QUESTION** What would the graphed results look like if the null hypothesis were correct?

1. Put your finger on the point on the *x*-axis that indicates that one minute has passed since the start of the experiment.

2. Read up until you hit the green line and the gray line. The green line represents data from proteins; the gray line represents data from tRNAs.

3. Check the *y*-axis—which indicates the amount of radioactive leucine present—at each point.

4. It should be clear that early in the experiment, almost all the radioactive leucine is attached to tRNA, not protein.

Next, do the same four steps at the point on the *x*-axis labeled 10 minutes (since the start of the experiment). Your conclusion now should be that late in the experiment, almost all the radioactive leucine is attached to proteins, not tRNA.

These results inspired the use of the word transfer in tRNA's name, because amino acids are transferred from the RNA to a growing polypeptide. The experiment also confirmed that tRNAs act as the interpreter during translation: tRNAs are Crick's adapter molecules.

What Do tRNAs Look Like?

Transfer RNAs serve as chemical go-betweens that allow amino acids to interact with an mRNA template. But precisely how does the connection occur?

This question was answered by research on tRNA's molecular structure. The initial studies established the sequence of nucleotides in various tRNAs, or what is termed their primary structure. Transfer RNA sequences are relatively short, ranging from 75 to 85 nucleotides in length.

When biologists studied the primary sequence closely, they noticed that certain parts of the molecules can form secondary structures. Specifically, some sequences of bases in the tRNA molecule can form hydrogen bonds with complementary base sequences elsewhere in the same molecule. As a result, portions of the molecule form stem-and-loop structures (introduced in Chapter 4). The stems are short stretches of double-stranded RNA; the loops are single stranded.

Two aspects of tRNA's secondary structure proved especially important. A CCA sequence at the 3′ end of each tRNA molecule offered a site for amino acid attachment, while a triplet on the loop at the other end of the structure could serve as an anticodon. An **anticodon** is a set of three ribonucleotides that forms base pairs with the mRNA codon.

Later, X-ray crystallography studies revealed the tertiary structure of tRNAs. Recall that the tertiary structure of a molecule is the three-dimensional arrangement of its atoms (Chapter 3). As **FIGURE 17.12** shows, tRNAs fold into an L-shaped molecule. The anticodon is at one end of the structure; the CCA sequence and attached amino acid are at the other end.

All the tRNAs in a cell have the same general structure, shaped like an upside-down L. They vary at the anticodon and attached amino acid. The tertiary structure of tRNAs is important because it maintains a precise physical distance between the anticodon and amino acid. As it turns out, this separation is important in positioning the amino acid and the anticodon within the ribosome.

✔ If you understand the structure of tRNAs, you should be able to (1) describe where on the L-shaped structure the amino acid attaches; and (2) explain the relationship between the anticodon of a tRNA and a codon in an mRNA.

FIGURE 17.12 The Structure of an Aminoacyl Transfer RNA. The anticodon forms complementary base pairs with an mRNA codon.

How Are Amino Acids Attached to tRNAs?

How are amino acids linked to tRNAs? Just as important, what allows the right amino acid for a particular tRNA to be attached?

- An input of energy, in the form of ATP, is required to attach an amino acid to a tRNA.

- Enzymes called **aminoacyl-tRNA synthetases** catalyze the addition of amino acids to tRNAs—what biologists call "charging" a tRNA.

- For each of the 20 major amino acids, there is a different aminoacyl-tRNA synthetase and one or more tRNAs.

Each aminoacyl-tRNA synthetase has a binding site for a particular amino acid and a particular tRNA. Subtle differences in tRNA shape and base sequence allow the enzymes to recognize the correct tRNA for the correct amino acid. The combination of a tRNA molecule covalently linked to an amino acid is called an **aminoacyl tRNA. FIGURE 17.13** (see page 328) shows an aminoacyl-tRNA synthetase bound to a tRNA that has just been charged with an amino acid. Note how tightly the two structures fit together—making it possible for the enzyme and its tRNA and amino acid substrates to interact in a precise way.

How Many tRNAs Are There?

After characterizing all the different types of tRNAs, biologists encountered a paradox. According to the genetic code (Chapter 16), the 20 most common amino acids found in proteins are specified by 61 different mRNA codons. Instead of containing 61 different tRNAs with 61 different anticodons, though, most cells contain only about 40. How can all 61 codons be translated with only two-thirds that number of tRNAs?

Amino acid

Aminoacyl
tRNA

Aminoacyl-
tRNA
synthetase

tRNA

FIGURE 17.13 Aminoacyl-tRNA Synthetases Couple the Appropriate Amino Acid to the Appropriate tRNA.

To resolve this paradox, Francis Crick proposed what is known as the **wobble hypothesis.** Recall that:

1. Many amino acids are specified by more than one codon.

2. Codons for the same amino acid tend to have the same nucleotides at the first and second positions but a different nucleotide at the third position.

For example, both of the codons CAA and CAG code for the amino acid glutamine. Surprisingly, experimental data have shown that a tRNA with an anticodon of GUU can base-pair with both CAA and CAG in mRNA. The GUU anticodon matches the first two bases (C and A) in both codons, but the U in the anticodon's third position forms a nonstandard base pair with a G in the CAG codon.

Crick proposed that inside the ribosome, certain bases in the third position of tRNA anticodons can bind to bases in the third position of a codon in a manner that does not match Watson–Crick base pairing. If so, this would allow a limited flexibility, or "wobble," in the base pairing.

According to the wobble hypothesis, particular nonstandard base pairs—such as G-U—are acceptable in the third position of a codon and do not change the amino acid that the codon specifies. In this way, wobble in the third position of a codon allows just 40 or so tRNAs to bind to all 61 mRNA codons.

17.5 The Structure and Function of Ribosomes

Recall that protein synthesis occurs when the sequence of bases in an mRNA is translated into a sequence of amino acids in a polypeptide. The translation of each mRNA codon begins when the anticodon of an aminoacyl tRNA binds to the codon. Translation of a codon is complete when a peptide bond forms between the tRNA's amino acid and the growing polypeptide chain.

Both of these events take place inside a ribosome. Biologists have known since the 1930s that ribosomes contain many proteins and **ribosomal RNAs (rRNAs).** Later work showed that ribosomes can be separated into two major substructures, called the large subunit and small subunit. Each ribosome subunit consists of a complex of RNA molecules and proteins. The small subunit holds the mRNA in place during translation; the large subunit is where peptide-bond formation takes place.

FIGURE 17.14 shows two views of how the molecules required for translation fit together. Note that during protein synthesis, three distinct tRNAs are lined up inside the ribosome. All three are bound to their corresponding mRNA codons.

- The tRNA that is on the right in the figure, and colored red, carries an amino acid. This tRNA's position in the ribosome is called the A site—"A" for acceptor or aminoacyl.

- The tRNA that is in the middle (green) holds the growing polypeptide chain and occupies the P site, for peptidyl, inside the ribosome. (Think of "P" for peptide-bond formation.)

- The left-hand (blue) tRNA no longer has an amino acid attached and is about to leave the ribosome. It occupies the ribosome's E site—"E" for exit.

Because all tRNAs have similar secondary and tertiary structure, they all fit equally well in the A, P, and E sites. The ribosome is a macromolecular machine that synthesizes proteins in a three-step sequence:

1. An aminoacyl tRNA diffuses into the A site; if its anticodon matches a codon in mRNA, it stays in the ribosome.

2. A peptide bond forms between the amino acid held by the aminoacyl tRNA in the A site and the growing polypeptide, which was held by a tRNA in the P site.

3. The ribosome moves down the mRNA by one codon, and all three tRNAs move one position within the ribosome. The tRNA in the E site exits; the tRNA in the P site moves to the E site; and the tRNA in the A site switches to the P site.

The protein that is being synthesized grows by one amino acid each time this three-step sequence repeats. The process occurs up to 20 times per second in bacterial ribosomes and about 2 times per second in eukaryotic ribosomes. Protein synthesis starts at the amino end (N-terminus) of a polypeptide and proceeds to the carboxy end (C-terminus; see Chapter 3).

This introduction to how tRNAs, mRNAs, and ribosomes interact during protein synthesis leaves several key questions unanswered. How do mRNAs and ribosomes get together to start the process? Once protein synthesis is under way, how is peptide-bond formation catalyzed inside the ribosome? And how does protein synthesis conclude when the ribosome reaches the end of the message? Let's consider each question in turn.

(a) Diagram of ribosome during translation (interior view)

Peptide bond formation occurs here

Gly
Glu
His
Ser | Ala

The **P site** holds the tRNA with growing polypeptide attached

The **A site** holds an aminoacyl tRNA

The **E site** holds a tRNA that will exit

E   P   A

GUGAGU CGA — Anticodon

5′ UUAUUUCGGGGAA**CAC**UCAGCUGAGGAUACUAU 3′
mRNA

Codon

Large subunit

Small subunit

(b) Model of ribosome during translation (exterior view)

Ribosomal RNA (rRNA)

Protein

tRNA

FIGURE 17.14 The Structure of the Ribosome. Ribosomes have three distinct tRNA binding sites in their interior.

Initiating Translation

To translate an mRNA properly, a ribosome must begin at a specific point in the message, translate the mRNA up to the message's termination codon, and then stop. Using the same terminology that they apply to transcription, biologists call these three phases of protein synthesis initiation, elongation, and termination, respectively.

One key to understanding translation initiation is to recall that a start codon (usually AUG) is found near the 5′ end of all mRNAs and that it codes for the amino acid methionine (Chapter 16).

FIGURE 17.15 shows how translation gets under way in bacteria. The process begins when a section of rRNA in a small ribosomal subunit binds to a complementary sequence on an mRNA. The mRNA region is called the **ribosome binding site,** or **Shine–Dalgarno sequence,** after the biologists who discovered it. The site is about six nucleotides upstream from the start codon.

PROCESS: INITIATING TRANSLATION IN BACTERIA

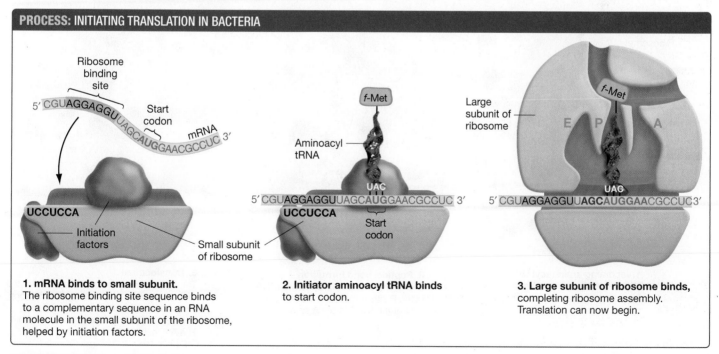

1. mRNA binds to small subunit.
The ribosome binding site sequence binds to a complementary sequence in an RNA molecule in the small subunit of the ribosome, helped by initiation factors.

2. Initiator aminoacyl tRNA binds to start codon.

3. Large subunit of ribosome binds, completing ribosome assembly. Translation can now begin.

FIGURE 17.15 Initiation of Translation.

The interactions between the small subunit, the message, and the tRNA are mediated by proteins called **initiation factors** (Figure 17.15, step 1). Initiation factors help in preparing the ribosome for translation, including binding the first aminoacyl tRNA to the ribosome. In bacteria this initiator tRNA bears a modified form of methionine called *N*-formylmethionine (abbreviated *f*-met) (Figure 17.15, step 2). In eukaryotes, this initiating tRNA carries a normal methionine. Initiation factors also prevent the small and large subunits of the ribosome from coming together until the initiator tRNA is in place at the AUG start codon, and they help bind the mRNA to the small ribosomal subunit.

Initiation is complete when the large subunit joins the complex (Figure 17.15, step 3). When the ribosome is completely assembled, the tRNA bearing *f*-met occupies the P site.

To summarize, translation initiation is a three-step process in bacteria: **(1)** The mRNA binds to a small ribosomal subunit, **(2)** the initiator aminoacyl tRNA bearing *f*-met binds to the start codon, and **(3)** the large ribosomal subunit binds, completing the complex.

In eukaryotes, the details of initiation are different, but they still involve recognition of a start codon, assembly of the ribosome, assistance from initiation factors, and the positioning of a methionine-carrying tRNA in the P site. The cap and poly(A) tail are also important in assembling the ribosome on the mRNA.

Elongation: Extending the Polypeptide

At the start of elongation, the E and A sites in the ribosome are empty of tRNAs. As a result, an mRNA codon is exposed in the A site. As step 1 in **FIGURE 17.16** illustrates, elongation proceeds when an aminoacyl tRNA binds to the codon in the A site by complementary base pairing between the anticodon and codon.

When both the P site and A site are occupied by tRNAs, the amino acids on the tRNAs are in the ribosome's active site. This is where peptide-bond formation—the essence of protein synthesis—occurs.

Peptide-bond formation is one of the most important reactions that take place in cells because manufacturing proteins is among the most fundamental of all cell processes. Biologists wondered, is it the ribosome's proteins or RNAs that catalyze this reaction?

Is the Ribosome an Enzyme or a Ribozyme? Because ribosomes contain both protein and RNA, researchers had argued for decades over whether the active site consisted of protein or RNA. The debate was not resolved until the year 2000, when researchers completed three-dimensional models that were detailed enough to reveal the structure of the active site. These models confirmed that the active site consists entirely of ribosomal RNA. Based on these results, biologists are now convinced that protein synthesis is catalyzed by RNA. The ribosome is a ribozyme—not a protein-based enzyme.

The observation that protein synthesis is catalyzed by RNA is important because it supports the RNA-world hypothesis (Chapter 4). Recall that proponents of this hypothesis claim that life began with RNA molecules and that the presence of DNA and proteins in cells evolved later. If the RNA-world hypothesis is correct, then it would make sense that the production of proteins is catalyzed by RNA.

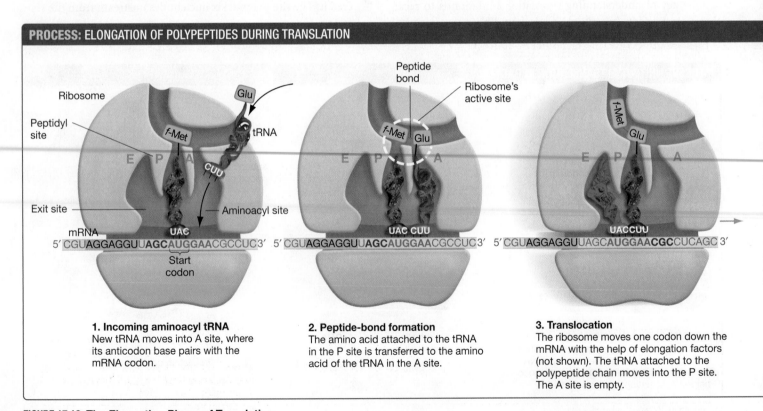

PROCESS: ELONGATION OF POLYPEPTIDES DURING TRANSLATION

1. Incoming aminoacyl tRNA
New tRNA moves into A site, where its anticodon base pairs with the mRNA codon.

2. Peptide-bond formation
The amino acid attached to the tRNA in the P site is transferred to the amino acid of the tRNA in the A site.

3. Translocation
The ribosome moves one codon down the mRNA with the help of elongation factors (not shown). The tRNA attached to the polypeptide chain moves into the P site. The A site is empty.

FIGURE 17.16 The Elongation Phase of Translation.

Moving Down the mRNA   What happens after a peptide bond forms? Step 2 in Figure 17.16 shows that when peptide-bond formation is complete, the polypeptide chain is transferred from the tRNA in the P site to the amino acid held by the tRNA in the A site. Step 3 shows the process called **translocation,** which occurs when proteins called **elongation factors** help move the ribosome relative to the mRNA so that translation occurs in the 5′ → 3′ direction. Translocation is an energy-demanding event that requires GTP.

Translocation does several things: It moves the uncharged RNA into the E site; it moves the tRNA containing the growing polypeptide into the P site; and it opens the A site and exposes a new mRNA codon. The empty tRNA that finds itself in the E site is ejected into the cytosol.

The three steps in elongation—(**1**) arrival of aminoacyl tRNA, (**2**) peptide-bond formation, and (**3**) translocation—repeat down the length of the mRNA. Recent three-dimensional models of ribosomes in various stages of translation show that the machine is highly dynamic. The ribosome constantly changes shape as tRNAs come and go and catalysis and translocation occur. The ribosome is a complex and dynamic macromolecular machine.

Terminating Translation

How does protein synthesis end? Recall that the genetic code includes three stop codons: UAA, UAG, and UGA (see Chapter 16). In most cells, no aminoacyl tRNA has an anticodon that binds to these sequences. When the translocating ribosome reaches one of the stop codons, a protein called a **release factor** recognizes the stop codon and fills the A site (**FIGURE 17.17**, see page 333).

Release factors fit tightly into the A site because they have the size and shape of a tRNA coming into the ribosome. However, release factors do not carry an amino acid. When a release factor occupies the A site, the protein's active site catalyzes the hydrolysis of the bond that links the tRNA in the P site to the polypeptide chain. This reaction frees the polypeptide.

The newly synthesized polypeptide and uncharged tRNAs are released from the ribosome, the ribosome separates from the mRNA, and the two ribosomal subunits dissociate. The subunits are ready to attach to the start codon of another message and start translation anew. Termination occurs in very similar ways in bacteria and eukaryotes.

Post-Translational Modifications

Proteins are not fully formed and functional when termination occurs. From earlier chapters, it should be clear that most proteins go through an extensive series of processing steps, collectively called post-translational modification, before they are completely functional. These steps require a wide array of molecules and events and take place in many different locations throughout the cell.

Folding   Recall that a protein's function depends on its shape, and that a protein's shape depends on how it folds (see Chapter 3). Although folding can occur spontaneously, it is frequently speeded up by proteins called **molecular chaperones.**

Recent data have shown that in some bacteria, chaperone proteins bind to the ribosome near the "tunnel" where the growing polypeptide emerges from the ribosome. This finding suggests that folding occurs as the polypeptide emerges from the ribosome.

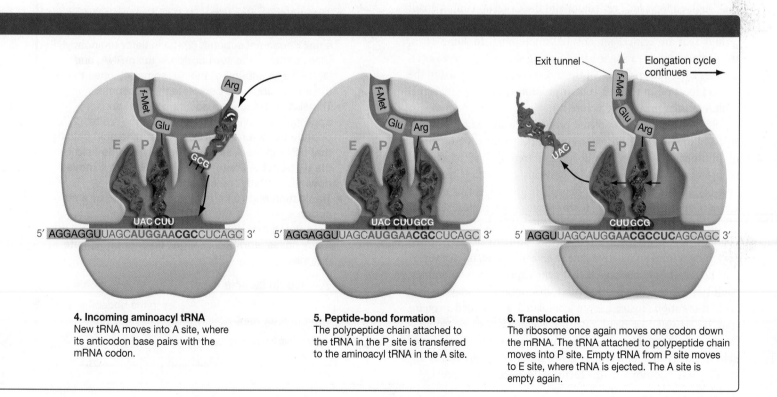

4. Incoming aminoacyl tRNA
New tRNA moves into A site, where its anticodon base pairs with the mRNA codon.

5. Peptide-bond formation
The polypeptide chain attached to the tRNA in the P site is transferred to the aminoacyl tRNA in the A site.

6. Translocation
The ribosome once again moves one codon down the mRNA. The tRNA attached to polypeptide chain moves into P site. Empty tRNA from P site moves to E site, where tRNA is ejected. The A site is empty again.

1. Release factor binds to stop codon.
When the translocating ribosome reaches a stop codon, a protein release factor fills the A site. The release factor breaks the bond linking the tRNA in the P site to the polypeptide chain.

2. Polypeptide and uncharged tRNAs are released.

3. Ribosome subunits separate.
The subunits are ready to attach to the start codon of another message.

FIGURE 17.17 Termination of Translation.

Chemical Modifications An earlier chapter pointed out that many eukaryotic proteins are extensively modified after they are synthesized (see Chapter 7). For example, in the organelles called the rough endoplasmic reticulum and the Golgi apparatus, small chemical groups may be added to proteins—often sugar or lipid groups that are critical for normal functioning. In some cases, the proteins receive a sugar-based sorting signal that serves as an address label and ensures that the molecule will be carried to the correct location in the cell. (Other proteins have a sorting signal built into their primary structure.)

In addition, many completed proteins are altered by enzymes that add or remove a phosphate group. Phosphorylation (addition of phosphate) and dephosphorylation (removal of phosphate) of proteins were introduced in previous chapters (Chapters 9 and 11). Recall that because a phosphate group has two negative charges, adding or removing a phosphate group can cause major changes in the shape and chemical reactivity of proteins. These changes have a dramatic effect on the protein's activity—often switching it from an inactive state to an active state, or vice versa.

The take-home message is that gene expression is a complex, multistep process that begins with transcription but may not end with translation. Instead, even completed and folded proteins may be activated or deactivated by events such as phosphorylation. In general, how are genes turned on or off? How does a cell "decide" which of its many genes should be expressed at any time? These questions are the focus of the next two chapters.

check your understanding

C Y U

If you understand that . . .

- Translation begins when (1) the ribosome binding site on an mRNA binds to an rRNA sequence in the small ribosomal subunit, (2) the initiator aminoacyl tRNA binds to the start codon in the mRNA, and (3) the large subunit of the ribosome attaches to the small subunit to complete the ribosome.
- Translation elongation occurs when (1) an appropriate aminoacyl tRNA enters the A site, (2) a peptide bond forms between the amino acid held by that tRNA in the A site and the polypeptide held by the tRNA in the P site, and (3) the ribosome moves down the mRNA one codon.
- Translation ends when the ribosome reaches a stop codon.
- Completed proteins are modified by folding and, in many cases, addition of sugar, lipid, or phosphate groups.

✔ You should be able to . . .

Explain why the E, P, and A sites in the ribosome are appropriately named.

Answers are available in Appendix A.

If you understand . . .

17.1 An Overview of Transcription

- RNA polymerase catalyzes the production of an RNA molecule whose base sequence is complementary to the base sequence of the DNA template strand.

- RNA polymerase binds DNA with the help of other proteins.

- RNA polymerase begins transcription by binding to promoter sequences in DNA.

- In bacteria, this binding occurs in conjunction with a protein called sigma. Sigma associates with RNA polymerase and then recognizes particular sequences within promoters that are centered 10 bases and 35 bases upstream from the start of the actual genetic message.

- Eukaryotic promoters vary more than bacterial promoters.

- In eukaryotes, transcription begins when a large array of proteins called basal transcription factors bind to a promoter. In response, RNA polymerase binds to the site.

- In both bacteria and eukaryotes, the RNA elongates in a $5' \rightarrow 3'$ direction.

- Transcription in bacteria ends when RNA polymerase encounters a stem-loop structure in the just transcribed RNA; in eukaryotes, transcription terminates after the RNA is cleaved downstream of the poly(A) sequence.

✔ You should be able to predict the consequences of a mutation in bacteria that inserts random nucleotides into the hairpin-coding region near the 3′ end of a transcribed region.

17.2 RNA Processing in Eukaryotes

- In eukaryotes, the primary (initial) transcript must be processed to produce a mature RNA.

- In primary transcripts, stretches of RNA called introns are spliced out and regions called exons are joined together.

- Complex macromolecular machines called spliceosomes splice introns out of pre-mRNA.

- A "cap" is added to the 5′ end of pre-mRNAs, and a poly(A) tail is added to their 3′ end.

- The cap and tail serve as recognition signals for translation and protect the message from degradation by ribonucleases.

- RNA processing occurs in the nucleus.

✔ You should be able to predict whether the protein-coding portion of a gene for an identical protein will be of the same or different lengths in a bacterium and a eukaryote.

17.3 An Introduction to Translation

- Ribosomes translate mRNAs into proteins with the help of adaptor molecules called transfer RNAs.

- In bacteria, an RNA is often transcribed and translated at the same time because there is no nucleus.

- In eukaryotes, transcription and translation of an RNA cannot occur together, because transcription and RNA processing occur in the nucleus and translation occurs in the cytoplasm.

- Experiments with radioactively labeled amino acids showed that transfer RNAs (tRNAs) serve as the chemical bridge between the RNA message and the polypeptide product.

✔ You should be able to explain why it is correct to say that transfer RNAs work as molecular adaptors.

17.4 The Structure and Function of Transfer RNA

- Each transfer RNA carries an amino acid corresponding to the tRNA's three-base-long anticodon.

- tRNAs have an L-shaped, tertiary structure. One leg of the L contains the anticodon, which forms complementary base pairs with the mRNA codon. The other leg holds the amino acid appropriate for that codon.

- Enzymes called aminoacyl-tRNA synthetases link the correct amino acid to the correct tRNA.

- Imprecise pairing—or "wobble pairing"—is allowed in the third position of the codon and anticodon, so only about 40 different tRNAs are required to translate the 61 codons that code for amino acids.

✔ You should be able to predict what would occur if a mutation caused an aminoacyl-tRNA synthetase to recognize two different amino acids.

17.5 The Structure and Function of Ribosomes

- Ribosomes are large macromolecular machines made of many proteins and RNAs.

- In the ribosome, the tRNA anticodon binds to a three-base-long mRNA codon to bring the correct amino acid into the ribosome.

- Peptide-bond formation by the ribosome is catalyzed by a ribozyme (RNA), not an enzyme (protein).

- Protein synthesis occurs in three steps: (1) an incoming aminoacyl tRNA occupies the A site; (2) the growing polypeptide chain is transferred from a peptidyl tRNA in the ribosome's P site to the amino acid bound to the tRNA in the A site, and a peptide bond is formed; and (3) the ribosome is translocated to the next codon on the mRNA, accompanied by ejection of the uncharged RNA from the E site.

- Chaperone proteins help fold newly synthesized proteins into their three-dimensional conformation (tertiary structure).

- Most proteins need to be modified after translation (post-translational modification) to activate them or target them to specific locations.

✔ You should be able to create a concept map (see BioSkills 15 in Appendix B) that describes the relationships among the following concepts and structures: translation, initiation, elongation, termination, growing polypeptide in P site, start codon, ribosome subunits.

(MB) MasteringBiology

1. MasteringBiology Assignments

Tutorials and Activities Chromosomal Mutations; Following the Instructions in DNA; Point Mutations Protein Synthesis (1 of 3): Overview; Protein Synthesis (2 of 3): Transcription and RNA Processing; Protein Synthesis (3 of 3): Translation and Protein Targeting Pathways; RNA Processing; RNA Synthesis; Synthesizing Proteins; Transcription; Translation; Types of RNA

Questions Reading Quizzes, Blue-Thread Questions, Test Bank

2. eText Read your book online, search, take notes, highlight text, and more.

3. The Study Area Practice Test, Cumulative Test, BioFlix® 3-D Animations, Videos, Activities, Audio Glossary, Word Study Tools, Art

You should be able to . . .

✔ TEST YOUR KNOWLEDGE

Answers are available in Appendix A

1. How did the A site of the ribosome get its name?
   a. It is where amino acids are joined to tRNAs, producing aminoacyl tRNAs.
   b. It is where the amino group on the growing polypeptide chain is available for peptide-bond formation.
   c. It is the site occupied by incoming aminoacyl tRNAs.
   d. It is surrounded by α-helices of ribosomal proteins.

2. Where is the start codon located?
   a. at the very start (5′ end) of the mRNA
   b. at the downstream end of the 3′ untranslated region (UTR)
   c. at the downstream end of the 5′ untranslated region (UTR)
   d. at the upstream end of the 3′ untranslated region (UTR)

3. What is the function of a molecular chaperone?

4. What does a bacterial RNA polymerase produce when it transcribes a protein-coding gene?

   a. rRNA
   b. tRNA
   c. mRNA
   d. pre-mRNA

5. Where is an amino acid attached to a tRNA?

6. Compared with mRNAs that have a cap and tail, what do researchers observe when eukaryotic mRNAs that lack a cap and poly(A) tail are translated within a cell?
   a. The primary transcript cannot be processed properly.
   b. Translation occurs inefficiently.
   c. Enzymes on the ribosome add back a cap and poly(A) tail.
   d. tRNAs become resistant to degradation (being broken down).

✔ TEST YOUR UNDERSTANDING

Answers are available in Appendix A

7. Explain the relationship between eukaryotic promoter sequences, basal transcription factors, and RNA polymerase. Explain the relationship between bacterial promoter sequences, sigma, and RNA polymerase.

8. According to the wobble rules, the correct amino acid can be added to a growing polypeptide chain even if the third base in the mRNA codon is not complementary to the corresponding base in the tRNA anticodon. How do the wobble rules relate to the redundancy of the genetic code?

9. RNases and proteases are enzymes that destroy RNAs and proteins, respectively. Which of the following enzymes when added to a spliceosome is predicted to prevent recognition of pre-mRNA regions critical for splicing?
   a. an RNase specific for tRNAs
   b. an RNase specific for snRNAs
   c. a protease specific for initiation factors
   d. a protease specific for a release factor

10. Describe the sequence of events that occurs during translation as a protein elongates by one amino acid and the ribosome moves down the mRNA. Your answer should specify what is happening in the ribosome's A site, P site, and E site.

11. **QUANTITATIVE** Controlling the rates of transcription and translation is important in bacteria to avoid collisions between ribosomes and RNA polymerases. Calculate the maximum rate of translation by a ribosome in a bacterial cell, provided in units of amino acids per second, so that the ribosome doesn't overtake an RNA polymerase that is transcribing mRNA at a rate of 60 nucleotides per second. How long would it take for this bacterial cell to translate an mRNA containing 1800 codons?

12. In an aminoacyl tRNA, why is the observed distance between the amino acid and the anticodon important?

✔ **TEST YOUR PROBLEM-SOLVING SKILLS**   *Answers are available in Appendix A*

13. The 5′ cap and poly(A) tail in eukaryotic mRNAs protect the message from degradation by ribonucleases. But why do ribonucleases exist? What function would an enzyme that destroys messages serve? Answer this question using the example of an mRNA for a hormone that causes human heart rate to increase.

14. The nucleotide shown below is called cordycepin triphosphate.

If cordycepin triphosphate is added to a cell-free transcription reaction, the nucleotide is added onto the growing RNA chain but no more nucleotides can be added. The added cordycepin is always found at the 3′ end of an RNA, confirming that synthesis occurs in the 5′ → 3′ direction. Why does cordycepin end transcription?

a. It prevents the association of RNA polymerase and sigma.
b. It irreversibly binds to the active site of RNA polymerase.
c. It cannot be recognized by RNA polymerase.
d. It lacks a 3′ OH.

15. Certain portions of the rRNAs in the large subunit of the ribosome are very similar in all organisms. To make sense of this finding, Carl Woese suggests that the conserved sequences have an important functional role. His logic is that these conserved sequences evolved in a common ancestor of all modern cells and are so important to cell function that any changes in the sequences cause death. In addition to rRNAs, which specific portions of the ribosome would you expect to be identical or nearly identical in all organisms? Explain your logic.

16. Recent structural models show that a poison called α-amanitin inhibits transcription by binding to a site inside eukaryotic RNA polymerase II but not to the active site itself. Based on the model of RNA polymerase in Figure 17.2, predict a place or places where α-amanitin might bind to inhibit transcription.

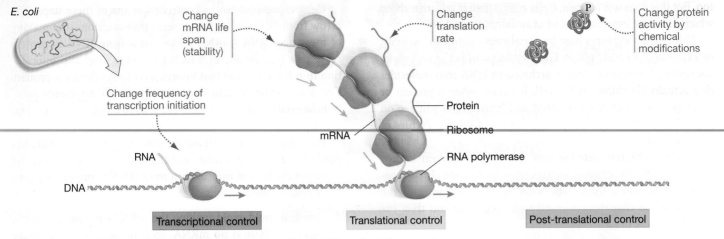

FIGURE 18.1 Gene Expression in Bacteria Can Be Regulated at Three Levels.

✔**EXERCISE** Label the mode of regulation that is slowest in response time and the mode that is fastest. Label the most efficient and least efficient mode in resource use.

controls. Just as important, some genes—such as those that code for the enzymes required for glycolysis—are transcribed all the time, or **constitutively.** Finally, it is critical to realize that gene expression is not an all-or-none proposition. Genes are not just "on" or "off"—instead, the level of expression can vary between these extremes.

The ability to regulate gene expression allows cells to respond to changes in their environment.

Metabolizing Lactose—A Model System

Many of the great advances in genetics have been achieved through the analysis of model systems (see Chapters 14–17). Mendel studied garden peas and discovered the fundamental patterns of gene transmission; Morgan studied fruit flies and confirmed the chromosome theory of inheritance; an array of researchers used *E. coli* and its viruses to work out the mechanisms of DNA synthesis, transcription, and translation. In early studies of gene regulation, a key model system was the metabolism of the sugar lactose in *E. coli.*

Jacques Monod, François Jacob, and many colleagues introduced lactose metabolism in *E. coli* as a model system during the 1950s and 1960s. Although they worked with a single species of bacteria, their results had a profound effect on thinking about gene regulation in all organisms.

E. coli can use a wide variety of sugars for ATP production, via cellular respiration or fermentation. These sugars also serve as raw material in the synthesis of amino acids, vitamins, and other complex compounds. Glucose, however, is *E. coli's* preferred carbon source—meaning that it is the source of energy and carbon atoms that the organism uses most efficiently.

A preference for glucose makes sense, because glycolysis begins with glucose and is the main pathway for the production of ATP. Lactose, the sugar found in milk, can also be used by *E. coli*, but it is not used until glucose supplies are depleted. Recall that

lactose is a disaccharide made up of one molecule of glucose and one molecule of galactose (see Chapter 5).

To use lactose, *E. coli* must first transport the sugar into the cell. Once lactose is inside the cell, the enzyme β-galactosidase catalyzes a reaction that breaks down the disaccharide into glucose and galactose. The glucose released by this reaction goes directly into the glycolytic pathway; other enzymes convert the galactose to a substance that can also be processed in the glycolytic pathway.

In the early 1950s, biologists discovered that *E. coli* produces high levels of β-galactosidase only when lactose is present in the environment. Based on this observation, researchers proposed that lactose itself regulates the gene for β-galactosidase—meaning that lactose acts as an inducer. An **inducer** is a small molecule that triggers transcription of a specific gene.

In the late 1950s, Jacques Monod wondered how the presence of glucose affects the regulation of the β-galactosidase gene. Would *E. coli* produce high levels of β-galactosidase when both glucose and lactose were present in the surrounding environment? As the experiment summarized in **FIGURE 18.2** shows, the answer was no. Significant amounts of β-galactosidase are produced only when lactose is present and glucose is not present.

Monod teamed up with François Jacob to investigate exactly how lactose and glucose regulate the genes responsible for lactose metabolism—the gene for the membrane protein that imports lactose and the gene for β-galactosidase. Discoveries about how these genes are regulated shed light on how genes in all organisms are controlled. Research on this system is still going strong, over 50 years later.

✔ You should be able to make a chart summarizing the molecules involved in regulating lactose use in *E. coli.* There should be 7 rows and 2 columns. Title the first column "Name" and the second column "Function." The rows are *lacZ, lacY,* operator, promoter, repressor, lactose, and glucose. As you read this chapter, fill in the "Function" column.

QUESTION: *E. coli* produces β-galactosidase when lactose is present. Does *E. coli* produce β-galactosidase when both glucose and lactose are present?

HYPOTHESIS: *E. coli* does not produce β-galactosidase when glucose is present, even if lactose is present. (Glucose is the preferred food source.)

NULL HYPOTHESIS: *E. coli* produces β-galactosidase whenever lactose is present, regardless of the presence or absence of glucose.

EXPERIMENTAL SETUP:

PREDICTION: β-Galactosidase will be produced only in treatment 3.

PREDICTION OF NULL HYPOTHESIS: β-Galactosidase will be produced in treatments 2 and 3.

RESULTS:

CONCLUSION: Glucose prevents expression of the gene for β-galactosidase. The presence of lactose without glucose stimulates expression of that gene.

FIGURE 18.2 Glucose Affects the Regulation of the β-Galactosidase Gene.

✔**QUESTION** How would you control growth conditions in the three treatments so that the results of this experiment are valid?

SOURCE: Pardee, A. B., F. Jacob, and J. Monod. 1959. The genetic control and cytoplasmic expression of "inducibility" in the synthesis of β-galactosidase by *E. coli. Journal of Molecular Biology* 1: 165–178.

18.2 Identifying Regulated Genes

To understand how *E. coli* controls production of β-galactosidase and the transport protein that brings lactose into the cell, Jacob and Monod first had to find the genes that code for these proteins. To do this, they employed the same tactic used in the pioneering studies of DNA replication, transcription, and translation reviewed in earlier chapters: They isolated and analyzed mutants. In this case, their goal was to find *E. coli* cells that could not metabolize lactose. Cells that can't use lactose

must lack either β-galactosidase or the lactose transporter protein.

To find mutants that are associated with a particular trait, a researcher has to complete two steps:

1. Generate a large number of individuals with mutations at random locations in their genomes. Monod and colleagues accomplished this step by exposing *E. coli* populations to **mutagens**—X-rays, UV light, or chemicals that damage DNA and increase mutation rates.

2. Screen the treated individuals for mutants with defects in the process or biochemical pathway in question—in this case, defects in lactose metabolism. Recall that a genetic screen is any technique for selecting individuals with certain types of mutations out of a large population, and that a mutant is an individual with a mutation (see Chapter 16).

The researchers were looking for cells that cannot grow in an environment that contains only lactose as an energy source. Normal cells grow well in this environment. How could the researchers select cells on the basis of *lack* of growth?

Replica Plating to Find Lactose Metabolism Mutants

Replica plating and growth on indicator plates were key techniques in the search for mutants with defects in lactose metabolism. **FIGURE 18.3** (see page 6) shows how **replica plating** works.

Step 1 When mutants with defects in lactose metabolism are desired, mutagenized bacteria are spread on a "master plate" filled with a gelatinous growth **medium** containing glucose but no lactose. Growth medium is any liquid or solid that supports the growth of cells. It is important that the mutant cells are capable of growing on the master plate. The bacteria are then allowed to grow, so that each cell produces a colony—a large number of identical cells descended from a single cell.

Step 2 A block covered with a piece of sterilized velvet is pressed onto the master plate. Some cells from each colony on the master plate are transferred to the velvet.

Step 3 The velvet is pressed onto a plate called a replica plate that contains medium that differs from the master plate by a single component. In this case, the second medium has only lactose and no glucose as the source of carbon and energy. Cells from the velvet stick to the replica plate's surface, producing an exact copy of the locations of the colonies on the master plate.

Step 4 After these transferred cells grow, compare the colonies on the replica plate with those on the master plate. In this example, colonies that grow on the master plate but are missing on the replica plate are mutants that cannot metabolize lactose. By picking cells from these colonies on the master plate, researchers build a collection of lactose metabolism mutants.

Several Genes Are Involved in Lactose Metabolism

The initial mutant screen yielded three types of mutants. In one class, the mutant cells were unable to cleave lactose—even when

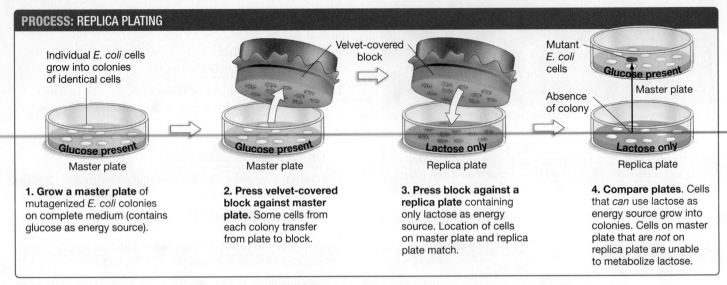

Individual *E. coli* cells grow into colonies of identical cells

Velvet-covered block

Mutant *E. coli* cells

Absence of colony

Glucose present
Master plate

Glucose present
Master plate

Lactose only
Replica plate

Glucose present
Master plate

Lactose only
Replica plate

1. Grow a master plate of mutagenized *E. coli* colonies on complete medium (contains glucose as energy source).

2. Press velvet-covered block against master plate. Some cells from each colony transfer from plate to block.

3. Press block against a replica plate containing only lactose as energy source. Location of cells on master plate and replica plate match.

4. Compare plates. Cells that *can* use lactose as energy source grow into colonies. Cells on master plate that are *not* on replica plate are unable to metabolize lactose.

FIGURE 18.3 Replica Plating Is a Technique for Identifying Mutants That Cannot Grow in Particular Conditions. Here, replica plating is used to isolate mutant *E. coli* cells with a deficiency in lactose metabolism.

✔**QUESTION** How would you alter this protocol to isolate mutant cells with a deficiency in the enzymes required to synthesize tryptophan?

lactose was in the medium and transported into cells to induce production of the β-galactosidase protein. Jacob and Monod concluded that these mutants must lack a functioning version of the β-galactosidase protein and, therefore, the gene that encodes β-galactosidase is defective. This gene was designated *lacZ*, and the mutant allele *lacZ⁻*.

In the second class of mutants, the cells failed to accumulate lactose inside the cell. To explain this result, Jacob and Monod hypothesized that the mutant cells had defective copies of the membrane protein responsible for transporting lactose into the cell. This protein was identified and named galactoside permease; the gene that encodes it was designated *lacY*. **FIGURE 18.4** summarizes the functions of β-galactosidase and galactoside permease.

The third and most surprising class of mutants did not show normal regulation of β-galactosidase and galactoside permease expression. Instead, these mutants made the proteins all the

time—even if no lactose was present. **TABLE 18.1** summarizes these three types of mutants.

Cells that are abnormal because they produce a product at all times are called **constitutive mutants.** The gene that was mutated to produce constitutive β-galactosidase and galactoside permease expression was named *lacI*. The letter I signified that these mutants did not need an inducer—lactose—to express β-galactosidase or galactoside permease.

To understand the significance of the *lacI* mutation, recall that in normal cells, the expression of the *lacZ* (β-galactosidase) and *lacY* (galactoside permease) genes is induced by lactose. But in

FIGURE 18.4 Two Proteins E. coli Needs for Using Lactose. The membrane protein galactoside permease brings lactose into the cell, and the enzyme β-galactosidase breaks lactose into its glucose and galactose subunits.

TABLE 18.1 Three Types of Lactose Metabolism Mutants in *E. coli*

Observed Phenotype	Interpretation	Genotype
1. Cells cannot cleave lactose, even in the presence of inducer.	No β-galactosidase; gene for β-galactosidase is defective. Call this gene *lacZ*.	*lacZ⁻*
2. Cells cannot accumulate lactose.	No membrane protein (galactoside permease) to import lactose; gene for galactoside permease is defective. Call this gene *lacY*.	*lacY⁻*
3. Cells cleave lactose even if lactose is absent as an inducer.	Constitutive (constant) expression of *lacZ* and *lacY*; gene for regulatory protein that shuts down *lacZ* and *lacY* is defective—it does not need to be induced by lactose. Call this gene *lacI*.	*lacI⁻*

FIGURE 18.5 The *lac* Genes Are in Close Physical Proximity. The associated proteins and their functions are shown above each gene.

cells with a mutant form of *lacI* (*lacI⁻* mutants), gene expression occurs with or without lactose. This means that *lacI⁻* mutants have a defect in gene regulation. In these mutants, the gene remains on when it should be turned off.

To pull these observations together, the researchers hypothesized that the normal product of the *lacI* gene prevents the transcription of *lacZ* and *lacY* when lactose is absent. Because lactose triggers production of β-galactosidase, it was reasonable to expect that the *lacI* gene or gene product interacts with lactose in some way. (Later work showed that the inducer is actually a derivative of lactose called *allolactose*. For historical accuracy and simplicity, however, this discussion refers to lactose itself as the inducer.)

Jacob and Monod had succeeded in identifying three genes involved in lactose metabolism: *lacZ*, *lacY*, and *lacI*. They concluded that *lacZ* and *lacY* code for proteins required for the metabolism and import of lactose, while *lacI* is responsible for some sort of regulatory function. When lactose is absent, the *lacI* gene or gene product shuts down the expression of *lacZ* and *lacY*. But when lactose is present, the opposite occurs—transcription of *lacZ* and *lacY* is induced.

✔ If you understand the genes involved in lactose metabolism, you should be able to describe the specific function of *lacZ* and *lacY*. You should also be able to describe the effect of the *lacI* gene product when lactose is present versus absent and explain why these effects are logical.

Jacob and Monod followed up on these experiments by mapping the location of the three genes on the *E. coli* chromosome (**FIGURE 18.5**). They discovered that the genes are close together. This was a crucial finding because it suggested that *lacZ* and *lacY* might be transcribed together. Could the *lacI* regulatory gene govern both the *lacZ* and *lacY* genes? How does *lacI* actually work? And why do lactose and glucose have opposite effects on gene expression?

18.3 Negative Control of Transcription

In principle, there are two general ways that transcription can be regulated: by negative control or positive control.

1. **Negative control** occurs when a regulatory protein called a **repressor** binds to DNA and shuts down transcription (**FIGURE 18.6a**).

2. **Positive control** occurs when a regulatory protein called an **activator** binds to DNA and triggers transcription (**FIGURE 18.6b**).

When you are driving a car, negative control is exerted by setting the parking brake; positive control occurs when you step on the gas pedal. It turned out that the *lacZ* and *lacY* genes in *E. coli* are controlled by engaging or releasing a parking brake—they are under negative control.

(a) Negative control: Regulatory protein *shuts down* transcription.

(b) Positive control: Regulatory protein *triggers* transcription.

FIGURE 18.6 Genes Are Regulated by Negative Control, Positive Control, or Both. (To review transcription initiation, see Figure 17.3.)

The hypothesis that the *lacZ* and *lacY* genes might be under negative control originated with Leo Szilard in the late 1950s. Szilard suggested to Monod that the *lacI* gene could code for a product that represses transcription of the *lacZ* and *lacY* genes. As it turned out, Szilard was right.

The *lacI* gene produces a repressor protein that exerts negative control over *lacZ* and *lacY* gene transcription. The repressor was proposed to bind directly to DNA at or near the promoter for the *lacZ* and *lacY* genes (**FIGURE 18.7a**).

To explain how lactose triggers transcription, Szilard and Monod proposed that lactose interacts with the repressor in a way that makes the repressor release from its binding site (**FIGURE 18.7b**). In negative control, the repressor is the parking brake; lactose releases the brake.

What about the constitutive mutants? **FIGURE 18.7c** shows that constitutive transcription is observed in *lacI⁻* mutants because a functional repressor is absent—the parking brake is broken.

To test the hypothesis of negative control by a repressor, Jacob, Monod, and co-workers added back a functioning copy of the *lacI* repressor gene to the *lacI⁻* mutants that made β-galactosidase all the time. If these cells were grown using glucose and no lactose, β-galactosidase production declined and then stopped.

This result supported the hypothesis that the repressor codes for a protein that shuts down transcription. Significantly, if the experimental cells were grown using lactose instead of glucose, β-galactosidase activity resumed. This result supported the hypothesis that lactose removes the repressor.

What's the take-home message? The *lacI* gene codes for a repressor protein that exerts negative control on *lacZ* and *lacY*. Lactose acts as an inducer by causing the repressor to release from DNA and ending negative control.

The Operon Model

Jacob and Monod summarized the results of their experiments with a comprehensive model of negative control that was published in 1961. One of their key conclusions was that the genes for β-galactosidase and galactoside permease are controlled together and transcribed into a single mRNA. To encapsulate this idea, they coined the term **operon** for a set of coordinately regulated bacterial genes that are transcribed together into one mRNA. Logically enough, the group of genes involved in lactose metabolism was termed the **lac operon.**

Later, a gene called *lacA* was found to be adjacent to *lacY* and *lacZ* and transcribed as part of the same operon. The *lacA* gene

FIGURE 18.7 The Hypothesis of Negative Control of the *lac* Operon. The plasma membrane and galactoside permease are shown as a reminder that lactose comes from outside the cell and controls genes within the *E. coli* chromosome. Repression of the *lac* operon is never complete, so there is always some galactoside permease to transport lactose into the cell and begin induction of the *lac* operon.

Regulatory sequence

lac operon

DNA

lacI promoter | lacI | Promoter of *lac* operon | Operator | lacZ | lacY | lacA

FIGURE 18.8 The *lac* Operon and *lacI* Gene. This view emphasizes the arrangement of genes and regulatory sequences and is not drawn to scale.

✔**EXERCISE** Using small, colored bits of candy or paper, add the repressor protein to the figure. Next, add RNA polymerase; then add lactose. At each step, explain what happens after the molecule is added.

codes for the enzyme transacetylase. This enzyme catalyzes reactions that allow certain types of sugars to be exported from the cell when they are too abundant and could harm the cell. The components of the *lac* operon are summarized in **FIGURE 18.8**.

Three hypotheses are central to the Jacob–Monod model of *lac* operon regulation:

1. The *lacZ*, *lacY*, and *lacA* genes are adjacent and are transcribed into one mRNA initiated from the single promoter of the *lac* operon. This is known as cotranscription, and it results in the coordinated expression of the three genes.

2. The repressor is a protein encoded by *lacI* that binds to DNA and prevents transcription of the *lac* operon genes (*lacZ*, *lacY*, and *lacA*). Jacob and Monod proposed that *lacI* is expressed constitutively, and that the repressor binds to a section of DNA in the *lac* operon called the **operator.**

3. The inducer (lactose) binds to the repressor. When it does, the repressor changes shape. The shape change causes the repressor to come off the DNA. Recall that this form of control over protein function is **allosteric regulation** (see Chapter 8). In allosteric regulation, a small molecule binds to a protein and causes it to change its shape and activity. When the inducer binds to the repressor, the repressor can no longer bind to DNA and transcription can proceed.

✔ If you understand negative control of the *lac* operon, you should be able to predict the effect of a mutation in the *lacI* gene that alters the repressor so it cannot bind to lactose, and the effect of a mutation in the operator that prevents repressor binding.

How Does Glucose Regulate the *lac* Operon?

The model of *lac* operon control, summarized in Figure 18.7, is elegant and successful in explaining experimental results. But it is not complete. After studying the model, you may think of an important question that it fails to answer: Where does glucose fit in?

Transcription of the *lac* operon is drastically reduced when glucose is present in the environment—even when lactose is available to induce β-galactosidase expression (see Figure 18.2). This makes sense, given that glucose is *E. coli*'s preferred carbon source. When glucose is already present, the cell doesn't need to cleave lactose as a way of acquiring glucose.

How can glucose prevent expression of the *lac* operon? Researchers recently discovered that glucose inhibits the lactose

transport activity of galactoside permease through a chain of molecular events. When both glucose and lactose are present in the environment, the transport of lactose into the cell is inhibited. Because lactose does not accumulate in the cytoplasm, the repressor remains bound to the operator. Negative control (as in Figure 18.7a) is in place. In contrast, when glucose levels outside the cell are low, galactoside permease is active. If lactose is present, it is transported into the cell and induces *lac* operon expression (as in Figure 18.7b).

The mechanism of glucose preventing the transport of inducer is known as inducer exclusion. Inducer exclusion affects the activity of many different sugar transporters in addition to galactoside permease. It allows *E. coli* to preferentially use glucose, even when other sugars are also present outside the cell.

This understanding of how glucose regulates the *lac* operon is relatively new. For decades, researchers thought that when glucose levels outside the cell declined, an activator protein called CAP bound to a regulatory sequence in DNA just upstream of the promoter to increase the frequency of transcription initiation. There is strong evidence that binding of CAP to the regulatory sequence is important for efficient transcription of the *lac* operon. However, recent results indicate that CAP is always bound to the regulatory sequence, even in the presence of glucose.

Why Has the *lac* Operon Model Been So Important?

The *lac* operon has been an immensely important model system for two reasons. First, follow-up work showed that many bacterial genes and operons are under negative control by repressor proteins. This means that the findings on the *lac* operon are general. Second, the *lac* operon model introduced a fundamentally important idea: Gene expression is regulated by physical contact between regulatory proteins and specific regulatory sites in DNA. Publication of the *lac* operon model was a watershed event in the history of biological science.

Work on the *lac* operon also offered an important example of post-translational control over gene expression. To understand why, you have to realize that the repressor protein is always present; it is transcribed and translated constitutively. When a rapid change in *lac* operon activity is needed, it does not require changes in the transcription or translation of new repressor proteins. Instead, the activity of *existing* repressor proteins is altered.

This is exactly the prediction made at the beginning of this chapter—post-translational control is best when a rapid response is needed. As it turns out, this is a common type of control. In most cases, the activity of key regulatory proteins is controlled by post-translational modifications.

check your understanding

If you understand that . . .

- Negative control occurs when something must be taken away for transcription to occur.
- The *lac* operon repressor exerts negative control over three protein-coding genes by binding to the operator site in DNA, near the promoter.
- For transcription to occur in the *lac* operon, an inducer molecule (a derivative of lactose) must bind to the repressor, causing it to change shape and release from the operator.
- The *lac* operon is not transcribed when glucose is available because glucose prevents lactose transport into the cell.

✔ **You should be able to . . .**

1. Explain why lactose should induce transcription of the *lac* operon.

2. Diagram the *lac* operon, showing the relative positions of the operator, the promoter, and the three protein-coding genes; indicate what is happening at the operon in the absence of lactose and in the presence of lactose.

Answers are available in Appendix A.

18.4 Positive Control of Transcription

Positive control is an important way of controlling transcription. In positive control, an activator protein binds to a regulatory sequence in DNA when genes are turned on. When bound to DNA, the activator interacts with RNA polymerase to increase the rate of initiating transcription (see Figure 18.6b).

The ***ara* operon** provides an important example of positive control and of the process of science. The *ara* operon wasn't discovered in the laboratory of a famous scientist. Instead, students working on a project for a laboratory course were the first to uncover it.

This operon contains three genes that allow *E. coli* to use the sugar arabinose. Arabinose is found in many plant cell walls. When you eat vegetables, arabinose is available to the bacteria that inhabit your gut. Without arabinose in the environment, the *ara* operon is not transcribed. But when arabinose is present, transcription of the *ara* operon is turned on by an activator protein called **AraC**. The *ara* operon and an adjacent gene, *araC*, that codes for the araC activator are shown in **FIGURE 18.9**.

FIGURE 18.10 outlines how AraC controls the *ara* operon. The AraC protein is allosterically regulated by arabinose. When bound to arabinose, two copies of the AraC protein attach to a regulatory sequence of DNA called the *ara* initiator that lies just upstream of the promoter (see Figure 18.10a). Once AraC is bound to DNA, it can also bind to RNA polymerase. This interaction between AraC and the RNA polymerase helps to dock the polymerase to the promoter and accelerate the initiation of transcription.

Continued work on the *ara* operon revealed a surprise—AraC is both an activator and a repressor. In the absence of arabinose, the two copies of the AraC protein remain together; but while one araC copy remains bound to the initiator, the other copy now binds to a different regulatory site in DNA, the *ara* operator, as shown in Figure 18.10b. In this configuration, AraC works as a repressor to prevent the transcription of both the *ara* operon and the *araC* gene.

✔ If you understand positive control by the AraC protein, you should be able to predict the effect of a mutation that removes the part of AraC that binds to RNA polymerase.

18.5 Global Gene Regulation

A theme of this chapter is that cells respond to changing environments. To compete for resources, bacteria must be able to coordinate the expression of large sets of genes. As you've seen for the *lac* and *ara* operons, an effective way to express sets of genes together is to group them into an operon and transcribe them into a single mRNA. But there are limits to the size of operons. How can bacterial cells manage responses that require the expression of dozens or even hundreds of genes?

FIGURE 18.9 The *ara* Operon, Regulatory Sequences, and *araC* Gene.

(a) AraC protein is an **activator** when bound to arabinose.

(b) AraC protein is a **repressor** when arabinose is absent.

FIGURE 18.10 Positive and Negative Regulation of the *ara* Operon.

Global gene regulation is the coordinated regulation of many genes. You already learned that alternative sigma proteins provide one way for bacteria to turn on large numbers of genes in response to environmental change (Chapter 17). But there are other means of global gene regulation, such as grouping genes into a **regulon**—a set of separate genes or operons that contain the same regulatory sequences and that are controlled by a single type of regulatory protein.

Regulons allow bacteria to respond to challenges that include shortages of nutrients, sudden changes in temperature, exposure to radiation, or shifts in habitat. Let's explore how regulons work in general, and then look at two specific examples.

A general strategy for controlling regulon genes is shown in **FIGURE 18.11**. In this example, the regulon consists of many genes that are scattered across the genome. All of these genes are controlled by the same type of repressor protein that binds to the same operator sequences near the promoter of each gene. When an environmental change triggers the removal of the repressor protein from all the operators, every gene in the regulon is transcribed.

Regulons can be under negative control by a repressor protein or positive control by an activator protein. The regulon in Figure 18.11 is under negative control. The SOS response regulon works exactly this way to allow bacterial cells to repair extensive damage to DNA that can occur when cells are exposed to ultraviolet light, other types of radiation, or some chemicals. Damaged DNA sets off an SOS signal that induces the transcription of more than 40 genes that code for DNA repair enzymes and for DNA polymerases that can use damaged DNA as a template. Without the SOS response, bacteria with massive DNA damage would face almost certain death.

The ToxR regulon of *Vibrio cholera*—the bacterium that causes cholera—is under positive control. This regulon allows *V. cholera* to colonize the human gut and to produce toxins that cause diarrhea. Cholera kills 120,000 people each year and sickens as many as 18 million. ToxR regulon genes are inactive when *V. cholera* lives outside a human host. When bacteria from contaminated drinking water encounter the environment of the human gut, this sets off a signal that activates an activator protein. The activator induces a response by binding to a regulatory DNA sequence near the promoters of all ToxR regulon genes to stimulate their transcription. The diarrhea induced by this regulon is adaptive for *V. cholera* because it spreads more bacteria into the environment to infect new hosts.

What are the general messages of this chapter? Interactions among protein regulators and the DNA sequences they bind produce finely tuned control over gene expression, regulating individual genes, operons, or large sets of genes. With these exquisite controls over gene expression, bacteria have been able to compete, grow, and reproduce for more than 3 billion years of life's history.

Do eukaryotes control their genes the same way as bacteria? If not, what are the differences? These questions are the focus of the next chapter.

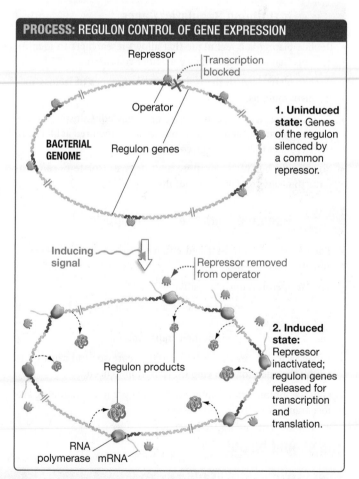

PROCESS: REGULON CONTROL OF GENE EXPRESSION

1. Uninduced state: Genes of the regulon silenced by a common repressor.

2. Induced state: Repressor inactivated; regulon genes released for transcription and translation.

FIGURE 18.11 Genes of a Regulon Are Expressed Together. The "| |" symbols indicate regions of the bacterial genome not shown between regulon genes.

If you understand . . .

18.1 An Overview of Gene Regulation and Information Flow

- Changes in gene expression allow bacterial cells to respond to environmental changes.
- Most gene products are produced or activated only when they are needed.
- Gene expression can be controlled at three levels: transcription, translation, or post-translation (protein activation).
- Transcriptional control can be negative or positive. Negative control occurs when a regulatory protein prevents transcription. Positive control occurs when a regulatory protein increases the frequency of initiating transcription.

✔ You should be able to describe one component of the *lac* operon that is under transcriptional control and one component that is under post-translational control.

18.2 Identifying Regulated Genes

- Replica plating is a technique that allows researchers to identify mutants that cannot grow in a particular condition.
- Replica plating led to the isolation of three types of lactose metabolism mutants.
- Transcription may be constitutive or regulated. Constitutive expression occurs in genes whose products are required at all times, such as genes that encode glycolytic enzymes.

✔ You should be able to propose a strategy to isolate *E. coli* mutants that can grow at 33°C, but not at 42°C.

18.3 Negative Control of Transcription

- The *lac* operon is transcribed efficiently when lactose is present and glucose is absent.
- The *lac* operon is under negative control.
- Negative control occurs because a repressor protein binds to an operator sequence in DNA near the promoter of the protein-encoding genes to prevent their transcription.
- When lactose is present, it binds to the repressor and causes it to fall off the operator, allowing transcription to occur.
- Glucose inhibits transcription of the *lac* operon by inhibiting lactose transport into the cell.

✔ You should be able explain how the operator, repressor, and inducer relate to a car's parking brake.

18.4 Positive Control of Transcription

- Positive control of transcription occurs when a regulatory protein called an activator binds to a regulatory sequence in DNA.
- Activator proteins bind to RNA polymerase in addition to DNA. Binding between the activator and RNA polymerase increases the rate of transcription initiation.
- The *ara* operon codes for genes required for metabolism of the sugar arabinose. The operon is controlled by the AraC regulatory protein. AraC is an activator when bound to arabinose and a repressor when the protein is not bound to arabinose.

✔ You should be able to predict if mutations in the *ara* initiator sequence of the *ara* operon are most likely to affect positive regulation, negative regulation, or both.

18.5 Global Gene Regulation

- Bacterial cells often need to coordinate the expression of large sets of genes in response to changing environments.
- Regulons coordinate the expression of different genes by using a shared regulator that acts on a regulatory sequence found in all genes of the regulon. Regulons can work through negative control using repressors, or through positive control using activators.

✔ You should be able to propose a method that would allow more genes to become part of the SOS regulon.

(MB) MasteringBiology

1. MasteringBiology Assignments

Tutorials and Activities The *lac* Operon

Questions Reading Quizzes, Blue-Thread Questions, Test Bank

2. eText Read your book online, search, take notes, highlight text, and more.

3. The Study Area Practice Test, Cumulative Test, BioFlix® 3-D animations, Videos, Activities, Audio Glossary, Word Study Tools, Art

You should be able to . . .

✔ TEST YOUR KNOWLEDGE

Answers are available in Appendix A

1. Replica plating is used to isolate mutants that
   a. can produce an enzyme.
   b. cannot grow in a particular condition.
   c. can utilize lactose.
   d. turn yellow when lactose is broken down.

2. Why are the genes involved in lactose metabolism considered to be an operon?
   a. They occupy adjacent locations on the *E. coli* chromosome.
   b. They have a similar function.
   c. They are all required for normal cell function.
   d. They are all controlled by the same promoter.

3. In the *lac* operon, the repressor inhibits transcription when
   a. the repressor is bound to the inducer.
   b. the repressor is not bound to the inducer.
   c. the repressor is bound to glucose.
   d. the repressor is not bound to the operator.

4. Activators bind to regulatory sequences in _____ and to _____ polymerase.

5. How does inducer exclusion control gene expression in the *lac* operon?

6. A regulon is a set of genes controlled by
   a. one type of regulator of transcription.
   b. two or more different alternative sigma proteins.
   c. many different types of promoters.
   d. glucose.

TEST YOUR UNDERSTANDING *Answers are available in Appendix A*

7. *E. coli* expresses genes for glycolytic enzymes constitutively. Why?

8. Explain the difference between positive and negative control over transcription.

9. Predict what would happen if the *lac* repressor protein were altered so it could not bind inducer.
   a. The repressor could not bind to DNA.
   b. The repressor would always be bound to DNA.
   c. The repressor could bind to DNA only when cells were grown with glucose.
   d. The repressor could bind to DNA only when cells were grown without glucose.

10. Predict what would happen to regulation of the *lac* operon if the *lacI* gene were moved 50,000 nucleotides upstream of its normal location.

11. If any of the following hypothetical drugs could be developed, which would be most effective in preventing cholera?
    a. a drug that increased the amount of the ToxR activator
    b. a drug that blocked the DNA-binding activity of the activator
    c. a drug that increased rates of transcription in *V. cholerae*
    d. a drug that increased rates of translation in *V. cholerae*

12. IPTG is a molecule with a structure very similar to lactose. IPTG can be transported into cells by galactoside permease and can bind to the *lac* repressor protein. However, unlike lactose, IPTG is not broken down by β-galactosidase. Predict what would occur regarding *lac* operon regulation if IPTG were added to *E. coli* growth medium containing arabinose and no glucose or lactose.

TEST YOUR PROBLEM-SOLVING SKILLS *Answers are available in Appendix A*

13. You are interested in using bacteria to metabolize wastes at an old chemical plant and convert them into harmless compounds. You find bacteria that are able to tolerate high levels of the toxic compounds toluene and benzene, and you suspect that this is because the bacteria can break down these compounds into less-toxic products. If that is true, these toluene- and benzene-resistant strains will be valuable for cleaning up toxic sites. How could you find out whether these bacteria are metabolizing toluene as a source of carbon compounds?

14. **QUANTITATIVE** Imagine that you are repeating the replica-plating procedure of Jacob and Monod to find mutants that can't grow using lactose. After treating cells with a mutagen, you anticipate a mutation rate of 1×10^{-4} lactose-nonutilizing mutants per mutagen-treated cell. Based on this estimate, how many cells should you replica-plate to have a good chance of finding one mutant?

15. A type of mutation in the *lac* operator known as *lacO^c* prevents repressor binding to DNA and causes constitutive transcription of the *lac* operon. Which of the following secondary mutations might restore normal regulation to the *lac* operon in a *lacO^c* mutant?
    a. a *lacI* mutation that decreases the ability of the repressor to bind the inducer
    b. a *lacI* mutation that produces a repressor than can recognize the mutated *lacO^c* DNA sequence
    c. a promoter mutation that prevents it from being recognized by sigma
    d. an RNA polymerase mutation that allows it to bind to the promoter without using sigma

16. X-gal is a colorless, lactose-like molecule that can be split into two fragments by β-galactosidase. One of these product molecules is blue. The following photograph shows *E. coli* colonies growing in a medium that contains X-gal.

Find three colonies whose cells have functioning copies of β-galactosidase. Find three colonies whose cells might have mutations in the *lacZ* or in the *lacY* genes. Suppose you analyze the protein-coding sequence of the *lacZ* and *lacY* genes of cells from the three mutant colonies and find that these sequences are wild type (normal). What other region of the *lac* operon might be altered to account for the mutant phenotype of these colonies?

19 Control of Gene Expression in Eukaryotes

In this chapter you will learn that

Eukaryotes control gene expression many different ways

beginning with

Gene regulation in eukaryotes
(an overview)   19.1

then explore

| Regulation via chromatin remodeling **19.2** | Regulation during transcription initiation **19.3** | Regulation following transcription **19.4** |

compare

Regulation in prokaryotes **19.5**

apply

Regulation defects and cancer **19.6**

A model of eukaryotic DNA in the condensed state. The DNA (shown in red and pink) is wrapped around proteins (in green). The DNA has to be uncoiled before transcription can take place.

This chapter is part of the Big Picture. See how on pages 366–367.

Bacteria regulate gene expression to respond to changes in their environment. *Escherichia coli* thrive best if the genes that are required to import and cleave lactose are expressed only when the cells are relying on lactose as a source of energy (see Chapter 18).

Unicellular eukaryotes face similar challenges. Consider the yeast *Saccharomyces cerevisiae*, which is used extensively in the production of beer, wine, and bread. In nature this species lives on the skins of grapes and other fruits, where the sugars that the cells use as food vary in type and concentration as the fruit ripens, falls, and rots. For yeast cells to grow and reproduce efficiently, gene expression has to be modified in response to these changes.

The cells that make up multicellular eukaryotes face additional challenges. Consider your body, which contains trillions of cells, each with a specialized structure and function. You have heart

✔ When you see this checkmark, stop and test yourself. Answers are available in Appendix A.

muscle cells, lung cells, nerve cells, skin cells, and so on. Even though these cells are different, they contain the same genes. Your bone cells and blood cells aren't different because of a difference in their genes but because they *express* different genes. Your bone cells have blood-cell genes—they just don't transcribe them.

Why not? The answer is that your cells respond to their environment, just as bacteria and unicellular eukaryotes do. But there's a key difference. The cells in a multicellular eukaryote express different genes in response to changes in the *internal* environment—specifically, to signals from other cells. As a human being or an oak tree develops, cells that are located in different parts of the organism are exposed to different cell–cell signals. As a result, they express different genes. **Differential gene expression** is responsible for creating different cell types, arranging them into tissues, and coordinating their activity to form the multicellular society we call an individual.

How does all of this regulation and differentiation happen? Later chapters introduce the signals that trigger the formation of muscle, bone, leaf, and flower cells (see Unit 4). In contrast, this chapter focuses on what happens after a eukaryotic cell receives such a signal. Let's start with an overview of how gene expression can be controlled, and close with a look at how defects in the process can trigger cancer.

19.1 Gene Regulation in Eukaryotes— An Overview

Like bacteria, eukaryotes can control gene expression at the levels of transcription, translation, and post-translation. But as **FIGURE 19.1** shows, three additional levels of control occur in eukaryotes as genetic information flows from DNA to proteins.

The first additional level of control involves the DNA–protein complex at the top of the figure. In eukaryotes, DNA is wrapped around proteins to create a structure called **chromatin.** Eukaryotic genes have promoters, just as bacterial genes do; but before transcription can begin in eukaryotes, the stretch of DNA containing the promoter must be released from tight interactions with proteins, so that RNA polymerase can make contact with the promoter. To capture this idea, biologists say that **chromatin remodeling** must occur before transcription.

The second level of regulation that is unique to eukaryotes is **RNA processing**—the steps required to produce a mature, processed mRNA from a primary RNA transcript. Recall that introns have to be spliced out of primary transcripts (see Chapter 17). In many cases, carefully orchestrated alternative splicing occurs—meaning that different combinations of exons are included in the mRNA. If different cells use different splicing patterns, different gene products result.

Third, mRNA life span is regulated in eukaryotes: mRNAs that remain in the cell for a long time tend to be translated more than mRNAs that have a short life span.

Each of the six potential control points shown in Figure 19.1 is employed in eukaryotic cells. This chapter explores all six—chromatin

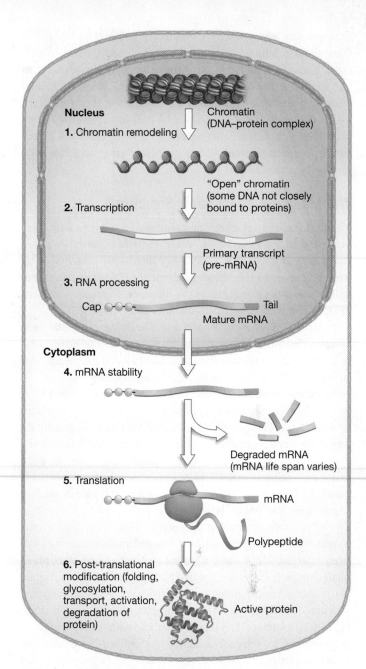

FIGURE 19.1 In Eukaryotes, Gene Expression Can Be Controlled at Many Different Levels.

remodeling, transcription, RNA processing, mRNA stability, translation, and post-translational modification of proteins.

To appreciate the breadth and complexity of gene regulation in eukaryotes, let's follow the series of events that occur as an embryonic cell responds to a developmental signal. Suppose a molecule arrives that specifies the production of a muscle-specific protein. What happens next?

19.2 Chromatin Remodeling

For a molecular signal to trigger the transcription of a particular gene, the chromatin around the target gene must be remodeled.

To appreciate why, consider that a typical cell in your body contains about 6 billion base pairs of DNA. Lined up end to end, these nucleotide pairs would form a double helix about 2 m (6.5 feet) long. But the nucleus that holds this DNA is only about 5 μm in diameter—far less than the thickness of this page. To fit inside the nucleus, the DNA must be packed tightly—so tightly that RNA polymerase can't access it. How is DNA packaged? And how can it be unpacked at particular genes so RNA polymerase can transcribe it?

What Is Chromatin's Basic Structure?

The first data on the chemical composition of chromatin were published in the early 1900s, when researchers established that eukaryotic DNA is intimately associated with proteins. Later work documented that the most abundant DNA-associated proteins belong to a group called the **histones.** Chromatin consists of DNA complexed with histones and other proteins.

In the 1970s electron micrographs like the one in **FIGURE 19.2a** revealed that chromatin has a regular structure. In some preparations for electron microscopy, chromatin looked like beads on a string. The "beads" came to be called **nucleosomes.**

More information emerged in 1984 when researchers determined the three-dimensional structure of eukaryotic chromatin by using X-ray crystallography (see **BioSkills 11** in Appendix B). The X-ray crystallographic data indicated that each nucleosome consists of DNA wrapped almost twice around a core of eight histone proteins. As **FIGURE 19.2b** indicates, a histone called H1 "seals" DNA to each nucleosome. Between each pair of nucleosomes there is a "linker" stretch of DNA.

The intimate association between DNA and histones occurs in part because DNA is negatively charged and histones are positively charged. DNA has a negative charge because of its phosphate groups; histones are positively charged because they contain many lysines and arginines, two positively charged amino acids.

There are additional layers of complexity in packaging DNA. H1 histones interact with one another and with histones in other nucleosomes to produce a tightly packed structure like that shown in Figure 19.2b. Based on its width, this structure is called the 30-nanometer fiber. (Recall that a nanometer is one-billionth of a meter and is abbreviated nm.)

Finally, the 30-nm fibers are attached at intervals along their length to proteins that form a scaffold or framework inside the nucleus. In this way, the entire chromosome is organized and held in place. When chromosomes condense before mitosis or meiosis, the scaffold proteins and 30-nm fibers are folded into still larger and more tightly packed structures.

A eukaryotic chromosome, then, is made up of chromatin that has several layers of organization: The DNA is wrapped around histones to form nucleosomes, nucleosomes are packed into 30-nm fibers, 30-nm fibers are attached to scaffold proteins, and the entire assembly can be folded into the highly condensed structure observed during cell division.

Although research has shown that bacterial DNA interacts with proteins that are organized similarly to nucleosomes,

(a) Nucleosomes in chromatin

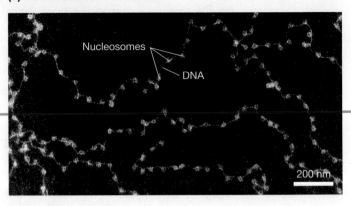

Nucleosomes

DNA

200 nm

(b) Nucleosome structure

Linker DNA

H1 protein attached to linker DNA and nucleosome

Nucleosome

Group of 8 histone proteins

30-nm fiber

30 nm

Protein scaffold

Chromosome in its condensed form (the form it's in during cell division)

FIGURE 19.2 Chromatin Has Several Levels of Organization.

nothing like the 30-nm fibers or higher-order arrangements has been observed in bacterial chromosomes.

The elaborate structure of eukaryotic chromatin does more than just package DNA so that it fits into the nucleus. Chromatin structure also has profound implications for the control of gene expression. To appreciate this point, consider the 30-nm fiber illustrated in Figure 19.2b. If this tightly packed stretch of DNA contains a promoter, how can RNA polymerase bind to it and initiate transcription?

Evidence that Chromatin Structure Is Altered in Active Genes

Once the nucleosome-based structure of chromatin was established, biologists hypothesized that the close physical interaction between DNA and histones must be altered for RNA polymerase to make contact with DNA. More specifically, biologists hypothesized that a gene could not be transcribed until the condensed chromatin near its promoter was remodeled.

The central idea is that chromatin must be decondensed, to expose the promoter so RNA polymerase can bind to it. If so, then chromatin remodeling would represent the first step in the control of eukaryotic gene expression. Two types of studies have provided strong support for this hypothesis.

DNA in Condensed Chromatin Is Protected from DNase DNases are enzymes that cut DNA. Some DNases cleave DNA at random locations, and these cannot cut efficiently if DNA is tightly wrapped with proteins. As **FIGURE 19.3** shows, this type of DNase works effectively only if DNA is in a decondensed configuration.

Harold Weintraub and Mark Groudine used this observation to test the hypothesis that the DNA of actively transcribed genes is in an open configuration. In chicken blood cells, they compared chromatin structure in two genes: β-globin and ovalbumin. β-globin is a protein that is part of the hemoglobin found in red blood cells; ovalbumin is a protein found in egg white. In blood cells, the β-globin gene is transcribed at high levels, but the ovalbumin gene is not transcribed at all.

After treating blood cells with DNase and then analyzing the state of chromatin at the β-globin and ovalbumin genes, the researchers found that DNase cut the β-globin gene DNA much more readily than DNA of the ovalbumin gene. They interpreted this finding as evidence that in blood cells, chromatin of the

actively transcribed β-globin gene was decondensed; and conversely, chromatin of the non-transcribed ovalbumin gene was condensed. Studies using DNase on different genes in different cell types yielded similar results.

Histone Mutants The second type of evidence in support of the chromatin-remodeling hypothesis comes from studies of mutant brewer's yeast cells that do not produce the usual complement of histones. In these mutant cells, many genes that are normally never transcribed are instead always transcribed at high levels.

To interpret this finding, biologists hypothesized that the lack of histone proteins prevented the assembly of normal chromatin. If the absence of normal histone–DNA interactions promotes transcription, then the presence of normal histone–DNA interactions must prevent it.

Taken together, the data suggest that in their normal, or default, state, eukaryotic genes are turned off. This is a new mechanism of negative control—different from repressor proteins (introduced in Chapter 18). When DNA is wrapped into a 30-nm fiber, the parking brake is on. If so, then gene expression depends on chromatin being opened up in the promoter region.

How Is Chromatin Altered?

Research on chromatin remodeling has been proceeding at a furious pace, and biologists have succeeded in identifying some of the key players that work to change the state of chromatin condensation. These include enzymes that add methyl groups to DNA, enzymes that chemically modify histones, and macromolecular machines that actively reshape chromatin. Let's examine each of these in turn.

DNA Methylation A group of enzymes known as **DNA methyltransferases** add methyl groups ($-CH_3$) to cytosine residues in DNA. In mammals, the sequence recognized by these enzymes is a C next to a G in one strand of the DNA. This sequence is abbreviated CpG and is shown below in methylated form within a stretch of DNA.

Why is **DNA methylation** important? Methylated CpG sequences are recognized by proteins that trigger chromatin condensation. Actively transcribed genes usually have low levels of methylated CpG near their promoters, and non-transcribed genes usually have high levels of methylated CpG.

Histone Modification DNA methylation is only one part of the chromatin alteration story. A large set of enzymes adds a variety of chemical groups to specific amino acids of histone proteins. These include phosphate groups, methyl groups, short polypeptide chains, and acetyl groups ($-COCH_3$). Addition of these groups to histones promotes condensed or decondensed chromatin depending on the specific set of modifications.

Condensed chromatin   Decondensed chromatin

Treatment with DNase I

Condensed chromatin   Degraded DNA

FIGURE 19.3 DNase Assay for Chromatin Structure. DNase is an enzyme that cuts DNA at random locations. It cannot cut condensed chromatin.

To account for these effects, researchers have proposed the existence of a **histone code.** The histone code hypothesis postulates that particular combinations of histone modifications set the state of chromatin condensation for a particular gene. In turn, this has an important role in regulating transcription. Let's take a closer look at one way histone modifications can control chromatin structure.

As shown in **FIGURE 19.4,** two different types of enzymes can add or remove acetyl groups from histones. **Histone acetyltransferases (HATs)** add acetyl groups to the positively charged lysine residues in histones, and **histone deacetylases (HDACs)** remove them. **Acetylation** of histones usually results in decondensed chromatin, a state associated with active transcription. How can acetylation of histones promote chromatin decondensation? When HATs add acetyl groups, the acetyl groups neutralize the positive charge on lysine residues and loosens the close association of nucleosomes with the negatively charged DNA. The addition of acetyl groups also creates a binding site for other proteins that help open the chromatin.

In contrast, when HDACs remove acetyl groups from histones, this process usually leads to condensed chromatin, a state associated with no transcription. HATs are an on switch for transcription, and HDACs are an off switch.

Chromatin-Remodeling Complexes

Other major players in chromatin alteration and gene regulation are enzymes that form macromolecular machines called **chromatin-remodeling complexes.** These machines harness the energy in ATP to reshape chromatin. Chromatin-remodeling complexes cause nucleosomes to slide along the DNA or, in some cases, knock the histones completely off the DNA to open up stretches of chromatin and allow gene transcription.

Condensed chromatin

FIGURE 19.4 Acetyl Groups Decondense Chromatin. Histone acetyltransferases (HATs) cause chromatin to decondense; histone deacetylases (HDACs) cause it to condense.

✔**QUESTION** Are HAT and HDAC elements in positive control or negative control? Explain your reasoning.

DNA methylation, histone modifications, and chromatin-remodeling complexes work together to fine-tune chromatin condensation at specific genes. The take-home message from work on chromatin remodeling is simple: The condensation state of chromatin is critical in determining whether transcription occurs.

Chromatin Modifications Can Be Inherited

The pattern of chromatin modifications varies from one cell type to another. For example, suppose within an individual you analyzed the same gene in a muscle cell and a brain cell. This and other genes would likely have a different pattern of DNA methylation and histone acetylation in the two cell types.

DNA methylation and histone modifications are an example of **epigenetic inheritance,** the collective term for patterns of inheritance that are due to something other than differences in DNA sequences. The *epi*– of epigenetics comes from the Greek word meaning "upon." It implies another level of inheritance that adds to standard DNA-based mechanisms to explain how different phenotypes are transmitted.

With epigenetic inheritance, if a cell received a "become muscle" signal early in development, it would modify its chromatin in distinctive ways and pass those modifications on to its descendants. Muscle cells are different from brain cells not because they contain different genes, but largely because they have inherited different patterns of DNA methylation and histone modifications during the course of their development.

But the story of epigenetic inheritance involves more than just differentiation of cell types during development. Evidence is emerging that epigenetic mechanisms can record early-life events and that this archive can be difficult to erase. This is the case when prenatal conditions alter the patterns of chromatin modification and the later-life phenotypes of a mother's offspring.

For example, biologists have long known that rats born to mothers fed low-protein diets during pregnancy and while nursing have a greatly increased risk of developing disorders similar to type 2 diabetes. Type 2 diabetes is a serious and increasingly common disease that alters the cellular uptake of glucose (Chapter 44). Both genetic factors and environmental factors, such as diet, play important roles in diabetes development.

One significant gene associated with diabetes is *Hnf4a. Hnf4a* codes for a regulator of genes involved in glucose uptake. Rats born to protein-deprived mothers develop symptoms of diabetes later in life, even when these rats are fed a normal, healthy diet from the time they are weaned. These diabetic rats also express the *Hnf4a* regulatory gene at lower levels than normal rats. Could epigenetic mechanisms be at work in silencing *Hnf4a* expression?

One team's approach to probing this question is shown in **FIGURE 19.5.** Using a treatment group and a control group, the researchers measured the types of histone modifications found at a key regulatory region of the *Hnf4a* gene. A regulatory region is a section of DNA that, like prokaryotic operators (Chapter 18), is involved in controlling the activity of a gene. The chromatin at this region has to be opened up for transcription to occur. Levels

QUESTION: Does poor nutrition in a mother produce epigenetic effects in offspring?

HYPOTHESIS: Protein-deprived mothers will produce offspring with abnormal histone modifications.

NULL HYPOTHESIS: Protein-deprived mothers will produce offspring with normal histone modifications.

EXPERIMENTAL SETUP:

1. Provide rat mothers with a normal or a low-protein diet during pregnancy and while nursing.

2. After weaning, feed rat pups a normal diet and raise to old age.

3. Determine types of histone modifications for a regulatory gene involved in diabetes. (Also measure gene transcription.)

TRANSCRIPTION

Hnf4a

Modifications promoting:

Modifications promoting:

condensed chromatin

decondensed chromatin

PREDICTION: Offspring of mothers fed a low-protein diet will have abnormal histone modifications.

PREDICTION OF NULL HYPOTHESIS: Offspring of mothers fed a low-protein diet will have normal histone modifications.

RESULTS:

CONCLUSION: A mother's diet influences chromatin modifications and gene expression patterns throughout her offspring's life.

FIGURE 19.5 Events in Early Life Can Be Recorded through Epigenetic Mechanisms.

SOURCE: Sandovici, I., N. H. Smith, and M. D. Nitert. 2011. Maternal diet and aging alter the epigenetic control of a promoter-enhancer interaction at the *Hnf4a* gene in rat pancreatic islets. *Proceedings of the National Academy of Sciences USA* 108: 5449–5454.

✔ **QUESTION** What could researchers do to prove that the histone modifications are causing reduced regulatory gene transcription?

of *Hnf4a* transcription in the control and treatment groups were also measured. What did the team learn?

As the graph on the left in the "Results" section of Figure 19.5 shows, they found that histone modifications that lead to condensed chromatin were *elevated* in rats born to malnourished mothers compared to control offspring. Conversely, histone modifications associated with decondensed chromatin were significantly *reduced* in the treatment group. The graph on the right confirms that transcription of *Hnf4a* was much lower in the treatment group than the control group. Together these results demonstrate a correlation between altered histone modifications and decreased levels of *Hnf4a* gene expression.

Remember that in this study, all rats were provided a healthy diet after weaning. This finding implies that a mother's nutritional status during pregnancy and nursing is responsible for the types of chromatin modifications seen in rats that develop diabetes.

Chromatin remodeling must occur before transcription. Now the question is, What happens once a section of chromatin is opened and DNA becomes accessible to RNA polymerase?

check your understanding

C Y U

If you understand that . . .

- Eukaryotic DNA is wrapped tightly around histones, forming nucleosomes, which are then coiled into structures called 30-nm chromatin fibers.
- Before transcription can begin, the DNA–protein complex of chromatin must be decondensed so that RNA polymerase can contact the promoter.
- Methylation of DNA and specific chemical modifications of histones play a key role in determining whether chromatin is opened and a gene is expressed.
- In many cases, the patterns of DNA methylation and histone modification in a cell are passed on to its daughter cells.

✔ **You should be able to . . .**

1. Predict how gene expression will be affected if a cell is grown with compounds that prevent DNA methylation.

2. Explain how certain patterns of histone acetylation or DNA methylation could influence whether a cell became a muscle cell or a brain cell.

Answers are available in Appendix A.

19.3 Initiating Transcription: Regulatory Sequences and Regulatory Proteins

As in bacteria, the **promoter** is a site in DNA where RNA polymerase binds to initiate transcription. Recent findings from genome sequencing projects (see Chapter 20) have shown that there is still much to be learned about eukaryotic promoters. What is known is that promoters in eukaryotes are more complex than bacterial promoters, often containing two or three conserved sequences that serve as binding sites for proteins needed to start transcription. The most intensively studied of these is a sequence known as the **TATA box.**

Once a promoter that contains a TATA box has been exposed by chromatin remodeling, the first step in initiating transcription is binding of the **TATA-binding protein (TBP).** Proteins related to TBP also work on promoters with other conserved sequences. But the binding of TBP or any of its relatives does not guarantee that a gene will be transcribed. In eukaryotes, a wide array of other DNA sequences and proteins work together to allow transcription.

Promoter-Proximal Elements Are Regulatory Sequences Near the Promoter

The first **regulatory sequences** in eukaryotic DNA were discovered in the late 1970s, when Yasuji Oshima and co-workers set out to understand how yeast cells control the metabolism of the sugar galactose.

When galactose is absent, *S. cerevisiae* cells produce only tiny quantities of the enzymes required to metabolize it. But when galactose is present, transcription of the genes encoding these enzymes increases by a factor of 1000.

The team's first major result was the discovery of mutant cells that failed to produce any of the five enzymes required for galactose metabolism, even if galactose was present. To interpret this observation, they hypothesized that

1. the five genes are regulated together, even though they are not on the same chromosome;

2. normal cells have an activator protein that exerts positive control over the five genes;

3. the mutant cells have a mutation that completely disables the activator protein.

Other researchers were able to isolate the regulatory protein and confirm that it binds to a short stretch of DNA located just upstream from the promoter for all five genes required for galactose use.

In bacteria, genes that need to be regulated together are often clustered into a single operon and transcribed into a single mRNA. In contrast, eukaryotes use the strategy uncovered by Oshima's group for the galactose-metabolizing genes in yeast—co-regulated genes are not clustered together, but instead share a regulatory DNA sequence that binds the same regulatory protein.

Regulatory DNA sequences similar to those first discovered in yeast have now been found in a wide array of eukaryotic genes and species. Regulatory sequences like these that are located close to the promoter and bind regulatory proteins are termed **promoter-proximal elements.**

Unlike the promoter itself, promoter-proximal elements have sequences that are unique to specific sets of genes. In this way, they furnish a mechanism for eukaryotic cells to express certain genes but not others.

The discovery of promoter-proximal elements and a mechanism of positive control suggested a satisfying parallel between gene regulation in bacteria and in eukaryotes. This picture changed, however, when researchers discovered a new class of eukaryotic DNA regulatory sequences—sequences unlike anything in bacteria.

Enhancers Are Regulatory Sequences Far from the Promoter

Susumu Tonegawa and colleagues made a startling discovery while exploring how human cells regulate gene expression. The gene studied by Tonegawa's group was broken into many introns and exons. Recall that introns are transcribed sequences that are spliced out of the primary transcript; exons are transcribed regions that are included in the mature RNA once splicing is complete (Chapter 17). The researchers discovered a regulatory sequence within one of the introns that was required for transcription of the gene.

This finding was remarkable for two reasons: (1) The regulatory sequence was thousands of bases away from the promoter, and (2) it was downstream of the promoter instead of upstream. Regulatory sequences that are far from the promoter and activate transcription are termed **enhancers.** Follow-up work has shown that enhancers occur in all eukaryotes and that they have several key characteristics:

- Enhancers can be more than 100,000 bases away from the promoter. They can be located in introns or in untranscribed sequences on either the 5′ or 3′ side of the gene (See **FIGURE 19.6**).

- Like promoter-proximal elements, many types of enhancers exist.

- Most genes have more than one enhancer.

- Enhancers usually have binding sites for more than one protein.

- Enhancers can work even if their normal 5′→3′ orientation is flipped, or if they are moved to a new location in the vicinity of the gene.

Enhancers are regulatory DNA sequences unique to eukaryotes. When regulatory proteins called **transcriptional activators** bind to enhancers, transcription begins. Thus, enhancers and activators are like a gas pedal—an element in positive control. Eukaryotes also possess regulatory sequences that are similar in structure and share key characteristics with enhancers but work to inhibit transcription. These DNA sequences are called **silencers.** When regulatory proteins called **repressors** bind to silencers, transcription is shut down. Silencers and repressors are like a brake—an element in negative control.

The Role of Transcription Factors in Differential Gene Expression

Follow-up work supported the hypothesis that enhancers and silencers are binding sites for activators and repressors that regulate transcription. Collectively, these proteins are termed

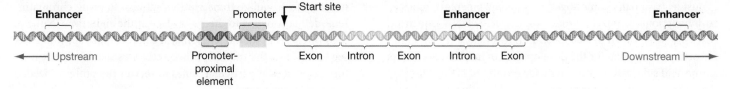

FIGURE 19.6 Enhancers and Promoter-Proximal Elements Regulate the Expression of Eukaryotic Genes. Promoter-proximal elements are near the promoter. Enhancers are located farther away, may be upstream or downstream of the promoter, and may even be within introns. Exons and introns are not drawn to scale. They are typically very large compared with regulatory sequences.

✔**EXERCISE** Compare and contrast the structure of this typical eukaryotic gene and the structure of a bacterial operon.

regulatory transcription factors, or often **transcription factors** for short. By analyzing mutant yeast, fruit flies, and roundworms that have defects in the expression of particular genes, biologists have identified a large number of transcription factors that bind to enhancers, silencers, and promoter-proximal elements.

These results support one of the most general statements researchers are able to make about gene regulation in eukaryotes: Different types of cells express different genes because they have different transcription factors. In multicellular species, the transcription factors, in turn, are produced in response to signals that arrive from other cells early in embryonic development.

For example, if a signal that says "become a muscle cell" reaches a cell in the early embryo, it triggers a signal transduction cascade (see Chapter 11) that leads to the production of transcription factors specific to muscle cells. Because different transcription factors bind to specific regulatory sequences, they turn on the production of muscle-specific proteins. But if no "become-a-muscle-cell" signal arrives, then no muscle-specific transcription factors are produced and no muscle-specific gene expression takes place.

Differential gene expression is a result of the production or activation of specific transcription factors. Eukaryotic genes are turned on when transcription factors bind to enhancers and promoter-proximal elements; the genes are turned off when transcription factors bind to silencers or when chromatin is condensed. Distinctive transcription factors are what make a muscle cell a muscle cell and a bone cell a bone cell.

How Do Transcription Factors Recognize Specific DNA Sequences?

Each transcription factor must be able to recognize and bind to a specific DNA sequence. How can it do this?

Recall that DNA bases are partially exposed in the major and minor grooves of the DNA double helix (see Figure 4.7 for a review). The edges of an AT base pair and a GC base pair that project into the grooves of the DNA helix contain different sets of atoms and have different surface shapes (**FIGURE 19.7a**). These differences in composition and shape can be recognized by transcription factors.

(a) AT and CG base pairs present different shapes and chemical groups in the grooves of DNA.

(b) Transcription factors recognize a specific sequence of bases in target DNA.

FIGURE 19.7 How Transcription Factors Bind to Regulatory Sequences. (a) The edges of base pairs that project into the major and minor grooves of the double helix present a different structure and set of atoms. Atoms that can participate in hydrogen bonding with amino acids of transcription factors are indicated by arrows. The methyl group on thymine (T) is also important in recognition. **(b)** A transcription factor (green) involved in muscle-cell differentiation binding to a regulatory sequence in DNA. The bases recognized by the protein are highlighted in red.

Just as base pairs come together by complementary molecular interactions, so too can proteins and specific DNA sequences. An example is shown in **FIGURE 19.7b**. In this case, a transcription factor that is essential for the development of muscle cells inserts amino acid side chains into two major grooves of DNA. This particular transcription factor binds to a specific enhancer sequence because of complementary interactions between base pairs and its amino acids. Without such specific interactions between transcription factors and DNA, the development of muscle cells—or any other specialized cell type—would not be possible.

A Model for Transcription Initiation

Although gene expression can be controlled at many levels, regulating the start of transcription is at center stage. For a process so important, many questions remain. What is clear is that transcription factors must interact with regulatory sequences to initiate transcription.

Besides the regulatory transcription factors you've learned about that bind to enhancers, silencers, and promoter-proximal elements, there is another type: **basal transcription factors.** These are proteins that interact with the promoter and are not restricted to particular genes or cell types. The term basal implies that these proteins are necessary for transcription to occur, but they do not provide much in the way of regulation. The promoter-recognized TATA-binding protein (TBP) that you learned about earlier is an example of a basal transcription factor that is common to many genes. ✔ If you understand this concept, you should be able to compare and contrast the regulatory and basal transcription factors found in muscle cells versus nerve cells.

In addition to transcription factors, a large complex of proteins called **Mediator** acts as a bridge between regulatory transcription factors, basal transcription factors, and RNA polymerase II.

FIGURE 19.8 summarizes one model for how transcription is initiated in eukaryotes.

Step 1 Transcriptional activators bind to DNA and recruit chromatin-remodeling complexes and histone acetyltransferases (HATs).

Step 2 The chromatin-remodeling complexes and HATs open a swath of chromatin that includes the promoter, promoter-proximal elements, and enhancers.

Step 3 Other transcriptional activators bind to the newly exposed enhancers and promoter-proximal elements; basal transcription factors bind to the promoter and recruit RNA polymerase II.

Step 4 Mediator connects the transcriptional activators and basal transcription factors that are bound to DNA. This step is made possible through DNA looping. RNA polymerase II can now begin transcription.

✔ If you understand this model, you should be able to explain why DNA forms loops near the promoter in order for transcription to begin.

An important point in this model of transcription initiation is the dual role of transcriptional activators. Activators work not only to stimulate transcription but also to bring chromatin-remodeling proteins to the right place at the right time. None of the proteins that remodel chromatin can recognize specific DNA sequences. It is the transcriptional activators that bind to regulatory sequences at particular genes to recruit the proteins needed to remodel chromatin.

The role of transcriptional activators in bringing in proteins that decondense chromatin leads to a chicken-and-egg paradox: How can an activator bind to DNA in the first place if chromatin is condensed? It turns out that except in its most highly condensed forms, chromatin is dynamic. DNA occasionally dissociates from nucleosomes, exposing regulatory sequences to activators that are present in a particular cell type.

Getting RNA polymerase to initiate transcription requires interactions between many proteins, including transcriptional activators that are bound to enhancers and promoter-proximal elements, Mediator, basal transcription factors, and RNA polymerase itself. The result is a large, macromolecular machine that is positioned at a gene's start site and capable of starting transcription.

Compared with what happens in bacteria, where just three to five proteins may interact at the promoter to initiate transcription, the process in eukaryotes is remarkably complicated.

check your understanding

If you understand that . . .

- Eukaryotic genes have regulatory sequences called promoter-proximal elements close to their promoters.
- Eukaryotic genes also have regulatory sequences called enhancers or silencers far from their promoters.
- Transcription initiation is a multistep process that begins when transcriptional activators bind to DNA and recruit proteins that open chromatin.
- Interactions between regulatory transcription factors, Mediator, and basal transcription factors position RNA polymerase II at the gene's start site.

✔ **You should be able to . . .**

1. Compare and contrast the nature of regulatory sequences and regulatory proteins in bacteria versus eukaryotes.

2. Explain why the presence of certain transcription factors could influence whether a cell becomes a muscle cell or a brain cell.

Answers are available in Appendix A.

19.4 Post-Transcriptional Control

Chromatin remodeling and transcription are just the opening to the story of gene regulation. Once a gene is transcribed, a series of events has to occur before a final product appears (see Figure 19.1). Each of these events offers an opportunity

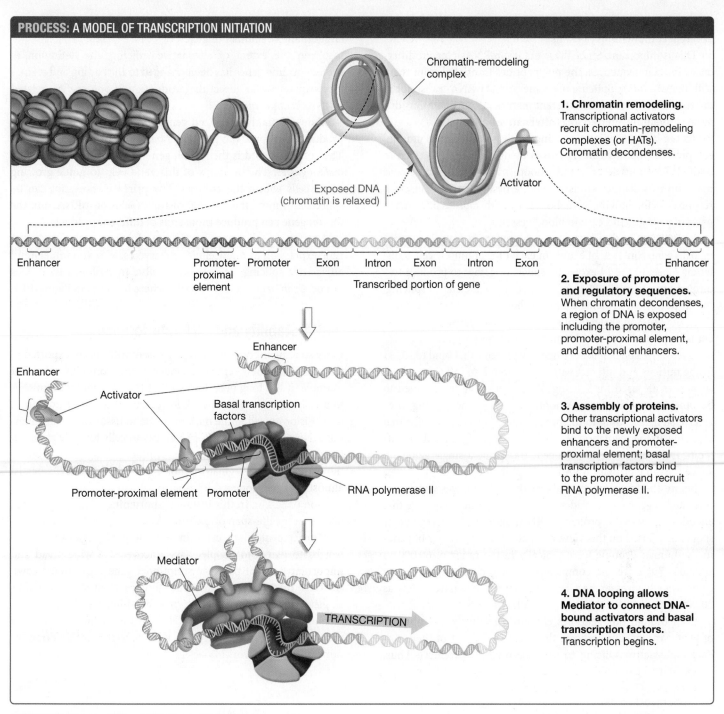

1. Chromatin remodeling. Transcriptional activators recruit chromatin-remodeling complexes (or HATs). Chromatin decondenses.

Chromatin-remodeling complex

Exposed DNA (chromatin is relaxed)

Activator

Enhancer

Promoter-proximal element

Promoter

Exon Intron Exon Intron Exon

Transcribed portion of gene

Enhancer

2. Exposure of promoter and regulatory sequences. When chromatin decondenses, a region of DNA is exposed including the promoter, promoter-proximal element, and additional enhancers.

Enhancer

Enhancer

Activator

Basal transcription factors

Promoter-proximal element

Promoter

RNA polymerase II

3. Assembly of proteins. Other transcriptional activators bind to the newly exposed enhancers and promoter-proximal element; basal transcription factors bind to the promoter and recruit RNA polymerase II.

Mediator

TRANSCRIPTION

4. DNA looping allows Mediator to connect DNA-bound activators and basal transcription factors. Transcription begins.

FIGURE 19.8 Transcription Initiation in Eukaryotes.

to regulate gene expression, and each is used in some cells at least some of the time. These control mechanisms include (1) splicing RNAs in various ways, (2) modifying the life span of mRNAs, (3) altering the rate at which translation is initiated, and (4) activating or inactivating proteins after translation has occurred. Let's consider each in turn.

Alternative Splicing of mRNAs

Introns are spliced out in the nucleus as the primary RNA is transcribed. Recall that the mRNA that results from splicing consists of sequences encoded by exons, and that it is protected by a cap on the 5′ end and a long poly(A) tail on the 3′ end (see Chapter 17). You may also recall that splicing is accomplished by macromolecular machines called **spliceosomes,** and that many primary transcripts can be spliced in more than one way. This turns out to be a major way of regulating eukaryotic gene expression.

During splicing, gene expression is regulated when selected exons are removed from the primary transcript along with the introns. As a result, the same primary RNA transcript can yield more than one kind of mature, processed mRNA, consisting of different combinations of exons.

This is important. Since these mature mRNAs contain differences in their sequences, the polypeptides translated from them will likewise differ. Splicing the same primary RNA transcript in different ways to produce different mature mRNAs and thus different proteins is referred to as **alternative splicing.**

To see how alternative splicing works, consider the muscle-cell protein tropomyosin. The tropomyosin gene is expressed in both skeletal muscle cells and smooth muscle cells. These cells make up two distinct kinds of muscle tissue. Skeletal muscle is responsible for moving your bones; smooth muscle lines many parts of your gut and certain blood vessels.

As **FIGURE 19.9a** shows, the primary transcript from the tropomyosin gene contains 14 exons. In each type of muscle cell, a different subset of the 14 exons are spliced together to produce two different mRNAs (**FIGURE 19.9b**). As a result of alternative splicing, the tropomyosin proteins found in these two cell types are distinct. One reason skeletal muscle and smooth muscle are different is that they contain different types of tropomyosin.

Alternative splicing is controlled by proteins that bind to RNAs in the nucleus and interact with spliceosomes to influence which sequences are used for splicing. When cells that are destined to become skeletal muscle or smooth muscle are developing, they receive signals leading to the production of specific proteins that are active in the regulation of splicing. Instead of transcribing different tropomyosin genes, the cells transcribe a single gene and splice the same primary RNA transcript in different ways.

Before the importance of alternative splicing was widely appreciated, a gene was considered to be a nucleotide sequence that encodes one specific protein or RNA, along with its regulatory sequences. Based on this view, estimates for the number of genes in the human genome were typically in the range of 60,000 to 100,000. But once the complete human genome sequence became available, researchers realized that we may have as few as 20,000 sequences for primary mRNA transcripts.

Even though our genomes contain a relatively low number of genes, recent data indicate that over 90 percent of them undergo alternative splicing to produce multiple products. Thus, the number of different proteins that your cells can produce is far larger than the number of genes.

Given the extent of alternative splicing, the definition of protein-coding genes has been changed to the coding and regulatory sequences that direct the production of one *or more* related mRNAs and polypeptides.

Many alternatively spliced genes produce just a few different products, but some can produce a bewildering array of mRNAs. The current record is the *Dscam* gene of the fruit fly *Drosophila melanogaster*. The products of this gene help to guide growing nerve cells within the embryo. The primary transcript can be spliced into more than 38,000 distinct forms of mRNA, and the *Dscam* gene can produce thousands of different products.

Alternative splicing is a major mechanism in the control of gene expression in multicellular eukaryotes. ✔ If you understand alternative splicing, you should be able to explain why it does not occur in bacteria and describe where it occurs in Figure 19.1.

mRNA Stability and RNA Interference

Once splicing is complete and processed mRNAs are exported to the cytoplasm, new regulatory mechanisms come into play. For example, it has long been known that the life span of an mRNA in a cell can vary. The mRNA for casein—the major protein in milk—is produced in the mammary gland tissue of female mammals. Normally, casein mRNA persists in cells for only about an hour, and little casein protein is produced. But when a female mouse is lactating, regulatory molecules help the mRNAs persist almost 30 times longer—leading to a huge increase in the production of casein. In this instance, shortening of the poly(A) tail decreases the life-span of the mRNA.

More recent discoveries in a variety of organisms, from worms to plants to people, have uncovered a widespread and important form of post-transcriptional gene regulation known as **RNA interference.** RNA interference occurs when a tiny, single-stranded RNA held by a protein complex binds to a complementary sequence in an mRNA. This event unleashes either the destruction of the mRNA or a block to the mRNA's translation. How does it work?

(a) Tropomyosin gene

(b) Alternative splicing produces more than one mature mRNA.

FIGURE 19.9 Alternative Splicing Produces More than One Mature mRNA from the Same Gene.

FIGURE 19.10 walks through the sequence of events.

Step 1 RNA interference begins when RNA polymerase transcribes genes that code for RNAs that double back on themselves to form a hairpin. Hairpin formation occurs because pairs of bases within the RNA transcript are complementary.

Step 2 Some of the RNA is trimmed by enzymes in the nucleus; then the double-stranded hairpin that remains is exported to the cytoplasm.

Step 3 In the cytoplasm, another enzyme cuts out the hairpin loop to form double-stranded RNA molecules that are only about 22 nucleotides long.

Step 4 One of the strands from this short RNA is taken up by a group of proteins called the *RNA-induced silencing complex*, or RISC. The RNA strand held by the RISC is a **microRNA (miRNA).**

Step 5 Once it is part of a RISC, the miRNA binds to its complementary sequences in a target mRNA.

Step 6 If the match between a miRNA and an mRNA is perfect, an enzyme in the RISC destroys the mRNA by cutting it in two. In effect, tight binding by a miRNA is a "kiss of death" for the mRNA. If the match isn't perfect, however, the mRNA is not destroyed. Instead, its translation is inhibited. Either way, miRNAs "interfere" with gene expression.

The first papers on RNA interference were published in the mid-1990s, and the first miRNAs were characterized in 2001. Since then, research on miRNAs and RNA interference has exploded. Current data suggest that a typical animal or plant species has at least 500 genes that code for miRNAs and that each miRNA regulates more than one mRNA. Because of this evidence, it is estimated that a large percentage of all animal and plant genes are controlled by these tiny molecules. miRNAs are critical for normal development, and mutations in miRNA genes are associated with many diseases. RNA interference is increasingly recognized as a key aspect of post-transcriptional control.

Researchers are currently testing whether certain miRNAs could be used to knock out specific genes associated with illness, or to destroy mRNAs produced by viruses during an infection. Research on RNA interference has quickly moved from an exciting new frontier in basic biology to possible applications in medicine.

✔ If you understand RNA interference, you should be able to describe where it occurs in Figure 19.1.

How Is Translation Controlled?

RNA interference is not the only mechanism of gene control that acts on mRNAs and affects whether translation occurs. For example, cells may slow or stop translation in response to a sudden increase in temperature or infection by a virus. The slowdown occurs because regulatory proteins that are activated by the temperature spike or viral invasion add a phosphate group to a protein that is part of the ribosome.

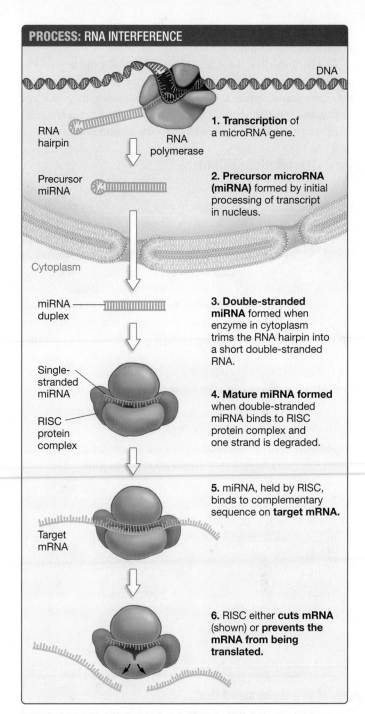

PROCESS: RNA INTERFERENCE

DNA

RNA hairpin

RNA polymerase

1. Transcription of a microRNA gene.

Precursor miRNA

2. Precursor microRNA (miRNA) formed by initial processing of transcript in nucleus.

Cytoplasm

miRNA duplex

3. Double-stranded miRNA formed when enzyme in cytoplasm trims the RNA hairpin into a short double-stranded RNA.

Single-stranded miRNA

RISC protein complex

4. Mature miRNA formed when double-stranded miRNA binds to RISC protein complex and one strand is degraded.

5. miRNA, held by RISC, binds to complementary sequence on **target mRNA**.

Target mRNA

6. RISC either **cuts mRNA** (shown) or **prevents the mRNA from being translated.**

FIGURE 19.10 MicroRNAs Either Target mRNAs for Destruction or Prevent Their Translation. MicroRNAs are held by the RISC protein complex and bind to target mRNAs by complementary base pairing.

You might recall that phosphorylation frequently leads to changes in the shape and chemical reactivity of proteins. In the case of the phosphorylated ribosomal protein, the shape change slows or prevents translation.

For the cell, this dramatic change in gene expression can mean the difference between life and death. High temperatures disrupt protein folding, so shutting down translation prevents the production of improperly folded polypeptides. If the problem is an invading virus, the cell stops the infection because it avoids manufacturing viral proteins.

Mechanisms like these are a reminder that gene expression can be regulated at multiple points: at the level of chromatin structure, transcription initiation, RNA processing, mRNA availability, and translation. But that's not all. Let's look at the last level possible: altering protein activity, after translation is complete.

Post-Translational Control

In bacteria, mechanisms of post-translational regulation are important because they allow cells to respond rapidly to new conditions (see Chapter 18). The same is true for eukaryotes. Instead of waiting for transcription, RNA processing, and translation to occur, the cell can keep an existing but inactive protein waiting in the wings and then quickly activate it in response to altered conditions. This is the essence of post-translational control.

There is a trade-off, however: Speed is gained at the expense of efficiency. Transcription, RNA processing, and translation use up energy and materials; it is wasteful to produce proteins that won't be used.

You have already encountered several important mechanisms of post-translational control over gene expression.

- Proteins are folded into their final, active conformation by chaperone proteins (see Chapters 3 and 17).

- Proteins may be modified by enzymes that add carbohydrate groups (see Chapter 7) or cleave off certain amino acids.

- Phosphorylation is an extremely common mechanism for activating or deactivating proteins. You just learned how a ribosomal protein is deactivated by adding a phosphate. You also might recall discussion of the activation of cyclin–Cdk complexes by phosphorylation and the subsequent entry into M phase of the cell cycle (see Chapter 12).

Yet another key mechanism of post-translational control—the targeted destruction of proteins—was first introduced by

check your understanding

If you understand that . . .

- Alternative splicing allows a single gene to code for many products.
- RNA interference is one of several mechanisms for controlling an mRNA's life span and translation rate.
- Ubiquitin tagging and destruction by proteasomes is one of many mechanisms for controlling a protein's life span.

✔ **You should be able to . . .**

1. Explain why the discovery of alternative splicing forced biologists to change their definition of the gene.
2. Explain why RNA interference is aptly named.

Answers are available in Appendix A.

describing the short life span of cyclin proteins. When a protein such as a cyclin needs to be destroyed, enzymes mark it by adding many copies of a small polypeptide called ubiquitin. Ubiquitin got its name because it is ubiquitous in cells. A macromolecular machine called the **proteasome** recognizes proteins that have a ubiquitin tag and cuts them into short segments.

As you can see, the regulation of gene expression in eukaryotes includes everything from opening chromatin in the nucleus to controlling the life span of proteins.

Do bacteria use the same range of regulatory mechanisms? Let's explore this question next.

19.5 How Does Gene Expression Compare in Bacteria and Eukaryotes?

Almost as soon as biologists knew that information in DNA is transcribed into RNA and then translated into proteins, they began asking questions about how that flow of information is regulated. **TABLE 19.1** summarizes what biologists have learned over the past half century about how bacterial and eukaryotic gene expression is controlled—organized by the six steps in gene expression introduced in Figure 19.1.

How does the regulation of gene expression differ in bacteria and eukaryotes? Biologists point to five fundamental differences in the control of gene expression in bacteria and eukaryotes:

1. **DNA Packaging** The chromatin of eukaryotic DNA must be decondensed for basal and regulatory transcription factors to gain access to genes and for RNA polymerase to initiate transcription. The tight packaging of eukaryotic DNA means that the default state of transcription in eukaryotes is "off." In contrast, the default state of transcription in bacteria, which lack histone proteins and have freely accessible promoters, is "on." Chromatin structure provides a mechanism of negative control that does not exist in bacteria.

2. **Complexity of transcription** Transcriptional control is much more complex in eukaryotes than in bacteria. The sheer number of eukaryotic proteins involved in regulating transcription dwarfs that in bacteria, as does the complexity of their interactions.

3. **Coordinated transcription** In bacteria, genes that take part in the same cellular response are often organized into operons and transcribed together from a single promoter. In contrast, operons are rare in eukaryotes. Instead, for coordinated gene expression, eukaryotes rely on the strategy used in bacterial regulons—physically scattered genes are expressed together when the same regulatory transcription factors trigger the transcription of genes with the same DNA regulatory sequences.

4. **Greater reliance on post-transcriptional control** Eukaryotes make greater use of post-transcriptional control, such as alternative splicing. Alternative splicing allows eukaryotes to regulate the production of many proteins from each gene.

Level of Regulation	Bacteria	Eukaryotes
Chromatin remodeling	• Limited packaging of DNA • Remodeling not a major issue in regulating gene expression.	• Extensive packaging of DNA • Chromatin must be decondensed for transcription to begin.
Transcription	• Positive and negative control by regulatory proteins that act at sites close to the promoter • Sigma interacts with promoter.	• Positive and negative control by regulatory proteins that act at sites close to *and* far from promoter • Large set of basal transcription factors interact with promoter. • Mediator required.
RNA processing	• Rare	• Extensive processing: alternative splicing of introns
mRNA stability	• Rarely used for control	• Commonly used: RNA interference limits life span or translation rate of many mRNAs.
Translation	• Regulatory proteins bind to mRNAs and ribosomes and affect translation rate.	• Regulatory proteins bind to mRNAs and ribosomes and affect translation rate.
Post-translational modification	• Folding by chaperone proteins • Chemical modification (e.g., phosphorylation) changes protein activity.	• Folding by chaperone proteins • Chemical modification (e.g., phosphorylation) changes protein activity. • Ubiquitination targets proteins for destruction by proteasome.

Alternative splicing, microRNAs, and regulation of mRNA stability are seldom seen in bacteria, but these constitute major elements of control in eukaryotes.

To date, biologists do not have a good explanation for why gene expression is so much more complex in eukaryotes than it is in bacteria. After decades of research, the debate continues.

For multicellular eukaryotes, one hypothesis is that the requirement for complex regulation of gene expression during development has driven the evolution of such complicated and multilayered means of controlling genes.

Normal regulation of gene expression results in the orderly development of an embryo and appropriate responses to environmental change in adults. What happens when gene expression goes awry? Unfortunately, one answer is uncontrolled cell growth and the set of diseases called cancer. Understanding how changes in gene expression can lead to cancer is one of today's great research frontiers.

19.6 Linking Cancer with Defects in Gene Regulation

All cancers involve uncontrolled cell division. What allows this unbridled increase in cell number? Each type of cancer is caused by a different set of mutations that lead to cancer when they affect one of two classes of genes: (1) genes that stop or slow the cell cycle, and (2) genes that trigger cell growth and division by initiating specific phases in the cell cycle. Many of the genes that are mutated in cancer influence gene regulation. Let's take a closer look at how altered gene regulation can cause uncontrolled cell growth.

The Genetic Basis of Uncontrolled Cell Growth

As you learned in the chapter on the cell cycle (Chapter 12), proteins that stop or slow the cell cycle when conditions are unfavorable for cell division are called tumor suppressors. Logically enough, the genes that code for these proteins are called **tumor suppressor** genes. If the function of a tumor suppressor gene is lost because of mutation, then a key brake on the cell cycle is eliminated.

Genes that stimulate cell division are called **proto-oncogenes** (literally, "first cancer genes"). In normal cells, the proteins produced from proto-oncogenes are active only when conditions are appropriate for growth. In cancerous cells, defects in the regulation of proto-oncogenes can cause these genes to stimulate growth at all times. (In the context of cancer, cell growth refers to an increase in cell numbers, not an increase in the size of individual cells.) In such cases, a mutation has converted the proto-oncogene into an **oncogene**—an allele that promotes cancer development.

For cancers to develop, many mutations are required within a single cell, and these alter both tumor suppressor genes and proto-oncogenes.

The *p53* Tumor Suppressor: A Case Study

To gain a deeper understanding of how defects in gene expression can lead to cancer, consider research on the gene that is most often defective in human cancers. The gene is called *p53* because

(a) Normal cell

UV light

Damaged DNA

p53 activation ◄┄┄ Promotes transcription of genes for cell cycle arrest, DNA repair, or apoptosis.

G_1 inhibition

DNA restored

G_0

M

G_2

S

(b) *p53* mutant cell

UV light

Damaged *p53* gene

No active p53 ◄┄┄ No transcription of genes for cell cycle arrest, DNA repair, or apoptisis.

No G_1 inhibition

Synthesis of damaged DNA

G_0

M

G_2

S

FIGURE 19.11 Consequences of *p53* Mutation.

the protein it codes for has a molecular weight of approximately 53 kilodaltons. Sequencing studies have revealed that mutant, nonfunctional forms of *p53* are found in over half of all human cancers. The *p53* gene codes for a regulatory transcription factor.

Researchers began to understand what *p53* does when they exposed normal, noncancerous human cells to UV radiation and noticed that levels of active *p53* protein increased markedly. Recall that UV radiation damages DNA (see Chapter 15). Follow-up studies confirmed that there is a close correlation between DNA damage and the amount of *p53* in a cell. In addition, analyses of the protein's primary structure suggested that it might contain a DNA-binding region similar to the one shown for the muscle-specific transcription factor in Figure 19.7.

These observations inspired the hypothesis that *p53* is a regulatory transcription factor that serves as a master brake on the cell cycle. In this model, shown in **FIGURE 19.11**, *p53* is activated by DNA damage. Activated *p53* binds to the enhancers of genes that arrest the cell cycle, repair DNA damage, and when all else fails, trigger apoptosis (cell death). Expression of these genes allows the cell to halt the cell cycle in order to repair its DNA, if this is possible, or commit suicide if the DNA damage is too severe. In mutant cells that lack a form of *p53* that can bind to enhancers, DNA damage cannot arrest the cell cycle, the cell cannot kill

check your understanding

C Y U

If you understand that . . .

- Many mutations associated with cancer alter gene regulation.
- Cancer is associated with mutations that lead to loss of control over the cell cycle.
- Uncontrolled cell growth may result when a mutation in a regulatory gene creates a protein that activates the cell cycle constitutively.
- Uncontrolled cell growth may result when a mutation prevents a tumor suppressor protein from shutting down the cell cycle in damaged cells.

✔ You should be able to . . .

1. Explain why cancer has a common pattern of uncontrolled cell growth, but not a common cause. Your answer should refer to the six levels of gene regulation outlined in Figure 19.1.

2. Explain why loss-of-function mutations in *p53* are observed in so many cancers.

Answers are available in Appendix A.

itself, and damaged DNA is replicated. This situation leads to mutations that can move the cell farther down the road to cancer. The *p53* protein is like a quality control officer. If it is missing, errors are made and things go downhill.

Here are some of the results that support this model of *p53* function:

- *p53* activates many different genes, including genes for cell cycle regulation, DNA repair, and apoptosis.

- X-ray crystallography studies show that *p53* binds directly to DNA regulatory sequences of the genes it controls.

- Virtually all the *p53* mutations associated with cancer are located in the protein's DNA-binding region and alter amino acids that interfere with *p53*'s ability to bind to regulatory DNA sequences.

The role of *p53* in preventing cancer is so important that biologists call this gene "the guardian of the genome." Today, biologists are searching for molecules that can restore *p53* activity to protect the genome and act against cancer.

If you understand . . .

19.1 Gene Regulation in Eukaryotes— An Overview

- Changes in gene expression allow eukaryotic cells to respond to changes in the environment and cause distinct cell types to develop.

- In a multicellular eukaryote, cells are different because they express different genes, not because they have different genes.

- Gene expression is regulated at six levels: Chromatin has to be remodeled, the transcription of specific genes may be initiated or repressed, mRNAs may be spliced in different ways to produce a different product, the life span of specific mRNAs may be extended or shortened, translation rate may be increased or decreased, and the life span or activity of particular proteins may be altered.

✔ You should be able to describe how the presence of the nuclear envelope, and the physical separation of transcription and translation, influences the levels of gene regulation observed in eukaryotes versus bacteria and archaea.

19.2 Chromatin Remodeling

- Eukaryotic DNA is packaged with proteins into chromatin that must be opened before transcription can occur.

- Eukaryotic DNA is wrapped around histone proteins to form bead-like nucleosomes that are then coiled into 30-nm fibers and higher-order chromatin structures.

- Transcription cannot be initiated until chromatin around regulatory regions is decondensed.

- The state of chromatin condensation depends on the methylation of cytosines in DNA, acetylation and other modifications of histones, and the action of molecular machines called chromatin-remodeling complexes.

- Patterns of DNA methylation and histone modifications can be passed from mother cells to daughter cells.

- Epigenetic inheritance is the inheritance of different phenotypes due to anything other than differences in alleles; transmitting patterns of chromatin condensation from mother to daughter cells or from parent to offspring is a mechanism of epigenetic inheritance.

✔ You should be able to explain why chromatin remodeling has to be the first step in gene activation.

19.3 Initiating Transcription: Regulatory Sequences and Regulatory Proteins

- In eukaryotes, transcription is triggered by regulatory proteins called transcription factors that bind to sequences both close to and far from the promoter.

- Regulatory transcription factors can be activators or repressors; these bind to regulatory sequences called (1) promoter-proximal sequences that are near promoters or (2) enhancers and silencers that are often located at a distance from gene promoters.

- Amino acids on regulatory transcription factors interact with the projections of base pairs in the grooves of the DNA helix to allow binding to specific regulatory sequences.

- The first regulatory transcription factors that bind to DNA recruit proteins that loosen the interaction between nucleosomes and DNA, making the promoter, promoter-proximal elements, and enhancers accessible to other transcription factors.

- Interactions between regulatory and basal transcription factors occur through a complex of proteins called Mediator and lead to the positioning of RNA polymerase at the promoter and the start of transcription.

✔ You should be able to draw a model of a eukaryotic gene undergoing transcription. Label enhancers, promoter-proximal elements, the promoter, activators, basal transcription factors, Mediator, and RNA polymerase.

19.4 Post-Transcriptional Control

- Once transcription is complete, gene expression is controlled by **(1)** alternative splicing, **(2)** RNA interference, and **(3)** activation or inactivation of protein products.

- Alternative splicing allows a single gene to produce more than one version of an mRNA and more than one kind of protein. It is regulated by proteins that interact with potential splice sites in the primary transcript to control which ones the spliceosome uses.

- RNA interference occurs when tiny strands of complementary RNA bind to mRNAs in association with the protein complex called RISC. This marks the mRNAs for destruction or prevents their translation. If the short RNA strands come from a transcribed cellular gene, they are known as microRNAs (miRNAs).

- Once translation occurs, proteins may be activated or inactivated by the addition or removal of chemical groups such as phosphates, or marked for destruction in the proteasome by adding polypeptides known as ubiquitin.

✔ You should be able to explain why humans can have so few genes.

19.5 How Does Gene Expression Compare in Bacteria and Eukaryotes?

- Examine Table 19.1 Regulating Gene Expression in Bacteria and Eukaryotes.

✔ You should be able to compare and contrast how bacteria and eukaryotes regulate the transcription of genes that need to be turned on together.

19.6 Linking Cancer with Defects in Gene Regulation

- If mutations alter regulatory proteins that promote or inhibit progression through the cell cycle, then uncontrolled cell growth and tumor formation may result.

✔ You should be able to explain why mutations of *p53* that prevent the protein from binding to DNA set the stage for cancer.

(MB) MasteringBiology

1. MasteringBiology Assignments

Tutorials and Activities Control of Gene Expression; Control of Transcription; DNA Packing; Overview: Control of Gene Expression; Post-Transcriptional Control Mechanisms; Regulation of Gene Expression in Eukaryotes; Transcription Initiation in Eukaryotes

Questions Reading Quizzes, Blue-Thread Questions, Test Bank

2. eText Read your book online, search, take notes, highlight text, and more.

3. The Study Area Practice Test, Cumulative Test, BioFlix® 3-D Animations, Videos, Activities, Audio Glossary, Word Study Tools, Art

---

You should be able to . . .

✔ TEST YOUR KNOWLEDGE

Answers are available in Appendix A

1. What is chromatin?
   a. the histone-containing protein core of the nucleosome
   b. the 30-nm fiber
   c. the complex of DNA and proteins found in eukaryotes
   d. the histone *and* non-histone proteins in eukaryotic nuclei

2. What is a tumor suppressor?

3. Which of the following statements about enhancers is correct?
   a. They contain a unique base sequence called a TATA box.
   b. They are located only in 5′-flanking regions.
   c. They are located only in introns.
   d. They are found in a variety of locations and are functional in any orientation.

4. In eukaryotes, what allows only certain genes to be expressed in certain types of cells?

5. What is alternative splicing?
   a. phosphorylation events that lead to different types of post-translational regulation
   b. mRNA processing events that lead to different combinations of exons being spliced together
   c. folding events that lead to proteins with alternative conformations
   d. action by regulatory proteins that leads to changes in the life span of an mRNA

6. What types of proteins bind to promoter-proximal elements?
   a. the basal transcription complex
   b. the basal transcription complex plus RNA polymerase
   c. basal transcription factors such as TBP
   d. regulatory transcription factors such as activators

7. Compare and contrast (a) enhancers and the *E. coli araC* binding site (see Chapter 18), (b) promoter-proximal elements and the operator of the *lac* operon, and (c) basal transcription factors and sigma.

8. Explain how alternative splicing could play a role in changing eukaryotic gene expression in response to changes in the environment.

9. Compare and contrast (a) enhancers and silencers; (b) promoter-proximal elements and enhancers; and (c) regulatory transcription factors and Mediator.

10. Predict how a drug that inhibits histone deacetylase will alter gene expression.

11. Relative to the genetic code, the histone code
   a. has more triplets.
   b. is much simpler to read because of complementary base pairing.
   c. does not depend on particular base sequences.
   d. requires methylated Cs rather than standard Cs in DNA.

12. Predict how a mutation that caused continuous production of active *p53* would affect the cell.

13. In the follow-up work to the experiment shown in Figure 19.5, the researchers used a technique that allowed them to see if two DNA sequences are in close physical proximity (association). They applied this method to examine how often an enhancer and the promoter of the *Hnf4a* regulatory gene were near each other. A logical prediction is that compared with rats born to mothers fed a healthy diet, the *Hnf4a* gene in rats born to mothers fed a protein poor diet would
   a. show no difference in how often the promoter and enhancer associated.
   b. never show any promoter–enhancer association.
   c. show a lower frequency of promoter–enhancer association.
   d. show a higher frequency of promoter–enhancer association.

14. **QUANTITATIVE** Imagine repeating the experiment on epigenetic inheritance that is shown in Figure 19.5. You measure the amount of radioactive uridine (U) incorporated into *Hnf4a* mRNA in counts per minute (cpm) to determine the level of *Hnf4a* gene transcription in rats born to mothers fed either a normal diet or a low protein diet. The results are 11,478 cpm for the normal diet

and 7368 cpm for the low-protein diet conditions. You should prepare a graph similar to the one at the bottom of Figure 19.5 that shows the normalized results relative to the normal diet condition. Normalizing values means that value obtained from one condition is expressed as 1.0 (the norm; the normal diet in this case) and the values obtained from any other conditions (low-protein diet in this case) are expressed as decimal values relative to the norm.

15. After DNA damage, levels of activated p53 protein in the cytoplasm increase. Design an experiment to determine whether this increase is due to increased transcription of the *p53* gene or to activation of preexisting p53 proteins by a post-translational mechanism such as phosphorylation.

16. Researchers have discovered that if a single-stranded mRNA produced by a virus is used as a template to create a double-stranded RNA, that this double-stranded RNA often blocks infection by the virus. Propose an explanation for how infection can be blocked by this type of double-stranded RNA.

The Big Picture

Copying, using, and transmitting genetic information is fundamental to life. Cells use the genetic information archived in their DNA to respond to changes in the environment and, in multicellular organisms, to develop into specific cell types.

Hereditary information is transmitted to offspring with random changes called mutations. Thus, genetic information is dynamic—both within generations and between generations.

Note that each box in the concept map indicates the chapter and section where you can go for review. Also, be sure to do the blue exercises in the Check Your Understanding box below.

check your understanding

If you understand the big picture . . .

✓ You should be able to . . .

1. Draw stars next to the three elements of the central dogma of molecular biology.

2. Add arrows and labels indicating what reverse transcriptase does.

3. Draw an E in the corners of boxes that refer only to eukaryotes, not prokaryotes.

4. Fill in the blue ovals with appropriate linking verbs or phrases.

Answers are available in Appendix A.

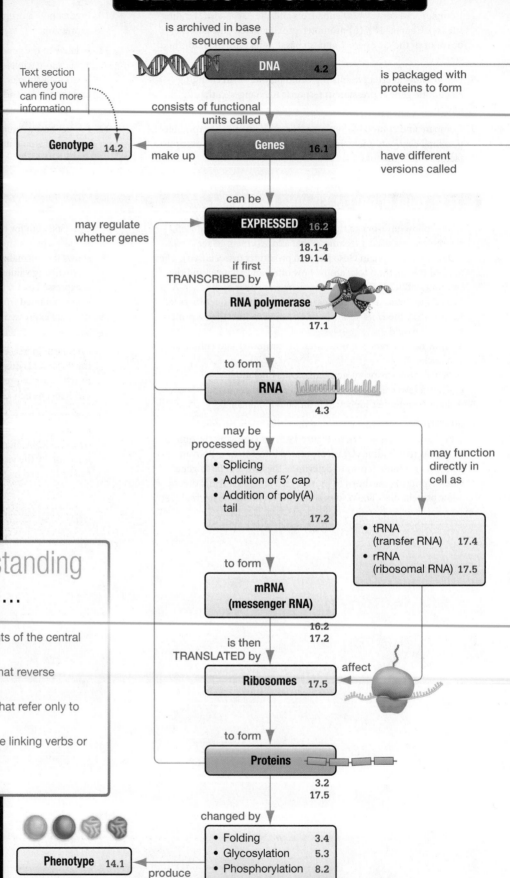

GENETIC INFORMATION

is archived in base sequences of

DNA 4.2

is packaged with proteins to form

Text section where you can find more information

consists of functional units called

Genotype 14.2

make up

Genes 16.1

have different versions called

can be

may regulate whether genes

EXPRESSED 16.2

18.1-4
19.1-4

if first TRANSCRIBED by

RNA polymerase

17.1

to form

RNA 4.3

may function directly in cell as

may be processed by

- Splicing
- Addition of 5′ cap
- Addition of poly(A) tail

17.2

- tRNA (transfer RNA) 17.4
- rRNA (ribosomal RNA) 17.5

to form

mRNA (messenger RNA)

16.2
17.2

is then TRANSLATED by

Ribosomes 17.5

affect

to form

Proteins

3.2
17.5

changed by

- Folding 3.4
- Glycosylation 5.3
- Phosphorylation 8.2
- Degradation 19.4

Phenotype 14.1

produce

Chromatin 19.2

Chromosomes 12.1 13.1

may change due to

- Breakage
- Duplication or deletion due to errors in meiosis
- Damage by radiation or other agents

13.3

15.5
16.4

causing

Mutation 16.4

can be

Alleles 14.2

are

COPIED 15.3

and

TRANSMITTED 12.1 13.1

can be

by

DNA polymerase 15.3

occasionally makes errors, causing

Mutation 16.4

to somatic cells by

MITOSIS 12.1

to germ cells by

MEIOSIS 13.1

includes

- Independent assortment
- Recombination

13.2

14.3–4

starts with

Parent cell

2n

starts with

Parent cell

2n

ends with

2n   2n

Two daughter cells with the same genetic information as the parent cell (unless mutation has occurred).

ends with

n   n

n   n

Four daughter cells with half the genetic information as the parent cell.

occurs during

GROWTH and ASEXUAL REPRODUCTION

12.1
12.2

result in

Low genetic diversity

occurs during

SEXUAL REPRODUCTION

13.4

results in

High genetic diversity

20 Analyzing and Engineering Genes

In this chapter you will learn that

Biotechnology depends on methods to analyze and alter genomes

by exploring ↓

Case studies of key genetic technologies

applied to

Recombinant DNA technologies **20.1**	Genome analysis	The polymerase chain reaction **20.2**
Finding genes by mapping **20.4**	Medicine	Dideoxy DNA sequencing **20.3**
Gene therapy **20.5**	Agriculture	Transgenic crops **20.6**

The rice plants in these bottles have been genetically engineered—using techniques introduced in this chapter—to produce a molecule needed for a key vitamin.

The molecular revolution in biological science got its start when researchers confirmed that DNA is the hereditary material and succeeded in describing the molecule's secondary structure. But when biologists discovered how to remove DNA sequences from an organism, manipulate them, and insert them into different individuals, the molecular revolution really took off.

Efforts to manipulate DNA sequences in organisms are often referred to as genetic engineering. Genetic engineering became possible with the discovery of enzymes that cut DNA at specific sites and of other enzymes that paste DNA sequences together. These new molecular tools were extremely powerful. Biologists no longer had to rely solely on controlled breeding experiments to change the genetic characteristics of individuals. Instead, they could mix and match specific DNA sequences in

✔ When you see this checkmark, stop and test yourself. Answers are available in Appendix A.

the lab. Because successful efforts to manipulate genes usually result in novel combinations of DNA, techniques used to engineer genes are often referred to as **recombinant DNA technology.**

This chapter uses a series of case histories to introduce basic molecular biology techniques in the context of solving problems. It also considers the ethical and economic issues raised by efforts to manipulate genes. What are the potential perils and benefits of introducing recombinant genes into human beings, food plants, and other organisms? This question, one of the great ethical challenges of the twenty-first century, is a recurrent theme in the following pages.

20.1 Case 1—The Effort to Cure Pituitary Dwarfism: Basic Recombinant DNA Technologies

To understand the basic techniques and tools of genetic engineering, let's consider the role they played in developing a treatment for pituitary dwarfism in humans.

The pituitary gland is a structure at the base of the mammalian brain that produces several important biomolecules, including a protein that stimulates growth. This protein, which was found to be just 191 amino acids long, was named human growth hormone (HGH). In humans, the gene that codes for it is called *GH1*.

The discovery of growth hormone led researchers immediately to suspect that at least some forms of inherited dwarfism might be caused by a defect in the *GH1* gene. This hypothesis was confirmed by studies showing that people with certain types of dwarfism produce little growth hormone or none at all. These people have defective copies of *GH1* and exhibit pituitary dwarfism, type I (**FIGURE 20.1a**).

By studying the pedigrees of families in which dwarfism was common, several teams of researchers established that pituitary dwarfism, type I, is an autosomal recessive trait (see Chapter 14). In other words, affected individuals have two copies of the defective allele. Individuals who are affected by pituitary dwarfism have normal body proportions but grow more slowly than average people, reach puberty from two to 10 years later than average, and are short in stature as adults—typically no more than 120 cm (4 feet) tall (**FIGURE 20.1b**).

Why Did Early Efforts to Treat the Disease Fail?

Once the molecular basis of pituitary dwarfism was understood, physicians began treating the disease with injections of naturally produced growth hormone. This approach was inspired by the spectacular success that had been achieved in treating type I diabetes mellitus. This form of diabetes is caused by a deficiency of the peptide hormone insulin, and clinicians had been able to alleviate the disease's symptoms by injecting patients with insulin from pigs.

Early trials showed that people with pituitary dwarfism could be treated successfully with growth hormone therapy, but only if the protein came from humans. Growth hormones isolated from pigs, cows, or other animals were ineffective. Until the 1980s, however, the only source of human growth hormone was pituitary glands from human cadavers. As a result, the drug was extremely scarce and expensive.

It turned out that meeting demand was the least of the problems with growth hormone therapy. To understand why, recall that infectious proteins called prions can cause degenerative brain disorders in mammals (see Chapter 3). When some of the children treated with human growth hormone developed a prion disease in their teens and twenties, physicians realized that the supply of growth hormone was contaminated with a prion protein from

(a) *GH1* codes for a pituitary growth hormone.

Normal *GH1* gene Defective *GH1* gene

Normal amount of GH1 protein produced

Little or no GH1 protein produced in pituitary gland

Pituitary dwarfism (slower growth, shorter stature)

(b) Normal versus GH1-deficient

FIGURE 20.1 Pituitary Dwarfism Is a Genetic Disease. (a) If mutations in the human *GH1* sequence are severe enough to inactivate the gene, pituitary dwarfism may result. **(b)** William Harrison and Charles Stratton, in a photo taken about 1860. Harrison and Stratton were both celebrated comedians and performers. Stratton, whose stage name was Tom Thumb, enjoyed audiences in the White House with Abraham Lincoln and Buckingham Palace with Queen Victoria. Stratton had pituitary dwarfism; Harrison had normal height.

some of the cadavers supplying the hormone. In 1984, the use of growth hormone isolated from cadavers was banned.

Steps in Engineering a Safe Supply of Growth Hormone

To replace natural sources of growth hormone, researchers turned to genetic engineering. Their plan was to insert fully functional copies of human *GH1* into the bacterium *Escherichia coli*, which they hoped would then produce huge quantities of recombinant progeny. If the plan worked, the recombinant cells would produce uncontaminated growth hormone in sufficient quantities to meet demand at an affordable price.

The plan required investigators to find *GH1*, obtain many copies of the gene, and insert them into *E. coli* cells. Their ability to do these things hinged on using basic tools of molecular biology.

Using Reverse Transcriptase to Produce cDNAs An enzyme called reverse transcriptase (Chapter 16) is responsible for an exception to the central dogma of molecular biology: It allows information to flow from RNA to DNA. More specifically, reverse transcriptase catalyzes the synthesis of DNA from an RNA template.

DNA that is produced from RNA is called **complementary DNA,** or **cDNA.** Although reverse transcriptase initially produces a single-stranded cDNA, it is also capable of synthesizing the complementary strand to yield a double-stranded DNA. In many cases, however, researchers add a chemically synthesized primer to single-stranded cDNAs and use DNA polymerase to synthesize the second strand (**FIGURE 20.2**).

Reverse transcriptase played a key role in the search for the growth hormone gene. Knowing that *GH1* is actively transcribed in the pituitary gland, researchers isolated mRNAs from pituitary-gland cells and used the enzyme to reverse-transcribe those mRNAs to cDNAs. These reaction products contained double-stranded cDNAs corresponding to each gene that is actively expressed in pituitary cells.

The next move? Isolating each of the cDNAs and making many identical copies of them.

Using Plasmids in Cloning Producing many copies of a gene is referred to as **DNA cloning.** If a researcher says that she has cloned a gene, it means that she has isolated it and then produced many identical copies.

In many cases, researchers can clone a gene by inserting it into a small, circular DNA molecule called a **plasmid.** You might recall that plasmids are common in bacterial cells (see Chapter 7). They are physically separate from the bacterial chromosome and are not required for normal growth and reproduction. Most replicate independently of the chromosome. Some plasmids carry genes for antibiotic resistance or other traits that increase the cell's ability to grow in a particular environment.

Researchers realized that if they could splice a loose piece of DNA into a plasmid and then insert the modified plasmid into a bacterial cell, the engineered plasmid would be replicated and passed on to daughter cells as the bacterium grew and divided. If this recombinant bacterium were then placed in a nutrient broth and allowed to grow and reproduce overnight, billions of copies of the original cell, each containing identical modified plasmid DNA, would result. When a plasmid is used in this way—to make copies of a foreign DNA sequence—it is called a **cloning vector,** or simply a **vector.**

Biologists harvest the recombinant genes by breaking the bacteria open, isolating all the DNA, and then separating the plasmids from the main chromosomes. But how do they insert a gene into a plasmid in the first place?

Using Restriction Endonucleases and DNA Ligase to Cut and Paste DNA To cut a gene out for later insertion into a cloning vector, researchers use enzymes called restriction endonucleases. A **restriction endonuclease** is a bacterial enzyme that cuts DNA molecules at specific base sequences. In bacterial cells, these enzymes cut up DNA from invading viruses and prevent a fatal infection.

FIGURE 20.2 Reverse Transcriptase Catalyzes the Synthesis of DNA from RNA. The single-stranded DNA produced by reverse transcriptase is complementary to the RNA template. The cDNA can be made double stranded by reverse transcriptase or DNA polymerase. DNA polymerase requires a primer.

More than 800 restriction endonucleases are known, and many of them cut DNA only at sequences that form palindromes. In English, a word or sentence is a palindrome if it reads the same way backward as it does forward. "Madam, I'm Adam" is an example. In biology, a stretch of double-stranded DNA forms a palindrome if the $5' \rightarrow 3'$ sequence of one strand is identical to the $5' \rightarrow 3'$ sequence on the antiparallel, complementary strand.

To insert the pituitary-gland cDNAs into plasmids, researchers performed the sequence of steps outlined in **FIGURE 20.3**.

Step 1 The left side of the figure shows a plasmid containing a palindromic sequence, or recognition site, that is cut by a specific restriction endonuclease. As the right side of the figure shows, the researchers attached the same palindromic sequence to the ends of each cDNA in their sample.

Step 2 They cut the recognition sites in each plasmid (left) and at the ends of each cDNA (right) with a restriction endonuclease called EcoRI. (The name comes from the fact that this

enzyme was the first (Roman numeral *I*) restriction endonuclease discovered in *E. coli* strain *R*Y13.)

Step 3 Like most restriction endonucleases, EcoRI makes a staggered cut in the palindrome. The resulting DNA fragments are described as having **sticky ends,** because the single-stranded bases on one fragment are complementary to the single-stranded bases on the other fragment. As a result, the two ends can pair up and hydrogen-bond to each other: The complementary sequences in the sticky ends of the plasmid (in grey) will bind to the sticky ends in the cDNA (in red) by complementary base pairing.

Step 4 Finally, researchers used **DNA ligase**—the enzyme that connects Okazaki fragments during DNA replication (see Chapter 15)—to seal the pieces of DNA together at the arrows marked in green.

The creation of sticky ends in DNA is important. If restriction sites in different DNA sequences are cut with the same restriction

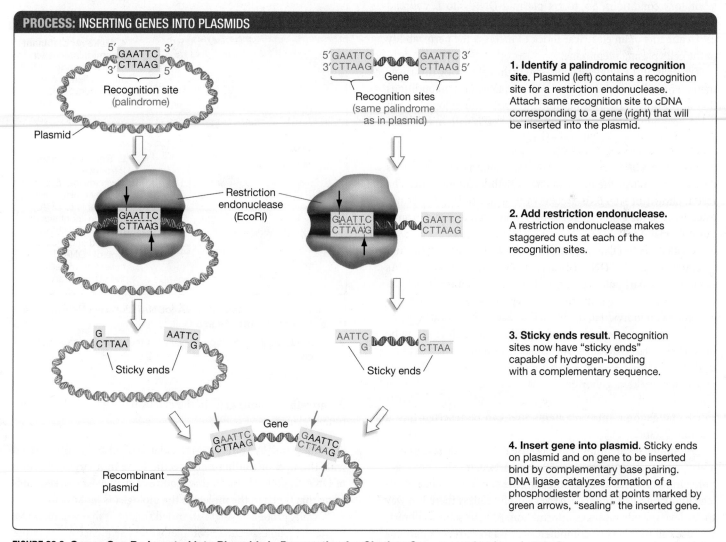

FIGURE 20.3 Genes Can Be Inserted into Plasmids in Preparation for Cloning. Once a gene has been inserted into a plasmid, the recombinant plasmid can be introduced into bacterial cells that grow and divide to produce many identical copies of the gene.

endonuclease, the presence of the same sticky ends in both samples of DNA promotes joining of the resulting fragments. This is the essence of recombinant DNA technology—the ability to create novel combinations of DNA sequences by cutting specific sequences and pasting them into new locations.

After performing this procedure, the researchers who were hunting for the growth hormone gene had a set of recombinant plasmids. Each contained a cDNA made from one of the many human pituitary-gland mRNAs. But the *GH1* gene was still not cloned.

Transformation: Introducing Recombinant Plasmids into Bacterial Cells
If a recombinant plasmid can be inserted into a bacterial or yeast cell, the foreign DNA will be copied and transmitted to new cells as the host cell grows and divides. In short, it can be cloned. In this way, researchers can obtain millions or billions of copies of specific genes. How is the insertion brought about?

Cells that take up DNA from the environment and incorporate it into their genomes are said to undergo **transformation.** Most bacterial cells do not take up DNA on their own under laboratory conditions. So, to get plasmid DNA into *E. coli* and other common laboratory species, researchers use simple chemical treatments or an electrical shock to increase the permeability of the cell's plasma membrane.

Typically, just a single plasmid enters the cell during this treatment. The cells are then spread out on plates under conditions that allow only cells with plasmids to grow into colonies. Each colony contains millions of identical cells, each with many identical copies of the recombinant plasmid.

Producing a cDNA Library
FIGURE 20.4 summarizes the steps covered thus far in the hunt for the growth hormone gene. The result, shown in step 5, is a collection of transformed bacterial cells. Each of the cells contains a plasmid with one cDNA from a pituitary gland mRNA.

A collection of DNA sequences, each of which is inserted into a vector, is called a **DNA library.** If the sequences are cDNAs made from a particular cell type or tissue, the library is called a **cDNA library.** If the sequences are fragments of DNA from the genome of an individual, the library is called a **genomic library.**

DNA libraries are made up of cloned genes. Each gene can be produced in large quantity and isolated in pure form. ✔ If you understand this concept, you should be able to describe how you could make a genomic library starting with DNA from your own cells and using the restriction endonuclease EcoRI to cut the genome into fragments that can be inserted into a plasmid vector.

DNA libraries are important because they provide researchers a way to store DNA fragments from a particular cell type or genome in a form that is accessible for gene cloning. But like a college library, a DNA library isn't very useful unless there is a way to retrieve specific pieces of information. At your school's library, you use call numbers or computer searches to retrieve a particular book or article. How do you go about retrieving a particular gene from a DNA library? For example, how did researchers find

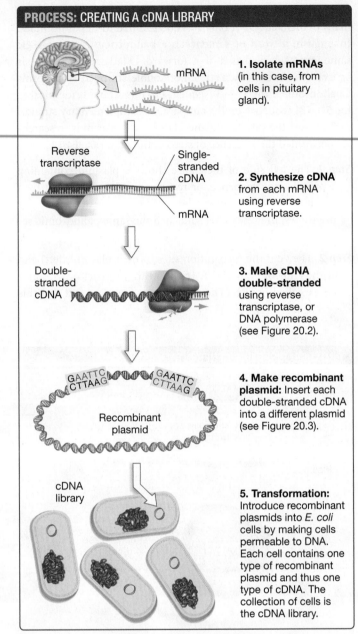

PROCESS: CREATING A cDNA LIBRARY

1. **Isolate mRNAs** (in this case, from cells in pituitary gland).

2. **Synthesize cDNA** from each mRNA using reverse transcriptase.

3. **Make cDNA double-stranded** using reverse transcriptase, or DNA polymerase (see Figure 20.2).

4. **Make recombinant plasmid:** Insert each double-stranded cDNA into a different plasmid (see Figure 20.3).

5. **Transformation:** Introduce recombinant plasmids into *E. coli* cells by making cells permeable to DNA. Each cell contains one type of recombinant plasmid and thus one type of cDNA. The collection of cells is the cDNA library.

FIGURE 20.4 Complementary DNA (cDNA) Libraries Represent a Collection of the mRNAs in a Cell.

✔**QUESTION** Would each type of cDNA in the library be represented just once? Why or why not?

the growth hormone gene in the cDNA library of the human pituitary gland?

Screening a DNA Library
Molecular biologists are often faced with the task of finding one specific gene in a large collection of DNA fragments. To do this requires a **probe**—a marked molecule that binds to the molecule the biologist is looking for.

A DNA probe is a single-stranded fragment that will bind to a particular single-stranded complementary sequence in a mixture of DNAs. By binding to the target sequence, the probe marks the fragment containing that sequence, distinguishing it from all the other DNA fragments in the sample. As **FIGURE 20.5** shows, a DNA probe must be labeled in some way so that it can be found after it has bound to the complementary sequence in the large sample of fragments.

✔ If you understand the concept of a DNA probe, you should be able to explain why the probe must be single stranded and labeled in order to work, and why it binds to just one specific fragment. You should also be able to indicate where a probe with the sequence AATCG (recall that sequences are always written 5′→3′) will bind to a target DNA with the sequence TTTTACCCATTTACGATTGGCCT (again written 5′→3′).

To find an appropriate probe for the human growth hormone gene, researchers began by using the genetic code to predict possible DNA sequences of *GH1*. They could do this because they knew the sequence of amino acids in the polypeptide, which they could use to infer the codons that coded for each amino acid. You made similar inferences in some of the exercises in earlier chapters. But recall that there is more than one codon for most amino acids (see Chapter 16). As a result, the researchers could not infer a unique sequence for the growth hormone gene. Instead, they deduced a set of possible sequences that could encode the *GH1* gene.

The next step was to chemically synthesize the set of short, single-stranded DNAs that were complementary to the possible *GH1* sequences. Because one of these sequences would bind to single-stranded fragments from the actual gene by complementary base pairing, it could act as a probe. In this case, the label the researchers attached to the probe was a radioactive atom.

FIGURE 20.6 shows how researchers used this probe to find the plasmid in the cDNA library that contained *GH1*. (For more

information on how to use probes, see **BioSkills 9** in Appendix B.) As predicted, the labeled probe bound to its complementary sequence in the cDNA library—identifying the recombinant cell that contained the human growth hormone cDNA.

Mass-Producing Growth Hormone To accomplish their goal of producing large quantities of the human growth hormone, the investigators used recombinant DNA techniques to transfer the growth hormone cDNA to a new plasmid. The plasmid in question contained a promoter sequence recognized by *E. coli*'s RNA polymerase holoenzyme (see Chapter 17). The recombinant plasmids were then introduced into *E. coli* cells.

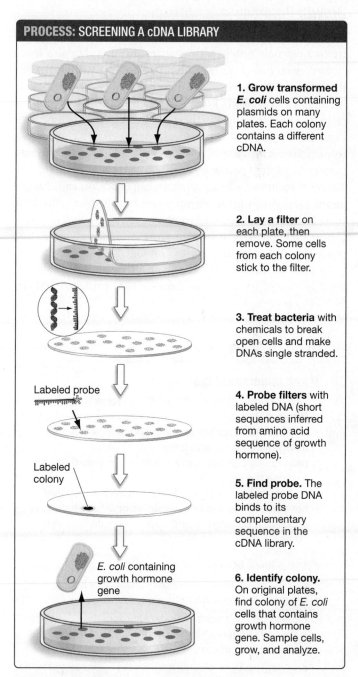

PROCESS: SCREENING A cDNA LIBRARY

1. **Grow transformed** *E. coli* cells containing plasmids on many plates. Each colony contains a different cDNA.

2. **Lay a filter** on each plate, then remove. Some cells from each colony stick to the filter.

3. **Treat bacteria** with chemicals to break open cells and make DNAs single stranded.

Labeled probe

4. **Probe filters** with labeled DNA (short sequences inferred from amino acid sequence of growth hormone).

Labeled colony

5. **Find probe.** The labeled probe DNA binds to its complementary sequence in the cDNA library.

E. coli containing growth hormone gene

6. **Identify colony.** On original plates, find colony of *E. coli* cells that contains growth hormone gene. Sample cells, grow, and analyze.

FIGURE 20.6 Finding Specific Genes by Probing a cDNA Library.

PROCESS: USING A DNA PROBE

Labeled probe

1. **Make probe.** Single-stranded DNA probe has a label that can be visualized.

2. **Expose probe** to collection of single-stranded DNA sequences.

3. **Find probe.** Probe binds to complementary sequences in target DNA—and only to that DNA. Target DNA is now labeled and can be isolated.

FIGURE 20.5 DNA Probes Bind to and Identify Specific Target Sequences.

The resulting transformed *E. coli* cells now contained a gene for human growth hormone attached to a promoter. These cells began to transcribe and translate the human growth hormone gene. Human growth hormone accumulated in the cells and was subsequently isolated and purified.

Today, bacterial cells containing the human growth hormone gene are grown in huge quantities. These cells have proved to be a safe and reliable source of the human growth hormone protein. The effort to cure pituitary dwarfism using recombinant DNA technology was a spectacular success—a triumph of applied biology, or **biotechnology.**

Ethical Concerns over Recombinant Growth Hormone

As supplies of growth hormone increased, physicians used it in treating not only people with pituitary dwarfism but also short children who had no growth hormone deficiency. Even though the treatment requires several injections per week until adult height is reached, growth hormone therapy was popular because it often increased the height of these children by a few centimeters.

In essence, growth hormone was being used as a cosmetic—a way to improve appearance in cultures where height is deemed attractive. But if short people are discriminated against in a culture, is a medical treatment a better solution than education and changes in attitudes? And what if parents wanted a tall child to be even taller, to enhance her potential success as, say, a basketball player?

Currently, the U.S. Food and Drug Administration has approved the use of human growth hormone for only the shortest 1.2 percent of children. These individuals are projected to reach adult heights of less than 160 cm (5′3″) in males and 150 cm (4′11″) in women.

check your understanding

If you understand that . . .

- **The essence of recombinant DNA technology is to cut DNA into fragments with a restriction endonuclease, paste specific sequences together using DNA ligase, and insert the resulting recombinant genes into a bacterial (or yeast) cell to express the genes.**
- **A DNA library consists of different sequences that have been inserted into plasmids or other vectors. A probe can be used to find specific sequences in the library.**

✔ **You should be able to . . .**

1. Explain why restriction endonucleases like EcoRI create DNA fragments with sticky ends.

2. Explain why the word probe is appropriate to describe a labeled sequence that is used to find a particular gene in a DNA library.

Answers are available in Appendix A.

Growth hormone has also become a popular performance-enhancing drug for athletes, because it improves the maintenance of bone density and muscle mass. Part of its popularity stems from the fact that it is difficult to detect in current drug tests.

Should athletes be able to enhance their physical skills by taking hormones or other types of drugs? Is the drug safe at the dosages athletes are using? These questions are being debated by physicians, researchers, agencies that govern sports, and legislative bodies.

In the meantime, it is clear that while solving one important problem, recombinant DNA technology created others. One of this chapter's recurring themes is that genetic engineering has costs that must be carefully weighed against its benefits.

20.2 Case 2–Amplification of Fossil DNA: The Polymerase Chain Reaction

Inserting a gene into a bacterial plasmid is one method for cloning DNA. The polymerase chain reaction is another.

The **polymerase chain reaction (PCR)** is an in vitro DNA synthesis reaction that uses DNA polymerase to replicate a specific section of DNA over and over. It generates many identical copies of a particular region of DNA.

Requirements of PCR

Although PCR is much faster and technologically easier than cloning genes into a DNA library, there is a catch: PCR is possible only when a researcher already has some information about DNA sequences near the gene in question. Sequence information is required because to do a polymerase chain reaction, you have to start by synthesizing short lengths of single-stranded DNA that match sequences on either side of the gene. These short segments act as primers for the DNA synthesis reaction.

As **FIGURE 20.7a** shows, the primer sequences must be complementary to bases on either side of the target gene—the DNA you want to copy. One primer is complementary to a sequence on one side of the target gene; the other primer is complementary to a sequence on the opposite strand of DNA, on the other side of the target gene. If the target DNA molecule is made single stranded, then the primers will bind, or anneal, to their complementary sequences, as shown in **FIGURE 20.7b**. You might recall that DNA polymerase cannot work without a primer. Once the primers are bound, DNA polymerase can extend each new strand of DNA in the 5′→3′ direction.

FIGURE 20.8 shows the sequence of the polymerase chain reaction.

Step 1 The researcher creates a reaction mix containing an abundant supply of the four deoxyribonucleoside triphosphates (dNTPs; see Chapter 15), a DNA sample that includes the gene of interest, many copies of the two primers, and a heat-resistant DNA polymerase called *Taq* polymerase.

(a) PCR primers must bind to sequences on either side of the target sequence, on opposite strands.

Primer binding site

Primer binding site

Region of DNA to be amplified by PCR

(b) When target DNA is made single stranded, primers bind and allow DNA polymerase to work.

Primer

Primer

FIGURE 20.7 The Polymerase Chain Reaction Requires Appropriate Primers. (a) To design an appropriate primer, the base sequence at the primer binding sites must be known. **(b)** The primers bind by complementary base pairing to single-stranded target DNA.

✔**EXERCISE** Indicate where DNA polymerase would begin to work on each strand; add an arrow indicating the direction of DNA synthesis.

Step 2 The reaction mix is heated to 95°C. At this temperature, the double-stranded template DNA denatures. This means that the two DNA strands separate, forming single-stranded templates.

Step 3 The mixture is allowed to cool to 50–60°C. In this temperature range, the primers bind, or anneal, to complementary portions of the single-stranded template DNA. This step is called primer annealing.

Step 4 The reaction mix is heated to 72°C. At this temperature, *Taq* polymerase efficiently synthesizes the complementary DNA strand from the dNTPs, starting at the primer. This step is called *extension*.

Step 5 Repeat steps 2 through 4.

Step 6 Continue repeating steps 2 through 4 until the necessary number of copies is obtained.

The temperature changes required in each step are controlled by automated PCR machines, and there is no need to add more components once the reaction starts.

Taq polymerase is a DNA polymerase found in the thermophilic ("heat-loving") bacterium *Thermus aquaticus*, which was discovered in a hot spring in Yellowstone National Park. Researchers use *Taq* polymerase because the PCR mixture has to be heated, and *Taq* polymerase is heat stable. Enzymes from most organisms are destroyed at high temperature, but *Taq* polymerase functions normally even when heated to 95°C.

The denaturation, primer annealing, and extension steps constitute a single PCR cycle. If one copy of the template sequence existed in the original sample, then two copies are present at the

PROCESS: THE POLYMERASE CHAIN REACTION

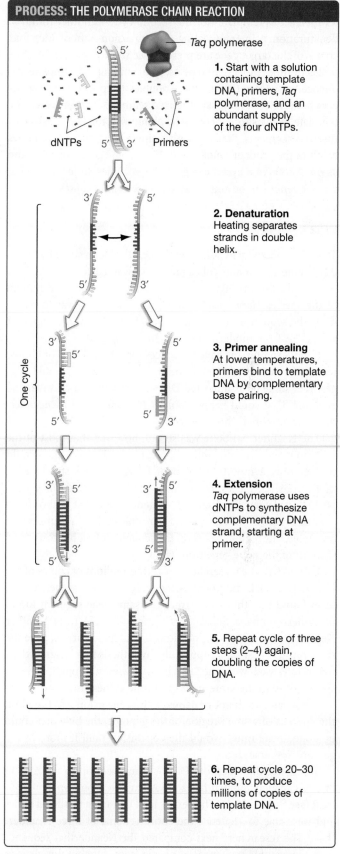

Taq polymerase

dNTPs   Primers

1. Start with a solution containing template DNA, primers, *Taq* polymerase, and an abundant supply of the four dNTPs.

2. Denaturation
Heating separates strands in double helix.

3. Primer annealing
At lower temperatures, primers bind to template DNA by complementary base pairing.

4. Extension
Taq polymerase uses dNTPs to synthesize complementary DNA strand, starting at primer.

One cycle

5. Repeat cycle of three steps (2–4) again, doubling the copies of DNA.

6. Repeat cycle 20–30 times, to produce millions of copies of template DNA.

FIGURE 20.8 The Polymerase Chain Reaction Produces Many Copies of a Specific Sequence. Each PCR cycle (denaturation, primer annealing, and extension) results in a doubling of the number of target sequences.

end of the first cycle (see step 4 in Figure 20.8). These two copies then act as templates for the second cycle—another round of denaturation, primer annealing, and extension—after which four copies of the target gene are present (see step 5).

Each time the cycle repeats, the amount of template sequence in the reaction mixture doubles (step 6). Doubling occurs because each newly synthesized segment of DNA serves as a template in the subsequent cycle, along with the previously synthesized segments. Starting with a single copy, successive cycles result in the production of 2, 4, 8, 16, 32, 64, 128, 256 copies, and so on. A total of n cycles can generate 2^n copies. In just 20 cycles, one sequence can be amplified to over a million copies.

PCR in Action

To understand why PCR is so valuable, consider a study by biologist Svante Pääbo and colleagues, who wanted to analyze DNA recovered from the 30,000-year-old bones of a fossilized human of the species *Homo neanderthalensis*. Their goal was to determine the sequence of bases in the ancient DNA and compare it with DNA from modern humans (*Homo sapiens*).

If modern humans have sequences that are identical or almost identical to the sequences found in Neanderthals, it would suggest that some of us inherited DNA directly from a Neanderthal ancestor. That could happen only if *H. sapiens* and *H. neanderthalensis* interbred while they coexisted in Europe.

The Neanderthal bone was so old, however, that most of the DNA in it had degraded into tiny fragments. The biologists could recover only a minute amount of DNA that was still in even moderate-sized pieces. Fortunately, the Neanderthal DNA sample included a few fragments of the gene region that Pääbo's team wanted to study. The researchers were able to design primers that bracketed this region, based on the sequence of highly conserved sections of the same gene from *H. sapiens*.

Using PCR, the researchers produced millions of copies of the Neanderthal DNA fragment. After analyzing these sequences, the team found that they differ from the same gene segment found in modern humans. Subsequent work with DNA from 14 other Neanderthal fossils, from locations throughout Europe, gave the same result. These data support the hypothesis that Neanderthals never interbred with modern humans—even though the two species lived in the same areas of Europe at the same time.

But a nagging doubt remained—the conclusion was based on the analysis of a small region of the genome, the best that could be done at the time. Would the conclusion hold if more of the genome was analyzed?

To answer this question, Pääbo's team went on to use a form of DNA sequencing that has a DNA amplification step similar to PCR (see Section 20.3). They extracted DNA from fossilized bones and were able to sequence the entire genome of three Neanderthals. The researchers next compared the Neanderthal sequence with the genome sequences of people from different populations living today.

Their conclusion? For African populations there was no evidence of inbreeding with Neanderthals. The story is different,

however, for non-African populations. People from these groups have a small amount of DNA, roughly 1 percent to 4 percent of each person's genome, that is derived from Neanderthal ancestors. So, many people living today can claim just a touch of Neanderthal in their family tree.

PCR has opened new research possibilities in countless areas beyond ancient DNA. For example:

- Forensic scientists, who use biological analyses to help solve crimes, clone DNA from tiny drops of blood or hair. The copied DNA can then be analyzed to identify victims, implicate perpetrators, or exonerate the falsely accused.

- Genetic counselors, who advise couples on how likely their offspring are to suffer from inherited diseases, can use PCR to find out if an embryo conceived by the couple has alleles associated with deadly illness.

Because the complete genomes of a wide array of organisms have now been sequenced, researchers can easily find appropriate primer sequences to use in cloning almost any target gene by PCR. The polymerase chain reaction is now one of the most basic and widely used techniques in biology.

check your understanding

If you understand that . . .

- PCR is a technique for amplifying a specific region of DNA into millions of copies, which can then be sequenced or used for other types of analyses.

✔ **You should be able to . . .**

1. Explain the purpose of the denaturation, annealing, and extension steps in a PCR cycle, and why "chain reaction" is an appropriate part of the term PCR.

2. Write down the sequence of a double-stranded DNA that is 50 base pairs long. Then design 21-base-pair-long primers that would allow you to amplify the segment by PCR.

Answers are available in Appendix A.

20.3 Case 3–Sanger's Breakthrough: Dideoxy DNA Sequencing

Once researchers have cloned a gene from a DNA library or by PCR, determining the gene's base sequence is usually one of the first things they want to do. Learning a gene's sequence is valuable for a variety of reasons. For example:

- Once a gene's sequence is known, the amino acid sequence of its product can be inferred from the genetic code. Knowing a protein's primary structure often provides clues to its function.

- Comparing sequences is fundamental to understanding why alleles vary in function—for example, why one allele causes disease and another doesn't.

- Researchers can infer evolutionary relationships by comparing the sequences of the same gene in different species (see Chapter 1). This information can be used to study an array of questions, ranging from how new traits evolve to where new diseases come from.

How do researchers sequence DNA? In 1977 Frederick Sanger published a technique called dideoxy sequencing that is still in use today.

The Logic of Dideoxy Sequencing

As **FIGURE 20.9** shows, **dideoxy sequencing** is a clever variation on the basic in vitro DNA synthesis reaction. But saying "clever" is an understatement. Sanger had to link three important insights to make his sequencing strategy work.

Dideoxynucleotides Terminate DNA Synthesis Sanger's first insight was to use monomers for DNA synthesis called dideoxyribonucleoside triphosphates (ddNTPs) along with the normal deoxyribonucleoside triphosphates (dNTPs) (Chapter 15) in the reaction mix. The ddNTPs are identical to dNTPs except that they lack a hydroxyl group at their 3′ carbon. Four types of ddNTPs are used in dideoxy sequencing, each named according to whether it contains adenine (ddATP), thymine (ddTTP), cytosine (ddCTP), or guanine (ddGTP). The use of ddNTPs inspired the name dideoxy sequencing.

Sanger realized that if a ddNTP were added to a growing DNA strand, it would terminate synthesis. Why? After a ddNTP is added, no hydroxyl group is available on a 3′ carbon to link to the 5′ carbon on an incoming dNTP monomer. As a result, DNA polymerization stops once a ddNTP is added.

Fragment Length Correlates with the Location of Each Base Sanger linked the ability of ddNTPs to stop DNA synthesis to a second fundamental insight: Every time a ddNTP is added to a growing strand, the result is a fragment with a length corresponding to the position in the template of a base complementary to the ddNTP. To produce these fragments, biologists create a reaction mix containing (1) many copies of the template DNA with (2) a primer, (3) DNA polymerase, (4) a large supply of the four dNTPs, and (5) a small amount of the four ddNTPs (Figure 20.9, step 1). Each of the four ddNTPs carries a different fluorescent tag. (In the figure, ddGTP is purple, ddCTP is blue, ddATP is green, and ddTTP is orange.) Fluorescent molecules absorb light at one wavelength and reemit the light at a longer wavelength. As described in **BioSkills 10** (see Appendix B), fluorescent tags provide a very sensitive way of detecting molecules.

Under these conditions, many daughter strands of different lengths are synthesized. All fragments that are the same length end in the same kind of ddNTP.

Step 2 in Figure 20.9 shows why:

- DNA polymerase synthesizes a complementary strand from each template in the reaction mix.

- The synthesis of each one of these complementary strands starts at the same point—the primer.

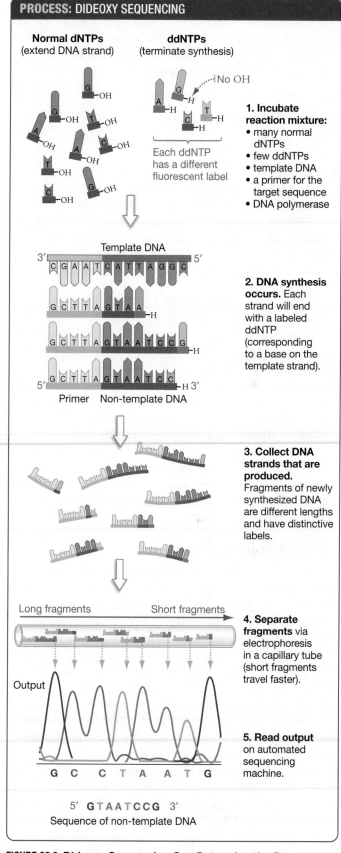

FIGURE 20.9 Dideoxy Sequencing Can Determine the Base Sequence of DNA.

- Because there are many dNTPs and relatively few ddNTPs in the reaction mix, dNTPs are usually incorporated opposite each complementary base on the template strand as DNA polymerase works its way along the template strand. Incorporating a dNTP allows DNA synthesis to continue.

- Occasionally, one of the few ddNTPs is incorporated into the growing strand, opposite the corresponding base in the template. The complementary base in the template strand pairs randomly with either a ddNTP or a dNTP.

- The addition of the ddNTP stops further elongation.

- "Stops" of this kind happen for each base in the template strand. As a result, the overall reaction produces a collection of newly synthesized strands (fragments) whose various lengths correspond to the location of each base in the template strand (see step 3 in Figure 20.9). Each fragment will fluoresce in the color of its ddNTP.

DNA Sequence Can Be Read from Fragments Lined Up by Size
Sanger's third insight? When the DNA fragments produced by the synthesis reactions are lined up by size, the dideoxy monomers on the successive fragments reveal the sequence of bases in the template DNA. To line up fragments in order of size, biologists separate them using gel electrophoresis (step 4 in Figure 20.9). As step 5 shows, a machine can read the pattern of fluorescence, indicating the sequence of bases in the newly synthesized strand.

Dideoxy sequencing ranks among the greatest of all technological advances in the history of biological science. Its impact is comparable to the development of light and electron microscopes, microelectrodes for recording membrane potentials, and recombinant gene technology.

"Next Generation" Sequencing

Sequencing technology is advancing at a blindingly fast pace. Dideoxy sequencing is still performed, particularly when relatively long sequences need to be read with great accuracy. But new, "next generation" sequencing approaches now make it possible to sequence much faster and more cheaply.

Most of the newer methods are based on amplification steps related to PCR. These create many copies of each template DNA molecule and allow sequencing of minute quantities of DNA. Even more important, millions of different DNAs can be amplified and sequenced in a single run. Methods that allow simultaneous sequencing or analysis of huge numbers of different molecules are called massively parallel approaches. These sequencing technologies are opening research possibilities that were barely imaginable only a few years ago. Today, instead of sequencing individual genes, researchers often obtain the sequence of entire genomes. A case in point is the sequencing of the Neanderthal genome—this was made possible using a massively parallel approach.

The project to sequence the human genome for the first time took 10 years and cost $3 billion U.S. dollars. Now researchers could sequence your genome in a day for less than $5,000.

If you understand that . . .
- Dideoxy DNA sequencing, which makes use of monomers that stop DNA synthesis in predictable ways, allows researchers to determine the sequence of bases present in a length of DNA.

✔ **You should be able to . . .**
Explain why labeled ddNTPs have to be present in small numbers relative to the number of unlabeled dNTPs.

Answers are available in Appendix A.

20.4 Case 4–The Huntington's Disease Story: Finding Genes by Mapping

Mendel had no idea what a "hereditary determinant" actually was. But now we know. Biology's molecular revolution has allowed researchers to find and characterize individual genes—to explore the connection between genotype and phenotype as explicitly and directly as possible. The question is, How do researchers find the genes associated with certain traits in the first place? How do you find the gene responsible for seed shape in peas, or white eyes in flies, or DNA polymerase III in *E. coli*?

One widely used approach is conceptually simple: You begin with a map of known sites in the genome and then look for an association between one of those known sites and the phenotype you're interested in. The gene that affects the phenotype is probably close to the known site.

In practice, the process is not so simple. As an example of how this type of gene hunt is done, let's consider one of the first successful searches ever conducted for a human gene—the gene associated with Huntington's disease.

How Was the Huntington's Disease Gene Found?

Huntington's disease is a rare but devastating illness. Typically, affected individuals first show symptoms between the ages of 35 and 45. At onset, an individual appears to be clumsier than normal and tends to develop small tics and abnormal movements. As the disease progresses, uncontrollable movements become more pronounced. Eventually the affected individual twists and writhes involuntarily. Personality and intelligence are also compromised—to the extent that the early stage of this disease is sometimes misdiagnosed as the brain disorder schizophrenia. The illness may continue to progress for 10 to 20 years and is eventually fatal.

Because Huntington's disease runs in families, physicians suspected that it was a genetic disease. An analysis of pedigrees from families affected by Huntington's disease suggested that the trait was due to a single, autosomal dominant allele (see Chapter 14). To understand the molecular basis for the disease, researchers used many of the tools and techniques of genetic engineering that are shown in **TABLES 20.1** and **20.2** (see pages 380–381) as they set out to locate and identify the gene or genes involved. It took over 10 years of intensive effort to reach this goal.

The search for the Huntington's disease gene was led by Nancy Wexler, whose mother had died of the disease. If the trait was indeed due to an autosomal dominant allele, it meant that Wexler had a 50 percent chance of receiving the allele from her mother and would begin to show symptoms when she reached middle age.

Using Genetic Markers To locate the gene or genes associated with a particular phenotype, such as a disease, researchers traditionally start with a **genetic map,** also known as a **linkage map** (see Chapter 14). Recall that a genetic map shows the relative positions of genes on the same chromosome, determined by analyzing the frequency of recombination between pairs of genes. Biologists also use **physical maps** of the genome. A physical map shows the absolute position of a gene—in numbers of base pairs—along a chromosome. Genome sequencing (see Chapter 21) has produced physical maps for a wide array of species.

A genetic map is valuable in gene hunts because it contains **genetic markers**—easily identified genes or DNA sequences that have known locations. Each genetic marker provides a landmark—a position along a chromosome that is known relative to other markers. The key is to find where the unknown gene lies relative to the established genetic markers.

To understand how genetic markers can be used, let's use a hypothetical example. Suppose that you knew the position of a hair-color gene in humans relative to other genetic markers. Suppose too that various alleles of this gene contributed to the development of black hair, red hair, blond hair, and brown hair in the group of people you were studying. This variation in phenotype associated with the marker is crucial. To be useful in a gene hunt, a genetic marker has to be **polymorphic,** meaning that the phenotype associated with the marker varies. In our hypothetical example, hair color is a polymorphic genetic marker.

Now suppose that the genetic disease called cystic fibrosis is common among the individuals you were studying and that your goal is to find the gene associated with cystic fibrosis. Further, suppose that people who have cystic fibrosis almost always have black hair—even though they are just as likely as unaffected individuals to have any other inherited trait observed in the study population, such as the presence or absence of a widow's peak or detached earlobes.

If you observe that a certain form of a marker and a certain disease are almost always inherited together, this means that the marker gene and disease gene are physically close to each other on the same chromosome—they are closely linked. If they were not closely linked, then crossing over in between them would be common and they would *not* be inherited together. In this hypothetical study, you could infer that the gene for cystic fibrosis is very close to the hair-color gene. ✔ If you understand this concept, you should be able to explain why it's helpful to hunt for genes using a genetic map with many genetic markers rather than only a few.

Gene hunts in humans boil down to this: Researchers have to find a large number of people who are affected and unaffected. Then they must attempt to locate a genetic marker that almost always occurs in one form in the affected individuals but only rarely in unaffected people. If such a marker is found, the disease gene is almost guaranteed to be nearby.

The types of genetic markers used in gene mapping have changed over time. Today, researchers often have a large catalog of polymorphic genetic markers available, including the particularly abundant markers known as **single nucleotide polymorphisms** (**SNPs,** pronounced *snips*). A SNP is a site in DNA that varies between alleles at a single base pair. Below is an example of a SNP:

To date, roughly 10 million human SNPs have been identified.

In the late 1970s and early 1980s, when biologists were searching for the Huntington's disease gene, SNPs were unknown. The best genetic markers available were restriction sites—short stretches of DNA where restriction endonucleases cut the double helix. These sequences are also known as restriction endonuclease recognition sites.

The restriction sites that Wexler's team used were polymorphic: Some alleles had a sequence that allowed cuts to occur; but in other alleles, the DNA sequence at the same site varied slightly, and no cuts occurred. Thus, just as an individual might have an A instead of a C at a certain SNP, an individual might have a restriction site allele that allowed cutting or not. Wexler's team was looking for restriction site alleles that were almost always present in diseased individuals but not found any more often than predicted by chance in healthy individuals (**FIGURE 20.10**).

FIGURE 20.10 Genetic Markers Can be Used to Locate Disease Alleles. Because of genetic recombination, genetic markers that are far from the gene of interest are equally likely to be found in both affected and unaffected individuals—there will be no association between particular forms of the marker and either the normal or the disease-causing allele.

Tool	Description	How Used	Illustration
Reverse transcriptase	Enzyme that catalyzes synthesis of a complementary DNA (cDNA) from an RNA template.	Many applications, including making cDNAs used in constructing a genetic library.	
Restriction endonucleases	Enzymes that cut DNA at a specific sequence—often a palindromic sequence that is six base pairs long.	Allows researchers to cut DNA at specific locations. Cuts in palindromic sites create "sticky ends."	
DNA ligase	An enzyme that catalyzes the formation of a phosphodiester bond between nucleotides on the same DNA strand.	Ligates (joins) sequences that were cut with a restriction endonuclease. Gives researchers the ability to splice fragments of DNA together.	
Plasmids	Small, extrachromosomal circles of DNA found in many bacteria and in some yeast.	After a target gene is inserted into a plasmid, the recombinant plasmid serves as a vector for transferring the gene into a bacterial or yeast cell, so the gene can be cloned.	
Taq polymerase	DNA polymerase from the bacterium *Thermus aquaticus*. Catalyzes synthesis of DNA from a primed DNA template; remains stable at 95°C.	Responsible for the "primer extension" step in the polymerase chain reaction. Heat stability allows enzyme to be active even after the 95°C denaturation step of PCR.	
Single nucleotide polymorphisms (SNPs)	Sites in DNA where the identity of a single base pair varies between alleles.	An important type of polymorphic DNA sequence that is useful in creating the genetic maps required for gene hunts.	

A Linkage Study   Once a genetic map containing many genetic markers has been assembled, to find the gene in question, biologists need help from groups of people that include individuals affected by an inherited disease. Recall that the fundamental goal is to find a genetic marker that is almost always inherited along with the disease-causing allele. Biologists call this a linkage study. Gene hunts based on linkage studies are more likely to be successful if large groups are involved. Large sample sizes reduce the possibility that researchers will observe an association between one or more markers and the disease just by chance, rather than because they are closely linked.

Huntington's disease is rare, but Wexler's team was fortunate to find a large, extended family affected with the disease living along the shores of Lake Maracaibo, Venezuela. The researchers followed the inheritance of Huntington's disease and various polymorphic genetic markers within this extended family.

From historical records, the researchers deduced that the Huntington's disease allele was introduced to this family by an English sailor who visited the area in the early 1800s. At the time of the study, over 3000 of his descendants were living in the area. Hundreds of these people had been diagnosed with Huntington's disease. To help in the search for the gene, family members agreed to donate skin or blood samples for DNA analysis.

When Wexler's team looked for associations between the presence or absence of the disease phenotype and genetic markers observed in each family member, they found one marker that

Technique	Description	How Used	Illustration
Recombinant DNA technology	Taking a copy of a gene from one individual and placing it in the genome of a different individual (usually of a different species).	Many applications, including DNA cloning, gene therapy (see Section 20.5), and biotechnology (see Sections 20.1 and 20.6).	Inserted gene
DNA libraries	A collection of all DNA sequences present in a particular source. The library consists of individual DNA fragments that are isolated and inserted into a plasmid or other vector, so they can be cloned.	cDNA libraries allow researchers to catalog the genes being expressed in a particular cell type and to work with coding sequences uninterrupted by introns. Genomic libraries allow researchers to archive all the DNA sequences present in a genome. Libraries can be screened to find a particular target gene.	Stored cDNA
Probing/screening a DNA library	Use of a labeled, known DNA fragment to hybridize (by complementary base pairing) with a collection of unlabeled, unknown fragments.	Allows a researcher to find a particular DNA sequence in a large collection of sequences.	Labeled probe
Polymerase chain reaction (PCR)	A DNA synthesis reaction that uses known primer sequences on either side of a target gene. Reaction is based on many cycles of DNA denaturation, primer annealing, and primer extension.	Produces many identical copies of a target sequence. A shortcut method for DNA cloning.	
Dideoxy sequencing	In vitro DNA synthesis reaction that includes dideoxyribonucleoside triphosphates (ddNTPs) as monomers.	Determining the base sequence of a gene or other section of DNA.	G C C T A A T G
Genetic mapping	Creation of a map showing the relative positions of genes or specific DNA sequences on chromosomes. Done by analyzing the frequency of recombination between sequences.	Many applications, including use of mapped genetic markers in genetic association studies to find unknown genes associated with diseases or other distinctive phenotypes.	Yellow body / White eyes / Ruby eyes 1.4 6.1

turned out to be especially important. Four different restriction site alleles (*A, B, C,* and *D*) were present at this location in the genome. The key finding was that the *C* form of the marker was almost always found in diseased individuals. Almost certainly, the English sailor who introduced the Huntington's disease allele had the *C* form of the marker in his DNA. The marker and the Huntington's disease gene are so close together that recombination in between them has been extremely rare. No other genetic marker showed this tight association with Huntington's disease.

From the human genetic map that was available at the time, Wexler's team knew that the marker they had identified was on chromosome 4. Eventually the team succeeded in narrowing down the location of the marker, and thus the Huntington's

disease gene, to a region about 500,000 base pairs long. Because the haploid human genome contains over 3 billion base pairs, this was a huge step in focusing the search for the gene.

Pinpointing the Defect   Once the general location of the Huntington's disease gene was known, biologists looked in that region for exons that encode an mRNA. Then they used dideoxy sequencing to determine the sequence of exons from diseased and normal individuals, compared the data, and pinpointed specific bases that differed between the two groups.

When this analysis was complete, the research team found that individuals with Huntington's disease have an unusually large number of CAG codons near the 5' end of one gene. CAG codes for glutamine. Healthy individuals have 11–25 copies of the CAG codon at that location, while affected individuals have 42 or more copies.

When the Huntington's disease research team confirmed that the increased number of CAG codons was always observed in affected individuals, they concluded that the long search for the Huntington's disease gene was over. They named the newly discovered gene *IT15* and its protein product huntingtin. The huntingtin protein is involved in the development of nerve cells. Only later in life do the mutant forms of the protein cause disease.

What Are the Benefits of Finding a Disease Gene?

How have efforts to find disease genes improved human health and welfare? Has the effort to locate the Huntington's disease gene helped researchers and physicians understand and treat the illness? Biomedical researchers point to three major benefits of disease-gene discovery.

Improved Understanding of the Phenotype   Once a disease gene is found and its sequence is known, researchers can usually figure out why its product causes disease. In the case of *IT15*, autopsies of Huntington's patients had shown that their brains decrease in size, and that the brain tissue contains clumps, or aggregates of the protein now called huntingtin.

Huntingtin aggregates are a direct consequence of the increased number of CAG repeats in the *IT15* gene. Long stretches of glutamine are known to promote protein aggregations. The leading hypothesis to explain Huntington's disease proposes that a gradual buildup of huntingtin aggregates triggers neurons to undergo apoptosis, or programmed cell death.

These results explained why Huntington's disease is pleiotropic (see Chapter 14). Patients suffer from abnormal movements *and* personality changes because neurons throughout the brain are killed. The results also help explain why the disease takes so long to appear, and why it is progressive: The defective huntingtin proteins take time to build up to harmful levels and then continue to increase over time. Finally, understanding the molecular mechanism responsible for the illness explained why the disease allele is dominant. One copy of the defective gene is enough to produce fatal concentrations of huntingtin aggregates.

Therapy   Once *IT15* was found, biologists began a search for new therapies for Huntington's disease by introducing the defective allele into mice, using the types of genetic engineering techniques discussed in Section 20.5. These mice with alleles that have been modified by genetic engineering are called **transgenic** (literally, "across-genes").

Transgenic mice that produce defective versions of the huntingtin protein develop a version of Huntington's disease, exhibiting tremors and abnormal movements, higher-than-normal levels of aggression toward litter and cage mates, and a loss of neurons in the brain. Laboratory animals with disease symptoms that parallel those of a human disease provide an **animal model** of the disease (see **BioSkills 13** in Appendix B).

Animal models are valuable in disease research because they can be used to test potential treatments. For example, research groups are using transgenic mice to test drugs that may prevent or reduce the aggregation of the huntingtin protein.

Genetic Testing   When the Huntington's gene was found and sequenced, biologists used the knowledge to develop a test for the presence of the defective allele. The test consists of obtaining a DNA sample from an individual and using the polymerase chain reaction to amplify the chromosome region that contains the CAG repeats responsible for the disease. If the number of CAG repeats is 35 or less, the individual is not considered at risk. Forty or more repeats results in a positive diagnosis for Huntington's.

Thanks to genetic maps based on SNPs, gene hunts are increasingly successful. Biologists have recently documented alleles associated with a predisposition to developing type I and type II diabetes, breast and ovarian cancer, obesity, coronary heart disease, bipolar disorder, Crohn's disease, and rheumatoid arthritis. Genetic testing for these alleles is now available for both prenatal and adult screens.

Ethical Concerns over Genetic Testing

Knowing the genetic basis of human diseases offers hope, but it also raises difficult ethical issues.

Genetic testing, for example, can create serious moral and legal dilemmas as well as harrowing personal choices. Consider that some people maintain that it is morally wrong to terminate any pregnancy, even if the fetus is certain to be born with a debilitating or fatal genetic disease. Think too about Nancy Wexler's position soon after the discovery of *IT15*: Would you choose to be tested for the defective allele and risk finding out that you were almost certain to develop an incurable disease such as Huntington's?

There are other, equally serious, questions. Should people be tested for any disease that has no cure? Should it be legal for insurance companies to test clients? If so, can companies refuse to insure people at risk for diseases that require expensive treatments? What about employers?

These questions are being debated by political and religious leaders, health-care workers, philosophers, and the public at large. In many cases, we've yet to find answers.

check your understanding

If you understand that . . .
- Genes for particular traits can be located if they are inherited together with a known genetic marker.

✔ **You should be able to . . .**
Describe how you would design a study aimed at identifying alleles associated with alcoholism.

Answers are available in Appendix A.

20.5 Case 5–Severe Immune Disorders: The Potential of Gene Therapy

For physicians who treat inherited disorders such as Huntington's disease, sickle-cell anemia, and cystic fibrosis, the ultimate goal is to cure the disease. This may be done by replacing or augmenting defective copies of the gene with normal alleles. This approach to treatment is called **gene therapy.**

For gene therapy to succeed, two crucial requirements must be met. First, the sequence of the allele associated with the healthy phenotype must be known. Second, a method must be available for introducing this allele into affected individuals and having it be expressed in the correct tissues, in the correct amount, and at the correct time. If the defective allele is dominant, then the introduction step may be even more complicated: In at least some cases, the introduced allele must physically replace or block the expression of the undesirable dominant allele.

How Can Genes Be Introduced into Human Cells?

Section 20.1 reviewed how recombinant DNA sequences are packaged into plasmids and taken up by *E. coli* cells. However, humans and other mammals lack plasmids. How can foreign genes be introduced into human cells?

Researchers have focused on packaging foreign genes into viruses for transport into human cells. These viruses have been engineered so they can deliver genes to cells but cannot replicate to produce new viruses. Viral infection begins when a virus particle attaches to a cell and delivers its genome into the cell (Chapter 36). For some viruses, the viral DNA becomes integrated into a host-cell chromosome. This trait makes it possible to use these viruses as vectors to carry engineered genes into the chromosomes of target cells. Potentially, the genes delivered by the virus could be expressed and produce a product capable of curing a genetic disease.

Vectors used today in gene therapy are often modified retroviruses. Retroviruses have genomes made of single-stranded RNA. When a **retrovirus** infects a human cell, a reverse transcriptase encoded by the virus catalyzes the production of a DNA copy of the virus's RNA genome. Other viral enzymes catalyze the insertion of the viral DNA into a host-cell chromosome. If an RNA version of a human gene can be packaged into a recombinant retrovirus, then the virus will insert a DNA copy of the human gene into a chromosome in a target cell (**FIGURE 20.11**).

Unfortunately, there are problems associated with using retroviruses as agents in gene therapy. For example, if the virus happens to insert the recombinant human gene in a position that disrupts the function of an important gene in the target cell, the consequences may be serious. Despite these risks, modified retroviruses are still among the best vectors currently available for human gene therapy.

Using Gene Therapy to Treat X-Linked Immune Deficiency

In 2000, a research team reported the successful use of gene therapy to treat an illness called severe combined immunodeficiency (SCID). Children who are born with SCID lack a normal immune system and are unable to fight off infections.

The type of SCID the team treated is designated SCID-X1, because it is caused by mutations in a gene on the X chromosome.

PROCESS: INTRODUCING A FOREIGN GENE INTO HUMAN CELLS

Human RNA   Retrovirus RNA

Reverse transcriptase

DNA complementary to introduced RNA

Reverse transcriptase

Human cell

Double-stranded DNA version of introduced gene

Host chromosome

1. Retrovirus engineered to contain recombinant RNA, which has both viral sequences and human sequences.

2. Target cell infected. RNA version of human gene enters host cell.

3. DNA produced. Viral enzymes make double-stranded DNA version of introduced gene.

4. DNA inserted. Human gene is inserted into host chromosome and transcribed.

FIGURE 20.11 Modified Retroviruses Can Insert a Foreign Gene into a Host-Cell Chromosome. (Many details have been omitted for conceptual clarity.)

✔**QUESTION** What happens if the recombinant DNA is inserted in the middle of a gene that is critical to normal cell function?

FIGURE 20.12 A "Bubble Child." Children with SCID cannot fight off bacterial or viral infections. As a result, such children must live in a sterile environment.

The gene codes for a receptor protein necessary for the development of immune system cells, called T cells, that develop in bone marrow (Chapter 51).

Traditionally, physicians have treated SCID-X1 by keeping the patient in a sterile environment, isolated from any direct human contact, until the person could receive a transplant of bone-marrow tissue from a close relative (**FIGURE 20.12**). In most cases, the T cells that the patient needs are produced by the transplanted bone-marrow cells and allow the individual to live normally. In some cases, though, no suitable donor is available. Could gene therapy cure this disease by furnishing functioning copies of the defective gene?

After extensive testing suggested that their treatment plan was safe and effective, the research team gained approval to treat 10 boys, each less than 1 year old, who had SCID-X1 but no suitable bone-marrow donor. The researchers removed bone marrow from each child, collected the stem cells that produce mature T cells, and infected those cells with an engineered retrovirus that delivered the normal receptor gene. Cells that began to produce normal receptor protein were then isolated and transferred back into the patients (**FIGURE 20.13**).

Within four months after reinsertion of the transformed marrow cells, nine of the boys had normal levels of functioning T cells. These patients were removed from germ-free isolation rooms and began residing at home, where they grew and developed normally.

Subsequently, however, four of the boys developed a cancer characterized by unchecked growth of T cells. Follow-up analyses of their bone-marrow cells showed that the normal receptor gene had been inserted either near a gene for a transcription factor that triggers T-cell growth, or near a gene for a cyclin that drives the cell cycle (see Chapter 12). The inserted receptor gene provided an enhancer that led to constitutive (constant) expression of the transcription factor or cyclin.

Three of the four boys responded to cancer chemotherapy and are healthy. The fourth did not respond to treatment and died of cancer.

The tenth boy to receive gene therapy never produced T cells. For unknown reasons, his recombinant stem cells failed to function normally when they were transplanted back into his bone marrow. Fortunately, physicians were later able to find a bone-marrow donor whose cells matched the boy's closely enough to make a successful transplant possible.

Ethical Concerns over Gene Therapy

Throughout the history of medicine, efforts to test new drugs, vaccines, and surgical protocols have always carried a risk for the patients involved. Gene therapy experiments are no different. The researchers who run gene therapy trials must explain the risks clearly and make every effort to minimize them.

PROCESS: ONE APPROACH TO GENE THERAPY

1. Stem cells are isolated from the patient's bone marrow and grown in vitro.

2. Engineered retroviruses carry the normal allele into host cells.

3. Cells that express normal alleles are isolated and implanted into patient.

SCID patient

FIGURE 20.13 Gene Therapy Can Cure a Genetic Disorder. For gene therapy to work in the case of a loss-of-function allele, copies of a normal allele have to be introduced into a patient's cells and be expressed.

The initial report on the development of cancer in the boys who received gene therapy for SCID-X1 concluded with the following statement: "We have proposed . . . a halt to our trial until further evaluation of the causes of this adverse event and a careful reassessment of the risks and benefits of continuing our study of gene therapy."

When recombinant DNA technology first became possible, many researchers thought they would live to see most or all of the serious inherited diseases caused by single-gene mutations cured by gene therapy. After several decades of rare successes punctuated by tragic failures, that optimism was tempered. In the past few years, however, renewed hope has emerged for gene therapy. Improved vectors have been used successfully to treat two different forms of blindness, a brain disorder, and another type of SCID. Perhaps gene therapy is finally poised to deliver on some of its promises.

20.6 Case 6—The Development of Golden Rice: Biotechnology in Agriculture

Progress in transforming crop plants with recombinant genes has been breathtaking. In 2010, a total 10 percent of all farmland worldwide was planted with transgenic crops—and this number is predicted to show double-digit growth over the foreseeable future. In the same year, roughly 90 percent of soybeans, cotton, and corn grown in the United States were genetically engineered. You almost certainly have eaten food from a genetically modified plant sometime, if not today.

Transgenic crops have been engineered largely to meet three objectives:

1. **Reducing losses from herbivore damage**   For instance, researchers have transferred a gene from the bacterium *Bacillus thuringiensis* into corn; the presence of the "Bt toxin" encoded by this gene protects the plant from corn borers and other caterpillar pests.

2. **Reducing competition with weeds**   An example is the genetic engineering of soybeans for resistance to an herbicide—a molecule that kills plants—called glyphosate. Soybean fields with the engineered strain can be sprayed with glyphosate to kill weeds without harming the soybeans.

3. **Improving food quality**   An important example is engineering soybeans and canola to produce a higher percentage of unsaturated fatty acids relative to saturated fatty acids (Chapter 6). Reducing the amounts of saturated fatty acids helps prevent heart disease, so these crops produce healthier vegetable oils.

How is this work done?

Rice as a Target Crop

Almost half the world's population depends on rice as its staple food. Unfortunately, rice is a poor source of some vitamins and essential nutrients—including vitamin A. Vitamin A deficiency causes blindness in 250,000 Southeast Asian children each year. It also increases susceptibility to diarrhea, respiratory infections, and childhood diseases such as measles.

Humans and other mammals synthesize vitamin A from a precursor molecule known as β-carotene (beta-carotene). β-carotene belongs to a family of orange, yellow, and red plant pigments called the carotenoids (Chapter 10).

Rice plants synthesize β-carotene in their chloroplasts but not in the carbohydrate-rich seed tissue called endosperm—the part of the rice seed that you eat. Could genetic engineering produce a strain of rice plants that synthesizes β-carotene in the endosperm?

Synthesizing β-Carotene in Rice

To explore the possibility of genetically engineering rice, a research team searched for compounds in rice endosperm that could serve as precursors for the synthesis of β-carotene. They found that maturing rice endosperm contains a molecule that could be converted to β-carotene in three enzyme-catalyzed reactions. The researchers reasoned that if genes that encode these enzymes could be introduced into rice plants along with regulatory sequences that would trigger their synthesis in endosperm, it should be possible to create a transgenic strain of rice that would contain β-carotene.

Fortunately, genes that encode two of the required enzymes had already been isolated from daffodils, and the gene for the third enzyme had been purified from a bacterium. These genes had been cloned in bacteria. To each of the coding sequences, biologists added regulatory sequences from an endosperm-specific protein. This segment would promote transcription of the recombinant sequences in endosperm cells.

Next, the three sets of sequences had to be inserted into rice plants. How are foreign genes introduced into plants?

The *Agrobacterium* Transformation System

Agrobacterium tumefaciens is a bacterium that infects plant tissues and triggers formation of tumorlike growths called galls. When researchers looked into how these infections occur, they found that a plasmid carried by the *Agrobacterium* cells, called a **Ti (tumor-inducing) plasmid,** plays a key role (**FIGURE 20.14**, see page 386).

Ti plasmids contain several sets of genes. One set encodes products that allow the bacterium to bind to the cell walls of a host. Another set, referred to as the virulence genes, encodes the proteins required to transfer part of the Ti DNA, called T-DNA (transferred DNA), into the plant cell. The T-DNA then travels to the nucleus and integrates into host-cell chromosomes (Figure 20.14, step 1). T-DNA genes are expressed and their products induce the infected cell to grow and divide. This results in the formation of a gall that houses a growing population of *Agrobacterium* cells (Figure 20.14, step 2).

Researchers soon realized that the Ti plasmid offers an efficient way to introduce recombinant genes into plant cells. Follow-up experiments confirmed that recombinant genes could be

1. Transfer of Ti genes. A section of DNA from the Ti plasmid, called T-DNA, integrates into the chromosomes of plant cells infected by the bacterium.

2. Transcription of Ti genes. When transcribed, Ti genes induce the infected plant cell to begin growing and dividing. The resulting gall encloses an increasing number of *Agrobacterium* cells.

FIGURE 20.14 *Agrobacterium* Infections Introduce Genes into a Plant Host-Cell Chromosome. Ti plasmids of *Agrobacterium* cells induce gall formation—a tumorlike growth.

1. Start with normal Ti plasmids.

2. Remove tumor-inducing genes.

3. Add genes for enzymes required for β-carotene synthesis along with regulatory sequences that will activate gene expression in endosperm.

(b) Golden rice (right) is engineered to synthesize β-carotene.

FIGURE 20.15 Constructing a Ti Plasmid to Produce "Golden Rice." Golden rice is a transgenic strain capable of synthesizing β-carotene in the endosperm of its seeds.

added to the T-DNA that integrates into the host chromosome, that the gall-inducing genes could be removed from the T-DNA, and that the resulting sequence is efficiently transferred and expressed in its new host plant.

Using the Ti Plasmid to Produce Golden Rice

To generate a strain of rice that produces all three enzymes needed to synthesize β-carotene in endosperm, the researchers exposed embryos to *Agrobacterium* cells containing genetically modified Ti plasmids (**FIGURE 20.15**). When the transgenic rice plants grew and produced seeds, the researchers found that some rice grains contained so much β-carotene that they were yellow. The biologists called the engineered plants "golden rice."

Follow-up experiments used gene sequences from corn, rather than daffodil, to produce 23 times more β-carotene in rice than the original transformants contained.

Will golden rice help solve a serious public health problem? The answer is not clear. Many environmental groups are strongly opposed to golden rice and any other engineered crops. Regulatory agencies in an array of countries would need to approve the use of golden rice—but there is strong resistance in some nations—and seed would have to be made available to farmers at an affordable price. The barriers to answering the question of whether golden rice can solve health problems are more societal than scientific.

It's important to recognize that each solution offered by genetic engineering introduces new issues to resolve. Biology students and others who are well informed about the techniques and issues involved will be important participants in this debate.

If you understand . . .

20.1 Basic Recombinant DNA Technologies

- In genetic engineering, DNA is added to a cell either to modify the cell's properties or to clone (obtain many identical copies of) the DNA.

- Restriction endonucleases cut DNA at specific locations. The resulting DNA fragments can be inserted into plasmids or other vectors with the help of DNA ligase.

- In many cases, the DNA fragments are inserted into vectors containing regulatory sequences that control expression of inserted genes.

✔ You should be able to explain why a plasmid is needed for gene cloning.

20.2 The Polymerase Chain Reaction

- The polymerase chain reaction (PCR) produces many identical copies of a gene without using cells for cloning.

- PCR depends on having primers that bracket a target stretch of DNA. These allow *Taq* polymerase, a heat-stable DNA polymerase, to amplify a single target DNA sequence to millions of identical copies.

✔ You should be able to list the advantages and disadvantages of cloning in cells versus using PCR to obtain many copies of genes.

20.3 Dideoxy DNA Sequencing

- Dideoxy sequencing determines the sequence of bases in DNA.

- Dideoxy sequencing is based on an in vitro DNA synthesis reaction in which dideoxyribonucleotides stop different DNA replication reactions at different bases in the sequence.

- The DNA fragments of different lengths that are generated by a dideoxy sequencing reaction are separated via gel electrophoresis to determine the sequence of bases.

✔ You should be able to explain how the newly synthesized DNA fragments—when they are lined up by size—can be used to determine the sequence of bases in the template DNA.

20.4 Finding Genes by Mapping

- Genetic maps are often used to find genes associated with phenotypes such as diseases.

- If individuals with a certain phenotype share a particular form of a polymorphic genetic marker (a mapped site with two or more forms in DNA that is unrelated to the phenotype), the gene responsible for the phenotype is likely to be near that marker.

- Once the general area of a gene is known, DNA in the region can be sequenced to determine exactly where the gene is located.

✔ You should be able to explain why genetic markers that are not polymorphic (come in only one form) are not useful in gene hunts.

20.5 The Potential of Gene Therapy

- Researchers are working to cure genetic diseases by gene therapy. This involves inserting normal copies of the defective gene into patients.

- In humans, genes used for gene therapy are often introduced using modified viruses.

- Gene therapy has faced many difficulties but recently has met with some notable successes.

✔ You should be able to describe what makes retroviruses well suited for gene therapy and what the concerns are about their use.

20.6 Biotechnology in Agriculture

- Many important crop plants are genetically engineered for traits that include pest and herbicide resistance, and improved food quality.

- Genes are often introduced into crops by infecting plant cells with bacteria that integrate their plasmid genes into the host-plant genome. By adding recombinant genes to these plasmids, researchers have been able to introduce genes that improve crops.

✔ You should be able to explain how genes are inserted into plants.

MasteringBiology

1. MasteringBiology Assignments

Tutorials and Activities Analyzing DNA Fragments Using Gel Electrophoresis; Cloning a Gene in Bacteria; Gel Electrophoresis of DNA; Making Decisions about DNA Technology: Golden Rice; Producing Human Growth Hormone; Restriction Enzymes; Restriction Enzymes, Recombinant DNA, and Gene Cloning; The Polymerase Chain Reaction

Questions Reading Quizzes, Blue-Thread Questions, Test Bank

2. eText Read your book online, search, take notes, highlight text, and more.

3. The Study Area Practice Test, Cumulative Test, BioFlix® 3-D Animations, Videos, Activities, Audio Glossary, Word Study Tools, Art

You should be able to . . .

1. What do restriction endonucleases do?

2. What is a plasmid?
   a. an organelle found in many bacteria and certain eukaryotes
   b. a circular DNA molecule that often replicates independently of the main chromosome(s)
   c. a type of virus that has a DNA genome and that infects certain types of human cells, including lung and respiratory tract tissue
   d. a type of virus that has an RNA genome, codes for reverse transcriptase, and inserts a cDNA copy of its genome into host cells

3. When present in a DNA synthesis reaction mixture, a ddNTP molecule is added to the growing chain of DNA. No further nucleotides can be added afterward. Why?

4. Once the gene that causes Huntington's disease was found, researchers introduced the defective allele into mice to create an animal model of the disease. Why was this model valuable?
   a. It allowed the testing of potential drug therapies without endangering human patients.
   b. It allowed the study of how the gene is regulated.

    c. It allowed the production of large quantities of the huntingtin protein.
    d. It allowed the study of how the gene was transmitted from parents to offspring.

5. To begin the hunt for the human growth hormone gene, researchers created a cDNA library from cells in the pituitary gland. What did this library contain?
   a. only the sequence encoding growth hormone
   b. DNA versions of all the mRNAs in the pituitary-gland cells
   c. all the coding sequences in the human genome, but no introns
   d. all the coding sequences in the human genome, including introns

6. What does it mean to say that a genetic marker and a disease gene are closely linked?
   a. The marker lies within the coding region for the disease gene.
   b. The sequence of the marker and the sequence of the disease gene are extremely similar.
   c. The marker and the disease gene are on different chromosomes.
   d. The marker and the disease gene are in close physical proximity and tend to be inherited together.

7. Explain how restriction endonucleases and DNA ligase are used to insert foreign genes into plasmids and create recombinant DNA. Make a drawing that shows why sticky ends are sticky and that identifies the location where DNA ligase catalyzes a key reaction.

8. **QUANTITATIVE** If a particular sequence of DNA were amplified using 25 PCR cycles, then the amount of this DNA would be predicted to increase by _____ -fold.

9. What is a cDNA library? Would you expect the cDNA library from a human muscle cell to be different from the cDNA library from a human nerve cell in the same individual? Explain why or why not.

10. What are genetic markers, and how are they used to create a genetic map?

11. Researchers added regulatory sequences from an endosperm-specific gene to the Ti plasmids used in creating golden rice. This was important to
    a. allow inserted genes to integrate into the plant genome.
    b. increase the endosperm growth rate.
    c. prevent the introduced plasmid from harming the endosperm.
    d. promote expression of introduced genes in the rice grain.

12. Compare and contrast PCR with the DNA synthesis that occurs in cells (see Chapter 15).

13. Suppose you had a large amount of sequence data, similar to the data that Nancy Wexler's team had in the region of the Huntington's disease gene, and that you knew that mRNAs of the species being studied typically contain protein-coding regions about 1500 bases long. How would you use the genetic code (see Chapter 16) and information on the structure of promoters (see Chapters 17 and 18) to find the precise location of one or more genes in your sequence?

14. Modifying germ-line or somatic cells for gene therapy involves the same ethical concerns. True or false?

15. Describe similarities between how researchers screen a DNA library and how they perform a genetic screen—for example, for mutant *E. coli* cells that cannot metabolize lactose (see Chapter 18).

16. A friend of yours is doing a series of PCRs and comes to you for advice. She purchased two sets of primers, hoping that one set

would amplify the template sequence shown here. (The dashed lines in the template sequence stand for a long sequence of unspecified bases.) Neither of the primer pairs produced any product DNA, however.

	Primer a		Primer b
Primer Pair 1:	5' CAAGTCC 3'	&	5' GCTGGAC 3'
Primer Pair 2:	5' GGACTTG 3'	&	5' GTCCAGC 3'
Template:	5' ATTCGGACTTG---GTCCAGCTAGAGG 3'		
	3' TAAGCCTGAAC---CAGGTCGATCTCC 5'		

a. Explain why each primer pair didn't work. Indicate whether both primers are at fault, or just one of them.
b. Your friend doesn't want to buy new primers. She asks you whether she can salvage this experiment. What do you tell your friend to do?

21 Genomics and Beyond

In this chapter you will learn that

Genomes are full of surprises

by exploring ↓

Whole-genome sequencing **21.1**

and what the data reveal about ↓

then asking ↓

Bacterial and Archaeal genomes 21.2

Eukaryotic genomes 21.3

What are functional genomics, proteomics, and systems biology? 21.4

A high-throughput robotic genome sequencer. Advances in DNA sequencing technologies are opening new questions to biologists and deepening the understanding of life and its evolution.

The first data sets describing the complete DNA sequence, or **genome,** of humans were published in February 2001. These papers were immediately hailed as a landmark in the history of science. In just 50 years, biologists had gone from not understanding the molecular nature of the gene to knowing the molecular makeup of every gene present in our species.

Years later, knowledge continues to stream from the multinational effort called the **Human Genome Project** and its many spinoffs. It's important to recognize, though, that research on *Homo sapiens* is part of a much larger, ongoing effort to gain insights from the genome sequences of an array of other eukaryotes, bacteria, and archaea. The pace of progress in this field is nothing short of explosive.

The effort to sequence, interpret, and compare whole genomes is **genomics.** While whole-genome sequencing supplies a list of the genes present in an organism, **functional genomics**

✔ When you see this checkmark, stop and test yourself. Answers are available in Appendix A.

answers questions about the functioning of that genome, such as what particular genes do and how they're expressed.

Genomics has spawned a host of related fields. These are often referred to as the *–omics*—proteomics, metabolomics, and transcriptomics—but also include emerging areas like systems biology. Like genomics, these fields take a holistic approach to learning about the entire set of proteins, metabolites, or RNA transcripts present in a given cell or tissue type at a given time.

As an introductory biology student, you are part of the first generation trained in the genome era. Genomics and the related fields it has generated are revolutionizing biological science. They will almost certainly be an important part of your personal and professional life. Let's delve in.

21.1 Whole-Genome Sequencing

Genomics has moved to the cutting edge of research in biology, largely because of technological advances. These began with the development of dideoxy sequencing and progressed to next-generation sequencing techniques (introduced in Chapter 20). These technical breakthroughs have enabled obtaining immense quantities of high-quality sequence data rapidly and at low cost.

As technology continues to become faster and less expensive, the pace of genome sequencing accelerates. The result is that an almost mind-boggling number of sequences of genes and whole genomes are now being generated. As this book goes to press, the primary international repository for DNA sequence data contained over 425 *billion* nucleotides. By way of comparison, a haploid human genome contains about 3 billion nucleotides on each strand of the DNA double helix.

FIGURE 21.1 gives a visual sense of the growth in sequence data by plotting the number of nucleotides, in billions, versus time. There are three large international online repositories for

FIGURE 21.1 The Total Number of Bases Sequenced Is Growing Rapidly. Data from the EMBL Nucleotide Sequence Database (also known as EMBL-Bank).

DATA: European Nucleotide Archive/EMBL-Bank Release Notes. Release 110, December 2011. www.ebi.ac.uk/embl/.

sequence data; the numbers plotted here were compiled from one of them. Over 15 years, this database has grown at a staggering average rate of about 46 percent per year.

How Are Complete Genomes Sequenced?

Genomes range in size from about a half million base pairs to several billion. But even under the best conditions, a single dideoxy sequencing reaction can analyze only about 1000 nucleotides. Reads from next-generation sequencing are even shorter. How do investigators break a genome into sequencing-sized pieces and then figure out how the thousands or millions of pieces go back together?

Shotgun Sequencing   When researchers first set out to sequence the genome of a species, they usually rely on an approach known as **shotgun sequencing.** In shotgun sequencing, a genome is broken up into a set of overlapping fragments that are small enough to be sequenced completely. The regions of overlap are then used as guides for putting the sequenced fragments back into the correct order (**FIGURE 21.2**).

Step 1 Application of high-frequency sound waves, or sonication, is used to break a genome randomly into pieces about 160 kilobases (kb) long (1 kb = 1000 bases).

Step 2 Each 160-kb piece is inserted into a type of cloning vector called a **bacterial artificial chromosome (BAC).** BACs are able to replicate large segments of DNA. Each BAC is then inserted into a different *Escherichia coli* cell (using techniques introduced in Chapter 20), creating a **BAC library.** A BAC library is a genomic library: a set of all the DNA sequences in a particular genome, split into small segments and inserted into cloning vectors. By allowing each cell to grow into a colony, researchers can isolate large numbers of each 160-kb fragment.

Step 3 After many copies of each 160-kb fragment have been produced, each cloned DNA is again broken into fragments—but this time, into segments about 1 kb long.

Step 4 These small fragments are then inserted into plasmids and placed inside bacterial cells. (Note that by this point the genome has been broken down twice, into increasingly manageable pieces: 160-kb fragments in BACs and 1-kb segments in plasmids.) The plasmids are copied many times as each bacterial cell grows into a large population. Cloned 1-kb fragments are then available for sequencing reactions.

Step 5 Next, the cloned 1-kb fragments from each 160-kb BAC clone are sequenced, and computer programs analyze regions where the ends of different 1-kb segments overlap. Overlaps occur because many copies were made of each 160-kb segment, and these copies were fragmented randomly by sonication.

Step 6 A computer program searches for overlaps between 1-kb fragments from a single BAC clone and stitches the sequences together until a continuous sequence across the BAC has been reconstructed.

Step 7 The ends of the reconstructed BACs are analyzed in a similar way. The goal is to link sequences from each 160-kb

Genomic DNA

~160-kb fragments

BAC library

BAC

Main bacterial chromosome

1-kb fragments

Many copies (three shown) of each 160-kb fragment, each broken differently

"Shotgun clones"

Shotgun sequences

TAGCGATCGATTTAGACTCGATAA
TAGACTCGATAAGGATGCGATACTACG

TAGCGATCGATTTAGACTCGATAAGGATGCGATACTACG

Draft sequence

1. **Break DNA at random locations into fragments of ~160 kb**, using sonication.

2. **Clone using BACs.** Insert fragments into bacterial artificial chromosomes; grow in *E. coli* cells to obtain large numbers of each fragment.

3. **Break into 1-kb fragments.** Purify many copies of each 160-kb fragment, then randomly break each into a set of 1-kb fragments, using sonication; if enough DNA is broken up, there will be overlap between many fragments.

4. **Clone using plasmids.** Insert 1-kb fragments into plasmids; grow in *E. coli* cells. Obtain many copies of each fragment.

5. **Sequence each fragment.** Find regions where different fragments overlap.

6. Assemble the sequence from all of the **1-kb fragments** from each original 160-kb fragment by matching overlapping ends.

7. Assemble the sequence from all of the **160-kb fragments** (from different BACs) by matching overlapping ends.

FIGURE 21.2 Shotgun Sequencing Breaks Large Genomes into Many Short Segments.

✔**QUESTION** A shotgun blast produces many small, scattered pieces of shot. Why is "shotgun" an appropriate way to describe this sequencing strategy?

segment based on regions of overlap until the sequence of an entire chromosome is assembled.

In essence, the shotgun strategy consists of breaking a genome into many small fragments, sequencing each fragment, and then putting the sequence data back in the correct order.

✔ If you understand shotgun sequencing, you should be able to explain why it is essential for fragments to have regions of overlap.

The Impact of Next-Generation Sequencing Today, there are approaches that are much faster and cheaper than dideoxy sequencing (see Chapter 20). These next-generation methods are massively parallel, meaning that millions of DNA fragments can be sequenced simultaneously in one run of a sequencing machine. The downside of these methods is that they produce sequence reads of only about 50–200 nucleotides, depending on the particular technology. This is in contrast to the roughly 1000 nucleotides obtained by dideoxy sequencing. For piecing together a whole genome, especially one with many repetitive sequences such as those present in most eukaryotes, these read lengths are too short.

But if a complete genome is already available for the organism, next-generation sequencing offers a remarkably quick and inexpensive way to sequence the entire genome from a particular individual—with all the tiny fragments arranged in the correct order by being compared with the "master genome." This sequencing power has opened up possibilities that were unimaginable even a few years ago.

Consider what was involved in sequencing the human genome. The Human Genome Project required more than 15 years and about $3 billion to assemble the first human genome sequence. In 2011, ten years after the first human genome sequence was available, more than 2700 individual human genomes had been sequenced at a cost as low as $5000 per genome. This is a 2700-fold increase in output and a 600,000-fold drop in price.

What's important is not only the numbers, but what can be done with the information. One illustration is an offshoot of the Human Genome Project, the 1000 Genomes Project. This effort has already sequenced the genomes of over 1000 people selected from diverse populations spread across the planet. An important goal is to assess the genetic similarities and differences among people in order to understand our own evolution. Many similarly ambitious and exciting genomics projects are under way.

Bioinformatics How do researchers piece together the millions of fragments produced by shotgun sequencing? Once a complete genome is assembled, how are the raw sequence data and information about genes and their products made available to the international community of researchers? How can genomes of different species be compared to learn about evolutionary relationships?

The answer is **bioinformatics**—a field that fuses mathematics, computer science, and biology in an effort to manage and analyze sequence data. Researchers in bioinformatics have created searchable databases. These vast repositories hold sequence information that allows investigators to evaluate the similarities between newly discovered genes and genes that have been studied previously in the same or other species.

The World Wide Web has put sequence databases at the fingertips of anyone with an Internet connection. For example, the U.S. National Center for Biotechnology Information (NCBI) is only a click away on your computer. At this free and publicly accessible site, you can search billions of nucleotides by using programs such as BLAST, which can quickly find DNA sequences related to any new gene uncovered in a genomics project.

The vast quantity of data generated by genome sequencing centers makes bioinformatics an indispensable element of genomics.

Which Genomes Are Being Sequenced, and Why?

The first genome to be sequenced from an organism—not a virus—came from a bacterium that lives in the human upper respiratory tract. *Haemophilus influenzae* has one circular chromosome and a total of 1.8 million base pairs of DNA. Its genome was small enough to sequence completely in a reasonable amount of time and within a reasonable budget, given the technology available in the early 1990s. *H. influenzae* was an important research subject because it causes earaches and respiratory tract infections in children. One strain is also capable of infecting the membranes surrounding the brain and spinal cord, causing meningitis.

Publication of the *H. influenzae* genome in 1995 was quickly followed by publication of complete genomes sequenced from an assortment of bacteria and archaea. Sequencing of the first eukaryotic genome, from the yeast *Saccharomyces cerevisiae*, was finished in 1996. Today, the genomes of more than 2000 species from all domains of life have been sequenced. That number is certain to continue climbing in the coming years.

Most of the organisms that have been selected for whole-genome sequencing have interesting biological properties, represent a particular branch of life informative for evolutionary investigations, or cause disease. For example:

- Genomes of bacteria and archaea from hot environments have been sequenced in the hopes of discovering enzymes useful for high-temperature industrial applications.

- A set of more than 50 genomes of diverse bacteria and archaea was sequenced to explore patterns of evolution that are impossible to study by other means.

- Genomes such as those of rice and maize (corn) have been sequenced for crop improvement applications.

- The fruit fly *Drosophila melanogaster*, the roundworm *Caenorhabditis elegans*, the house mouse *Mus musculus*, and the mustard plant *Arabidopsis thaliana* were analyzed because they serve as model organisms in biology (see **BioSkills 13** in Appendix B). Data from these and other well-studied organisms have helped researchers interpret the human genome.

- The platypus and African elephant genomes have been sequenced to reveal evolutionary relationships among mammals. Although the elephant genomes are not complete, the available data confirmed that there are two distinct species of African elephant. This information is vital to conservation plans.

Which Sequences Are Genes?

Obtaining raw sequence data is just the beginning of the effort to understand a genome. As researchers point out, raw sequence data are analogous to the parts list for a house. The list, however, would read something like "windowwabeborogovestaircasedoorjubjub" because it has no punctuation and contains portions that appear to have no meaning.

Where do the genes for "window," "staircase," and "door" start and end? Are the segments that read "wabeborogove" and "jubjub" important in gene regulation, or are they simply spacers or other types of sequences that have no function at all?

The most basic task in interpreting a genome is to identify which bases constitute genes. This task is called **genome annotation**. Recall the current definition of a gene: A segment of DNA that codes for a functional RNA or protein product—or a series of alternatively spliced products—and that regulates their production. In bacteria and archaea, identifying genes is relatively straightforward. The task is much more difficult in eukaryotes.

Identifying Genes in Bacterial and Archaeal Genomes To interpret bacterial and archaeal genomes, biologists begin with computer programs that scan the sequence of a genome in both directions. These programs identify each reading frame that is possible on the two strands of the DNA. Recall that a reading frame is a continuous sequence of non-overlapping codons (see Chapter 16).

Codons consist of three bases, so three reading frames are possible on each DNA strand, for a total of six possible reading frames (**FIGURE 21.3**). Because randomly generated sequences contain a stop codon at about 1 in every 20 codons on average, a long stretch of codons that lacks a stop codon is a good indication of a protein-coding sequence. The computer programs draw attention to any "gene-sized" stretches of sequence that lack an internal stop codon and are flanked by a stop codon and a start codon. Because polypeptides range in size from a few dozen amino acids to many hundreds of amino acids, gene-sized stretches of sequence range from several hundred bases to thousands of bases. In addition, the computer programs look for sequences typical of promoters, ribosome binding sites, or other regulatory sites. DNA segments that are identified in this way are called **open reading frames,** or **ORFs.**

Once an ORF is found, a computer program compares its sequence with the sequences of known genes from another species.

FIGURE 21.3 Open Reading Frames Can Identify Genes. Computer programs use the genetic code to translate the three possible reading frames on each strand of DNA. A long stretch of codons that lacks a stop codon may be an open reading frame (ORF) and identify a gene.

✔**QUESTION** To predict the mRNA codons that would be produced by a particular reading frame, a computer analyzes the DNA in the 3′-to-5′ direction. Why?

If the ORF is unlike any gene that has so far been described in any species, further research is required before it can actually be considered a gene. In contrast, if an ORF shares a significant amount of sequence with a known gene from another species, then it is very likely to be a gene.

Similarities between genes in different species are usually due to **homology.** If genes are homologous, it means they are similar because they are related by descent from a common ancestor. Homologous genes have similar base sequences and frequently the same or a similar function. For example, consider the genes that code for enzymes involved in repairing mismatches in DNA (introduced in Chapter 15). The mismatch-repair genes in *E. coli*, yeast, and humans are similar in DNA sequence and function. To explain this similarity, biologists hypothesize that the common ancestor of all cells living today had mismatch-repair genes—thus, the descendants of this ancestral species also have versions of these genes.

Identifying Genes in Eukaryotic Genomes Mining eukaryotic sequence data for protein-coding genes is complicated. For example, because coding regions are broken up by introns, it is not possible to scan for long ORFs. Instead, researchers combine an array of approaches.

Perhaps the most productive gene-finding strategy has been to isolate mRNAs from cells, use reverse transcriptase to produce a cDNA version of each mRNA, and sequence a portion of the resulting molecule to produce an **expressed sequence tag,** or **EST.** ESTs represent protein-coding genes. To locate the gene, researchers use the EST to find the matching sequence in genomic DNA.

Although ESTs and many other gene-finding strategies have been fruitful, it will likely be many years before biologists are convinced they have identified all the genes in even a single eukaryotic genome. As gene identification efforts continue, researchers are analyzing the data and making some remarkable discoveries. Let's first consider what genome sequencing has revealed about the nature of bacterial and archaeal genomes and then move on to eukaryotes. Is the effort to sequence whole genomes paying off?

21.2 Bacterial and Archaeal Genomes

Biologists have obtained the genome sequences of thousands of distinct bacterial and archaeal species or strains. For example, researchers have sequenced the genome of a laboratory population of *Escherichia coli*—derived from the harmless strain that lives in your gut—as well as the genome of a form that causes severe disease in humans. As a result, researchers can identify genes that differ between these strains and begin experiments to learn what accounts for the infectious properties of some strains.

What general observations have biologists been able to make about the nature of all these bacterial and archaeal genomes?

The Natural History of Prokaryotic Genomes

In a sense, biologists who are working in genomics can be compared to the naturalists of the eighteenth and nineteenth centuries. These early biologists explored the globe, collecting the plants and animals they encountered. Their goals were to describe what existed and identify any patterns. Similarly, the first task of a genome sequencer is to catalog what is in a genome—specifically, the number, type, and organization of genes—and then look for patterns within and between different genomes. Here are some principles that have emerged from analysis of prokaryotic genomes:

- Bacterial and archaeal genomes are compact. They have uninterrupted coding sequences, little space between genes, extensive use of operons, and relatively few regulatory sequences. This structure leads to a linear relationship between genome size and gene number. Look at the graph of bacterial and archaeal genomes in **FIGURE 21.4a**, and notice how genome size and the number of genes increase together in a nearly straight line. In contrast, this simple relationship does not hold for eukaryotic genomes (**FIGURE 21.4b**). Section 21.3 explores these features of eukaryotic genome organization in detail.

- In bacteria, there is a correlation between the size of a genome and the metabolic capabilities of the organism. Species that live in a variety of habitats and use a wide array of molecules for food have large genomes; parasites—species that make use of a host's biochemical machinery rather than synthesizing their own molecules—have small genomes.

- The function of many bacterial and archaeal genes is still unknown. Across a wide range of species, a function cannot yet be assigned to 15 to 30 percent of genes.

- Most of the genes found in one species are not shared widely with others. In a study that sampled the genomes of diverse prokaryotes, every time a new genome was sequenced, more than 1000 new genes were discovered. Only a small set of genes involved in processes such as DNA replication, transcription, and translation are similar across a wide range of bacteria or archaea.

- The content of genomes varies widely even within species. For example, sequencing of 17 different *E. coli* strains revealed a total of about 13,000 different genes. The genome of any one strain has only about 4400 genes, and roughly 2200 of these are shared by all strains. The remaining genes in each genome are found only in some strains but not in others.

- Prokaryotic genomes are frequently rearranged during evolution. Even closely related species show little similarity in gene order.

Perhaps the most surprising observation of all is that in many bacterial and archaeal species, a significant proportion of the genome appears to have been acquired from other, often distantly related, species. The movement of DNA from one species to another is called **lateral gene transfer.** As you'll learn (Chapter 27), there are many ways to define a species. But in all traditional definitions, members of one species cannot exchange genes with members of another species. Lateral gene transfer counters this view: Instead of moving vertically from generation to generation within a species, in lateral gene transfer, genes move "laterally" between different coexisting species.

Genomics has shown that lateral gene transfer is more common than ever imagined. This finding is causing many biologists to wonder if conventional evolutionary trees capture the way bacteria and archaea have evolved.

Lateral Gene Transfer in Bacteria and Archaea

The extent of lateral gene transfer is still debated, but genomic data indicate that all bacterial and archaeal species have experienced lateral gene transfer. Lateral gene transfer is a major force in the evolution of bacteria and archaea.

One illustration is the bacterium *Thermotoga maritima*. This species thrives in high-temperature environments near deep-sea vents. Almost 25 percent of the genes in this species are closely related to genes found in archaea that live in the same habitats. The archaea-like genes occur in well-defined clusters within the *T. maritima* genome. These observations support the hypothesis that these sequences were transferred in large pieces from an archaean to a bacterium—organisms in two different domains of life.

How are laterally transferred genes identified? Biologists primarily use two criteria: (**1**) A gene is much more similar to genes in distantly related species than to those in closely related species,

(a) In bacteria (◉) and archaea (●), genome size and number of genes increase together in a linear relationship.

(b) The same relationship does not hold true for eukaryotes (●).

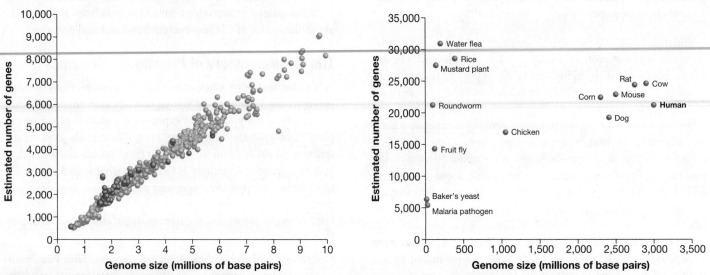

FIGURE 21.4 Relationship Between Genome Size and Gene Number.

DATA: (a) Hou, Y., and S. Lin. 2009. *PLoS ONE* 4(9): e6978, Supplemental Table S1. (b) KEGG: Kyoto Encyclopedia of Genes and Genomes, KEGG Organisms: Complete Genomes. www.genome.jp/kegg/.

and (2) the proportion of G-C base pairs to A-T base pairs in a particular gene or series of genes is markedly different from the base composition of the rest of the genome. This second criterion works because the proportion of G-C base pairs in a genome is characteristic of the particular genus or species.

How can genes move from one species to another? In some cases, plasmids are responsible. For example, most of the genes that are responsible for conferring resistance to antibiotics are found on plasmids. Researchers have documented the transfer of plasmid-borne antibiotic-resistance genes between distantly related bacteria. In many cases of lateral gene transfer, genes from plasmids become integrated into the main chromosome of a bacterium through genetic recombination.

Lateral gene transfer may also occur by transformation—when bacteria and archaea take up raw pieces of DNA from the environment—and by viruses that pick up DNA from one cell and transfer it to another cell.

There is no doubt that lateral gene transfer occurs even between distantly related organisms. What is still debated, however, is how much this shakes the tree of life. If lateral gene transfer is rare, then evolutionary paths in the bacteria and archaea form a set of branches that begin at common ancestors and spread out to descendants. If lateral gene transfer is as widespread as some biologists believe, then evolutionary paths must form a complex interconnected network that links species in a web of vertical and lateral gene transfers. These alternative views are shown in **FIGURE 21.5**.

In light of new genomics findings, will a tree of life stand for bacteria and archaea? Or are the evolutionary relationships in these domains of life best viewed as a web? Stay tuned.

Metagenomics

Biologists continue to gain important insights from sequencing the genomes of individual species and strains. But more recently, some research groups have taken a different approach: cataloging all the genes present in a community of bacteria and archaea. This type of research is called **metagenomics,** or **environmental sequencing.** The subject of these studies is genes—not organisms.

The first environmental sequencing study was conducted in the Sargasso Sea, near the Caribbean island of Bermuda. Researchers chose the spot because it is extremely nutrient poor and species poor—a desert in the ocean. This is a common strategy

in biological science: Start by studying a simple system, then go on to more complex situations. To inventory the complete array of bacterial genes present, the research group collected cells from different water depths and locations, isolated DNA from the samples, and sequenced the DNA.

After analyzing over 1 billion nucleotides, the team concluded that at least 1800 bacterial species were present, of which 148 were previously undiscovered. They also identified more than 1.2 million genes that had never before been characterized. These genes included over 780 sequences that code for proteins similar to the rhodopsin found in the cells of your retina—a molecule that is absorbing the light entering your eye right now. Follow-up work suggests that most of the Sargasso Sea rhodopsin-like molecules are also absorbing light, and that bacterial cells use the energy of the light to pump protons across their plasma membranes—creating a chemiosmotic gradient that can synthesize ATP (see Chapter 9).

Many metagenomics studies are going on today, including one that examined microbes of the human gut (Chapter 29). All these investigations are providing new insights about the living world, from how rhodopsin-like proteins help bacteria thrive in a desert to how your lunch is being digested.

21.3 Eukaryotic Genomes

DNA sequencing has revealed some extraordinary features of eukaryotic genome organization. Genome size varies widely, but the number of genes is much more similar. For example, the genomes shown in Figure 21.4b vary roughly 250-fold in size (from 12 to 3000 million base pairs) but less than sixfold in the number of genes (from 5400 to 30,900). How can gene number be so similar among organisms with vastly different genome sizes and that range in complexity from single-celled parasites to large multicellular plants and animals?

Tree of life   **Web** of life

FIGURE 21.5 Tree of Life and Web of Life Views of Evolution. The broken horizontal arrows indicate lateral gene transfer events.

Before exploring possible answers, let's consider two daunting challenges in sequencing eukaryotic genomes. The first is size. The largest bacterial genome, that of *Sorangium cellulosum*, is slightly more than 13 million base pairs. However, as Figure 21.4b shows, even modest-sized eukaryotic genomes are much larger. The 3-billion-base-pair human genome dwarfs those of bacteria but is miniscule compared with the genome of the Japanese flower *Paris japonica*. This plant's genome contains 149 billion base pairs of DNA—that's 50 times the size of the human genome!

The second challenge in sequencing eukaryotic genes is coping with sequences that are repeated many times. Many eukaryotic genomes are dominated by repeated DNA sequences that occur between genes or inside introns and do not code for products used by the organism. These repeated sequences greatly complicate the work of aligning and interpreting sequence data. They also explain some of the paradox of the immense variation in eukaryotic genome sizes. If repeated sequences don't code for products needed by cells, what do they do?

Transposable Elements and Other Repeated Sequences

In many eukaryotes, the exons and regulatory sequences associated with genes make up a relatively small percentage of the genome. Over 90 percent of a bacterial or archaeal genome consists of genes, but about 50 percent of an average eukaryotic genome consists of repeated sequences that do not code for a product used by the cell.

When repeated sequences were discovered, they were initially considered "junk DNA" that was nonfunctional and probably unimportant and uninteresting. But subsequent work has shown that many of the repeated sequences observed in eukaryotes are actually derived from sequences known as transposable elements.

Transposable elements are segments of DNA that are capable of being inserted into new locations, or transposing, in a genome. They were first discovered in corn by Barbara McClintock and later shown to be present in organisms from every domain of life. Transposable elements behave similarly to some viruses that insert into the genome. In contrast to viruses, however, transposable elements seldom leave their host cell—instead, they make copies of themselves that become inserted in new locations. Transposable elements are passed from mother to daughter cell and from parents to offspring, generation after generation, because they are part of the genome.

A transposable element is an example of what biologists call selfish DNA: a DNA sequence that has invaded a host and persists and reproduces using the resources of the host. Selfish or not, transposable elements play a big role in a species as they move from place to place and cause mutation. Like any mutation, a transposable element insertion into a new site in the genome can have negative, neutral, or positive effects on fitness. Transposable elements are a significant part of almost all cellular genomes and play a major role in evolution. Transposable elements have many effects, disrupting the coding sequence of genes, changing patterns of gene regulation, and promoting gene

duplication and loss. They are genome invaders that also shape the structure and function of genomes in profound ways.

How Do Transposable Elements Work? Transposable elements come in a wide variety of types and spread through genomes in a variety of ways. Different organisms—*E. coli*, fruit flies, yeast, and humans, for example—contain distinct types of transposable elements. Bacterial and archaeal genomes, however, have far fewer transposable elements compared with most eukaryotes. This observation has inspired the hypothesis that bacteria and archaea either have efficient means of removing parasitic sequences or can somehow thwart insertion events.

As an example of how these selfish DNA sequences work, consider a well-studied type of transposable element called a **long interspersed nuclear element (LINE).** LINEs are found in humans and other animals. Because LINEs have a reverse transcriptase like retroviruses (introduced in Chapter 20), biologists hypothesize that they are derived from them evolutionarily. Your genome contains nearly 1 million LINEs, each between 1000 and 5000 bases long. Transposable elements of different types make up over 45 percent of the human genome and 85 percent of the corn (maize) genome. **FIGURE 21.6** illustrates the steps that allow an active LINE to transpose.

Most of the LINEs observed in the human genome do not actually function, however, because they don't contain a promoter or the genes for either reverse transcriptase or integrase. To make sense of this observation, researchers hypothesize that the insertion process illustrated in steps 6 and 7 of Figure 21.6 is often disrupted in some way, leaving the inserted replica of the original LINE incomplete.

Research on transposable elements and lateral gene transfer has revolutionized how biologists view the genome. Many genomes are riddled with transposable elements, and others have undergone radical change in response to lateral gene transfer events. In other words, genomes are much more dynamic and complex than previously thought. Their size and composition can change dramatically over time.

Repeated Sequences and DNA Fingerprinting In addition to containing repeated sequences from transposable elements, many eukaryotic genomes have several thousand loci of relatively short DNA sequences. They are repeated in tandem, one after another, contiguously along part of a chromosome.

These tandem repetitive DNA sequences fall into two major classes:

1. Repeating units that are just 2 to about 6 bases long. These are **microsatellites,** also known as **short tandem repeats (STRs)** or **simple sequence repeats.** The most common type of microsatellite in humans is a repeated stretch of the dinucleotide AC, giving the sequence ACACACAC . . . .

2. Repeating units that are longer, from about 6 to 100 bases long. These are **minisatellites,** or **variable number tandem repeats (VNTRs).**

Microsatellite sequences are thought to originate when DNA polymerase skips or mistakenly adds extra bases during replication;

Gene for reverse transcriptase

Gene for integrase

DNA

Original location of LINE

1. A long interspersed nuclear element (LINE) exists in the genome.

Nuclear envelope

LINE mRNA

RNA polymerase

2. RNA polymerase transcribes the LINE, producing LINE mRNA.

LINE protein

Ribosome

3. LINE mRNA exits nucleus and is translated.

LINE mRNA and LINE proteins

Reverse transcriptase

Integrase

4. LINE mRNA and proteins enter nucleus.

cDNA

mRNA

Reverse transcriptase

5. Reverse transcriptase makes LINE cDNA from mRNA, then makes cDNA double stranded.

Integrase

Cytoplasm

6. Integrase cuts chromosomal DNA and inserts LINE cDNA.

Original copy

New copy

7. New copy of the LINE is integrated into a new position in genome.

FIGURE 21.6 Transposable Elements Spread within a Genome.

the origin of minisatellites is still unclear. Together, the two types of repeated sequences make up 3 percent of the human genome.

Soon after these sequences were first characterized, Alec Jeffreys and co-workers established that microsatellite and minisatellite loci are "hypervariable," meaning that they vary among individuals much more than any other type of sequence does.

FIGURE 21.7 (see page 398) illustrates one mechanism to explain why microsatellites and minisatellites have so many different alleles: a process called **unequal crossover.** Here's how it works: Homologous chromosomes sometimes align incorrectly during prophase of meiosis I. Instead of lining up in exactly the same location, the two chromosomes pair in a way that matches up bases in different DNA repeats. When crossover occurs, the resulting chromosomes have different numbers of repeats.

Repeated sequences are particularly prone to unequal crossover, because their homologs are so similar that they are likely to misalign. If the region in question has a unique number of repeats, it represents a unique allele. Like any other alleles, microsatellite and minisatellite alleles are transmitted from parents to offspring.

The variation in repeat number among individuals is more than a curious feature of genome organization—it is the basis

of most DNA fingerprinting. **DNA fingerprinting** refers to any technique for identifying individuals based on the unique features of their genomes. Because microsatellite and minisatellite sequences vary so much, they are now the sequences of choice for DNA fingerprinting. How can these sequences be used for DNA fingerprinting?

Investigators obtain a DNA sample and perform the polymerase chain reaction (PCR), using primers that flank a region containing an STR (**FIGURE 21.8**; see page 398). Once the region has been amplified, it can be analyzed to determine the number of repeats present. Primers are now available that allow the analysis of many different STR loci.

These advances have profound impacts on society. Police use DNA fingerprinting to put people behind bars, and DNA fingerprinting has been used to show that people who were accused of crimes were actually innocent. Beyond criminal investigations, DNA fingerprinting has also been used to assign paternity and to identify remains.

Now that some characteristics of eukaryotic genomes have been reviewed, let's consider the genes they contain. Let's start with the most basic question of all: Where do eukaryotic genes come from?

8 repeats

8 repeats

1. **Homologs pair up.** At first these two homologous chromosome sections contain the same number of microsatellite or minisatellite repeats.

2. **Repeats misalign.** The repeats misalign during meiosis I. Crossing over and recombination occur.

Chromosomes break and exchange here

10 repeats

6 repeats

3. **New repeat numbers are created.** Products of meiosis have additional or deleted repeats; the 10-repeat region and 6-repeat region represent new alleles.

FIGURE 21.7 Unequal Crossover Adds or Deletes DNA Repeats.

Gene Families

In eukaryotes, the major source of new genes is the duplication of existing genes. Biologists infer that genes have been duplicated recently when they find groups of genes that are similar in such features as the arrangement of exons and introns, and their base sequence. Within a species, genes that are similar to each other in structure and function are considered to be part of the same **gene family.**

The degree of sequence similarity among members of a gene family varies. In the genes that code for ribosomal RNAs (rRNAs) in vertebrates, the sequences are virtually identical—meaning that each individual has many exact copies of the same gene. In other cases, though, 50 percent or less of the bases are identical.

How Do Gene Families Arise? Genes that make up gene families are hypothesized to have arisen from a common ancestral sequence through gene duplication. When **gene duplication** occurs, an extra copy of a gene is added to the genome.

The most common type of gene duplication results from unequal crossover during meiosis—the same process that resulted in extra microsatellite and minisatellite repeats in Figure 21.7. Gene-sized segments of chromosomes can be deleted or duplicated if homologous chromosomes misalign during prophase of meiosis I and an unequal crossover occurs. Like microsatellites or minisatellites, the duplicated segments are arranged in tandem—one after the other.

New Genes—New Functions? Gene duplication is important because the original gene is still functional and produces a normal product. As a result, the new, duplicated stretches of sequence are redundant. If mutations in the duplicated sequence alter the protein product so that it performs a valuable new function, then an important new gene has been created.

(a) Use PCR to amplify STR loci.

STR locus

5′ Primer

3′

Primer

5′

Individual 1:

4 repeats

Individual 2:

8 repeats

Individual 3:

2 repeats

Different individuals may have different numbers of repeats at this locus

(b) Compare number of STR repeats in alleles to test paternity.

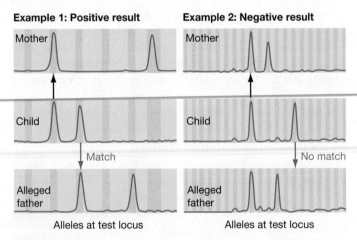

Example 1: Positive result

Example 2: Negative result

Mother

Child

Match

Alleles at test locus

Mother

Child

No match

Alleles at test locus

Alleged father

Alleged father

FIGURE 21.8 DNA Fingerprinting Can Be Used to Identify Parents. **(a)** The lengths of STR loci vary. Only one allele is shown for each individual. Individuals are often heterozygous, so the repeat number varies within and between individuals. **(b)** Here, the position of each peak indicates the number of repeats at a particular locus. Each individual is heterozygous and thus shows two peaks. One of the peaks from the mother and one of the peaks from the father should line up with a peak in the child. Typically 6 to 16 loci are tested to determine paternity.

Human β-globin gene family

ψβ2   ε   Gγ   Aγ   ψβ1   δ   β

Pseudogene   Coding gene

FIGURE 21.9 Gene Families Are Closely Related Genes. The β-globin gene family is shown with coding genes in red and pseudogenes in yellow.

✔**EXERCISE** Suppose that during prophase of meiosis I, the β locus on one chromosome aligned with the ψβ2 locus on another chromosome. Then crossing over occurred in the noncoding sequences just to the left (as oriented in the figure) of this β-ψβ2 pairing. List the order of the β-globin-family genes that would result on each chromosome.

As an example of a gene family, consider the human globin genes diagrammed in **FIGURE 21.9**. Collectively, this set of genes is known as the β-globin gene family, and they code for proteins that form part of hemoglobin—the oxygen-carrying molecule in your red blood cells. Each coding gene in the family serves a slightly different function. For example, some genes are transcribed only in the fetus or the adult. The product of the fetal gene binds oxygen more tightly than the proteins expressed in adults. Consequently, oxygen moves from the mother's blood, where it is not as tightly bound to hemoglobin, to the fetus's blood (see Chapter 45).

In addition to creating genes with new functions, mutations in duplicated regions often make gene expression impossible. For example, a mutation could produce a stop codon in the middle of an exon. A member of a gene family that resembles a working gene but does not code for a functional product because of a mutation is a **pseudogene.** Pseudogenes do not function. Note that the β-globin gene family contains pseudogenes along with several genes that code for oxygen-transporting proteins. The number of pseudogenes is remarkable. In the human genome, for example, there are roughly as many pseudogenes as functional genes.

Lateral Gene Transfer in Eukaryotes Genome sequencing projects are revealing more and more instances of lateral gene transfer in eukaryotes. What is the role of lateral gene transfer in the evolution of eukaryotes?

There are some clear examples of lateral gene transfer playing pivotal roles in eukaryote evolution. One key example is the capture of bacterial cells that were predecessors of today's mitochondria and chloroplasts and the subsequent transfer of many bacterial genes to the host genome (see Chapter 30). But in general, lateral gene transfer seems to be relatively rare in eukaryotes compared with prokaryotes. While biologists debate whether a tree or a web of life best describes evolution in bacteria and archaea, the eukaryotic tree remains rooted in what appear to be more modest rates of lateral gene transfer.

Insights from the Human Genome Project

More than 10 years ago, President Clinton announced the completion of the human genome sequence as "the most wondrous map ever produced by humankind." What has analysis of human genome sequence revealed? Let's first explore what types of DNA sequences make up the human genome and then examine some questions raised by this wondrous map.

Given biologists' focus on protein-coding genes, the composition of the human genome was unexpected. As **FIGURE 21.10** reveals, less than 2 percent of the genome consists of protein-coding exons, and nearly half is made of transposable elements. Introns make up over one-quarter of the genome and are 17 times more abundant than protein-coding exons.

Of all the observations stemming from the human genome, perhaps the most striking is that organisms with complex morphology, biochemistry, and behavior do not have particularly large numbers of genes. Notice in Figure 21.4b that the total number of genes in humans is only 50 percent more than the number in fruit flies, is about the same as in roundworms, and is substantially lower than the number of genes in water fleas, rice, and the mustard plant *Arabidopsis thaliana*.

Before the human genome was sequenced, many biologists expected that humans would have at least 100,000 genes. But we may only have a fifth of that number. How can this be?

In prokaryotes there is a correlation between genome size, gene number, a cell's metabolic capabilities, and the cell's ability to live in a variety of habitats. But why isn't there a stronger correlation between gene number and morphological, biochemical, and behavioral complexity in eukaryotes?

FIGURE 21.10 Composition of the Human Genome.

DATA: Gregory, T. R. 2005. *Nature Reviews Genetics* 6: 699–708.

One hypothesis to explain this observation is based on **alternative splicing.** Recall that the exons of a particular gene can be spliced in ways that produce distinct, mature mRNAs. As a result, a single eukaryotic gene can code for multiple transcripts and thus multiple proteins (see Chapter 19). The alternative-splicing hypothesis claims that multicellular eukaryotes do not need enormous numbers of distinct genes. Instead, more extensive use of alternative splicing in these organisms creates different proteins from the same gene.

In support of the alternative-splicing hypothesis, researchers have analyzed the mRNAs produced by human genes. They estimate that at least 90 percent of genes produce transcripts that are alternatively spliced, with an average of more than three distinct transcripts per gene. This means that the number of different proteins that can be produced is more than triple the gene number. Because of extensive alternative splicing, the number of genes and number of proteins do not have to be tightly linked.

There is likely more to the story, however, than extensive alternative splicing. Accompanying the rapid advances in sequencing technology have been equally dramatic advances in methods for studying how the genome functions. Some of these methods are described later, in Section 21.4. But what's important here are the findings.

Roughly 90 percent of the human genome is transcribed. This is far more than believed even a few years ago. Some of these transcripts code for regulatory RNAs with known roles, such as microRNAs (Chapter 19). Many of these transcripts, however, have no currently known roles. If many of these newly discovered RNAs have a function, then the human genome and the genomes of other complex organisms may not be so gene poor after all. If so,

this would require an adjustment in how we view a typical gene. Instead of considering most genes to code for proteins, perhaps most genes produce regulatory RNAs that are never translated. As you read this, biologists are working hard on this central question that has been opened by functional studies of the human genome.

What are these functional studies, and how are they carried out? The following section considers these questions.

21.4 Functional Genomics, Proteomics, and Systems Biology

Genomics researcher Eric Lander has compared the sequencing of the human genome to the establishment of the periodic table of the elements in chemistry. Once the periodic table was validated, chemists focused on understanding how the elements combine to form molecules. Similarly, biologists now want to understand how the elements of the human genome combine to produce an individual.

Remember that a genome sequence is essentially a parts list. Once that list is assembled, researchers delve deeper to understand how genes interact to produce an organism.

What Is Functional Genomics?

For decades, biologists have worked at understanding how and when individual genes are expressed. Research on the *lac* operon is typical of this type of study (see Chapter 18). But now researchers can ask how and when *all* the genes in an organism are expressed.

Large-scale analysis of gene expression is part of functional genomics—research on how genes work together to produce a phenotype. The effort is motivated by the realization that gene products do not exist in a vacuum. Instead, groups of RNAs and proteins act together to respond to environmental challenges such as extreme heat or drought. Similarly, distinct groups of genes are transcribed at different stages as a multicellular eukaryote grows and develops.

A basic tool of functional genomics is a microarray. A **DNA microarray** consists of as many as 1 million different single-stranded DNA segments that are permanently attached at one end to a glass slide or silicon chip. Each DNA sequence is linked to the slide or chip in a known location and serves as a probe for a specific transcript. For example, the slide pictured in **FIGURE 21.11** contains thousands of spots, each one containing single-stranded DNA from a unique exon found in the human genome.

Microarrays can be used for many applications. The most common use is to learn which genes are expressed as RNAs in a particular cell type under particular conditions. A typical experiment done with a DNA microarray would follow the protocol outlined in **FIGURE 21.12.** For example, if the researchers' goal was to learn how a cell alters its gene expression to meet the challenges of heat stress, the first step would be to isolate mRNAs produced in control cells functioning at normal temperature and in cells of the same kind exposed to high temperatures.

check your understanding

C Y U

If you understand that . . .

- Eukaryotic genomes are riddled with transposable elements.
- Relatively short repeated sequences are common in eukaryotic genomes.
- In eukaryotes, many coding sequences are organized into families of genes with related functions.
- Much of the human genome does not code for proteins.
- The recent discovery that much of the genome is transcribed suggests that eukaryotic gene expression may be much more complex than previously thought.

✔ You should be able to . . .

1. Explain why there is no simple relationship between the size of a eukaryotic genome and the complexity of the organism.
2. Estimate the ratio of protein-coding portions of the human genome to all transcribed regions of the genome.

Answers are available in Appendix A.

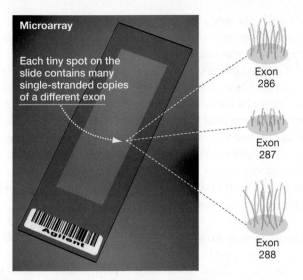

FIGURE 21.11 A DNA Microarray. To create this microarray, thousands of short, single-stranded DNA sequences were synthesized in defined positions on a glass plate. In this microarray, the synthesized DNAs represent portions of all exons from a particular species.

Once they had purified mRNAs from the two populations of cells (step 1), investigators would use reverse transcriptase to make a single-stranded cDNA version of each RNA in each of the two samples (see Chapter 20). In addition to the four standard dNTPs, one of the DNA building blocks used in synthesizing the cDNA would carry a fluorescent label (step 2). The label used for the cDNA of the control cells would glow one color (let's say green), while the label chosen for the cDNA of the heat-stressed cells would glow another color (let's say red). The labeled cDNAs of both colors would then be used to bind to the complementary DNA probes on the microarray (step 3). This step is called hybridization because hybrids between probe DNAs and cDNAs will form.

Out of all the exons in the genome, then, only the exons that are being expressed by the two populations of cells will be labeled on the microarray. In this example, genes that are expressed by the control cells under normal conditions will be labeled green, while those expressed by the cells during heat stress will be labeled red. If an exon in the microarray is expressed under both sets of conditions, then both green- and red-labeled cDNAs will bind to the DNA in that spot and make it appear yellow (step 4).

A microarray lets researchers study the expression of thousands of genes at a time. As a result, they can identify which sets of genes are expressed in concert under specific sets of conditions.

Researchers can use microarrays to establish which genes are transcribed in different organs and tissues, in cancers, or—as you saw in Figure 21.12—in response to changes in environmental conditions such as heat stress. ✔ **If you understand how microarrays are used, you should be able to explain how you would use a DNA microarray to compare the genes expressed in brain cells versus liver cells of an adult human.**

Besides using microarrays, investigators are now able to assess gene expression by directly sequencing cDNAs using next-generation sequencing technologies. For example, if biologists

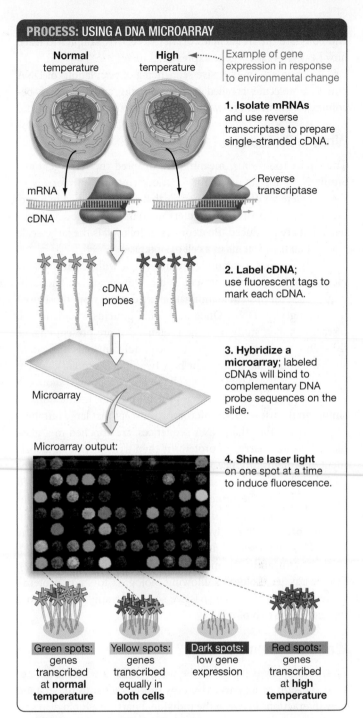

FIGURE 21.12 DNA Microarrays Can Be Used to Study Changes in Gene Expression. By probing a microarray with labeled cDNAs synthesized from mRNAs, researchers can identify which sequences are being transcribed. Here mRNAs from cells growing at normal temperature are detected by green color, while mRNAs from cells growing at high temperature are detected by red color.

wanted to learn about gene expression changes induced by heat stress, they would start as they did for the microarray by preparing two different sets of cDNAs, one from control and one from heat-stressed cells. But instead of using these cDNAs to hybridize with microarray probes, biologists would sequence millions of cDNAs from each treatment type. Using bioinformatics tools, they would then determine the frequency of each type of cDNA and compare the frequencies of the two cDNA samples to learn how heat stress alters gene expression.

This approach of extensive sequencing of populations of DNA or cDNA molecules is called **deep sequencing,** and it is quickly becoming an important research approach for functional genomics.

What Is Proteomics?

The Greek root *–ome*, meaning all, inspired the term genome. Similarly, biologists use the term **transcriptome** in referring to the complete set of DNA sequences that are transcribed in a particular cell, and **proteome** in referring to the complete set of proteins that are produced. **Proteomics,** it follows, is the large-scale study of all the proteins in a cell or organism.

Like genomics, proteomic studies begin with a parts list by identifying the proteins present in a cell or organelle. The techniques used for protein identification are distinct from those used in working with DNA. Once individual proteins are identified, researchers then study how the proteins that are present change through time, interact, or vary between different cells. Instead of studying individual proteins or how two proteins might interact, proteomics is based on studying all the proteins present at once.

One approach to studying protein–protein interactions is similar to the use of DNA microarrays, except that large numbers of proteins, rather than DNA sequences, are attached to a glass plate. This microarray of proteins is then treated with an assortment of proteins produced by the same organism. These proteins are labeled with a fluorescent or radioactive tag. If any labeled proteins bind to the proteins in the microarray, the two molecules may also interact in the cell. In this way, researchers can identify proteins that physically bind to one another—like the G proteins and associated enzymes (introduced in Chapter 11), or the cyclin and Cdk molecules (introduced in Chapter 12). Microarray technology is allowing biologists to study protein-protein interactions on a massive scale, opening the door to a new approach to biology.

What Is Systems Biology?

Systems biology is based on the premise that a whole is greater than the sum of its parts. **Systems biology** aims to understand how interactions between the individual parts of a biological system create new properties. For example, how does metabolism come about from the interaction of proteins within a cell? How does cancer arise from the interplay of individual genes? Complex properties that arise from the interaction of simpler elements are **emergent properties.**

DNA replication, metabolism, cancer, and the development of an organism all are emergent properties because they come about from the interactions of simpler elements—genes and proteins. ✔ If you understand this concept, you should be able to explain why cell replication is an emergent property.

Many systems biology investigations focus on mapping the interactions between genes or proteins. Genomics and proteomics provide the parts list needed to start these systems biology studies. It's the job of a systems biologist to learn how these parts are linked together into networks and how new properties emerge from these interactions.

Let's look at how a systems biology approach was taken to predict all possible interactions between proteins in *Schizosaccharomyces pombe*, a species of yeast. An interaction can mean either that two proteins bind to one another to form a stable complex or that one protein acts on another protein. An example of the first type of interaction is cyclin binding to a cyclin-dependent kinase (Cdk); an example of the second interaction type is when the cyclin–Cdk complex acts as a kinase to add a phosphate to a target protein (see Chapter 12).

To begin their study, the researchers used existing databases of known protein–protein interactions to make predictions about the fission yeast protein interaction network. Then they verified some of the predicted interactions experimentally to show that their prediction method worked. Once they had shown that their prediction method was accurate, they mapped 37,325 possible interactions between 3438 different proteins. If you look at the interaction network in **FIGURE 21.13**, you will see how clusters of interacting proteins emerge from the tangle. These clusters reveal highly connected portions of the network.

The significance? Some of the predicted interactions in this yeast may have implications for human disease. For example, some of the key players in the cluster associated with signal transduction point to previously unexplored relationships that could be important in understanding cancer and neurodegenerative disorders, such as Huntington's disease (see Chapter 20), and for targeted drug design.

Work in genomics, proteomics, and systems biology is opening the door to knowing how cells work in health and disease. These new areas of biology also lay the foundation for understanding one of the most wondrous aspects of life—how a single cell develops into a complete organism. This topic is explored in the unit ahead.

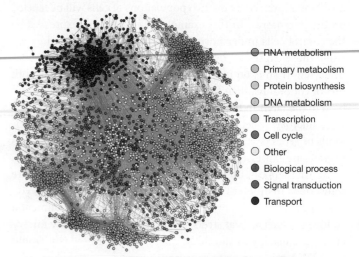

- RNA metabolism
- Primary metabolism
- Protein biosynthesis
- DNA metabolism
- Transcription
- Cell cycle
- Other
- Biological process
- Signal transduction
- Transport

FIGURE 21.13 Yeast Protein Interaction Network. Each circle represents a different protein, with each protein color-coded according to its cellular function. Lines that connect a pair of proteins indicate an interaction.

DATA: V. Pancaldi et al. 2012. *G3: Genes, Genomes, Genetics* 2: 453–467.

If you understand . . .

21.1 Whole-Genome Sequencing

- Advances in DNA sequencing technologies have allowed investigators to sequence DNA more rapidly and cheaply, resulting in a flood of genome data.

- Thousands of genomes have been sequenced to date for many different purposes.

- Bioinformatics is the application of computer science to genome analysis and is essential for genome research.

- Researchers annotate genome sequences by finding genes and determining their function.

- To identify genes in bacteria and archaea, researchers use computers to scan the genome for start and stop codons that are in the same reading frame and that are separated by gene-sized stretches of sequence.

- Genes are also identified based on their homology (similarity due to evolutionary relatedness) to previously identified genes in other species.

- To find genes in eukaryotes, researchers couple the use of computers and study of RNAs to identify transcribed sequences.

✔ You should be able to describe how a research group that discovered a gene for coat color in mice would determine whether a homologous gene exists in the human genome.

21.2 Bacterial and Archaeal Genomes

- Bacterial and archaeal genomes are small relative to many eukaryotic genomes and have tightly spaced genes.

- There is a linear relationship between bacterial and archaeal gene number and genome size, and a similar correspondence in the relationship between metabolic capacity and genome size.

- The function of many of the genes identified in bacteria and archaea is still unknown.

- There is a huge amount of genetic diversity in bacterial and archaeal genomes, even among different strains of the same species.

- Lateral gene transfer is common in bacteria and archaea. It is an important source of new genes in many species.

- Metagenomics allows the analysis of all the genes of all the bacteria or archaea in a community.

✔ You should be able to propose genome characteristics you would look for to distinguish a bacterial parasite from a nonparasitic bacterial species that is found in a range of environments.

21.3 Eukaryotic Genomes

- Eukaryotic genomes tend to be large and complex. They include many sequences that have little to no effect on the fitness of the organism, and many transcribed sequences whose function is not known.

- There is no correlation between morphological complexity and gene number in eukaryotes.

- Because of alternative splicing, the number of distinct transcripts produced in many eukaryotes is much larger than the gene number.

- Gene duplication has been an important source of new genes in eukaryotes.

- Lateral gene transfer occurs in eukaryotes, but appears to play a smaller role than in bacteria and archaea.

- Recent findings show that much more of the eukaryotic genome is transcribed than previously believed, and the function of many noncoding transcripts is unknown.

✔ You should be able to explain what features of eukaryotic genomes result in a lack of correspondence between genome size, gene number, and morphological complexity.

21.4 Functional Genomics, Proteomics, and Systems Biology

- Functional genomics uses tools such as DNA microarrays to learn patterns of gene expression and gene function.

- Proteomics is similar to genomics but works to identify the complete set of proteins expressed in a cell, how this set changes under different conditions, and how it relates to phenotype.

- Systems biology starts with genomics and proteomics data and studies the set of interactions between different genes or proteins to understand how biological systems work.

✔ You should be able to explain how Figure 21.13 indicates that there are networks within networks for interacting proteins associated with particular functions.

MB MasteringBiology

1. **MasteringBiology Assignments**

   Tutorials and Activities DNA Fingerprinting; Human Genome Project: Genes on Human Chromosome 17; Human Genome Sequencing Strategies; Shotgun Approach to Whole-Genome Sequencing

   Questions Reading Quizzes, Blue-Thread Questions, Test Bank

2. **eText** Read your book online, search, take notes, highlight text, and more.

3. **The Study Area** Practice Test, Cumulative Test, BioFlix® 3-D Animations, Videos, Activities, Audio Glossary, Word Study Tools, Art

You should be able to . . .

1. What is an open reading frame in bacteria?
   a. a gene whose function is already known
   b. a DNA section that is thought to code for a protein because it is similar to a complementary DNA (cDNA)
   c. a DNA section that is thought to code for a protein because it has a start codon and a stop codon flanking hundreds of nucleotides
   d. any member of a gene family

2. What best describes the logic behind shotgun sequencing?
   a. Break the genome into tiny pieces. Sequence each piece. Use overlapping ends to assemble the pieces in the correct order.
   b. Start with one end of each chromosome. Sequence straight through to the other end of the chromosome.
   c. Use a variety of techniques to identify genes and ORFs. Sequence these segments—not the noncoding and repeated sequences.
   d. Break the genome into pieces. Map the location of each piece. Then sequence each piece.

3. A _____ is a 2- to 6-base-pair repeated sequence in DNA.
   a. LINE
   b. restriction site
   c. gene duplication
   d. microsatellite

4. What is a leading hypothesis to explain the paradox that large, morphologically complex eukaryotes such as humans have relatively small numbers of genes?
   a. lateral transfer of genes from other species
   b. alternative splicing of mRNAs
   c. polyploidy, or the doubling of the genome's entire chromosome complement
   d. expansion of gene families through gene duplication

5. What evidence do biologists use to infer that a gene is part of a gene family?

6. What are some characteristics of a pseudogene?

7. Explain how open reading frames are identified in the genomes of bacteria and archaea. Why is it more difficult to find open reading frames in eukaryotes?

8. In a genomics-based search for mutations that caused a patient's cancer, which of the following would provide the most informative comparison with the cancer cell?
   a. the average human DNA sequence available from a database
   b. the DNA sequence of a noncancerous cell from another person
   c. the DNA sequence of a noncancerous cell from the patient
   d. the DNA sequence of a different tumor type from another cancer patient

9. QUANTITATIVE Gene density is the number of genes per unit length of DNA. Most often, gene density is expressed as the number of genes per million base pairs (Mbp). Go to Figure 21.4b and find the approximate number of genes estimated in water fleas and in humans and the size of each genome. What is the gene density per Mbp in water fleas? What is the gene density per Mbp in humans? How much greater is the gene density in water fleas relative to humans?

10. In DNA fingerprinting, why is it an advantage to analyze an STR locus with many different repeat length alleles versus an STR locus with only a few different repeat length alleles?

11. Explain how microarrays of short, single-stranded DNAs that represent many or all of the exons in a genome are used to document changes in the transcription of genes over time or in response to environmental challenges.

12. Explain the concept of homology and how identifying homologous genes helps researchers identify the function of unknown genes.

13. Parasites lack genes for many of the enzymes found in their hosts. Most parasites, however, have evolved from free-living ancestors that had larger genomes. Based on these observations, W. Ford Doolittle claims that the loss of genes in parasites represents an evolutionary trend. He summarizes his hypothesis with the quip "use it or lose it." What does he mean?

14. According to eyewitness accounts, communist revolutionaries executed Nicholas II, the last czar of Russia, along with his wife and five children, the family physician, and about a dozen servants. Many decades after this event, a grave purported to hold the remains of the royal family was discovered. Biologists were asked to analyze DNA from each adult and juvenile skeleton and determine whether the bodies were indeed those of several young siblings, two parents, and several unrelated adults. If the remains of the family were in this grave, predict how similar the DNA fingerprints would be between the parents, the children, and the unrelated individuals in the grave.

15. Pleiotropy (a concept presented in Chapter 14) occurs when a mutation in one gene results in many different phenotypes. In a study of a gene interaction network, similar to the protein interaction network shown in Figure 21.13, which type of gene would you predict to exhibit the greatest degree of pleiotropy when mutated?
   a. a gene that interacts with one neighbor
   b. a gene that has a duplicate copy
   c. a gene that is the center of interactions with functionally related genes
   d. a gene that is the center of interactions with genes of many different functions

16. One hypothesis for differences between humans and chimps involves differences in gene regulation. A recent study used microarrays to compare the patterns of expression of genes that are active in the brain, liver, and blood of chimpanzees and humans. The overall patterns of gene expression were similar in the liver and blood of the two species, but the expression patterns were strikingly different in the brain. How do these results relate to the hypothesis?

22 Principles of Development

In this chapter you will learn that

A common set of principles guides development

first surveying

Developmental processes shared by all multicellular organisms **22.1**

then asking

How do cells containing the same genes express different specialized subsets? **22.2**

via

Chemical signals and regulatory genes **22.3**

which, if mutated, can produce

Evolutionary change **22.4**

An embryonic fish, still attached to the nutrient-filled yolk in the egg. This chapter introduces the processes responsible for transforming a fertilized egg into an individual that has specialized cells, tissues, and organs.

What is today's greatest challenge in biological science? Although there are many candidates, one of the most compelling is the question addressed in Unit 4: How does a multicellular individual develop from a single cell—a fertilized egg?

It's important to pause for a moment and think about the magnitude of this problem. Consider that once you were a single cell. If you could have watched your own development, you would have seen that single cell divide rapidly and form a ball of tiny, identical-looking cells. At that point, the fertilized egg had become an **embryo**—a young, developing organism. After continued cell division, large groups of cells suddenly began moving into the interior of the embryo. Cell division continued at a dizzying pace. After a week or two the embryo elongated, and a recognizable head and tail portion appeared. Tiny precursors of vertebrae became visible, along with rudimentary eyes. Eventually

✔ When you see this checkmark, stop and test yourself. Answers are available in Appendix A.

buds emerged and went on to form your limbs. As development continued, the embryo became you.

Biologists who have watched this process in humans or other organisms never cease to marvel at it. How does all the growth and formation of body parts happen?

To understand how researchers are answering this question, you'll need to draw on what you've already learned about cell–cell interactions, the regulation of gene expression, and a host of other topics described in previous units. Part of the excitement surrounding developmental biology is that it draws on insights from biochemistry, cell biology, genetics, and evolution. It is one of the most interdisciplinary fields in all of biology. Let's delve in.

22.1 Shared Developmental Processes

Over one hundred years of research has culminated in one of the great insights of contemporary biology: A few fundamental principles are common to all developmental processes in every multicellular organism. This discovery has brought a unified understanding to how the embryos of fruit flies, oak trees, and humans each grow and develop from a single cell into complex, multicellular individuals.

An individual develops as cells divide; signal to one another about where they are, what they are doing, and what type of cell they are becoming; begin to express certain genes rather than others; move or expand in specific directions; and, in the case of some cells, die (**TABLE 22.1**). Let's first consider these processes in turn, while keeping in mind that in the embryo they are interdependent and occur together. Then, let's go on to consider more specific questions about how cells interact and specialize.

Cell Division

For an embryo to grow and develop, its cells have to proliferate—they must divide and make more cells. This statement is obvious, but less obvious and equally important is the following point: The location, timing, and extent of cell division have to be tightly controlled.

How is cell division controlled? You might recall that mitosis and cytokinesis are responsible for cell proliferation in eukaryotes (Chapter 12) and that cells initiate mitosis in response to a regulatory protein complex called M phase–promoting factor (MPF). You also may remember that each stage of the cell cycle has carefully regulated checkpoints, and that cells control their division in response to what biologists call "social controls"—meaning, signals from other cells.

SUMMARY TABLE 22.1 **Essential Developmental Processes**

Cell proliferation		Cells divide by mitosis and cytokinesis. The timing, location, and amount of cell division are regulated.
Cell–cell interactions		Signals that are produced by cells influence their neighbors to divide, die, move, or differentiate.
Cell differentiation		Undifferentiated cells specialize at specific times and places in a stepwise fashion.
Cell movement and expansion		Cells can move past one another within a block of animal cells, causing drastic shape changes in the embryo.
		Cells can break away from a block of animal cells and migrate to new locations.
		Plant cells can regulate the plane of cell division and expand in specific directions, causing dramatic changes in shape.
Programmed cell death		The timing, location, and amount of cell death are regulated.

As development proceeds in both plants and animals, most cells stop dividing when they mature. But in later stages of development and in the adult, populations of undifferentiated cells keep proliferating throughout the individual's life. These are **stem cells.** Stem cells divide to produce a daughter cell that remains a stem cell and another daughter cell that ultimately differentiates into a specialized cell type. By producing one daughter that stays a stem cell, populations of stem cells are maintained.

- In plants, stem cells are located in **meristems.** Meristems are present in embryos and adults and produce the stems, roots, leaves, flowers, and other structures that develop throughout a plant's life (see Chapter 24).

- In animals, stem cells exist in specific locations in the body. In adults, stem cells proliferate to replace skin, blood, and gut cells that die; repair wounds; and create a constant supply of disease-fighting cells in the immune system.

Cell–Cell Interactions

Cells interact constantly during development. They engage in "conversations" with their neighbors through a diverse set of chemical signals. This ongoing communication drives most of the cellular behaviors that are the foundation of development.

You might recall that when a signal arrives at the surface of a cell that possesses the appropriate receptor, the message is received and processed (Chapter 11). Signal molecules used in development may diffuse in the watery environment that surrounds cells, but they also can be present on the surface of other cells or bound to the extracellular matrix. These nondiffusible signals are detected by developing cells as they expand or move within the embryo.

In most cases in development, when a signal is bound by a receptor, it activates a signal transduction cascade that triggers the production of transcription factors (Chapter 19). As a result, patterns of gene expression change, and this alters the cell's activity. In response to a chemical signal, embryonic cells may divide, differentiate, move, expand, or die. In this way, the fate of a cell inside an embryo hinges on the signals it receives from other cells.

Cell Differentiation

As development progresses, most cells undergo **differentiation**— the process of becoming a specialized type of cell. Differentiation is a progressive, step-by-step process. As cells differentiate, decisions are made at branch points to follow one developmental path or another. As a result, a fertilized egg may give rise to hundreds of distinctive cell types.

In the case of your own development, some of your embryonic cells differentiated to form muscle cells that contract and relax, while others became nerve cells that conduct electrical signals throughout the body. As an oak tree develops, some cells secrete thickened walls and transport water as part of a woody stem; other cells become flattened and secrete the waxy coating found on the surface of leaves.

Remarkably, sometimes the process of differentiation can be reversed. For example, some plant cells can "de-differentiate" after they have become specialized. When a branch of a western red cedar tree droops low enough to make contact with the soil, cells in the branch de-differentiate. These cells then re-differentiate to form root cells, resulting in the growth of a fully formed root where the branch came to rest on the ground.

For decades, biologists thought that animal cells could not de-differentiate. As you'll see in Section 22.2, however, the cloning of frogs and mammals reversed this view. In a spectacular recent study, scientists found that it is even possible to de-differentiate many adult human cells. Adding as few as three transcription factors to adult cells can cause them to de-differentiate into cells that resemble those of the early embryo. The hope is that these de-differentiated cells can be coaxed to re-differentiate in tissue culture to create any desired cell type that might be needed to replace cells damaged by disease. If this dream becomes reality, it will largely be through understanding cell differentiation during normal development.

Cell Movement and Cell Expansion

Many animal cells have to move to new locations for normal development to occur. In plants, cells cannot move because of the cell wall. However, plants are masters at controlling both the direction of cell expansion—the growth of an individual cell—and the orientation of cell division to determine where cells are added.

Some of the most dramatic cell movements in animals occur early in development once rapid cell divisions have produced a mass of cells. During a sequence of events called **gastrulation,** cells in different parts of the mass rearrange themselves into three distinctive layers, which then give rise to the skin, gut, and other basic parts of the body (gastrulation is described in Chapter 23).

Later in development, some animal cells break away from their original sites and migrate to new locations in the embryo. There they give rise to germ cells (cells that produce sperm or eggs), pigment-containing cells, precursors of blood cells, or certain nerve cells. If these cell movements go awry, the embryo will be deformed or die.

Plant cells, in contrast, are encased in stiff cell walls and do not move. Instead, the direction of cell division and cell expansion is carefully regulated to form stems, leaves, roots, and all other parts of the plant. Differential cell expansion is a key part of plant development, just as regulated cell movement is a key part of animal development.

Cell Death

Cell division is controlled by signals from other cells. But sometimes, signals trigger cell death. Cell death is a highly regulated aspect of plant and animal development. **Programmed cell death** occurs as certain tissues and organs take shape.

There are a few different forms of programmed cell death. In animals, the predominant one used in development is **apoptosis** (literally, "falling away"). Although biologists debate whether to call the form of programmed cell death that occurs in plants apoptosis, an important role for cell death in plant development is undisputed.

Apoptosis plays fascinating roles. For example, the feet of a chicken embryo initially develop with webs. Cells that are present between the toes must die to form separate toes (**FIGURE 22.1a**). The same is true for you—without apoptosis, your hands and feet would be webbed like a duck's feet. A duck has webbed feet because cells between the toes do not undergo apoptosis. Apoptosis doesn't end with digits. Vertebrate nerve cells are overproduced in the embryo. As a normal part of "wiring" an effective nervous system, at least half of embryonic nerve cells are destroyed by apoptosis.

The initial studies on apoptosis were done on the roundworm *Caenorhabditis* (pronounced *see-no-rab-DIE-tiss*) *elegans*. This species is a popular model organism in developmental biology because the adult contains only about a thousand cells and because the embryonic worm is transparent, so biologists can follow individual cells throughout development. (For more on *C. elegans*, see **BioSkills 13** in Appendix B.)

As a *C. elegans* worm matures, exactly 131 of its cells undergo apoptosis. To explore how this happens, Hillary Ellis and Robert Horvitz studied mutant embryos that do *not* exhibit the normal pattern of cell death and then used techniques (introduced in Chapter 20) to locate the mutated genes responsible for the defect. Their early work uncovered two genes required for apoptosis. The researchers proposed that the genes are part of a genetic program for apoptosis—in short, they are cell-death genes.

In follow-up work, researchers searched databases of DNA sequences and discovered that mice have similar genes. When a research team used genetic engineering techniques to produce mice in which both copies of one of their cell-death genes were knocked out, the embryos had severe malformation of the brain (**FIGURE 22.1b**). The defect occurred because nerve cells that normally die during development survived.

The same genes exist in humans and other mammals. Abnormal apoptosis—either too much or too little—occurs in certain diseases. For example, inappropriate activation of programmed cell death is proposed to be involved in some devastating neurodegenerative diseases such as ALS (Lou Gehrig's disease).

(a) Chicken embryo with normal (left) and defective (right) cell-death genes.

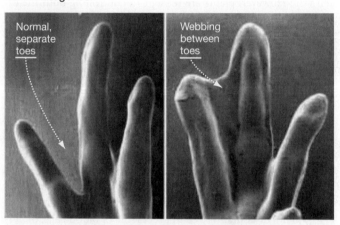

(b) Mouse embryo with normal (left) and defective (right) cell-death genes.

FIGURE 22.1 Programmed Cell Death Is a Normal Part of Development.

22.2 Genetic Equivalence and Differential Gene Expression in Development

The differentiation of a cell occurs through differential gene expression. The muscle cells in your body are different from your nerve cells because they express different genes and therefore produce different proteins. The water-transport cells in an oak tree are different from its leaf-surface cells for the same reason.

If you think about these statements, you'll realize that they have to be true. The only way that cells can have different structures and functions is if they contain different molecules. What is less obvious is whether cells express different genes because they contain different genes, or whether all the cells in a body contain the same genes but express only a specialized subset.

Evidence that Differentiated Plant Cells Are Genetically Equivalent

If cells from the branch of a cedar tree can de-differentiate to form roots, the branch cells must contain the genes required by root cells. Gardeners and farmers have known for centuries that in many plant species, complete new plants can be produced from a small section of a root or shoot.

From observations like this, researchers strongly suspected that all plant cells contain the same genes—meaning that they show **genetic equivalence**. This notion of genetically equivalent cells was confirmed in the 1950s when biologists were able to grow entire tobacco plants or carrots from a single, differentiated cell taken from an adult. (For more information on how plant cells are grown in culture, see **BioSkills 12** in Appendix B.) These experiments confirmed that differentiated plant cells are genetically equivalent.

Evidence that Differentiated Animal Cells Are Genetically Equivalent

The issue of genetic equivalence proved more difficult to resolve in animals than plants. Serious work began in the 1950s, but the question wasn't settled until the late 1990s.

Early experiments were based on transferring nuclei from differentiated frog cells into unfertilized egg cells whose nuclei had been removed. Some of these transplanted nuclei were able to direct the development of tadpoles. These results provided strong evidence that all cells in the same individual are genetically equivalent.

This conclusion was confirmed in a stunning way in 1997 when Ian Wilmut and colleagues reported their results of nuclear transfer experiments in sheep. As **FIGURE 22.2** shows, the researchers removed mammary-gland cells from a female, grew them in culture, and fused them with eggs whose nuclei had been removed. As the upper drawings show, the eggs came from a black-faced breed of sheep, while the donor nuclei came from a white-faced breed. After developing in culture, the embryos were implanted into surrogate mothers. In one of several hundred such transfer attempts, an apparently normal white-faced lamb named Dolly was born.

Genetic tests showed that Dolly was a **clone**—a genetically identical copy—of the white-faced donor of the mammary-gland cell. Dolly grew into a fertile adult. Soon after, other research groups reported similar results in mice and cows. More recently, a menagerie of animals, including horses, cows, monkeys, and dogs, have been cloned. The success rate of the procedure remains low, however, with the vast majority of nuclear transplants failing to produce healthy newborns.

Taken together, research on cloning plants and animals has shown that cellular differentiation typically does not involve changes in the genetic makeup of cells. Instead, it results from differential gene expression.

How Does Differential Gene Expression Occur?

Eukaryotic cells control gene expression at several different levels: chromatin remodeling and modification, transcriptional regulation, alternative splicing of mRNAs, selective destruction of mRNAs, translation rate, and activation and deactivation of proteins after they are translated (see Chapter 19). All of these processes occur during development. Which is most important in differentiation?

The answer is transcriptional control. To understand why, ask yourself whether a muscle cell should produce mRNAs and proteins that are specifically required by nerve cells. The answer is no. If it did, it would also have to either produce microRNAs that disable nerve-cell mRNAs or expend the energy to translate the mRNAs into nerve-cell proteins that would then have to be maintained in an inactive state. It is more logical to expect that muscle cells transcribe genes required by muscle cells and not genes needed for specialized functions in other types of cells. This is exactly what researchers have found.

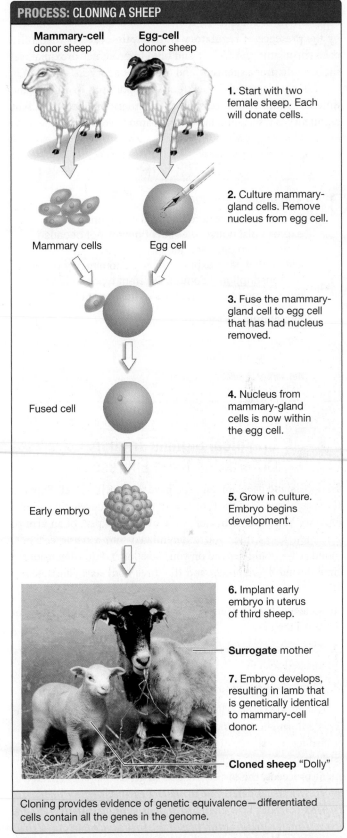

PROCESS: CLONING A SHEEP

Mammary-cell donor sheep   **Egg-cell** donor sheep

1. Start with two female sheep. Each will donate cells.

Mammary cells   Egg cell

2. Culture mammary-gland cells. Remove nucleus from egg cell.

Fused cell

3. Fuse the mammary-gland cell to egg cell that has had nucleus removed.

4. Nucleus from mammary-gland cells is now within the egg cell.

Early embryo

5. Grow in culture. Embryo begins development.

6. Implant early embryo in uterus of third sheep.

Surrogate mother

7. Embryo develops, resulting in lamb that is genetically identical to mammary-cell donor.

Cloned sheep "Dolly"

Cloning provides evidence of genetic equivalence—differentiated cells contain all the genes in the genome.

FIGURE 22.2 Mammals Can Be Cloned. The lamb that resulted from this experiment was genetically identical to the white-faced individual that donated the nucleus, not the black-faced egg donor or surrogate mother.

Transcription is the fundamental level of control during differentiation. In eukaryotes, transcription is controlled primarily by the presence of **regulatory transcription factors** that influence chromatin remodeling and bind to promoter-proximal elements, enhancers, silencers, and other regulatory sites in DNA.

This simple insight is extremely important. To understand differentiation, researchers have to understand how and why regulatory transcription factors vary among cells.

22.3 Chemical Signals Trigger Differential Gene Expression

To understand development, you have to think like a cell. Suppose that you were one of the thousands of cells in a developing animal embryo. Your fate—whether you ended up as part of an arm or a kidney, or whether you differentiated into a nerve cell or a blood cell—would depend on your position in four dimensions—time during development and the three body axes illustrated in **FIGURE 22.3**.

1. One axis runs **anterior** (toward the head) to **posterior** (toward the tail).

2. One axis runs **dorsal** (toward the back) to **ventral** (toward the belly).

3. One axis runs left to right.

Chemical signals tell cells where they are in time and space. This information works through signal receptors and signal transduction cascades to activate transcription factors that turn specific genes on or off, resulting in differentiation. As development proceeds, the distinctive sets of genes that are activated at successive stages determine the fate of each cell.

Let's consider how chemical signals activate differential gene expression, beginning with one of the first developmental signals ever discovered. Although you'll be analyzing what happens as a fruit fly embryo develops, keep an important point in mind: Principles that were discovered in fruit flies are relevant to all multicellular organisms—from mustard plants to humans.

(a) The three body axes observed in humans and other animals...

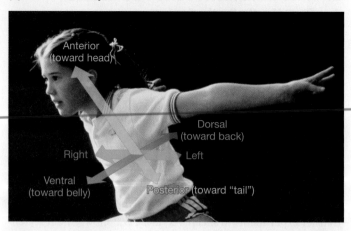

Anterior (toward head)

Dorsal (toward back)

Right

Left

Ventral (toward belly)

Posterior (toward "tail")

(b) ...are initially established in embryos.

Mouse

Anterior (toward head)

Dorsal (toward back)

Right

Left

Ventral (toward belly)

Posterior (toward "tail")

1 mm

FIGURE 22.3 Most Animals Have Three Major Body Axes.

Morphogens Set Up the Major Body Axes

Biologists use the term **pattern formation** to describe the events that determine the spatial organization of an embryo. If a molecule signals that a target cell is in the head or tail, or dorsal or ventral side of an embryo, then that molecule is involved in pattern formation.

Many molecules that work in pattern formation are present in a concentration gradient, with high concentrations near the source of the molecule and lower concentrations farther away. If cells detect the different concentrations to learn their position, then this molecule is a **morphogen.** Morphogens play central roles in setting up the body axes and at many later stages of development.

Complex patterns don't develop all at once. Instead, pattern formation is progressive. Morphogens set up the major elements of the body—the anterior–posterior, dorsal–ventral, and left–right axes. Genes activated by these morphogens generate signals of their own with more specific information about the location of cells. As development continues, the process repeats: New signals arrive and activate genes that specify finer and finer control over

(a) A normal fruit fly embryo

Anterior structures

Posterior structures

(b) A mutant that does not express *bicoid*

Posterior structures

Posterior structures

FIGURE 22.4 Pattern-Formation Mutants Have Misshapen Bodies.

✔ **QUESTION** Predict the phenotype of a mutant that expresses twice the normal levels of Bicoid instead of no Bicoid at all.

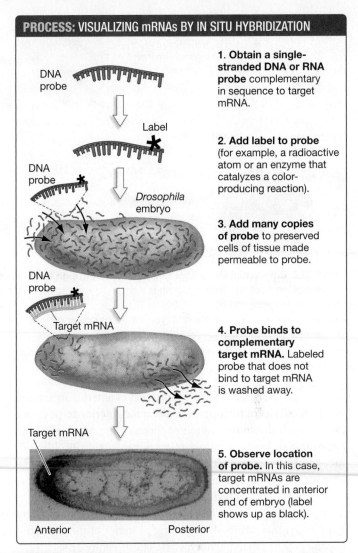

PROCESS: VISUALIZING mRNAs BY IN SITU HYBRIDIZATION

DNA probe

1. Obtain a single-stranded DNA or RNA probe complementary in sequence to target mRNA.

Label

2. Add label to probe (for example, a radioactive atom or an enzyme that catalyzes a color-producing reaction).

DNA probe

Drosophila embryo

3. Add many copies of probe to preserved cells of tissue made permeable to probe.

DNA probe

Target mRNA

4. Probe binds to complementary target mRNA. Labeled probe that does not bind to target mRNA is washed away.

Target mRNA

5. Observe location of probe. In this case, target mRNAs are concentrated in anterior end of embryo (label shows up as black).

Anterior   Posterior

FIGURE 22.5 In Situ Hybridization Allows Researchers to Pinpoint the Location of Specific mRNAs. The micrograph in the last step shows the location of the *bicoid* mRNA in a fruit fly embryo.

what a cell becomes. Development unfolds in a cascade of events unleashed by a few key signals.

The Discovery of Bicoid   Morphogens had long been proposed on theoretical grounds, but discovery of the first morphogen emerged from work on the fruit fly *Drosophila melanogaster*—a key model organism in genetics and development (see **BioSkills 13** in Appendix B).

Christiane Nüsslein-Volhard and Eric Wieschaus started this work in the 1970s, taking an approach based on genetics. They began by exposing adult flies to chemicals that cause mutations and examining their offspring for defects in development. One of the most dramatic mutations affected the anterior–posterior axis of the embryos. As **FIGURE 22.4** shows, the mutant embryos were missing all the structures normally found in the anterior end. Instead, the anterior end contained some structures normally found in the posterior.

The mutated gene responsible for this phenotype was dubbed *bicoid*, meaning two tailed. Based on its phenotype, Nüsslein-Volhard and Wieschaus suspected that the product of the normal *bicoid* gene must provide positional information. In other words, they hypothesized that the *bicoid* gene coded for a morphogen that tells cells where they are located along the anterior–posterior body axis.

The Importance of Morphogen Concentration Gradients   To test their hypothesis that *bicoid* encodes a morphogen, Nüsslein-Volhard and colleagues cloned and sequenced the *bicoid* gene (using techniques introduced in Chapter 20). They then used **in situ** (literally, "in place") **hybridization** to find where *bicoid* mRNAs are located. As **FIGURE 22.5** shows, in situ hybridization works by adding a label to single-stranded DNA or RNA molecules to create a probe that is complementary in sequence to the

mRNA of interest. In this case, the probe sequence was designed to bind to *bicoid* mRNA inside the embryo. As a result, the labeled probes marked the location of the *bicoid* mRNAs.

When Nüsslein-Volhard's group treated eggs and early embryos with labeled probes, they found the *bicoid* mRNA was tightly localized to the anterior end (see step 5 in Figure 22.5). In related studies they also made another surprising discovery: the *bicoid* gene is not expressed in the egg or embryo—it is only transcribed in special cells of the mother fruit fly that are near the egg. The *bicoid* mRNA is then deposited by these maternal cells into the developing egg at the anterior end. Follow-up work showed that when these mRNAs are translated, the protein product forms a steep concentration gradient: Bicoid protein is abundant in the anterior end but declines to progressively lower concentrations in the posterior end.

Later work showed that the Bicoid morphogen is a regulatory transcription factor (see Chapter 19). It binds to DNA and

High levels of Bicoid protein result in high transcription of genes that trigger formation of anterior structures

Drosophila embryo

Anterior

Posterior

Little or no Bicoid protein results in little or no transcription of genes that trigger formation of anterior structures

Enhancer sequence

Bicoid proteins (transcription factors)

High level of transcription

Promoter sequence

Genes required for anterior structures

No transcription

Genes required for anterior structures

FIGURE 22.6 Differential Gene Expression Occurs through the Presence or Absence of Regulatory Transcription Factors. Bicoid protein is a regulatory transcription factor that triggers the formation of anterior structures. Because the concentration of Bicoid decreases toward the embryo's posterior, different genes are expressed in the anterior of the embryo than in the posterior.

✔**QUESTION** Predict the effect on development if some cytoplasm from the anterior of a normal early *Drosophila* embryo were added to the central region of a mutant embryo that lacked *bicoid* mRNA.

activates genes required for the formation of anterior structures. In effect, cells learn their position along the anterior-to-posterior axis through the concentration of Bicoid. High levels of Bicoid result in high levels of transcription of genes with Bicoid binding sites. In this way, high concentrations tell a cell that it's in the anterior, and lower concentrations tell the cell that it's more posterior (**FIGURE 22.6**). When Bicoid is lacking, cells throughout the embryo get a "you're in the posterior" message from a different morphogen that has its source in the posterior of the embryo—leading to the mutant phenotype you saw in Figure 22.4.

Morphogens in Plant Development Plants also use morphogens to provide positional information. One important example is auxin, a small molecule that is used first in early development and then repeatedly and in different ways throughout the plant's life. In the embryo, auxin enters cells and triggers the production of transcription factors that affect differentiation.

Auxin is produced in the embryo's meristem cells that lie at the tip of what will become the stem. Auxin is actively transported toward the base of the embryo—what will become the root. In the process, an auxin concentration gradient forms. A high concentration of auxin signals, "You're near the top of the stem"; lower auxin concentrations that accumulate at the base of the embryo signal, "You're the root."

Regulatory Genes Provide Increasingly Specific Positional Information

The initial work on *bicoid* illustrated the importance of morphogens as a general theme in development. A second fundamental developmental principle common to both plants and animals soon emerged: Differentiation is a progressive, step-by-step process.

Dividing Up the Embryo Along with the "two-tailed monster" *bicoid* mutants, Nüsslein-Volhard and Wieschaus found an array of embryos that had defects in smaller regions of their body. Flies and many other animals are constructed from a set of **segments**—portions of the body that are repeated along its length. The mutants with altered segments had defective **segmentation genes.**

Researchers have now identified four classes of segmentation genes that act in a cascade:

1. First, maternally expressed genes deposit morphogen mRNAs in the egg. Morphogens, such as *bicoid*, control the formation of large groups of segments that span the anterior or the posterior halves of the embryo.

2. Gap genes are expressed second, in broad regions along the head-to-tail axis (**FIGURE 22.7a**). Gap genes control the formation of groups of segments that define large body regions.

3. Pair-rule genes are expressed next, in alternating bands along the embryo (**FIGURE 22.7b**). This pattern suggests that these genes control the formation of individual segments.

4. Segment polarity genes are expressed later, in a restricted band within every segment (**FIGURE 22.7c**). This pattern and order of expression imply that segment polarity genes create specific regions within each segment.

Specifying Segment Identity Once segmentation gene products create segments along the anterior–posterior axis of a fly embryo, development continues with the activation of **Hox genes.** These genes are turned on by products of the segmentation genes. *Hox* genes have spectacular and far-reaching effects. While segmentation gene products establish each segment, *Hox* gene products specify each segment's identity by activating effector genes.

(a) Gap genes

Anterior   Posterior

Early in development, gap genes define broad regions that often span several segments.

(b) Pair-rule genes

Later in development, pair-rule genes demarcate individual segments.

(c) Segment polarity genes

Still later, segment polarity genes delineate regions within individual segments.

FIGURE 22.7 Domains of Gene Expression Demarcate Body Regions in Fruit Flies. The embryos in **(a)** and **(b)** were stained for two different segmentation gene products. The embryo in **(c)** was stained for one segmentation gene product.

Effector gene expression leads to the production of the specialized proteins and structures of differentiated cells.

The *Hox* genes were discovered when researchers found adult fruit flies with body parts in the wrong place. This bizarre outcome is called homeosis, and the types of mutations that cause it are termed **homeotic mutations.** For example, homeotic mutations in *Hox* genes can turn a segment in the middle part of the body into a segment just like the one that lies in front of it. Instead of bearing the pair of small balancing structures called halteres, the transformed segment now bears a pair of wings—as does the segment in front of it. The mutant has four wings instead of the normal two (**FIGURE 22.8**).

Hox genes code for transcription factors that trigger the development of the structures appropriate to each type of segment, such as antennae, wings, or legs. The *Hox* genes are distinguished by a sequence called the **homeobox** that encodes a DNA-binding domain.

To put all this information into perspective, it helps to realize that the genes involved in early development form a regulatory cascade (**FIGURE 22.9**; see page 414). Morphogens—produced by maternally expressed genes and deposited in the egg—trigger the production of other regulatory signals and transcription factors, which initiate the production of another set of signals and regulatory proteins, and so on down the chain. Ultimately, the genes needed to form antennae, wings, and legs are turned on in the right places and times within the developing embryo. Similar types of regulatory cascades exist in all animal and plant species.

The unfolding regulatory cascades provide progressively detailed information about where cells are located. This positional information, in turn, causes changes in cell proliferation, death, movement, interactions, and differentiation.

- Because chemical signals vary in identity and concentration along the three major body axes, cells in different locations receive unique positional information.

Normal fruit fly   Homeotic mutant   Homeotic mutant

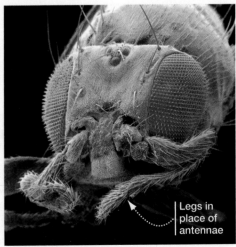

FIGURE 22.8 Homeotic Mutants in *Drosophila* Have Structures in the Wrong Locations. These colorized scanning electron micrographs show homeotic mutants in fruit flies. Wings are growing where small, stabilizing structures called halteres should be, or legs are growing where antennae should be.

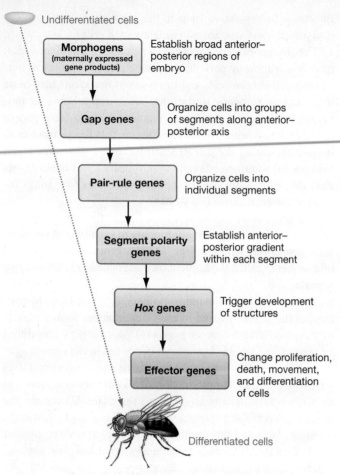

FIGURE 22.9 A Regulatory Gene Cascade in Fruit Flies. This cascade establishes the positions and identities of structures along the anterior-to-posterior axis. Morphogens—products of maternally expressed genes that are deposited in the egg—unleash the sequential expression of genes that work at each level of the cascade. Genes at one level of the cascade control the expression of genes at lower levels. The *Hox* genes are controlled by gap, pair-rule, and segment polarity genes. The effector genes are controlled by both segment polarity and *Hox* genes.

FIGURE 22.10 Organization and Expression of *Hox* Genes. The location of *Hox* genes on the chromosome correlates with their pattern of expression in the embryos of flies and mice. Matching colors of genes and regions in the embryo indicate where each gene is expressed at its highest levels. The genes represented by same-color boxes in the fly **(a)** and mouse **(b)** are considered homologous. For simplicity, three features are not shown in this figure: the fly genes are present in two separate groups that were created by a relatively recent chromosome breakage in the fly lineage, the mouse has three similar *Hox* gene clusters in addition to the one shown, and there is overlap in the regions in which *Hox* genes are expressed.

✔QUESTION What evidence would support the claim that these genes are homologous?

- Each level in a regulatory cascade provides more specific information about where a cell is located.

- As regulatory cascades are deployed over time, a cell's identity becomes more and more finely specified.

Chemical Signals and Regulatory Genes Are Evolutionarily Conserved

Early genetic mapping studies revealed that the *Drosophila Hox* genes were present in clusters as shown in **FIGURE 22.10a**. Once methods to isolate and sequence DNA were available (see Chapter 20), researchers soon applied them to the *Hox* genes. The results of molecular studies were striking—the order of the *Hox* genes along the chromosome corresponds to the order of where the genes are expressed along the anterior-to-posterior axis.

Conservation of *Hox* Gene Organization With information about fruit fly *Hox* genes in hand, a few biologists thought it was worth looking for similar genes in other animals. This was a long shot, because it seemed unlikely that creatures as different as flies, sea urchins, mice, and humans would use the same genes to build such vastly different bodies. The gamble paid off. Virtually all animals were found to possess related sets of *Hox* genes. Incredibly, these *Hox* genes were organized in clusters in an order that aligned with the *Drosophila* genes. Even more amazing, the curious relationship first seen in *Drosophila* between *Hox* gene order along the chromosome and where a particular *Hox* gene is expressed was shared by all these animals.

A comparison of *Hox* genes in flies and mice (**FIGURE 22.10b**) reveals that there are more *Hox* genes in the mouse, but their chromosomal organization is similar. Notice how genes at the "left-hand" side of the cluster are expressed in the anterior end of both embryos, while genes at the opposite end of the cluster are expressed in posterior regions. The discovery that similar regulatory genes are used to build such a diverse array of animals ranks as one of the greatest surprises in modern biology.

Conservation of *Hox* Gene Function Is there evidence that *Hox* genes in organisms other than *Drosophila* actually specify parts of the animal? Could it be that these genes are organized and expressed in similar ways but aren't used to specify pattern except in flies? Mutations discovered in many animals, including humans, have shown that when *Hox* genes are altered, defects in pattern formation result. Based on these data, biologists conclude that in flies, mice, humans, and most other animals, *Hox* genes play a key role in specifying which body structures to build.

This conclusion was supported in dramatic fashion when researchers in William McGinnis's lab introduced the *Hoxb6* gene from mice into fruit fly eggs. The *Hoxb6* gene in mice is similar in structure and sequence to the *Antp* gene of flies (Figure 22.10). Because it was introduced without its normal regulatory sequences, the *Hoxb6* gene was expressed throughout the treated fly embryos. The resulting larvae had defects identical to those observed in naturally occurring fly mutants in which the *Antp* gene is improperly expressed throughout the embryo. This is a stunning result: A mouse gene not only affected the development of a fly, but even mimicked the effect of the related fly gene.

To interpret these observations, biologists hypothesize that the genes in *Hox* complexes of animals are **homologous**—meaning that they are similar because they are descended from genes in a common ancestor. This hypothesis implies that the first *Hox* genes arose very early in animal evolution. For the past roughly 600 million years of animal evolution, *Hox* gene products have been helping to direct the development of animals.

The take-home message from these studies is that key molecular mechanisms of pattern formation have been highly conserved during animal evolution. The discovery of these shared mechanisms is one of the most significant results to have emerged from studies of animal development. Although animal bodies are spectacularly diverse in size and shape, the underlying mechanisms responsible for their development are similar.

One Regulator Can Be Used Many Different Ways

Regulatory gene cascades and the evolutionary conservation of key developmental genes are general features of animal and plant development. But biologists have articulated another organizing principle as well: During development, the same regulatory genes, proteins, and chemical signals are used over and over in a variety of contexts.

The Wnt signal protein offers an impressive example of repeated uses of the same molecule to achieve different ends. *Drosophila* has a single *Wnt* gene known as *wingless*. From the name, you can guess that *wingless* is involved in wing development—flies that have a mutated *wingless* gene lack wings. The first role of *wingless*, however, is not in controlling wing formation, but in specifying boundaries between segments. The list doesn't stop here—it turns out that *wingless* is also a regulator of many other

developmental decisions. The *wingless* gene produces a signal protein that binds to specific receptors on different target cells and controls development of the leg, head, and digestive system as well as specifying particular nerve cells.

An analogy for this developmental principle of one signal, many messages is the signal called a pinch. When you were little, a pinch on the cheek from your grandmother indicated something very different from a pinch on the arm from an older sibling. It also means something different from a pinch you might receive now from someone you are romantically attracted to. In development, as in human communication, the context in which a signal is sent and received—its location, timing, and intensity—has a major effect on the signal's meaning and consequences.

In contrast to single genes such as *wingless* that are used in many different ways during the development of one organism, in many different organisms there are genes that control similar events in development. Animals and plants each have a conserved set of **tool-kit genes** that code for signals, signal-transduction pathways, and regulatory proteins that are used to direct related aspects of development in many different species.

Hox genes are an important example of tool-kit genes because they are present in a wide array of animals and are used to specify body structures. Similar tool-kit genes can, however, produce dramatically different structures such as the wing of a bat or the flipper of a whale if they are used at different times, in different places, and in different developmental contexts.

What determines the ways tool-kit genes are used? The answer lies largely in the regulatory sequences of DNA. Mutation of these sequences allows tool-kit genes to be used in new ways in different species. As you'll see in the next section, changes in regulatory sequence have profound effects on evolution.

22.4 Changes in Developmental Gene Expression Underlie Evolutionary Change

This chapter has shown that for an embryo to develop, cells have to proliferate, move or expand, differentiate, interact in specific ways, and sometimes die. Differentiation is caused by differential gene expression. It results from signals that tell cells where they are in time and space, triggering deployment of a complex cascade of transcription factors.

If any of these processes are disrupted, the embryo is likely to die. But if one of these processes is modified, the effect may be a structure with a different size, shape, or activity. As a result, the embryo will develop new features, and the adult will have a novel phenotype.

Once biologists began working out the regulatory signals and cascades introduced earlier in the chapter, they realized that the genetic changes altering these developmental processes must be the foundation of evolutionary change. The increase in body size that has occurred during human evolution, for example, must have resulted from mutations that altered the signals, regulatory sequences in DNA, or transcription factors that are involved in the amount and timing of cell proliferation throughout the body.

An emerging research field called evolutionary developmental biology, or **evo-devo,** focuses on understanding how changes in developmentally important genes have led to the evolution of new forms such as the flower, the leaf, and animal limbs. As an example of this exciting work, let's consider how snakes *lost* their limbs.

Although some snakes do not develop any sort of forelimb or hindlimb, boas and pythons have tiny pelvic (hip) bones and a rudimentary femur (thigh bone). The fossil record shows that the ancestor of all snakes had four functional legs. How did snakes lose their legs?

Researchers were able to answer this question because the signals and genes responsible for limb development are well understood. In chicken embryos, scientists had learned that *Hox* genes called *Hoxc6* and *Hoxc8* are expressed together in cells where ribs form, but *Hoxc6* is expressed alone in the region that gives rise to the forelimbs (**FIGURE 22.11a**).

Could modifications in where homologous *Hox* genes are expressed in snakes explain their limb loss? As **FIGURE 22.11b** shows, *Hoxc6* and *Hoxc8* are expressed together throughout the snake embryo—including in the region where forelimbs should form. These data suggest that a change in the regulation of where *Hoxc8* is expressed led to the evolutionary loss of the forelimb. Snakes make ribs instead of forelimbs.

It turns out that there's more than one way to lose a limb, even in the same animal. Hindlimb loss in snakes is due to defects in production of a signaling molecule encoded by the gene *sonic hedgehog.* This gene is homologous to a *Drosophila* segment polarity gene, *hedgehog,* which also produces a signal protein. Recent work has shown that defects in *sonic hedgehog* signaling affect animals other than snakes, as well—loss of *sonic hedgehog* signals in the pelvic region of whales led to the disappearance of their hindlimbs. Changes in the regulation of one widespread tool-kit gene—*sonic hedgehog*—have achieved extraordinary outcomes in animals as different as snakes and whales.

In limb loss, biologists can point to alterations in gene expression or in the production of specific signals that explain why an important evolutionary change occurred. Changes in the regulation of tool-kit genes led to developmental changes that build a snake or whale. These creatures have been legless ever since.

(a) Pattern of gene expression in tetrapods.

(b) Pattern of gene expression in snakes.

In the area where *Hoxc6* is expressed by itself, the forelimb forms

In the areas where *Hoxc6* and *Hoxc8* are expressed together, ribs form

Hoxc6 and *Hoxc8* are always expressed together, so no forelimbs form

Chick embryo — Forelimb — *Hoxc8* — *Hoxc6*

Snake embryo — *Hoxc8* — *Hoxc6*

FIGURE 22.11 Changes in Homeotic Gene Expression Led to Limb Loss in Snakes. *Hoxc6* and *Hoxc8* code for transcription factors. If both transcription factors are present in a population of cells along the anterior–posterior axis of a tetrapod (a four-limbed vertebrate), genes that lead to the formation of vertebrae and ribs are activated. But if only *Hoxc6* is expressed, then genes that lead to the formation of forelimbs are activated.

If you understand . . .

22.1 Shared Developmental Processes

- During development in all organisms, cells divide, interact, differentiate, move or expand in a directed manner, and sometimes die.

- Cells have to proliferate in a regulated manner to promote growth of the body.

- Chemical signals sent between cells provide a constant flow of information about where cells are in space and time.

- Cells undergo a step-by-step process of differentiation that leads to specialized cell types.

- Development requires the controlled movement and expansion of cells. In animals, many cells move in a coordinated way throughout development. Plant cells cannot move because of the cell wall, but plant cells precisely control the plane of cell cleavage and the direction of cell expansion to shape the embryo and the adult.

- Programmed cell death of some cells is an essential part of normal development.

✔ You should be able to predict how a mutation that disrupts control of the plane of cell cleavage would affect the development of an oak seedling.

22.2 Genetic Equivalence and Differential Gene Expression in Development

- Cells become specialized because they express different genes, not because they contain different genes.

- Cloning of plants and animals from differentiated adult cells indicates the genetic equivalence—that is, all cells of an individual contain the same genetic information.

- An implication of genetic equivalence is that cells are different because they express a common set of genes differently. Creating different cell types by expression of different genes from the same genome is differential gene expression.

- Stem cells in animals and plants do not fully differentiate, but proliferate throughout life to provide a source of cells that can specialize to replace damaged cells or promote growth.

- Transcriptional regulation through the production of different sets of transcription factors in different types of cells is the most important form of gene control during development.

You should be able to explain why it's possible to propagate banana trees from cuttings. Your answer should cover how cells in a stem cutting can give rise to roots.

22.3 Chemical Signals Trigger Differential Gene Expression

- Cells "know" where they are in the body and how far along development has progressed because they receive a steady stream of signals. These signals are produced by other cells and diffuse or are transported throughout the body.

- Morphogens are signals present in a gradient. Cells can learn where they are along a body axis by "reading" the concentration of morphogen.

- Cells respond to chemical signals by activating signal transduction pathways that lead to expression of a distinctive set of proteins. These proteins are expressed because new groups of genes are transcribed in response to the signal.

- Cells gradually acquire positional information, first learning where they are in broad regions and later in smaller parts of each region.

- In some organisms, products of maternally expressed genes (morphogens) are deposited in the egg, and these activate the sequential expression of many other segmentation genes that are linked in regulatory cascades.

- Chemical signals are conserved across species.

- Each specific signaling pathway is used over and over in development in different contexts to achieved different outcomes.

- A set of conserved tool-kit genes is used to control core developmental processes in a wide array of species.

✔ You should be able to explain why a mutation in a *Drosophila* pair-rule gene affects the expression of segment polarity genes but has no effect on the expression of gap genes.

22.4 Changes in Developmental Gene Expression Underlie Evolutionary Change

- Mutations in genes responsible for development can lead to the evolution of new body sizes, shapes, and structures. These changes are important components of evolution.

- Mutations in regulatory sequences of DNA are especially important because they can change when, where, and to what levels key regulators are expressed.

✔ You should be able to use the information in Figure 22.11 to predict the effect on chick development of replacing the *Hoxc8* gene regulatory sequence with the regulatory sequences of *Hoxc6* gene.

MasteringBiology

1. MasteringBiology Assignments

Tutorials and Activities Early Pattern Formation in *Drosophila*; Pattern Formation; Role of *bicoid* Gene in *Drosophila* Development

Questions Reading Quizzes, Blue-Thread Questions, Test Bank

2. eText Read your book online, search, take notes, highlight text, and more.

3. The Study Area Practice Test, Cumulative Test, BioFlix® 3-D Animations, Videos, Activities, Audio Glossary, Word Study Tools, Art

You should be able to . . .

TEST YOUR KNOWLEDGE · Answers are available in Appendix A

TEST YOUR KNOWLEDGE

1. What is apoptosis?
   a. an experimental technique used to kill specific cells
   b. programmed cell death that is required for normal development
   c. a pathological condition observed only in damaged or diseased organisms
   d. a developmental mechanism unique to the roundworm *C. elegans*

2. In adult animals, _____ are a source of undifferentiated cells that can divide to produce cells that can specialize.

3. Why is in situ hybridization such a valuable tool for studying development?
   a. It can identify the location of specific mRNAs to provide information on where particular genes are being expressed.
   b. It allows researchers to understand how cell–cell signals and regulatory transcription factors interact.

   c. It provides information on gene evolution.
   d. It can locate promoters and enhancers of important developmental genes.

4. Embryonic cells can learn their position by "reading" the concentration of a morphogen.
   a. True
   b. False

5. What is a homeotic mutant?
   a. an individual with a structure located in the wrong place
   b. an individual with an abnormal head-to-tail axis
   c. an individual that is missing segments
   d. An individual with double the normal number of structures

6. A tool-kit gene is _____.

TEST YOUR UNDERSTANDING · Answers are available in Appendix A

TEST YOUR UNDERSTANDING

7. What does it mean to say that cell proliferation, death, movement, or expansion; differentiation; and interaction are shared developmental processes?

8. Explain the logic of cloning animals to test the hypothesis that all cells in the animal body are genetically equivalent.

9. How did researchers go about looking for genes that have important roles in establishing the anterior–posterior body axis and body segments in *Drosophila*?

10. Why is it significant that many of the genes involved in development encode regulatory transcription factors?

11. Explain the connection between the existence of regulatory cascades and the observation that differentiation is a step-by-step process.

12. Which of the following provides the strongest evidence for the conservation of tool-kit genes?
    a. Bicoid moved from one fly embryo into the posterior of another fly embryo causes the formation of two head regions.
    b. Mutation of an unrelated gene in another species of fly has a similar effect to mutation of *bicoid* in *Drosophila*.
    c. A mouse *Hox* gene can be used to take over the function of a mutated *Drosophila Hox* gene.
    d. Sheep can be cloned by fusing a differentiated adult cell with an enucleated egg.

TEST YOUR PROBLEM-SOLVING SKILLS · Answers are available in Appendix A

TEST YOUR PROBLEM-SOLVING SKILLS

13. Biologists found that the products of two different *Drosophila* genes are required to keep *bicoid* mRNA concentrated at the anterior end of the egg. In individuals with mutant forms of these proteins, *bicoid* mRNA spreads farther toward the posterior pole than it normally does. Predict what effect these mutations will have on segmentation of the larva.

14. In 1992 David Vaux and colleagues used genetic engineering technology to introduce a human gene for a protein that inhibits apoptosis into embryos of the roundworm *C. elegans*. When the team examined the embryos, they found that cells that normally undergo programmed cell death survived. What is the significance of this observation?

15. **QUANTITATIVE** Imagine a situation in which a morphogen has its source at the posterior end of a *Drosophila* embryo. Every 100 μm from the posterior pole, the morphogen concentration decreases by half. If a cell required 1/16th the amount of morphogen found at the posterior pole to form part of a leg, how far from the posterior pole would the leg form?
    a. 100 μm
    b. 160 μm
    c. 400 μm
    d. 1600 μm

16. Fruit flies have 6 legs; spiders have 8 legs; centipedes have many; earthworms have none. All these species are segmented and have *Hox* genes. Generate a hypothesis to explain how variation in leg number evolved in these species.

23 An Introduction to Animal Development

In this chapter you will learn how

An overall sequence of events drives animal development

by examining

| Fertilization 23.1 | produces → | Single-celled zygote |

and

| Cleavage 23.2 | produces → | Multicellular embryo |

and

| Gastrulation 23.3 | produces → | Germ layers and body axes |

and

| Organogenesis 23.4 | produces → | Tissues and organs |

A human embryo, about 3 days old, on the tip of a pin.

"Making Babies: 25,000 Genes, Some Assembly Required." This wry chapter title written by evolutionary-developmental biologist Sean Carroll highlights what is surely one of the most remarkable events of life. How *is* a baby assembled from a single cell using instructions in the genome?

In a previous chapter (Chapter 22), you learned about the basic cellular processes of development. The central message was that cell–cell signals cause different sets of transcription factors to be produced in various cells throughout the embryo, resulting in differential gene expression and differentiation. For example, muscle cells "know" that they are muscle cells because they have received specific types and amounts of cell–cell signals. In response, they produce muscle-specific proteins.

✔ When you see this checkmark, stop and test yourself. Answers are available in Appendix A.

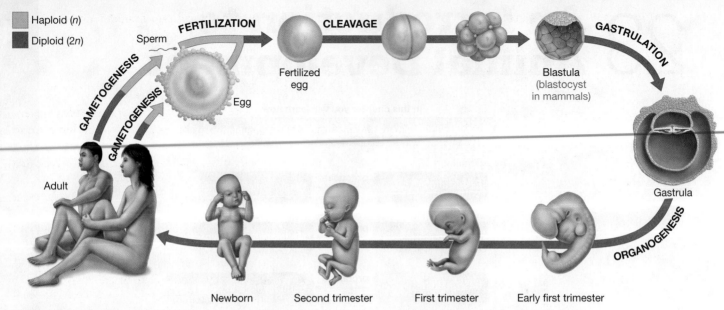

FIGURE 23.1 Development Proceeds in Ordered Phases. In animals, the development of a new individual starts with fertilization and continues with early cell divisions (cleavage) that result in a blastula. Gastrulation then rearranges the blastula into a multilayered structure called a gastrula. Organogenesis follows and leads to the formation of adult tissues and structures. The cycle is continued into a new generation with gametogenesis, the formation of sperm and egg, and their fusion in fertilization.

This chapter's goal is to apply these general developmental principles to the events that are needed to develop an animal; in essence, what's involved in "some assembly required." The major stages in this assembly process for a human are summarized in **FIGURE 23.1**.

Let's begin with the union of sperm and egg, and end with formation of differentiated cells within organs. At each step, remember that it is the readout of the genes that produces the cell–cell signals and regulatory proteins needed to direct developing cells to divide, move, interact, and differentiate—to do the things required to form a new individual.

23.1 Fertilization

Sperm and egg come together to create the **zygote,** the first cell of the new individual. What could be simpler? When this process called **fertilization** is examined closely, however, the question is quickly turned on its head. What could be more complex than two cells finding each other and doing something as odd as fusing together?

Sperm and eggs are among the most specialized of all cells. Recall that sperm and egg are gametes, haploid reproductive cells that are produced by meiosis and differentiation (Chapter 13). Sperm, the male gametes, are stripped-down speedsters designed to deliver the paternal (the father's) haploid genome to the egg. Eggs are huge and specialized to support early—and in some organisms, virtually all—development. A frog egg is more than 400,000 times larger in volume, and an ostrich egg is over

a billion times larger, than other cells in the body. Human eggs, in comparison, are relatively small, but they still contain more than 100 times the volume of a "typical" cell. A later chapter (Chapter 50) explores the process of gametogenesis, how eggs and sperm are formed.

What are the challenges of fertilization? First, sperm need to find the egg (**FIGURE 23.2a**). In many species, this is accomplished by following chemical signals secreted by the egg. Next, sperm and egg must recognize one another. In organisms that use external fertilization—that is, fertilization outside the body—there's an additional challenge. For these organisms, there needs to be a mechanism to ensure that only sperm and egg from the same species come together. Once in contact, sperm and egg have to fuse, something that few other cells in the body ever do. What's more, in most species, fusion must be limited to a single sperm so the egg does not receive extra chromosomes. Finally, the fusion of the two gametes has to trigger the onset of development. There's little simple about fertilization.

Research on fertilization began in earnest early in the twentieth century, when biologists started to study the sperm–egg interaction in sea urchins. Like most aquatic animals, sea urchins shed their gametes into the surrounding water, and fertilization occurs externally. **FIGURE 23.2b** outlines major events of this process.

Sea urchin sperm are drawn to the egg by following a gradient of a chemical secreted by the egg's jelly layer (step 1). The head of the sperm initially binds to the jelly layer that surrounds the egg cell (step 2). This triggers the acrosome reaction, a release of the contents of the sperm's acrosome (step 3). The acrosome is a large

(a) Many sperm compete to fertilize an egg.

(b) Steps in fertilization

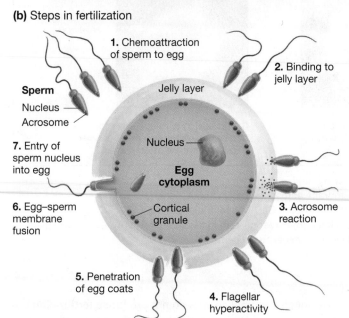

1. Chemoattraction of sperm to egg
2. Binding to jelly layer
3. Acrosome reaction
4. Flagellar hyperactivity
5. Penetration of egg coats
6. Egg–sperm membrane fusion
7. Entry of sperm nucleus into egg

Sperm
Nucleus
Acrosome

Jelly layer

Nucleus

Egg cytoplasm

Cortical granule

FIGURE 23.2 Fertilization Is a Complex Process. (a) Many sperm competing to fertilize an egg. **(b)** Steps of fertilization.

vesicle containing digestive enzymes that bore a hole through the jelly layer and through another protective coat called the vitelline envelope. The initial contact with the egg's jelly layer ramps up the beating of the flagellum to power the sperm through the egg layers (steps 4 and 5).

Once the sperm head contacts the egg-cell surface, specific proteins in the plasma membranes of the egg and sperm induce membrane fusion (step 6). The sperm nucleus then enters the egg (step 7). Contact between the sperm and egg plasma membranes initiates many responses in the egg. Sperm and egg nuclei fuse, and the egg is now an activated machine ready for development. Fertilization is complete.

But for species like the sea urchin that cast their sperm indiscriminately, what prevents cross-species fertilization and the consequent production of dysfunctional hybrid offspring? After all, in many habitats sperm and eggs from different species are mixed in the same sea water.

How Do Gametes from the Same Species Recognize Each Other?

In the 1970s, Victor Vacquier and co-workers succeeded in identifying a protein on the head of sea urchin sperm that binds to the plasma membrane of sea urchin eggs in a species-specific manner. They called this protein bindin. Follow-up work showed that bindin proteins from even very closely related species are distinct. This led to the proposal that bindin ensures that a sperm binds only to eggs from the same species.

But what binds to bindin? If bindin is a key, what's the lock? Charles Glabe and Vacquier hypothesized that egg-cell membranes contain a receptor for bindin. To test this hypothesis, they set out to isolate the receptor molecule from sea urchin eggs.

The researchers began by treating sea urchin eggs with a set of chemicals that released proteins from the egg cell surface. When the investigators separated the proteins that were released from the egg surface, they found one that bound to isolated bindin molecules. Significantly, further testing by the researchers showed that this binding occurred in a species-specific manner. The egg-cell receptor molecule only bound to bindin isolated from sperm of the same species, not to bindin isolated from sperm of different species. Based on these observations, Glabe and Vacquier concluded that they had found the egg-cell receptor protein for sperm.

During sea urchin fertilization, species-specific bindin molecules on sperm interact with species-specific receptors on the surface of the egg. Similar types of specific protein–protein interactions occur between the sperm and egg cells of mammals.

✔ If you understand the importance of protein–protein interactions for binding sperm and egg, you should be able to explain why adding a molecule that bound to a bindin-like protein on the human sperm head would be an effective contraceptive.

What Prevents More Than One Sperm from Entering the Egg?

In early studies on sea urchin fertilization, researchers noticed that only one sperm fertilized an egg, even when hundreds of sperm were clustered around it. This observation makes sense: Multiple fertilization, or polyspermy, would result in a zygote that had more than two copies of each chromosome. Embryos with more than two copies of each chromosome die. How do animals avoid polyspermy?

In sea urchins, there are two mechanisms that block polyspermy. One works immediately after sperm–egg contact and is only briefly effective. A second mechanism, seen in almost all species, including mammals, results in the erection of a physical barrier to sperm entry. This second process begins when sperm entry causes calcium ions (Ca^{2+}) to be released from storage

(a) PROCESS: A WAVE OF Ca²⁺ RELEASE SPREADS FROM THE SITE OF SPERM ENTRY

Sperm enters egg here

Ca²⁺

Ca²⁺

A wave of calcium ion release starts at the point of sperm entry and spreads throughout the egg

(b) PROCESS: FERTILIZATION ENVELOPE PREVENTS ENTRY OF MORE THAN ONE SPERM

Sperm enters egg here

Excess sperm

Fertilization envelope

1. Egg is covered with sperm. One sperm enters.

2. Fertilization envelope begins to lift and clear away excess sperm.

3. Fertilization envelope expands across egg. When complete, all excess sperm are cleared away.

FIGURE 23.3 A Physical Barrier Erected after Fertilization Prevents Polyspermy. (a) During fertilization, a wave of Ca²⁺ begins at the point of sperm entry and spreads under the plasma membrane across the egg in about 30 seconds. Colors represent Ca²⁺ concentration, where white is highest and blue lowest. Ca²⁺ concentrations were measured using an injected reagent that fluoresces in the presence of calcium ions. **(b)** In response to increased Ca²⁺ concentrations, a fertilization envelope arises in about 40 seconds and clears away excess sperm.

areas inside the egg. As **FIGURE 23.3a** shows, a wave of calcium ion release starts at the point of sperm entry and spreads throughout the egg.

The egg contains thousands of vesicles called cortical granules that lie just beneath the plasma membrane (see **FIGURE 23.2b**). Ca²⁺ causes the cortical granules to fuse with the plasma membrane and release their contents by exocytosis to the exterior. As you'll see in a later chapter, this same mechanism of Ca²⁺-mediated exocytosis is used to allow nerve cells to communicate with other cells (Chapter 46).

The egg cortical granules release proteases that digest the exterior-facing portion of the receptor for sperm. This prevents any new sperm from binding to the egg surface. In addition, ions and other compounds released by the cortical granules accumulate between the egg cell's plasma membrane and the vitelline envelope. These concentrated solutes cause water to flow into the space between the plasma membrane and vitelline envelope by osmosis (see Chapter 6). The influx of water lifts the vitelline envelope away from the cell. Compounds from the cortical granules cross-link molecules in the vitelline envelope together to form a tough **fertilization envelope** (**FIGURE 23.3b**). This impenetrable barrier keeps additional sperm from reaching the sea urchin egg.

Although mammalian eggs do not produce a fertilization envelope, a wave of Ca²⁺ ions also trigger the release of enzymes from cortical granules that perform a function similar to the sea

check your understanding

C Y U

If you understand that . . .

- Fertilization depends on specific interactions between proteins on the plasma membranes of animal sperm and egg cells.
- Enzymes that digest the egg-cell receptor for sperm help to prevent polyspermy.

✔ **You should be able to . . .**

1. Predict the consequences of mutations that change the structure of bindin of sea urchin species A to the bindin structure of sea urchin species B.

2. Describe the role of calcium signaling in processes that block polyspermy.

Answers are available in Appendix A.

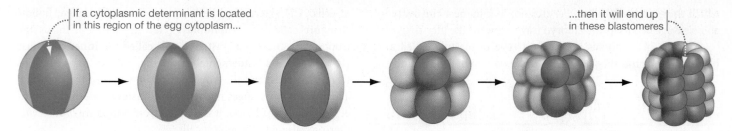

If a cytoplasmic determinant is located in this region of the egg cytoplasm...

...then it will end up in these blastomeres

FIGURE 23.4 Cytoplasmic Determinants Are Partitioned into Certain Blastomeres during Cleavage.

urchin enzymes that destroy the egg's receptor for sperm. These enzymes modify proteins on the egg-cell surface and on a structure similar to the vitelline envelope. This process blocks polyspermy by preventing the binding of any additional sperm once the egg is fertilized.

Once a sperm nucleus enters a sea urchin or human egg, the chromosomes of the two haploid nuclei come together to form a diploid genome. Development is off and running.

23.2 Cleavage

Fertilization activates development. One cell becomes two, two become four, and the beginnings of a multicellular organism are in place. The stage of rapid cell divisions that follows fertilization is called **cleavage.**

In most animals, cleavage divides the egg cytoplasm without any overall growth of the embryo. The zygote simply divides without increasing in size. The key feature of cleavage is that the cytoplasm present in the egg is divided into a larger and larger number of smaller and smaller daughter cells.

Cleavage is fast—the fastest cell divisions recorded over an individual's lifetime. For example, in fruit flies, mitosis can occur every 10 minutes, producing about 50,000 cells in half a day.

The cells that are created during cleavage are called **blastomeres** (literally, "bud-part"). When cleavage is complete, the embryo in many animals consists of a mass of blastomeres called a **blastula** ("little-sprout," "bud").

Partitioning Cytoplasmic Determinants

A key to understanding cleavage is to analyze what is happening to the egg cytoplasm as it is divided into many blastomeres. Many embryos contain **cytoplasmic determinants.** Cytoplasmic determinants are regulatory molecules that are located in specific regions of the egg cytoplasm. Through cleavage, they end up in specific populations of blastomeres (**FIGURE 23.4**). As a result, certain regulatory gene cascades are triggered only in certain blastomeres.

By precisely dividing the egg cytoplasm to distribute (partition) cytoplasmic determinants to certain cells, cleavage initiates the step-by-step process that, in combination with signals received from other cells, results in cell differentiation.

Cleavage in Mammals

In humans and other mammals, cleavage occurs as the embryo moves down the fallopian tube, or oviduct, a structure that connects the ovary and the uterus (**FIGURE 23.5**). The ovary is the organ in which the egg matures, and the uterus is the organ in

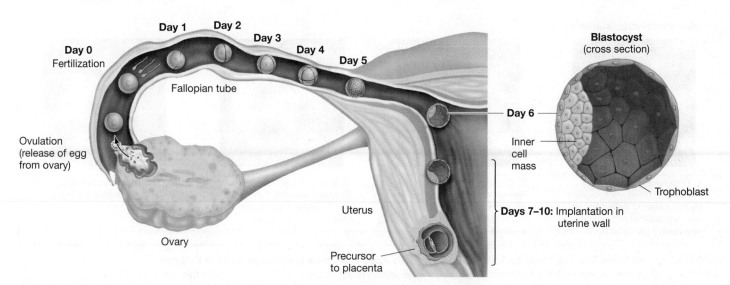

FIGURE 23.5 In Mammals, Cleavage Occurs before Implantation into the Uterus. Human cleavage and the human blastocyst are shown here.

which the embryo develops. Fertilization occurs near the ovary, and cleavage occurs as the embryo travels toward the uterus.

Cleavage in mammals results in a type of blastula called a **blastocyst** ("sprout-bag"), which has two major populations of cells. The exterior of the blastocyst is a thin-walled, hollow structure called the **trophoblast** ("feeding-sprout"). Inside the trophoblast, there is a cluster of cells called the **inner cell mass (ICM).** This is a fundamental distinction—the embryo develops from ICM cells, and the trophoblast forms part of the **placenta.** The placenta exchanges nutrients and wastes between the mother's and embryo's blood. It is derived from a mixture of maternal uterine cells and trophoblast cells.

The placenta is a key evolutionary innovation of mammals that allows internal development—but it isn't part of a newborn or adult.

23.3 Gastrulation

As cleavage nears completion, cell division slows. During the next phase of development, cell proliferation is no longer the most important developmental process. Instead, cell movement takes the leading role. In this stage of development, called **gastrulation,** extensive and highly organized cell movements radically rearrange the embryonic cells into a structure called the gastrula. **FIGURE 23.6** shows the remarkable reshaping of a frog embryo during gastrulation.

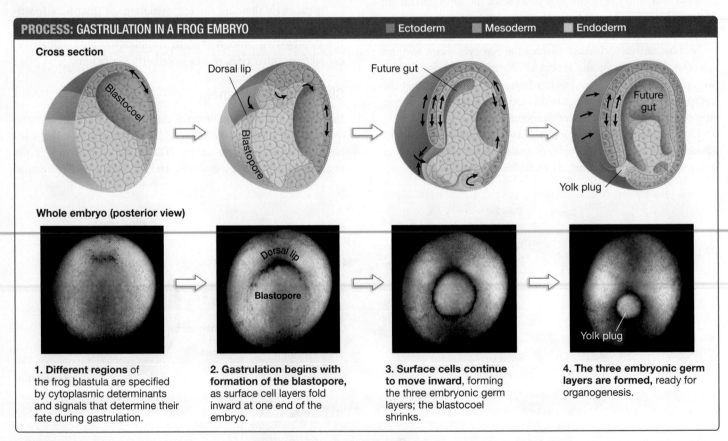

PROCESS: GASTRULATION IN A FROG EMBRYO ☐ Ectoderm ☐ Mesoderm ☐ Endoderm

Cross section

Dorsal lip · Blastocoel · Blastopore · Future gut · Future gut · Yolk plug

Whole embryo (posterior view)

Dorsal lip · Blastopore · Yolk plug

1. Different regions of the frog blastula are specified by cytoplasmic determinants and signals that determine their fate during gastrulation.

2. Gastrulation begins with formation of the blastopore, as surface cell layers fold inward at one end of the embryo.

3. Surface cells continue to move inward, forming the three embryonic germ layers; the blastocoel shrinks.

4. The three embryonic germ layers are formed, ready for organogenesis.

FIGURE 23.6 Gastrulation Creates Head-to-Tail and Back-to-Belly Axes and Three Germ Layers in the Embryo. Gastrulation in the frog is shown here. In all animal embryos, gastrulation requires coordination of a remarkable set of cell movements.

Research on this phase of development started in the 1920s with efforts to follow the movement of individual cells during newt and frog gastrulation. In these early experiments, tiny blocks of agar (a gelatinous compound) were soaked with a nontoxic dye. The dyed blocks were then pressed against the surface of blastula-stage embryos so that a small number of blastomeres became marked with dye. By allowing marked embryos to develop and then examining them at intervals during gastrulation, researchers were able to follow the movement of cells. Today, research on gastrulation continues, using new ways to mark individual cells and track them throughout the developing embryo.

Formation of Germ Layers

The pattern of gastrulation varies widely among animal species, but the general outcome is the same: Gastrulation forms three embryonic tissue layers. A **tissue** is an integrated set of cells that function as a unit.

Most animal embryos have three primary tissue layers: **(1)** ectoderm ("outside skin"), **(2)** mesoderm ("middle skin"), and **(3)** endoderm ("inner skin"). These embryonic tissues are called **germ layers** because they give rise to all the organs and tissues of the adult. It is important to note that each germ layer forms certain tissues and organs, and the correspondence between germ layer and organ type is the same in most animals.

Figure 23.6 shows two views of how the cell movements of gastrulation form the three germ layers in a frog embryo. The cell movements of gastrulation are remarkably complex. Cells that will become ectoderm are shown in blue; cells destined to form mesoderm are shown in pink, and cells that will form endoderm are colored yellow. These cell fates are set by signals between blastomeres and cytoplasmic determinants within blastomeres.

Step 1 The frog blastula contains a fluid-filled interior space called the **blastocoel.** The blastula of most animals contains a blastocoel.

Step 2 As gastrulation begins, an indentation (invagination) forms on the outer surface. In frogs this invagination starts out as a slit that eventually forms a circular opening known as the **blastopore.**

Step 3 Cells from the surface move into the interior of the embryo through the blastopore, forming a tube that will become the gut or digestive tract. The blastocoel is displaced and eventually disappears.

Step 4 The movement of cells into the embryo ultimately results in the formation of three layers—an inside, a middle, and an outside. These are the endoderm, mesoderm, and ectoderm.

FIGURE 23.7 shows what the germ layers produce in a human. Each germ layer gives rise to the same types of organs in virtually all animals. **Ectoderm** forms the outer covering of the body and the nervous system; **mesoderm** produces muscle, most internal organs, and connective tissues such as bone and cartilage; **endoderm** produces the lining of the digestive tract and the many organs that develop from the gut, such as the liver and lungs.

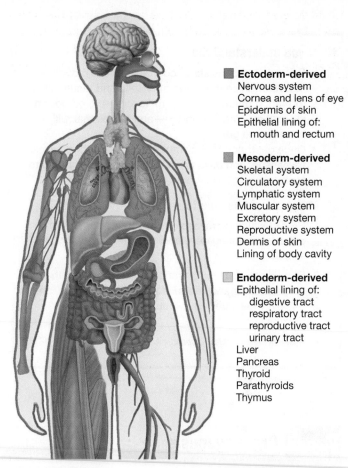

■ **Ectoderm-derived**
Nervous system
Cornea and lens of eye
Epidermis of skin
Epithelial lining of:
    mouth and rectum

■ **Mesoderm-derived**
Skeletal system
Circulatory system
Lymphatic system
Muscular system
Excretory system
Reproductive system
Dermis of skin
Lining of body cavity

■ **Endoderm-derived**
Epithelial lining of:
    digestive tract
    respiratory tract
    reproductive tract
    urinary tract
Liver
Pancreas
Thyroid
Parathyroids
Thymus

FIGURE 23.7 The Three Embryonic Germ Layers Give Rise to Different Adult Tissues and Organs. Each germ layer forms the same types of organs in all animals.

Creating Body Axes

In addition to establishing the germ layers, gastrulation has another major role: Creating the body axes. In frogs, for example, the blastopore becomes the anus (posterior), the region at the opposite end of the gut tube becomes the mouth (anterior), the region where cells first move into the blastopore (as drawn in Figure 23.6) defines the dorsal, or back, side of the embryo, and the region opposite this becomes the ventral region. In this way, the anterior–posterior and dorsal–ventral axes of the body become apparent.

Note, however, that in frogs, the major body axes are partially determined early in development by cytoplasmic determinants that were stored in the egg and partitioned into blastomeres during cleavage. These master regulators start regulatory gene cascades that begin the step-by-step process of differentiation (see Chapter 22).

Current research on gastrulation is focused on understanding how cells can coordinate their behaviors—especially, the mechanisms responsible for cell movement and navigation. Meanwhile, like the embryo itself, let's move along to the next phase in development—the formation of tissues and organs as the body takes shape.

23.4 Organogenesis

Gastrulation creates the rudiments of an animal. There are outside, inside, and middle layers; a head and a tail; a back and a belly—but much remains to be done. For one thing, there are no organs within the embryo. The heart, brain, liver, and lungs all need to be formed from the germ layers, properly positioned, and connected with other organs. **Organogenesis** ("organ-origin") is the process of tissue and organ formation. During organogenesis, cells proliferate, move, differentiate, and assemble into tissues and organs.

To explore how organogenesis takes place, let's consider one example—how muscle tissue forms in the embryo. Muscle comes from mesoderm, the middle germ layer. To form muscle, mesodermal cells must specialize and organize into muscle tissue. What causes undifferentiated mesodermal cells to form muscle tissue and to begin producing the muscle-specific proteins that make breathing, walking, and turning the pages of textbooks possible?

Organizing Mesoderm into Somites

To understand the formation of any tissue or organ, it is necessary to trace back through many earlier structures, some of which exist only in the embryo. Let's begin with the assembly of key embryonic structures called somites.

Somites are paired blocks of mesodermal tissue that extend along either side of the dorsal midline (top middle) of the embryo. In other words, they run along the "back" of the

(a) Surface view of somites

(b) Cross section of somites

FIGURE 23.8 Somites Are Made of Mesodermal Cells. Somites (color-coded pink in these chick-embryo micrographs) form on either side of the neural tube, a structure that will develop into the brain and spinal cord and that runs along the middle of the embryo's back.

embryo. **FIGURE 23.8** shows two views of somites and surrounding embryonic structures early in organogenesis of a chick embryo, where somites are easily studied. Note that at this stage of development, a new structure has appeared in the dorsal mesoderm—a rod-like element called the **notochord.** This structure is shown in cross section in the figure and runs the length of the anterior–posterior axis of the embryo, just under the dorsal surface.

The notochord is unique to the group of animals called the **chordates,** which includes humans and all other vertebrates. In some species of chordates, the notochord is a long-lasting structure that functions as a simple internal skeleton—it stiffens the body and makes efficient swimming movements possible. But in vertebrates, such as chicks, frogs, and humans, the notochord is transient. It appears only in embryos. As organogenesis proceeds, many of the cells in the notochord undergo programmed cell death by apoptosis.

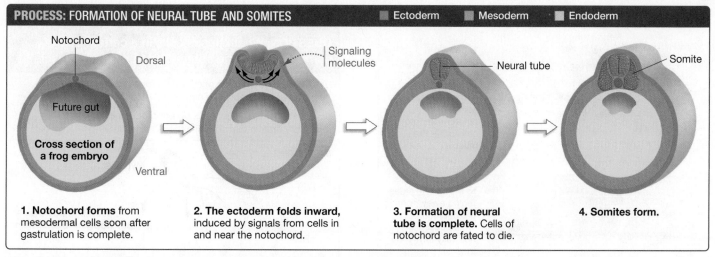

PROCESS: FORMATION OF NEURAL TUBE AND SOMITES ☐ Ectoderm ☐ Mesoderm ☐ Endoderm

Notochord

Dorsal

Future gut

Cross section of a frog embryo

Ventral

Signaling molecules

Neural tube

Somite

1. Notochord forms from mesodermal cells soon after gastrulation is complete.

2. The ectoderm folds inward, induced by signals from cells in and near the notochord.

3. Formation of neural tube is complete. Cells of notochord are fated to die.

4. Somites form.

FIGURE 23.9 The Notochord, Neural Tube, and Somites Form Early in Organogenesis. In vertebrates, the notochord forms from mesodermal cells soon after gastrulation is complete. Molecules produced in the notochord induce the formation of the neural tube.

✔**QUESTION** Predict how the formation of the neural tube would be affected by a drug that locks the cytoskeleton in place.

But where do somites come from? **FIGURE 23.9** illustrates the developmental path to somite formation in a frog embryo. What's shown here differs in detail but not in substance from what occurs in a chick, human, or any other vertebrate embryo. From your study of gastrulation, you should already be familiar with the ectoderm, mesoderm, endoderm, and future gut cavity shown in step 1 of Figure 23.9. The somites form from the mesoderm near the notochord.

Formation of Somites and the Neural Tube Somite formation, the stepping stone to formation of muscle, is coupled to the creation of another major structure in the embryo—the **neural tube.** The neural tube is the precursor of the brain and spinal cord. As step 2 in Figure 23.9 shows, the notochord produces signaling molecules that induce the ectoderm on the dorsal (back) side of the embryo to fold. The notochord is a key organizing element during organogenesis.

Folding of the dorsal ectoderm results from changes in cell shape—analogous to the differential cell expansion that occurs in plant cell development (see Chapter 22). The folding begins when signals from the notochord trigger massive changes in cytoskeletal elements inside each of the dorsal ectodermal cells. As the cytoskeleton is reorganized, the ectodermal cells grow taller and then constrict at their dorsal end and expand at their ventral end. The constriction above and expansion below makes the sheet of cells fold upward. When the folds of dorsal ectoderm come together, they form the neural tube (step 3). This is a monumental event in the development of chordates, because the newly formed neural tube is the beginning of the brain and spinal cord.

As organogenesis continues, cells near the neural tube and notochord becomes organized into blocks of mesodermal tissue. These are the somites (step 4). Somite formation is a response

to changes in the cell adhesion molecules that keep mesodermal cells attached to each other (see Chapter 11). Look back at Figure 23.8 to see how the somites, notochord, and neural tube appear in an actual embryo (in this case, chick).

Somites Are Precursors to Skin, Bone, and Muscle Somites are transient structures, just like the notochord is. But unlike the vertebrate notochord, somites produce many important structures of the adult. By marking somite cells and following them over time, researchers discovered that somites give rise to the lower (dermal) layer of the skin, to much of the skeleton, and to skeletal muscles.

As organogenesis proceeds, somite cells break away in distinct groups that migrate to their final location in the developing embryo. There they continue to proliferate and eventually differentiate. Cell movements like these are critical to organogenesis. As **FIGURE 23.10** shows, the fate of a somite cell depends on its position within the somite. When is the fate determined?

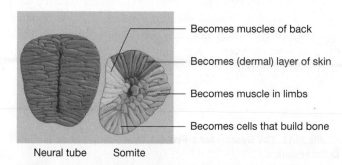

Becomes muscles of back

Becomes (dermal) layer of skin

Becomes muscle in limbs

Becomes cells that build bone

Neural tube   Somite

FIGURE 23.10 A Cell's Position within a Somite Determines Its Fate. Each somite eventually breaks up into distinct populations of cells, each of which gives rise to distinct tissues.

Neural tube

Somite

Somite

1. Early in development, remove cells from various locations in somites that normally differentiate into cells other than muscle.

2. Transplant cells to the somite sides facing away from the neural tube, where cells normally form limb muscle.

Future limb muscle

Future limb muscle

3. Cells from the outer side of somites will migrate away and become limb muscle, regardless of their normal fate and original location.

FIGURE 23.11 Evidence for Progressive Determination of Mesodermal Cells in Somites. By transplanting somite cells early in somite formation to new locations in another developing somite, researchers showed that in the early somites, the fate of cells is not yet determined.

Cell Determination By transplanting cells from one location to another within the developing somite, researchers found that initially any of those cells can become any of the somite-derived elements. For example, biologists transplanted cells from various parts of the early somite to the somite region farthest from the neural tube—the region that produces limb muscle. All the transplanted cells eventually formed limb muscles (**FIGURE 23.11**), even though these same cells normally would have formed other structures. Cells transplanted later in development, however, failed to become the cell type associated with their new position. Instead, the transplanted cells differentiated into the cell types they normally would have formed in their original position, even though they were now in a new and inappropriate location.

FIGURE 23.12 The Search for a Protein That Causes Muscle-Cell Differentiation.

Weintraub, H., S. J. Tapscott, R. L. Davis, et al. 1989. Activation of muscle-specific genes in pigment, nerve, fat, liver, and fibroblast cell lines by forced expression of *MyoD*. *Proceedings of the National Academy of Sciences, USA* 86: 5434–5438.

✔ **QUESTION** Why did the researchers have to attach a "general purpose" promoter to the cDNAs?

QUESTION: What causes cells in a certain part of somites to become committed to produce muscle?

HYPOTHESIS: Production of an mRNA for a regulatory protein (or proteins) by myoblasts commits them to their fate.

NULL HYPOTHESIS: Myoblasts do not produce an mRNA for a regulatory protein that commits them to their fate.

EXPERIMENTAL SETUP:

Myoblast

mRNA

1. Isolate mRNAs from myoblasts.

cDNA

2. Convert mRNAs to cDNAs using reverse transcriptase.

Promoter

cDNA

3. Attach "general purpose" promoters to cDNAs.

4. Introduce cDNAs into non-muscle cells.

PREDICTION: One of the myoblast-derived cDNAs will convert non-muscle cells into cells that produce muscle-specific proteins.

PREDICTION OF NULL HYPOTHESIS: None of the myoblast-derived cDNAs will convert non-muscle cells into cells that produce muscle-specific proteins.

RESULTS:

Muscle-like cell (produces muscle-specific proteins)

CONCLUSION: Certain somite cells produce an mRNA for a regulatory protein (later called MyoD) that commits them to differentiate into muscle.

These transplantation studies reveal an important aspect of development: Cells gradually become determined. **Determination** is the irreversible commitment of a cell to a particular developmental fate. Before determination, cells transplanted to a new location can take on the fate appropriate for that location. But once determined, cells do not alter their course, even if they are placed in a new environment and receive signals to become a different type of tissue.

Determination comes about through epigenetic mechanisms that were discussed in the chapter on eukaryotic gene expression (Chapter 19). Signals in the embryo induce the production of new sets of transcription factors. In turn, these transcription factors trigger new and long-lasting patterns of chromatin condensation. Gene expression patterns appropriate for a particular cell type are set in place, and switching to different patterns becomes nearly impossible.

In short, the determination of a cell's fate as the somite matures **(1)** is a step-by-step process guided by regulatory gene cascades and **(2)** is a function of the cell's location in the embryo. Location is important because different positions within the embryo have different molecular signals.

Signals coming from cells in the notochord, the neural tube, and nearby ectoderm and mesoderm set up divisions within the somite. Different combinations of these signals inform cells where they are within the somite and what fate they should adopt. These signals induce new sets of transcription factors and new patterns of gene expression that determine whether a cell becomes muscle, bone, or skin.

Differentiation of Muscle Cells

What makes a cell become muscle? Harold Weintraub and colleagues answered this question in the late 1980s by experimenting with cells called myoblasts. A myoblast is derived from cells in the somite. Myoblasts are destined to become muscle cells but have not yet begun producing muscle-specific proteins.

Weintraub and co-workers hypothesized that myoblasts must express at least one regulatory protein that commits them to their fate. The idea was that myoblasts begin producing this muscle-determining protein after they receive an appropriate set of signals from nearby tissues. In effect, the researchers were looking for a transcription factor in myoblasts that dictates, "Become a muscle cell."

FIGURE 23.12 outlines how the biologists went about searching for this hypothetical protein.

1. They began by isolating mRNAs from myoblasts.

2. They used reverse transcriptase to convert the mRNAs to cDNAs. Because myoblasts transcribe genes required for a muscle-cell fate, they reasoned that the cDNAs must include a muscle-determining gene transcript.

3. They attached a type of promoter to the cDNAs that would ensure expression of the cDNA in any type of cell.

4. They introduced the modified cDNAs into non-muscle cells called fibroblasts and monitored the development of the cells.

As predicted, one of the myoblast-derived cDNAs converted fibroblasts into muscle-like cells. Follow-up experiments showed that the same gene product could convert pigment cells, nerve cells, fat cells, and liver cells into cells that produced muscle-specific proteins.

Weintraub's group called the protein product of this gene **MyoD**, for *myo*blast *d*etermination. Subsequent work showed that the *MyoD* gene encodes a regulatory transcription factor and that the MyoD protein binds to enhancer elements located upstream of muscle-specific genes (see Chapter 19). MyoD is a master regulator of muscle differentiation.

In addition, researchers found that the MyoD protein activates further expression of the *MyoD* gene. This was a key observation because it meant that once *MyoD* is turned on, it triggers its own expression—meaning that the gene continues to be transcribed.

To put this specific example of muscle differentiation into context, think back to the sequence of events occurring in early development and to the principles of development (see Chapter 22). Differentiation is a step-by-step process that is complete when cells begin producing proteins that are specific to a particular cell type. By what path did the cells in your bicep become muscle cells?

Step 1 Fertilization of the egg inside your mother triggered the onset of cleavage, resulting in a blastocyst.

Step 2 Certain cells in the blastocyst began producing signals that triggered regulatory gene cascades that led specific cells during gastrulation to become mesoderm in your back.

Step 3 During organogenesis, signals from cells of the notochord and nearby structures induced the production of MyoD and other muscle-determining proteins in particular somite cells. In response, these target cells became committed to developing into muscle and moved into the regions that formed your arms.

Step 4 Later, these MyoD-expressing cells began producing muscle-specific proteins.

All of these steps made it possible for you to move your arms after birth. The rest, as they say, is history.

If you understand . . .

23.1 Fertilization

- Fertilization creates the zygote, the first cell of the animal.

- Fertilization involves mechanisms to attract the sperm to the egg; bind sperm to egg in a species-specific manner; prevent the entry of more than one sperm; fuse the sperm- and egg-cell plasma membranes; bring the chromosomes of egg and sperm together; and activate metabolism in the egg to support development.

✔ You should be able to explain why bindin and the egg-cell receptor for sperm are said to interact like a lock and key.

23.2 Cleavage

- Cleavage is the set of rapid, early cell divisions that follow fertilization.

- Cleavage divides the fertilized egg into a mass of cells.

- In many species, an array of signals and regulatory transcription factors (cytoplasmic determinants) present in the egg are distributed to different blastomeres. As a result, different blastomeres have different fates.

- Once cleavage is complete, the embryo consists of a mass of cells.

✔ You should be able to describe how cleavage in a frog embryo can result in certain cells containing a cytoplasmic determinant.

23.3 Gastrulation

- Gastrulation is the dramatic, coordinated movement of cells that transform the cleavage-stage embryo into the gastrula.

- Cell movements of gastrulation require changes in cell shape and size and changes in the affinity of one cell for another.

- The gastrula contains the three germ layers of ectoderm, mesoderm, and endoderm and has recognizable embryonic axes (anterior–posterior, dorsal–ventral, right–left).

- The germ layers give rise to distinct organs of the adult, and each germ layer produces organs of similar types in all animals.

✔ You should be able to compare and contrast, in broad terms, a mesodermal cell in the anterior end of an embryo and a mesodermal cell in the posterior end, regarding the cell-cell signals they receive and the regulatory transcription factors they contain.

23.4 Organogenesis

- Organogenesis is the formation of tissues and organs.

- Organogenesis occurs through the coordination of cell–cell signals, cell proliferation, cell movements, and differentiation. Differentiation is complete when cells express tissue-specific proteins and the specialized properties of adult cells.

- The creation of muscle is one example of organogenesis.

- Early in vertebrate organogenesis, cells in the notochord release cell–cell signals that induce the formation of a neural tube—precursor to the brain and spinal cord—from ectoderm that overlies the notochord.

- Blocks of mesodermal cells, called somites, form next to the neural tube.

- In response to cell–cell signals from the notochord, neural tube, and other nearby tissues, cells in the somites are determined, move to new positions, proliferate, and begin expressing tissue-specific proteins.

- Determination of cells occurs during organogenesis and is the generally irreversible commitment of a cell to follow a particular pathway of differentiation.

✔ You should be able to give examples of cell–cell signals, expansion, movement, proliferation, and differentiation during neural tube, somite, and muscle development in mammals.

MB **MasteringBiology**

1. MasteringBiology Assignments

Tutorials and Activities Early Stages of Animal Development; Embryonic Development; Frog Development; How Do Calcium Ions Help to Prevent Polyspermy During Egg Fertilization?; Sea Urchin Development

Questions Reading Quizzes, Blue-Thread Questions, Test Bank

2. eText Read your book online, search, take notes, highlight text, and more.

3. The Study Area Practice Test, Cumulative Test, BioFlix® 3-D Animations, Videos, Activities, Audio Glossary, Word Study Tools, Art

You should be able to . . .

1. What is bindin, and what does it do?

2. How is calcium signaling involved in blocking polyspermy?
   a. It triggers a regulatory gene cascade.
   b. Accumulation of calcium ions causes an outward flow of water, raising the fertilization envelope.
   c. It triggers fusion of cortical granules with the plasma membrane that leads to formation of the fertilization shell.
   d. Calcium ions bind to egg-cell receptors for sperm, blocking binding sites.

3. What happens during cleavage?
   a. The neural tube—precursor of the spinal cord and brain—forms.
   b. The inner cell mass of the blastocyst begins dividing rapidly to produce the placenta.
   c. The fertilized egg divides without growth occurring, forming a mass of cells.
   d. Massive movements of cells make the primary body axes visible and organize the three embryonic tissues.

4. What happens during gastrulation?
   a. The neural tube—precursor of the spinal cord and brain—forms.
   b. The inner cell mass of the blastocyst begins dividing rapidly to produce the placenta.
   c. The fertilized egg divides without growth occurring, forming a ball of cells.
   d. Massive movements of cells make the primary body axes visible and organize the three embryonic tissues.

5. Which of the embryonic germ layers gives rise to the brain?

6. Which of the following does *not* occur during organogenesis?
   a. establishment of the anterior–posterior axis
   b. differentiation of cells
   c. movement of cells into new positions
   d. extensive cell–cell signaling

7. Many frogs and mice are similar in size, yet a frog egg is vastly larger than a mouse egg. Propose a plausible explanation for the difference in the size of these eggs. Explain your reasoning.

8. A variety of agents are known that bind tightly to Ca^{2+} inside of cells. This binding prevents Ca^{2+} from exerting any of its normal effects. Imagine a situation in which a Ca^{2+}-binding agent was injected into a sea urchin egg before thousands of sperm were added to the seawater surrounding the egg. Predict how this treatment would affect fertilization, and provide reasons why.

9. Blastomeres look identical. Describe how they are not.

10. Plants depend on directional cell expansion to create structures. Which of the following events in animal development relies on a mechanism related to directional cell expansion in plants?
    a. formation of the neural tube from a sheet of ectoderm
    b. migration of myoblasts to developing limbs
    c. determination of cells in different regions of the somite
    d. formation of the inner cell mass within the blastocyst

11. Explain how cell-transplantation experiments provided evidence that cells in a somite are determined in a step-by-step fashion.

12. **QUANTITATIVE** In the study that discovered MyoD, Weintraub and his colleagues reasoned that MyoD was likely to be a rare mRNA. If they assumed that MyoD mRNA comprised 2×10^{-5} (0.00002) of all mRNAs, what is the smallest number of fibroblast cells into which they would have needed to introduce cDNAs so they would have a reasonable chance of finding one that expressed MyoD?

13. At the molecular level, explain why bindin on the sperm of the sea urchin *Strongylocentrotus purpuratus* attaches to the egg-cell receptor for sperm on eggs of *S. purpuratus*, but why bindin from other species of *Strongylocentrotus* cannot.

14. A molecule called G418 inhibits translation. When researchers treat fruit fly eggs with G418, the early stages of cleavage proceed normally after fertilization. What can you conclude from this experiment?

15. In the marine organisms called sea squirts, eggs contain a yellow pigment. Blastomeres that contain this yellow pigment become muscle cells. Is the yellow pigment a cytoplasmic determinant? Why or why not?

16. One reason chicks are such a valuable model organism for studying development is that their embryos can be surgically manipulated. Researchers have done experiments that involve transplanting an additional piece of notochord near the upper portion of the neural tube in an embryo that has just formed somites. A reasonable prediction for how this manipulation would affect somite development is that
    a. more bone than normal will be produced from the somite.
    b. more limb muscle than normal will be produced from the somite.
    c. more dermis than normal will be produced from the somite.
    d. somite development will be unchanged, because somite cells are determined as soon as the somite forms.

24 An Introduction to Plant Development

In this chapter you will learn how

Plant development is both similar to and distinct from animal development

by asking what happens during three key stages

Embryogenesis 24.1	Vegetative development 24.2	Reproductive development 24.3
including	including	including
What genes and proteins set up body axes?	What genes and proteins determine leaf shape?	What genes and proteins control organ patterns in flowers?

The 4700-year old bristlecone pine is Earth's oldest known living organism. This chapter explores how plant embryos can develop into adults that grow and reproduce throughout their lives.

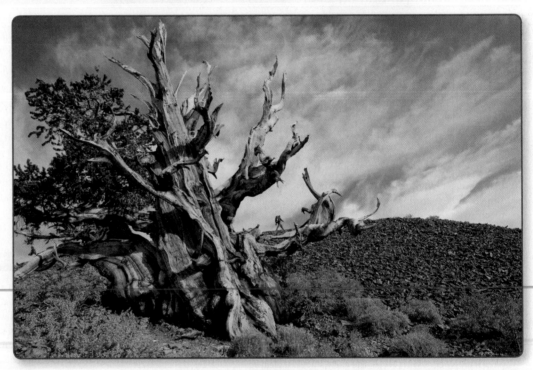

About the year 2750 B.C., perhaps more than a century before the first pyramids were constructed in ancient Egypt, a sperm cell and an egg cell from a bristlecone pine tree fused on a mountainside in today's California. A zygote was created that produced a tree that has been growing and developing ever since—for over 4700 years.

Unlike many animals, plants continue to grow and develop throughout their lives, whether that life lasts two weeks or thousands of years. Continuous development is a hallmark of plant biology. Why the difference?

To begin to answer this question, consider the distinct lifestyles that have evolved in plants and animals. Most animals move to find food and mates, while plants stand their ground, make their own food, and use water, wind, or insects to transport their gametes. Animal cells lack a rigid cell

✔ When you see this checkmark, stop and test yourself. Answers are available in Appendix A.

wall, while plant cells are encased in a protective but constraining wall. These very different ways of life account for many of the contrasts in plant and animal development.

Animals, land plants, two groups of fungi, and the red and brown algae all have evolved independent solutions to the common problem of how to build a multicellular organism. However, despite their separate evolutionary paths, there are abundant similarities in plant and animal development. How can this be explained? Recall that a few fundamental principles are common to the developmental events observed in all multicellular organisms (Chapter 22)—there are a limited number of ways to build a plant or animal.

This chapter is about how plants develop, and how plant development is both similar to and distinct from the development of animals. Let's begin with embryogenesis in plants, the formation of a multicellular embryo from a single-celled zygote.

24.1 Embryogenesis

FIGURE 24.1 provides an overview of development in a flowering plant—the most familiar, species-rich, and abundant group of plants (see Chapter 31). The development of flowering plants is the primary focus of this chapter. In this figure and in many other places in this chapter, a small species called *Arabidopsis thaliana*—the best-studied plant—will serve as a model organism (see **BioSkills 13** in Appendix B).

Note that the sequence of developmental events, or life cycle, diagrammed in Figure 24.1 begins with gametogenesis, or gamete formation. Next come pollination and fertilization. In *Arabidopsis* and other flowering plants, an egg is fertilized inside a protective, womb-like structure called the ovule. You'll learn more about these processes in a later chapter (Chapter 41).

Development continues inside the ovule with **embryogenesis**— literally, "embryo-origin." In flowering plants, embryogenesis ends with the maturation of the ovule into a **seed**—a structure that contains the embryo and a supply of nutrients surrounded by a protective maternally-provided coat.

An embryo may remain in a nongrowing state inside the seed for months, years, or in some cases, centuries. This state of arrested development is rare in animals. When conditions are favorable the seed undergoes **germination**—meaning that it resumes growth— to form a seedling, beginning the process of organogenesis. Here is another contrast with animal development: organogenesis—the development of organs—continues throughout the plant's life.

Vegetative development produces the nonreproductive portions of the plant body—the roots, leaves, and stems. As the plant matures, some cells in the stem will become converted to reproductive structures, a process known as **reproductive development.**

Reproductive development is a distinguishing feature of plant development. In animals, determination of cells that have the potential to form sperm or egg—the **germ line**—is one of the earliest events in development. In plants, there is no determined germ line. A small group of cells in each stem have the potential to switch from vegetative to reproductive development in response to environmental conditions.

What Happens during Plant Embryogenesis?

The sequence of events in plant embryogenesis was worked out through careful observation of developing embryos. Embryogenesis is shown in **FIGURE 24.2** (see page 434) within the overall scheme for development of the model plant *Arabidopsis*.

After fertilization, the zygote undergoes an asymmetric cell division (see Figure 24.2 steps 1 and 2). The cells resulting from this initial division are different in size, content, and fate. Similar asymmetric cell divisions are seen often in animal development.

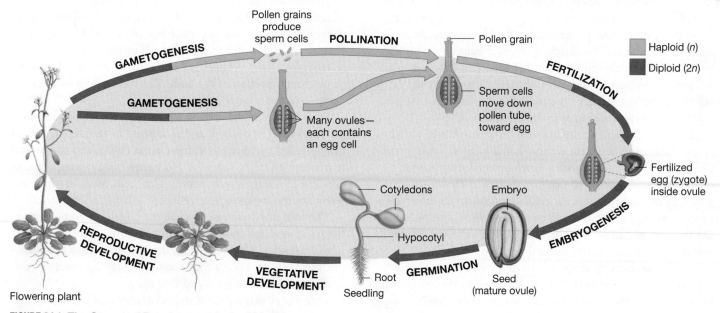

FIGURE 24.1 The Stages of Development in *Arabidopsis*.

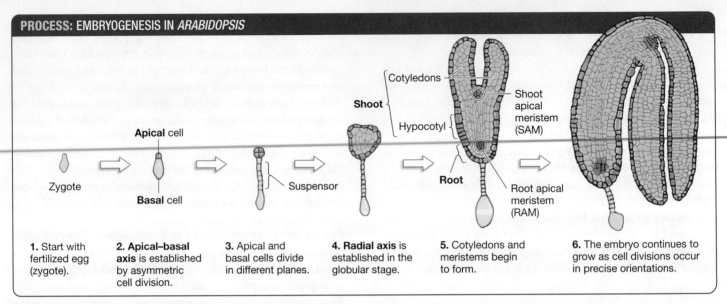

Cotyledons

Shoot

Hypocotyl

Shoot apical meristem (SAM)

Root

Root apical meristem (RAM)

Apical cell

Basal cell

Zygote

Suspensor

1. Start with fertilized egg (zygote).

2. **Apical–basal axis** is established by asymmetric cell division.

3. Apical and basal cells divide in different planes.

4. **Radial axis** is established in the globular stage.

5. Cotyledons and meristems begin to form.

6. The embryo continues to grow as cell divisions occur in precise orientations.

FIGURE 24.2 Embryogenesis Takes Place Inside the Developing Seed. Only the embryo is shown in these panels.

✓**EXERCISE** Label the apical–basal and radial axes on the globular-stage embryo.

(For example, think of the distribution of cytoplasmic determinants in Figure 23.4.) The bottom, or basal, cell is large, and it gives rise to a column of cells called the suspensor. The suspensor connects the embryo to surrounding tissues within the seed. The small cell above the basal cell, the apical cell, gives rise to most of the embryo.

Apical refers to the tip; **basal** refers to the base. The asymmetry between the basal and apical cells helps establish one of the primary axes of the plant body: the **apical–basal axis.**

As steps 3 and 4 in Figure 24.2 show, the basal cell divides perpendicularly to the apical–basal axis to produce the suspensor. Only one cell in the suspensor—the one closest to the apical cell—contributes cells to the embryo and thus to the mature adult. The apical cell, in contrast, divides both perpendicularly to the apical–basal axis and parallel to it.

As you learned in an earlier chapter, the regulation of cleavage plane orientation and the direction of cell expansion are critical for creating structures in the plant (Chapter 22). Perpendicular and parallel divisions in the cells at the apical tip of the plant embryo give rise to a ball of cells attached to the suspensor. At this point, the embryo is said to be in the globular stage.

You should be able to see how the division between an apical cell that forms the plant embryo and a basal cell that forms a supportive structure parallels a developmental strategy used by mammalian embryos. Recall that in mammals, the inner cell mass produces the embryo, and the cells surrounding the inner cell mass create part of the placenta (see Figure 23.5). The placenta, like the plant suspensor, is a structure that is essential for development but that doesn't contribute to the embryo.

Another asymmetry arises as cells of the globular-stage plant embryo continue to divide. Cells in the interior are completely surrounded by other cells of the embryo. In contrast, cells in the outermost layer contact surrounding tissues in the seed in addition to underlying embryo cells. Interior and exterior cells become visibly different. This creates the second major body axis, the radial axis. The radial axis extends from the interior of the plant body out to the exterior. Note that the animals examined in the animal development chapter (Chapter 23) do not have a radial axis.

The initial events in embryogenesis illustrate a general point: Just as for most animal cells, the fate of a plant cell can be summed up in the old quip about the three keys to success in real estate—"location, location, location." Starting with the initial division that creates the apical and basal cells, plant cells differentiate based on where they are in the body.

As the ball of cells continues to grow and develop, embryonic leaves, or **cotyledons,** begin to take shape (step 5). The cotyledons are connected to the developing root by a stem-like structure called the **hypocotyl.** Together, the cotyledons and hypocotyl make up the **shoot,** which will become the aboveground portion of the body. The shoot system functions in photosynthesis, support, and reproduction. The **root,** in contrast, forms the belowground portion of the body. The root system anchors the plant and gathers water and nutrients.

Once the apical–basal and radial axes are established, and as the cotyledons, hypocotyl, and root begin to take shape, groups of cells called the **shoot apical meristem (SAM)** and **root apical meristem (RAM)** are specified. **Meristems** contain groups of stem cells that divide repeatedly into daughter cells, some of which differentiate into the specialized cells of the plant (see Chapter 22).

The root meristem can form all the underground portions of the plant, and the shoot meristem can form all the aerial portions, including reproductive structures. Throughout a plant's life, meristematic tissues continue to produce cells that can differentiate into adult tissues and structures.

Note that all of this growth and development takes place without any of the cell migration seen in gastrulation in animals. Because plant cells have stiff cell walls, they do not move. This is one of the

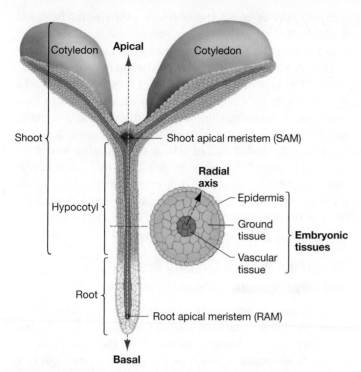

FIGURE 24.3 Embryogenesis Establishes the Two Body Axes and Three Embryonic Tissues.

fundamental differences between plant and animal development. For the cotyledons and other embryonic structures to take shape, it is important that cell divisions occur in precise orientations (step 6 of Figure 24.2). What's more, the resulting cells exhibit differential growth. Some cells grow larger than others, and even more important, the direction of cell expansion along the apical–basal or radial axes is tightly controlled and often radically different.

During embryogenesis, three embryonic tissues are produced and arranged along the radial axis (**FIGURE 24.3**).

1. The **epidermis** (literally, "over-skin") is an outer covering of specialized cells that protects the individual.

2. Inside the epidermal layer of cells is **ground tissue,** a mass of cells that may later differentiate into cells that are specialized for photosynthesis, food storage, or other functions.

3. The **vascular tissue** in the center of the plant will eventually differentiate into specialized cells that transport food and water between root and shoot.

These embryonic tissues are analogous to the ectoderm, mesoderm, and endoderm of developing animals (see Chapter 23). Although the apical–basal and radial axes of plants do not directly correspond to animal axes, note the parallel of creating axes during early embryogenesis in both groups of organisms.

Which Genes and Proteins Set Up Body Axes?

The genetic approach to development that was pioneered with research on *Drosophila melanogaster* has also been a powerful way of studying plant embryogenesis. Although the specific genes involved are different in plants than in animals, the basic mechanism by which genes direct the earliest events in development is similar.

Research on the genetics of early development in plants was pioneered by Gerd Jürgens and colleagues in the 1990s. This research group set out to identify genes that are transcribed in the zygote or embryo of *Arabidopsis* and that are responsible for establishing the apical–basal axis of the plant body. It's no coincidence that this effort was similar to the work on anterior–posterior pattern formation in *Drosophila* (see Chapter 22)—Jürgens had participated in the work with flies.

The biologists' initial goal was to identify individuals with developmental defects at the seedling stage. More specifically, they were looking for patterning mutants that lacked particular regions along the apical–basal axis of the body. The team succeeded in finding several bizarre-looking mutants (**FIGURE 24.4**). Apical mutants lacked the first leaves, or cotyledons. Some mutants, called central mutants, lacked the embryonic stem, or hypocotyl. Other plants, dubbed basal mutants, lacked both hypocotyls and roots.

To interpret these results, the researchers suggested that each type of *Arabidopsis* mutant had a defect in a different gene and that each gene played a role in specifying the position of cells along the apical–basal axis of the body. They hypothesized that these genes are analogous to the segmentation genes of fruit flies, which specify the fate of cells within well-defined regions along the anterior–posterior axis of insects.

What are these *Arabidopsis* genes, and how do they exert their effects? To answer these questions, consider the gene responsible for the basal mutants lacking hypocotyls and roots. This gene has been cloned and sequenced and named *MONOPTEROS*. Because its DNA sequence indicated that its protein product has a DNA-binding domain, *MONOPTEROS* was hypothesized to encode a transcription factor that regulates the activity of target genes. This hypothesis was later shown to be correct.

The *MONOPTEROS* gene is activated in response to signals from auxin—a cell-to-cell signal molecule. Auxin is produced in the shoot apical meristem and transported toward the basal parts of the individual. This results in an auxin concentration gradient along the apical–basal axis of a plant. Much like the Bicoid concentration

FIGURE 24.4 *Arabidopsis* **Developmental Mutations Can Result in Misshapen Embryos.** Researchers have identified *Arabidopsis* patterning mutants that are missing specific sections of the body along the apical–basal axis.

gradient in fruit fly embryos (see Chapter 22), the auxin concentration gradient provides positional information.

Together these observations indicate that auxin acts as a morphogen to trigger the production of the regulatory transcription factor MONOPTEROS. In turn, MONOPTEROS unleashes a regulatory cascade that determines which cells in the basal portion of the embryo will form hypocotyl and roots.

The take-home message? Although the specific genes and proteins differ in animals and plants, independent evolution in the two groups has either converged or been constrained to adopt solutions for development that involve common principles. Cell-to-cell signals and regulatory cascades result in the step-by-step specification of a cell's position and fate.

check your understanding

If you understand that . . .

C Y U

- Early embryonic development results in the formation of the apical–basal and radial axes of the plant body and three embryonic tissues.
- The early structures of the shoot and root systems form along the body's apical–basal axis.
- The genes responsible for setting up the body axes are currently the focus of intense research.

✔ **You should be able to . . .**

1. Relate the "location, location, location" quip to the differentiation of epidermal, ground, and vascular tissue.

2. Predict the effect on the *MONOPTEROS* gene and on the embryo of adding auxin to embryonic root cells.

Answers are available in Appendix A.

24.2 Vegetative Development

Whether a plant lives for one season or the 4700 seasons of the oldest bristlecone pine, it continues to develop throughout its life. New roots, stems, and leaves appear and grow. This is vegetative development.

Vegetative development is finely tuned to the environment. Consider just one environmental condition—the availability of light.

You might recall that plants primarily use wavelengths in the blue and red portions of the spectrum to drive photosynthesis (see Chapter 10). Now suppose that you are an oak tree with a life expectancy of 300 years. The quality and quantity of light that your leaves receive depends on where you happen to germinate. Are you growing on flat ground? With a southern exposure in full sun? With a northern exposure rarely exposed to full sun?

In addition, from the time you emerge from an acorn to the time of your death, the light you receive will depend on changes in climate and weather as well as on your size relative to the size

and proximity of plants that compete with you for light and shade. Finally, the leaves in your bottommost branches experience a different light regime from the leaves at your apex.

How do plants cope with all this variation in their living conditions? Unlike most animals, they don't move around to find a place that suits their requirements. Instead, they adjust to their immediate surroundings, in large part through the continuous growth and development of roots, stems, and leaves. If an oak tree is heavily shaded on one side, it stops growing in that direction and extends branches on the other side. If it is heavily shaded on all sides, its growth is directed upward. This constant adjustment to changing environmental conditions is possible because of the meristems that are located at the tips of shoots and roots.

Meristems Provide Lifelong Growth and Development

Just as in animals, when early embryonic development in a plant is complete, the basic body axes are established and the initial structures have formed. For the rest of the individual's life, further development is driven by the groups of stem cells held within meristems (**FIGURE 24.5a**). Meristematic tissue is located at each tip in the shoot and at the tips of the root system. As a result, the individual is capable of growing in any direction aboveground or belowground, depending on conditions.

FIGURE 24.5b provides a close-up view of a shoot apical meristem, or SAM. The cells within the meristem are small and undifferentiated. Within each meristem, the rate of cell division is dictated by cell–cell signals produced in response to environmental cues, such as the arrival of spring, the presence of abundant water, or the amount of light striking the plant. Just below and at the periphery of the meristem, daughter cells produced by mitosis and cytokinesis in the meristem grow in specific directions and initially differentiate into epidermal, ground, or vascular tissue. Eventually these cells will differentiate into more specialized cell types.

Careful microscopy allowed biologists to tease out the sequence of events that occur as meristems grow, and intense research continues to explore how interactions between auxin and other cell–cell signals influence the fate of cells produced by meristems.

Recently researchers have also taken up the question of which genes respond to these cell–cell signals to direct the development of specific structures. Let's consider one example—the genetic control of leaf shape.

Which Genes and Proteins Determine Leaf Shape?

Once a leaf begins to grow, three different axes must be formed: the proximal–distal, mediolateral (middle-to-side), and adaxial–abaxial (upper–lower; **FIGURE 24.6**). Proximal is toward the main body; distal is away from the main body. The mediolateral axis runs from the middle of a leaf toward its margin. Unlike the shoot, but similar to the body of most animals, most leaves are bilaterally symmetric—they have mirror-image left and right

(a) Meristems are located at the tips of shoots and roots.

(b) Cross section of a shoot apical meristem.

Shoot meristems

Root meristems

Developing leaves

Dividing, undifferentiated **meristematic cells**

Cells differentiating into **vascular tissue**

Cells differentiating into **ground tissue**

Cells differentiating into **epidermal tissue**

50 μm

FIGURE 24.5 Development Takes Place in Meristems. (a) Each tip in the root and shoot systems contains a meristem. The individual can grow in any direction to which a meristem is oriented. **(b)** When stem cells in the meristem divide, the daughter cells either remain undifferentiated and continue to function as stem cells or differentiate into new epidermal, ground, or vascular cells.

sides. The amount of growth along these axes determines the shape of the leaf.

Recent research has begun to identify the genes responsible for specifying the three leaf axes. For example, analyses of mutant snapdragons—a flowering plant you may have seen growing in a garden—and other species has shown that a gene called

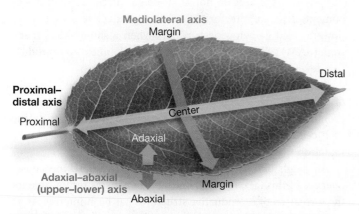

Mediolateral axis
Margin

Proximal–distal axis

Distal

Proximal

Center

Adaxial

Adaxial–abaxial (upper–lower) axis

Margin

Abaxial

FIGURE 24.6 Leaves Have Three Axes. Overall, the plant body has just two axes: apical–basal and radial. Individually, however, every leaf has the three axes shown here. (To remember the difference between adaxial and abaxial, pretend that the *b* in a*b*axial stands for "below.")

✔ **QUESTION** How do the body axes in plants compare and contrast with the body axes in animals?

PHANTASTICA (abbreviated *PHAN*) may play a critical role in setting up the adaxial–abaxial axis of leaves.

The protein product of *PHAN* is a regulatory transcription factor. PHAN controls the expression of genes that cause cells to form the upper surface of leaves. It is part of a regulatory cascade that begins with auxin and other cell–cell signals and ends with the growth of a normal-shaped leaf.

Research on tomatoes suggests that changes in *PHAN* expression may also underlie some of the evolutionary changes observed in leaf shape. Leaf shape varies widely among species (see Chapter 37). Simple leaves consist of a single blade, but as **FIGURE 24.7a** (see page 438) shows, tomatoes have compound leaves—each leaf blade is divided into smaller units called leaflets. Other species have palmately compound leaves, meaning that leaflets radiate from a single point.

To explore whether changes in *PHAN* might have a role in the evolution of various leaf shapes, a team of biologists produced transgenic tomato plants using techniques like those that created "golden rice" (Chapter 20). In these tomato plants, the PHAN transcription factor was expressed at reduced levels in the developing leaf.

As **FIGURE 24.7b** shows, leaf shape in the transgenic tomato plants was dramatically altered. Some had simple leaves that were cup shaped, while others had several leaflets emerging from the same point (palmate leaves). Although it is still uncertain exactly how PHAN affects leaf shape, it is clear that *PHAN* expression plays a role.

(a) Normal tomato leaf

(b) Transgenic leaves (reduced *PHAN* expression)

Simple, cup-shaped

1 cm

Palmately compound

1 cm

Compound

1 cm

FIGURE 24.7 Changes in the Expression of Genes That Establish Leaf Axes Can Change Leaf Shape.

These results have inspired the hypothesis that at least some evolutionary changes in leaf size and shape are due to mutations that alter *PHAN* expression. Alleles that result in lowered *PHAN* expression might lead to simple leaves with a single blade; alleles that increase the extent of *PHAN* expression might result in compound leaves like those of normal tomatoes. By experimentally altering the genes that regulate development, researchers are beginning to understand the genetic changes leading to different types of leaves. Biologists now have another example of how changes in regulatory pathways that direct development could underlie evolutionary change (see Chapter 22). Clearly, evo-devo is not just for animals.

Do Plant Cells Become Irreversibly Determined?

Normal animal cells are irreversibly determined and do not de-differentiate without considerable coaxing. You may recall that producing Dolly, the first cloned mammal, wasn't easy (see Chapter 22). More than 300 embryos were generated from differentiated adult cells, but only one healthy lamb—Dolly—developed. In plants, however, the situation is different.

For millennia, people have grown plants from cuttings. Removing just part of a stem and planting it can produce a new plant with a complete set of roots. What this requires is either **(1)** for plant cells to de-differentiate into an unspecialized cell type and then for the descendants of these cells to repeat development, or **(2)** for one type of specialized cell to differentiate directly into another cell type.

Unlike animal cells, determination in plants either doesn't occur or is readily reversible in at least some cells. This extraordinary flexibility is another hallmark of plant development.

24.3 Reproductive Development

An important difference between animal and plant development is how reproductive tissues and organs form. In animals, the cells that give rise to sperm and egg cells are set aside early in development. These **germ cells** migrate to the ovaries or testes as the reproductive organs begin to develop. As a result, the cells that give rise to animal gametes undergo relatively few rounds of mitosis—perhaps 20 to 50—before meiosis and gametogenesis.

In contrast, plants do not set aside germ cells early in development. Instead, the formation of reproductive structures—flowers—and gametogenesis occurs when a shoot apical meristem (SAM) switches from vegetative development to reproductive development.

As a result, meristematic cells that have divided hundreds of times can give rise to the reproductive organs of plants, and eventually to sperm and eggs. Because mutations occur during each cell cycle, plants generate much more genetic variation by mutation than animals do.

Although biologists have just begun to explore the consequences of this fact, research on the mechanisms responsible for the formation of reproductive structures has been intense. As an introduction to this work, let's consider some highlights of research on *Arabidopsis*.

The Floral Meristem and the Flower

When specialized proteins in *Arabidopsis* sense that nights are getting shorter and the temperature is favorable, they trigger the

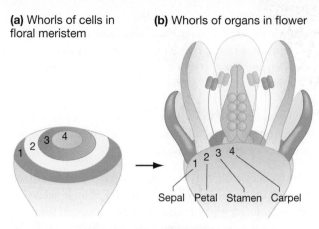

(a) Whorls of cells in floral meristem

(b) Whorls of organs in flower

Sepal Petal Stamen Carpel

FIGURE 24.8 Flowers Are Composed of Four Organs.

production of signals that convert SAMs from vegetative to reproductive development. Through a series of steps, a SAM can produce a **floral meristem;** instead of vegetative structures, such as roots, stems, or leaves, it produces flowers, which contain the plant's reproductive organs. The genes that take part in the regulatory cascade responsible for the maturation of a floral meristem are now well known.

As an *Arabidopsis* flower develops, the floral meristem produces four kinds of organs: **(1)** sepals, **(2)** petals, **(3)** stamens, and **(4)** carpels. Each of these organs is a modified leaf (**FIGURE 24.8**), and all are arranged in a characteristic pattern of whorls within whorls (a whorl is a circular arrangement).

- **Sepals** are located around the outside of the flower and protect the flower as it develops.

- Inside the sepals is a whorl of **petals,** which enclose the male and female reproductive organs.

- The pollen-producing organs, or **stamens,** are located in a whorl inside the petals.

- In the center of the entire structure are egg-producing reproductive organs, or **carpels.** (Ovules most often are located at the base of carpels.)

How does the floral meristem produce these four organs in their characteristic arrangement?

The first hint of an answer came over 100 years ago, when researchers discovered several types of homeotic mutant flowers in popular garden plants. In the mutants, one kind of floral organ was replaced by another. For example, one homeotic mutant had flowers with sepals, petals, another ring of petals, and carpels instead of having sepals, petals, stamens, and carpels. These mutants are similar to *Drosophila* homeotic mutants (see Chapter 22), where individuals have legs or antennae growing in the wrong location in place of the appropriate structure.

Just as an analysis of homeotic mutants in fruit flies triggered a breakthrough in understanding the genetic control of the body plan in animals, an analysis of homeotic flower mutants in *Arabidopsis* triggered a breakthrough in understanding the genetic control of flower structure.

The Genetic Control of Flower Structures

Over 100 years after floral homeotic mutants were first described, Elliot Meyerowitz and colleagues assembled a large collection of *Arabidopsis* homeotic flower-structure mutants. The researchers' goal was to identify and characterize the genes responsible for specifying the four floral organs.

The group found three general classes of mutants that are shown in **FIGURE 24.9**. Some mutants had only carpels and stamens; others had only sepals and carpels; still others had only petals and sepals. The key observation was that each type of mutant lacked the elements found in *two* of the four whorls.

What was going on? The biologists hypothesized that each class of homeotic mutant was caused by a defect in a single gene. They reasoned that if three genes set up the pattern of a flower, the

| **Wild type** Normal arrangement of organs | **Mutant class 1** Only carpels and stamens | **Mutant class 2** Only sepals and carpels | **Mutant class 3** Only petals and sepals |

Whorl 4: Carpel
Whorl 3: Stamen
Whorl 2: Petal
Whorl 1: Sepal

Sepal Petal Stamen Carpel   Carpel Stamen Stamen Carpel   Sepal Sepal Carpel Carpel   Sepal Petal Petal Sepal

FIGURE 24.9 Homeotic Mutants of *Arabidopsis* Can Have Flower Organs in the Wrong Locations.
Orange labels in the mutants indicate a homeotic transformation (misplaced structure).

mutant phenotypes suggested a hypothesis for how the three gene products interact. Because they referred to the three hypothetical genes as *A*, *B*, and *C*, the hypothesis is called the ABC model.

The ABC Model Three basic ideas underlie the ABC model (**FIGURE 24.10a**):

- Each of the three genes is expressed in two adjacent whorls.

- Because each gene is expressed in two adjacent whorls, a total of four different combinations of gene products can occur.

- Each of these four combinations of gene products triggers the development of a different floral organ.

Specifically, the Meyerowitz group proposed that (**1**) the A protein alone causes cells to form sepals, (**2**) a combination of A and B proteins sets up the formation of petals, (**3**) B and C combined specify stamens, and (**4**) the C protein alone designates cells as the precursors of carpels.

Does this model explain how the three classes of homeotic mutants occur? The answer is yes, if two additional elements are added to the model:

- The A protein inhibits production of the C protein.
- The C protein inhibits production of the A protein.

With these ideas in mind, patterns of gene expression can be predicted that correspond to the mutant phenotypes. These are shown in **FIGURE 24.10B**.

For example, if the *A* gene is disabled by mutation, then it no longer inhibits the expression of the *C* gene, and all cells produce the C protein. As a result, cells in the outermost whorl express only C protein and develop into carpels, while cells in the whorl just to the inside produce B and C proteins and develop into stamens.

✔ If you understand the ABC model, you should be able to explain why in mutants that lack C gene expression, *A* genes become expressed across all whorls.

Testing the Model Although the ABC model is plausible, elegant, and most important, capable of explaining the data, it needed to be tested. To do so, Meyerowitz and co-workers mapped the genes responsible for the mutant phenotypes and cloned the genes. Once they had isolated the genes, they were able to obtain probes to perform in situ hybridizations (see Chapter 22). The goal was to learn the pattern of expression of the *A*, *B*, and *C* genes and see if that pattern corresponded to the model's predictions.

The mRNAs for the *A*, *B*, and *C* genes showed up in the sets of whorls predicted by the model. The *A* gene is expressed in the outer two whorls, the *B* gene is expressed in the middle two whorls, and the *C* gene is expressed in the inner two whorls. Later work showed that there weren't single *A*, *B*, and *C* genes, but groups of *A*, *B*, and *C* class genes. However, this finding didn't change the outline of the ABC model.

The result that *A*, *B*, and *C* class genes were expressed in the predicted regions supported the ABC model. Just as different

(a) The ABC model

Idea 1: The products of three genes pattern the flower; each gene is expressed in two adjacent whorls.

Idea 2: Four different combinations of proteins occur.

Idea 3: Each protein combination triggers development of a different floral organ.

(b) Predictions of the ABC model

FIGURE 24.10 The ABC Model for Genetic Control of Flower Development. The ABC model is a hypothesis to explain the existence of three types of homeotic mutants in *Arabidopsis* flowers. The regions of gene expression were determined by in situ hybridization.

combinations of *Hox* gene products specify the identity of fly segments, different combinations of floral identity genes specify the parts of a flower.

MADS-Box Genes When Meyerowitz and others analyzed the DNA sequences of the floral organ identity genes, they discovered that all three classes of genes contained a segment called the **MADS box.** The MADS box encodes the MADS domain—a sequence of about 60 amino acids that binds to DNA. Based on this observation, the researchers hypothesized that the floral genes encode transcription factors that bind to regulatory sequences to trigger the expression of genes required for sepal, petal, carpel, and stamen formation.

The researchers found many remarkable similarities between the *MADS-box* genes and the *Hox* genes present in *Drosophila* and other animals.

- *MADS-box* and *Hox* genes encode DNA-binding transcription factors.

- *MADS-box* and *Hox* gene products are both part of regulatory cascades that lead to the specification of structures. *Hox* genes regulate the expression of genes responsible for forming limbs or specific parts of the body; *MADS-box* genes regulate the expression of genes responsible for forming flowers.

- Both sets of genes tell cells where they are in the body and can produce homeotic mutants if they do not work properly.

Biologists are currently working to identify the genes targeted by the ABC proteins, just as they are working to identify the genes controlled by the *Hox* gene products in animals. Eventually, they hope to understand the complete regulatory cascade from initiation of the floral meristem to the expression of proteins that are specific to petals or other floral organs.

Ultimately, the goal is to explore how changes in this cascade could have led to the evolution of the spectacular diversity of flowers observed today. Were mutations in *MADS-box* genes, or the effector genes that they control, responsible for the flamboyant petals of the lady slipper orchid, the bright red sepals of fuchsia, and the elaborate stamens of a sugar maple tree?

Although different genes are involved in plant and animal development, the underlying genetic principles for producing a multicellular body are similar.

Like animals, plants have a genetic toolkit for development. The tools have turned out to be different from those of the animal kit. The differences, however, are not in the types of tools—hammer versus saw—but in their particular forms: nail hammer versus finishing hammer; circular saw versus band saw. For example, plants use MADS-domain transcription factors to specify structures, and animals use Hox transcription factors; plants use graded concentrations of auxin to specify position within the embryo, and (some) animals use graded concentrations of Bicoid. The details are different, but the broad mechanisms are the same.

Why should this be? To return to the opening theme of this chapter, the answer must lie in there being a limited number of ways to build a multicellular organism, coupled with the different evolutionary paths taken in plants and animals to achieve this end. Understanding evolution is the key to understanding biology—and evolution is what you'll explore in the next unit.

check your understanding

If you understand that . . .

- Reproductive development in plants begins when SAMs are converted to floral meristems.
- In *Arabidopsis*, development of the four floral organs depends on the expression of regulatory transcription factors encoded by the *A*, *B*, and *C* genes.

✔ **You should be able to . . .**

1. Describe fundamental differences in the generation of plant and animal reproductive cells and structures.

2. Compare and contrast the *MADS-box* genes of plants with the *Hox* genes of animals.

Answers are available in Appendix A.

CHAPTER 24 REVIEW

For media, go to MasteringBiology

If you understand . . .

24.1 Embryogenesis

- Embryogenesis is the early stage of development following fertilization. Embryogenesis occurs in the seed and usually includes a period of arrested development that often lasts from weeks to months.

- Body axes are established early in development.

- Three types of embryonic tissues (vascular, ground, epidermis) are formed along the radial axis during embryogenesis.

- Cell migration is not possible in plants. In its absence, tight control of the orientation of the cleavage plane during cell division and differential growth of cells along a particular body axis create the different plant structures.

- A shoot and root meristem arise early in embryogenesis. Meristems contain stem cells that produce all cells for the remainder of the plant's life.

- Master regulatory genes unleash gene-regulatory cascades. These regulatory cascades pattern different regions of the early embryo along the apical–basal and radial axes.

✓ You should be able to predict how the types of patterning mutants obtained by Gerd Jürgens and colleagues (see Figure 24.4) would be different if there were only one master regulatory gene that determined pattern along the apical–basal axis.

24.2 Vegetative Development

- Vegetative development is the development and growth of all plant structures except flowers.

- Vegetative development occurs throughout the plant's life.

- Vegetative development depends on meristems present at the tips of roots and shoots.

- The growth that occurs in vegetative development is finely tuned to the environment, allowing plants to adopt forms that maximize their survival and reproduction.

- Master regulators that control vegetative development are being discovered.

✓ You should be able to describe what an imaginary fly would look like if it underwent the continuous vegetative development found in plants.

24.3 Reproductive Development

- Reproductive development is initiated when a shoot meristem switches from producing cells that will form shoots and leaves to producing cells that will form flowers, the plant's reproductive structure.

- The switch between vegetative and reproductive development in shoot meristems occurs in response to environmental conditions.

- The ability to switch from vegetative to reproductive development means that plants do not have a devoted germ line—any shoot meristem cell has the potential to produce reproductive cells.

- Once a floral meristem is established, combinations of regulatory transcription factors encoded by *A*, *B*, and *C* classes of genes interact to specify the flower's sepals, petals, stamens, and carpels.

✓ You should be able to predict according to the ABC model the phenotype of a mutant that lacked both the *B* and *C* genes.

(MB) MasteringBiology

1. MasteringBiology Assignments

Tutorials and Activities The ABC Model of Flowering; Developmental Biology of Plants

Questions Reading Quizzes, Blue-Thread Questions, Test Bank

2. eText Read your book online, search, take notes, highlight text, and more.

3. The Study Area Practice Test, Cumulative Test, BioFlix® 3-D Animations, Videos, Activities, Audio Glossary, Word Study Tools, Art

You should be able to . . .

✓ TEST YOUR KNOWLEDGE

Answers are available in Appendix A

1. From the standpoint of evolution, why are differences in the development of plants and animals expected?

2. Which of the following is *not* a contrast between plant and animal development?
   a. Under certain conditions, plant cells can "de-differentiate" readily.
   b. The fate of a cell is determined in part by its location in the embryo.
   c. Germ cells are not set aside early in plant development.
   d. Plant cells do not move.

3. The MADS box is
   a. part of a *Hox* gene.
   b. a DNA sequence that encodes the DNA-binding portion of a type of transcription factor.
   c. a DNA regulatory sequence for genes involved in flower patterning.
   d. a micro-RNA that controls the timing of flowering.

4. Which of the following does *not* occur during embryogenesis?
   a. formation of the radial axis
   b. production of the suspensor
   c. formation of the cotyledons and hypocotyl
   d. formation of the leaf lateral and proximal–distal axes

5. When does the apical–basal axis first become apparent?
   a. when the epidermal, ground, and vascular tissues form
   b. when the cotyledons, hypocotyl, and root form
   c. when the first cell division produces the apical cell and basal cell
   d. during the globular stage, when the suspensor is complete

6. What evidence suggests that changes in the way the *PHAN* gene is expressed could be partly responsible for evolutionary changes in leaf shape?
   a. If *PHAN* gene expression is manipulated experimentally, individuals produce leaf types found in different species.
   b. Experiments have shown that *PHAN* plays a role in establishing the upper–lower surface axis in leaves.
   c. Sequencing studies and other data have shown that *PHAN* encodes a regulatory transcription factor.
   d. All plant species surveyed to date have a gene homologous to *PHAN*.

7. If auxin acts as a morphogen for determining position along the apical–basal axis, then auxin must be
   a. a transcription factor.
   b. a hormone.
   c. present in a concentration gradient.
   d. synthesized faster than it is degraded.

8. In plants, reproductive tissues may develop late in life, from cells that have undergone mitosis hundreds or thousands of times. How does this differ from animals, and what are the consequences?

9. What qualities of particular meristem cells make them stem cells?

10. In what sense are the epidermal, ground, and vascular tissues produced in the SAMs and RAMs of a 300-year-old oak tree "embryonic"?

11. When in situ hybridization experiments documented where *A*, *B*, and *C* genes were expressed in developing *Arabidopsis* flowers, it was considered strong support for the ABC model. Explain why.

12. Give an example of how each of the fundamental developmental processes—cell proliferation, expansion, interaction, and differentiation—plays a role in plant development.

13. **QUANTITATIVE** Imagine a species that made use of two genes, *D* and *E*, to pattern its flowers. Using only these two genes, how many different structures could be specified? In this hypothetical situation—which is not unlike the real situation of the *ABC* genes in flower patterning—consider that the only thing that's important is whether a gene is on or off (the amount of gene product isn't a factor) and that a default structure can be specified when neither gene is active.

14. Indeterminate growth is the type that occurs when growth is unlimited in duration. In contrast, determinate growth is of limited duration and then stops. Which form of growth is observed in vegetative development and which in reproductive development? Explain your logic.

15. A mutation that causes the *B* gene to be expressed across the floral meristem (without altering *B* or *C* gene expression) would be predicted to lead to

   a. petals in the two outer whorls and stamens in the two inner whorls.
   b. a pattern of sepals, petals, stamens, and carpels moving from the outer whorl to the center.
   c. sepals in the two outer whorls and carpals in the two inner whorls.
   d. a pattern of petals, sepals, carpels and stamens moving from the outer whorl to the center.

16. **QUANTITATIVE** Assume that a gene shared by animals and plants mutates at a rate of 1×10^{-6} per cell division. If a germ-line cell in a female mouse undergoes 50 cell divisions before forming an egg, and a cell in an old oak tree undergoes 5000 cell divisions before forming an egg, how much more likely is it to find a mutation in this gene in the oak tree egg relative to the mouse egg?

APPENDIX A Answers

CHAPTER 1

IN-TEXT QUESTIONS AND EXERCISES

p. 4 Fig. 1.2 analyze If Pasteur had done any of the things listed, he would have had more than one variable in his experiment. This would allow critics to claim that he got different results because of the differences in broth types, heating, or flask types—not the difference in exposure to preexisting cells. The results would not be definitive.
p. 6 apply The average kernel protein content would decline, from 11 percent to a much lower value over time.
p. 6 CYU apply The data points would all be about 11 percent, indicating no change in average kernel protein content over time.
p. 7 Fig. 1.4 apply Molds and other fungi are more closely related to green algae because they differ from plants at two positions (5 and 8 from left) but differ from green algae at only one position (8).
p. 8 Fig. 1.6 apply The eukaryotic cell is roughly 10 times the size of the prokaryotic cell.
p. 9 CYU evaluate From the sequence data provided, species A and B differ only in one ribonucleotide of the rRNA sequence (position 10 from left). Species C differs from species A and B in four ribonucleotides (positions 1, 2, 9, and 10). A correctly drawn phylogenetic tree would indicate that species A and B appear to be closely related, and species C is more distantly related. See **FIGURE A1.1**.
p. 13 (1) analyze You could conclude that the ants weren't navigating normally, because they had been caught and released and transferred to a new channel. **(2)** analyze You could conclude that the ants can't navigate normally on their manipulated legs.
p. 13 CYU The key here is to test predation rates during the hottest part of the day (when desert ants actually feed) versus other parts of the day. The experiment would best be done in the field, where natural predators are present. One approach would be to capture a large number of ants, divide the group in two, and measure predation rates (number of ants killed per hour) when they are placed in normal habitat during the hottest part of the day versus an hour before (or after). **(1)** analyze The control group here is the normal condition—ants out during the hottest part of the day. If you didn't include a control, a critic could argue that predation did or did not occur because of your experimental setup or manipulation, not because of differences in temperature. **(2)** analyze You would need to make sure that there is no difference in body size, walking speed, how they were captured and maintained, or other traits that might make the ants in the two groups more or less susceptible to predators. They should also be put out in the same habitat, so the presence of predators is the same in the two treatments.
p. 13 Fig. 1.10 analyze The interpretation of the experiment would not likely change, but your confidence in the conclusions drawn would be reduced if you used just one ant.

IF YOU UNDERSTAND . . .

1.1. understand Dead cells cannot regulate the passage of materials between exterior and interior spaces, replicate, use energy, or process information. **1.2** understand Observations on thousands of diverse species supported the claim that all organisms consist of cells. The hypothesis that all cells come from preexisting cells was supported when Pasteur showed that new cells do not arise and grow in a boiled liquid unless they are introduced from the air. **1.3** understand If seeds with higher protein content leave the most

offspring, then individuals with low protein in their seeds will become rare over time. **1.4** understand A newly discovered species can be classified as a member of the Bacteria if the sequence of its rRNA contains some features found only in Bacteria. The same logic applies to classifying a new species in the Archaea or Eukarya. **1.5** understand (1) A hypothesis is an explanation of how the world works; a prediction is an outcome you should observe if the hypothesis is correct. (2) Experiments are convincing because they measure predictions from two opposing hypotheses. Both predicted actions cannot occur, so one hypothesis will be supported while the other will not.

YOU SHOULD BE ABLE TO . . .

✔ Test Your Knowledge

1. remember d **2.** understand d **3.** remember populations **4.** understand b **5.** understand An individual's ability to survive and reproduce **6.** understand c

✔ Test Your Understanding

7. evaluate That the entity they discovered replicates, processes information, acquires and uses energy, is cellular, and that its populations evolve. **8.** understand a **9.** understand Over time, traits that increased the fitness of individuals in this habitat became increasingly frequent in the population. **10.** understand Individuals with certain traits are selected, in the sense that they produce the most offspring. **11.** analyze Yes. If evolution is defined as "change in the characteristics of a population over time," then those organisms that are most closely related should have experienced less change over time. On a phylogenetic tree, species with substantially similar rRNA sequences would be diagrammed with a closer common ancestor—one that had the sequences they inherited—than the ancestors shared between species with dissimilar rRNA sequences. **12.** understand A null hypothesis specifies what a researcher should observe when the hypothesis being tested isn't correct.

✔ Test Your Problem-Solving Skills

13. analyze A scientific theory is not a guess—it is an idea whose validity can be tested with data. Both the cell theory and the theory of evolution have been validated by large bodies of observational and experimental data. **14.** apply If all eukaryotes living today have a nucleus, then it is logical to conclude that the nucleus arose in a common ancestor of all eukaryotes, indicated by the arrow you should have added to the figure. See **FIGURE A1.2**. If it had arisen in a common ancestor of Bacteria or Archaea, then species in those groups would have had to lose the trait—an unlikely event. **15.** evaluate The data set was so large and diverse that it was no longer reasonable to argue that noncellular lifeforms would be discovered. **16.** apply b

BIG PICTURE Doing Biology

p. 16 CYU (1) understand Biologists design and carry out a study, either observational or experimental, to test their ideas. As part of this process, they state their ideas as a hypothesis and null hypothesis and make predictions. They analyze and interpret the data they have gathered, and determine whether the data support their ideas. If not, they revisit their ideas and come up with an alternative hypothesis and design another study to test these new predictions. **(2)** understand There are many

possible examples. Consider, for example, the experiment on navigation in foraging desert ants (Chapter 1). In addition to testing how the ants use information on stride length and number to calculate how far they are from the nest (multicellular organism and population levels), researchers also could test how the "pedometer" works at the level of cells and molecules. **(3)** analyze A hypothesis is a testable statement to explain a specific phenomenon or a set of observations. The word theory refers to proposed explanations for very broad patterns in nature that are supported by a wide body of evidence. A theory serves as a framework for the development of new hypotheses. **(4)** analyze The next step is to relate your findings to existing theories and the current scientific literature, and then to communicate your findings to colleagues through informal conversations, presentations at scientific meetings, and eventually publication in peer-reviewed journals.

CHAPTER 2

IN-TEXT QUESTIONS AND EXERCISES

p. 20 Fig. 2.3 apply There are 15 electrons in phosphorus, so there must be 15 protons, which is the atomic number. Since the mass number is 31, then the number of neutrons is 16.
p. 23 understand *Water:* arrows pointing from hydrogens to oxygen atom; *ammonia:* arrows pointing from hydrogens to nitrogen atom; *methane:* double arrows between carbons and hydrogens; *carbon dioxide:* arrows pointing from carbon to oxygens; *molecular nitrogen:* double arrows between nitrogens.
p. 23 Fig. 2.7 understand Oxygen and nitrogen have high electronegativities. They hold shared electrons more tightly than C, H, and many other atoms, resulting in polar bonds.
p. 24 CYU evaluate See **FIGURE A2.1**.

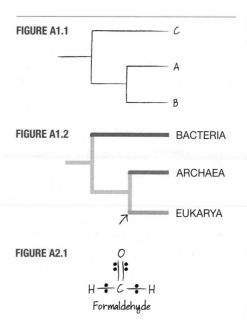

FIGURE A1.1

FIGURE A1.2

FIGURE A2.1

Formaldehyde

p. 25 (1) [evaluate] $\delta^+H-O^{\delta^-}-H^{\delta^+}$ (2) [apply] If water were linear, the partial negative charge on oxygen would have partial positive charges on either side. Compared to the actual, bent molecule, the partial negative charge would be much less exposed and less able to participate in hydrogen bonding.

p. 26 Fig. 2.14 [understand] Oils are nonpolar. They have long chains of carbon atoms bonded to hydrogen atoms, which share electrons evenly because their electronegativities are similar. When an oil and water are mixed, the polar water molecules interact with each other via hydrogen bonding rather than with the nonpolar oil molecules, which interact with themselves instead.

p. 28 Table 2.2 [understand] "Cause" (Row 1): electrostatic attractions between partial charges on water and opposite charges on ions; hydrogen bonds; water and other polar molecules. "Biological Consequences" (Row 2): ice to float; freezing solid. "Cause" (Row 4): lots of heat energy; break hydrogen bonds and change water to a gas.

p. 29 [apply] The proton concentration would be 3.2×10^{-9} M.

p. 29 Fig. 2.17 [apply] The concentration of protons would decrease because milk is more basic (pH 6.5) than black coffee (pH 5).

p. 30 [apply] The bicarbonate concentration would increase. The protons (H^+) released from carbonic acid would react with the hydroxide ions (OH^-) dissociated from NaOH to form H_2O, leaving fewer protons free to react with bicarbonate to reform carbonic acid.

p. 32 CYU (1) [apply] The reaction would be spontaneous based on the change in potential energy; the reactants have higher chemical energy than the products. The entropy, however, is not increased based on the number of molecules, although heat given off from the reaction still results in increased entropy. (2) [understand] The electrons are shifted farther from the nuclei of the carbon and hydrogen atoms and closer to the nuclei of the more electronegative oxygen atoms.

p. 32 Fig. 2.19 [remember] See FIGURE A2.2.

p. 33 Fig. 2.21 [analyze] The water-filled flask is the ocean; the gas-filled flask is the atmosphere; the condensed water droplets are rain; the electrical sparks are lightning.

p. 37 Table 2.3 [understand] All the functional groups in Table 2.3, except the sulfhydryl group (−SH), are highly polar. The sulfhydryl group is only very slightly polar.

IF YOU UNDERSTAND . . .

2.1. [understand] The bonds in methane and ammonia are all covalent, but differ in polarity: Methane has nonpolar covalent bonds while ammonia has polar covalent bonds. Sodium chloride does not have covalent bonds; instead, ionic bonds hold the ionized sodium and chloride together. 2.2 [understand] Assuming neutral pH, amino and hydroxyl groups would interact with the partial negative charge on water's oxygen, because they both carry a positive charge (partial for the hydrogen in the hydroxyl). The carboxyl group would interact with the partial positive charges on water's hydrogens, since it would carry a negative charge after losing the proton from its hydroxyl. 2.3 [understand] Like solar radiation, the energy in electricity generates free radicals that would promote the reaction. 2.4 [understand] "Top-down" approach: The reaction responsible for synthesizing acetic acid is observed in cells and can serve as an intermediate for the formation of a more complex molecule (acetyl CoA) that is used by cells throughout the tree of life. "Bottom-up" approach: This reaction can also occur under conditions that mimic the early Earth environment in deep-sea vents. 2.5 [apply] The hydroxyls would increase the solubility of octane by introducing polar covalent bonds, which would make the molecule more hydrophilic. The high electronegativity of oxygen would decrease the potential energy of the modified molecule.

YOU SHOULD BE ABLE TO . . .

✔ **Test Your Knowledge**

1. [understand] b 2. [remember] a 3. [remember] c 4. [remember] d 5. [remember] Potential energy and entropy. 6. [remember] The prebiotic soup model and the surface metabolism model.

✔ **Test Your Understanding**

7. [apply] c. Acetic acid has more highly electronegative oxygen atoms than the other molecules. When bonded to carbon or hydrogen, each oxygen will result in a polar covalent bond. 8. [apply] Relative electronegativities would be F > O > H > Na. One bond would form with sodium, and it would be ionic. 9. [understand] When oxygen is covalently bonded to hydrogen, the difference in electronegativities between the atoms causes the electrons to spend more time near the oxygen. In contrast, the atoms in H_2 and O_2 have the same electronegativities, so they equally share electrons in their covalent bonds. 10. [apply] See FIGURE A2.3. 11. [apply] The dissociation reaction of carbonic acid lowers the pH of the solution by releasing extra H^+ into the solution. If additional CO_2 is added, the sequence of reactions would be driven to the right, which would make the ocean more acidic. 12. [understand] The carbon framework determines the overall shape of an organic molecule. The functional groups attached to the carbons determine the molecule's chemical behavior, because these groups are likely to interact with other molecules.

✔ **Test Your Problem-Solving Skills**

13. [analyze] b. 14. [analyze] No, they don't conflict. Shells that are farther from the protons (positive charges) in the nucleus house electrons that have greater potential energy than shells closer to the nucleus. 15. [analyze] One possible concept map relating the structure of water to its properties is shown below (see FIGURE A2.4). 16. [create] In hot weather, water absorbs large amounts of heat due to its high specific heat and high heat of vaporization. In cold weather, water releases the large amount of heat that it has absorbed.

FIGURE A2.2

FIGURE A2.3

FIGURE A2.4

CHAPTER 3

IN-TEXT QUESTIONS AND EXERCISES

p. 43 Fig. 3.2 [understand] The green R-groups contain mostly C and H, which have roughly equal electronegativities. Electrons are evenly shared in C–H bonds and C–S bonds, so the groups are nonpolar. Cysteine has a sulfur that is slightly more electronegative than hydrogen, so it will be less nonpolar than the other green groups. All of the pink R-groups have a highly electronegative oxygen atom with a partial negative charge, making them polar. **p. 44** [apply] From most hydrophilic to most hydrophobic: (1) aspartate, (2) asparagine, (3) tyrosine, (4) valine. The most hydrophilic amino acids will have side chains with full charges (ionized), like aspartate, followed by those with the largest number of highly electronegative atoms, like oxygen or nitrogen. Highly electronegative atoms produce polar covalent bonds with carbon or hydrogen. The most hydrophobic will not have oxygen or nitrogen in their side chains, but instead will have the largest number of C–H bonds, which are nonpolar covalent. **p. 47 CYU** [understand] See **FIGURE A3.1**. **p. 52 CYU (1)** [understand] Secondary, tertiary, and quaternary structure all depend on bonds and other interactions between amino acids that are linked in a chain in a specific order (primary structure). **(2)** [apply] There would be 20^5 different peptides, or 3.2×10^6 different primary sequences. **p. 55 CYU** [apply] Amino acid changes would be expected to be in the active site or in regions that affect the folded structure of this site. Either of these changes could result in a different active site that either binds a new substrate or catalyzes a different reaction.

IF YOU UNDERSTAND . . .

3.1. [understand] Look at the R-group of the amino acid. If there is a positive charge, then it is basic. If there is a negative charge, then it is acidic. If there is not a charge, but there is an oxygen atom, then it is polar uncharged. If there is no charge or oxygen, then it is nonpolar. **3.2** [apply] Nonpolar amino acid residues would be found in the interior of a globular protein, grouped with other nonpolar residues due to hydrophobic interactions. **3.3** [analyze] Both calmodulin and infectious prions require some form of induction to achieve their active conformations. Calcium ions are required for calmodulin to fold into its functional structure while prions are induced to change their shape by other, improperly folded prion proteins. **3.4** [analyze] *Catalysis:* Proteins are made of amino acids, which have many reactive functional groups, and can fold into different shapes that allow the formation of active sites. *Defense:* Similar to catalysis, the chemical properties and capacity for different shapes allows proteins to be made that can attach to virtually any type of invading virus or cell. *Signaling:* The flexibility in protein structure allows protein activities to be quickly turned on or off based on binding to signal molecules or ions.

YOU SHOULD BE ABLE TO . . .

✔ Test Your Knowledge

1. [remember] d **2.** [remember] The atoms and functional groups found in the side chains. **3.** [remember] b **4.** [remember] b **5.** [understand] The order and type of amino acids (i.e., the primary structure) contains the information that directs folding. **6.** [understand] a

✔ Test Your Understanding

7. [understand] Because the nonpolar amino acid residues are not able to interact with the water solvent, they are crowded together in the interior of a protein and surrounded by a network of hydrogen-bonded water molecules. This crowding leads to the development of van der Waals interactions that help glue the nonpolar side chains together. **8.** [understand] No, polymerization is a nonspontaneous reaction because the product molecules have lower entropy than the free form of the reactants and there would be nothing to prevent hydrolysis from reversing the reaction. **9.** [understand] Many possible correct answers, including (1) the presence of an active site in an enzyme that is precisely shaped to fit a substrate or substrates in the correct orientation for a reaction to occur; (2) the doughnut shape of porin that allows certain substances to pass through it; (3) the cable shape of collagen to provide structural support for cells and tissues. **10.** [create] Proteins are highly variable in overall shape and chemical properties due to variation in the composition of R-groups and the array of secondary through quaternary structures that are possible. This variation allows them to fulfill many different roles in the cell. Diversity in the shape and reactivity of active sites also makes them effective catalysts. **11.** [understand] c **12.** [create] In many proteins, especially those involved in cell signaling, their structure is affected by binding to other molecules or ions. Since the shape of the protein is directly involved in its function, the protein's activity is regulated by controlling how it is folded. If proteins were inflexible, this type of control could not occur.

✔ Test Your Problem-Solving Skills

13. [analyze] The side chain of proline is a cyclic structure that is covalently bonded to the nitrogen in the core amino group. This restricts the movement of the side chain relative to the core nitrogen, which further restricts the backbone when the nitrogen participates in a peptide bond with a neighboring amino acid. **14.** [create] See **FIGURE A3.2 15.** [apply] b. Phosphates have a negative charge, so they are most likely to form ionic bonds with the positively-charged side chains of basic amino acid residues. **16.** [apply] The inherited forms likely have some alteration in the primary structure such that the infectious form is spontaneously generated at a higher rate than normal. The amino acid sequence in these prions would likely differ from those transmitted between animals.

CHAPTER 4

IN-TEXT QUESTIONS AND EXERCISES

p. 59 [understand] See **FIGURE A4.1**.

FIGURE A3.1

FIGURE A3.2

FIGURE A4.1

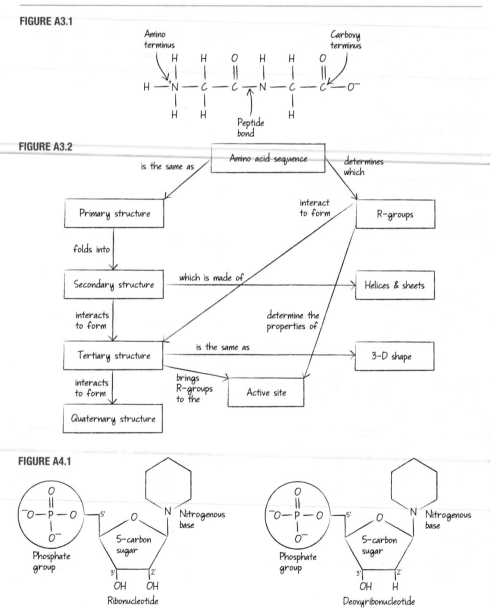

should get smaller and lighter in color since each succeeding generation has relatively less heavy DNA.

p. 290 apply See **FIGURE A15.1**. The new strands grow in opposite directions, each in the 5′ → 3′ direction.

p. 292 apply Helicase, topoisomerase, single-strand DNA-binding proteins, primase, and DNA polymerase are all required for leading-strand synthesis. If any one of these proteins is nonfunctional, DNA replication will not occur.

p. 293 apply See **FIGURE A15.2**. If DNA ligase were defective, then the leading strand would be continuous, and the lagging strand would have gaps in it where the Okazaki fragments had not been joined.

p. 294 CYU (1) understand DNA polymerase adds nucleotides only to the free 3′ −OH on a strand. Primase synthesizes a short RNA sequence that provides the free 3′ end necessary for DNA polymerase to start working. **(2)** understand The need to begin DNA synthesis many times on the lagging strand requires many new primers. Since DNA polymerase I is needed to remove primers, it is required predominantly on the primer-rich lagging strand.

p. 296 Fig. 15.13 apply As long as the RNA template could bind to the "overhanging" section of single-stranded DNA, any sequence could produce a longer strand. For example, 5′-CCCAUUCCC-3′ would work just as well.

p. 297 CYU (1) understand This is because telomerase is needed only to replicate one end of a linear DNA and bacterial DNAs lack ends because they are circular. **(2)** understand Since telomerase works by extending one strand of DNA without any external template and because DNA synthesis requires a template, telomerase must contain an internal template to allow it to extend a DNA chain.

p. 299 Fig. 15.16 analyze They are lower in energy and not absorbed effectively by the DNA bases.

p. 300 CYU (1) apply The mutation rate would be predicted to rise because differences in base-pair stability and shape make it possible for DNA polymerase to distinguish correct from incorrect base pairs during DNA replication. **(2)** apply The mutation rate should increase because without a way to distinguish which strand to use as a template for repair, about half of mismatches on average would be repaired using the incorrect strand as a template. **(3)** remember The enzyme that removes the dimer and surrounding DNA is specific to nucleotide excision repair. DNA polymerase and DNA ligase work in both nucleotide excision repair and normal DNA synthesis.

p. 300 Fig. 15.18 apply Exposure to UV radiation can cause formation of thymine dimers. If thymine dimers are not repaired, they represent mutations. If such mutations occur in genes controlling the cell cycle, cells can grow abnormally, resulting in cancers.

IF YOU UNDERSTAND . . .

15.1 apply These results would not allow distinguishing whether DNA or protein was the genetic material.

15.2 understand The bases added during DNA replication are shown in red type.

> Original DNA: **CAATTACGGA**
> **GTTAATGCCT**
> Replicated DNA: **CAATTACGGA**
> GTTAATGCCT
> CAATTACGGA
> **GTTAATGCCT**

15.3 understand See **FIGURE A15.3**. **15.4** understand Because cancer cells divide nearly without limit, it's important for these cells to have active telomerase so that chromosomes don't shorten to the point where cell division becomes impossible. **15.5** understand If errors in DNA aren't corrected, they represent mutations. When DNA repair systems fail, the mutation rate increases. As the mutation rate increases, the chance that one or more cell-cycle genes will be mutated increases. Mutations in these genes often result in uncontrolled cell division, ultimately leading to cancer.

YOU SHOULD BE ABLE TO . . .

✔ Test Your Knowledge

1. remember d **2.** understand a **3.** remember topoisomerase
4. remember b **5.** remember c **6.** remember telomerase

✔ Test Your Understanding

7. remember Labeling DNA or labeling proteins. **8.** understand DNA is constantly damaged, and many pathways have evolved to repair this onslaught of damage. If a DNA repair pathway is inactivated by mutation, damage is inefficiently repaired. Consequently mutation rates increase, and the increased number of mutations increases the probability that cancer-causing mutations will occur. **9.** understand On the lagging strand, DNA polymerase moves away from the replication fork. When helicase unwinds a new section of DNA, primase must build a new primer on the template for the lagging strand (closer to the fork) and another polymerase molecule must begin synthesis at this point. This makes the lagging-strand synthesis discontinuous. On the leading strand, DNA polymerase moves in the same direction as helicase, so synthesis can continue, without interruption, from a single primer (at the origin of replication). **10.** understand Telomerase binds to the overhang at the end of a chromosome. Once bound, it begins catalyzing the addition of deoxyribonucleotides to the overhang in the

5′ → 3′ direction, lengthening the overhang. This allows primase, DNA polymerase, and ligase to catalyze the addition of deoxyribonucleotides to the lagging strand in the 5′ → 3′ direction, restoring the lagging strand to its original length. **11.** apply a (Because the ability to distinguish which strand contains the incorrect base would be lost.) **12.** analyze d (The regularity of DNA's structure allows enzymes to recognize any type of damage that distorts this regular structure.)

✔ Test Your Problem-Solving Skills

13. analyze c (If DNA polymerase could synthesize DNA 3′ → 5′ as well as the normal 5′ → 3′, then both newly synthesized DNA strands could be extended to follow the replication fork.) **14.** apply (a) In **FIGURE A15.4**, the gray lines represent DNA strands containing radioactivity. (b) After one round of replication in radioactive solution, one double-stranded DNA would be radioactive in both strands and the other would not be radioactive in either strand. After another round of DNA synthesis, this time in nonradioactive solution, one of the four DNA molecules would be radioactive in both strands and the other three DNA molecules would contain no radioactivity in any strand. **15.** analyze (a) The double mutant of both *uvrA* and *recA* is most sensitive to UV light; the single mutants are in between; and the wild type is

FIGURE A15.1

Direction of new strand synthesis

Direction of new strand synthesis

FIGURE A15.2

Okazaki fragments

FIGURE A15.3

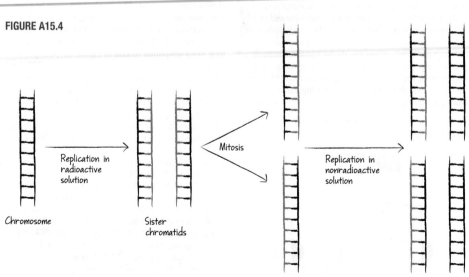

Lagging   Leading

Leading   Lagging

FIGURE A15.4

Chromosome

Replication in radioactive solution

Sister chromatids

Mitosis

Replication in nonradioactive solution

least sensitive. (b) The *recA* gene contributes more to UV repair through most of the UV dose levels. But at very high UV doses, the *uvrA* gene is somewhat more important than the *recA* gene. **16.** `apply` About 4600 seconds or 77 minutes. This answer comes from knowing that replication proceeds bidirectionally, so replication from each fork is predicted to replicate half the chromosome. This is 4.6 million base pairs/2 = 2.3 million base pairs. At 500 base pairs per second, this requires 2.3 million base pairs/500 base pairs per second = 4600 seconds. To obtain the time in minutes, divide 4600 seconds by 60 seconds per minute.

CHAPTER 16

IN-TEXT QUESTIONS AND EXERCISES

p. 305 Fig. 16.1 `apply` No, it could not make citrulline from ornithine without enzyme 2. Yes, it would no longer need enzyme 2 to make citrulline.

p. 306 Fig. 16.2 `create` Many possibilities: strain of fungi used, exact method for creating mutants and harvesting spores to grow, exact growing conditions (temperature, light, recipe for growth medium—including concentrations of supplements), objective criteria for determining growth or no growth.

p. 309 CYU `remember` Change in DNA sequence, change in sequence of transcribed mRNA, potential change in amino acid sequence of protein, likely altered protein function (if amino acid sequence was altered), likely change in phenotype.

p. 312 `analyze` (1) The codons in Figure 16.4 are translated correctly. (2) See **FIGURE A16.1**. (3) There are many possibilities (just pick alternative codons for one or more of the amino acids); one is an mRNA sequence (running 5′ → 3′): 5′ GCG-AAC-GAU-UUC-CAG 3′. To get the corresponding DNA sequence, write this sequence but substitute Ts for Us: 5′ GCG-AAC-GAT-TTC-CAG 3′. Now write the complementary bases, which will be in the 3′ → 5′ direction: 3′ CGC-TTG-CTA-AAG-GTC 5′. When this second strand is transcribed by RNA polymerase, it will produce the mRNA given with the proper 5′ → 3′ orientation.

p. 312 CYU (1) `apply` Note the 3′ → 5′ polarity of the DNA sequences in the accompanying table, and in the subsequent answer. This means that the complementary mRNA codon will read 5′ → 3′. U (rather than T) is the base transcribed from A.

DNA	mRNA Codon	Amino Acid
ATA	UAU	Tyrosine
ATG	UAC	Tyrosine
ATT	UAA	Stop
GCA	UGC	Cysteine

(2) `understand` The ATA → ATG mutation would have no effect on the protein. The ATA → ATT mutation introduces a stop codon, so the resulting polypeptide would be shortened. This would result in synthesis of a mutant protein much shorter than the original protein. The ATA → GCA mutation might have a profound effect on the protein's conformation because cysteine's structure is different from tyrosine's.

p. 312 Fig. 16.7 `apply` See **FIGURE A16.1**.

p. 314 Fig. 16.9 `analyze` Chromosomes 2, 3, 6, 10, 13, 14, 15, 18, 19, 21, 22, and the X chromosome show aneuploidy. Virtually every chromosome has structural rearrangements, and translocations are the most obvious. These are seen when two or more different colors occur on the same chromosome.

IF YOU UNDERSTAND . . .

16.1 `understand` Ornithine, citrulline or arginine could be added to allow growth. As Figure 16.1 shows, these compounds are made after the steps catalyzed by the enzymes 1, 2, and 3 that are needed to produce arginine.

16.2 `understand` An inhibitor of RNA synthesis will eventually prevent the synthesis of new proteins because newly synthesized mRNA is needed for translation. **16.3** `apply` AUG UGG AAA/AAG CAA/CAG **16.4** `understand` Since redundancy is having more than one codon specify a particular amino acid, redundancy makes it possible for there to be a point mutation without altering the amino acid. This is a silent mutation. A silent mutation is likely to be neutral because there is no change in amino acid sequence.

YOU SHOULD BE ABLE TO . . .

✔ Test Your Knowledge

1. `remember` d **2.** `understand` a **3.** `understand` d **4.** `remember` Because there is no chemical complementarity between nucleotides and amino acids; and because in eukaryotes, DNA is in the nucleus but translation occurs in the cytoplasm. **5.** `understand` d **6.** `remember` A codon that signals the end of translation.

✔ Test Your Understanding

7. `understand` Because the Morse code and genetic code both use simple elements (dots and dashes; 4 different bases) in different orders to encode complex information (words; amino acid sequences).

8.

Substrate 1 $\xrightarrow{A}$ Substrate 2 $\xrightarrow{B}$ Substrate 3 $\xrightarrow{C}$
Substrate 4 $\xrightarrow{D}$ Substrate 5 $\xrightarrow{E}$ Biological Sciazine

`apply` Substrate 3 would accumulate. Hypothesis: The individuals have a mutation in the gene for enzyme D.

9. `understand` They supported an important prediction of the hypothesis: Losing a gene (via mutation) resulted in loss of an enzyme. **10.** `understand` c **11.** `understand` In a triplet code, addition or deletion of 1–2 bases disrupts the reading frame "downstream" of the mutation site(s), resulting in a dysfunctional protein. But addition or deletion of 3 bases restores the reading frame—the normal sequence is disrupted only between the first and third mutation. The resulting protein is altered but may still be able to function normally. Only a triplet code would show these patterns. **12.** `understand` A point mutation changes the nucleotide sequence of an existing allele, creating a new one, so it always changes the genotype. But because the genetic code is redundant, and because point mutations can occur in DNA sequences that do not code for amino acids, these point mutations do not change the protein product and therefore do not change the phenotype.

✔ Test Your Problem-Solving Skills

13. `apply` See **FIGURE A16.2**. **14.** `analyze` Every copying error would result in a mutation that would change the amino acid sequence of the protein and would likely affect its function. **15.** `analyze` Before the central dogma was understood, DNA was known to be the hereditary material, but no one knew how particular sequences of bases resulted in the production of RNA and protein products. The central dogma clarified how genotypes produce phenotypes. **16.** `analyze` c

CHAPTER 17

IN-TEXT QUESTIONS AND EXERCISES

p. 318 Fig. 17.1 `remember` RNA is synthesized in the 5′ → 3′ direction; the DNA template is "read" 3′ → 5′.

p. 321 CYU (1) `apply` Transcription would be reduced or absent because the missing nucleotides are in the −10 region, one of the two critical parts of the promoter. **(2)** `understand` NTPs are required because the three phosphate groups raise the monomer's potential energy enough to make the polymerization reaction exergonic.

p. 322 Fig. 17.5 `apply` There would be no loops—the molecules would match up exactly.

p. 323 CYU (1) `understand` The subunits contain both RNA (the *ribonucleo*– in the name) and proteins. **(2)** `understand` The cap and tail protect mRNAs from degradation and facilitate translation.

p. 326 Fig. 17.11 `analyze` If the amino acids stayed attached to the tRNAs, the gray line in the graph would stay high and the green line low. If the amino acids were transferred to some other cell component, the gray line would decline but the green line would be low.

p. 327 `understand` (1) The amino acid attaches on the top right of the L-shaped structure. (2) The anticodon is antiparallel in orientation to the mRNA codon, and it contains the complementary bases.

p. 332 CYU `understand` E is for exit—the site where uncharged tRNAs are ejected; P is for peptidyl (or peptide bond)—the site where peptide bond formation takes place; A is for aminoacyl—the site where aminoacyl tRNAs enter.

IF YOU UNDERSTAND . . .

17.1 `apply` Transcription would continue past the normal point because the insertion of nucleotides would disrupt the structure of the RNA hairpin that functions as a terminator. **17.2** `apply` The protein-coding segment of the gene is predicted to be longer in eukaryotes because of the presence of introns. **17.3** `understand` The tRNAs act as adaptors because they couple the information contained in the nucleotides of mRNA to that contained in the amino acid sequence of proteins. **17.4** `apply` An incorrect amino acid would appear often in proteins. This is because the altered synthetase would sometimes add the correct amino acid for a particular tRNA and at other

---

FIGURE A16.1

mRNA sequence:
5′ AUG-CUG-GAG-GGG-GUU-AGA-CAU 3′

Amino acid sequence:
Met–Leu–Glu–Gly–Val–Arg–His

FIGURE A16.2

Bottom DNA strand:
3′ AACTT-TAC(start)-GGG-CAA-ACC-TCT-AGC-CCA-ATG-TCG-ATC(stop)-AGTTTC 5′

mRNA sequence:
5′ AUG-CCC-GUU-UGG-AGA-UCG-GGU-UAC-AGC-UAG 3′

Amino acid sequence:
Met–Pro–Val–Trp–Arg–Ser–Gly–Tyr–Ser

times add the incorrect amino acid. **17.5** create One possible concept map is shown in **FIGURE A17.1**.

YOU SHOULD BE ABLE TO . . .

✔ Test Your Knowledge

1. understand c **2.** remember c **3.** understand To speed the correct folding of newly synthesized proteins **4.** remember d **5.** remember At the 3′ end **6.** apply b

✔ Test Your Understanding

7. understand Basal transcription factors bind to promoter sequences in eukaryotic DNA and facilitate the binding of RNA polymerase. As part of the RNA polymerase holoenzyme, sigma binds to a promoter sequence in bacterial DNA and to allow RNA polymerase to initiate at the start of genes. **8.** analyze The wobble rules allow a single tRNA to pair with more than one type of mRNA codon. This is distinct from redundancy, in which more than one codon can specify a single amino acid. If the wobble rules did not exist, there would need to be one tRNA for each amino-acid-specifying codon in the redundant genetic code. **9.** apply b **10.** apply After a peptide bond forms between the polypeptide and the amino acid held by the tRNA in the A site, the ribosome moves down the mRNA. As it does, an uncharged tRNA leaves the E site. The now-uncharged tRNA that was in the P site enters the E site; the tRNA holding the polypeptide chain moves from the A site to the P site, and a new aminoacyl tRNA enters the A site. **11.** apply The ribosome's active site is made up of RNA, not protein. **12.** understand The separation allows the aminoacyl tRNA to place the amino acid into the ribosome's active site while reaching to the distant codon on the mRNA.

✔ Test Your Problem-Solving Skills

13. create Ribonucleases degrade mRNAs that are no longer needed by the cell. If an mRNA for a hormone that increased heart rate were never degraded, the hormone would be produced continuously and heart rate would stay elevated—a dangerous situation. **14.** apply d **15.** analyze The regions most crucial to the ribosome's function should be the most highly conserved: the active site, the E, P, and A sites, and the site where mRNAs initially bind. **16.** analyze The most likely locations are one of the grooves or channels where RNA, DNA, and ribonucleotides move through the enzyme—plugging one of them would prevent transcription.

CHAPTER 18

IN-TEXT QUESTIONS AND EXERCISES

p. 338 understand See **TABLE A18.1**.
p. 338 Fig. 18.1 analyze Write "Slowest response, most efficient resource use" next to the transcriptional control label. Write "Fastest response, least efficient resource use" next to the post-translational control label.
p. 339 Fig. 18.2 apply Plates from all three treatments must be identical and contain identical growth medium, except for the presence of the sugars labeled in the figure. Also, all plates must be grown under the same physical conditions (temperature, light) for the same time.
p. 340 Fig. 18.3 apply Use a medium with all 20 amino acids when producing a master plate of mutagenized *E. coli* colonies; then use a replica plate that contains all the amino acids except tryptophan. Choose cells from the master plate that did *not* grow on the replica plate.
p. 341 understand *lacZ* codes for the β-galactosidase enzyme, which breaks the disaccharide lactose into glucose and galactose. *lacY* codes for the galactoside permease enzyme, which transports lactose into the bacterial cell. *lacI* codes for a protein that shuts down production of the other *lac* products. When lactose is absent, the *lacI* product prevents transcription. This is logical because there is no reason for the cell to make β-galactosidase

and galactoside permease if there is no lactose to metabolize. But when lactose is present, it interacts with *lacI* in some way so that *lacZ* and *lacY* are induced (their transcription can occur). When lactose is present, the enzymes that metabolize it are expressed.
p. 343 apply A mutation that prevents lactose binding to repressor is predicted to prevent transcription of the *lac* operon under any condition. This is because the repressor would never come off the operator. A mutation in the operator that prevents repressor binding is predicted to lead to constitutive expression of the *lac* operon.
p. 343 Fig. 18.8 understand Put the "Repressor protein" on the operator. No transcription will take place. Then put the "RNA polymerase" on the promoter. No transcription will take place. Finally, put "lactose" on the repressor protein and then remove the resulting lactose–repressor complex from the operon. Transcription will begin.
p. 344 CYU (1) understand It is logical that the genes for metabolizing lactose should be expressed only when lactose is available. **(2)** understand See **FIGURE A18.1**.
p. 344 apply This mutation is predicted to lower rates of transcription initiation because AraC's binding to

RNA polymerase is essential for AraC to work as an activator.

IF YOU UNDERSTAND . . .

18.1 understand Production of β-galactosidase and galactosidase permease are under transcriptional control—transcription depends on the action of regulatory proteins. The activity of the repressor is under post-translational control. **18.2** apply Treat *E. coli* cells with a mutagen, and create a master plate that is grown at 33°C. Replica-plate this master plate and grow the replica plate at 42°C. Look for colonies that are on the master plate at 33°C but not on the replica plate at 42°C. **18.3** understand The operator is the parking brake; the repressor locks it in place, and the inducer releases it. **18.4** apply *ara* initiator mutations are likely to affect positive and negative control because AraC must bind to the *ara* initiator sequence for both forms of control. **18.5** create Mutations that create operators for the SOS regulon repressor protein would put new genes under control of the repressor and incorporate them into the regulon.

FIGURE A17.1

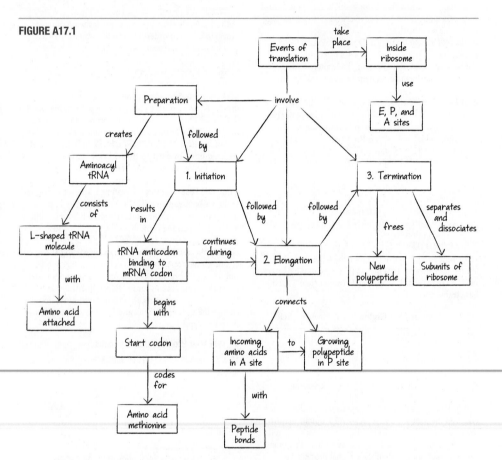

TABLE A18.1

Name	Function
lacZ	Gene for β-galactosidase
lacY	Gene for galactoside permease
Operator	Binding site for repressor
Promoter	Binding site for RNA polymerase
Repressor	Shuts down transcription
Lactose	Binds to repressor and stimulates transcription (removes negative control)
Glucose	At low concentration, increases transcription

YOU SHOULD BE ABLE TO . . .

✔ Test Your Knowledge

1. `remember` b **2.** `understand` d **3.** `apply` b **4.** `remember` DNA; RNA;
5. `understand` When glucose and another sugar are present in the environment, inducer exclusion prevents the use of the other sugar and allows only use of glucose.
6. `remember` a

✔ Test Your Understanding

7. `understand` The glycolytic enzymes are always needed in the cell because they are required to produce ATP, and ATP is always needed. **8.** `understand` Positive control means that a regulatory protein, when present, causes transcription to increase. Negative control means that a regulatory protein, when present, prevents transcription. **9.** `apply` b. **10.** `apply` Regulation of the *lac* operon should be normal. The location of the *lacI* gene isn't important, because the gene produces a protein that diffuses within the cell to the operator. **11.** `analyze` b (since the activator needs to bind to regulatory sequences to activate gene expression, preventing DNA binding would cripple the regulon and prevent cholera).**12.** `analyze` The *lac* operon would be strongly induced. Once inside the cell, the IPTG will bind to the repressor, causing it to release from DNA. IPTG cannot be broken down, so its concentration will remain high. Finally, since glucose is absent, there will be no inducer exclusion to inhibit IPTG transport through the galactoside permease transporter.

✔ Test Your Problem-Solving Skills

13. `create` Set up cultures with individuals that all come from the same colony of toluene-tolerating bacteria. Half the cultures should have toluene as the only source of carbon; half should have glucose or another common source of carbon. The glucose-containing medium serves as a control to ensure that cells can be grown in the lab. Cells will grow in both cultures if they are able to use toluene as a source of carbon; they will grow only in glucose-containing medium if toluene cannot be used as a carbon source. **14.** `apply` At a rate of 1×10^{-4} mutants per cell, you would on average find one mutant in every $10,000$ (1×10^4) cells. Therefore, you should screen a bit more (~ 2–3 times more) than 10,000 cells to be reasonably sure of finding at least one mutant. **15.** `apply` b
16. `analyze` Cells with functioning β-galactosidase will produce blue colonies; cells with *lacZ* mutations or *lacY* mutations will not produce β-galactosidase and will produce white colonies. The *lac* promoter could be mutated so that RNA polymerase cannot bind.

CHAPTER 19

IN-TEXT QUESTIONS AND EXERCISES

p. 352 Fig. 19.4 `analyze` Acetylation of histones decondenses chromatin and allows transcription to begin, so HATs are elements in positive control. Deacetylation condenses chromatin and inactivates transcription, so HDACs are elements in negative control.
p. 353 CYU (1) `apply` Many more genes than normal are predicted to be expressed because the inability to methylate DNA would lead to more decondensed chromatin.
(2) `understand` Addition of acetyl groups to histones or methyl groups to DNA can cause chromatin to decondense or condense, respectively. Different patterns of acetylation or methylation will determine which genes in muscle cells versus brain cells can be transcribed and which are not available for transcription.
p. 353 Fig. 19.5 `create` They could do something to change histone modifications to see how this affects gene transcription instead of just making the observation that certain histone modifications and low rates of transcription go together.
p. 355 Fig. 19.6 `analyze` A typical eukaryotic gene usually contains introns and is regulated by multiple enhancers. Bacterial operons lack introns and enhancers. The

promoter-proximal element found in some eukaryotic genes is comparable to the *araC* binding site in the *ara* operon of bacteria. Bacterial operons have a single promoter but code for more than one protein; eukaryotic genes code for a single product.
p. 356 `analyze` The basal transcription factors found in muscle and nerve cells are similar or identical; the regulatory transcription factors found in each cell type are different.
p. 356 `understand` DNA forms loops when distant regulatory regions, such as silencers and enhancers, are brought close to the promoter through binding of regulatory transcription factors to mediator.
p. 356 CYU (1) `analyze` Bacterial regulatory sequences are found close to the promoter; eukaryotic regulatory sequences can be close to the promoter or far from it. Bacterial regulatory proteins interact directly with RNA polymerase to initiate or prevent transcription; eukaryotic regulatory proteins influence transcription by altering chromatin structure or binding to the basal transcription complex through mediator proteins.
(2) `understand` Certain regulatory proteins decondense chromatin at muscle- or brain-specific genes and then activate or repress the transcription of cell-type-specific genes. Muscle-specific genes are expressed only if muscle-specific regulatory proteins are produced and activated.
p. 358 `understand` Alternative splicing does not occur in bacteria because bacterial genes do not contain introns—in bacteria, each gene codes for a single product. Alternative splicing is part of step 3 in Figure 19.1.

p. 359 `remember` Step 4 and step 5. RNA interference either (1) decreases the life span of mRNAs or (2) inhibits translation.
p. 360 CYU (1) `understand` It became clear that a single gene can code for multiple products instead of a single one.
(2) `understand` The miRNAs interfere with mRNAs by targeting them for destruction or preventing them from being translated.
p. 362 CYU (1) `understand` Many different types of mutations can disrupt control of the cell cycle and initiate cancer. These mutations can affect any of the six levels of control over gene regulation outlined in Figure 19.1.
(2) `understand` The p53 protein is responsible for shutting down the cell cycle in cells with damaged DNA. If the protein does not function, then cells with damaged DNA—and thus many mutations—continue to divide. If these cells have mutations in genes that regulate the cell cycle, then they may continue to divide in an uncontrolled fashion.

IF YOU UNDERSTAND . . .

19.1 `understand` Because eukaryotic RNAs are not translated as soon as transcription occurs, it is possible for RNA processing to occur, which creates variation in the mRNAs produced from a primary RNA transcript and in their life spans. **19.2** `understand` The default state of eukaryotic genes is "off," because the highly condensed state of the chromatin makes DNA unavailable to RNA polymerase. **19.3** `understand` See **FIGURE A19.1**. **19.4** `understand` Alternative splicing makes it possible for a single gene to

FIGURE A18.1

FIGURE A19.1

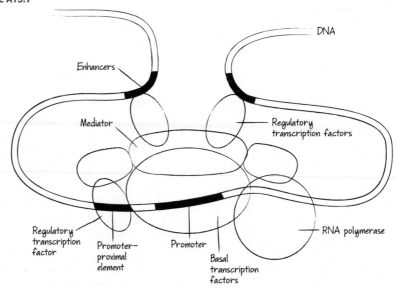

code for multiple products. **19.5** analyze One difference is that, in bacteria, genes that need to be turned on together are often clustered together in operons. Eukaryotes do not use operons. A similarity is in the way eukaryotes turn on many genes together and the strategy of bacterial regulons. Here, genes that are in many different locations share a DNA regulatory sequence and are activated when a regulatory transcription factor binds to the regulatory sequence. **19.6** understand In mutant cells that lack a form of p53 that can bind to enhancers, DNA damage cannot arrest the cell cycle, the cell cannot kill itself, and damaged DNA is replicated. This leads to mutations that can move the cell farther down the road to cancer.

YOU SHOULD BE ABLE TO . . .

✔ Test Your Knowledge

1. remember c **2.** remember A tumor suppressor is a gene or protein that holds cell division in check unless conditions are right for the cell to divide. **3.** understand d **4.** remember The set of regulatory transcription factors present in a particular cell, not differences in DNA sequence, are largely responsible for which genes are expressed. **5.** remember b **6.** remember d

✔ Test Your Understanding

7. analyze (a) Enhancers and the *araC* site are similar because both are sites in DNA where regulatory proteins bind. They are different because enhancers generally are located at great distances from the promoter, whereas the *araC* site is located nearer the promoter. (b) Promoter-proximal elements and the *lac* operon operator are both regulatory sites in DNA located close to the promoter. (c) Basal transcription factors and sigma are proteins that must bind to the promoter before RNA polymerase can initiate transcription. They differ because sigma is part of the RNA polymerase holoenzyme, while the basal transcription complex recruits RNA polymerase to the promoter. **8.** understand If changes in the environment cause changes in how spliceosomes function, then the RNAs and proteins produced from a particular gene could change in a way that helps the cell cope with the new environmental conditions. **9.** analyze (a) Enhancers and silencers are both regulatory sequences located at a distance from the promoter. Enhancers bind regulatory transcription factors that activate transcription; silencers bind regulatory proteins that shut down transcription. (b) Promoter-proximal elements and enhancers are both regulatory sequences that bind positive regulatory transcription factors. Promoter-proximal elements are located close to the promoter; enhancers are far from the promoter. (c) Transcription factors bind to regulatory sites in DNA; mediator does not bind to DNA but instead forms a bridge between regulatory transcription factors and

FIGURE A19.2

basal transcription factors. **10.** apply Inhibition of a histone deacetylase is predicted to leave acetyl groups on histones longer than normal. This is predicted to keep gene transcription going longer than normal, leading to higher levels of particular proteins and a change in the phenotype of the cell. **11.** analyze c; this is because there are many more modifications in the histone code relative to the genetic code. **12.** apply The cell is predicted to arrest in the cell cycle and most likely will commit suicide through activation of apoptosis genes because of the continually active *p53*.

✔ Test Your Problem-Solving Skills

13. apply c; This is because promoters and enhancers are brought into close physical proximity when transcription begins (see Figure 19.8). Since rats of malnourished mothers initiate *Hnf4a* gene transcription infrequently, the promoter and enhancer will be together less often in these animals compared with rats born to well-nourished mothers. **14.** understand See **FIGURE A19.2**. The value for the normal diet should be shown as 1.0 and the value for the low-protein diet should be shown as 0.64 (0.64 comes from the ratio of the cpm of the low-protein diet divided by the cpm of the normal diet, or 7368/11,478). **15.** create You could treat a culture of DNA-damaged cells with a drug that stops transcription and then compare them with untreated DNA-damaged cells. If transcriptional control regulates p53 levels, then the p53 level would be lower in the treated cells versus control cells. If control of p53 levels is post-translational, then in both cultures the p53 level would be the same. *Other approaches:* Add labeled NTPs to damaged cells and see if they are incorporated into mRNAs for p53; or add labeled amino acids to damaged cells and see if they are incorporated into completed p53 proteins; or add labeled phosphate groups and see if they are added to p53 proteins. **16.** create The double-stranded RNA could be cut by the same enzyme that creates double-stranded miRNAs from miRNA precursors. This double-stranded RNA may be incorporated into RISC and converted into a single-stranded RNA. The single-stranded RNA held by RISC would work in RNA interference to trigger the destruction of the complementary viral mRNA.

BIG PICTURE Genetic Information

p. 366 CYU (1) remember Star = DNA, mRNA, proteins. **(2)** understand RNA "is reverse transcribed by" reverse transcriptase "to form" DNA. **(3)** analyze E = splicing, etc.; E = meiosis and sexual reproduction (along with their links). **(4)** analyze Chromatin "makes up" chromosomes; independent assortment and recombination "contribute to" high genetic diversity.

CHAPTER 20

IN-TEXT QUESTIONS AND EXERCISES

p. 372 understand Isolate the DNA and cut it into small fragments with EcoRI, which leaves sticky ends. Cut copies of a plasmid or other vector with EcoRI. Mix the fragments and plasmids under conditions that promote complementary base pairing by sticky ends of fragments and plasmids. Use DNA ligase to catalyze formation of phosphodiester bonds and seal the sequences.
p. 372 Fig. 20.4 apply No—many times, because many copies of each type of mRNA were present in the pituitary cells, and many pituitary cells were used to prepare the library.
p. 373 apply The probe must be single stranded so that it will bind by complementary base pairing to the target DNA, and it must be labeled so that it can be detected. The probe will base-pair only with fragments that include a sequence complementary to the probe's sequence. A probe with the sequence 5′ AATCG 3′ will

bind to the region of the target DNA that has the sequence 5′ CGATT 3′ as shown here:

$$5'\ \text{AATCG}\ 3'$$
$$3'\ \text{TCCGGTTAGCATTACCATTTT}\ 5'$$

p. 374 CYU (1) understand When the endonuclease makes a staggered cut in a palindromic sequence and the strands separate, the single-stranded bases that are left will bind ("stick") to the single-stranded bases left where the endonuclease cut the same palindrome at a different location. **(2)** understand The word probe means to examine thoroughly. A DNA probe "examines" a large set of sequences thoroughly and binds to one—the one that has a complementary base sequence.
p. 375 Fig. 20.7 analyze The polymerase will begin at the 3′ end of each primer. On the top strand in part (b), it will move to the left; on the bottom strand, it will move to the right. As always, synthesis is in the 5′-to-3′ direction.
p. 376 CYU (1) understand Denaturation makes DNA single stranded so the primer can bind to the sequence during the annealing step. Once the primer is in place, *Taq* polymerase can synthesize the rest of the strand during the extension step. It is a "chain reaction" because the products of each reaction cycle are used in the next reaction cycle—this is why the number of copies doubles in each cycle. **(2)** apply One of many possible answers is shown in **FIGURE A20.1**.
p. 378 CYU understand If ddNTPs were present at high concentration, they would almost always be incorporated—meaning that only fragments from the first complementary base in the sequence would be produced.
p. 379 understand Using a map with many markers makes it more likely that there will be one marker that is very tightly linked to the gene of interest—meaning that a form of the marker will almost always be associated with the phenotype you are tracking.
p. 382 CYU create Start with a genetic map with as many polymorphic markers as possible. Determine the genotype at these markers for a large number of individuals who have the same type of alcoholism—one that is thought to have a genetic component—as well as a large number of unaffected individuals. Look for particular versions of a marker that is almost always found in affected individuals. Genes that contribute to a predisposition to alcoholism will be near that marker.
p. 383 Fig. 20.11 apply The insertion will probably disrupt the gene and have serious consequences for the cell and potentially the individual.

IF YOU UNDERSTAND . . .

20.1 understand Special features of DNA are needed to allow replication in a cell. It is unlikely that any DNA fragment generated by cutting the DNA of one species with a restriction enzyme would replicate when inserted into a bacterial cell. By placing DNA fragments within plasmids that normally replicate in a bacterial cell, the inserted DNA can be replicated along with the plasmid.
20.2 analyze Advantages of cloning in cells: No knowledge of the sequence is required. Disadvantages of cloning in cells: It is slower and technically more difficult than PCR. PCR advantages: It is fast and easy, and it can amplify a DNA sequence that is rare in the sample. PCR disadvantage: It requires knowledge of sequences on either side of the target gene, so primers can be designed.
20.3 understand The length of each fragment is dictated by where a ddNTP was incorporated into the growing strand, and each ddNTP corresponds to a base on the template strand. Thus the sequence of fragment sizes corresponds to the sequence on the template DNA.
20.4 understand In this case, both affected and unaffected individuals would have the same marker, so there would be no way to associate a particular form of the marker with the gene and phenotype you are interested in.
20.5 understand An advantage of using retrovirus vectors is that any foreign genes they carry are integrated into

human chromosomes, so delivered genes become a permanent feature of the cell's genome. The concern is that they may integrate at sites that alter gene function in ways that harm cell function, or even worse, lead to cancer. **20.6** understand Genes can be inserted into the Ti plasmids carried by *Agrobacterium*, and the recombinant plasmids transferred to plant cells infected by the bacterium.

YOU SHOULD BE ABLE TO . . .

✔ Test Your Knowledge

1. understand They cut DNA at specific sites, known as recognition sites, to produce DNA fragments useful for cloning. **2.** remember b **3.** understand ddNTPs lack the —OH (hydroxyl) group on the 3′ carbon of deoxyribose sugar that is required to extend the DNA chain during synthesis. **4.** understand a **5.** remember b **6.** understand d

✔ Test Your Understanding

7. understand When a restriction endonuclease cuts a "foreign gene" sequence and a plasmid, the same sticky ends are created on the excised foreign gene and the cut plasmid. After the sticky ends on the foreign gene and the plasmid anneal, DNA ligase catalyzes closure of the DNA backbone, sealing the foreign gene into the plasmid DNA (see **FIGURE A20.2**). **8.** apply If the DNA at each cycle steadily doubles, it is predicted to yield a 33.6 million-fold increase (a 2^{25}-fold increase). **9.** understand A cDNA library is a collection of complementary DNAs made from all the mRNAs present in a certain group of cells. A cDNA library from a human nerve cell would be different from one made from a human muscle cell, because nerve cells and muscle cells express many different genes that are specific to their cell type. **10.** remember Genetic markers are genes or other loci

that have known locations in the genome. When these locations are diagrammed, they represent the physical relationships between landmarks—in other words, they form a map. **11.** understand d; these regulatory sequences promote expression of introduced genes in the endosperm—the rice grain eaten as food. **12.** analyze PCR and cellular DNA synthesis are similar in the sense of producing copies of a template DNA. Both rely on primers and DNA polymerase. The major difference between the two is that PCR copies only a specific target sequence, but the entire genome is copied during cellular DNA synthesis.

✔ Test Your Problem-Solving Skills

13. create You could use a computer program to identify likely promoter sequences in the sequence data and then look for sequences just downstream that have an AUG start codon and codons that could be part of the protein-coding exons of a potential protein about 500 amino acids (500 codons or 1500 bases) long. **14.** analyze False. Since somatic-cell modifications cannot be passed on to future generations, but germ-line modifications may be passed on to offspring, germ-line modification opens a new set of ethical questions. **15.** analyze In both techniques, researchers use an indicator to identify either a gene of interest or a colony of bacteria with a particular trait. The problem is the same—picking one particular thing (a certain gene or a cell with a particular mutation) out of a large collection. **16.** analyze (a) Primer 1b binds to the top right strand and would allow DNA polymerase to synthesize the top strand across the target gene. Primer 1a, however, binds to the upper left strand and would allow DNA polymerase to synthesize the upper strand *away* from the target gene. Primer 2a binds to the bottom left strand and would allow DNA polymerase to synthesize the bottom strand across the target gene. Primer 2b, however, binds

to the bottom right strand and would allow DNA polymerase to synthesize away from the target gene. (b) She could use primer 1b with primer 2a.

CHAPTER 21

IN-TEXT QUESTIONS AND EXERCISES

p. 391 understand If no overlap occurred, there would be no way of ordering the fragments correctly. You would be able to put fragments only in random order—not the correct order.

p. 391 Fig. 21.2 understand Shotgun sequencing is based on fragmenting the genome into many small pieces.

p. 393 Fig. 21.3 understand Because the mRNA codons that match up with each strand are oriented in the 5′-to-3′ direction.

p. 395 CYU (1) understand Parasites don't need genes that code for enzymes required to synthesize molecules they acquire from their hosts. **(2)** understand If two closely related species inherited the same gene from a common ancestor, the genes should be similar. But if one species acquired the same gene from a distantly related species via lateral gene transfer, then the genes should be much less similar.

p. 399 Fig. 21.9 understand

Chromosome 1:
ψβ2-ε-Gγ-Aγ-ψβ1-δ-ψβ2-ε-Gγ-Aγ-ψβ1-δ-β

Chromosome 2: β

p. 400 CYU (1) understand Eukaryotic genomes have vastly different numbers of transposable elements and other repeated sequences that don't directly contribute to phenotype. So genome size can vary widely without changing the number of coding genes or the complexity of the organism. **(2)** apply Because most of the genome is transcribed, this number is very close to the percentage of the genome that codes for exons. In humans, the ratio would be roughly 2%.

p. 401 understand Start with a microarray containing exons from a large number of human genes. Isolate mRNAs from brain tissue and liver tissue, and make labeled cDNAs from each. Probe the microarray with cDNA made from each type of tissue, and record where binding occurs. Binding events identify exons that are transcribed in each type of tissue. Compare the results to identify genes that are expressed in brain but not liver, or in liver but not brain.

p. 402 understand Cell replication is an emergent property because it is due to the interaction of proteins working in a network, yet is a property that could not be predicted from the analysis of any one of these proteins.

IF YOU UNDERSTAND . . .

21.1 understand If a search of human gene sequence databases revealed a gene that was similar in base sequence, and if follow-up work confirmed that the mouse and human genes were similar in their pattern of exons and introns and regulatory sequences, then the researchers could claim that they are homologous. **21.2** understand The genome of a parasite is predicted to be smaller and to have fewer genes, particularly for metabolism, than the genome of a nonparasite. **21.3** understand These features include variable numbers of transposable elements; noncoding repeated sequences; and noncoding, non-repetitive DNA. These add to genome size without adding genes that directly influence phenotype. Additionally, mechanisms such as alternative splicing and the possibility of many noncoding regulatory RNAs can create situations in which few protein-coding genes but many different proteins are expressed in intricately regulated ways—features that would increase morphological complexity without increasing gene number. **21.4** understand Color-coded clusters shown in Figure 21.13 represent networks of interacting proteins that work together to carry out a particular cellular function. But it's clear from the figure that smaller, clusters are also connected into larger networks.

FIGURE A20.1

5′ CATGACTATTACGTATCGGGTACTATGCTATCGATCTAGCTACGCTAGCT 3′
3′ GTACTGATAATGCATAGCCCATGATACGATAGCTAGATCGATGCGATCGA 5′

Primer #1, which will anneal to the 3′ end of the top strand:

5′ AGCTAGCGTAGCTAGATCGAT 3′

Primer #2, which will anneal to the 3′ end of the bottom strand:

5′ CATGACTATTACGTATCGGGT 3′

The primers will bind to the separated strands of the parent DNA sequence as follows:

5′ CATGACTATTACGTATCGGGTACTATGCTATCGATCTAGCTACGCTAGCT 3′
                                 3′ TAGCTAGATCGATGCGATCGA 5′

5′ CATGACTATTACGTATCGGGT 3′
3′ GTACTGATAATGCATAGCCCATGATACGATAGCTAGATCGATGCGATCGA 5′

FIGURE A20.2

Sticky ends on cut plasmid and gene to be inserted are complementary and can base pair

DNA ligase forms four phosphodiester bonds between gene and plasmid DNA at the points indicated by arrows

YOU SHOULD BE ABLE TO . . .

✔ Test Your Knowledge

1. `remember` c **2.** `understand` a **3.** `remember` d **4.** `remember` b **5.** `understand` Finding two or more genes of similar sequence in the genome. **6.** `remember` Having a sequence that is clearly related to a functional gene but with a crippling mutation such as a stop codon or deletion.

✔ Test Your Understanding

7. `analyze` Computer programs are used to scan sequences in both directions to find ATG start codons, a gene-sized logical sequence with recognizable codons, and then a stop codon. One can also look for characteristic promoter, operator, and other regulatory sites. It is more difficult to identify open reading frames in eukaryotes because their genomes are so much larger and because of the presence of introns and repeated sequences. **8.** `evaluate` c **9.** `apply` Water flea gene density is about 31,000 genes/200 Mbp = 155 genes/Mbp. In humans, gene density is about 21,000 genes/3000 Mbp = 7 genes/Mbp. The relative gene density of water flea/human = 155/7 = 22. **10.** `analyze` Because with many different alleles, the chance that a match is coincidental is low relative to the chance that they match by identity (they come from the same person). **11.** `understand` A DNA microarray experiment identifies which genes are being expressed in a particular cell at a particular time. If a series of experiments shows that different genes are expressed in cells at different times or under different conditions, it implies that expression was turned on or off in response to changes in age or changes in conditions. **12.** `understand` Homology is a similarity among different species that is due to their inheritance from a common ancestor. If a newly sequenced gene is found to be homologous with a known gene of a different species, it is assumed that the gene products have similar function.

✔ Test Your Problem-Solving Skills

13. `analyze` If "gene A" is not necessary for existence, it can be lost by an event like unequal crossing over (on the chromosome with deleted segments) with no ill effects on the organism. In fact, individuals who have lost unnecessary genes are probably at a competitive advantage, because they no longer have to spend time and energy copying and repairing unused genes. **14.** `apply` If the grave were authentic, it might include two very different parental patterns along with five children whose patterns each represented a mix between the two parents. The other unrelated individuals would have patterns not shared by anyone else in the grave. **15.** `apply` d; mutation of such a central gene is likely to influence the phenotypes associated with all the genes it interacts with. If these interacting genes are involved in many different functions, the effects of mutation of a central gene are likely to be widespread, or pleiotropic. **16.** `apply` You would expect that the livers and blood of chimps and humans would function similarly, but that strong differences occur in brain function. The microarray data support this prediction and suggest chimp and human brains are different because certain genes are turned on or off at different times and expressed in different amounts.

CHAPTER 22

IN-TEXT QUESTIONS AND EXERCISES

p. 410 CYU `understand` Biologists have been able to grow entire plants from a single differentiated cell taken from an adult. They have also succeeded in producing animals by transferring the nucleus of a fully differentiated cell to an egg whose nucleus has been removed. **p. 411 Fig. 22.4** `apply` Anterior structures are predicted to be shifted farther back in the embryo, since higher-than-normal levels of the anterior-specifying Bicoid protein would be present in posterior regions. **p. 412 Fig. 22.6** `apply` Since Bicoid is located in the anterior cytoplasm and works to turn on genes that produce anterior structures, a prediction is that this manipulation would create a concentration gradient of Bicoid with its highest point in the middle of the mutant embryo. This would lead to the most anterior-like structures in the center of the embryo and more posterior structures forming on either side. **p. 414 Fig. 22.10** `create` Genes are usually considered homologous if they have similar DNA sequence and exon–intron structure. **p. 415 CYU (1)** `understand` Bicoid is a transcription factor, and cells that experience a high versus medium versus low concentrations of Bicoid express Bicoid-regulated genes at different levels. **(2)** `understand` Bicoid proteins control the expression of gap genes, which—in effect—tell cells which third of the embryo they are in along the anterior–posterior axis. Gap genes control the expression of pair-rule genes, which organize the embryo into individual segments. Pair-rule genes control expression of segment polarity genes, which establish an anterior–posterior polarity to each individual segment. Segment polarity genes turn on homeotic genes, which then trigger genes for producing segment-specific structures like wings or legs.

IF YOU UNDERSTAND . . .

22.1 `apply` The seedling would be misshapen because controlling the plane of cleavage is essential for formation of normal structures in plants. **22.2.** `understand` Cells in the stem contain a complete set of genes, so if they can be de-differentiated and receive appropriate signals from other cells, they could differentiate into any cell type and cooperate to form any structure of the plant, including roots. **22.3** `analyze` Because segmentation genes work in a cascade, mutation of a pair-rule gene is expected to influence the expression of genes in the next level, segment polarity genes, but not affect the expression of earlier acting genes, such as gap genes. **22.4** `apply` This procedure is predicted to lead to the loss of limbs because the regulatory sequence replacement will cause the regions of expression of *Hoxc8* and *Hoxc6* to overlap.

YOU SHOULD BE ABLE TO . . .

✔ Test Your Knowledge

1. `remember` b **2.** `remember` stem cells **3.** `understand` a **4.** `remember` True **5.** `remember` a **6.** `remember` a conserved gene that can be expressed at different times and places during development to produce different types of structures in different organisms.

✔ Test Your Understanding

7. `understand` They occur in both plants and animals, and are responsible for the changes that occur as an embryo develops. **8.** `understand` If an adult cell used for cloning had some permanent change or loss of genetic information during development, it should not be able to direct the development of a viable adult. But if the adult cell is genetically equivalent to a fertilized egg, then it should be capable of directing the development of a new individual. **9.** `understand` The researchers exposed adult flies to treatments that induce mutations and then looked for embryos with defects in the anterior–posterior body axis or body segmentation. The embryos had mutations in genes required for body axis formation and segmentation. **10.** `understand` Development—specifically differentiation—depends on changes in gene expression. Changes in gene expression depend on differences in regulatory transcription factors. **11.** `create` Differentiation is triggered by the production of transcription factors, which induce other transcription factors, and so on—a sequence that constitutes a regulatory cascade—as development progresses. At each step in the cascade, a new subset of genes is activated—resulting in a step-by-step progression from undifferentiated to fully differentiated cells. **12.** `analyze` c. This result shows that a mouse tool-kit gene is similar enough (conserved) to a fly gene to be able to take over its function.

✔ Test Your Problem-Solving Skills

13. `apply` There would be more Bicoid protein farther toward the posterior and less in the anterior—meaning that the anterior segments would be "less anterior" in their characteristics and the posterior segments would be "more anterior." **14.** `analyze` The result means that the human gene was expressed in worms and that its protein product has performed the same function in worms as in humans. This is strong evidence that the genes are homologous and that their function has been evolutionarily conserved. **15.** `apply` c. Since the morphogen concentration is halved every 100 μm away from the posterior pole, dropping to 1/16 the highest concentration would require 4 "halvings" of the initial concentration (1/2 × 1/2 × 1/2 × 1/2 = 1/16), and this would take place in 4 steps of 100 μm each—or 400 μm from the morphogen source. This is where the gene is predicted to be formed. **16.** `create` Tool-kit genes—such as *Hox* genes—are responsible for triggering the production of structures like wings or legs in particular segments. One hypothesis to explain the variation is that changes in gene expression of tool-kit genes led to different numbers of segments that express leg-forming *Hox* genes.

CHAPTER 23

IN-TEXT QUESTIONS AND EXERCISES

p. 421 `understand` If the bindin-like protein were blocked, it would not be able to bind to the egg-cell receptor for sperm. Fertilization would not take place. **p. 422 CYU (1)** `apply` This change is predicted to allow the sperm of species A to bind to—and likely fertilize—the eggs of species B. **(2)** `understand` The entry of a sperm triggers a wave of calcium ion release around the egg-cell membrane. Thus, calcium ions act as a signal that indicates "A sperm has entered the egg." In response to the increase in calcium ion concentration, cortical granules fuse with the egg-cell membrane and release their contents to the exterior, causing the destruction of the sperm receptor and, in sea urchins, the formation of a fertilization envelope. **p. 424 CYU** `understand` If cytoplasmic determinants could not be localized, they would be distributed to all blastomeres. This would specify the same fate for all blastomeres and their descendants instead of different fates. **p. 426 CYU (1)** `understand` Ectoderm and endoderm, because these are the inside and outside layers. **(2)** `apply` The future anterior–posterior and dorsal–ventral region of the embryo could first be identified as the blastopore forms and cells begin migrating into the embryo. The blastopore marks the future posterior of the embryo, and the region opposite the blastopore is where the anterior will form. Cells that first migrate into the blastopore come from the dorsal side, so this is the future dorsal region; the ventral region is on the opposite side. **p. 427 Fig. 23.9** `understand` Because neural tube formation depends on rearrangements of the cytoskeleton, a drug that locked the cytoskeleton in place is predicted to prevent formation of the neural tube. **p. 428 Fig. 23.12** `analyze` It made gene expression possible in non-muscle cells. A "general purpose" promoter can lead to transcription of an inserted cDNA in any type of cell, including fibroblasts.

IF YOU UNDERSTAND . . .

23.1 `understand` Bindin and the egg-cell receptor for sperm bind to each other. For this to happen, their tertiary structures must fit together somewhat like the structures of a lock and key. **23.2** understand If the cytoplasmic determinant is localized to a certain part of the egg cytoplasm, blastomeres will either contain or not contain the determinant, because the egg's cytoplasm becomes divided between blastomeres. **23.3** analyze Both cells would contain cell–cell signals and transcription factors specific to mesodermal cells, but the anterior cell would contain anterior-specific signals and regulatory transcription factors while the posterior cell would contain posterior-specific signals and regulators. **23.4** understand Cell–cell signals from the notochord direct the formation of the neural tube; subsequently, cell–cell signals from the notochord, neural tube, and ectoderm direct the differentiation of somite cells. Cells expand during neural tube formation. Somite cells move to new positions, proliferate, and differentiate.

YOU SHOULD BE ABLE TO . . .

✔ Test Your Knowledge

1. remember Bindin is a protein on the sperm-cell head that binds to a receptor on the egg plasma membrane to allow species-specific attachment of sperm and egg. **2.** understand c **3.** understand c **4.** understand d **5.** remember ectoderm; **6.** remember a

✔ Test Your Understanding

7. evaluate As mammals, mice form a placenta that allows the embryo to obtain nutrients from the mother. There's no need for lots of nutrient reserves to support development. Frogs develop outside the mother's body, and the embryo is on its own. There is a need for nutrient reserves, and the large egg provides this. Eggs contain stores of nutrients. **8.** apply This treatment is predicted to promote polyspermy, because there would be no release of Ca^{2+} inside the cell to trigger a block to polyspermy. **9.** understand They contain different types and/or concentrations of cytoplasmic determinants. **10.** analyze a; this is because ectodermal cells have to constrict on their dorsal margins and expand on their ventral margin to fold ectoderm into a neural tube. **11.** understand When transplanted early in development, somite cells become the cell type associated with their new location. But when transplanted later in development, somite cells become the cell type associated with the original location. These observations indicate that the same cell is not committed to a particular fate until later in development. **12.** apply More than 50,000 fibroblasts would need to be screened. This is because if 2×10^{-5} of all mRNAs and cDNAs coded for MyoD, then 50,000 cells would need to be screened on average to find one with a MyoD cDNA (50,000 comes from the quotient of $1/2 \times 10^{-5}$ MyoD cDNAs per cDNA). Since this is an average, in some groups of 50,000 cells, there may not be a MyoD cDNA. Therefore, to be reasonably sure of finding at least one MyoD cDNA, more than 50,000 cells should be screened.

✔ Test Your Problem-Solving Skills

13. analyze The interactions that allow proteins to bind to each other are extremely specific—certain amino acids have to line up in certain positions with specific charges and/or chemical groups exposed. Sperm–egg binding is species specific because each form of bindin binds to a different egg-cell receptor for sperm. **14.** evaluate Translation is not needed for cleavage to occur normally—meaning that all the proteins needed for early cleavage are present in the egg. **15.** create The yellow pigment may be a cytoplasmic determinant, or it may be only a molecule that is found in the same cells as the cytoplasmic determinant. Without isolating the pigment and adding it to cells to see if it acts as a cytoplasmic determinant, these two possibilities cannot be distinguished. **16.** apply a (See Figure 23.10 and note how the notochord is closest

to the somite region that forms bone. If an additional notochord were transplanted near the top of the neural tube, signals from the notochord would be expected to convert the nearby somite into bone-producing somite, adding to the amount of bone.)

CHAPTER 24

IN-TEXT QUESTIONS AND EXERCISES

p. 434 Fig. 24.2 remember See **FIGURE A24.1**.
p. 436 CYU (1) understand Cells located on the outside of the embryo become epidermal tissue; cells in the interior become vascular tissue; cells in between become ground tissue. **(2)** apply The *MONOPTEROS* gene would be overexpressed, producing higher levels of the MONOPTEROUS transcription factor. The embryonic cells would get information indicating that they are in the shoot apex, where MONOPTEROS levels are high. As a result, the root would not develop normally.
p. 437 Fig. 24.6 analyze Many animals have three body axes analogous to the three axes of leaves: anterior–posterior (like the leaf's proximal–distal), dorsal–ventral (like the leaf's adaxial–abaxial), and a left–right axis (like the leaf's mediolateral axis). The main plant body, in contrast, has only two axes: apical–basal and radial.
p. 438 CYU (1) create Simply planting a cut stem will often produce a new root meristem. This occurs whenever the cutting produces a new plant. **(2)** create If the plant could be genetically engineered to express higher levels of the PHAN transcription factor in leaves, the likely result would be a shift in leaf type from simple to compound.
p. 440 apply Mutual inhibition between the A and C genes helps create separate regions of gene expression. If C gene expression is lost, then the inhibition of A gene expression is also lost, and A genes are expressed in their normal location and where they are normally inhibited by C gene expression.
p. 441 CYU (1) analyze Plants generate many reproductive structures—flowers—many times and in many places during the plant's life. Plant reproductive cells are created from vegetative cells, not cells that are set aside early in development as reproductive cells. In contrast, animals form reproductive structures (ovaries and testes) in set locations, produce reproductive structures only once during development and set aside specialized reproductive cells (the germ line) early in their development. **(2)** analyze Like *Hox* genes, *MADS-box* genes code for DNA-binding proteins that function as transcriptional regulators. Both Hox proteins and MADS-domain proteins are involved in regulatory transcription cascades that lead to development of specific structures in specific locations. Although they perform similar functions, *Hox* genes and *MADS-box* genes evolved independently and differ in their base sequences. Thus, these genes and their protein products are not homologous.

IF YOU UNDERSTAND . . .

24.1 apply If there were only one master regulatory gene for patterning, the predicted result of loss of that gene's function by mutation would be disorganization of the entire apical–basal axis, not just removal of parts of it as seen in the actual mutants. **24.2** apply Many things could be imagined, but all should point in the direction of many repeated body parts (for example, 21 wings in a variety of shapes and positions on the body) and a very large overall size (if the fly is old) due to continuous growth. **24.3** apply The prediction is that sepals would develop from every whorl. This is because when the C gene product is absent, A gene expression spreads across the flower. Since there is no B gene expression, only sepals would form—the structure associated with only A gene expression.

YOU SHOULD BE ABLE TO . . .

✔ Test Your Knowledge

1. understand This is because mechanisms for development evolved independently in plants and animals. **2.** understand b **3.** remember b **4.** remember d **5.** remember c **6.** understand a.

✔ Test Your Understanding

7. understand c. **8.** understand Animal cells that lead to sperm or egg formation are "sequestered" early in development, so that they are involved only in reproduction and undergo few rounds of mitosis. Because mutations occur in every round of mitosis, plant eggs and sperm should contain many more mutations than animal eggs and sperm. **9.** understand Meristem cells are undifferentiated and divide to produce a cell that remains in the meristem and a cell that can commit to a particular pathway of differentiation. These are properties of stem cells. **10.** understand Just like the three tissue layers in plant embryos, the tissues produced in the SAMs and RAMs of a 300-year-old oak tree can differentiate into all of the specialized cell types found in a mature plant. **11.** evaluate The ABC model predicted that the A, B, and C genes would each be expressed in specific whorls in the developing flower. The experimental data showed that the prediction was correct. **12.** understand Cell proliferation is responsible for expansion of meristems and growth in size. Asymmetric cell division along with cell expansion is responsible for giving the adult plant a particular shape. Differentiation is responsible for generating functional tissues and organs in the adult plant. Cell–cell signals are responsible for setting up the apical–basal axis of the embryo and for guiding the pollen tube to the egg.

✔ Test Your Problem-Solving Skills

13. create Four. This is because there are four states of gene activity (in any cell, D can be on or off and E can be on or off) and therefore four structures that could be specified using this code. Sets of on–off switches are known as Boolean systems and are the basis for information storage in computers and (sometimes) in cells. **14.** analyze Because vegetative growth is continuous and occurs in all directions (unrestricted), it could be considered indeterminate. In contrast, reproductive growth produces mature reproductive organs and stops when those organs are complete. Because its duration is limited, it could be considered determinate. **15.** apply a; see Figure 24.10 to understand the reasoning. A and B together create petals, and B and C together create stamens. **16.** analyze One hundred times more likely. This is because 100 times more cell divisions were required to create the oak egg, and the rate of mutation per cell division is constant. (You don't need to consider the absolute rate of 1×10^{-6}/division, since the question asks about relative chances only).

FIGURE A24.1

Apical
Radial ←
Basal

CHAPTER 25

IN-TEXT QUESTIONS AND ANSWERS

p. 448 Fig. 25.4 analyze Under special creation, fossils with transitional features would be explained as separate types, unrelated to other organisms and just coincidentally having intermediate features.

p. 448 Figs. 25.5 analyze If vestigial traits result from inheritance of acquired characteristics, some individuals must have lost the traits during their own lifetimes and passed the reduced traits on to their offspring. For example, a certain monkey's long tail might have been bitten off by a predator. The new traits would then somehow have passed to the individual's eggs or sperm, resulting in shorter-tailed offspring, until humans with a coccyx resulted.

p. 453 CYU (1) evaluate If birds evolved from dinosaurs you would expect to find transitional fossil dinosaurs with feathers. Such fossils have been found. **(2)** analyze The DNA sequences of chimpanzees and humans are so similar because we share a recent common ancestor.

p. 455 (1) understand Relapse occurred because the few bacteria remaining after drug therapy were not eliminated by the patient's weakened immune system and began to reproduce quickly. **(2)** understand No—almost all of the cells present at the start of step 3 would have been resistant to the drug.

p. 456 understand When antibiotics are overused, some bacteria will be killed, including the bacteria that are harmless or beneficial to us. However, bacteria that are resistant to these antibiotics will flourish and multiply, reducing the likelihood that antibiotic treatment will be effective in the future.

p. 457 Fig. 25.16 analyze "Prediction": Beak measurements were different before and after the drought. "Prediction of null hypothesis": No difference was found in beak measurements before and after the drought.

p. 459 CYU (1) understand *Postulate 1:* Traits vary within a population. *Postulate 2:* Some of the trait variation is heritable. *Postulate 3:* There is variation in reproductive success (some individuals produce more offspring than others). *Postulate 4:* Individuals with certain heritable traits produce the most offspring. The first two postulates describe heritable variation; the second two describe differential reproductive success. **(2)** understand Beak size and shape and body size vary among individual finches, in part because of differences in their genotypes (some alleles lead to deeper or shallower beaks, for example). When a drought hit, individuals with deep beaks survived better and produced more offspring than individuals with shallow beaks.

p. 460 analyze In biology, an adaptation is any heritable trait that increases an individual's ability to produce offspring in a particular environment. In everyday English, adaptation is often used to refer to an individual's nonheritable adjustment to meet an environmental challenge, a phenomenon that biologists call acclimatization. The phenotypic changes resulting from acclimatization are not passed on to offspring.

IF YOU UNDERSTAND . . .

25.1 analyze Typological thinking is based on the idea that species are unchanging types and that variations within species are unimportant or even misleading. Population thinking recognizes the variation that occurs within a species as critically important. **25.2** remember Many answers are possible. For example, the evolution of an antibiotic-resistant TB bacterium demonstrates that species change through time. The homologous sequences in human and fruit fly genes demonstrate that these species are related by common ancestry. **25.3** analyze In biology, fitness is the ability of an individual to produce viable, fertile offspring, relative to that ability in other individuals in the population. In everyday English, fitness is a physical attribute

that is acquired as a result of practice or exercise. **25.4** apply Under special creation, changes in *Mycobacterium tuberculosis* populations would be explained as individual creative events governed by an intelligent creator. Under the theory of evolution by inheritance of acquired characters, changes in *M. tuberculosis* populations would be explained by the cells trying to transcribe genes in the presence of the drug, and their *rpoB* gene becoming altered as a result. **25.5** understand Brain size in *H. sapiens* might be constrained by the need for babies to pass through the mother's birth canal, by the energy required to maintain a large brain as an adult, or by lack of genetic variation for even larger brain size. Flying speed in falcons might be constrained by loss of maneuverability (and thus less success in hunting), the energy demands of extremely rapid flight, or lack of genetic variation for even faster flight.

YOU SHOULD BE ABLE TO . . .

✔ Test Your Knowledge

1. remember b **2.** remember False **3.** remember a **4.** understand d **5.** remember homologous **6.** remember c

✔ Test Your Understanding

1. understand Mutation produces new genetic variations, at random, without any forethought as to which variations might prove adaptive in the future. Individuals with mutations that are disadvantageous won't produce many offspring, but individuals with beneficial mutations will produce many offspring. The beneficial mutations will thus increase in frequency through selection. **2.** analyze Artificial selection is determined by human choice and is goal directed, whereas natural selection is the unplanned differential reproductive success of individuals that vary in their heritable traits. **3.** analyze b (In some environments, being big and strong lowers fitness.) **4.** analyze There were 751 finches before the drought and 90 finches afterward, so only 90/751 = 0.12, or 12 percent, survived. There were more finches with deep beaks before the drought only because the population was much larger, but the *average* beak depth increased after the drought. **5.** analyze The evidence for within-patient evolution is that DNA sequences at the start and end of treatment were identical except for a single nucleotide change in the *rpoB* gene. If rifampin were banned, *rpoB* mutant strains likely would have had lower fitness in the drug-free environment and would not continue to increase in frequency in *M. tuberculosis* populations. **6.** apply (1) Traits vary within rabbit populations, such as fur color and hearing ability. (2) Some of this variation is heritable. (3) More rabbits are produced than can survive. (4) Rabbits with certain heritable traits, such as the ability to camouflage in their environment or hear the approach of foxes, produce more offspring.

✔ Test Your Problem-Solving Skills

1. analyze The theory of evolution by natural selection predicts that white and brown deer mice are descendants of an ancestral population that varied in color. The white mice had higher fitness in the beach environment, where they were more likely to escape the notice of predators. Likewise, the brown mice had a higher fitness in the shady forest environment. Over time, the average color of each population changed. Special creation claims that the two colors of mice do not change and were created by divine intervention. Evolution by inheritance of acquired characteristic predicts that the mice in the different environments needed to change color, so they did so and then passed their traits onto their offspring. **2.** analyze b (No, because changes in height due to nutrition and reduced incidence of disease are not heritable.) **3.** evaluate Evidence to support a common ancestor of humans and chimpanzees about 6–7 million years ago includes the following: fossils from Africa that date from 6–7 million years ago with primate characteristics that

are common to chimps and humans, and DNA sequence comparisons showing homologies between chimps and humans (the rate of change of some genes can be used to estimate time of divergence). **4.** evaluate The theory of evolution fits the six criteria as follows: (1) and (2): It provides a common underlying mechanism responsible for puzzling observations such as homology, geographic proximity of similar species, the law of succession in the fossil record, vestigial traits, and extinctions. (3) and (4): It suggests new lines of research to test predictions about the outcome of changing environmental conditions in populations, about the presence of transitional forms in the fossil record, and so on. (5): It is a simple idea that explains the tremendous diversity of living and fossil organisms and why species continue to change today. (6): The realization that all organisms are related by common descent and that none are higher or lower than others was a surprise.

CHAPTER 26

IN-TEXT QUESTIONS AND EXERCISES

p. 467 (1) apply The frequencies of the three genotypes are shown by $p^2 + 2pq + q^2 = 1$.

$$\text{Freq}(A_1A_1) = p^2 = (0.6)^2 = 0.36.$$
$$\text{Freq}(A_1A_2) = 2pq = 2(0.6)(0.4) = 0.48.$$
$$\text{Freq}(A_2A_2) = q^2 = (0.4)^2 = 0.16.$$

(2) apply Allele frequencies in the offspring gene pool:

$$\text{Freq}(A_1) = 0.36 + 1/2(0.48) = 0.60.$$
$$\text{Freq}(A_2) = 1/2(0.48) + 0.16 = 0.40.$$

(3) apply Evolution has not occurred, because the allele frequencies have not changed.

p. 469 Table 26.1 apply The observed allele frequencies, calculated from the observed genotype frequencies, are 0.43 for *M* and 0.57 for *N*. The expected genotypes, calculated from the observed allele frequencies under the Hardy–Weinberg principle, are 0.185 for *MM*, 0.49 for *MN*, 0.325 for *NN*.

p. 470 CYU apply Given the observed genotype frequencies, the observed allele frequencies are

$$\text{Freq}(A_1) = 0.574 + 1/2(0.339) = 0.744.$$
$$\text{Freq}(A_2) = 1/2(0.339) + 0.087 = 0.256.$$

Given these allele frequencies, the genotype frequencies expected under the Hardy–Weinberg principle are A_1A_1: $0.744^2 = 0.554$; A_1A_2: $2(0.744 \times 0.256) = 0.381$; A_2A_2: $0.256^2 = 0.066$. There are 4 percent too few heterozygotes observed, relative to the expected proportion. One of the assumptions of the Hardy–Weinberg principle is not met at this gene in this population, at this time.

p. 470 apply The proportions of homozygotes should increase, and the proportions of heterozygotes should decrease, but allele frequencies will not change if no selection is occurring at this locus.

p. 474 apply There will be an excess of observed genotypes containing the favored allele compared to the proportion expected under Hardy–Weinberg proportions.

p. 478 (1) understand Sperm are small and hence relatively cheap to produce, whereas eggs are large and require a large investment of resources to produce. **(2)** understand Sperm are inexpensive to produce, so reproductive success for males depends on their ability to find mates—not on their ability to find resources to produce sperm. The opposite pattern holds for females. Sexual selection is based on variation in ability to find mates, so is more intense in males—leading to more exaggerated traits in males.

p. 479 apply If allele frequencies are changing due to drift, the populations in Table 26.1 would behave like the simulated populations in Figure 26.12—frequencies would drift up and down over time, and diverge.

p. 479 Fig. 26.12 analyze See **FIGURE A26.1**.

p. 480 Fig. 26.13 analyze The smaller the population, the greater the chance that genetic drift will cause alleles to be fixed or lost in the population in a short time. Starting with a smaller population makes the experiment easier.

p. 481 Fig. 26.14 analyze Original population: freq(A_1) = $(9 + 9 + 11)/54 = 0.54$; New population: freq(A_1) = $(2 + 2 + 1)/6 = 0.83$. The frequency of A_1 has increased dramatically.

p. 482 CYU (1) understand When allele frequencies fluctuate randomly up and down, sooner or later the frequency of an allele will hit 0. That allele thus is lost from the population, and the other allele at that locus is fixed. **(2)** understand In small populations, sampling error is large. For example, the accidental death of a few individuals would have a large impact on allele frequencies.

IF YOU UNDERSTAND . . .

26.1 apply The frequencies of two alleles are shown by $p + q = 1$. The frequencies of the three genotypes are shown by $p^2 + 2pq + q^2 = 1$.

If freq(A_1) = 0.2, then freq(A_2) = 0.8.
Freq(A_1A_1) = $p^2 = (0.2)^2 = 0.04$.
Freq(A_1A_2) = $2pq = 2(0.2)(0.8) = 0.32$.
Freq(A_2A_2) = $q^2 = (0.8)^2 = 0.64$.

26.2 apply It increased homozygosity of recessive alleles associated with diseases such as hemophilia. As a result, the royal families were plagued by genetic diseases. **26.3** understand The Hardy–Weinberg principle predicts that allele frequencies will stay the same over time. Natural selection favors some alleles over others, causing the frequency of those alleles to increase. **26.4** apply Genetic drift will rapidly reduce genetic variation in small populations. Thus, it is a high priority to keep the populations of endangered animals as large as possible. **26.5** apply In some cases, gene flow could be used to increase genetic variation and reduce inbreeding depression. However, gene flow can also be deleterious if the new alleles reduce the fitness of the endangered population. **26.6** apply Mutation rates themselves are too slow to affect an endangered population in a short time—unless a deleterious or beneficial mutation exposes the population to selection. (In instances where chemicals cause mutations in organisms, such as pesticides causing abnormal development in frogs, actions could be taken to remove the source of chemicals.)

YOU SHOULD BE ABLE TO . . .

✔ Test Your Knowledge

1. understand It defines what genotype and allele frequencies should be expected if evolutionary processes and nonrandom mating are *not* occurring. **2.** remember b **3.** understand a **4.** remember b **5.** remember True **6.** understand c

✔ Test Your Understanding

1. apply d; freq(A_1) = 0.9, so freq(A_2) = 0.1. Freq(A_2A_2) = $q^2 = (0.1)^2 = 0.01$. The number of babies with cystic

---

FIGURE A26.1

fibrosis = $2500(0.01) = 25$. **2.** apply Inbreeding causes an increase in the number of homozygotes. Individuals who are homozygous for recessive deleterious alleles are exposed to—and eliminated by—directional natural selection. **3.** evaluate This statement is false. Mutations do not occur because an organism wants or needs them. They just happen by accident and can be beneficial, neutral, or deleterious. **4.** understand Sexual selection is most intense on the sex that makes the least investment in the offspring, so that sex tends to have exaggerated traits that make individuals successful in competition for mates. In this way, sexual dimorphism evolves. **5.** analyze An allele in males, because a successful male can have as many as 100 offspring—10 times more than a successful female. **6.** create Your concept map should have linking verbs that relate the following information: Selection may decrease genetic variation or maintain it (e.g., if heterozygotes are favored). Genetic drift reduces it. Gene flow may increase or decrease it (depending on whether immigrants bring new alleles or emigrants remove alleles). Mutation increases it.

✔ Test Your Problem-Solving Skills

1. apply If we let p stand for the frequency of the loss-of-function allele, we know that $p^2 = 0.0001$; therefore $p = \sqrt{0.0001} = 0.01$. By subtraction, the frequency of normal alleles is 0.99. Under the Hardy–Weinberg principle, the frequency of heterozygotes is $2pq$, or $2 \times 0.01 \times 0.99$, which is 0.0198. **2.** apply c **3.** analyze If males never lift a finger to help females raise children, the fundamental asymmetry of sex is pronounced and sexual dimorphism should be high. If males invest a great deal in raising offspring, then the fundamental asymmetry of sex is small and sexual dimorphism should be low. **4.** apply Captive-bred young, transferred adults, and habitat corridors could be introduced to counteract the damaging effects of drift and inbreeding in the small, isolated populations. It is important, though, to introduce individuals only from similar habitats and connect patches of similar habitat with corridors—to avoid introducing alleles that lower fitness.

CHAPTER 27

IN-TEXT QUESTIONS AND EXERCISES

p. 493 CYU (1) understand Biological species concept: Reproductive isolation means that no gene flow is occurring. Morphological species concept: If populations are evolving independently, they may have evolved morphological differences. Phylogenetic species concept: If populations are evolving independently, they should have synapomorphies that identify them as independent twigs on the tree of life. **(2)** understand Biological species concept: It cannot be used to evaluate fossils, species that reproduce asexually, or species that do not occur in the same area and therefore never have the opportunity to mate. Morphospecies concept: It cannot identify species that differ in traits other than morphology and is subjective—it can lead to differences of opinion that cannot be resolved by data. Phylogenetic species concept: Reliable phylogenetic information exists only for a small number of organisms.

p. 494 Fig. 27.4 analyze The disruption of the salt marsh was a vicariance event that split the range of seaside sparrows into two areas (Gulf and Atlantic coasts), limiting gene flow and eventually resulting in two monophyletic groups.

p. 496 Fig. 27.7 analyze It creates genetic isolation—the precondition for divergence—by separating apple maggot flies from hawthorn flies, which mate on hawthorn fruits.

p. 497 create The difference in timing of ripening of apples and hawthorns causes divergence by separating apple flies and hawthorn flies not only in space (apple fruits versus hawthorn fruits) but also in time. The

---

earlier ripening of apples means that apple maggots develop in relatively warm conditions in the fall compared to hawthorn maggots. Natural selection may favor traits in apple maggots that make them adapted to warmer temperatures and traits in hawthorn maggots that make them adapted to cooler temperatures.

p. 498 apply A diploid grape plant experiences a defect in meiosis resulting in the formation of diploid gametes. The individual self-fertilizes, producing tetraploid offspring. The tetraploid grape plants self-fertilize or mate with other tetraploid individuals, producing a tetraploid population.

p. 499 apply A tetraploid ($4n$) species (in this case, wheat) gives rise to diploid ($2n$) gametes, and a diploid wheat species gives rise to haploid (n) gametes. When a haploid gamete (n) fertilizes a diploid gamete ($2n$), a sterile triploid offspring results. If an error in mitosis occurs and chromosome number doubles before meiosis, then hexaploid wheat is formed.

p. 499 CYU understand Use one of the cases given in the chapter to illustrate dispersal (Galápagos finches), vicariance (snapping shrimp), or habitat specialization (apple maggot flies). In each case, drift will cause allele frequencies to change randomly. Differences in the habitats occupied by the isolated populations will cause allele frequencies to change under natural selection. Any mutation that occurs in one population will not occur in the other due to the lack, or low level, of gene flow. Or use polyploidization in maidenhair fern or *Tragopogon* to illustrate how a massive mutation can itself create immediate genetic isolation and divergence.

p. 501 Fig. 27.12 evaluate If the same processes were at work in the past that are at work in an experiment with living organisms, then the outcome of the experiment should be a valid replication of the outcome that occurred in the past.

p. 502 analyze Gene regions in the experimental hybrid would not align with those of *H. annuus* and *H. petiolaris*, meaning that the gray bar representing the DNA of the experimental hybrid could have had no orange or red regions corresponding to *H. annuus* and *H. petiolaris*.

IF YOU UNDERSTAND THAT . . .

27.1 apply Human populations would not be considered separate species under the biological concept, because human populations can successfully interbreed. They would not be considered separate species under the morphological species concept, because all human populations have the same basic morphology. Although human races differ in minor superficial attributes such as skin color and hair texture, they have virtually identical anatomy and physiology in all other regards. Nor would human populations be separate species under the phylogenetic species concept, because they all arose from a very recent common ancestor. (DNA comparisons have revealed that human races are remarkably similar genetically and do not differ enough genetically to qualify even for subspecies status.) **27.2** create (Many possibilities) Collect a group of fruit flies that are living in your kitchen, separate the individuals into different cages, and expose the two new populations to dramatically different environmental conditions—heat, lighting, food sources. In essence: Create genetic isolation and then conditions for divergence due to drift, mutation, and selection. **27.3** evaluate The sample experiment described above is an example of vicariance, because the experimenter fragmented the habitat into isolated cages. (A different mechanism of speciation may have been involved in your example.) **27.4** apply Hybrid individuals will increase in frequency, and natural selection will favor parents that hybridize. The hybrid zone will increase. The two parent populations may go extinct if hybrids live in the same environment.

✔ **Test Your Knowledge**

1. `remember` a 2. `remember` b 3. `understand` a 4. `understand` a
5. `understand` d 6. `remember` False

✔ **Test Your Understanding**

1. `analyze` Direct observation: Galápagos finch colonization, apple maggot flies, *Tragopogon* allopolyploids, maidenhair ferns. Indirect studies: snapping shrimp, sunflowers. 2. `understand` a 3. `analyze` Based on biological and morphospecies criteria (breeding range and morphological traits, respectively), one species with six subspecies of seaside sparrows were recognized. However, based on phylogenetic criteria (comparison of gene sequences), biologists concluded that seaside sparrows represent only two monophyletic groups. 4. `evaluate` Some flies breed on apple fruits, others breed on hawthorn fruits. This reduces gene flow. Because apple fruits and hawthorn fruits have different scents and other characteristics, selection is causing the populations to diverge. 5. `apply` Decreasing. Gene flow tends to equalize allele frequencies among populations. 6. `apply` c

✔ **Test Your Problem-Solving Skills**

1. `analyze` b 2. `create` Sexual selection can promote divergence because it affects gene flow directly. For example, if blond women began mating only with blond men, and dark-haired women began mating only with dark-haired men, gene flow would decrease between these two populations even though they live in the same geographic area. Given enough time, these two populations would diverge due to mutation, genetic drift, and natural selection. 3. `evaluate` Two things: (1) Some flies happen to have alleles that allow them to respond to apple scents; others happen to have alleles that allow them to respond to hawthorn scents. There is no "need" involved. (2) The alleles for scent response exist; they are not acquired by spending time on the fruit. 4. `analyze` If the isolated populations and habitat fragments are small enough, the species is likely to dwindle to extinction due to inbreeding and loss of genetic variation or catastrophes like a severe storm or a disease outbreak. If the isolated populations survive, they are likely to diverge into new species because they are genetically isolated and because the habitats may differ.

CHAPTER 28

IN-TEXT QUESTIONS AND EXERCISES

p. 507 Fig. 28.1 `apply` Many trees are possible. See **FIGURE A28.1**.

p. 510 `understand` Because whales and hippos share SINEs 4–7 and no other species has these four, they are synapomorphies that define them as a monophyletic group. The similarity is unlikely to be due to homoplasy because the chance of four SINEs inserting in exactly the same place in two different species, independently, is very small.

p. 510 CYU `analyze` Hair and limb structures in humans and whales are examples of homology because they are traits that can be traced to a common ancestor. All mammals have hair and similar limb bone structure. However, extensive hair loss and advanced social behavior in whales and humans are examples of homoplasy. These traits are not common to all mammalian species and likely arose independently during the evolution of specific mammalian lineages.

p. 511 Fig. 28.4 `apply` See **FIGURE A28.2**. SINEs group (19, 20) identifies peccaries and pigs as a monophyletic group.

p. 513 `apply` Since some organisms fossilize better than others—such as mollusks with hard shells—a paleontologist would be mistaken to conclude that mollusks were more abundant than other organisms at a certain time just because mollusks are more abundant in the fossil

record. Similarly, a paleontologist would be mistaken to conclude that no organisms occupied an ancient desert based on the lack of fossil evidence; organisms in deserts do not fossilize well. Several other examples are possible.

p. 517 `create` In habitats on the mainland, complex communities have been established for a long time, so tarweeds in California and *Anolis* lizards on the mainland experience greater competition for resources and are limited to specific niches.

p. 520 CYU `understand` In terms of ecological opportunity, many resources were available because no other animals (or other types of organisms) existed to exploit them. Also, the evolution of new species in new niches made new niches available for predators. Morphological innovations like limbs and complex mouthparts were important because they made it possible for animals to live in habitats other than the benthic area, and to consume new types of food.

p. 520 Fig. 28.14 `analyze` The dinosaurs went extinct during the end-Cretaceous. About 15 percent of families went extinct.

IF YOU UNDERSTAND . . .

28.1 `apply` Humans are the only mammal that walks upright on two legs, so this character is a synapomorphy that defines the hominins. **28.2** `apply` Snail shells are hard structures, and snails often live in marine environments that fossilize readily, so snail fossils are relatively abundant. Abundant fossils are more helpful than rare fossils for estimating changes in relative abundance over time. **28.3** `analyze` They could eat food that was available off the substrate. And once animals lived off (and away from) the bottom, it created an opportunity for other animals who could eat them to evolve. **28.4** `evaluate` This statement is false. Human-induced extinctions today are not due to poor adaptation to the normal environment. Humans are changing habitats rapidly and in unusual ways (such as clear-cutting forests), so most species cannot adapt to these changes rapidly enough to avoid going extinct. In this way, human-induced extinctions are more similar to mass extinctions than to background extinctions.

YOU SHOULD BE ABLE TO . . .

✔ **Test Your Knowledge**

1. `remember` c 2. `understand` a 3. `understand` b 4. `understand` False
5. `understand` d 6. `remember` True

✔ **Test Your Understanding**

1. `evaluate` The fossil record is biased because recent, abundant organisms with hard parts that live underground or in environments where sediments are being deposited are most likely to fossilize. Even so, it is the only data available on what organisms that lived in the past looked like, and where they lived. 2. `create` An enormous amount of animal diversity appeared in a time frame that was short compared to the sweep of earlier Earth history. The most common analogy is to an explosion, referring to the sudden and spectacular radiation of lineages. Another analogy is your hand, where one line (your arm) comes into a node (your palm) and several lineages (your fingers) come out. Many other answers are possible. 3. `understand` Homoplasy is rare relative to homology, so phylogenetic trees that minimize the total change required are usually more accurate. In the case of the artiodactyl astragalus, parsimony was misleading because the astragalus was lost when whales evolved limblessness—creating two changes (a loss following a gain) instead of one (a gain). 4. `create` (Many answers are possible.) Adaptive radiation of *Anolis* lizards after they colonized new islands in the Caribbean. *Hypothesis:* After lizards arrived on each new island, where there were no predators or competitors, they rapidly diversified to occupy four distinct types of habitats on each island. Adaptive radiation during the Cambrian period

following a morphological innovation. *Hypothesis:* Additional *Hox* genes made it possible to organize a large, complex body; the evolution of complex mouthparts and limbs made it possible for animals to move and find food in new ways. 5. `apply` About 90 percent of species went extinct during the end-Permian, so about 10 percent survived, written as 0.1 in decimals. $1880 \times 0.1 = 188$ would make it to graduation. 6. `apply` d

✔ **Test Your Problem-Solving Skills**

1. `understand` b 2. `evaluate` Many trees are possible. See **FIGURE A28.3**. Sponges are the outgroup. The most parsimonious tree has the fewest number of changes (five in this case). Adding more characters to the matrix can improve the reliability of the results. 3. `understand` It would be helpful to have (1) a large crater or other physical evidence from a major impact dated at 251 mya, (2) shocked quartz and microtektites in rock layers dating to the end-Permian era, and (3) high levels of iridium or other elements common in space rocks and rare on Earth, dated to the time of the extinctions. 4. `create` Many concept map topologies are possible. Maps should follow this logic: Oxygen is an effective final electron acceptor in cellular respiration because of its high electronegativity. Organisms that use it as a final electron acceptor can produce more usable energy than organisms that do not use oxygen, but only if it is available. With more available energy, aerobic organisms can grow larger and move faster.

BIG PICTURE Evolution

p. 526 CYU (1) `understand` Circle = inbreeding, sexual selection, nonrandom mating, natural selection, genetic drift,

FIGURE A28.1

FIGURE A28.2

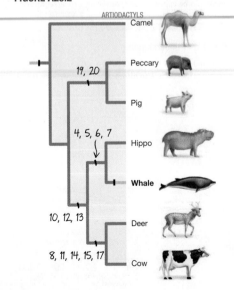

ARTIODACTYLS

mutation, and gene flow. **(2)** [analyze] Adaptation "increases" fitness; synapomorphies "identify branches on" the tree of life. **(3)** [analyze] Several answers possible, e.g., adaptive radiations and mass extinctions → can be studied using → the fossil record. **(4)** [analyze] Genetic drift, mutation, and gene flow "are random with respect to" fitness.

CHAPTER 29

IN-TEXT QUESTIONS AND EXERCISES

p. 529 Fig. 29.1 [analyze] Prokaryotic—it would require just one evolutionary change, the origin of the nuclear envelope in Eukarya. If it were eukaryotic, it would require that the nuclear envelope was lost in both Bacteria and Archaea—two changes are less parsimonious (see Chapter 28).

p. 529 Table 29.1 [apply] See **FIGURE A29.2**.

p. 534 Fig. 29.5 [analyze] Weak—different culture conditions may have revealed different species.

p. 535 CYU (1) [create] Conditions should mimic a spill—sand or stones with a layer of crude oil, or seawater with oil floating on top. Add samples, from sites contaminated with oil, that might contain cells capable of using molecules in oil as electron donors or electron acceptors. Other conditions (temperature, pH, etc.) should be realistic. **(2)** [create] Use metagenomic analysis. After isolating DNA from a soil sample, fragment and sequence the DNA. Compare the sequences to those from known organisms and use the data to place the species on the tree of life.

p. 538 [apply] cyanobacteria—photoautotrophs; *Clostridium aceticum*—chemoorganoautotroph; *Nitrosopumilus* sp.—chemolithoautotrophs; helicobacteria— photoheterotrophs; *Escherichia coli*—chemoorganoheterotroph; *Beggitoa*—chemolithotrophic heterotrophs.

p. 539 Fig. 29.10 [apply] Table 29.5 contains the answers. For example, for organisms called sulfate reducers you would have H_2 as the electron donor, SO_4^{2-} as the electron acceptor, and H_2S as the reduced by-product. For humans the electron donor is glucose ($C_6H_{12}O_6$), the electron acceptor is O_2, and the reduced by-product is water (H_2O).

p. 541 [understand] Different species of bacteria that have different forms of bacterial chlorophyll will not absorb the same wavelengths of light and will therefore not compete.

p. 542 Fig. 29.12 [apply] Aerobic respiration. More free energy is released when oxygen is the final electron acceptor than when any other molecule is used, so more ATP can be produced and used for growth.

p. 543 [apply] If bacteria and archaea did not exist, then (1) the atmosphere would have little or no oxygen, and (2) almost all nitrogen would exist in molecular form (the gas N_2).

p. 543 Fig. 29.13 [apply] Add a label for "animals" in the upper right quarter of the circle, and draw arrows leading from "Organic compounds with amino groups" to "animals" and from "animals" to "NH_3."

p. 544 CYU [evaluate] Eukaryotes can only (1) fix carbon via the Calvin–Benson pathway, (2) use aerobic respiration with organic compounds as electron donors, and (3) perform oxygenic photosynthesis. Among bacteria and archaea, there is much more diversity in pathways for carbon fixation, respiration, and photosynthesis, along with many more fermentation pathways.

p. 545a, p. 545b, p. 546a, p. 546b, p. 547 [apply] See **FIGURE A29.1**.

p. 548 Fig. 29.21 [apply] About 600,000 *N. maritimus* cells would fit end to end on a meter stick.

IF YOU UNDERSTAND . . .

29.1 [analyze] Eukaryotes have a nuclear envelope that encloses their chromosomes; bacteria and archaea do not. Bacteria have cell walls that contain peptidoglycan, and archaea have phospholipids containing isoprene subunits in their plasma membranes. Thus, the exteriors of a bacterium and archaeon are radically different. Archaea and eukaryotes also have similar machinery for processing genetic information. **29.2** [apply] After extracting and sequencing DNA from bacteria that live in an insect gut, analyze the resulting sequences for genes that are similar to known genes involved in nitrogen fixation. **29.3** [understand] They are compatible, and thrive in each others' presence—one species' waste product is the other species' food.

YOU SHOULD BE ABLE TO . . .

✔ Test Your Knowledge

1. [understand] c **2.** [remember] b **3.** [remember] d **4.** [remember] d **5.** [remember] False—the enzyme nitrogenase is poisoned by oxygen. **6.** [remember] peptidoglycan

✔ Test Your Understanding

1. [understand] c **2.** [understand] An electron donor provides the potential energy required to produce ATP. **3.** [evaluate] Yes. The array of substances that bacteria and archaea can use as electron donors, electron acceptors, and fermentation substrates, along with the diversity of ways that they can fix carbon and perform photosynthesis, allows them to live just about anywhere. **4.** [apply] They should get energy from reduced organic compounds "stolen" from their hosts. **5.** [understand] Large amounts of potential energy are released and ATP produced when oxygen is the electron acceptor, because oxygen is so electronegative.

FIGURE A28.3

FIGURE A29.1

FIGURE A29.2

Large body size and high growth rates are not possible without large amounts of ATP. **6.** understand They are paraphyletic because the prokaryotes include some (bacteria and archaea) but not all (eukaryotes) groups derived from the common ancestor of all organisms living today.

✓ Test Your Problem-Solving Skills

1. understand c (It is pathogenic on a wide variety of organisms). **2.** analyze This result supports their hypothesis because the drug poisons the enzymes of the electron transport chain and prevents electron transfer to Fe^{3-}, which is required to drive magnetite synthesis. If magnetite had still formed, another explanation would have been needed. **3.** create Hypothesis: A high rate of tooth cavities in Western children is due to an excess of sucrose in the diet, which is absent from the diets of East African children. To test this hypothesis, switch a group of East African children to a diet that contains sucrose, switch a group of Western children to a diet lacking sucrose, and monitor the presence of *S. mutans* in both groups of children. **4.** create Look in waters or soil polluted with benzene-containing compounds. Put samples from these environments in culture tubes where benzene is the only source of carbon. Monitor the cultures and study the cells that grow efficiently.

CHAPTER 30

IN-TEXT QUESTIONS AND EXERCISES

p. 529 CYU (1) understand Opisthokonts have a flagellum at the base or back of the cell; alveolate cells contain unique support structures called alveoli; stramenopiles—meaning "straw hairs"—have straw-like hairs on their flagella. **(2)** understand In direct sequencing, DNA is isolated directly from the environment and analyzed to place species on the tree of life. It is not necessary to actually see the species being studied.

p. 557 CYU (1) understand *Plasmodium* species are transmitted to humans by mosquitoes. If mosquitoes can be prevented from biting people, they cannot spread the disease. **(2)** apply Iron added → primary producers (photosynthetic protists and bacteria) bloom → more carbon dioxide taken up from atmosphere during photosynthesis → consumers bloom, eat primary producers → bodies of primary producers and consumers fall to bottom of ocean → large deposits of carbon-containing compounds form on ocean floor.

p. 560 apply See FIGURE A30.1.

p. 560 Fig. 30.8 analyze Two—one derived from the original bacterium and one derived from the eukaryotic cell that engulfed the bacterium.

p. 561 apply A photosynthetic bacterium (e.g., a cyanobacterium) could have been engulfed by a larger eukaryotic cell. If it was not digested, it could continue to photosynthesize and supply sugars to the host cell.

p. 562 apply Membrane infoldings observed in bacterial species today support the hypothesis's plausibility—they confirm that the initial steps actually could have happened. The continuity of the nuclear envelope and ER are consistent with the hypothesis, which predicts that the two structures are derived from the same source (infolded membranes).

p. 564 apply The chloroplast genes label should come off of the branch that leads to cyanobacteria. (If you had a phylogeny of just the cyanobacteria, the chloroplast branch would be located somewhere inside.)

p. 565 analyze The acquisition of the mitochondrion and the chloroplast represent the transfer of entire genomes, and not just single genes, to a new organism.

p. 566 apply Yes—when food is scarce or population density is high, the environment is changing rapidly (deteriorating). Offspring that are genetically unlike their parents may be better able to cope with the new and challenging environment.

p. 569a understand **(1)** Alternation of generations refers to a life cycle in which there are multicellular haploid phases and multicellular diploid phases. A gametophyte is the multicellular haploid phase that produces gametes; the sporophyte is the multicellular diploid phase that produces spores. A spore is a cell that grows into a multicellular individual but is not produced by fusion of two cells. A zygote is a cell that grows into a multicellular individual but *is* produced by fusion of two cells (gametes). Gametes are haploid cells that fuse to form a zygote. **(2)** See FIGURE A30.2.

p. 569b apply Most likely, the two types of amoebae evolved independently. The alternative hypothesis is that the common ancestor of alveolates, stramenopiles, rhizarians, plants, opisthokonts, and amoebozoa were amoeboid and that this growth form was lost many times. See FIGURE A30.1.

p. 569c, p. 569d, p. 569e, p. 570a, p. 570b, p. 570c apply See FIGURE A30.1.

p. 571 analyze If euglenids could take in food via phagocytosis (ingestive feeding), then that would have provided a mechanism by which a smaller photosynthetic protist could have been engulfed and incorporated into the cell via secondary endosymbiosis.

p. 572a, p. 572b, p. 573, p 574. apply See FIGURE A30.1.

IF YOU UNDERSTAND . . .

30.1 understand Photosynthetic protists use CO_2 and light to produce sugars and other organic compounds, so they furnish the first or primary source of organic material in an ecosystem. **30.2** remember Eukaryotic cells contain many organelles, including a nucleus, endomembrane system, and an extensive cytoskeleton. Most prokaryotic cells contain few or no organelles, and no nucleus or endomembrane system and lack a complex cytoskeleton. **30.3** understand **(1)** Outside membrane was from host eukaryote; inside from engulfed cyanobacterium. **(2)** From the outside in, the four membranes are derived from the eukaryote that engulfed a chloroplast-containing eukaryote, the plasma membrane of the eukaryote that was engulfed, the outer membrane of the engulfed cell's chloroplast, and the inner membrane of its chloroplast.

YOU SHOULD BE ABLE TO . . .

✓ Test Your Knowledge

1. understand b **2.** remember b **3.** remember a **4.** remember b **5.** remember False. **6.** remember diatoms.

✓ Test Your Understanding

1. understand a **2.** understand Because all eukaryotes living today have cells with a nuclear envelope, it is valid to infer that their common ancestor also had a nuclear envelope. Because bacteria and archaea do not have a nuclear envelope, it is valid to infer that the trait arose in the common ancestor of eukaryotes. **3.** understand The host cell provided a protected environment and carbon compounds for the endosymbiont; the endosymbiont provided increased ATP from the carbon compounds **4.** understand It confirmed a fundamental prediction made by the hypothesis and could not be explained by any alternative hypothesis. **5.** understand All alveolates have alveoli, which are unique structures that function in supporting the cell. Among alveolates there are species that are (1) ingestive feeders, photosynthetic, or parasitic, and that (2) move using cilia, flagella, or a type of amoeboid movement.

✓ Test Your Problem-Solving Skills

1. analyze d **2.** analyze If the apicoplast that is found in *Plasmodium* (the organism that causes malaria) is genetically similar to chloroplasts, and if glyphosate poisons chloroplasts, it is reasonable to hypothesize that glyphosate will poison the apicoplast and potentially kill the *Plasmodium*. This would be a good treatment strategy

for malaria because humans have no chloroplasts, provided that the glyphosate produces no other effects that would be detrimental to humans. **3.** evaluate Primary producers usually grow faster when CO_2 concentration increases, but to date they have not grown fast enough to make CO_2 levels drop—CO_2 levels have been increasing steadily over decades. **4.** analyze Given that lateral gene transfer can occur at different points in a phylogenetic history, specific genes can become part of a lineage by a different route from that taken by other genes of the organism. In the case of chlorophyll *a*, its history traces back to a bacterium being engulfed by a protist and forming a chloroplast.

FIGURE A30.1

FIGURE A30.2

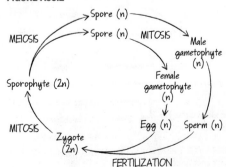

CHAPTER 31

IN-TEXT QUESTIONS AND EXERCISES

p. 582 CYU [understand] (1) Green algae and land plants share an array of morphological traits that are synapomorphies, including the chlorophylls they contain; (2) green algae appear before land plants in the fossil record; and (3) on phylogenetic trees estimated from DNA sequence data, green algae and land plants share a most recent common ancestor—green algae are the initial groups to diverge, and land plants diverge subsequently.

p. 583 Fig. 31.6 [remember] If you find the common ancestor of all the green algae (at the base of the Ulvophyceae), the lineages that are collectively called green algae don't include that common ancestor and all of its descendants—only some of its descendants. The same is true for non-vascular plants (the common ancestor here is at the base of the Hepaticophyta) and seedless vascular plants (the common ancestor here is at the base of the Lycophyta).

p. 586 Fig. 31.8 [understand] Water flows more easily through a short, wide pipe than through a long, skinny one because there is less resistance from the walls of the pipe.

p. 587 Fig. 31.9 [understand] To map an innovation on a phylogenetic tree, biologists determine the location(s) on the tree that is (are) consistent with all descendants from that point having the innovation (unless, in rare cases, the innovation was lost in a few descendants).

p. 589 [understand] Alternation of generations occurs when there are multicellular haploid individuals and multicellular diploid individuals in a life cycle. One hypothesis is that this life cycle evolved because the plants with a large sporophyte were more successful at using the wind to disperse many spores, while a small gametophyte produces gametes that can swim.

p. 590 [apply] In the hornwort photo, the sporophyte is the spike-like green and yellow structure; the gametophyte is the leafy-looking structure underneath. The horsetail gametophyte is the microscopic individual on the left; the sporophyte is the much larger individual on the right.

p. 590 Fig. 31.14 [understand] There are multicellular haploid stages and multicellular diploid stages in these plants.

p. 592 Fig. 31.18 [analyze] Gymnosperm gametophytes are microscopic, so they are even smaller than fern gametophytes. The gymnosperm gametophyte is completely dependent on the sporophyte for nutrition, while fern gametophytes are not. Fern gametophytes are photosynthetic and even supply nutrition to the young sporophytes.

p. 593 Fig. 31.19 [analyze] Consistent—the fossil data suggest that gymnosperms evolved earlier, and gymnosperms have larger gametophytes than angiosperms.

p. 595 Fig. 31.21 [analyze] To eliminate all other variables except for the one being tested, such as the possibility of pollination by other insects in the field. Scales differ because the two insect species visited flowers at very different frequencies.

p. 596 CYU (1) [understand] Cuticle prevents water loss from the plant; UV-absorbing compounds allow plants to be exposed to high light intensities without damage to their DNA; vascular tissue moves water up from the soil and moves photosynthetic products down to the roots.
(2) [apply] See **FIGURE A31.1**.
p. 600a, p. 600b, p. 601, p. 603a, p. 603b, p. 604, p. 605, p. 606 [apply] See **FIGURE A31.1**.
p. 596 Fig. 31.23 [apply] See **FIGURE A31.1**.

IF YOU UNDERSTAND . . .

31.1 [apply] Rates of soil formation will decline; rates of soil loss will increase. **31.2** [understand] All green algae and land plants contain chlorophyll *a* and *b*, have similar arrangements of thylakoid membranes, and store starch. **31.3** [analyze] Both spores and seeds have a tough, protective coat, so they can survive while being dispersed to

a new location. Seeds have the advantage of carrying a store of nutrients with them—when a spore germinates, it must make its own food via photosynthesis right away.

YOU SHOULD BE ABLE TO . . .

✔ Test Your Knowledge

1. [remember] c **2.** [remember] c **3.** [remember] a **4.** [remember] c **5.** [remember] (a) Gametangia are found in all land plant groups except angiosperms; (b) pollen is found in gymnosperms and angiosperms; (c) seeds are found in gymnosperms and angiosperms, (d) fruit is found in angiosperms.
6. [understand] In a gametophyte-dominant life cycle, the gametophyte is larger and longer lived than the sporophyte and produces most of the nutrition. In a sporophyte-dominant life cycle, the sporophyte generation is the larger, longer-lived, and photosynthetic phase of the life cycle.

✔ Test Your Understanding

1. [remember] c **2.** [understand] Plants build and hold soils required for human agriculture and forestry, and they increase water supplies that humans can use for drinking, irrigation, or industrial use. Plants release oxygen that we breathe. **3.** [analyze] Cuticle prevents water loss from leaves but also prevents entry of CO_2 required for photosynthesis. Stomata allow CO_2 to diffuse but can close to minimize water loss. Liverwort pores allow gas exchange but cannot be closed if conditions become dry. **4.** [understand] They provided the support needed for plants to grow upright and not fall over in response to wind or gravity. Erect growth allowed plants to compete for light. **5.** [apply] Homosporous plants produce a single type of spore that develops into a gametophyte that produces both egg and sperm. Heterosporous plants produce two different types of spores that develop into two different gametophytes that produce either egg or sperm. In a tulip, the microsporangium is found within the stamen,

and the megasporangium is found within the ovule. Microspores divide by mitosis to form male gametophytes (pollen grains); megaspores divide by mitosis to form the female gametophyte. **6.** [understand] Animal pollinators increase the efficiency of pollination so plants can save energy by making less pollen.

✔ Test Your Problem-Solving Skills

1. [understand] d **2.** [analyze] A "reversion" to wind pollination might be favored by natural selection because it is costly to produce a flower that can attract animal pollinators. Because wind-pollinated species grow in dense clusters, they can maximize the chance that the wind will carry pollen from one individual to another (less likely if the individuals are far apart). Wind-pollinated deciduous trees flower in early spring before their developing leaves begin to block the wind. **3.** [create] The combination represents a compromise between efficiency and safety. Tracheids can still transport water if vessels become blocked by air bubbles. **4.** [create] Alter one characteristic of a flower, and present the flower to the normal pollinator. As a control, present the normal (unaltered) flower to the normal pollinator. Record the amount of time the pollinator spends in the flower, the amount of pollen removed, or some other measure of pollination success. Repeat for other altered characteristics. Analyze the data to determine which altered characteristic affects pollination success the most.

CHAPTER 32

IN-TEXT QUESTIONS AND EXERCISES

p. 619 CYU (1) [understand] Because they are made up of a network of thin, branching hyphae, mycelia have a large surface area, which makes absorption efficient.
(2) [remember] Swimming spores and gametes, zygosporangia, basidia, asci.

FIGURE A31.1

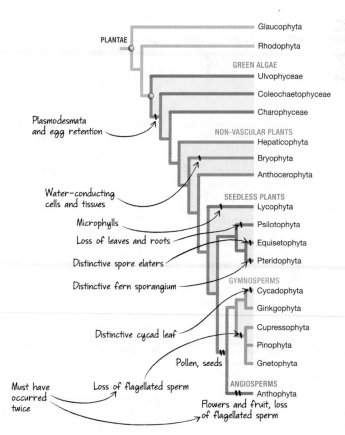

p. 622 Fig. 32.10 analyze Labeled nutrient experiments identify which nutrients are exchanged and in which direction. They explain *why* plants do better in the presence of mycorrhizae, and why the fungus also benefits.

p. 625 analyze Human sperm and egg undergo plasmogamy followed by karyogamy during fertilization, but heterokaryotic cells do not occur in humans or other eukaryotes besides fungi.

p. 625 Fig. 32.12 apply The haploid mycelium.

p. 626 Fig. 32.13 analyze Mycelia produced by asexual reproduction are genetically identical to their parent; mycelia produced by sexual reproduction are genetically different from both parents (each spore has a unique genotype).

p. 628 CYU (1) understand When birch tree seedlings are grown in the presence and absence of EMF, individuals denied their normal EMF are not able to acquire sufficient nitrogen and phosphorus. **(2)** analyze Each type of fungus involves meiosis and the production of haploid spores. In a zygosporangium, meiosis occurs in a multinucleate cell; in basidia and asci, this cell has a single diploid nucleus. In asci, meiosis is followed by one round of mitosis.

p. 628, p. 630a, p. 630b, p. 630c, p. 631, p. 632a, p. 632b apply See **FIGURE A32.1**.

IF YOU UNDERSTAND . . .

32.1 apply Loss of mycorrhizal fungi would decrease nutrient delivery to plants and reduce their growth inside the experimental plots, compared to control plots with intact fungi. Also, lack of fungi would slow decay of dead plant material, causing a dramatic buildup of dead organic material. **32.2** understand Like humans, fungi are composed of eukaryotic cells. Bacteria are prokaryotic and thus have many more unique targets for antibiotics. **32.3** understand Haploid hyphae fuse, forming a heterokaryotic mycelium. When karyogamy occurs, a diploid nucleus forms—just as when gametes fuse. **32.4** understand See the diversity boxes.

YOU SHOULD BE ABLE TO . . .

✔ **Test Your Knowledge**

1. remember a **2.** remember d **3.** remember b **4.** remember b **5.** understand Pores in septa allow nutrients to move from regions of acquisition to regions of mycelial growth; **6.** remember commensal.

✔ **Test Your Understanding**

7. understand c **8.** evaluate Along with a few bacteria, fungi are the only organisms that can digest wood completely. If the wood is not digested, carbon remains trapped in wood. Without fungi, CO_2 would be tied up and unavailable for photosynthesis, and the presence of undecayed organic matter would reduce the space available for plants to grow. **9.** analyze Fungi produce enzymes that degrade cellulose and lignin. **10.** analyze Plant roots have much smaller surface area than EMF or AMF. Hyphae are much smaller than the smallest portions of plant roots so can penetrate dead material more efficiently. Extracellular digestion—which plant roots cannot do—allows fungi to break large molecules into small compounds that can be absorbed. **11.** apply DNA and RNA contain phosphorus. Mycorrhizal fungi effectively increase the volume of soil mined by root systems and can make adequate amounts of phosphorus available for supporting rapid cell division. **12.** analyze Both compounds are processed via extracellular digestion. Different enzymes are involved, however. The degradation of lignin is uncontrolled and does not yield useful products; digestion of cellulose is controlled and produces useful glucose molecules.

✔ **Test Your Problem-Solving Skills**

13. apply c **14.** create (1) Confirm that the chytrid fungus is found only in sick frogs and not healthy frogs.

(2) Isolate the chytrid fungus and grow it in a pure culture. (3) Expose healthy frogs to the cultured fungus and see if they become sick. (4) Isolate the fungus from the experimental frogs, grow it in culture, and test whether it is the same as the original fungus. **15.** create Each of the different cellulase enzymes attacks cellulose in a different way, so producing all the enzymes together increases the efficiency of the fungus in breaking down cellulose completely. It is likely that lignin peroxidase is produced along with cellulases, so they can act in concert to degrade wood. To test this idea, you could harvest enzymes secreted from a mycelium before and after it contacts wood in a culture dish, and see if the cellulases and lignin peroxidase appear together once the mycelium begins growing on the wood. **16.** create Collect a large array of colorful mushrooms that are poisonous; also capture mushroom-eating animals, such as squirrels. Present a hungry squirrel with a choice of mushrooms that have been dyed or painted a drab color versus others treated with a solution that is identical to the dye or paint used but uncolored. Record which mushrooms the squirrel eats. Repeat the test with many squirrels and many mushrooms.

CHAPTER 33

IN-TEXT QUESTIONS AND EXERCISES

p. 639 Fig. 33.2 apply The label and bar for "Segmentation?" should go on the branch to the left of the clam shell and "Mollusca" label. The label and bar for "Toolkit genes for segmentation?" should go on the same branch as cephalization, CNS, and coelom.

p. 640 apply The genetic tool kit required for multicellularity includes genes that regulate the cell cycle, and thus cell division. When something goes wrong in this process, uncontrolled growth called cancer can result. Thus, cancer may be as ancient as multicellularity itself.

p. 642 Fig. 33.7 analyze Stained *Hox* and *dpp* gene products would either not be found in *Nematostella* at all or would not occur in the same anterior–posterior and dorsal–ventral pattern as observed in bilaterians.

p. 646 CYU (1) apply Yes, your body has a tube-within-a-tube design. Humans are vertebrates, vertebrates are chordates, chordates are bilaterians, and all bilaterians have a tube-within-a-tube design, consisting of a bilaterally symmetrical, slender body with an inner tube (gut), an outer tube (mostly skin), muscles and organs derived from mesoderm in between, and one end (head) cephalized with sensory organs and brain. **(2)** understand When an animal moves through an environment in one direction, its ability to acquire food and perceive and respond to threats is greater if its feeding, sensing, and information-processing structures are at the leading end.

p. 651 Fig. 33.14 evaluate The larva and metamorphosis arrow should be circled in the life cycle. It might be adaptive to skip the feeding larval stage if more food resources are available for the mother (who must expend energy to produce the yolk) than for the larva, if predation pressure on larvae is high, or if other environmental factors for larvae are unfavorable, resulting in high mortality of the larvae.

p. 652 CYU (1) understand The mouthparts of deposit feeders are relatively simple because these animals simply gulp relatively soft material. The mouthparts of mass feeders are more complex because they have to tear off and process chunks of relatively hard material.

(2) create Gametes that are shed into aquatic environments can float or swim. This cannot happen on land, so internal fertilization is more common there. Also, gametes are more at risk of drying out on land, and internal fertilization eliminates that risk.

p. 653 apply The origin of cilia-powered swimming should be marked on the Ctenophora branch.

p. 654 apply The origin of cnidocytes should be marked on the Cnidaria branch.

IF YOU UNDERSTAND . . .

33.1 apply It should increase dramatically—animals would no longer be consuming plant material. **33.2** understand Your drawing should show the tube-within-a-tube design of a worm, with one end labeled head and containing the mouth and brain. From the brain, one or two major nerve tracts should run the length of the body. The gut should go from the mouth to the other end, ending in the anus. In between the gut and outer body wall, there should be blocks or layers of muscle. **33.3** apply There are many possible answers. For example, within the phylum Arthropoda: The sensory systems have diversified (e.g., eyes are important for sight in flies, antennae are important for smelling in moths); feeding strategies have diversified (e.g., butterflies are fluid feeders but caterpillars are mass feeders); ecological roles have diversified (e.g., butterflies are herbivores but spiders are carnivores); limbs have diversified (e.g., butterflies have legs and wings but caterpillars have lobe-like limbs); and some insects, like grasshoppers, have direct development whereas others, like mosquitoes, have indirect development. **33.4** remember If you see a

FIGURE A32.1

sessile, multicellular animal with no symmetry or radial symmetry, with outcurrent holes, and choanocytes, it is a sponge. If you see a radially symmetric, transparent, planktonic organism moving via rows of cilia, it is a ctenophore. If you see a diploblastic, outwardly radially symmetric animal that is either a single polyp or colony of multiple polyps, or it is a medusa, then the animal is a cnidarian.

YOU SHOULD BE ABLE TO . . .

✔ Test Your Knowledge

1. `remember` a 2. `remember` True 3. `remember` protostomes, deuterostomes 4. `understand` a 5. `remember` d 6. `remember` d

✔ Test Your Understanding

1. `understand` Diploblasts have two types of embryonic tissue (ectoderm, endoderm); triploblasts have three (ectoderm, mesoderm, endoderm). Mesoderm made it possible for an enclosed, mesoderm-lined cavity to develop, creating a coelom. 2. `analyze` Many unicellular organisms are heterotrophic, but they can consume only small bits of food. Animals are multicellular, so they are larger and can consume larger amounts of food—making them important consumers in food webs. 3. `apply` There are about 18.4 named arthropods for every named species of chordate (1,160,000/63,000), about 1.3 named mollusks for every named species of chordate (85,000/63,000), and about 0.4 named species of nematode worm for every named species of chordate (25,000/63,000). These numbers are likely to be underestimates because there are probably many more arthropods, mollusks, and nematode worms than have been described, yet relatively few of the larger, more familiar chordates (including vertebrates) left to be described. 4. `understand` In oviparous species, the mother adds nutrient-rich yolk to the egg that nourishes the developing embryo. In viviparous species, the mother transfers nutrients directly from her body to the growing embryo. 5. `evaluate` You could stain the gene products of the segmentation genes in chordates (such as a mouse embryo) and observe whether the expression pattern correlates with the expression pattern observed in arthropods (such as a fly embryo). 6. `apply` c

✔ Test Your Problem-Solving Skills

1. `analyze` Yes—if the same gene is found in nematodes and humans, it was likely found in the common ancestor of protostomes and deuterostomes. If so, then fruit flies should also have this gene. 2. `apply` Asexual reproduction is likely to be favored when the sea anemones have plenty of resources such as space and food. Sexual reproduction is likely to be favored when the sea anemones are stressed due to lack of resources, predation, infection, or other factors. In this case, it is adaptive to reproduce in a way that enables dispersal to a new site and/or genetically diverse offspring, some of which will be more likely to survive selection than a clone of identical animals would. 3. `apply` Since echinoderms evolved from bilaterally symmetric ancestors with nerve cords, they should have nerve cords too. However, the organization of the nerve cords should be more diffuse, because echinoderms need to take in and process information from multiple directions—not just one. 4. `apply` d

CHAPTER 34

IN-TEXT QUESTIONS AND EXERCISES

p. 659 Fig. 34.3 `apply` We know that aquatic living was the ancestral trait because the outgroups (e.g., Cnidaria, Porifera) were aquatic. We also have fossil evidence that the aquatic lifestyle was ancestral.
p. 660, p. 662, p. 665, p. 666, p. 667, p. 671, p. 674 `apply` See **FIGURE A34.1**.

p. 662 Fig. 34.6 `evaluate` Different phyla of worms have different feeding strategies and mouthparts, and they eat different foods, so they are not all in direct competition for food.
p. 664 CYU (1) `apply` Snails on land would be expected to have gills or lungs inside the body or otherwise protected from drying out, thinner shells so that they are not as heavy, and desiccation-resistant eggs. **(2)** `evaluate` The tentacles of the octopus could be probed during development for *Hox* genes known to be important in establishing the foot of snails. If the expression pattern is similar, homology would be supported.
p. 664 Fig. 34.10 `apply` You should have drawn a circle around all the tentacles.

IF YOU UNDERSTAND . . .

34.1 `remember` Ecological opportunity was important in the diversification of protostomes because many lineages made the transition from water to land, radiating into new niches. Genetic opportunity was important because changes in the expression of developmental tool-kit genes enabled diverse modifications of body plans based on existing genes. **34.2** `understand` One source of diversification is the loss of ancestral traits; just because a species does not possess a trait does not mean that its ancestor didn't. **34.3** `analyze` Both arthropods and vertebrates have segmented bodies, which are especially amenable to modularity since a change in the expression pattern of developmental tool-kit genes can change the number, size, and type of segments, exposing morphological diversity to the adaptive forces of natural selection.

YOU SHOULD BE ABLE TO . . .

✔ Test Your Knowledge

1. `remember` False 2. `remember` a 3. `remember` c 4. `understand` c 5. `remember` Both groups are bilaterally symmetric triploblasts with the protostome pattern of development. 6. `remember` b

✔ Test Your Understanding

1. `remember` d 2. `apply` Both annelids and arthropods have segmented bodies, unlike other protostomes such as flatworms and nematodes. 3. `evaluate` Clams are relatively sessile as adults. Because larvae can swim, they provide a mechanism to disperse to new locations. To test this idea, you could measure the distance traveled by clam larvae and measure their success in recruiting to new habitats compared to larvae that are artificially limited to stay near their parents. 4. `apply` Land plants would require cuticles or other mechanisms to keep their tissues from drying out, a mechanism of keeping gametes from

drying out, and structural support to retain body shape despite the effects of gravity. 5. `evaluate` The ability to fly allowed insects to disperse to new habitats and find new food sources efficiently. 6. `analyze` It might be adaptive for juveniles of hemimetabolous individuals to use the same food source as the adults when the food source is plentiful.

✔ Test Your Problem-Solving Skills

1. `analyze` See **FIGURE A34.2**. 2. `analyze` If the ancestors of brachiopods and mollusks lived in similar habitats and experienced natural selection that favored similar traits, then they would have evolved to have similar forms and habitats. This is called convergent evolution (see Chapter 28). 3. `evaluate` This hypothesis is reasonable because simple characteristics often occur before more elaborate ones; this could be true for the radula and foot in aplacophorans. An alternate hypothesis would be that these structures' simple forms are derived characteristics that are adaptations for parasitism, but the evidence suggests that aplacophorans are carnivores and detritivores, not parasites. The position of the aplacophorans in the phylogeny could be tested by using DNA sequence data rather than morphological characters in a phylogenetic analysis. 4. `apply` b

CHAPTER 35

IN-TEXT QUESTIONS AND EXERCISES

p. 682 Fig. 35.1 `apply` Invertebrates are a paraphyletic group. The group does not include all the descendants of the common ancestor because vertebrates are excluded.
p. 684 CYU `understand` If it's an echinoderm, it should have five-part radial symmetry, a calcium carbonate endoskeleton just underneath the skin, and a water vascular system (e.g., visible tube feet).
p. 684, p. 685a, p. 685b `apply` See **FIGURE A35.1**.

FIGURE A34.2

FIGURE A34.1

FIGURE A35.1

p. 690 Fig. 35.12 `apply` Mammals and birds are equally related to amphibians, because birds and mammals share a common ancestor, and this ancestor shares a common ancestor with amphibians.

p. 691 `apply` See **FIGURE A35.2**

p. 693 Fig. 35.16 `analyze` If this phylogeny were estimated on the basis of limb traits, it would be based on the *assumption* that the limb evolved from fish fins in a series of steps. But because the phylogeny was estimated from other types of data, it is legitimate to use it to analyze the evolution of the limb without making a circular argument.

p. 694 Fig. 35.18 `analyze` The yolk sac is smaller here—their function in an amniotic egg has been taken over by the placenta.

p. 695 CYU (1) `understand` Jaws allow animals to capture food efficiently and process it by crushing or tearing. The ability to process food more efficiently increased the importance of jawed vertebrates as herbivores and carnivores. The tetrapod limb enables vertebrates to locomote on land. The amniotic egg enables vertebrates to lay eggs on land; amniotic eggs are resistant to drying out. The placenta protects the growing fetus and enables the mother to be mobile. Parental care reduces the number of offspring that can be raised, but increases their survivorship. Flight enabled pterosaurs, bats, and birds to expand into new habitats. **(2)** `analyze` Protostomes and deuterostomes

living on land both have mechanisms for preventing their eggs from drying out (e.g., thick membranes in snail and insect eggs; amniotic egg in tetrapods), preventing their skin from drying out (e.g., waxy cuticle in insects, scaly skin in reptiles), and keeping their respiratory surfaces from drying out (e.g., lungs in land snails and tracheal system in insects; lungs in tetrapods).

p. 696 `apply` See **FIGURE A35.3**.

p. 698a, p. 698b, p. 699, p. 700 `apply` See **FIGURE A35.4**.

p. 701a, p. 701b, p. 702a, p. 702b, p. 703a, p. 703b `apply` See **FIGURE A35.3**.

p. 705 Table 35.1 `analyze` 44 percent, 26 percent; relative to body size, *H. floresiensis* has a smaller brain than *H. sapiens*.

p. 706 Fig. 35.37 `apply` Five hominin species existed 1.8 mya, and three existed 100,000 years ago.

p. 706 Fig. 35.38 `analyze` The forehead became much larger, and the face became "flatter"; the brow ridges are less prominent in later skulls than in earlier skulls.

IF YOU UNDERSTAND . . .

35.1 `apply` (1) Mussels and clams will increase dramatically; (2) kelp density will increase.

35.2 `understand` Your cartoon should have a dorsal hollow nerve chord along the top, dorsal edge; another line representing the notochord below the nerve cord; a tail

labeled at one end; and a series of lines or bumps representing pharyngeal gill slits near the other end.

35.3 `understand` Like today's lungfish, the earliest tetrapods could have used their limbs for pulling themselves along the substrate in shallow-water habitats. **35.4** `understand` The earliest fossils of *H. sapiens* are found in Africa, and phylogenetic analyses of living human populations indicate that the most basal groups are all African.

YOU SHOULD BE ABLE TO . . .

✔ Test Your Knowledge

1. `remember` True **2.** `remember` a **3.** `understand` d **4.** `remember` mammals **5.** `understand` d **6.** `remember` b

✔ Test Your Understanding

1. `understand` c **2.** `understand` Pharyngeal gill slits function in suspension feeding. The notochord furnishes a simple endoskeleton that stiffens the body; electrical signals that coordinate movement are carried by the dorsal hollow nerve cord to the muscles in the post-anal tail, which beats back and forth to make swimming possible. Cephalochordates and Urochordates are chordates, but they are not vertebrates. **3.** `analyze` If jaws are derived forms of gill arches, then the same genes and the same cells should be involved in the development of the jaw and the gill arches. **4.** `understand` Increased parental care allows

FIGURE A35.2

FIGURE A35.4

FIGURE A35.3

offspring to be better developed, and thus have increased chances of survival, before they have to live on their own. **5.** evaluate Yes: Over a short time interval, many species that occupy an array of foods and habitats evolved. Changes in tooth and jaw structure, tool use, and body size suggest that different hominin species exploited different types of food. **6.** evaluate A phylogeny of modern humans would reject the out-of-Africa hypothesis if the most ancestral lineages in the phylogeny were from any other region but Africa, such as Asia or Europe.

✔ Test Your Problem-Solving Skills

1. apply b **2.** evaluate Xenoturbellidans either retain traits that were present in the common ancestor of all deuterostomes, or they have lost many complex morphological characteristics. **3.** evaluate Vertebrate limbs are diverse in form and function (they can serve as arms, flippers, wings, etc.) but have homologous bones within. Molecular probes could be used to visualize the expression patterns of tool-kit genes like *Hox* in the embryonic limbs of different vertebrate species to show how changes in the expression of homologous genes can cause changes to the structure and function of limbs. **4.** analyze See **FIGURE A35.5.** It is important to use indigenous people for this study to minimize the amount of gene flow (transfer of alleles from one population to another) that occurred during the history of the study populations due to immigration and interbreeding.

CHAPTER 36

IN-TEXT QUESTIONS AND EXERCISES

p. 717 Fig. 36.7 apply Starting with 1 virion, approximately 80 virions were produced after 40 minutes. If all of these virions infected new cells, you expect each of the 80 infected cells to produce 80 virions by 90 minutes, or a total of 6400 virions. The cells appear to replicate once every 30 minutes, which means that they will have replicated 3 times during this period. The total number of cells at 90 minutes would be approximately 2^3, or 8 cells. There would be 800 times as many virions as cells.
p. 717 Fig. 36.8 analyze No—it only shows that CD4 is required. (Subsequent work showed that other proteins are involved as well.)
p. 721 analyze In T4, the viral genome is injected into the host while HIV inserts its genome via membrane fusion. HIV must first convert its genome into double-stranded DNA before replication, which is not necessary for T4, which has a DNA genome. Both viruses use double-stranded DNA to make mRNA that is translated into viral proteins. T4 virions are made in the cytosol of the cell and are released when the cell bursts. HIV virions bud from the cell surface, which does not require cell death.
p. 721 Fig. 36.12 create Budding may disrupt the integrity of the host-cell plasma membrane enough to kill the cell.
p. 722 CYU analyze Like cars in the assembly line of an automobile plant, new virions are assembled from premanufactured parts, and large numbers of progeny are produced per generation. Cells, on the other hand, reproduce by the division of a single integrated unit, resulting in only two progeny per generation.
p. 724 Fig. 36.15 apply You should have the following bars and labels: on branch to HIV-2 (sooty mangabey to human); on branch to HIV-1 strain O (chimp to human); on branch to HIV-1 strain N (chimp to human); on branch to HIV-1 strain M (chimp to human).

p. 725 CYU (1) analyze The escaped-genes and degenerate-cell hypotheses state that viruses originated from cells, while the RNA-world hypothesis suggests that viruses originated in parallel with, maybe even influenced, the origin of cells. The escaped-gene and RNA-world hypotheses state that the viruses originated from parasitic molecules. The degenerate-cell hypothesis states that viruses originated from parasitic organisms. **(2)** apply A mutation that made transmission between humans more efficient would make it more dangerous. Such mutation could occur via genomic reassortment in pigs, where an avian virus and a human virus could co-infect cells and produce recombinants.

IF YOU UNDERSTAND . . .

36.1 understand HIV infects helper T cells. These cells play a central role in the human immune response. As latently infected cells are activated by the immune response, HIV replication causes cell death. Over several years, steady decline in helper T cells leads to a compromised immune response that allows other infections to eventually kill the infected person. **36.2** understand (1) Fusion inhibitors block viral envelope proteins or host cell receptors. (2) Protease inhibitors prevent processing/assembly of viral proteins. (3) Reverse transcriptase inhibitors block reverse transcriptase, preventing replication of the genome. **36.3** apply There is heritable variation among virions, due to random changes that occur as their genomes are copied. There is also differential reproductive success among virions in their ability to successfully infect host cells. This differential success is due to the presence of certain heritable traits. **36.4** analyze Both classes have positive-sense single-stranded RNA genomes. Class IV viruses use an RNA replicase to make a negative-sense copy of the genome, which is then used as a template for producing mRNA and genomic RNAs. Class VI viruses use reverse transcriptase to convert their RNA genome into dsDNA, which is then integrated into the host genome. There it serves as a template to for transcribing viral mRNA and genomic RNAs by host-cell RNA polymerase.

YOU SHOULD BE ABLE TO . . .

✔ Test Your Knowledge

1. remember d **2.** remember b **3.** understand In class VI and VII viruses, reverse transcriptase converts positive-sense single-stranded RNA to double-stranded DNA.
4. remember b **5.** remember c **6.** remember The unique type of genetic material included in the virion (classes I–V) in addition to the manner in which the genome is replicated (compare classes VI and VII with classes IV and I, respectively).

✔ Test Your Understanding

7. analyze A virus with an envelope exits host cells by budding. A virus that lacks an envelope exits host cells by lysis (or other mechanisms that don't involve budding). **8.** analyze (1) Rate of viral genome replication is much higher via the lytic cycle. In lysogeny, the viral genome can replicate only when the host cell replicates. (2) Only the lytic cycle produces virions. (3) The lytic

FIGURE A35.5

cycle results in host cell death, while the host cell continues to survive during lysogeny. **9.** create Viruses rely on host-cell enzymes to replicate, whereas bacteria do not. Therefore, many drugs designed to disrupt the virus life cycle cannot be used, because they would kill host cells as well. Only viral-specific proteins are good targets for drug design. **10.** evaluate Each major hypothesis to explain the origin of viruses is associated with a different type of genome. Escaped genes: single- and double-stranded DNA or possibly single-stranded RNA. Degenerate cells: double-stranded DNA. RNA world: single- or double-stranded RNA. **11.** understand The phylogenies of SIVs and HIVs show that the two shared common ancestors, but that SIVs are ancestral to the HIVs. Also, there are plausible mechanisms for SIVs to be transmitted to humans through butchering or contact with pets, but fewer or no plausible mechanisms for HIVs to be transmitted to monkeys or chimps. **12.** apply a; the single-stranded positive-sense RNA genome of this virus can serve as an mRNA to produce viral proteins, while all of the other possibilities require viral enzymes to transcribe the genome into mRNA.

✔ Test Your Problem-Solving Skills

13. create Culture the *Staphylococcus* strain outside the human host and then add the virus to determine whether the virus kills the bacterium efficiently. Then test the virus on cultured human cells to determine whether the virus harms human cells. Then test the virus on monkeys or other animals to determine if it is safe. Finally, you could test the virus on human volunteers.
14. evaluate Prevention is currently the most cost-effective program, but it does not help people who are already infected. Treatment with effective drugs not only prolongs lives but also reduces virus loads in infected people, so that they have less chance of infecting others. **15.** apply b; by prolonging the infection in this manner, these viruses are more likely to be transmitted to a new host. **16.** evaluate Viruses cannot be considered to be alive by the definition given in (1), because viruses are not capable of replicating by themselves. By the definition given in (2), it can be argued that viruses are alive because they store, maintain, replicate, and use genetic information—although they cannot perform all these tasks on their own.

CHAPTER 37

IN-TEXT QUESTIONS AND EXERCISES

p. 732 Fig. 37.1 apply New branches and leaves should develop to the right (the plant will also lean that way); new lateral roots will develop to the left.
p. 734 Fig. 37.3 analyze Lawn grasses are shallow rooted, so when the top part of soil dries out, the plants cannot grow.
p. 735 Table 37.1 understand For aerobic respiration, which generates ATP needed to keep cells alive.
p. 736 Fig. 37.5 evaluate The individuals are genetically identical—thus, any differences in size or shape in the different habitats are due to phenotypic plasticity and not genetic differences.
p. 738 Fig. 37.6 analyze Large surface area provides more area to capture photons, but also more area from which to lose water and be exposed to potentially damaging tearing forces from wind.

p. 739 CYU (1) remember See **FIGURE A37.1 (2)** understand The generalized body of a plant has a taproot with many lateral roots and broad leaves. Examples of deviations from this generalized body structure follow. Modified roots: Unlike taproots, *fibrous roots* do not have one central root, and *adventitious roots* arise from stems. Modified stems: *Stolons* grow along the soil and grow roots and leaves at each node, and *rhizomes* grow horizontally underground. Both of these modified stems function in asexual reproduction. Modified leaves: The *needle-shaped leaves* of cacti don't lose water to transpiration compared with typical broad leaves, and they protect the plant from herbivory. *Tendrils* on climbing plants are modified leaves that do not photosynthesize, but wrap around trees or other substrates to facilitate climbing.

p. 745 Fig. 37.16 evaluate Many possible answers; for example, in leaf cells, genes encoding proteins involved in photosynthesis would be expressed. These genes wouldn't be expressed in the root cells.

p. 747 CYU (1) understand The apical meristem gives rise to the three primary meristems. The apical meristem is a single mass of cells localized at the tip of a root or shoot; the primary meristems are localized in distinctive locations behind the apical meristem. **(2)** understand Epidermal cells are flattened and lack chloroplasts; they secrete the cuticle (in shoots) or extend water and nutrient-absorbing root hairs (in roots) and protect the plant. Parenchyma cells are "metabolic workhorse cells" found throughout the plant body; they perform photosynthesis and synthesize, store, and/or transport materials. Tracheids are long, slim cells with pits in their secondary cell walls; they are found in xylem and are dead when mature, and they conduct water and solutes up the plant. Sieve-tube elements are long, thin cells found in phloem; they are alive when mature and conduct sugars and other solutes up and down the plant.

p. 750 understand Secondary growth adds cells only to the inside and outside of the vascular cambium, so the nail remains the same distance above the ground as the tree grows.

FIGURE A37.1

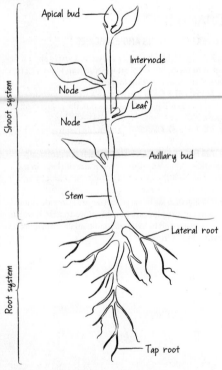

Apical bud
Internode
Node
Leaf
Node
Axillary bud
Stem
Lateral root
Tap root
Shoot system
Root system

p. 750 Fig. 37.24 remember You should have labeled a light band as early wood, and the dark band to its right as late wood; both bands together comprise one growth ring.

p. 751 CYU (1) apply The rings are small on the shaded side and large on the sunny side. **(2)** remember See **FIGURE A37.2**.

IF YOU UNDERSTAND . . .

37.1 analyze Phenotypic plasticity is more important in (1) environments where conditions vary because it gives individuals the ability to change the growth pattern of their roots, shoots, and stems to access sunlight, water, and other nutrients as the environment changes; and (2) in long-lived species because it gives individuals a mechanism to change their growth pattern as the environment changes throughout their lifetime.

37.2 analyze The zone of cellular maturation is where root hairs emerge. Root hairs are the site where most of the water and ions are absorbed in the root system. **37.3** apply (1) In both cases, the structure would stop growing; (2) the shoot would lack epidermal cells and would die. **37.4** apply (1) The plant would not produce secondary xylem and phloem, so its girth and transport ability would be reduced, though it would still produce bark. (2) The trunk would have much wider tree rings on one side than the other side.

YOU SHOULD BE ABLE TO . . .

✔ Test Your Knowledge

1. remember b **2.** understand a **3.** remember c **4.** remember a **5.** remember Cactus spines are modified leaves; thorns are modified stems. **6.** analyze Collenchyma and sclerenchyma cells both have thickened cell walls and function in providing structural support. Unlike collenchyma cells, sclerenchyma cells have lignified cell walls and are dead at maturity.

✔ Test Your Understanding

7. remember b **8.** evaluate The general function of both systems is to acquire resources: The shoot system captures light and carbon dioxide; the root system absorbs water and nutrients. Vascular tissue is continuous throughout both the shoot and root systems. Diversity in roots and shoots enables plants of different species to live together in the same environment without directly competing for resources. **9.** analyze Continuous growth enhances phenotypic plasticity because it allows plants to grow and respond to changes or challenges in their environment (such as changes in light and water availability). **10.** apply Cuticle reduces water loss; stomata facilitate gas exchange. Plants from wet habitats should have a relatively large number of stomata and thin cuticle. Plants

FIGURE A37.2

Cork
Cork cambium
Secondary phloem
Vascular cambium
Secondary xylem

living in dry habitats should have relatively few stomata and thick cuticle. **11.** analyze Parenchyma cells in the ground tissue perform photosynthesis and/or synthesize and store materials. In vascular tissue, parenchyma cells in rays conduct water and solutes across the stem; other parenchyma cells differentiate into the sieve-tube elements and companion cells in phloem. **12.** understand Cells produced to the inside of the vascular cambium differentiate into secondary xylem; cells produced to the outside of the vascular cambium differentiate into secondary phloem.

✔ Test Your Problem-Solving Skills

13. apply c **14.** remember Asparagus—stem; Brussels sprouts—lateral buds; celery—petiole; spinach—leaf (petiole and blade); carrots—taproot; potato—modified stem. **15.** create They grow continuously, so do not have alternating groups of large and small cells that form rings. **16.** analyze Girdling disrupts transport of solutes in secondary phloem. The tree's root system starves.

CHAPTER 38

IN-TEXT QUESTIONS AND EXERCISES

p. 756 apply (1) Add a solute (e.g., salt or sugar) to the left side of the U-tube, so that solute concentration is higher than on the right side. (2) Water would move from the right side of the U-tube into the left side.

p. 757 apply Pull the plunger up.

p. 757 Fig. 38.2 apply It increases (becomes less negative), which reduces the solute potential gradient between the two sides.

p. 758 CYU (1) apply When a cell is in equilibrium with pure water, the negative solute potential of the cell will be balanced by the positive pressure potential (turgor pressure). **(2)** analyze Irrigation water contains low concentrations of salts so the solute potential will be near zero. As water evaporates, salts remain in the soil, lowering the solute potential.

p. 760 Fig. 38.7 understand See **FIGURE A38.1**.

p. 761 apply The individual would not be able to exclude toxic solutes from the xylem. If these solutes were transported throughout the plant in high enough concentrations, they could damage tissues. The mutant could also not maintain ions in the xylem and so could not generate root pressure.

p. 764 Fig. 38.12 analyze The line on the graph would go up, and large jumps would occur each time the light level was turned up, because increased transpiration rates would increase negative pressure (tension) at the leaf surface and thus increase the water-potential gradient.

p. 765 CYU (1) understand At the air–water interface under a stoma, water molecules cannot hydrogen-bond in all directions—they can bond only with water molecules below them. This creates tension at the surface. When transpiration occurs, the curvature of the meniscus increases—increasing tension. **(2)** apply When a nearby stoma closes, the humidity inside the air space increases, transpiration decreases, and surface tension on menisci decreases. The same changes occur when a rain shower starts. When air outside the leaf dries, though, the opposite occurs: The humidity inside the air space decreases, transpiration increases, and surface tension on menisci increases.

p. 768 Fig. 38.17 understand See **FIGURE A38.2**.

p. 771 CYU (1) understand In early spring, growing leaves are sinks and phloem sap moves toward them. In midsummer, leaves are sources and phloem sap moves away from them. **(2)** create The movement of water and solutes in sieve elements is by bulk flow, so stationary organelles would provide resistance to the flow.

p. 771 Fig. 38.22 create With time the plant will metabolize the radioactive sucrose, and levels of radioactivity will decline.

38.1 analyze Because xylem cells are dead at maturity, there are no plasma membranes for water to cross and thus no solute potential. The only significant force acting on water in xylem is pressure. **38.2** evaluate It drops, because transpiration rates from stomata decrease, reducing the water potential gradient between roots and leaves. **38.3** understand CAM plants open their stomata only at night, and thus they transpire when temperatures are cooler. **38.4** analyze To create high pressure in phloem near sources and low pressure near sinks, water has to move from xylem to phloem near sources and from phloem back to xylem near sinks. If xylem were not close to phloem, this water movement could not occur, and the pressure gradient in phloem wouldn't exist.

YOU SHOULD BE ABLE TO . . .

✔ **Test Your Knowledge**

1. understand d 2. apply b 3. understand a 4. remember c 5. remember Endodermal cells. 6. remember Companion cells.

✔ **Test Your Understanding**

7. understand c 8. apply See **FIGURE A38.3**. 9. analyze The force responsible for root pressure is the high solute potential of root cells at night, when they continue to accumulate ions from soil and water follows by osmosis. The mechanism is active, in the sense that ions are imported

against their concentration gradient. Capillarity is driven by adhesion of water molecules to the sides of xylem cells that creates a pull upward, and by cohesion with water molecules below. Cohesion-tension, generated by transpiration, is driven by the loss of water molecules from menisci in leaves, which creates a large negative pressure (tension). Both capillarity and cohesion-tension are passive processes. **10.** understand Water moves up xylem because of transpiration-induced tension created at the air–water interface under stomata, which is communicated down to the roots by the cohesion of water molecules. Sap flows in phloem because of a pressure gradient that exists between source and sink cells, driven by differences in sucrose concentration and flows of water into or out of nearby xylem. **11.** understand By pumping protons out, companion cells create a strong electrochemical gradient favoring entry of protons. A cotransport protein (a symporter) uses this proton gradient to import sucrose molecules *against* their concentration gradient. **12.** apply When it germinates—sucrose stored inside the seed is released to the growing embryo, which cannot yet make enough sucrose to feed itself.

✔ **Test Your Problem-Solving Skills**

13. apply b 14. create Plants would not grow as quickly or as tall, because the taller they grew, the more energy they would need to use to transport water and thus the less they would have available for growth. **15.** analyze Because

they cannot readily replace water that is lost to transpiration, they have to close stomata. During a heat wave, transpiration cannot cool the plants' tissues. The trees bake to death. **16.** create Closing aquaporins will slow or stop movement of water from cells into the xylem and out of the plant via transpiration. Because water does not leave the cells, they are able to maintain turgor (normal solute potentials).

CHAPTER 39

IN-TEXT QUESTIONS AND EXERCISES

p. 776 Fig. 39.1 evaluate To test the hypothesis that most of a plant's mass comes from CO_2 in the atmosphere, design an experiment where plants are grown in the presence or absence of CO_2 and compare growth, and/or grow plants in the presence of labeled CO_2 and document the presence of labeled carbon in plant tissues. **p. 777 Fig. 39.3** evaluate Because soil is so complex, it would be difficult to defend the assumption that the soils with and without added copper are identical except for the difference in copper concentrations. Also, it would be better to compare treatments that had no copper versus normal amounts of copper, instead of comparing with and without added copper. **p. 783 Fig. 39.10** analyze The proton gradient arrow in (b) should begin above the membrane and cross the

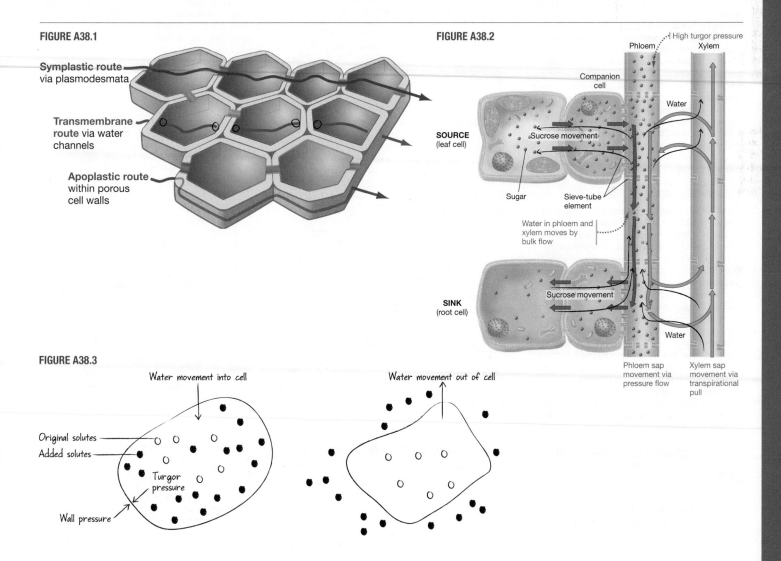

FIGURE A38.1

Symplastic route via plasmodesmata

Transmembrane route via water channels

Apoplastic route within porous cell walls

FIGURE A38.2

High turgor pressure
Phloem · Xylem
Companion cell
Water
SOURCE (leaf cell)
Sucrose movement
Sugar · Sieve-tube element
Water in phloem and xylem moves by bulk flow
SINK (root cell)
Sucrose movement
Water
Phloem sap movement via pressure flow
Xylem sap movement via transpirational pull

FIGURE A38.3

Water movement into cell
Water movement out of cell
Original solutes
Added solutes
Turgor pressure
Wall pressure

membrane pointing down. The electrical gradient arrow in (c) should also begin above the membrane (marked positive) and cross the membrane pointing down (toward negative).

p. 784a apply (1) If there is no voltage across the root-hair membrane, there is no electrical gradient favoring entry of cations, and absorption stops. (2) There is no route for cations to cross the root hair membrane, so absorption stops.

p. 784b apply (1) If there is no proton gradient across the root-hair membrane, there is no gradient favoring entry of protons, so symport of anions stops. (2) There is no route for anions to cross the root-hair membrane, so absorption stops.

p. 786 CYU (1) understand Proton pumps establish an electrical gradient across root-hair membranes, and the inside of the membrane is much more negative than the outside. Cations follow this electrical gradient into the cell. Anions are transported with protons via symporters.

(2) create See **FIGURE A39.1**.

p. 788 CYU (1) analyze Many possible answers, including the following: If N-containing ions are abundant in the soil, the energetic cost of maintaining nitrogen-fixing bacteria may outweigh the benefits in terms of increased nitrogen availability. If mycorrhizal fungi provide a plant with nitrogen-containing ions, the energetic cost of maintaining nitrogen-fixing bacteria may outweigh the benefits in terms of increased nitrogen availability. Plants that grow near species with N-fixing bacteria might gain nitrogen by absorbing nitrogen-containing ions after root nodules die, or by "stealing it" (roots of different species will often grow together). There could be a genetic constraint (see Chapter 25): Plant species may simply lack the alleles required to manage the relationship. Note: To be considered correct, your hypothesis should be plausible (based on the underlying biology) and testable. **(2)** analyze Many possible answers, including: Because the bacterial cell enters root cells, it is important for the plant to have reliable signals that it is not a parasitic bacterium. It is advantageous for the plant to be able to reject nitrogen-fixing bacteria if usable nitrogen is already abundant in the soil.

p. 790 Fig. 39.17 analyze Mistletoe is parasitic (it extracts water and certain nutrients from host trees); bromeliads are not (they are epiphytes and don't affect the fitness of their host plants).

IF YOU UNDERSTAND . . .

39.1 evaluate Add nitrogen to an experimental plot and compare growth within the plot to growth of a similar plot nearby. Do many replicates of the comparison, to convince yourself and others that the results are not due to unusual circumstances in one or a few plots. **39.2** understand Loam is a mixture of sand, clay, and organic matter so it can retain water and nutrients, and provide aeration, all properties that benefit plants. **39.3** apply A higher membrane voltage would allow cations in soil to cross root-hair membranes more readily through membrane channels and would allow anions to enter more readily via symporters. Nutrient absorption should increase. **39.4** evaluate Grow pea plants in the presence of rhizobia, exposed to air with (1) N_2 containing the heavy isotope of nitrogen and (2) radioactive carbon dioxide. Allow the plant to grow and then analyze the rhizobia and plant tissues. If the rhizobia–pea plant interaction is mutualistic, then ^{15}N-containing compounds and radioactive-carbon-containing compounds should be observed in both rhizobia and plant tissues. As a control, grow pea plants in the presence of the labeled compounds but without rhizobia. If the mutualism hypothesis is correct, the plant should contain labeled carbon but no heavy nitrogen. **39.5** analyze Despite the abundance of rain in tropical rain forests, many epiphytes there have limited access to water because their roots don't reach the soil. Like desert plants, these epiphytes are adapted to conserve water.

YOU SHOULD BE ABLE TO . . .

✔ Test Your Knowledge

1. remember b **2.** understand d **3.** remember d **4.** understand c **5.** remember Mycorrhizal fungi obtain most of their nitrogen from proteins in soil organic matter. Rhizobia obtain nitrogen from N_2 gas. **6.** remember Plant

✔ Test Your Understanding

7. remember b **8.** evaluate (1) Grow a large sample of corn plants with a solution containing all essential nutrients needed for normal growth and reproduction. Grow another set of genetically identical (or similar) corn plants in a solution containing all essential nutrients except iron. Compare growth, color, and seed production (and/or other attributes) in the two treatments. (2) Grow corn plants in solutions containing all essential nutrients in normal amounts, but with varying concentrations of iron. Determine the lowest iron concentration at which plants exhibit normal growth rates. **9.** understand If nutrients in soil are scarce, there is intense natural selection favoring alternative ways of obtaining nutrients—for example, by digesting insects or stealing nutrients.
10. apply Higher in soils with both sand and clay. Clay fills spaces between sand grains and slows (1) passage of water through soil and away from roots, and (2) leaching of anions in soil water. Clay particles also provide negative charges that retain cations so they are available to plants. The presence of sand keeps soil loose enough to allow roots to penetrate. **11.** understand Some metal ions are poisonous to plants, and high levels of other ions are toxic. Passive mechanisms of ion exclusion—such as a lack of ion channels that allow passage into root cells—do not require an expenditure of ATP. Active mechanisms of exclusion—such as production of metallothioneins—require an expenditure of ATP.
12. analyze (1) The amounts required and the mechanisms of absorption are different. Much larger amounts of C, H, and O are needed than mineral nutrients. CO_2 enters

plants via stomata, and water is absorbed passively along a water potential gradient; mineral nutrients enter via active uptake in root hairs or via mycorrhizae. (2) Macronutrients are required in relatively large quantities and usually function as components of macromolecules; micronutrients are required in relatively small quantities and usually function as enzyme cofactors.

✔ Test Your Problem-Solving Skills

13. apply The soil lost 60 g while the tree gained 74,000 g. Dividing 60 by 74,000 and converting to a percentage equals 0.08%, far less than 4%. The water that van Helmont used may have contained many of the macro- and micronutrients that were incorporated into plant mass. To test this hypothesis, conduct an experiment of the same design but have one treatment with pure water and one treatment with water containing solutes ("hard water"). Compare the percentage of soil mass incorporated into the willow tree in both treatments. Another explanation: Willow may be unusual. To test this hypothesis, repeat the experiment with willow and several other species under identical conditions, and compare the percentage of soil mass incorporated into the plants. A third hypothesis: Van Helmont's measurements were inaccurate. Repeat the experiment. **14.** analyze Acid rain inundates the soil with protons, which bind with the negatively charged clay particles—displacing the cations that are normally bound there. Cations may then be leached away. **15.** evaluate If the hypothesis is correct, vanadate should inhibit phosphorus uptake. To test this hypothesis, grow plants in solution with radioactively labeled phosphorus without vanadate and in solution with vanadate. Measure the amount of labeled phosphorus in root cells with and without vanadate. **16.** apply Alder should decompose much faster, because it provides more nitrogen to the fungi and bacteria that are responsible for decomposition—meaning that they can grow faster.

FIGURE A39.1

1. Passive exclusion: Root hairs have no membrane proteins to transport Cd^{2+}

2. Active exclusion: Metallothionein proteins bind to Cd^{2+} that enter cells, making them harmless

3. Active exclusion: Proteins in tonoplast move Cd^{2+} into vacuole for safe storage

4. Passive exclusion: Endodermal cells have no membrane proteins to transport Cd^{2+} past the Casparian strip

CHAPTER 40

IN-TEXT QUESTIONS AND EXERCISES

p. 796 CYU (1) [understand] The only cells that can respond to a specific environmental signal or hormone are cells that have an appropriate receptor for that signal or hormone. **(2)** [remember] Once a receptor is activated, it triggers production of many second messengers or the activation of many proteins in a phosphorylation cascade.

p. 797 Fig. 40.4 [evaluate] Wild-type plants already have a normal *PHOT1* gene and can respond to directional blue light. Inserting a defective gene into a wild-type plant would have no effect.

p. 798 Fig. 40.5 [evaluate] For "Tip removed": Cut the tip and replace it, to control for the hypothesis that the wound itself influences bending (not the loss of the tip). For "Tip covered": Cover the tip with a transparent cover instead of opaque one, to control for the hypothesis that the cover itself affects bending (not excluding light).

p. 799 CYU (1) [apply] Using a syringe or other device, apply auxin to the west side of each stem, just below the tip. **(2)** [apply] Stems would become very long and would not respond to directional light.

p. 801 Table 40.1 [analyze] The average germination rate of lettuce seeds last exposed to red light is about 99 percent. The average for seeds last exposed to far-red light is about 50 percent. Both of these values are much higher than the germination rate of buried seeds that receive no light at all; their germination rate is only 9 percent.

p. 803 CYU (1) [understand] Lettuce grows best in full sunlight. If a seed receives red light, it indicates that the seed is in full sunlight—meaning that conditions for growth are good. But if a seed receives far-red light, it indicates that the seed is in shade—meaning that conditions for growth are poor. **(2)** [understand] Night length changes very predictably throughout the year, so it is a reliable signal for plants to perceive.

p. 803 Fig. 40.10 [apply] An appropriate control treatment would be to graft a leaf from a plant that had never been exposed to a short-night photoperiod onto a new plant that also had never been exposed to the correct conditions for flowering. If the hypothesis is correct, the grafted leaf would not induce flowering.

p. 804 Fig. 40.12 [evaluate] (One possibility) Direct pressure from a statolith changes the shape of the receptor protein. The shape change triggers a signal transduction cascade, leading to a gravitropic cell response.

p. 805 CYU (1) [analyze] In both roots and shoots, auxin is redistributed asymmetrically in response to a signal. In phototropism, auxin is redistributed to the shaded side of the shoot tip in response to blue light. In gravitropism, auxin is redistributed to the lower part of the root or shoot in response to gravity. **(2)** [understand] In root cells, auxin concentrations indicate the direction of gravity. The gravitropic response is triggered by changes in the distribution of auxin in root-tip cells.

p. 809 Fig. 40.19 [analyze] The dwarfed individuals can put more of their available energy into reproduction because they are using less energy for growth than taller individuals.

p. 810 Fig. 40.21 [analyze] Without these data, a critic could argue that the stomata closed in response to water potentials in the leaf, not a signal from the roots.

p. 813 CYU (1) [analyze] ABA: Stomata should close and photosynthesis should stop compared to individuals that do not receive extra ABA. GA: Their stems should elongate rapidly, compared to plants that do not receive extra GA. **(2)** [understand] The rotten apple produces ethylene, which, being a gas, stimulates all of the other apples in the bushel basket to ripen too fast.

p. 816 [analyze] Sacrificing a small number of cells due to the hypersensitive response is far less costly for the plant than death.

p. 819 CYU (1) [analyze] Because specific receptor proteins recognize and match specific proteins produced by pathogens, having a wide array of these proteins allows individuals to recognize and respond to a wide array of pathogens. **(2)** [understand] If herbivores are not present, synthesizing large quantities of proteinase inhibitors wastes resources (ATP and substrates) that could be used for growth and reproduction.

IF YOU UNDERSTAND . . .

40.1 [analyze] An external signal arrives at a receptor cell/protein (a person sees a barn on fire); the person calls (the receptor transduces the signal); the dispatcher rings a siren (a second messenger is produced or phosphorylation cascade occurs); the fire department members get to the pumper truck (gene expression or some other cell activity changes). **40.2** [apply] Sun-adapted (high-light) plants are likely to have a significant positive phototropic response (i.e., growth toward light), whereas shade-adapted (low-light) plants are less likely to respond phototropically. **40.3** [evaluate] They do not perform photosynthesis and therefore do not need to switch behavior based on being located in full sunlight versus shade. **40.4** [understand] Starch is denser than water, so it falls quickly to the bottom of cells. **40.5** [understand] A reduction in plant height in a windy environment might keep them upright. **40.6** [evaluate] Those hormones need to be transported throughout the plant, because cells throughout the body need to respond to the signal—not in a single direction. **40.7** [understand] In the hypersensitive response, a few cells are sacrificed to prevent a pathogen from spreading throughout the plant.

YOU SHOULD BE ABLE TO . . .

✓ Test Your Knowledge

1. [remember] d **2.** [remember] b **3.** [remember] d **4.** [understand] b **5.** [remember] hypersensitive response **6.** [remember] plasma membrane

✓ Test Your Understanding

7. [remember] a **8.** [remember] The P_{fr}–P_r switch in phytochromes lets plants know whether they are in shade or sunlight, and triggers responses that allow sun-adapted plants to avoid shade. Phototropins are activated by blue light, which indicates full sunlight, and trigger responses that allow plants to photosynthesize at full capacity. **9.** [analyze] Transduce means to convert energy from one form to another. There are many examples: A touch may be converted to an electrochemical potential, light energy or pressure into a shape change in a protein, and so on. **10.** [understand] (Many possibilities) Cells or tissue involved: Auxin directs gravitropism in roots but phototropism in shoots. Developmental stage or age: GA breaks dormancy in seeds but extends stems in older individuals. Concentration: Large concentrations of auxin in stems promote cell elongation; small concentrations do not. Other hormones: The concentration of GA relative to ABA influences whether germination proceeds. **11.** [apply] Ethylene promotes senescence; increased ethylene sensitivity in leaves (combined with reduced auxin concentrations) causes abscission; fruits exposed to ethylene ripen faster than fruits that are not exposed to ethylene. Produce dealers can maximize profits by delaying ripening during transit of fruit by removing ethylene and then adding ethylene just before fruits are displayed. **12.** [apply] When one leaf on the mutant plant is chewed on by a herbivore, the damaged leaf would not be able to signal to the rest of the plant, so it would not be able to fend off future herbivore attacks.

✓ Test Your Problem-Solving Skills

13. [apply] a **14.** [evaluate] Small-seeded plants need to perform photosynthesis early in seedling development or they will starve. Therefore, they need to germinate in direct sunlight. The food reserves in large-seeded plants can support seedling growth for a relatively long time without photosynthesis. Therefore, they do not need to germinate in direct sunlight. **15.** [evaluate] In these seeds, (a) little or no ABA is present, or (b) ABA is easily leached from the seeds, so its inhibitory effects are eliminated. Test hypothesis (a) by determining whether ABA is present in the seeds. Test hypothesis (b) by comparing the amount of ABA present before and after running water—enough to mimic the amount of rain that falls in a tropical rain forest over a few days or weeks—over the seeds. **16.** [apply] New types of plant resistance provide a selective pressure for pathogens to evolve new mechanisms to evade the resistance.

CHAPTER 41

IN-TEXT QUESTIONS AND EXERCISES

p. 825 [understand] (1) The microsporocyte divides by meiosis to generate male spores; the megasporocyte divides by meiosis to generate female spores. (2) In angiosperms, the female gametophyte stays within the ovary at the base of the flower even after it matures and produces an egg cell. Fertilization and seed development take place in the same location.

p. 829 CYU (1) [analyze] Both are diploid cells that divide by meiosis to produce spores. The megasporocyte produces a megaspore (female spore); a microsporocyte produces a microspore (male spore). **(2)** [analyze] Both are multicellular individuals that produce gametes by mitosis. The female gametophyte is larger than the male gametophyte and produces an egg; the male gametophyte produces sperm.

p. 829 Fig. 41.8 [understand] A gametophyte is the multicellular individual that produces gametes by mitosis. The pollen grain is a gametophyte because it is multicellular and produces sperm by mitosis.

p. 832 CYU (1) [understand] Insects feed on nectar and/or pollen in flowers. Flowers provide food rewards to insects to encourage visitation. The individuals that attract the most pollinators produce the most offspring. **(2)** [understand] One product of double fertilization is the diploid zygote, which will eventually grow by mitosis into a mature sporophyte. The other product is the triploid endosperm nucleus, which will grow by mitosis to form a source of nutrients for the embryo.

p. 833 [remember] The endosperm nucleus in the central cell is triploid; the zygote is diploid; the synergid and other cells remaining from the female gametophyte are haploid.

p. 834 [analyze] A cotyledon and a radicle develop into organs with distinct functions, but both are composed of the same tissues: Epidermis is derived from protoderm, ground tissue is derived from ground meristem, and vascular tissue is derived from the procambium.

p. 835 Fig. 41.14 [evaluate] Hypothesis: Fruit changes color when it ripens as a signal to fruit eaters. The color change is advantageous because seeds are not mature in unripe fruit and are unlikely to survive. Fruit eaters disperse mature seeds in ripe fruit, which are more likely to survive.

p. 836 Fig. 41.16 [evaluate] Hackberries served as a control to demonstrate that animals that avoided chilies were hungry,

p. 837 Fig. 41.17 [understand] Beans—their cotyledons are aboveground.

IF YOU UNDERSTAND . . .

41.1 [remember] The sporophyte is diploid, the gametophyte is haploid, spores and gametes are haploid, and zygotes are diploid. **41.2** [apply] There would be four embryo sacs in each ovule, and thus four eggs. **41.3** [apply] Endosperm tissue is generated when two haploid, female nuclei fuse with one haploid sperm nucleus, forming a cell with three sets of chromosomes. In contrast, a zygote is formed by the fusion of one haploid sperm nucleus and one haploid egg, forming a cell with two sets of chromosomes. **41.4** analyze A cold requirement for germination ensures that seeds from plants native to habitats that experience cold winters will not germinate in the fall and fail to become established before winter.

YOU SHOULD BE ABLE TO . . .

✔ Test Your Knowledge

1. remember b **2.** understand c **3.** remember a **4.** understand d
5. remember Megasporocytes (female) and microsporocytes (male) undergo meiosis to generate megaspores and microspores, respectively. These divide mitotically to give rise to female and male gametophytes—the embryo sac and pollen grain, respectively. **6.** remember Pollination is the transfer of pollen from an anther to a stigma. Fertilization is the fusion of haploid gametes.

✔ Test Your Understanding

7. understand c **8.** analyze Asexual reproduction is a quick, efficient way for a plant to produce large numbers of offspring. The disadvantage is that offspring are genetically identical to the parent and thus are vulnerable to the same diseases and able to thrive only in habitats similar to those inhabited by the parent. **9.** analyze Sepals are the outermost structures in a flower and usually protect the structures within; petals are located just inside the sepals and usually function to advertise the flower to pollinators. In wind- and bee-pollinated flowers, sepals probably do not vary much in overall function or general structure. Petals are often lacking in wind-pollinated species; they tend to be broad and flat (to provide a landing site) and colored purple, blue, or yellow in bee-pollinated species (often with ultraviolet markings). **10.** analyze Outcrossing increases genetic diversity among offspring, making them more likely to thrive if environmental conditions change from the parental generation. However, outcrossing requires cross-pollination to be successful. Self-fertilization results in relatively low genetic diversity among offspring but ensures that pollination succeeds. **11.** remember The female gametophyte develops inside the ovule; the ovary develops into the pericarp of the fruit after fertilization. A mature ovule contains both gametophyte and sporophyte tissue; the ovary and carpel are all sporophyte tissue. **12.** analyze The endosperm of a corn seed and the cotyledons of a bean seed both contain nutrients needed for the seed to germinate. The bean cotyledons have absorbed the nutrients of the endosperm.

✔ Test Your Problem-Solving Skills

13. remember c **14.** apply Subtracting the number of used pollen grains per plant (600) from the number of pollen grains generated per plant (10 million) divided by 10 million reveals that 99.994% of the pollen is wasted. **15.** evaluate In response to "cheater" plants, insects should be under intense selection pressure to detect and avoid species that do not offer a food reward. Individuals would have to be able to distinguish scents, colors, or other traits that identify cheaters. In response to cheater insects, plants possessing unusually thick or toxic petals or sepals that prevented insects from bypassing the anthers on their way to the nectaries would be more successful at producing seed in the next generation. **16.** evaluate Acorns have a large, edible mass and are usually animal dispersed (e.g., by squirrels that store them and forget some). Cherries have an edible fruit and are animal dispersed. Burrs stick to animals and are dispersed as the animals move around. Dandelion seeds float in wind. To estimate the distance that each type of seed is dispersed from the parent, (1) set up "seed traps" to capture seeds at various distances from the parent plant; (2) sample locations at various distances from the parent and analyze young individuals—using techniques introduced in Chapter 21 to determine if they are offspring from the parent being studied; (3) mark seeds, if possible, and find them again after dispersal.

BIG PICTURE Plant and Animal Form and Function

p. 840 CYU (1) understand Plants and animals may accumulate compounds in the epidermis and skin (such as waxes and other lipids) that reduce water loss. Both plants and animals evolved closable openings in this barrier to allow regulated gas exchange. Plants and animals may also change the amount of the body exposed to the sun (in plants by moving leaves, in animals by changing posture or taking shelter) to reduce evaporative water loss.
(2) understand Many plants and animals keep their eggs inside their bodies and depend on sperm being deposited near them rather than releasing both to the environment where they could dry out. **(3)** understand Very large animals and plants need to build strong supporting structures such as thick bones or trunks to withstand the force of gravity. They also need elaborate internal transport systems to bring nutrients to each cell and carry wastes away. **(4)** analyze Cellular respiration fits into one of the key functions for survival—Nutrition. Both plants and animals depend on sugar (produced by plants via photosynthesis) as their energy source to make ATP via cellular respiration.

CHAPTER 42

IN-TEXT QUESTIONS AND EXERCISES

p. 844 Fig. 42.2 create There would be no difference between experimental and control males.
p. 847 Fig. 42.4 analyze Based on the projections called axons that go to specific cells and tissues, neurons transmit signals to specific tissues rather than broadly throughout the body.
p. 850 CYU analyze *Connective tissues* consist of cells embedded in a matrix. The density of the matrix determines the rigidity of the connective tissue and its function in padding/protection, structural support, or transport. *Nervous tissue* is composed of neurons and support cells. Neurons have long projections that function as "biological wires" for transmitting electrical signals. *Muscle tissue* has cells that can contract and functions in movement. *Epithelial tissue* has tightly packed layers of cells with distinct apical and basal surfaces. They protect underlying tissues and regulate the entry and exit of materials into these tissues.
p. 850 apply A newborn.
p. 851 Fig. 42.10 apply The puppy has to eat relatively more because it has a higher mass-specific metabolic rate than an adult dog.
p. 852 Fig. 42.11 apply There should be no oxygen uptake on either side of the chamber.
p. 853 CYU (1) understand As three-dimensional size increases, volume increases more quickly than does surface area, so the surface area/volume ratio decreases.
(2) analyze The squirrel has a higher mass-specific metabolic rate because it is a smaller animal. It therefore needs more energy per unit body mass than a bear.
p. 858 CYU (1) analyze Endotherms can remain active during the winter and at night and sustain high levels of aerobic activities such as running or flying, but they require large amounts of food energy. Ectotherms need much less food and can devote a larger proportion of their food intake to reproduction, but they have a hard time maintaining high activity levels at night or in cold weather. **(2)** apply The whale's tongue would become colder because not as much heat would be transferred from warm arterial blood to cool venous blood with concurrent flow as with countercurrent flow, and as a result more heat from the arterial blood would be lost to the environment.

IF YOU UNDERSTAND . . .

42.1 understand Some adaptations to cold environments in mammals include the evolution of thick coats of hair,

---

countercurrent heat exchangers in the extremities, and dark coloration to absorb radiation. Acclimatization may include further thickening and darkening of hair upon exposure to cold. **42.2** apply Loosening tight junctions would compromise the barrier between the external and internal environments. It would allow water and other substances (including toxins) to pass more readily across the epithelial boundary. **42.3** create King Kong is endothermic, and his huge mass would generate a great deal of heat. His relatively small surface area, especially as compared to a normal-sized gorilla, would not be able to dissipate the heat—especially if it were covered with fur. **42.4** create Individuals could acclimatize by normal homeostatic mechanisms (increased sweating, seeking shade, etc.). Individuals with alleles that allowed them to function better at higher temperatures would produce more offspring than individuals without those alleles. Over time, this would lead to adaptation via natural selection. **42.5** create *Apatosaurus* was likely a homeotherm because of its huge size. Because of its low surface area/volume ratio, metabolically produced heat would be trapped inside the dinosaur's body. Even if exposed to cold or warm temperatures, the dinosaur's temperature would not change much because it takes a long time for it to lose or gain heat across such a relatively small surface area.

YOU SHOULD BE ABLE TO . . .

✔ Test Your Knowledge

1. understand True **2.** remember Epithelial **3.** understand b
4. understand d **5.** understand a **6.** analyze a

✔ Test Your Understanding

7. understand It is where environmental changes are sensed first. As an effector, it has a large surface area for losing or gaining heat. **8.** analyze A warm frog can move, breathe, and digest much faster than a cold frog, because enzymes work faster (rates of chemical reactions increase) at 35°C than at 5°C. A warm frog needs much more food, though, to support this high metabolic rate. **9.** understand *Absorptive sections* have many folds and projections that increase their surface area for absorption. *Capillaries* have a high surface area because they are thin and highly branched, making exchange of fluids and gases efficient. *Beaks of Galápagos finches* have sizes and shapes that correlate with the type of food each species eats. Large beaks are used to crack large seeds; long, thin beaks are used to pick insects off surfaces; etc. *Fish gills* are thin, flattened structures with a large surface area, which facilitates the efficient exchange of gases and wastes. **10.** analyze A larger sphere has a relatively smaller surface area for its interior volume than does a smaller sphere, because surface area increases with size at a lower rate than volume does. Diffusion will be more efficient in the smaller sphere. **11.** apply See **FIGURE A42.1**. The graph shows that as size (length) increases, the volume increases more rapidly than the surface area. **12.** understand c. This is an example of negative feedback because insulin is released in response to an increase in a parameter (blood sugar concentration)

FIGURE A42.1

and acts to reduce that parameter so that homeostasis is maintained. In the other examples, homeostasis is not maintained.

✔ Test Your Problem-Solving Skills

13. `create` Large individuals require more food, so are more susceptible to starvation if food sources can't be defended (e.g., when seeds are scattered around). They may also be slower and/or more visible to predators. **14.** `create` You would have to show changes in the frequencies of alleles whose products are affected by temperature—for example, increases in the frequencies of alleles for enzymes that operate best at higher temperature. **15.** `create` The heat-dissipating system should have a very high surface area/volume ratio. Based on biological structures, it should be flattened (thin) and highly folded and branched, and/or contain tubelike projections. **16.** `understand` a. Turtles are ectothermic, so they do not expend much energy to produce body heat. Mice are endothermic, and expend energy to regulate body temperature. Mice will therefore have a higher basal metabolic rate and consume more food.

CHAPTER 43

IN-TEXT QUESTIONS AND EXERCISES

p. 863 Fig. 43.1 `understand` The black and white molecules are moving down their concentration gradients, and the concentration of red molecules does not affect the concentration of black or white molecules.

p. 867 Fig. 43.6 `understand` The sodium–potassium ATPase performs primary active transport; the sodium–chloride–potassium cotransporter performs secondary active transport; the chloride and potassium channels are involved in passive transport.

p. 869 CYU `apply` The osmolarity of the fish's tissue would decrease because the pumps would no longer be able to import sodium.

p. 869 Fig. 43.8 `analyze` The underside of the abdomen is shaded by the grasshopper's body. This location is relatively cool and should reduce the loss of water during respiration.

p. 871 CYU (1) `understand` Uric acid is insoluble in water and can be excreted without much water loss. **(2)** `understand` When electrolytes are reabsorbed in the hindgut epithelia, water follows along an osmotic gradient.

p. 873 `understand` Blood contains cells and large molecules as well as ions, nutrients, and wastes; the filtrate contains only ions, nutrients, and wastes.

p. 876 `apply` (1) Less water reabsorption will occur because the osmotic gradient is not as steep. (2) Filtrate osmolarity will be lower because less water has been reabsorbed. (3) Salt reabsorption will be reduced because the concentration of NaCl in the filtrate is lower.

p. 878 `apply` Ethanol consumption inhibits water reabsorption, leading to a larger volume of less concentrated urine. Nicotine consumption increases water reabsorption, leading to a smaller volume of more concentrated urine.

p. 878 CYU `apply` Water intake increases blood pressure and filtration rate in the renal corpuscle. It leads to lowered electrolyte concentration in the blood and filtrate and the production of large volumes of dilute urine. Eating large amounts of salt results in concentrated, hyperosmotic urine. Water deprivation triggers ADH release and the production of concentrated, hyperosmotic urine.

p. 879 Fig. 43.19 `create` Carrying a large volume of water in the bladder may be energetically costly to the animal. The animal also might not be able to run as quickly and may have an increased predation risk.

IF YOU UNDERSTAND . . .

43.1 `understand` A sodium gradient establishes an osmotic gradient that moves water. It also sets up an electrochemical gradient that can move ions against their electrochemical gradient by secondary active transport through a cotransporter, or with their electrochemical gradient by passive diffusion through a channel. **43.2** `analyze` Cartilaginous fishes such as sharks maintain a high concentration of urea in their blood, which raises their osmolarity to about the same as seawater. Therefore, they lose very little water to the environment. Bony fishes drink seawater and maintain their osmolarity lower than that of seawater by actively transporting excess ions out of the gills. Sharks actively excrete sodium chloride through a rectal gland. **43.3** `understand` When comparing chloride cells in gills of salmon in the ocean versus freshwater, the cells are in different locations on the gills, different forms of Na^+/K^+-ATPase exist, and $Na^+K^+Cl^-$ transporters are located on the basolateral side of the membrane in ocean salmon and on the apical side of the membrane in freshwater salmon. **43.4** `apply` Aquatic insects produce more urea, and especially more ammonia, than terrestrial insects because these nitrogenous wastes are energetically less expensive to produce than uric acid. Ammonia and urea require more water to be excreted, but aquatic insects have access to plenty of water. **43.5** `create` The longer the loop of Henle, the steeper the osmotic gradient in the interstitial fluid in the medulla. Steep osmotic gradients allow a great deal of water to be reabsorbed as urine passes through the collecting duct, resulting in highly concentrated urine.

YOU SHOULD BE ABLE TO . . .

✔ Test Your Knowledge

1. `remember` True **2.** `understand` c **3.** `apply` a **4.** `remember` d **5.** `understand` c **6.** `remember` bladder

✔ Test Your Understanding

7. `create` b. Fishes excrete nitrogenous wastes in the form of ammonia. Although the ammonia can easily diffuse across the fish's gills into the water, freshwater ponds are often small and ammonia can build up to potentially toxic levels. **8.** `analyze` Ammonia is toxic and must be diluted with large amounts of water to be excreted safely. Urea and uric acid are safer and do not have to be excreted with large amounts of water but are more expensive to produce in terms of energy expenditure. Fish excrete ammonia; mammals excrete urea; insects excrete uric acid. You would expect the embryos inside terrestrial eggs to excrete uric acid, since it is the least toxic and is insoluble in water. **9.** `understand` Mitochondria are needed to produce ATP. A key component of salt regulation in fish is the Na^+/K^+-ATPase, which requires ATP to function. Because ATP fuels establishment of the ion gradients necessary for the cotransport mechanisms used by chloride cells as well as other epithelial cells, an abundance of mitochondria would be expected in a chloride cell. **10.** `create` The wax layer should be thicker in insects inhabiting dry habitats, because the animals there are under more osmotic stress (more evaporation due to low humidity; less available drinking water). Insects that live in extremely humid habitats may lack the ability to close their spiracles; animals that live in humid habitats may excrete primarily ammonia or urea and have relatively short loops of Henle. In each case, you could test these predictions by comparing individuals of closely related species that live in dry versus humid habitats. **11.** `create` Without a loop of Henle, there would be no concentration gradient in the medullary interstitial fluid, so water would not be absorbed from the pre-urine in the collecting duct; concentrated urine could not be formed. **12.** `create` Freshwater has such a low osmolarity that if invertebrates osmoconformed to it, they would lack sufficient quantities of ions to conduct electrical currents in their excitable tissues.

✔ Test Your Problem-Solving Skills

13. `apply` c. Without aquaporins in the collecting duct, water cannot be reabsorbed, which would result in increased urine volume and decreased urine osmolarity. **14.** `create` The mammal's urine volume would increase dramatically. Ouabain would poison Na^+/K^+-ATPase pumps in the nephron, which would prevent reabsorption of ions and nutrients. Without the osmotic gradient established by reabsorption of ions and nutrients, the water would remain in the nephron and exit as urine. **15.** `analyze` The observation that the Na^+ concentration decreased by 30 percent, even as volume also decreased, indicates that Na^+ ions were reabsorbed to a greater degree than water. The observation that the urea concentration increased by 50 percent indicates that urea was not reabsorbed. **16.** `analyze` The hypothesis that the mussels are osmoconformers is supported because their hemolymph is very close in osmolarity to that of the wide range of water osmolarities into which they are placed.

CHAPTER 44

IN-TEXT QUESTIONS AND EXERCISES

p. 891 `understand` Nutrient absorption occurs in the small intestine, but not in the esophagus or stomach, and the rate of absorption increases with available surface area.

p. 894 Fig. 44.14 `understand` Frog eggs don't normally make the sodium-glucose cotransporter protein, so the researchers could be confident that if the protein appeared, it was from the injected RNA. They would not be confident of this if the RNAs were injected into rabbit epithelial cells, where the protein was probably already present.

p. 896 `apply` (1) Decreased energy yield from foods due to reduced glucose absorption, (2) increased fecal glucose content due to the passing of unabsorbed glucose into the colon, (3) watery feces due to decreased water reabsorption in the large intestine (lower osmotic gradient). (An additional effect would be increased flatulence due to the metabolism of unabsorbed glucose by bacteria that produce gases as waste products.)

p. 896 CYU (1) `understand` *Mouth:* Food is taken in; teeth physically break down food into smaller particles; salivary amylase begins to break down carbohydrates; lingual lipase initiates the digestion of fats. *Esophagus:* Food is moved to the stomach via peristaltic contractions. *Stomach:* HCl denatures proteins; pepsin begins to digest them. *Small intestine:* Pancreatic enzymes complete the digestion of carbohydrates, proteins, lipids, and nucleic acids. Most of the water and all of the nutrients are absorbed here. *Large intestine:* Water is reabsorbed and feces are formed. *Anus:* Feces accumulate in the rectum and are expelled out the anus. **(2)** `apply` If the release of bile salts is inhibited, fats would not be digested and absorbed efficiently, and they would pass into the large intestine. The individual would likely produce fatty feces and lose weight over time. Trypsin inactivation would inhibit protein digestion and reduce absorption of amino acids, likely leading to weight loss and muscle atrophy due to protein deprivation. If the Na^+-glucose cotransporter is blocked, then glucose would not be absorbed into the bloodstream. The individual would likely become sluggish from lack of energy.

p. 897 Fig. 44.16 `understand` In type 2 diabetes mellitus, the top, black arrow from glucose (in the blood) to glycogen (inside liver and muscle cells) is disrupted. In type 1 diabetes mellitus, the green arrow on the top left, indicating insulin produced by the pancreas, is disrupted.

p. 898 `understand` Individuals with type 1 diabetes mellitus have normal insulin receptors on their liver cells but do not produce and release insulin from the pancreas. Individuals with type 2 diabetes mellitus produce and release insulin, but they have fewer insulin receptors and/or these receptors do not respond as well to insulin. In both types of diseased individuals, glucose concentrations in the blood remain high. In individuals without disease, both insulin production and insulin receptors are normal, so liver cells respond to insulin by taking up glucose from the blood and storing it.

p. 898 Fig. 44.17 analyze The data do not prove that obesity causes type 2 diabetes mellitus. The study provides correlational data and cannot directly assign causation. However, the study provides strong evidence of an association between obesity and type 2 diabetes.

p. 899 CYU (1) understand Type 1 diabetes mellitus is an autoimmune disease, meaning that the body's immune system mistakenly kills the insulin-producing cells of the pancreas. Type 2 diabetes mellitus is caused by a combination of factors, including genetic predisposition, obesity, and poor diet. **(2)** apply When the blood sugar of individuals with type 1 diabetes mellitus gets too high, they should inject themselves with insulin, which will trigger the absorption and storage of glucose by their cells. When their blood glucose gets too low, they should eat something with a high concentration of sugar (such as orange juice or a candy bar) to increase their blood glucose levels quickly. (If glucose is dangerously low, a glucagon shot may be administered.)

IF YOU UNDERSTAND . . .

44.1 analyze One serving of skim milk has approximately 80 kcal, while whole milk has about 152 kcal. **44.2** apply Its mouth or tongue would be modified to make probing of flowers and sucking nectar efficient. It would not require teeth, and its digestive tract would be relatively simple. For example, the stomach would not have to pulverize food or digest large amounts of protein, and the large intestine would not have to process and store large amounts of bulky waste. **44.3** analyze Amylase will work best at a neutral or slightly alkaline pH (typical of mouth and small intestine), whereas pepsin will work best at a low pH (typical of the stomach). **44.4** understand Type 1 diabetes mellitus leads to high glucose because insulin is not produced, and therefore glucose cannot be moved from blood into cells. Type 2 diabetes mellitus leads to high blood glucose because insulin does not bind in sufficient amounts to its receptor, so glucose cannot be transported from blood into cells.

YOU SHOULD BE ABLE TO . . .

✔ Test Your Knowledge

1. remember a **2.** understand d **3.** remember d **4.** remember b **5.** remember insulin, glucagon **6.** remember False

✔ Test Your Understanding

7. understand The bird crop is an enlarged sac that can hold quickly ingested food; in leaf-eating species it is filled with symbiotic organisms and functions as a fermentation vessel. The cow rumen is an enlarged portion of the stomach; the elephant cecum is enlarged relative to other species. Both structures are filled with symbiotic organisms and function as fermentation vessels. **8.** analyze Digestive enzymes break down macromolecules (proteins, carbohydrates, nucleic acids, lipids). If they weren't produced in an inactive form, they would destroy the cells that produce and secrete them. **9.** understand b **10.** understand Pima living in the United States are genetically similar to Pima living in Mexico, but they have higher rates of type 2 diabetes. This suggests that the difference in diabetes rates is due to environmental factors such as diet and exercise. **11.** understand Humans have sharp canines and incisors, which may help to bite and tear meat. Humans also have flat molars, which help in grinding grains and other plant material. **12.** apply In a type 1 diabetic who took too much insulin, the blood glucose would become very low. The person would become sluggish and disoriented, and might slip into a diabetic coma.

✔ Test Your Problem-Solving Skills

13. create A proton pump inhibitor reduces activity of the proton pumps in parietal cells, reducing the amount of HCl produced. The contents of the stomach would therefore become less acidic, helping to prevent acid reflux. **14.** analyze b **15.** evaluate The result is still valid because the injection caused secretion—if it is also true that lack of injection led to no secretion. The criticism is also somewhat valid, however, because the researchers couldn't rule out the hypothesis that signaling from nerves plays some sort of role, too. **16.** create Terrestrial animals are exposed to increased risk of water loss, and the large intestine is where water reabsorption occurs. In most cases, fish do not need to reabsorb large amounts of water from their feces.

CHAPTER 45

IN-TEXT QUESTIONS AND EXERCISES

p. 905 CYU (1) analyze Oxygen partial pressure is high in mountain streams because the water is cold, mixes constantly, and has a high surface area (due to white water). Oxygen partial pressure is low at the ocean bottom because that is far from the surface, where gas exchange takes place, and there is relatively little mixing. **(2)** create *Warm-water species:* Large amount of air, because the oxygen-carrying capacity of warm water is low. *Vigorous algal growth:* Small amount of air, because algae contribute oxygen to the water through photosynthesis. *Sedentary animals:* Small amount of air, because sedentary animals require relatively little oxygen.

p. 906 Fig. 45.4 analyze External gills are ventilated passively and are efficient because they are in direct contact with water. They are exposed to predators and mechanical damage, however. Internal gills are protected but have to be ventilated by some type of active mechanism for water flow.

p. 907 Fig. 45.6 apply If "concurrent flow" occurred, oxygen transfer would stop, because the partial pressure gradient driving diffusion would fall to zero partway along the length of the capillary.

p. 909 Fig. 45.7 analyze Using many animals increases confidence that the results are true for most or all members of the population and are not due to one or a few unusual individuals or circumstances.

p. 912 CYU (1) analyze Common features include large surface area, short diffusion distance (a thin gas exchange membrane), and a mechanism that keeps fresh air or water moving over the gas exchange surface. Only fish gills use a countercurrent exchange mechanism; only tracheae deliver oxygen directly to cells without using a circulatory system; only lungs (mammalian) contain "dead space"—areas that are not involved in gas exchange. **(2)** The P_{O_2} would go down as oxygen is used up, the P_{CO_2} would go up as CO_2 diffuses into the blood but cannot be exhaled, and pH would drop as the CO_2 dissolves in blood to form bicarbonate and H^+ ions.

p. 913 create There would be an even larger change in the oxygen saturation of hemoglobin in response to an even smaller change in the partial pressure of oxygen.

p. 914 Fig. 45.16 analyze According to the data in the figure, when the tissue P_{O_2} is 30 mm Hg, the oxygen saturation of hemoglobin is about 15 percent for blood at pH 7.2 and about 25 percent at pH 7.4. Therefore, about 85 percent of the oxygen is released from hemoglobin at pH 7.2, but only about 75 percent of the oxygen is released at pH 7.4.

p. 915 Fig. 45.18 analyze In the lungs, a strong partial pressure gradient favors diffusion of dissolved CO_2 from blood into the alveoli. As the partial pressure of CO_2 in the blood declines, hydrogen ions leave hemoglobin and react with bicarbonate to form more CO_2, which then diffuses into the alveoli and is exhaled from the lungs.

p. 916 CYU analyze The curve for Tibetans should be shifted to the left relative to people adapted to sea level—meaning that their hemoglobin has a higher affinity for oxygen at all partial pressures.

p. 921 Fig. 45.24 analyze Air from the alveoli mixes with air in the "dead space" in the bronchi and trachea on its way out of the body. This dead space air is from the previous inhalation (P_{O_2} = 160 mm Hg; P_{CO_2} = 0.3), so when the alveolar air mixes with the dead space air, the partial pressures in the exhaled air achieve levels intermediate between that of inhaled and alveolar air.

p. 922 apply Without the delay at the AV node, the ventricles would not have the chance to fully fill with blood from the atria. As a consequence, the volume of blood ejected from the ventricles would decline.

p. 924 CYU (1) understand Blood plasma that leaks out of the capillaries and is not reabsorbed into capillaries enters lymphatic vessels that eventually merge with blood vessels. **(2)** understand See FIGURE A45.1.

IF YOU UNDERSTAND . . .

45.1 understand Cellular respiration. **45.2** create It is harder to extract oxygen from water than it is to extract it from air. This limits the metabolic rate of water breathers. **45.3** analyze Large animals have a relatively small body surface area relative to their volume. If they had to rely solely on gas exchange across their skin, the skin surface area would not be large enough to exchange the volume of oxygen needed to meet their metabolic needs. **45.4** create Since O_2 cannot compete as well as CO for binding sites in hemoglobin, O_2 transport decreases. As oxygen levels in the blood drop, tissues (most crucially in the brain) become deprived of oxygen, and suffocation occurs. **45.5** apply Their circulatory systems should have relatively high pressures—achieved by independent systemic and pulmonary circulations powered by a four-chambered heart—to maximize the delivery of oxygenated blood to metabolically active tissues.

YOU SHOULD BE ABLE TO . . .

✔ Test Your Knowledge

1. understand b **2.** remember c **3.** understand d **4.** understand d **5.** remember c **6.** understand An open circulatory system is less efficient than a closed circulatory system because it is not capable of directing the hemolymph toward specific organs, and because the pressure is lower, resulting in slower flow of hemolymph.

✔ Test Your Understanding

7. understand During exercise, P_{O_2} decreases in the tissues and P_{CO_2} increases. The increase in P_{CO_2} lowers tissue pH. The drop in P_{O_2} and pH causes more oxygen to be released from hemoglobin. **8.** understand In blood, CO_2 is converted to carbonic acid, which dissociates into a proton and a bicarbonate ion. Because the protons bind to deoxygenated hemoglobin, they do not cause a dramatic drop in blood pH (hemoglobin acts as a buffer). Also, small decreases in blood pH trigger a homeostatic response that increases breathing rate and expulsion of CO_2. **9.** understand Because airflow through bird lungs is unidirectional, the gas exchange surfaces are

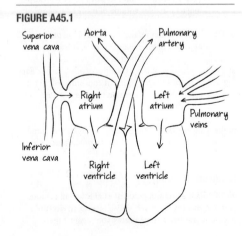

FIGURE A45.1

continuously ventilated with fresh, oxygenated air. The trachea and bronchi in a mammalian lung do not have a gas exchange surface, and bidirectional airflow in these species means that "stale" air has to be expelled before "fresh" air can be inhaled. As a result, the alveoli are not ventilated continuously. **10.** (understand) Lungs increase the temperature of the air and are moist to allow greater solubility of gases (increasing k); alveoli present a large surface area (large A); the epithelium of alveoli is thin (small D); and constant delivery of deoxygenated blood to alveoli maintains a steep partial pressure gradient, favoring diffusion of oxygen into the body ($P_2 - P_1$ is high). **11.** (analyze) Yes—the trait compensates for the small P_{O_2} gradient between stagnant water and the blood of the carp. **12.** (analyze) c

✔ Test Your Problem-Solving Skills

13. (create) Cold water carries more oxygen than warm water does, so icefish blood can carry enough oxygen to supply the tissues with oxygen even in the absence of hemoglobin. The oxygen and carbon dioxide are simply dissolved in the blood. **14.** (analyze) The insect tracheal system delivers O_2 directly to respiring cells; in humans, O_2 is first taken up by the blood of the circulatory system, which then delivers it to the cells. The human respiratory system also has specialized muscles devoted to ventilating the lungs. The open circulatory system of the insect is a low-pressure system, which makes use of pumps (hearts) and body movements to circulate hemolymph. The closed circulatory system of humans is a high-pressure system, which can respond to rapid changes in O_2 demand by tissues. **15.** (understand) If the pulmonary circulation were under pressure as high as that found in the systemic circulation, large amounts of fluid would be forced out of capillaries in the lungs. There is a conflict between the thin surface required for efficient gas exchange and the thickness needed in blood vessels to withstand high pressure. **16.** (analyze) b

CHAPTER 46

IN-TEXT QUESTIONS AND EXERCISES

p. 931 Fig. 46.3 (apply) No—as K^+ leaves the cell along its concentration gradient, the interior of the cell becomes more negative. As a result, an electrical gradient favoring movement of K^+ into the cell begins to counteract the concentration gradient favoring movement of K^+ out of the cell. Eventually, the two opposing forces balance out, and there is no net movement of K^+.

p. 932a (apply) $E_{Ca2+} = 58$ mV $\times \log(1 \text{ mM}/0.0001 \text{ mM})/(+2) = 116$ mV.

p. 932b (apply) The membrane potential would be near zero because the Na^+ and K^+ would diffuse across the membrane along their concentration gradients.

p. 934 CYU (1) (apply) The resting potential would fall (be much less negative) because K^+ could no longer leak out of the cell. **(2)** (understand) The size of an action potential from a particular neuron does not vary, so it cannot contain information. Only the frequency of action potentials from the same neuron varies.

p. 934 (understand) See **FIGURE A46.1.**

p. 935 (1) (understand) Na^+ channels open when the cell membrane depolarizes, and as more channels open, more Na^+ flows into the cell, further depolarizing the cell, causing more voltage-gated Na^+ channels to open. **(2)** (apply) Sodium would flow inward as potassium flowed outward, and their charges would cancel each other out. The membrane would not depolarize.

p. 938 CYU (1) (understand) Once the threshold level of depolarization is attained, the probability that the voltage-gated sodium channels will open approaches 100 percent. But below threshold, the massive opening of Na^+ channels does not occur. This is why the action potential is all or none. **(2)** (apply) The membrane would depolarize and stay depolarized.

p. 938 Fig. 46.10 (understand) Get a solution taken from the synapse between the heart muscle and the vagus nerve *without* the nerve being stimulated. Expose a second heart to this solution. There should be no change in heart rate.

p. 940 (analyze) The inside of the cell becomes more positive (depolarized). This shifts the membrane potential closer to the threshold, making it more likely that the postsynaptic cell will fire an action potential.

p. 942 CYU (apply) The EPSP would not be as strong if an enzyme that broke down excitatory neurotransmitter increased in concentration.

p. 943 Fig. 46.16 (analyze) In the rest-and-digest mode, pupils take in less light stimulation, the heartbeat slows to conserve energy, liver conserves glucose, and digestion is promoted by release of gallbladder products. In the fight-or-flight mode, the pupils open to take in more light, the heartbeat increases to support muscle activity, and the liver releases glucose into the blood to fuel the fight or the flight (for example, from a predator).

p. 944 Fig. 46.18 (understand) Point to your forehead and top of your head for the frontal lobe, the top and top rear of your head for the parietal lobe, the back of your head for the occipital lobe, and the sides of your head (just above your ear openings) for the temporal lobes.

p. 946 Fig. 46.21 (analyze) No—for example, part (b) indicates that the size of the brain area devoted to the trunk is no bigger than the size of the brain area devoted to the thumb.

p. 948 CYU (1) (understand) One method is to study individuals with known brain lesions and correlate the location of the defect with a deficit in mental or physical function. Another is to directly stimulate brain areas in conscious patients during brain surgery and record the response. **(2)** (remember) Learning and memory involve changes in the number, sensitivity, and placement of synapses and neurons.

IF YOU UNDERSTAND . . .

46.1 (understand) See **FIGURE A46.2. 46.2** (understand) Because the behavior of voltage-gated Na^+ and K^+ channels does not change. Every time a membrane depolarizes past the threshold and all of the voltage-gated Na^+ channels are open, the same amount of Na^+ will enter the cell, and the cell will depolarize to the same extent. In addition, K^+ channels open at the same time and allow the same amount of K^+ to leave the cell, so that it repolarizes to the same extent. **46.3** (understand) Summation means that EPSPs and IPSPs produced at various sites combine to determine the response or lack of response in a postsynaptic cell. Thus the behavior of a postsynaptic cell depends on the integration of input from many synapses. **46.4** (apply) The animal would likely die because its basic bodily functions (circulation, breathing, digestion, etc.) would be disrupted.

FIGURE A46.1

FIGURE A46.2

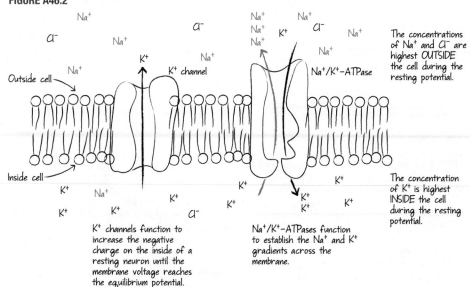

The concentrations of Na^+ and Cl^- are highest OUTSIDE the cell during the resting potential.

The concentration of K^+ is highest INSIDE the cell during the resting potential.

K^+ channels function to increase the negative charge on the inside of a resting neuron until the membrane voltage reaches the equilibrium potential.

Na^+/K^+-ATPases function to establish the Na^+ and K^+ gradients across the membrane.

YOU SHOULD BE ABLE TO . . .

✔ Test Your Knowledge

1. `remember` b 2. `remember` d 3. `understand` c 4. `remember` a 5. `remember` b 6. `understand` See FIGURE A46.3.

✔ Test Your Understanding

7. `understand` The Na+/K+-ATPase pumps 3 Na+ out of the cell for every 2 K+ it brings in. Since more positive charges leave the cell than enter it, there is a difference in charge on the two sides of the membrane and thus a voltage. 8. `analyze` The ligand-gated channel opens or closes in response to binding by a small molecule (for example, a neurotransmitter); the voltage-gated channel opens or closes in response to changes in membrane potential. 9. `understand` EPSPs and IPSPs are produced by flow of ions through channels in the membrane, which change the membrane potential in the postsynaptic cell. If a flow of ions at one point depolarizes the cell but a nearby flow of ions hyperpolarizes the cell, then in combination the flows of ions cancel each other out. But if two adjacent flows of ions depolarize the cell, then their effects are additive and sum to produce a total that is more likely to reach threshold. 10. `analyze` The somatic system responds to external stimuli and controls voluntary skeletal muscle activity, such as movement of arms and legs. The autonomic system controls internal involuntary activities, such as digestion, heart rate, and gland activities. 11. `apply` A drug that activates the sympathetic nervous system would increase heart rate. 12. `understand` a

✔ Test Your Problem-Solving Skills

13. `apply` The neurotransmitters will stay in the synaptic cleft longer, so the amount of binding to ligand-gated channels will increase. Ion flows into the postsynaptic cell will increase dramatically, affecting its membrane potential and likelihood of firing action potentials. 14. `analyze` The benefit of these approaches is that they allow the specific role of the brain region to be assessed in a living person. However, diseased or damaged brains may not respond like healthy and undamaged brains. Also, the extent of lesions and the exact location of electrical stimulation may be difficult to determine, making correlations between the regions affected and the response imprecise. 15. `analyze` c (Because the current is reduced from normal levels, but not eliminated completely, there must be more than one type of potassium channel present. The venom probably knocks out just one specific type of channel.) 16. `create` A reasonable hypothesis is that the new neurons arise from stem cells in the brain.

CHAPTER 47

IN-TEXT QUESTIONS AND EXERCISES

p. 959 CYU (1) `apply` A punctured eardrum wouldn't vibrate correctly and would result in hearing loss at all frequencies in the affected ear. If the stereocilia are too short to come into contact with the tectorial membrane, vibration of the basilar membrane will not cause them to bend, and sound will not be detected. A loss in basilar membrane flexibility would result in the inability to hear lower-pitched sounds, such as human speech. (2) `understand` The lateral line system senses pressure waves in the water, allowing fishes to detect the presence of such objects as prey, predators, or potential mates. p. 959 Fig. 47.9 `create` Vision could potentially be used to detect prey in diurnal catfish. To control for this, catfish could be blinded or "blindfolded" with an opaque material over their eyes to permit study of the lateral line system. p. 961 Fig. 47.12 `create` The change in retinal's shape would induce a change in opsin's shape. A protein's

function correlates with its shape, so opsin's function is likely to change as well. p. 962 `analyze` cGMP is a second messenger in the sense that it signals the sodium channel that transducing is inactive, and it is also a ligand that opens sodium channels. Transducin is like a G protein because it switches from "off" to "on" in response to a receptor protein, activating a key protein as a result. Closing Na+ channels stops the entry of positive charges into the cell, making the inside of the cell more negative relative to the outside. p. 962 Fig. 47.13 `understand` This ion flow occurs when the photoreceptor cell is not receiving light—when the cell is in the dark. p. 963 Fig. 47.15 `create` The S, M, and L opsin proteins are each different in structure. These structural differences affect the ability of retinal to absorb specific frequencies of light and change shape. p. 964 CYU (1) `understand` Retinal acts like an on–off switch that indicates whether light has fallen on a rod cell. When retinal absorbs light, it changes shape. The shape change triggers events that result in a change in action potentials, signaling that light has been absorbed. (2) `apply` Retina detachment from a jarring car accident would separate the retina from the optic nerve, resulting in blindness. The mutation would produce blue–purple color blindness, because the S opsin responds to wavelengths in that region of the spectrum. A clouded lens would reduce the amount of light that reaches the retina, reducing visual sensitivity. p. 967 CYU (1) `understand` Extremely hot food damages taste receptors, so those proteins would no longer be able to respond to their chemical triggers. (2) `create` One hypothesis to explain why the vomeronasal organ is vestigial in most primates is that primates have evolved other complex methods of communication, including gestures, facial expressions, and language, that render sensation with the vomeronasal organ less important. p. 967 `apply` (1) A rattlesnake with its eyes covered will strike effectively because its pits will sense the prey. (2) A rattlesnake with its eyes and pits covered with cotton cloth will strike effectively. The cloth will block vision but the heat from the animal will penetrate the cloth to stimulate the thermoreceptors in the pits. (3) A rattlesnake with its eyes and pits covered with an opaque heat-blocking material will not be able to strike effectively, because the material blocks vision and heat energy.

p. 969 CYU (1) `create` Electrogenic fishes may have a survival advantage over other fishes in murky water, where vision would be compromised. (2) `create` Female sea turtles likely use magnetic cues to locate their natal beaches.

IF YOU UNDERSTAND . . .

47.1 `create` The sensory neurons that used to serve that limb are cut, but may still fire action potentials that stimulate brain areas associated with awareness of pain in the missing limb. 47.2 `analyze` Each of these senses involves ion channels that open in response to application of pressure to cell membranes. 47.3 `understand` Because different animal species have different sets of opsin molecules that respond to different wavelengths of light absorbed by retinal they see different colors. 47.4 `create` Smells are part of the sensation of flavor, and people with nasal congestion cannot smell. In addition to smell being important, the small number of taste receptors can send many different frequencies of action potentials to the brain, and the taste receptors can be stimulated in many different combinations. 47.5 `understand` Capsaicin binds to thermoreceptors that are also activated at high temperatures, leading to the perception of heat. It can also bind to some nociceptors.

YOU SHOULD BE ABLE TO . . .

✔ Test Your Knowledge

1. `understand` a 2. `remember` d 3. `understand` c 4. `remember` b 5. `remember` false 6. `remember` magnetoreception

✔ Test Your Understanding

7. `understand` b 8. `understand` The many possible examples include several blue opsins allowing coelacanths to distinguish the blue wavelengths present in their deep-sea habitat, infrasound hearing allowing elephants to hear over long distances, ultrasonic hearing allowing bats to hunt by echolocation and moths to avoid bat predators, red and yellow opsins allowing fruit-eating mammals to distinguish ripe from unripe fruit. 9. `analyze` The data provide evidence that a pheromone acted to synchronize the women's menstrual cycles because the average time between onset of a woman's cycle and that of the rest of the women declined over time following move-in to the dorms in fall. It declined most rapidly after they first moved in together. The leveling of the graph over time

FIGURE A46.3

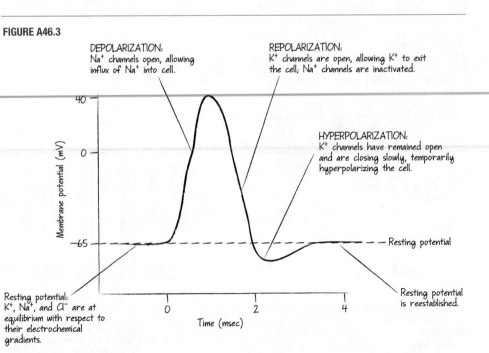

DEPOLARIZATION:
Na+ channels open, allowing influx of Na+ into cell.

REPOLARIZATION:
K+ channels are open, allowing K+ to exit the cell; Na+ channels are inactivated.

HYPERPOLARIZATION:
K+ channels have remained open and are closing slowly, temporarily hyperpolarizing the cell.

Resting potential

Membrane potential (mV)

Resting potential:
K+, Na+, and Cl− are at equilibrium with respect to their electrochemical gradients.

Resting potential is reestablished.

Time (msec)

shows that the onset dates became more synchronized. **10.** `understand` Ion channels in hair cells are thought to open in response to physical distortion caused by the bending of stereocilia. Ion channels in taste receptors, in contrast, change shape and open when certain bitter-tasting molecules bind to them. **11.** `evaluate` Dalton's hypothesis is reasonable because blue fluid absorbs red light and would prevent it from reaching the retina, and this could be tested by dissecting his eyes. Although the hypothesis was rejected, it inspired a rigorous test and required researchers to think of alternative explanations. **12.** `create` To determine whether electric eels are electrogenic, place an eel in a tank and measure the electricity of water to determine if the eel is emitting an electric field. To determine whether they are electroreceptive, place an eel in a tank in the dark and add an object the size and shape of a prey item that emits no scent or taste. Block the lateral line system with a chemical as in Figure 47.9. If the eel locates the object, it must do so using electroreception.

✔ Test Your Problem-Solving Skills

13. `understand` a; vomeronasal organs evolved in tetrapods, presumably as an adaptation for communication in terrestrial habitats. **14.** `analyze` Both the lateral line system and electroreceptors use hair cells to sense changes in the water. Hair cells in the lateral line system contain mechanoreceptors that are depolarized in response to changes in water pressure. In contrast, electroreceptors cause hair cells to depolarize in response to changes in the electric field. **15.** `evaluate` It is unlikely that a perfume contains human pheromones because they have never actually been characterized, although their presence is suspected. **16.** `create` Block all but a few hundred ommatidia in the center of each compound eye of many dragonflies with an opaque material, and observe any differences that might occur in their ability to detect and pursue prey compared to individuals whose ommatidia had been covered by a transparent material.

CHAPTER 48

IN-TEXT QUESTIONS AND EXERCISES

p. 974 `understand` The trucks are the Z discs, the ropes are the thin filaments, and the burly weightlifters are the thick filaments.
p. 974 Fig. 48.2 `analyze` The dark band includes thin filaments and a dense concentration of bulbous structures extending from the thick filament; the light band consists of thin filaments only.
p. 977 CYU (1) `understand` In a sarcomere, thick myosin filaments are sandwiched between thin actin filaments. When the heads on myosin contact actin and change conformation, they pull the actin filaments toward one another, shortening the whole sarcomere. **(2)** `apply` Increased acetylcholine release would result in an increased rate of muscle-cell contraction. Preventing conformational changes in troponin would prevent muscle contraction. Blocking the uptake of calcium ions into the sarcoplasmic reticulum would lead to sustained muscle contraction.
p. 980 `apply` The breast meat of pigeons is composed of dark meat because these muscles are specialized for endurance. Their dark color is due to a high concentration of myoglobin.
p. 981 `create` When earthworms shorten their longitudinal muscles, the segment containing these muscles shortens and squeezes the internal fluid, which becomes pressurized and pushes in all directions, expanding the circumference of the segment and pushing laterally against the burrow.
p. 983 CYU (1) `analyze` Exoskeletons are made of chitin, proteins, and other substances like calcium carbonate. Endoskeletons are made of calcium phosphate, calcium

carbonate, and proteins. Exoskeletons occur on the outsides of animals, whereas endoskeletons occur on the insides. Both types of skeletons are composed of rigid levers separated by joints such that motions are a result of changes in joint angles. Both types of skeletons attach to skeletal muscle and serve to transmit muscle forces. Hydrostatic skeletons also transmit muscle forces but are made of soft tissues that vary from animal to animal. Shape changes in hydrostatic skeletons occur not from changes in joint angles, but from shape changes of the bodies themselves. **(2)** `apply` No movement of the arm would occur, because for these antagonistic muscles to produce movement, one of them must be contracted while the other is relaxed.
p. 986 Fig. 48.16 `apply` Galloping at 3.5 m/s would cost about 75 percent more energy.
p. 987 Fig. 48.18 `apply` About 10 times more costly.

IF YOU UNDERSTAND . . .

48.1 `apply` Paralysis, because acetylcholine has to bind to its receptors on the membrane of postsynaptic muscle fibers for action potentials to propagate in the muscle and cause contraction. **48.2** `analyze` Skeletal muscle is multinucleate, unbranched, striated, and voluntary. Cardiac muscle is branched, striated, and involuntary. Smooth muscle is smooth and involuntary. **48.3** `analyze` When muscles contract in hydrostatic skeletons, the body deforms by becoming longer and thinner, or shorter and narrower, or by bending side to side. In vertebrates and arthropods, muscle contractions cause changes in the joint angles between rigid segments rather than shape changes in the segments themselves. **48.4** `understand` On land, an animal is constrained by its weight; larger animals cannot move as freely and are in greater danger of breaking than are smaller animals. In water, animals are constrained by their shapes. Some shapes would cause high drag, which would inhibit locomotion. In air, animals are constrained by their weight. Larger animals must create more lift to overcome gravity.

YOU SHOULD BE ABLE TO . . .

✔ Test Your Knowledge

1. `remember` a **2.** `remember` c **3.** `understand` d **4.** `remember` True **5.** `understand` a **6.** `understand` False.

✔ Test Your Understanding

7. `create` The key observation was that the banding pattern of sarcomeres changed during contraction. Even though the entire unit became shorter, only some portions moved relative to each other. This observation suggested that some portions of the structure slid past other portions. Muscle fibers have to have many mitochondria because large amounts of ATP are needed to power myosin heads to move along actin filaments; large amounts of calcium stored in smooth ER are needed to initiate contraction by binding to troponin. **8.** `apply` c **9.** `analyze` Acetylcholine reduces the rate and force of contraction of cardiac muscle, whereas it stimulates skeletal muscle to contract. **10.** `create` ATP is no longer produced after death, so calcium cannot be actively transported from the cytosol into the sarcoplasmic reticulum to enable the binding of myosin and actin. The proteins do not unbind, because this process requires ATP. **11.** `apply` The oxygen consumption of the runner would increase because the arches of his or her feet would no longer store as much elastic energy, which normally reduces the energetic cost of running. **12.** `create` Because diverse animals experience the same physical constraints where they live, natural selection favors the same kind of adaptations to those constraints, resulting in convergent evolution. For example,

dolphins and ichthyosaurs are distantly related but have a similar body shape due to their similar ability to swim rapidly through water.

✔ Test Your Problem-Solving Skills

13. `apply` In cardiac muscle, the binding of acetylcholine to its receptors causes the heart rate to slow. Ingestion of nightshade would increase heart rate because it blocks acetylcholine receptors. **14.** `apply` If the sarcomere is in a stretched state before stimulation, the initial force production would be reduced because less overlap would occur between actin and myosin; thus, fewer myosin heads could engage in the pull. **15.** `apply` c:

$$0.5 = v^2/(9.8 \text{ m/s}^2 \times 0.9 \text{ m})$$
$$0.5 = v^2/8.8 \text{ m}^2/\text{s}^2$$
$$v^2 = 0.5 \times 8.8 \text{ m}^2/\text{s}^2$$
$$v = \text{square root of } 4.4 \text{ m}^2/\text{s}^2$$
$$v = 2.1 \text{ m/s}$$

16. `evaluate` The bones of *T. rex* are available and can be analyzed to determine their structural properties, such as the mechanical advantage of the leg joints and the ability of the bones to withstand the different forces that the skeleton would experience at different speeds based on data from living animals. A very close estimate could be made, but the exact speed would be hard to determine without observing the dinosaur in action, since behavior affects speed.

CHAPTER 49

IN-TEXT QUESTIONS AND EXERCISES

p. 994 `apply` The current temperature is analogous to the sensory input; the thermostat, to the CNS; the thermostat signal, to the cell-to-cell signal; and the furnace, to the effector. If feedback inhibition fails, hormone production will continually increase.
p. 996 Fig. 49.4 `understand` Injecting the dog with a liquid extract from a different part of the body, or with a solution that is similar in chemical composition to the extract from the pancreas (e.g., in pH or ions present) but lacking molecules produced by the pancreas.
p. 997 CYU (1) `understand` It is difficult to differentiate between the nervous and endocrine systems because some neurons secrete hormones and some endocrine glands respond to neural signals. **(2)** `apply` Thyroid hormone receptors are intracellular because thyroid hormones are lipid soluble.
p. 998 `create` The cells could differ in receptors, signal transduction systems, or response systems (genes or proteins that can be activated).
p. 1001 `understand` Stressed individuals would be less able to heal or to fight pathogens.
p. 1001 Fig. 49.9 `understand` The saline injection controlled for any stress induced by the injection procedure and for introducing additional fluids.
p. 1002 Fig. 49.11 `apply` Same—a defect in both signal and receptor has the same effect—no response—as a defective signal alone or a defective receptor alone.
p. 1003 CYU (1) `analyze` In amphibians, an increase in thyroid hormones stimulates metamorphosis from a tadpole to an adult without a resting stage. In insects, a decrease in juvenile hormone and increase in ecdysone stimulate metamorphosis during a resting (pupal) stage. **(2)** `understand` By reducing immune system function, promoting release of fatty acids from storage cells and use of amino acids from muscles for energy production, and preventing release of glucose in response to signals from insulin, cortisol conserves glucose supplies for use by the brain.
p. 1004 Fig. 49.12 `create` Inject ACTH and monitor cortisol levels. If cortisol does not increase, adrenal failure is likely.

p. 1006 CYU (1) understand ACTH triggers the release of cortisol, but cortisol inhibits ACTH release by blocking the release of CRH from the hypothalamus and suppressing ACTH production. High levels of cortisol tend to lower cortisol levels in the future. **(2)** understand Processing centers in the brain are responsible for synthesizing a wide array of sensory input. To start a response to this sensory input, they stimulate neurosecretory cells in the hypothalamus. Neurohormones from these cells travel to the anterior pituitary, where they trigger the production and release of other hormones.

p. 1010 CYU (1) apply The steroid-hormone–receptor complex would probably fail to bind to the hormone-response element. Then gene expression would not change in response to the hormone—the arrival of the hormone would have little or no effect on the target cell. **(2)** analyze The mode of action would be similar to that of a peptide—hormone binding would have to trigger a signal transduction event that resulted in production of a second messenger or a phosphorylation cascade.

IF YOU UNDERSTAND . . .

49.1 understand The production and release of hormones are often controlled, directly or indirectly, by electrical or chemical signals from the nervous system. A good example is the hypothalamic–pituitary axis, where neuroendocrine signals regulate endocrine signals. **49.2** analyze Electrical signals are short lived and localized. All of the processes listed in the question are long-term changes in the organism that require responses by tissues and organs throughout the body. **49.3** apply The hypothalamus would continue to release CRH, stimulating the pituitary to release ACTH, which in turn would stimulate continued release of cortisol from the adrenal cortex. Sustained, high circulating levels of cortisol have negative effects. **49.4** apply A cell can respond to more than one hormone at a time, because cells produce different receptors for each hormone.

YOU SHOULD BE ABLE TO . . .

✔ Test Your Knowledge

1. understand c **2.** understand a **3.** remember a **4.** remember c **5.** remember Mammals—eye and pineal gland; lizards—pineal gland; birds—diffuse in brain **6.** remember True

✔ Test Your Understanding

7. understand b **8.** create This is one reason that the same hormone can trigger different effects in different tissues. For example, epinephrine binds to four different types of receptors in different tissues—eliciting a different response from each. **9.** analyze The posterior pituitary is an extension of the hypothalamus; the anterior pituitary is independent—it communicates with the hypothalamus via chemical signals in blood vessels. The posterior pituitary is a storage area for hypothalamic hormones; the anterior pituitary synthesizes and releases an array of hormones in response to releasing hormones from the hypothalamus. **10.** analyze Steroid hormones act directly, and alter gene expression. They usually bind to receptors inside the cell, forming a complex that binds to DNA and activates transcription. Nonsteroid hormones act indirectly, and activate proteins. They bind to receptors on the cell surface and trigger production of a second messenger or a phosphorylation cascade, ending in activation of proteins already present in the cell. **11.** create Subject volunteers to travel and jet lag, measure ACTH and cortisol levels, and correlate these results with their perceived level of jet lag symptoms. **12.** create Studies have shown that treatment with leptin is largely ineffective in reducing food intake in obese humans.

✔ Test Your Problem-Solving Skills

13. apply c (dehydration would result, because ADH would not be present) **14.** create There are two basic strategies: (1) remove the structure from some individuals and compare their behavior and condition to untreated individuals in the same environment, or (2) make a liquid extract from the structure, inject it into some individuals, and compare their behavior and condition to individuals in the same environment that were injected with water or a saline solution. **15.** evaluate The researcher's hypothesis initially sounds unreasonable, because peptide hormones cannot enter cells to initiate gene transcription. However, a second messenger activated by the hormone could possibly lead to changes in gene transcription. **16.** evaluate The graph shows that atrazine treatment reduces the testosterone concentrations in male frogs to near the levels of females. This result supports the hypothesis that atrazine is an endocrine disruptor that feminizes male amphibians.

CHAPTER 50

IN-TEXT QUESTIONS AND EXERCISES

p. 1015 Fig. 50.3 create Isolate and identify the molecules found in "crowded" water. Test each molecule by adding it, at the same concentration found in crowded water, to clean water occupied by a single *Daphnia* and recording whether the female produces a male-containing brood. Repeat with many test females, and for each molecule identified. As a control, record the number of male-containing broods produced in clean water.

p. 1016 apply Asexual reproduction would be expected in environments where conditions change little over time.

p. 1019 Fig. 50.7 evaluate No—the data are consistent with the displacement hypothesis, but there is no direct evidence for it. There are other plausible explanations for the data.

p. 1021 CYU (1) analyze Oviparity usually requires less energy input from the mother after egg laying, and mothers do not have to carry eggs around as long—meaning that they can lay more eggs and be more mobile. However, before egg laying, mothers have to produce all the nutrition the embryo will require in the egg, and eggs may not be well protected after laying. Viviparity usually increases the likelihood that the developing offspring will survive until birth, but it limits the number of young that can be produced to the space available in the mother's reproductive tract. If viviparous young can be nourished longer than oviparous young, then they may be larger and more capable of fending for themselves. **(2)** create Divide a population of sperm into two groups. Subject one group of sperm to spermkillerene at concentrations observed in the female reproductive tract (the experimental group) but not the other group (the control). Document the number of sperm that are still alive over time in both groups.

p. 1025 Fig. 50.13 create (Note: There is more than one possible answer—this is an example.) Having two gonads is an "insurance policy" against loss or damage. To test this idea, surgically remove one gonad from a large number of male and female rats. Do a similar operation on a large number of similar male and female rats, but do not remove either gonad. Once the animals have recovered, place them in a barn or other "natural" setting and let them breed. Compare reproductive success of individuals with paired versus unpaired gonads.

p. 1026 Fig. 50.15 analyze It is similar. In both cases, the hypothalamus produces a releasing factor (GnRH or CRH) that acts on the pituitary. The releasing factor stimulates release of regulatory hormones from the anterior pituitary (LH and FSH or ACTH). These hormones travel via the bloodstream and act on the gonads or adrenals to induce the release of hormones from these glands. In both cases, the hormones are involved in negative feedback control of the regulatory hormones from the pituitary.

p. 1029 apply (1) There should be an LH spike early in the cycle, followed by early ovulation. (This might not actually occur, though, if the follicle is not yet ready to rupture.) (2) LH levels should remain low—no mid-cycle spike and no ovulation.

p. 1030 CYU (1) understand FSH triggers maturation of an ovarian follicle. Its level rises at the end of a cycle because it is no longer inhibited by progesterone, which stops being produced at high levels when the corpus luteum degenerates. **(2)** apply The drug would keep FSH levels low, meaning that follicles would not mature and would not begin producing estradiol and progesterone. The uterine lining would not thicken.

p. 1032 analyze (1) The uterine lining may not be maintained adequately—if it degenerates, a miscarriage is likely. (2) Pregnancy tests detect hCG in urine. By the third trimester, hCG is no longer being produced by the placenta.

p. 1034 Fig. 50.23 apply The birth will be more difficult, as limbs or other body parts can get caught.

p. 1034 Fig. 50.24 apply A mother's chance of dying in childbirth in 1760 was a little over 1000 in 100,000 live births, or close to 1.0 percent. If she gave birth 10 times, she would have a 10 percent chance of dying in childbirth.

IF YOU UNDERSTAND . . .

50.1 create Parasites could wipe out a population of asexually produced animals because the animals do not vary genetically. However, a population of sexually produced animals would have enough genetic variability that some individuals could present better resistance to the parasites. **50.2** create Parental care demands resources (time, nutrients) that cannot be used to produce more eggs. Sperm competition is not likely when external fertilization occurs, and most fishes use external fertilization. **50.3** create The most common methods are surgical ligation (tying off) of the fallopian tubes in a woman to stop the delivery of the egg to the uterus and cutting the vas deferens (vasectomy) in a man to stop the delivery of sperm into the semen. **50.4** apply FSH and LH levels would increase following ovariectomy, because sex steroids would no longer be present to exert negative feedback upon the anterior pituitary. **50.5** understand The placenta is highly vascularized to provide a large surface area for transfer of oxygen and nutrients from the mother's circulation to the fetus's, and of wastes from the fetus's circulation to the mother's.

YOU SHOULD BE ABLE TO . . .

✔ Test Your Knowledge

1. remember a **2.** understand b **3.** remember a **4.** remember b **5.** remember Luteinizing hormone (LH) and follicle-stimulating hormone (FSH) **6.** remember True

✔ Test Your Understanding

7. remember Every offspring that is produced sexually is genetically unique; offspring that are produced asexually have DNA from their mother only and often are genetically identical to her. **8.** create *Daphnia* females produced males only when they were exposed to short day lengths *and* water from crowded populations *and* low food levels. These conditions are likely to occur in the fall. Sexual reproduction could be adaptive in fall if genetically variable offspring are better able to thrive in conditions that occur the following spring. **9.** analyze Spermatogenesis generates four haploid sperm cells from each primary spermatocyte; oogenesis produces only one haploid egg cell from each primary oocyte. Egg cells are much larger than sperm cells because they contain more cytoplasm. In males, the second meiotic division occurs right after the first meiotic division, but in females it is delayed until fertilization. **10.** understand LH triggers release of estradiol, but at low levels estradiol inhibits further release of LH. LH and FSH trigger release of progesterone, but progesterone inhibits further release of LH and FSH. High levels of estradiol trigger release of more LH. The follicle can produce high levels of estradiol only if it has grown and matured—meaning that it is ready for ovulation to occur. **11.** remember c **12.** apply Females should be choosier about their second mate because the sperm of the second male has an advantage in sperm competition.

✔ Test Your Problem-Solving Skills

13. create If all sheep were genetically identical, it is less likely that any individuals could survive a major adverse event (e.g., a disease outbreak or other environmental challenge) than if the population were composed of genetically diverse individuals, which are likely to vary in their ability to cope with the new conditions. **14.** apply Oviparous populations should produce larger eggs than viviparous populations. Because oviparous species deposit eggs in the environment, the eggs must contain all the nutrients and water required for the entire period of embryonic development. But because embryos in viviparous species receive nourishment directly from the mother, it is likely that their eggs are smaller. **15.** create One hypothesis might be that the shell membrane is a vestigial trait (see Chapter 25)—specifically, an "evolutionary holdover" from an ancestor of today's marsupials that laid eggs. **16.** analyze b

CHAPTER 51

IN-TEXT QUESTIONS AND EXERCISES

p. 1040 apply See **FIGURE A51.1**. Each type of Toll protein is a pattern-recognition receptor that responds to a different group of pathogens. Since bacteria and fungi are very different pathogens, you would not expect a receptor that recognizes fungi to be required for a response against bacteria, so flies lacking Toll would survive as well as the wild-type control.

p. 1041 CYU analyze (1) Applying direct pressure constricts blood vessels, mimicking the effect of histamine released from mast cells. (2) Cleaning the wound to remove dirt and debris mimics the phagocytic activity of macrophages and neutrophils to eliminate foreign cells and material. (3) Applying bandages impregnated with platelet-recruiting compounds mimics the effect of chemokines released from injured tissues and macrophages, which attract circulating leukocytes and platelets to facilitate blood clotting and wound repair.

p. 1041 Fig. 51.3 understand Rubor (reddening) is due to increased blood flow to the infected or injured area. Mast cells trigger dilation of vessels by releasing signaling molecules. Calor (heat) is the result of fever, which is activated by the release of cytokines from macrophages that are in the area of infection. Tumor (swelling) occurs

due to dilation of blood vessels triggered by mast cells and macrophages.

p. 1046 apply (1) In a human, 200 different light chains are possible (see text). The recombination of the *V*, *D*, and *J* segments results in $51 \times 27 \times 6 = 8262$ possible heavy chains. Since these chains combine to produce BCRs, a total of $200 \times 8262 = 1{,}652{,}400$ are possible within a human. (2) Only one type of antibody can be produced from a single human B cell.

p. 1047 CYU (1) understand The production of BCRs and TCRs via DNA recombination is required for an effective response. Selection against self-reactive lymphocytes is required for a safe response. **(2)** understand The cell is able to make a larger number of differing variable regions per unit of DNA when the variable region is segmented. For example, 3 segments of *V* and 3 segments of *J* can randomly assemble into 8 different variable light chains, whereas only 6 light chains would be possible with the same amount of DNA if the *V* regions were preassembled.

p. 1048 understand "Clonal" refers to the cloning—producing many exact copies—of cells that are "selected" by the binding of their receptor to a specific antigen (or MHC–peptide complex, which is introduced later).

p. 1050 understand See **TABLE A51.1**.

p. 1051 CYU create Individuals who are heterozygous for MHC genes have greater variability in their MHC proteins than do individuals who are homozygous for these genes. The increased variability in MHC proteins allows a greater variety of peptides to be presented to T cells, which thus would be able to recognize, attack, and eliminate a wider range of pathogens compared with T cells in homozygotes. As a result, heterozygous individuals would likely suffer fewer infections than homozygous individuals.

p. 1054 Fig. 51.17 apply See **FIGURE A51.2**. The response to the new antigen would resemble the primary response that occurred after the first injection of the original antigen.

p. 1055 CYU analyze Attenuated viruses are able to infect the cells of a vaccinated host, but not well enough to cause disease. Inactivated viruses are not able to infect cells at all, and instead remain as extracellular particles. Since attenuated viruses can infect cells, their antigens will be processed and presented by class I MHC complexes, which will promote a stronger cell-mediated response than the inactivated viruses.

IF YOU UNDERSTAND . . .

51.1 understand Toll-like receptors recognize molecular patterns of broad groups of pathogens, such as viruses, fungi, and bacteria. Upon binding to these molecules, the receptors cause signals to be transduced that activate cellular responses directed at eliminating these types of pathogens. **51.2** analyze The BCR locus would be smaller in mature B cells compared to immature cells, since recombination between the gene segments resulted in removing sections of DNA. **51.3** understand Antigen-presenting cells activate the CD4$^+$ T cells that are required for the full activation of B cells to differentiate into plasma cells. **51.4** understand Extracellular viruses are agglutinated and inactivated by antibodies produced in the humoral response. Intracellular viruses are eliminated by the destruction of their host via cytotoxic T cells in the cell-mediated response. **51.5** apply The IgE antibodies generate hypersensitive responses. When these are severe, they can lead to anaphylactic shock. If IgE antibodies were generated against self molecules, you would expect a strong response that will lead to either anaphylactic shock or chronic inflammation.

YOU SHOULD BE ABLE TO . . .

✔ Test Your Knowledge

1. remember c **2.** understand a **3.** remember b **4.** understand They are identical except that a B-cell receptor has a transmembrane domain that allows it to be located in the B-cell

TABLE A51.1

MHC type	Origin of Peptide	Type of T cell that binds	Activity stimulated
Class I MHC	Host cell cytosol	CD8+	Activates T cell to secrete molecules that kill infected cell
Class II MHC	Extracellular environment	CD4+	Activates T cell to secrete cytokines that support the immune response of other cells

FIGURE A51.2

FIGURE A51.1

plasma membrane. **5.** `remember` c **6.** `understand` The hygiene hypothesis claims that immune disorders such as allergies and autoimmunity have increased in frequency due to the implementation of hygienic practices that decrease common bacterial, viral, and parasitic infections.

✓ Test Your Understanding

7. `understand` B cells are first activated to divide via interactions between the BCRs and their complementary antigens. They are then fully activated to differentiate and continue to divide when their class II MCH–peptide complex interacts with an activated CD4$^+$ T cell that has a complementary TCR. **8.** `analyze` Both the BCR and TCR interact with epitopes of antigens via binding sites located in the variable regions of their polypeptide chains. A TCR is like one "arm" of a BCR. The BCR can interact with an antigen directly, while the TCR interacts with its epitope in the context of an MHC protein presented by another cell. **9.** `analyze` Vaccines have to contain an antigen that can stimulate an appropriate primary immune response. Vaccines have not worked for HIV, because the antigens on this virus are constantly changed through mutation, rendering it unrecognizable to memory cells generated following vaccination. **10.** `analyze` b **11.** `analyze` Pattern-recognition receptors on leukocytes bind to surface molecules that are present on many pathogens. BCRs and TCRs bind to particular epitopes of antigens that are pathogen specific. Pattern-recognition-receptor binding directly activates the cell, while BCRs and TCRs often require additional signals to become fully activated. **12.** `understand` The mixing and matching of different combinations of gene segments from the variable and joining regions of the light-chain immunoglobulin gene, along with diversity regions of the heavy-chain gene, gives lymphocytes a unique sequence in both chains of the BCR. The variable regions in the alpha and beta chains of the TCR are assembled from segments in a similar manner.

✓ Test Your Problem-Solving Skills

13. `apply` d **14.** `create` If B cells have receptors that bind to self molecules during maturation, they would become activated during the maturation process. Instead of the signal causing the cells to divide, these immature B cells might respond to the signal by becoming permanently inactivated or might be induced to undergo apoptosis. **15.** `analyze` Natural selection favors individuals that can create a large array of antibodies, because the high mutation rates and rapid evolution observed in pathogens means that they will constantly present the immune system with new antigens. If these antigens were not recognized by the immune system, the pathogens would multiply freely and kill the individual. **16.** `analyze` Tissue from the patient's own body is marked with the major histocompatibility (MHC) proteins that the patient's immune system recognizes as self. This results in the preservation, rather than the destruction, of the transplanted tissue. Tissue from a different person is marked with MHC proteins that are unique to that individual. The body of the transplant recipient will recognize the MHC proteins (and other molecules) of the donor as foreign, resulting in a full immune response and rejection of the grafted tissue.

CHAPTER 52

IN-TEXT QUESTIONS AND EXERCISES

p. 1062 Fig. 52.1 `apply` It would increase, because cattle would no longer succumb to the disease carried by tsetse flies. (In Africa, cattle are limited more by biotic conditions than abiotic conditions.)

p. 1064 Fig. 52.4 `apply` Since there were 28 plots and four treatments, there were probably about seven replicate plots per treatment. This is important because the abiotic and biotic characteristics of individual plots are likely to vary in natural landscapes.

p. 1066 `apply` See **FIGURE A52.1**.

p. 1067 CYU (1) `apply` There is little rain at the North Pole because dense, dry air descends and absorbs moisture. **(2)** `apply` The highest risk of skin cancer is in the tropics where the Sun is often directly overhead, resulting in a large amount of solar radiation per unit area.

p. 1069 Fig. 52.10 `analyze` Most of the remaining wildlands occur in Canada, at the top of the image.

p. 1070a `understand` Vines and epiphytes increase productivity because they are photosynthetic organisms that fill space between small trees and large trees—they capture light and use nutrients that might not be used in a forest that lacked vines and epiphytes.

p. 1070b `understand` Most leaves have a large surface area to capture light. Light is abundant in deserts, however, and plants with a large surface area would be susceptible to high water loss and/or overheating.

p. 1071a `understand` Crop grains like wheat and corn are grasses, so they thrive in the grassland biome.

p. 1071b `apply` There is a continuous grass cover and scattered trees. (A biome like this is called a savanna.)

p. 1072a `apply` Boreal forests should move north.

p. 1072b `understand` High elevations present an abiotic environment (precipitation and temperature) that is similar to the conditions in arctic tundra. As a result, the plant species present will have similar adaptations to cope with these physical conditions. These similarities are the result of convergent evolution.

p. 1073 Fig. 52.17 `analyze` Visible wavelengths enter the chamber from the top and through the glass, warming the plants and ground inside the chamber. Some of this heat energy is retained within the mini-greenhouse.

p. 1074 CYU (1) `analyze` Tropical dry forests are probably less productive than tropical wet forests because they have less water available to support photosynthesis during some periods of the year. **(2)** `apply` Your answer will depend on where you live. For example, if global warming continues at the current projected rates, several effects can be expected. If you live in a coastal area, you can expect water levels to rise. In general, plant communities will change because average temperature and moisture—and variation in temperature and moisture—will change. Also, development is likely to increase in most areas, transforming the landscape.

p. 1077a `apply` The littoral zone, because light is abundant and nutrients are available from the substrate.

p. 1077b `apply` Bogs are nitrogen poor, so plants that are able to capture and digest insects have a large advantage. This advantage does not exist in marshes and swamps, where nutrients are more readily available.

p. 1078a `apply` Cold water contains more oxygen than warm water, so it can support a higher rate of cellular respiration.

p. 1078b `apply` Species that live in estuaries must be able to tolerate variable salinity; freshwater marsh species do not. The abiotic environment (salt concentration) is so different that few species grow well in both habitats.

p. 1079 `apply` No—the aphotic zone is lightless, so natural selection favors individuals that do not invest energy in developing and maintaining eyes.

IF YOU UNDERSTAND . . .

52.1 `understand` Populations are made up of individual organisms; communities are made up of populations of different species; ecosystems are made up of groups of communities along with the abiotic environment; the biome is made up of all the world's ecosystems. **52.2** `apply` The Argentine ants would likely increase their range into the moist agricultural fields, displacing native ants. **52.3** `apply` All latitudes would get equal amounts of sunlight all year round. There would be no seasons and no changes in climate with latitude. **52.4** `apply` If precipitation does not increase, then increased transpiration rates will put plants under water stress and reduce productivity. **52.5** `apply` Like deserts, open oceans have plenty of sunlight but very low productivity. Unlike deserts, which are water limited, open oceans are nutrient limited.

YOU SHOULD BE ABLE TO . . .

✓ Test Your Knowledge

1. `remember` Organismal, population, community, ecosystem, and biosphere/global **2.** `remember` c **3.** `remember` b **4.** `remember` d **5.** `remember` c **6.** `remember` a

✓ Test Your Understanding

7. `create` Examples of possible questions: *Organismal:* How do humans cope with extremely hot (or cold) weather conditions? *Population:* How large will the total human population be in 50 years? *Community:* How are humans affecting ocean species such as cod and tuna? *Ecosystem:* How does nitrogen runoff from agricultural fields affect the availability of oxygen in the Gulf of Mexico? *Global:* How will human-induced changes in global temperature affect sea level? **8.** `apply` *Biotic:* These ants compete with native ants. *Abiotic:* Argentine ants require moist conditions, not too cold, not too hot. These factors interact; Argentine ants outcompete native ants in moist environments, but competition slows down their dispersal. **9.** `apply` b **10.** `analyze` Productivity in intertidal and neritic zones is high because sunlight is readily available and because nutrients are available from estuaries and deep-ocean currents. Productivity in the oceanic zone is extremely low—even though light is available at the surface—because nutrients are scarce. The deepest part of the oceanic zone may have nutrients available from the substrate but lacks light to support photosynthesis. **11.** `analyze` As elevation increases, biomes often change in a similar way as increasing in latitude (e.g., you might go from temperate forest to a boreal-type forest to tundra). **12.** `analyze` The natural biome of Eastern North America is primarily temperate forest. However, most of this area is now occupied by human-influenced biomes such as cities, suburbs, and farmland.

FIGURE A52.1

13. `apply` a **14.** `create` A single population of lizards was split when South America split from Africa due to continental drift. Fossils are too old to represent animals dispersed by human travel. **15.** `evaluate` Yes. As average global temperatures increase, some organisms may be able to adapt by moving their range. As mountaintops warm, however, organisms adapted to cold conditions will have nowhere to go. Ecologists have good reason to be concerned about these possible extinctions. **16.** `create` Temperature is relatively constant in the tropics all year, whereas temperature fluctuates dramatically at higher latitudes. Organisms that are physiologically adapted to constant temperatures may not have as much genetic variation for traits that enable adaptation to higher temperatures, compared to organisms at higher latitudes, which are adapted to tolerate different temperatures at different times of the year.

CHAPTER 53

IN-TEXT QUESTIONS AND EXERCISES

p. 1086 Fig. 53.4 `analyze` Even though half of each plot serves as a control, the natural landscape varies within plots, so there is a chance that foraging in the two subplots could be different due to factors not measured in the experiment. By measuring foraging activity before adding owls and seeds, the researchers established an additional control and a baseline.

p. 1087 CYU `apply` If owls were allowed to hunt within the treatment subplots, the threat of predation would increase. A larger amount of supplementary seed would be required to compensate for the increased risk of foraging. Gerbils do not consciously calculate costs and benefits. One possible proximate explanation is that the presence of hunting owls would increase the stress level in the gerbils (e.g., the hormone cortisol), which could suppress the hunger of the gerbils or reduce their drive to find food.

p. 1088 Fig. 53.6 `analyze` To form a better control, bring females into the lab and give them the same food and housing conditions as the treatment groups, but expose them to artificial lighting that simulates the short daylight conditions of winter.

p. 1089 CYU (1) `analyze` As in lizards, a rise in sex hormones in humans is a proximate cause of sexual readiness. **(2)** `analyze` Humans, like lizards, tend to choose a mate who will increase their fitness. For example, women often prefer men who will be able to help provide for offspring.

p. 1090 Fig. 53.7 `apply` The data would either be distributed randomly around the circles, showing no preference for direction, or the turtles would swim the same direction in both cases, suggesting that they are using compass orientation rather than map orientation.

p. 1093 Fig. 53.10 `apply` See FIGURE A53.1.

p. 1095 CYU (1) `analyze` Auditory communication allows individuals to communicate over long distances but can be heard by predators. Olfactory communication is effective in the dark, and scents can continue to carry information long after the signaler has left, but scents do not carry long distances. Visual communication is effective during the day but can be seen by predators. **(2)** `apply` The ability to detect deceit protects the individual from fitness costs (e.g., being eaten, having a mate's eggs fertilized by someone else); avoiding or punishing "liars" should also lower the frequency of deceit (because it becomes less successful). Alleles associated with detecting and avoiding or punishing liars should be favored by natural selection and increase in frequency.

p. 1096 Fig. 53.14 `apply` Between first cousins, $r = 1/2n \times 1/2n \times 1/2n = 1/8$.

p. 1097 `apply` Humans often live near kin, are good at recognizing kin, and in many cases can give kin resources for protection that increase fitness.

p. 1097 Fig. 53.15 `create` The control would be to drag an object of similar size through the colony, at similar times of day. This would test the hypothesis that prairie dogs are reacting to the presence of the experimenter and the disturbance—not to a predator.

IF YOU UNDERSTAND . . .

53.1 `apply` At the proximate level of causation, human teens experience an increase in the sex hormones testosterone and estrogen, which increases sex drive. At the ultimate level of causation, teens with a higher sex drive will, on average, have more offspring, which will increase their fitness and pass along their alleles at a higher rate than individuals with a lower sex drive. **53.2** `apply` You would have to consider variables that provide costs and benefits to the web-building spiders. For example, spinning silk and building webs uses up energy, so the size of the web, the amount of material used to build the web, and the frequency of building webs should be considered, as well as the effect of web-building on vulnerability to predation. But large, well-built webs can increase prey capture, so the rate of capture of prey of various types of webs also should be considered. **53.3** `create` Only a bowerbird with plenty of resources could afford the time and energy to build a highly decorated bower. Thus, the decoration behavior is likely an honest signal of male fitness. **53.4** `create` Migration is condition-dependent in some birds. If enough food is available locally, the costs of migration outweigh the benefits, so the birds' fitness will be higher if they stay put. **53.5** `apply` Deceitful communication should decrease. If large males are gone, few nests are available and more female-mimic males would compete at them. Most female-mimic males would have no (or fewer) eggs to fertilize and no male to take care of the eggs they did fertilize. There would be less natural selection pressure favoring deceitful communication. **53.6** `understand` (1) Long-lived, so that extensive interactions with kin and nonrelatives are possible; (2) good memory, to record reciprocal interactions; (3) kin nearby, to make inclusive fitness gains possible; (4) ability to share fitness benefits (e.g., food) between individuals.

YOU SHOULD BE ABLE TO . . .

✔ Test Your Knowledge

1. `understand` c **2.** `understand` Adaptive significance of behavior; how behavior affects fitness **3.** `remember` d **4.** `understand` c **5.** `understand` d **6.** `understand` c

✔ Test Your Understanding

7. `understand` The fitness of honeybees depends on how much energy they can bring into the hive by foraging. Hives with honeybees that successfully communicate locations of food resources will be favored by natural

FIGURE A53.1

The straight run of the upward waggle dance is twice as long as in Fig. 53.10

selection such that the frequency of alleles that determine the communication behavior will increase in the population. **8.** `evaluate` Evolution favors organisms that forage optimally—maximizing benefits and minimizing costs—increasing the frequency of optimal foraging in populations over time. However, individuals in populations vary and environmental circumstances vary. Some individuals will forage suboptimally and suffer reduced fitness. **9.** `apply` Only males that are in good shape (well nourished and free of disease) have the resources required to perform their displays. **10.** `understand` Longer day lengths indicate the arrival of spring, when renewed plant growth and insect activity make more food available. Courtship displays from males indicate the presence of males that can fertilize eggs. **11.** `analyze` This statement is true; sea turtles need both a map and a compass. A map provides information about the spatial relationships of places on a landscape; a compass allows sea turtles to orient the map—so that it aligns with the actual landscape. **12.** `apply` c

✔ Test Your Problem-Solving Skills

13. `create` One possibility would be to set up a testing arena with food supplies at various distances from the food source where adult flies are originally released. Identify rover and sitter adults by genotype and test them individually; then record how far each travels during a set time interval. If the rover and sitter alleles affect foraging movements in adults, then rovers should be more likely to move (and to move farther) than sitters. **14.** `analyze` a, b, c, d; the howler monkeys make condition-dependent decisions about what to eat and are likely to prefer high-energy foods, when available, or foods with nutrients or compounds that would increase their fitness. Hamilton's rule could be considered if some monkeys give the highest-quality foods to close relatives rather than consuming it themselves. **15.** `analyze` Individuals have an r of 1/2 with full siblings and an r of 1/8 with first cousins. If the biologist will lose his life, he needs to save two siblings, or eight first cousins, to keep the "lost copy" of his altruism alleles in the population. **16.** `apply` People would be expected to donate blood under two conditions: (1) They either received blood before or expect to have a transfusion in the future, and/or (2) they receive some other benefit in return, such as a good reputation among people who might be able to help them in other ways.

CHAPTER 54

IN-TEXT QUESTIONS AND EXERCISES

p. 1105 Fig. 54.2 `apply` See FIGURE A54.1 (see A:48)

p. 1106 (1) `apply` Compared to northern populations, fecundity should be high and survivorship low in southern populations. **(2)** `apply` Fecundity can be much higher if females lay eggs instead of retaining them in their bodies and giving birth to live young.

p. 1109 Fig. 54.6 `create` At high population density, competition for food limits the amount of energy available to female song sparrows for egg production.

p. 1112 CYU `create` One possibility is competition for food. To test this idea, compare carrying capacity in identical fenced-in areas that differ only in the amount of food added. (You could also test a hypothesis of space limitation by doubling the size of the enclosure but keeping the amount of food the same.)

p. 1114 Fig. 54.10 `apply` When the population of hares was low, the rate of increase was not density dependent on food. That is, plenty of food was available for the few rabbits, so competition for this resource was low.

p. 1117 CYU (1) `understand` Developed nations have roughly equal numbers of individuals in each age class because the fertility rate has been constant and survivorship high for many years. Developing countries have many more children and young people than older people, because the fertility rate and survivorship have been increasing. **(2)** [understand] Because survivorship is high in most human populations, changes in overall growth rate depend almost entirely on fertility rates.

p. 1118 Fig. 54.14 [apply] The total population size of first-generation females in the third year = 1000; in the fourth year = 240. The total population size of first- and second-generation females in the third year = 2308; in the fourth year = 1230. The number of 1-year-olds created by third-generation females (blue) is $981 \times 0.33 = 324$. The number of newborns (green) created by these 1-year-olds is $324 \times 4.0 = 1296$. Thus, the total population size at the end of the fourth year is expected to be $200 + 792 + 1296 + 324 + 198 + 40 = 2850$.

p. 1119 [apply] The population appears to be increasing over time: From the original 1000 females, there are 2308 females in the third year.

IF YOU UNDERSTAND . . .

54.1 [create] First you would need to observe snowshoe hares to determine how many pellets one individual can produce in one day, on average. Then you could use a quadrat to measure the number of fresh pellets produced in a known area of habitat in one day and use your pellets/hare number to convert this value to hares/area. **54.2** [analyze] In ancient Rome, survivorship was probably low and fecundity high—women started reproducing at a young age and did not live long, on average. In Rome today, the population has high survivorship and low fecundity. **54.3** [apply] See **FIGURE A54.2 54.4** [apply] The population density of mosquitoes would not be constant over time. Ignoring the number of predators that might be present, you would expect high densities of these short-lived insects after rains, followed by low densities during extended dry periods. **54.5** [understand] Fewer children are being born per female, but there are many more females of reproductive age due to high fecundity rates in the previous generation. **54.6** [apply] Small, isolated populations are likely to be wiped out by bad weather, a disease outbreak, or changes in the habitat. But in a metapopulation, migration between the individual small populations helps reestablish populations and maintain the overall size of the metapopulation.

YOU SHOULD BE ABLE TO . . .

✔ Test Your Knowledge

1. [remember] b **2.** [remember] births, deaths, immigration, emigration **3.** [remember] a **4.** [remember] True **5.** [understand] d **6.** [understand] c

✔ Test Your Understanding

7. [remember] c **8.** [apply] See **TABLE A54.1 9.** [create] The population has undergone near-exponential growth recently because advances in nutrition, sanitation, and medicine have allowed humans to live at high density without suffering from decreased survivorship and fecundity. Eventually, however, growth rates must slow as density-dependent effects such as disease and famine increase death rates and lower birth rates. **10.** [understand] (a) Corridors allow individuals to move between populations, increasing gene flow and making it possible to recolonize habitats where populations have been lost. (b) Maintaining unoccupied habitat makes it possible for the habitat to be recolonized. **11.** [apply] *Lacerta vivipara* is most vulnerable at the southern edge of its ranges, such as in the mountains of Spain. These lizards are cold adapted and may become locally extinct in areas that become too warm **12.** [analyze] r_{max} gives the population growth rate in the absence of density-dependent limitation; r is the actual growth rate, which is usually affected by density-dependent factors.

✔ Test Your Problem-Solving Skills

13. [apply] Fewer older individuals will be left in the population; there will be relatively more young individuals. If too many older individuals are taken, population growth rate may decline sharply as reproduction stops or slows. (But if relatively few older individuals are taken, more resources are available to younger individuals and their survivorship and fecundity, and the population's overall growth rate, may increase.) **14.** [create] The sunflowers and beetles are both metapopulations. To preserve them, you must preserve as many of the sunflower patches as possible (or plant more) and maintain corridors (that may be smaller sunflower patches) along which the beetles can migrate between the patches. **15.** [apply] As a sexually transmitted disease, AIDS will reduce the number of sexually active adults. If the epidemic continues unabated, the numbers of both reproductive-age adults and children will decline, causing a top-heavy age distribution dominated by older adults and the elderly. See **FIGURE A54.3. 16.** [analyze] c

CHAPTER 55

IN-TEXT QUESTIONS AND EXERCISES

p. 1126 [apply] Your graph should look like Figure 55.4, with *Semibalanus* on the left, *Chthamalus* is on the right, and the *x*-axis indicating height in the intertidal zone. The realized niche of *Chthamalus* is in the upper intertidal zone.

p. 1127 Fig. 55.5 [apply] If Connell had done the treatments on different rocks, a critic could argue that differences in survival were due to differences in the nature of the rocks—not differences in competition.

p. 1128 [apply] Your graph should look like the bottom graph in Figure 55.6, showing that the niches of *Semibalanus* and *Chthamalus* are nonoverlapping with respect to height in the intertidal zone.

p. 1131 Fig. 55.10 [apply] This experimental design tested the hypothesis that mussels can sense the presence of crabs. If the crabs had been fed mussels, a critic could argue that the mussels detect the presence of damaged mussels—not crabs. (As it turns out, the mussels can do both. The experimenters did the same experiment with broken mussel shells in the chamber instead of a fish-fed crab.)

FIGURE A54.1

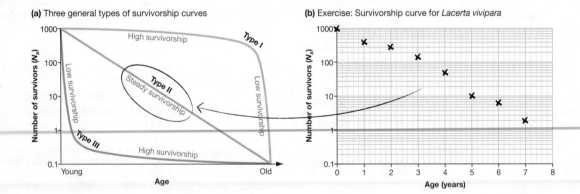

(a) Three general types of survivorship curves

High survivorship • Type I • Low survivorship • Type II • Steady survivorship • Low survivorship • Type III • High survivorship

Number of survivors (N_x) • Age • Young • Old

(b) Exercise: Survivorship curve for *Lacerta vivipara*

Number of survivors (N_x) • Age (years)

TABLE A54.1

Trait	Life-History Continuum		
	Left	Middle	Right
	High fecundity Low survivorship	Intermediate	Low fecundity High survivorship
Growth habit	Herbaceous	Shrub	Tree
Disease- and predator-fighting ability	Low	Medium	High
Seed size	Small	Medium	Large
Seed number	Many	Moderate	Few
Body size	Small	Medium	Large

FIGURE A54.2

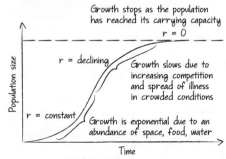

Growth stops as the population has reached its carrying capacity $r = 0$

$r =$ declining • Growth slows due to increasing competition and spread of illness in crowded conditions

$r =$ constant • Growth is exponential due to an abundance of space, food, water

Population size • Time

---

p. 1133 Fig. 55.12 `create` Put equal numbers of infected and uninfected ants in a pen that includes a bird predator. Count how many of each type are eaten over time.

p. 1134 Fig. 55.13 `apply` Many possible answers. For example, acacia trees may expend energy in producing the large bulbs the ants live in and the food they eat. The *Crematogaster* ants may expend energy and risk injury or death in repelling herbivores. The cleaner shrimp may occasionally get injured or eaten by the fish they are cleaning, or their diets may be somewhat restricted by the association. The host fish may miss meals or be at greater risk of predation when they are undergoing cleaning by the shrimp. In both examples of mutualism, however, the overall associations are positive for both parties.

p. 1135 CYU (1) `understand` The individuals do not choose or try to have traits that reduce competition—they simply have those traits (or not). Resource partitioning just happens, because individuals with traits that allow them to exploit different resources produce more offspring, which also have those traits. (Recall from Chapter 25 that natural selection occurs on individuals, but adaptive responses such as resource partitioning are properties of populations.) **(2)** `understand` When species interact via consumption, a trait that gives one species an advantage will exert natural selection on individuals of the other species who have traits that reduce that advantage. This reciprocal adaptation will continue indefinitely. An example is the interaction of *Plasmodium* with the human immune system: The human immune system has evolved the ability to detect proteins from the *Plasmodium* and kill infected cells; in response, *Plasmodium* has evolved different proteins that the immune system does not detect.

p. 1135 Fig. 55.14 `apply` These steps controlled for the alternative hypothesis that differences in treehopper survival were due to differences in the plants they occupied—not the presence or absence of ants.

p. 1137 Fig. 55.15 `analyze` Predictable, at least to a degree.

p. 1142 CYU (1) `understand` The shade provided by early successional species increases humidity, and decomposition of their tissues adds nutrients and organic material to the soil. These conditions favor growth by later successional species, which can outcompete the early successional species. **(2)** `understand` The presence or absence of a plant species where nitrogen fixation occurs would dramatically alter nutrient conditions, and thus the speed of succession and the types of species that could become established. For example, species that require high nitrogen would be favored on sites where alder grew, and species that can tolerate low nitrogen would thrive on sites where alder is absent.

p. 1142 Fig. 55.22 `apply` The effect would be to make the island more remote, which would lower the rate of immigration and move the whole immigration curve downward. The rate of extinction would increase, shifting the curve upward. The overall effect would be to decrease the number of species, because the island effectively would have become more remote.

p. 1143 `analyze` The calculations for community 1 are as follows. (Note that your final answer is sensitive to the number of digits you use in your calculations; see BioSkills 1 in Appendix B).

Species A	Species B
$p_A = 10/18 = 0.555$	$p_B = 1/18 = 0.0555$
$\ln 0.555 = -0.589$	$\ln 0.0555 = -2.891$
$-0.589 \times 0.555 = -0.327$	$-2.891 \times 0.0555 = -0.160$

Species C	Species D
$p_C = 1/18 = 0.0555$	$p_D = 3/18 = 0.167$
$\ln 0.0555 = -2.891$	$\ln 0.167 = -1.790$
$-2.891 \times 0.0555 = -0.160$	$-1.790 \times 0.167 = -0.299$

Species E	Species F
$p_E = 2/18 = 0.111$	$p_F = 1/18 = 0.0555$
$\ln 0.111 = -2.198$	$\ln 0.0555 = -2.891$
$-2.198 \times 0.111 = -0.244$	$-2.891 \times 0.0555 = -0.160$

Summing the values of $p \times \ln p$ for each species and multiplying by -1 gives the Shannon index of species diversity for community 1:

$$(-1)[(-0.327) + (-0.160) + (-0.160) + (-0.299) + (-0.244) + (-0.160)] = 1.350$$

Similar calculations would give species diversity values of 1.794 for community 2 and 1.610 for community 3.

IF YOU UNDERSTAND . . .

55.1 `understand` A mutualistic relationship becomes a parasitic one if one of the species stops receiving a benefit. The treehopper–ant mutualism becomes parasitic in years when spiders are rare, because the treehoppers no longer derive a benefit but pay a fitness cost (producing honeydew that the ants eat). **55.2** `apply` An invasive species could have a large impact on community structure by replacing the dominant plant species that creates physical structure, which in turn affects other members of the community; by replacing a keystone species that has an indirect effect on many other species; or by outcompeting many species, causing a direct reduction in species richness. **55.3** `apply` All species at Glacier Bay must be able to survive in a cold climate with the local amount of precipitation. The species earliest in the succession must also be able to grow on rock exposed as the glacier melts. But chance, historical differences in seed sources, and presence or absence of alder (and nitrogen fixation) created differences in the species present. **55.4** `apply` Many national parks are surrounded by altered habitats, so they are functionally similar to islands. Species richness in the park is more likely to be preserved if it is large and located nearby other wilderness areas.

YOU SHOULD BE ABLE TO . . .

✔ **Test Your Knowledge**

1. `remember` True **2.** `remember` a **3.** `remember` d **4.** `understand` d **5.** `remember` fecundity **6.** `remember` c

✔ **Test Your Understanding**

7. `evaluate` Yes—the treehopper–ant mutualism is parasitic or mutualistic, depending on conditions; competition can evolve into no competition over time if niche differentiation occurs; arms races mean that the outcome of host–parasite interactions can change over time, and so forth (many other examples). **8.** `understand` b **9.** `evaluate` (a) If community composition is predictable, then the species present should not change over time. But if composition is not predictable, then that species should undergo significant changes over time. (b) If community composition is predictable, then two sites with identical abiotic factors should develop identical communities. If community composition is not predictable, then sites with identical abiotic factors should develop variable communities. In most tests, the data best match the predictions of the "not predictable" hypothesis, though communities show elements of both. **10.** `analyze` Disturbance is any short-lived event that changes the distribution of resources. Compared to low-frequency fires, high-frequency fires would tend to be less severe (less fuel builds up) and would tend to exert more intense natural selection for adaptations to resist the effects of fire. Compared to low-severity fires, high-severity fires would open up more space for pioneering species and would tend to exert more intense natural selection for adaptations to resist the effects of fire. **11.** `apply` Early successional species are adapted to disperse to new environments (small seeds) and grow and reproduce quickly (reproduce at an early age, grow quickly). They can tolerate severe abiotic conditions (high temperature, low humidity, low nutrient availability) but have little competitive ability. These species are able to enter a new environment (with no competitors) and thrive. These attributes are considered adaptations because they increase the fitness of these species. **12.** `analyze` The idea is that high productivity will lead to high population density of consumers, leading to competition and intense natural selection favoring niche differentiation that leads to speciation.

✔ **Test Your Problem-Solving Skills**

13. `analyze` Natural selection will favor orchid individuals that have traits that resist bee attack: thicker flower walls, nectar storage in a different position, a toxin in the flower walls, and so on. Individuals could also be favored if their anthers were in a position that accomplished pollination even if bees eat through the walls of the nectar-storage structure. **14.** `apply` b **15.** `apply` The exact answer will depend on the location of the campus. The first species to appear must possess good dispersal ability, rapid growth, quick reproductive periods, and tolerance for very harsh and severe conditions. The two-acre plot is likely to be colonized first by pioneer species that have very "weedy" characteristics. But once colonization is under way, the course of succession will depend more on how the various species interact with each other. The presence of one species can inhibit or facilitate the arrival and establishment of another. For example, an early-arriving species might provide the shade and nutrients required by a late-arriving species. The site's history and nearby ecosystems may influence which species appear at each stage; for instance, an undisturbed ecosystem nearby could be a source for native species. The pattern and rate of this succession is also influenced by the overall environmental conditions affecting it. Only species with traits appropriate to the local climate are likely to colonize the site. **16.** `create` One reasonable experiment would involve constructing artificial ponds and introducing different numbers of plankton species

FIGURE A54.3

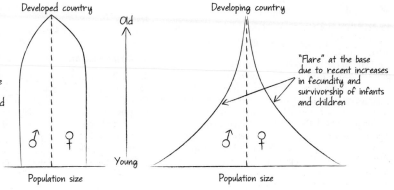

ANSWERS **A:49**

to different ponds, but the same total number of individuals. (Any natural immigration to the ponds would have to be prevented.) After a period of time, remove all of the plankton and measure the biomass present. Make a graph with number of species on the *x*-axis and total biomass on the *y*-axis. If the hypothesis is correct, the line of best fit through the data should have a positive slope.

CHAPTER 56

IN-TEXT QUESTIONS AND EXERCISES

p. 1152 `apply` To grow a kilogram of beef, first you have to grow 10 kg of grain or grass and feed it to the cow. Only 10 percent of this 10 kg will be used for growth and reproduction—the other 9 kg is used for cellular respiration or lost as heat.

p. 1152 `apply` Crustaceans and fish are ectothermic, so they are much more efficient at converting primary production into the biomass in their bodies than are endothermic birds and mammals.

p. 1153 `apply` Tuna are top predators in marine food webs. Mercury occurs in low concentrations in the water, but accumulates in higher concentrations at each trophic level. A person who eats a lot of tuna will accumulate mercury over time, potentially causing mercury poisoning. Sardines are primary consumers, a low trophic level, and therefore are not in danger of biomagnification of mercury.

p. 1155 Fig. 56.10 `apply` Average NPP = 125 g/m²/year. Area = 65%. Although NPP in the ocean is very small, the ocean is so vast that it ends up being the largest contributor to total NPP.

p. 1156 CYU (1) `understand` At each trophic level, most of the energy that is consumed is lost to cellular respiration and ultimately heat, metabolism, or other maintenance activities, which leaves only a small percentage of energy for biomass production (growth and reproduction). **(2)** `apply` Since NPP depends on temperature, it is logical to predict that NPP will increase with global warming. (However, Section 56.3 explains that global warming changes precipitation patterns, ocean stratification, and other factors that can limit NPP).

p. 1157 `understand` When the trees of tropical rain forests are hauled away and the remnants of the forest are burned, the nutrients are exported from the site, leaving only poor soil to nourish the crops.

p. 1158 Fig. 56.13 `create` One logical hypothesis is that the total amount exported increases as tree roots and other belowground organic material decay and begin to wash into the stream; then the amount exported should begin to decline as nitrate reserves become exhausted—eventually, there is no more nitrate to wash away.

p. 1159 Fig. 56.14 `apply` Much more water should evaporate. One possibility is that more water vapor will be blown over land and increase precipitation over land.

p. 1162 CYU `apply` **Water:** perhaps the most direct impacts concern rates of groundwater replenishment. Converting biomes into farms or suburbs decreases groundwater recharge and increases runoff. Irrigation pumps groundwater to the surface, where much of it runs off into streams. **Nitrogen:** Humans introduce huge quantities of nitrogen into ecosystems as fertilizer. **Carbon:** Humans are releasing huge quantities of stored carbon, in the form of fossil fuels, into the atmosphere as CO_2. Whereas humans are depleting supplies of freshwater below natural levels, we are increasing quantities of N and C above natural levels.

p. 1164 Fig. 56.22 `analyze` Photosynthesis increases in summer, resulting in removal of CO_2 from the atmosphere.

p. 1165 Fig. 56.23 `apply` $2 \times 7000 = 14{,}000$ million metric tons of CO_2, double the current total.

p. 1168 `apply` The pH of the water will fall (acidity will increase) because CO_2 from your breath reacts with water to form carbonic acid.

IF YOU UNDERSTAND . . .

56.1 `apply` Tofu is a plant, a primary producer at the base of the productivity pyramid. Cows are primary consumers, one level up. It would be about 10 times more efficient to eat tofu, because 10 times the quantity of plants would be necessary to provide one pound of beef.

56.2 `oral` Decomposition is slowest in cold, wet temperatures on land and in anoxic conditions in the ocean (or in freshwater). Stagnant water often becomes anoxic as decomposers use up available oxygen that is not replenished by diffusion from the atmosphere. **56.3** `apply` Many possible answers, including (1) reducing carbon dioxide emissions by finding alternatives to fossil fuels should decrease the amount of carbon dioxide released into the atmosphere, (2) recycling programs decrease carbon dioxide emissions from manufacturing plants and power plants, and (3) reforestation can tie up carbon dioxide in wood.

YOU SHOULD BE ABLE TO . . .

✔ Test Your Knowledge

1. `remember` True **2.** `understand` d **3.** `understand` a **4.** `remember` c **5.** `remember` c **6.** `remember` b

✔ Test Your Understanding

7. `apply` a **8.** `analyze` Sharks normally eat fish that eat the herbivorous fish that keep the growth of algae on corals in check. When sharks are absent, more of the intermediate fish survive, which decreases the number of herbivorous fish, causing algae to thrive on the coral reef. **9.** `analyze` Warmer, wetter climates speed decomposition; cool temperatures slow it. Decomposition regulates nutrient availability because it releases nutrients from detritus and allows them to reenter the food web. **10.** `evaluate` Yes, corals would be a good choice. Corals are important ecologically because they have a high NPP, providing energy to diverse animals in reef food webs. Corals are also vulnerable to climate change from the increase in water temperature, which can cause bleaching (loss of photosynthetic algae), as well as from the effects of ocean acidification, which slows the rate of coral growth. **11.** `analyze` Forest fires are an example of positive feedback; drier, hotter summers increase the risk of fires, and fires release CO_2 to the atmosphere, increasing global warming and the risk of fires. Forest growth is an example of negative feedback; as CO_2 increases in the atmosphere, the rate of forest growth in some areas may increase (due to increased availability of CO_2, increased temperature, and/or increased precipitation), reducing the concentration of CO_2 in the atmosphere. **12.** `analyze` The open ocean has almost no nutrient input from the land and has little upwelling to supply nutrients from the deep ocean. In contrast, intertidal and coastal areas receive large inputs of nutrients from rivers as well as from upwellings from ocean depths.

✔ Test Your Problem-Solving Skills

13. `create` One possibility is to add radioactive nitrogen to the water of the ponds in natural concentrations (control ponds) and elevated nitrogen concentrations (treatment ponds) and then follow this isotope through the primary producers, primary consumers, and secondary consumers in the system by measuring the amount of radioisotope in the tissues of organisms, and quantity of biomass, at each trophic level in the treatment versus control ponds **14.** `evaluate` c **15.** `apply` Without herbivores, there is no link in the nitrogen cycle between primary producers and secondary consumers. All of the plant nitrogen would go to the primary decomposers and back

into the soil. If decomposition is rapid enough, nitrogen would cycle quickly between primary producers, decomposers, the soil, and back to primary producers.
16. `apply` Atmospheric oxygen would increase due to extensive photosynthesis, but carbon dioxide levels would decrease because little decomposition was occurring. The temperature would drop because fewer greenhouse gases would be trapping heat reflected from the Earth's surface.

CHAPTER 57

IN-TEXT QUESTIONS AND EXERCISES

p. 1177 CYU `create` Do an all-taxon survey: organize experts and volunteers to collect, examine, and identify all species present. This could include direct sequencing studies (see Chapter 29) to document the bacteria and archaea present in different habitats on campus.

p. 1178 Fig. 57.7 `apply` Nearly all endangered species are threatened by more than one factor and therefore appear on the graph more than once, leading to species totals greater than 100 percent.

p. 1180 Fig. 57.9 `analyze` Because the treatments and study areas were assigned randomly, there is no bias involved in picking certain areas that are unusual in terms of their biomass or species diversity. They should represent a random sample of biomass and species diversity in fragments of various sizes versus intact forest.

p. 1183 Using the equation $S = (18.9)A^{0.15}$, plug in 5000 km² for *A* and solve for *S* such that $S = 18.9 \times 5000^{0.15} = 68$ species.

p. 1184 CYU (1) `understand` Fragmentation reduces habitat quality by creating edges that are susceptible to invasion and loss of species, due to changed abiotic conditions. Genetic problems occur inside fragments as species become inbred and/or lose genetic diversity via genetic drift. **(2)** `create` Habitat destruction and fragmentation in Borneo are obvious in the clear-cut areas. Overexploitation could be assessed by interviewing local citizens and law enforcement officers to the history of poaching orangutans for the pet trade or for other uses. Effects from invasive species could be assessed by published or local accounts of community structure over time. Pollution could be assessed by direct measurement of soil, water, air, and possibly tissue samples (in dead orangutans, feces, or blood samples). The effects of climate change could be assessed based on local weather records.

p. 1185 Fig. 57.12 `analyze` *Null hypothesis:* NPP is not affected by species richness or functional diversity of species. *Prediction:* NPP will be greater in plots with more species and more functional groups. *Prediction of null:* There will be no difference in NPP based on species richness or number of functional groups.

p. 1188 CYU `create` Establish several study plots in an area of Borneo where some of the forest has been converted to palm plantation but the adjacent forest (of similar elevation and topography) has rain forest still intact—half of the plots in the plantation and half in the rain forest. Monitor soil depth, soil nutrient levels, precipitation, and stream sediment levels regularly, and compare after several years.

p. 1191 Fig. 57.16 `apply` See **FIGURE A57.1**.

IF YOU UNDERSTAND . . .

57.1 `apply` At the genetic level, you could measure the genetic diversity of the orangutan population by comparing DNA sequences obtained from blood or feces samples. This information could help you to determine if a genetic bottleneck had occurred during the clear-cutting period. At the species level, you could monitor the increase in species richness of trees, birds, mammals, and other groups as the restoration progressed. At the ecosystem level, you could measure the ability of the growing forest to retain nutrients in the soil, prevent erosion, trap carbon, increase the prevalence of rain, and other functions **57.2** apply Small, geographically isolated populations of orangutans have low genetic diversity due to lack of gene flow and genetic drift and can suffer from inbreeding depression. Low genetic diversity makes them vulnerable to changes in their habitat, and small population size makes them susceptible to catastrophic events such as storms, disease outbreaks, or fires. **57.3** apply If many species are present, and each uses the available resources in a unique way, then a larger proportion of all available resources should be used—leading to higher biomass production. **57.4** evaluate Twelve percent of land area is already protected, but a mass extinction event is still going on. This is because 12 percent is not sufficient to save most species, and because many or most protected areas are not located in biodiversity hotspots.

YOU SHOULD BE ABLE TO . . .

✔ Test Your Knowledge

1. remember d **2.** remember a **3.** remember b **4.** remember d
5. understand c **6.** understand overexploitation

✔ Test Your Understanding

7. analyze b **8.** analyze By comparing the number of species estimated to reside in the park before and after the survey, biologists will have an idea of how much actual species diversity is being underestimated in other parts of the world—where research was at the level of Great Smoky Mountains National Park before the survey. The limitation is that it may be difficult to extrapolate from the data—no one knows if the situation at this park is typical of other habitats. **9.** evaluate No one correct answer; there are good arguments for both. In a case like the orangutans of Borneo, it is not possible to save the orangutans without saving their ecosystem. On the other hand, it is much easier to attract money and volunteers to help save a charismatic primate like an orangutan than to save a patch of forest. Both types of efforts are needed, their relative importance depends on the circumstance. **10.** evaluate In experiments with species native to North American grasslands, study plots that have more species produce more biomass, change less during a disturbance (drought), and recover from a disturbance faster than study plots with fewer species. **11.** understand Wildlife corridors facilitate the movement of individuals. Corridors allow areas to be recolonized if a species is lost and enable the introduction of new alleles that can counteract genetic drift and inbreeding in small, isolated populations. **12.** apply Ecosystem services are beneficial effects that ecosystems have on humans. It is rare to own or pay for these services, so no one has a vested interest in maintaining them. This is a primary reason that ecosystems are destroyed, even though the services they offer are valuable. The exception would be landowners who are paid *not* to log their forests, to maintain ecosystem functions like carbon storage.

✔ Test Your Problem-Solving Skills

13. apply d **14.** create Catalog existing biodiversity at the genetic, species, and ecosystem levels. Using these data, find areas that have the highest concentration of biodiversity at each level. Protect as many of these areas as possible, and connect them with corridors of habitat. **15.** evaluate There is no "correct" answer. One argument is that conservationists could lobby officials in Brazil and Indonesia to learn from the mistakes made in developed nations, and preserve enough forests to maintain biodiversity and keep ecosystem services intact—avoiding the expenditures that developed countries had to make to clean up pollution and restore ecosystems and endangered species. **16.** analyze Species with specialized food or habitat requirements, large size (and thus large requirements for land area), small population size, and low reproductive rate are vulnerable to extinction. A good example is the koala bear. Species that are particularly resistant to pressure from humans possess the opposite of many of these traits. A good example of a resistant species is the Norway rat.

BIG PICTURE Ecology

p. 1196 CYU (1) analyze Species interactions "are agents of" natural selection. **(2)** analyze As the human population climbs, more resources are needed to support the population. The harvesting of resources affects other species both directly (e.g. overfishing) and indirectly (e.g. climate change), resulting in decreased biodiversity. **(3)** analyze Both energy and nutrients flow through trophic levels in food webs. However, energy flows in one direction, ultimately dissipating as heat, whereas nutrients continue to cycle through ecosystems. **(4)** apply Human behavior "increases emissions of" CO_2 "causes" greenhouse effect "causes" climate change.

Bioskills

BIOSKILLS 1; p. B:2 CYU (1) apply 3.1 miles **(2)** apply 37°C **(3)** apply Multiply your weight in pounds by 1/2.2 (0.45).

(4) apply 4 **(5)** apply 4 **(6)** Two significant figures. When you multiply, the answer can have no more significant figures than the least accurate measurement—in this case, 1.6.

BIOSKILLS 2; p. B:3 CYU (1) apply "different-yoked-together" **(2)** apply "sugary-loosened" **(3)** apply "study-of-form" **(4)** apply "three-bodies"

BIOSKILLS 3; p. B:6 CYU (1) apply about 18% **(2)** apply a dramatic drop (almost 10%) **(3)** understand No—the order of presentation in a bar chart does not matter (though it's convenient to arrange the bars in a way that reinforces the overall message). **(4)** apply 11 **(5)** apply 68 inches

BIOSKILLS 4; p. B:7 CYU apply Test 2, the estimate based on the larger sample—the more replicates or observations you have, the more precise your estimate of the average should be.

BIOSKILLS 5; p. B:8 CYU (1) apply $1/2 \times 1/2 \times 1/2 \times 1/2 = 1/16$ **(2)** apply $1/6 + 1/6 + 1/6 = 1/2$

BIOSKILLS 6; p. B:9 CYU (1) understand exponential **(2)** apply $\ln N_t = \ln N_0 + rt$

BIOSKILLS 7; p. B:9 Fig. B7.1. analyze See **FIGURE BA.1**. **p. B:10 Fig. B7.2.** analyze See **FIGURE BA.2**. **p. B:11 Fig. B7.3.** analyze Figure B7.3d is different.

FIGURE BA.1

FIGURE BA.2

FIGURE A57.1

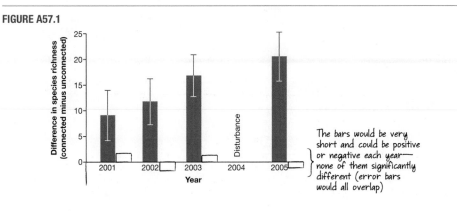

The bars would be very short and could be positive or negative each year—none of them significantly different (error bars would all overlap)

BIOSKILLS 8; p. B:12 Fig. B8.1. apply See **FIGURE BA.3.**

BIOSKILLS 9; p. B:13 Fig. B9.1 understand DNA and RNA are acids that tend to drop a proton in solution, giving them a negative charge.

p. B:15 CYU analyze The lane with no band comes from a sample where RNA X is not present. The same size RNA X is present in the next two lanes, but the faint band has very few copies while the dark band has many. In the fourth lane, the band is formed by a smaller version of RNA X, and relatively few copies are present.

BIOSKILLS 10; p. B:17 CYU (1) understand size and/or density. **(2)** apply Mitochondria, because they are larger in size compared with ribosomes.

BIOSKILLS 11; p. B:19 CYU (1) analyze No—it's just that no mitochondria happened to be present in this section

sliced through the cell. **(2)** explain Understanding a molecule's structure is often critical to understanding how it functions in cells.

BIOSKILLS 12; p. B:21 CYU (1) analyze It may not be clear that the results are relevant to noncancerous cells that are not growing in cell culture—that is, that the artificial conditions mimic natural conditions. **(2)** analyze It may not be clear that the results are relevant to individuals that developed normally, from an embryo—that is, that the artificial conditions mimic natural conditions.

BIOSKILLS 13; p. B:23 Fig. B13.1 analyze This is human body temperature—the natural habitat of *E. coli.*
p. B:24 CYU (1) analyze *Caenorhabditis elegans* would be a good possibility, because the cells that normally die have already been identified. You could find mutant individuals that lacked normal cell death; you could compare

the resulting embryos to normal embryos and be able to identify exactly which cells change as a result. **(2)** analyze Any of the multicellular organisms in the list would be a candidate, but *Dictyostelium discoideum* might be particularly interesting because cells stick to each other only during certain points in the life cycle. **(3)** analyze *Mus musculus*—as the only mammal in the list, it is the organism most likely to have a gene similar to the one you want to study.

BIOSKILLS 14; p. B:26 CYU synthesize Many examples are possible. See Figure 1.9 in Chapter 1, as an example of the format to use for your Research Box.

BIOSKILLS 15; p. B:27 CYU (1) analyze See **FIGURE BA.4.**
(2) analyze See **FIGURE BA.4.**

FIGURE BA.3

Molecular formula: CO_2

Structural formula: $O=C=O$

Ball-and-stick model:

Space-filling model:

FIGURE BA.4

BIOSKILL 1 the metric system and significant figures

Scientists ask questions that can be answered by observing or measuring things—by collecting data. What units are used to make measurements? When measurements are reported, how can you tell how reliable the data are?

The Metric System

The metric system is the system of units of measure used in every country of the world but three (Liberia, Myanmar, and the United States). It is also the basis of the SI system—the International System of Units (abbreviated from the French, *Système international d'unités*)—used in scientific publications.

The popularity of the metric system is based on its consistency and ease of use. These attributes, in turn, arise from the system's use of the base 10. For example, each unit of length in the system is related to all other measures of length in the system by a multiple of 10. There are 10 millimeters in a centimeter; 100 centimeters in a meter; 1000 meters in a kilometer.

Measures of length in the English system, in contrast, do not relate to each other in a regular way. Inches are routinely divided into 16ths; there are 12 inches in a foot; 3 feet in a yard; 5280 feet (or 1760 yards) in a mile.

If you have grown up in the United States and are accustomed to using the English system, it is extremely important to begin developing a working familiarity with metric units and values. **Tables B1.1** and **B1.2** (see B:2) should help you get started with this process.

As an example, consider the following question: An American football field is 120 yards long, while rugby fields are 144 meters

TABLE B1.1 Metric System Units and Conversions

Measurement	Unit of Measurement and Abbreviation	Metric System Equivalent	Converting Metric Units to English Units
Length	kilometer (km)	1 km = 1000 m = 10^3 m	1 km = 0.62 mile
	meter (m)	1 m = 100 cm	1 m = 1.09 yards = 3.28 feet = 39.37 inches
	centimeter (cm)	1 cm = 0.01 m = 10^{-2} m	1 cm = 0.3937 inch
	millimeter (mm)	1 mm = 0.001 m = 10^{-3} m	1 mm = 0.039 inch
	micrometer (μm)	1 μm = 10^{-6} m = 10^{-3} mm	
	nanometer (nm)	1 nm = 10^{-9} m = 10^{-3} μm	
Area	hectare (ha)	1 ha = 10,000 m^2	1 ha = 2.47 acres
	square meter (m^2)	1 m^2 = 10,000 cm^2	1 m^2 = 1.196 square yards
	square centimeter (cm^2)	1 cm^2 = 100 mm^2 = 10^{-4} m^2	1 cm^2 = 0.155 square inch
Volume	liter (L)	1 L = 1000 mL	1 L = 1.06 quarts
	milliliter (mL)	1 mL = 1000 μL = 10^{-3} L	1 mL = 0.034 fluid ounce
	microliter (μL)	1 μL = 10^{-6} L	
Mass	kilogram (kg)	1 kg = 1000 g	1 kg = 2.20 pounds
	gram (g)	1 g = 1000 mg	1 g = 0.035 ounce
	milligram (mg)	1 mg = 1000 μg = 10^{-3} g	
	microgram (μg)	1 μg = 10^{-6} g	
Temperature	Kelvin (K)*		K = °C + 273.15
	degrees Celsius (°C)		$°C = \frac{5}{9}(°F - 32)$
	degrees Fahrenheit (°F)		$°F = \frac{9}{5}°C + 32$

*Absolute zero is −273.15 °C = 0 K.

TABLE B1.2 Prefixes Used in the Metric System

Prefix	Abbreviation	Definition
nano–	n	$0.000\ 000\ 001 = 10^{-9}$
micro–	μ	$0.000\ 001 = 10^{-6}$
milli–	m	$0.001 = 10^{-3}$
centi–	c	$0.01 = 10^{-2}$
deci–	d	$0.1 = 10^{-1}$
–	–	$1 = 10^{0}$
kilo–	k	$1000 = 10^{3}$
mega–	M	$1\ 000\ 000 = 10^{6}$
giga–	G	$1\ 000\ 000\ 000 = 10^{9}$

long. In yards, how much longer is a rugby field than an American football field? To solve this problem, first convert meters to yards: 144 m × 1.09 yards/m = 157 yards (note that the unit "m" cancels out). The difference in yards is thus: 157 – 120 = 37 yards. If you did these calculations on a calculator, you might have come up with 36.96 yards. Why has the number of yards been rounded off? The answer lies in significant figures. Let's take a closer look.

Significant Figures

Significant figures or "sig figs"—the number of digits used to report the measurement—are critical when reporting scientific data. The number of significant figures in a measurement, such as 3.524, is the number of digits that are known with some degree of confidence (3, 5, and 2) plus the last digit (4), which is an estimate or approximation. How do scientists know how many digits to report?

Rules for Working with Significant Figures

The rules for counting significant figures are summarized here:

- All nonzero numbers are always significant.
- Leading zeros are never significant; these zeros do nothing but set the decimal point.
- Embedded zeros are always significant.

- Trailing zeros are significant *only* if the decimal point is specified (Hint: Change the number to scientific notation. It is easier to see the "trailing" zeros.)

Table B1.3 provides examples of how to apply these rules. The bottom line is that significant figures indicate the precision of measurements.

Precision versus Accuracy

If biologists count the number of bird eggs in a nest, they report the data as an exact number—say, 3 eggs. But if the same biologists are measuring the diameter of the eggs, the numbers will be inexact. Just how inexact they are depends on the equipment used to make the measurements. For example, if you measure the width of your textbook with a ruler several times, you'll get essentially the same measurement again and again. Precision refers to how closely individual measurements agree with each other. So, you have determined the length with precision, but how do you know if the ruler was accurate to begin with?

Accuracy refers to how closely a measured value agrees with the correct value. You don't know the accuracy of a measuring device unless you calibrate it, by comparing it against a ruler that is known to be accurate. As the sensitivity of equipment used to

check your understanding

If you understand BioSkill 1

✔ **You should be able to . . .**

1. **QUANTITATIVE** Calculate how many miles a runner completes in a 5.0-kilometer run.

2. **QUANTITATIVE** Calculate your normal body temperature in degrees Celsius (Normal body temperature is 98.6°F.).

3. **QUANTITATIVE** Calculate your current weight in kilograms.

4. **QUANTITATIVE** Calculate how many liters of milk you would need to buy to get approximately the same volume as a gallon of milk.

5. **QUANTITATIVE** Multiply the measurements 2.8723 and 1.6. How many significant figures does your answer have? Why?

TABLE B1.3 Rules for Working with Significant Figures

Example	Number of Significant Figures	Scientific Notation	Rule
35,200	5	3.52×10^{4}	All nonzero numbers are always significant
0.00352	3	3.52×10^{-3}	Leading zeros are not significant
1.035	4	$1.035\ (\times 10^{0})$	Imbedded zeros are always significant
200	1	2×10^{2}	Trailing zeros are significant only if the decimal point is specified
200.0	4	2.000×10^{2}	Trailing zeros are significant only if the decimal point is specified

make a measurement increases, the number of significant figures increases. For example, if you used a kitchen scale to weigh out some sodium chloride, it might be accurate to 3 ± 1 g (1 significant figure); but an analytical balance in the lab might be accurate to 3.524 ± 0.001 g (4 significant figures).

In science, only the numbers that have significance—that are obtained from measurement—are reported. It is important to follow the "sig fig rules" when reporting a measurement, so that data do not appear to be more accurate than the equipment allows.

Combining Measurements

How do you deal with combining measurements with different degrees of accuracy and precision? The simple rule to follow is that the accuracy of the final answer can be no greater than the least accurate measurement. So, when you multiply or divide measurements, the answer can have no more significant figures than the least accurate measurement. When you add or subtract measurements, the answer can have no more decimal places than the least accurate measurement.

As an example, consider that you are adding the following measurements: 5.9522, 2.065, and 1.06. If you plug these numbers into your calculator, the answer your calculator will give you is 9.0772. However, this is incorrect—you must round your answer off to the nearest value, 9.08, to the least number of decimal places in your data.

It is important to nail down the concept of significant figures and to practice working with metric units and values. The Check Your Understanding questions in this BioSkill should help you get started with this process.

BIOSKILL 2 · some common Latin and Greek roots used in biology

Greek or Latin Root	English Translation	Example Term
a, an	not	anaerobic
aero	air	aerobic
allo	other	allopatric
amphi	on both sides	amphipathic
anti	against	antibody
auto	self	autotroph
bi	two	bilateral symmetry
bio	life, living	bioinformatics
blast	bud, sprout	blastula
co	with	cofactor
cyto	cell	cytoplasm
di	two	diploid
ecto	outer	ectoparasite
endo	inner, within	endoparasite
epi	outer, upon	epidermis
exo	outside	exothermic
glyco	sugary	glycolysis
hetero	different	heterozygous
homo	alike	homozygous
hydro	water	hydrolysis
hyper	over, more than	hypertonic
hypo	under, less than	hypotonic
inter	between	interspecific
intra	within	intraspecific
iso	same	isotonic
logo, logy	study of	morphology
lyse, lysis	loosen, burst	glycolysis
macro	large	macromolecule

Greek or Latin Root	English Translation	Example Term
meta	change, turning point	metamorphosis
micro	small	microfilament
morph	form	morphology
oligo	few	oligopeptide
para	beside	parathyroid gland
photo	light	photosynthesis
poly	many	polymer
soma	body	somatic cells
sym, syn	together	symbiotic, synapsis
trans	across	translation
tri	three	trisomy
zygo	yoked together	zygote

check your understanding

If you understand BioSkill 2

✔ **You should be able to . . .**

Provide literal translations of the following terms:

1. heterozygote
2. glycolysis
3. morphology
4. trisomy

Graphs are the most common way to report data, for a simple reason. Compared to reading raw numerical values in a table or list, a graph makes it much easier to understand what the data mean.

Learning how to read and interpret graphs is one of the most basic skills you'll need to acquire as a biology student. As when learning piano or soccer or anything else, you need to understand a few key ideas to get started and then have a chance to practice—a lot—with some guidance and feedback.

Getting Started

To start reading a graph, you need to do three things: read the axes, figure out what the data points represent—that is, where they came from—and think about the overall message of the data. Let's consider each step in turn.

What Do the Axes Represent?

Graphs have two axes: one horizontal and one vertical. The horizontal axis of a graph is also called the x-axis or the abscissa. The vertical axis of a graph is also called the y-axis or the ordinate. Each axis represents a variable that takes on a range of values. These values are indicated by the ticks and labels on the axis. Note that each axis should *always* be clearly labeled with the unit or treatment it represents.

FIGURE B3.1 shows a scatterplot—a type of graph where continuous data are graphed on each axis. Continuous data can take an array of values over a range. In contrast, discrete data can take only a restricted set of values. If you were graphing the average height of men and women in your class, height is a continuous variable, but gender is a discrete variable.

For the example in this figure, the x-axis represents time in units of generations of maize; the y-axis represents the average percentage of the dry weight of a maize kernel that is protein.

To create a graph, researchers plot the independent variable on the x-axis and the dependent variable on the y-axis (Figure B3.1a). The terms independent and dependent are used because the values on the y-axis depend on the x-axis values. In our example, the researchers wanted to show how the protein content of maize kernels in a study population changed over time. Thus, the protein concentration plotted on the y-axis depended on the year (generation) plotted on the x-axis. The value on the y-axis always depends on the value on the x-axis, but not vice versa.

In many graphs in biology, the independent variable is either time or the various treatments used in an experiment. In these cases, the y-axis records how some quantity changes as a function of time or as the outcome of the treatments applied to the experimental cells or organisms.

(a) Read the axes—what is being plotted?

(b) Look at the bars or data points—what do they represent?

(c) What's the punchline?

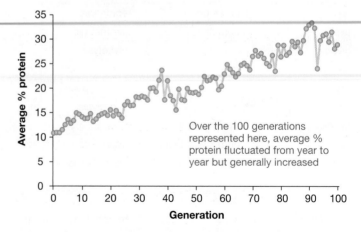

FIGURE B3.1 Scatterplots Are Used to Graph Continuous Data.

What Do the Data Points Represent?

Once you've read the axes, you need to figure out what each data point is. In our maize kernel example, the data point in Figure B3.1b represents the average percentage of protein found in a sample of kernels from a study population in a particular generation.

If it's difficult to figure out what the data points are, ask yourself where they came from—meaning, how the researchers got them. You can do this by understanding how the study was done and by understanding what is being plotted on each axis. The y-axis will tell you what they measured; the x-axis will usually tell you when they measured it or what group was measured. In some cases—for example, in a plot of average body size versus average brain size in primates—the x-axis will report a second variable that was measured.

In other cases, a data point on a graph may represent a relative or arbitrary unit of measurement. The data point shows the ratio of the amount of a substance, intensity, or other quantities, relative to a predetermined reference measurement. For example, the y-axis might show the percentage of relative activity of an enzyme—the rate of the enzyme-catalyzed reaction, scaled to the highest rate of activity observed (100 percent)—in experiments conducted under conditions that are identical except for one variable, such as pH or temperature (see Figure 8.14).

What Is the Overall Trend or Message?

Look at the data as a whole, and figure out what they mean. Figure B3.1c suggests an interpretation of the maize kernel example. If the graph shows how some quantity changes over time, ask yourself if that quantity is increasing, decreasing, fluctuating up and down, or staying the same. Then ask whether the pattern is the same over time or whether it changes over time.

When you're interpreting a graph, it's extremely important to limit your conclusions to the data presented. Don't extrapolate beyond the data, unless you are explicitly making a prediction based on the assumption that present trends will continue. For example, you can't say that the average percentage of protein content was increasing in the population before the experiment started, or that it will continue to increase in the future. You can say only what the data tell you.

Types of Graphs

Many of the graphs in this text are scatterplots like the one shown in Figure B3.1c, where individual data points are plotted. But you will also come across other types of graphs in this text.

Scatterplots, Lines, and Curves

Scatterplots sometimes have data points that are by themselves, but at other times data points will be connected by dot-to-dot lines to help make the overall trend clearer, as in Figure B3.1c, or may have a smooth line through them.

A *smooth line* through data points—sometimes straight, sometimes curved—is a mathematical "line of best fit." A line of best fit represents a mathematical function that summarizes the relationship between the x and y variables. It is "best" in the sense of fitting the data points most precisely. The line may pass through some of the points, none of the points, or all of the points.

Curved lines often take on characteristic shapes depending on the relationships between the x and y variable. For example, a bell-shaped curve depicts a normal distribution in which most data points are clumped near the middle, while a sigmoid or S-shaped curve exhibits small changes at first, which then accelerate and approach maximal value over time. Data from studies on population growth (see Chapter 54), enzyme kinetics (see Chapter 8), and oxygen–hemoglobin dissociation (see Chapter 45) typically fall on a curved line.

Bar Charts, Histograms, and Box-and-Whisker Plots

Scatterplots, or line-of-best-fit graphs, are the most appropriate type of graph when the data have a continuous range of values and you want to show individual data points. But other types of graphs are used to represent different types of distributions:

- *Bar charts* plot data that have discrete or categorical values instead of a continuous range of values. In many cases the bars might represent different treatment groups in an experiment, as in **FIGURE B3.2a** (see B:6). In this graph, the height of the bar indicates the average value. Statistical tests can be used to determine whether a difference between treatment groups is significant (see **BIOSKILLS 4**).

- *Histograms* illustrate frequency data and can be plotted as numbers or percentages. **FIGURE B3.2b** shows an example where height is plotted on the x-axis, and the number of students in a population is plotted on the y-axis. Each rectangle indicates the number of individuals in each interval of height, which reflects the relative frequency, in this population, of people whose heights are in that interval. The measurements could also be recalculated so that the y-axis would report the proportion of people in each interval. Then the sum of all the bars would equal 100 percent. Note that if you were to draw a smooth curve connecting the top of the bars on this histogram, the smooth curve would represent the shape of a bell.

- *Box-and-whisker plots* allow you to easily see where most of the data fall. Each box indicates where half of the data numbers are. The whiskers indicate the lower extreme and the upper extreme of the data. The vertical line inside each box indicates the median—meaning that half of the data are above this value and half are below (see Figure 1.9 for an example).

When you are looking at a bar chart that plots values from different treatments in an experiment, ask yourself if these values are the same or different. If the bar chart reports averages over discrete ranges of values, ask what trend is implied—as you would for a scatterplot.

(a) Bar chart

(b) Histogram

FIGURE B3.2 **Bar Charts and Histograms. (a)** Bar charts are used to graph data that are discontinuous or categorical. **(b)** Histograms show the distribution of frequencies or values in a population.

When you are looking at a histogram, ask whether there is a "hump" in the data—indicating a group of values that are more frequent than others. Is the hump in the center of the distribution of values, toward the left, or toward the right? If so, what does it mean?

Similarly, when you are looking at a box-and-whisker plot, ask yourself what information the graph gives you. What is the range of values for the data? Where are half the data points? Below what value is three quarters of the data?

Getting Practice

Working with this text will give you lots of practice with reading graphs—they appear in almost every chapter. In many cases we've inserted an arrow to represent your instructor's hand at the whiteboard, with a label that suggests an interpretation or draws your attention to an important point on the graph. In other cases, you should be able to figure out what the data mean on your own or with the help of other students or your instructor.

check your understanding

If you understand BioSkill 3
✔ **You should be able to . . .**

1. **QUANTITATIVE** Determine the total change in average percentage of protein in maize kernels, from the start of the experiment until the end.

2. **QUANTITATIVE** Determine the trend in average percentage of protein in maize kernels between generation 37 and generation 42.

3. Explain whether the conclusions from the bar chart in Figure B3.2a would be different if the data and label for Treatment 3 were put on the far left and the data and label for Treatment 1 on the far right.

4. **QUANTITATIVE** Determine approximately how many students in this class are 70 inches tall, by using Figure B3.2b.

5. **QUANTITATIVE** Determine the most common height in the class graphed in Figure B3.2b.

BIOSKILL 4 using statistical tests and interpreting standard error bars

When biologists do an experiment, they collect data on individuals in a treatment group and a control group, or several such comparison groups. Then they want to know whether the individuals in the two (or more) groups are different. For example, in one experiment student researchers measured how fast a product formed when they set up a reaction with three different concentrations of reactants (introduced in Chapter 8). Each treatment—meaning, each combination of reactant concentrations—was replicated many times.

FIGURE B4.1 graphs the average reaction rate for each of the three treatments in the experiment. Note that Treatments 1, 2, and 3 represent increasing concentrations of reactants. The thin

"I-beams" on each bar indicate the standard error of each average. The standard error is a quantity that indicates the uncertainty in the calculation of an average.

For example, if two trials with the same concentration of reactants had a reaction rate of 0.075 and two trials had a reaction rate of 0.025, then the average reaction rate would be 0.050. In this case, the standard error would be large. But if two trials had a reaction rate of 0.051 and two had a reaction rate of 0.049, the average would still be 0.050, but the standard error would be small.

In effect, the standard error quantifies how confident you are that the average you've calculated is the average you'd observe if

FIGURE B4.1 **Standard Error Bars Indicate the Uncertainty in an Average.**

you did the experiment under the same conditions an extremely large number of times. It is a measure of precision (see **BIOSKILLS 1**).

Once they had calculated these averages and standard errors, the students wanted to answer a question: Does reaction rate increase when reactant concentration increases?

After looking at the data, you might conclude that the answer is yes. But how could you come to a conclusion like this objectively, instead of subjectively?

The answer is to use a statistical test. This can be thought of as a three-step process.

1. Specify the null hypothesis, which is that reactant concentration has no effect on reaction rate.

2. Calculate a test statistic, which is a number that characterizes the size of the difference among the treatments. In this case, the test statistic compares the actual differences in reaction rates among treatments to the difference predicted by the null hypothesis. The null hypothesis predicts that there should be no difference.

3. The third step is to determine the probability of getting a test statistic at least as large as the one calculated just by chance. The answer comes from a reference distribution—a mathematical function that specifies the probability of getting various values of the test statistic if the null hypothesis is correct. (If you take a statistics course, you'll learn which test

statistics and reference distributions are relevant to different types of data.)

You are very likely to see small differences among treatment groups just by chance—even if no differences actually exist. If you flipped a coin 10 times, for example, you are unlikely to get exactly five heads and five tails, even if the coin is fair. A reference distribution tells you how likely you are to get each of the possible outcomes of the 10 flips if the coin is fair, just by chance.

In this case, the reference distribution indicated that if the null hypothesis of no actual difference in reaction rates is correct, you would see differences at least as large as those observed only 0.01 percent of the time just by chance. By convention, biologists consider a difference among treatment groups to be statistically significant if there is less than a 5 percent probability of observing it just by chance. Based on this convention, the student researchers were able to claim that the null hypothesis is not correct for reactant concentration. According to their data, the reaction they studied really does happen faster when reactant concentration increases.

You'll likely be doing actual statistical tests early in your undergraduate career. To use this text, though, you only need to know what statistical testing does. And you should take care to inspect the standard error bars on graphs in this book. As a *very* rough rule of thumb, averages often turn out to be significantly different, according to an appropriate statistical test, if there is no overlap between two times the standard errors.

check your understanding

If you understand BioSkill 4
✔ **You should be able to . . .**

QUANTITATIVE Determine which of the following tests used to estimate the average height of individuals in a class is likely to have the smallest standard error, and why.

- Measuring the height of two individuals chosen at random to estimate the average.

- Measuring the height of every student who showed up for class on a particular day to estimate the average.

Phylogenetic trees show the evolutionary relationships among species, just as a genealogy shows the relationships among people in your family. They are unusual diagrams, however, and it can take practice to interpret them correctly.

To understand how evolutionary trees work, consider **FIGURE B7.1**. Notice that a phylogenetic tree consists of a root (the most ancestral branch in the tree), branches, nodes, and tips.

- Branches represent populations through time. In this text, branches are drawn as horizontal lines. In most cases the length of the branch is arbitrary and has no meaning, but in some cases branch lengths are proportional to time or the extent of genetic difference among populations (if so, there will be a scale at the bottom of the tree). The vertical lines on the tree represent splitting events, where one group broke into two independent groups. Their length is arbitrary—chosen simply to make the tree more readable.

- Nodes (also called forks) occur where an ancestral group splits into two or more descendant groups (see point A in Figure B7.1). Thus, each node represents the most recent common ancestor of the two or more descendant populations that emerge from it. If more than two descendant groups emerge from a node, the node is called a polytomy (see node C). A polytomy usually means that the populations split from one another so quickly that it is not possible to tell which split off earlier or later.

- Tips (also called terminal nodes) are the tree's endpoints, which represent groups living today or a dead end—a branch

ending in extinction. The names at the tips can represent species or larger groups such as mammals or conifers.

Recall that a taxon (plural: taxa) is any named group of organisms (see Chapter 1). A taxon could be a single species, such as *Homo sapiens*, or a large group of species, such as Primates. Tips connected by a single node on a tree are called sister taxa.

The phylogenetic trees used in this text are all rooted. This means that the first, or most basal, node on the tree—the one on the far left in this book—is the most ancient. To determine where the root on a tree occurs, biologists include one or more outgroup species when they are collecting data to estimate a particular phylogeny. An outgroup is a taxonomic group that is known to have diverged before the rest of the taxa in the study. Outgroups are used to establish whether a trait is ancestral or derived. An ancestral trait is a characteristic that existed in an ancestor; a derived trait is a characteristic that is a modified form of the ancestral trait, found in a descendant.

In Figure B7.1, Taxon 1 is an outgroup to the monophyletic group consisting of taxa 2–6. A monophyletic group consists of an ancestral species and all of its descendants. The root of a tree is placed between the outgroup and the monophyletic group being studied. This position in Figure B7.1 is node A. Note that black hash marks are used to indicate a derived trait that is shared among the red branches, and another derived trait that is shared among the orange branches.

Understanding monophyletic groups is fundamental to reading and estimating phylogenetic trees. Monophyletic groups may also be called lineages or clades and can be identified using the "one-snip test": If you cut any branch on a phylogenetic tree, all of the branches and tips that fall off represent a monophyletic group. Using the one-snip test, you should be able to convince yourself that the monophyletic groups on a tree are nested. In Figure B7.1, for example, the monophyletic group comprising node A and taxa 1–6 contains a monophyletic group consisting of node B and taxa 2–6, which includes the monophyletic group represented by node C and taxa 4–6.

To put all these new terms and concepts to work, consider the phylogenetic tree in **FIGURE B7.2**, which shows the relationships between common chimpanzees and six human and humanlike species that lived over the past 5–6 million years. Chimps functioned as an outgroup in the analysis that led to this tree, so the root was placed at node A. The branches marked in red identify a monophyletic group called the hominins.

To practice how to read a tree, put your finger at the tree's root, at the far left, and work your way to the right. At node A, the ancestral population split into two descendant populations. One of these populations eventually evolved into today's chimps; the other gave rise to the six species of hominins pictured. Now

FIGURE B7.1 Phylogenetic Trees Have Roots, Branches, Nodes, and Tips.

✔**EXERCISE** Circle all four monophyletic groups present.

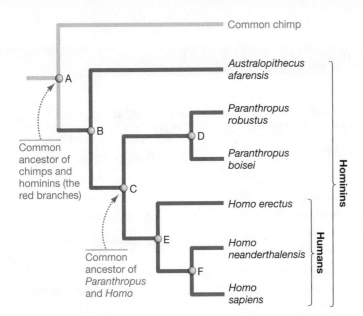

FIGURE B7.2 An Example of a Phylogenetic Tree. A phylogenetic tree showing the relationships of species in the monophyletic group called hominins.

✔**EXERCISE** All of the hominins walked on two legs—unlike chimps and all of the other primates. Add a mark on the phylogeny to show where upright posture evolved, and label it "origin of walking on two legs." Circle and label a pair of sister species. Label an outgroup to the monophyletic group called humans (species in the genus *Homo*).

continue moving your finger toward the tips of the tree until you hit node C. It should make sense to you that at this splitting event, one descendant population eventually gave rise to two *Paranthropus* species, while the other became the ancestor of humans—species in the genus *Homo*. As you study Figure B7.2, consider these two important points:

1. There are many equivalent ways of drawing this tree. For example, this version shows *Homo sapiens* on the bottom. But the tree would be identical if the two branches emerging from node E were rotated 180°, so that the species appeared in the order *Homo sapiens*, *Homo neanderthalensis*, *Homo erectus*. Trees are read from root to tips, not from top to bottom or bottom to top.

2. No species on any tree is any higher or lower than any other. Chimps and *Homo sapiens* have been evolving exactly the same amount of time since their divergence from a common ancestor—neither species is higher or lower than the other. It is legitimate to say that more ancient groups like *Australopithecus afarensis* have traits that are ancestral or more basal—meaning, that appeared earlier in evolution—compared to traits that appear in *Homo sapiens*, which are referred to as more derived.

FIGURE B7.3 presents a chance to test your tree-reading ability. Five of the six trees shown in this diagram are identical in terms of the evolutionary relationships they represent. One differs. The key to understanding the difference is to recognize that the ordering of tips does not matter in a tree—only the ordering of nodes (branch points) matters. You can think of a tree as being like a mobile: The tips can rotate without changing the underlying relationships.

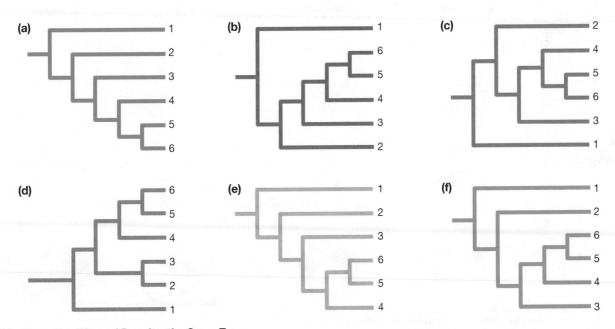

FIGURE B7.3 Alternative Ways of Drawing the Same Tree.

✔**QUESTION** Five of these six trees describe exactly the same relationships among taxa 1 through 6. Identify the tree that is different from the other five.

If you haven't had much chemistry yet, learning basic biological chemistry can be a challenge. One stumbling block is simply being able to read chemical structures efficiently and understand what they mean. This skill will come much easier once you have a little notation under your belt and you understand some basic symbols.

Atoms are the basic building blocks of everything in the universe, just as cells are the basic building blocks of your body. Every atom has a one- or two-letter symbol. **Table B8.1** shows the symbols for most of the atoms you'll encounter in this book. You should memorize these. The table also offers details on how the atoms form bonds as well as how they are represented in some visual models.

When atoms attach to each other by covalent bonding, a molecule forms. Biologists have a couple of different ways of representing molecules—you'll see each of these in the book and in class.

- Molecular formulas like those in **FIGURE B8.1a** simply list the atoms present in a molecule. Subscripts indicate how many of each atom are present. If the formula has no subscript, only one atom of that type is present. A methane (natural gas) molecule, for example, can be written as CH_4. It consists of one carbon atom and four hydrogen atoms.

TABLE B8.1 Some Attributes of Atoms Found in Organisms

Atom	Symbol	Number of Bonds It Can Form	Standard Color Code*
Hydrogen	H	1	white
Carbon	C	4	black
Nitrogen	N	3	blue
Oxygen	O	2	red
Sodium	Na	1	—
Magnesium	Mg	2	—
Phosphorus	P	5	orange or purple
Sulfur	S	2	yellow
Chlorine	Cl	1	—
Potassium	K	1	—
Calcium	Ca	2	—

*In ball-and-stick or space-filling models.

- Structural formulas like those in **FIGURE B8.1b** show which atoms in the molecule are bonded to each other. Each bond is indicated by a dash. The structural formula for methane in-

	Methane	Ammonia	Water	Oxygen
(a) Molecular formulas:	CH_4	NH_3	H_2O	O_2
(b) Structural formulas:	$H-\overset{\displaystyle H}{\underset{\displaystyle H}{C}}-H$	$H-\overset{}{\underset{\displaystyle H}{N}}-H$	$\overset{\displaystyle O}{H \quad H}$	$O=O$
(c) Ball-and-stick models:				
(d) Space-filling models:				

FIGURE B8.1 Molecules Can Be Represented in Several Different Ways.

✔**EXERCISE** Carbon dioxide consists of a carbon atom that forms a double bond with each of two oxygen atoms, for a total of four bonds. It is a linear molecule. Write carbon dioxide's molecular formula and then draw its structural formula, a ball-and-stick model, and a space-filling model.

dicates that each of the four hydrogen atoms forms one covalent bond with carbon, and that carbon makes a total of four covalent bonds. Single covalent bonds are symbolized by a single dash; double bonds are indicated by two dashes.

Even simple molecules have distinctive shapes, because different atoms make covalent bonds at different angles. Ball-and-stick and space-filling models show the geometry of the bonds accurately.

- In a ball-and-stick model, a stick is used to represent each covalent bond (see **FIGURE B8.1c**).

- In space-filling models, the atoms are simply stuck onto each other in their proper places (see **FIGURE B8.1d**).

To learn more about a molecule when you look at a chemical structure, ask yourself three questions:

1. *Is the molecule polar—meaning that some parts are more negatively or positively charged than others?* Molecules that contain nitrogen or oxygen atoms are often polar, because these atoms have such high electronegativity (see Chapter 2). This trait is important because polar molecules dissolve in water.

2. *Does the structural formula show atoms that might participate in chemical reactions?* For example, are there charged atoms or amino or carboxyl ($-COOH$) groups that might act as a base or an acid?

3. *In ball-and-stick and especially space-filling models of large molecules, are there interesting aspects of overall shape?* For example, is there a groove where a protein might bind to DNA, or a cleft where a substrate might undergo a reaction in an enzyme?

BIOSKILLS 9 separating and visualizing molecules

To study a molecule, you have to be able to isolate it. Isolating a molecule is a two-step process: the molecule has to be separated from other molecules in a mixture and then physically picked out or located in a purified form. **BIOSKILLS 9** focuses on the techniques that biologists use to separate nucleic acids and proteins and then find the particular one they are interested in.

Using Electrophoresis to Separate Molecules

In molecular biology, the standard technique for separating proteins and nucleic acids is called gel electrophoresis or, simply, electrophoresis (literally, "electricity-moving"). You may be using electrophoresis in a lab for this course, and you will certainly be analyzing data derived from electrophoresis in this text.

The principle behind electrophoresis is simple. Proteins (when denatured and coated with a special detergent) and nucleic acids carry a charge. As a result, these molecules move when placed in an electric field. Negatively charged molecules move toward the positive electrode; positively charged molecules move toward the negative electrode.

To separate a mixture of macromolecules so that each one can be isolated and analyzed, researchers place the sample in a gelatinous substance. More specifically, the sample is placed in a "well"—a slot in a sheet or slab of the gelatinous substance. The "gel" itself consists of long molecules that form a matrix of fibers. The gelatinous matrix has pores that act like a sieve through which the molecules can pass.

When an electrical field is applied across the gel, the molecules in the well move through the gel toward an electrode. Molecules that are smaller or more highly charged for their size move faster than do larger or less highly charged molecules. As they move, then, the molecules separate by size and by charge. Small and highly charged molecules end up at the bottom of the gel; large, less-charged molecules remain near the top.

An Example "Run"

FIGURE B9.1 (see B:14) shows the electrophoresis setup used in an experiment investigating how RNA molecules polymerize. In this case, the investigators wanted to document how long RNA molecules became over time, when ribonucleoside triphosphates were present in a particular type of solution.

Step 1 shows how they loaded samples of macromolecules, taken on different days during the experiment, into wells at the top of the gel slab. This is a general observation: Each well holds a different sample. In this and many other cases, the researchers also filled a well with a sample containing fragments of known size, called a size standard or "ladder."

In step 2, the researchers immersed the gel in a solution that conducts electricity and applied a voltage across the gel. The molecules in each well started to run down the gel, forming a lane. After several hours of allowing the molecules to move, the researchers removed the electric field (step 3). By then, molecules of different size and charge had separated from one another. In this case, small RNA molecules had reached the bottom of the gel. Above them were larger RNA molecules, which had run more slowly.

PROCESS: GEL ELECTROPHORESIS

Samples of macromolecules collected on different days

2 4 6 8 14

Fragments of known size

Wells —

Gel —

Power supply

Molecules that are smaller and more highly charged run farther than molecules that are larger and less highly charged

1. Load cavities ("wells") in gel with samples.

2. Hook up power supply and run gel. Molecules separate over time as some migrate faster than others.

3. Remove gel after samples have run its length.

FIGURE B9.1 **Macromolecules Can Be Separated via Gel Electrophoresis.**

✔QUESTION DNA and RNA run toward the positive electrode. Why are these molecules negatively charged?

Why Do Separated Molecules Form Bands?

When researchers visualize a particular molecule on a gel, using techniques described in this section, the image that results consists of bands: shallow lines that are as wide as a lane in the gel. Why?

To understand the answer, study **FIGURE B9.2**. The left panel shows the original mixture of molecules. In this cartoon, the size of each dot represents the size of each molecule. The key is to realize that the original sample contains many copies of each specific molecule, and that these copies run down the length of the gel together—meaning, at the same rate—because they have the same size and charge.

It's that simple: Molecules that are alike form a band because they stay together.

Using Thin Layer Chromatography to Separate Molecules

Gel electrophoresis is not the only way to separate molecules. Researchers also use a method called thin layer chromatography. This method was developed in the early 1900s by botanists who were analyzing the different-colored pigments from leaves of a plant (see Chapter 10), hence the name chromatography from the Greek words khroma for "color" and graphein, "to write."

In this method, rather than loading the sample into the well of a gel, the samples are deposited or "spotted" near the bottom of a stiff support, either glass or plastic, that is coated with a thin layer of silica gel, cellulose, or a similar porous material. The coated support is placed in a solvent solution. As the solvent

PROCESS: FORMATION OF BANDS ON GELS

Well

1. Start with a mixture of molecules in a well.

2. As electrophoresis starts, molecules begin to separate by size and charge.

3. As electrophoresis continues, separation increases. Molecules with the same size and charge "run" at the same rate.

4. If each molecule is visualized, the result is a set of bands.

FIGURE B9.2 **On a Gel, Molecules That Are Alike Form Bands.**

wicks upward through the coating by capillary action, it carries the molecules in the mixture with it. Molecules are carried at different rates, based on their size and solubility in the solvent.

Visualizing Molecules

Once molecules have been separated using electrophoresis or thin layer chromatography, they have to be detected. Unfortunately, although plant pigments are colored, proteins and nucleic acids are invisible unless they are tagged in some way. Let's first look at two of the most common tagging systems and then consider how researchers can tag and visualize specific molecules of interest and not others.

Using Radioactive Isotopes and Autoradiography

When molecular biology was getting under way, the first types of tags in common use were radioactive isotopes—forms of atoms that are unstable and release energy in the form of radiation.

In the polymerization experiment diagrammed in Figure B9.1, for example, the researchers had attached a radioactive phosphorus atom to the monomers—ribonucleoside triphosphates—used in the original reaction mix. Once polymers formed, they contained radioactive atoms. When electrophoresis was complete, the investigators visualized the polymers by laying X-ray film over the gel. Because radioactive emissions expose film, a black dot appears wherever a radioactive atom is located in the gel. So many black dots occur so close together that the collection forms a dark band.

This technique for visualizing macromolecules is called autoradiography. The autoradiograph that resulted from the polymerization experiment is shown in **FIGURE B9.3**. The samples, taken on days 2, 4, 6, 8, and 14 of the experiment, are noted along the bottom. The far right lane contains macromolecules of known size; this lane is used to estimate the size of the molecules in the experimental samples. The bands that appear in each sample lane represent the different polymers that had formed.

Reading a Gel

One of the keys to interpreting or "reading" a gel, or the corresponding autoradiograph, is to realize that darker bands contain more radioactive markers, indicating the presence of many radioactive molecules. Lighter bands contain fewer molecules.

To read a gel, then, you look for (1) the presence or absence of bands in some lanes—meaning, some experimental samples—versus others, and (2) contrasts in the darkness of the bands—meaning, differences in the number of molecules present.

For example, several conclusions can be drawn from the data in Figure B9.3. First, a variety of polymers formed at each stage. After the second day, for example, polymers from 12 to 18 monomers long had formed. Second, the overall length of polymers produced increased with time. At the end of the fourteenth day, most of the RNA molecules were between 20 and 40 monomers long.

FIGURE B9.3 Autoradiography Is a Technique for Visualizing Macromolecules. The molecules in a gel can be visualized in a number of ways. In this case, the RNA molecules in the gel exposed an X-ray film because they had radioactive atoms attached. When developed, the film is called an autoradiograph.

Starting in the late 1990s and early 2000s, it became much more common to tag nucleic acids with fluorescent tags. Once electrophoresis is complete, fluorescence can be detected by exposing the gel to an appropriate wavelength of light; the fluorescent tag fluoresces or glows in response (fluorescence is explained in Chapter 10).

Fluorescent tags have important advantages over radioactive isotopes: (1) They are safer to handle. (2) They are faster—you don't have to wait hours or days for the radioactive isotope to expose a film. (3) They come in multiple colors, so you can tag several different molecules in the same experiment and detect them independently.

Using Nucleic Acid Probes

In many cases, researchers want to find one specific molecule—a certain DNA sequence, for example—in the collection of molecules on a gel. How is this possible? The answer hinges on using a particular molecule as a probe.

You'll learn in more detail about how probes work in this text (Chapter 20). Here it's enough to get the general idea: A probe is a marked molecule that binds specifically to your molecule of interest. The "mark" is often a radioactive atom, a fluorescent tag, or an enzyme that catalyzes a color-forming or light-emitting reaction.

If you are looking for a particular DNA or RNA sequence on a gel, for example, you can expose the gel to a single-stranded probe that binds to the target sequence by complementary base pairing. Once it has bound, you can detect the band through autoradiography or fluorescence.

- *Southern blotting* is a technique for making DNA fragments that have been run out on a gel single stranded, transferring them from the gel to a nylon membrane, and then probing them to identify segments of interest. The technique was named after its inventor, Edwin Southern.

- *Northern blotting* is a technique for transferring RNA fragments from a gel to a nylon membrane and then probing them to detect target segments. The name is a lighthearted play on Southern blotting—the protocol from which it was derived.

Using Antibody Probes

How can researchers find a particular protein out of a large collection of different proteins? The answer is to use an antibody. An antibody is a protein that binds specifically to a section of a different protein (see Chapter 51 if you want more detail on the structure of antibodies and their function in the immune system).

To use an antibody as a probe, investigators attach a tag molecule—often an enzyme that catalyzes a color-forming reaction—to the antibody and allow it to react with proteins in a mixture. The antibody will stick to the specific protein that it binds to and then can be visualized thanks to the tag it carries.

If the proteins in question have been separated by gel electrophoresis and transferred to a membrane, the result is called a western blot. The name western is an extension of the Southern and northern patterns.

Using Radioimmunoassay and ELISA to Measure Amounts of Molecules

Another important method that makes use of antibodies is called a radioimmunoassay. This method is used when investigators want to measure tiny amounts of a molecule, such as a hormone in the blood. In this case, a known quantity of a hormone is labeled with a radioactive tag. This tagged hormone is then mixed with a known amount of antibody, and the two bind to one another. Next, a sample of blood, containing an unknown quantity of that same hormone, is added. The hormone from the blood and the radiolabeled hormone compete for antibody binding sites. As the concentration of unlabeled hormone increases, more of it binds to the antibody, displacing more of the radiolabeled hormone. The amount of unbound radiolabeled hormone is then measured. Using known standards as a reference, the amount of hormone in the blood can be determined.

Another commonly used technique based on similar principles is called ELISA (enzyme-linked immunosorbent assay). In this case, the amount of a particular molecule is measured using colorimetric signals instead of a radioactive signal.

check your understanding

If you understand BioSkill 9
✔ **You should be able to . . .**

Interpret a gel that has been stained for "RNA X." One lane contains no bands. Two lanes have a band in the same location, even though one of the bands is barely visible and the other is extremely dark. The fourth lane has a faint band located below the bands in the other lanes.

Biologists use a technique called differential centrifugation to isolate specific cell components. Differential centrifugation is based on breaking cells apart to create a complex mixture and then separating components in a centrifuge. A centrifuge accomplishes this task by spinning cells in a solution that allows molecules and other cell components to separate according to their density or size and shape. The individual parts of the cell can then be purified and studied in detail, in isolation from other parts of the cell.

The first step in preparing a cell sample for centrifugation is to release the cell components by breaking the cells apart. This can be done by putting them in a hypotonic solution, by exposing them to high-frequency vibration, by treating cells with a detergent, or by grinding them up. Each of these methods breaks apart plasma membranes and releases the contents of the cells.

The resulting pieces of plasma membrane quickly reseal to form small vesicles, often trapping cell components inside. The solution that results from the homogenization step is a mixture of these vesicles, free-floating macromolecules released from the cells, and organelles. A solution like this is called a cell extract or cell homogenate.

When a cell homogenate is placed in a centrifuge tube and spun at high speed, the components that are in solution tend to move outward, along the red arrow in **FIGURE B10.1a**. The effect is similar to a merry-go-round, which seems to push you outward in a straight line away from the spinning platform. In response to this outward-directed force, the cell homogenate exerts a centripetal (literally, "center-seeking") force that pushes the homogenate away from the bottom of the tube. Larger, denser molecules or particles resist this inward force more readily than do smaller, less dense ones and so reach the bottom of the centrifuge tube faster.

To separate the components of a cell extract, researchers often perform a series of centrifuge runs. Steps 1 and 2 of

(a) How a centrifuge works

When the centrifuge spins, the macromolecules tend to move toward the bottom of the centrifuge tube (red arrow)

The solution in the tube exerts a centripetal force, which resists movement of the molecules to the bottom of the tube (blue arrow)

Motor

Very large or dense molecules overcome the centripetal force more readily than smaller, less dense ones. As a result, larger, denser molecules move toward the bottom of the tube faster.

(b) PROCESS: DIFFERENTIAL CENTRIFUGATION

Low-speed centrifugation

Medium-speed centrifugation

High-speed centrifugation

Supernatant

Pellet

1. Start with uniform cell homogenate in centrifuge tube.

2. After low-speed spin, pellet contains large components. Transfer supernatant to new tube.

3. After medium-speed spin, pellet contains medium components. Transfer supernatant to new tube.

4. After high-speed spin, pellet contains small components.

(c) PROCESS: SUCROSE DENSITY–GRADIENT CENTRIFUGATION

Lower-density solution

Higher-density solution

Sample

1. Add sample to tube of variable-density solution.

2. Run centrifuge. Cell components are separated into distinct bands based on size and/or density.

3. To extract specific cell components for analysis, puncture bottom of tube with needle and collect drops from specific bands.

FIGURE B10.1 Cell Components Can Be Separated by Centrifugation. (a) The forces inside a centrifuge tube allow cell components to be separated. **(b)** Through a series of centrifuge runs made at increasingly higher speeds, an investigator can separate fractions of a cell homogenate by size via differential centrifugation. **(c)** A high-speed centrifuge run can achieve extremely fine separation among cell components by sucrose density–gradient centrifugation.

individuals. For example, when researchers examine a 33-cell *C. elegans* embryo, they know exactly which of the 959 cells in the adult will be derived from each of those 33 embryonic cells.

In addition, *C. elegans* are small (less than 1 mm long), are able to self-fertilize or cross-fertilize, and undergo early development in just 16 hours. The entire genome of *C. elegans* has now been sequenced.

Mus musculus

The house mouse *Mus musculus* is the most important model organism among mammals. For this reason, it is especially prominent in biomedical research, where researchers need to work on individuals with strong genetic and developmental similarities to humans.

The house mouse was an intelligent choice of model organism in mammals because it is small and thus relatively inexpensive to maintain in captivity, and because it breeds rapidly. A litter can contain 10 offspring, and generation time is only 12 weeks—meaning that several generations can be produced in a year. Descendants of wild house mice have been selected for docility and other traits that make them easy to handle and rear; these populations are referred to as laboratory mice (**FIGURE B13.1g**).

Some of the most valuable laboratory mice are strains with distinctive, well-characterized genotypes. Inbred strains are virtually homogenous genetically (see Chapter 26) and are useful in experiments where gene-by-gene or gene-by-environment interactions have to be controlled. Other populations carry mutations that knock out genes and cause diseases similar to those observed in humans. These individuals are useful for identifying the cause of genetic diseases and testing drugs or other types of therapies.

check your understanding

If you understand BioSkill 13

✔ **You should be able to . . .**

Determine which model organisms described here would be the best choice for the following studies. In each case, explain your reasoning.

1. A study of why specific cells in an embryo die at certain points in normal development. One goal is to understand the consequences for the individual when programmed cell death does not occur when it should.

2. A study of proteins that are required for cell–cell adhesion.

3. Research on a gene suspected to be involved in the formation of breast cancer in humans.

BIOSKILL 14   primary literature and peer review

As part of the process of doing science, biologists communicate their results to the scientific community through publications in scientific journals that report on their original research discoveries (see Chapter 1). These published reports are referred to, interchangeably, as the primary literature, research papers, or primary research articles.

What Is the Primary Literature?

Scientists publish "peer-reviewed" papers. This means that several experts in the field have carefully read the paper and considered its strengths and weaknesses. Reviewers write a critique of the paper and make a recommendation to the journal editor as to whether the paper should be published. Often reviewers will suggest additional experiments that need to be completed before a paper is considered acceptable for publication. The peer review process means that research discoveries are carefully vetted before they go to press.

A primary research paper can be distinguished from secondary sources—such as review articles, textbooks, and magazine articles—by looking for key characteristics. A primary research paper includes a detailed description of methods and results, written by the researchers who did the work. A typical paper contains a Title, Abstract, Introduction, Materials and Methods (or Experimental Design), Results and Discussion (**Table B14.1**), although the order and name of the sections varies among journals.

Getting Started

At first, trying to read the primary literature may seem like a daunting task. A paper may be peppered with unfamiliar terms and acronyms. If you tried to read a research paper from start to finish, like you might read a chapter in this textbook, it would be a frustrating experience. But, with practice, the scientific literature becomes approachable, and it is well worth the effort. The primary literature is the cutting edge, the place to read firsthand about the process of doing science. Becoming skilled at reading and evaluating scientific reports is a powerful way to learn how to think critically—to think like a biologist.

To get started, try breaking down reading the primary research article into a series of steps:

1. Read the authors' names. Where are they from? Are they working as a team or alone? After delving into the literature,

TABLE B14.1 Sections of a Primary Research Paper

Section	Characteristics
Title	Short, succinct, eye-catching
Abstract	Summary of Methods, Results, Discussion. Explains why the research was done and why the results are significant.
Introduction	Background information (what past work was done, why the work was important). States the objectives and hypotheses of the study and explains why the study is important.
Materials and Methods	Explains how the work was done and where it was done.
Results	Explains what the data show.
Discussion	Explains why the data show what they show, how the analysis relates to the objectives from the Introduction, and the significance of findings and how they advance the field.

certain familiar names will crop up again and again. You'll begin to recognize the experts in a particular field.

2. Read the title. It should summarize the key finding of the paper and tell you what you can expect to learn from the paper.

3. Read the abstract. The abstract summarizes the entire paper in a short paragraph. At this point, it might be tempting to stop reading. But sometimes authors understate or overstate the significance and conclusions of their work. You should never cite an article as a reference after having read only the abstract.

4. Read the Introduction. The first couple of paragraphs should make it clear what the objectives or hypotheses of the paper are; the remaining paragraphs will give you the background information you need to understand the point of the paper.

5. Flip through the article and look at the figures, illustrations, and tables, including reading the legends.

6. Read the Results section carefully. Ask yourself these questions: Do the results accurately describe the data presented in the paper? Were all the appropriate controls carried out in an experiment? Are there additional experiments that you think should have been performed? Are the figures and tables clearly labeled?

7. Consult the Materials and Methods section to help understand the research design and the techniques used.

8. Read the Discussion. The first and last paragraphs usually summarize the key findings and state their significance. The Discussion is the part of the paper where the results are explained in the context of the scientific literature. The authors should explain what their results mean.

Getting Practice

The best way to get practice is to start reading the scientific literature as often as possible. You could begin by reading some of the references cited in this textbook. You can get an electronic copy of most articles through online databases such as PubMed, ScienceDirect, or Google Scholar, or through your institution's library.

After reading a primary research paper, you should be able to paraphrase the significance of the paper in a few sentences, free of technical jargon. You should also be able to both praise and criticize several points of the paper. As you become more familiar with reading the scientific literature, you're likely to start thinking about what questions remain to be answered. And, you may even come up with "the next experiment."

check your understanding

If you understand BioSkill 14

✔ **You should be able to . . .**

After choosing a primary research paper on a topic in biology that you would like to know more about, select one figure in the Results section that reports on the experiment and construct a Research box (like the ones in this textbook) that depicts this experiment.

Glossary

5′ cap A modified guanine (G) nucleotide (7-methylguanylate) added to the 5′ end of eukaryotic mRNAs. Helps protect the mRNA from being degraded and promotes efficient initiation of translation.

abdomen A region of the body; in insects, one of the three prominent body regions called tagmata.

abiotic Not alive (e.g., air, water, and some components of soil). Compare with **biotic**.

aboveground biomass The total mass of living plants in an area, excluding roots.

abscisic acid (ABA) A plant hormone that inhibits growth; it stimulates stomatal closure and triggers dormancy.

abscission In plants, the normal (often seasonal) shedding of leaves, fruits, or flowers.

abscission zone The region at the base of a petiole where cell wall degradation occurs; results in the dropping of leaves.

absorption In animals, the uptake of ions and small molecules, derived from food, across the lining of the intestine and into the bloodstream.

absorption spectrum The amount of light of different wavelengths absorbed by a pigment. Usually depicted as a graph of light absorbed versus wavelength. Compare with **action spectrum**.

acclimation A change in a study organism's phenotype that occurs in response to laboratory conditions.

acclimatization A change in an individual's phenotype that occurs in response to a change in natural environmental conditions.

acetyl CoA A molecule produced by oxidation of pyruvate (the final product of glycolysis) in a reaction catalyzed by pyruvate dehydrogenase. Can enter the citric acid cycle and is used as a carbon source in the synthesis of fatty acids, steroids, and other compounds.

acetylation Addition of an acetyl group ($-COCH_3$) to a molecule. Acetylation of histone proteins is important in controlling chromatin condensation.

acetylcholine (Ach) A neurotransmitter, released by nerve cells at neuromuscular junctions, that triggers contraction of skeletal muscle cells but slows the rate of contraction in cardiac muscle cells. Also used as a neurotransmitter between neurons.

acid Any compound that gives up protons or accepts electrons during a chemical reaction or that releases hydrogen ions when dissolved in water.

acid-growth hypothesis The hypothesis that auxin triggers elongation of plant cells by increasing the activity of proton pumps, making the cell wall more acidic and leading to expansion of the cell wall and an influx of water.

acoelomate A bilaterian animal that lacks an internal body cavity (coelom). Compare with **coelomate** and **pseudocoelomate**.

acquired immune deficiency syndrome (AIDS) A human disease characterized by death of immune system cells (in particular helper T cells) and subsequent vulnerability to other infections. Caused by the human immunodeficiency virus (HIV).

acrosome A caplike structure, located on the head of a sperm cell, that contains enzymes capable of dissolving the outer coverings of an egg.

ACTH See **adrenocorticotropic hormone**.

actin A globular protein that can be polymerized to form filaments. Actin filaments are part of the cytoskeleton and constitute the thin filaments in skeletal muscle cells.

actin filament A long fiber, about 7 nm in diameter, composed of two intertwined strands of polymerized actin protein; one of the three types of cytoskeletal fibers. Involved in cell movement. Also called a *microfilament*. Compare with **intermediate filament** and **microtubule**.

action potential A rapid, temporary change in electrical potential across a membrane, from negative to positive and back to negative. Occurs in cells, such as neurons and muscle cells, that have an excitable membrane.

action spectrum The relative effectiveness of different wavelengths of light in driving a light-dependent process such as photosynthesis. Usually depicted as a graph of some measure of the process versus wavelength. Compare with **absorption spectrum**.

activation energy The amount of energy required to initiate a chemical reaction; specifically, the energy required to reach the transition state.

activator A protein that binds to a DNA regulatory sequence to increase the frequency of transcription initiation by RNA polymerase.

active site The location in an enzyme molecule where substrates (reactant molecules) bind and react.

active transport The movement of ions or molecules across a membrane against an electrochemical gradient. Requires energy (e.g., from hydrolysis of ATP) and assistance of a transport protein (e.g., pump).

adaptation Any heritable trait that increases the fitness of an individual with that trait, compared with individuals without that trait, in a particular environment.

adaptive immune response See **adaptive immunity**.

adaptive immunity Immunity to a particular pathogen or other antigen conferred by activated B and T cells in vertebrates. Characterized by specificity, diversity, memory, and self–nonself recognition. Also called *adaptive immune response*. Compare with **innate immunity**.

adaptive radiation Rapid evolutionary diversification within one lineage, producing many descendant species with a wide range of adaptive forms.

adenosine triphosphate (ATP) A molecule consisting of an adenine base, a sugar, and three phosphate groups that can be hydrolyzed to release energy. Universally used by cells to store and transfer energy.

adhesion The tendency of certain dissimilar molecules to cling together due to attractive forces. Compare with **cohesion**.

adipocyte A fat cell.

adrenal gland Either of two small endocrine glands, one above each kidney. The outer portion (cortex) secretes several steroid hormones; the inner portion (medulla) secretes epinephrine and norepinephrine.

adrenaline See **epinephrine**.

adrenocorticotropic hormone (ACTH) A peptide hormone, produced and secreted by the anterior pituitary, that stimulates release of steroid hormones (e.g., cortisol, aldosterone) from the adrenal cortex.

adult A sexually mature individual.

adventitious root A root that develops from a plant's shoot system instead of from the plant's root system.

aerobic Referring to any metabolic process, cell, or organism that uses oxygen as an electron acceptor. Compare with **anaerobic**.

afferent division The part of the nervous system that transmits information about the internal and external environment to the central nervous system. Consists mainly of sensory neurons. Compare with **efferent division**.

age class All the individuals of a specific age in a population.

age-specific fecundity The average number of female offspring produced by a female in a certain age class.

age structure The proportion of individuals in a population that are of each possible age.

agglutination Clumping together of cells or viruses by antibodies or other cross-linking molecules.

aggregate fruit A fruit (e.g., raspberry) that develops from a single flower that has many separate carpels. Compare with **multiple** and **simple fruit**.

AIDS See **acquired immune deficiency syndrome**.

albumen A solution of water and protein (particularly albumins), found in amniotic eggs, that nourishes the growing embryo. Also called *egg white*.

alcohol fermentation Catabolic pathway in which pyruvate produced by glycolysis is converted to ethanol in the absence of oxygen.

aldosterone A hormone that stimulates the kidney to conserve salt and water and promotes retention of sodium; produced in the adrenal cortex.

allele A particular version of a gene.

allergen Any molecule (antigen) that triggers an allergic response (an allergy).

allergy An IgE-mediated abnormal response to an antigen, usually characterized by dilation of blood vessels, contraction of smooth muscle cells, and increased activity of mucus-secreting cells.

allopatric speciation Speciation that occurs when populations of the same species become geographically isolated, often due to dispersal or vicariance. Compare with **sympatric speciation**.

allopatry Condition in which two or more populations live in different geographic areas. Compare with **sympatry**.

allopolyploidy (adjective: allopolyploid) The state of having more than two full sets of chromosomes (polyploidy) due to hybridization between different species. Compare with **autopolyploidy**.

allosteric regulation Regulation of a protein's function by binding of a regulatory molecule, usually to a specific site distinct from the active site, that causes a change in the protein's shape.

α-amylase See **amylase**.

α-helix (alpha-helix) A protein secondary structure in which the polypeptide backbone coils into a spiral shape stabilized by hydrogen bonding.

alternation of generations A life cycle involving alternation of a multicellular haploid stage (gametophyte) with a multicellular diploid stage (sporophyte). Occurs in most plants and some protists.

alternative splicing In eukaryotes, the splicing of primary RNA transcripts from a single gene in different ways to produce different mature mRNAs and thus different polypeptides.

altruism Any behavior that has a fitness cost to the individual (lowered survival and/or reproduction) and a fitness benefit to the recipient. See **reciprocal altruism**.

alveolus (plural: alveoli) Any of the tiny air-filled sacs of a mammalian lung.

ambisense RNA virus A ssRNA virus whose genome consists of at least one strand that contains both positive-sense and negative-sense regions.

aminoacyl tRNA A transfer RNA molecule that is covalently bound to an amino acid.

aminoacyl-tRNA synthetase An enzyme that catalyzes the addition of a particular amino acid to its corresponding tRNA molecule.

ammonia (NH$_3$) A small molecule, produced by the breakdown of proteins and nucleic acids, that is very toxic to cells. Is a strong base that gains a proton to form the ammonium ion (NH$_4^+$).

amnion The innermost of the membranes surrounding the embryo in an amniotic egg.

amniotes A major lineage of vertebrates (Amniota) that reproduce with amniotic eggs. Includes all reptiles (including birds) and mammals—all tetrapods except amphibians.

amniotic egg An egg that has a watertight shell or case enclosing a membrane-bound water supply (the amnion), food supply (yolk sac), and waste sac (allantois).

amoeboid motion See **cell crawling**.

amphibians A lineage of vertebrates, many of which breathe through their skin and feed on land but lay their eggs in water. Represent the earliest tetrapods; include frogs, salamanders, and caecilians.

amphipathic Containing hydrophilic and hydrophobic regions.

ampullae of Lorenzini Structures on the heads of sharks that contain cells with electroreceptors.

amylase Any enzyme that can break down starch by catalyzing hydrolysis of the glycosidic linkages between the glucose residues.

amyloplasts Starch-storing organelles (plastids) in plants. In root cap cells, they settle to the bottom of the cell and may be used as gravity detectors.

anabolic pathway Any set of chemical reactions that synthesizes large molecules from smaller ones. Generally requires an input of energy. Compare with **catabolic pathway**.

anadromous Having a life cycle in which adults live in the ocean (or large lakes) but migrate up freshwater streams to breed and lay eggs.

anaerobic Referring to any metabolic process, cell, or organism that uses an electron acceptor other than oxygen, including fermentation or anaerobic respiration. Compare with **aerobic**.

anaphase A stage in mitosis or meiosis during which chromosomes are moved to opposite poles of the spindle apparatus.

anatomy The study of the physical structure of organisms.

ancestral trait A trait found in the ancestors of a particular group.

aneuploidy (adjective: aneuploid) The state of having an abnormal number of copies of a certain chromosome.

angiosperm A flowering vascular plant that produces seeds within mature ovaries (fruits). The angiosperms form a single lineage. Compare with **gymnosperm**.

animal A member of a major lineage of eukaryotes (Animalia) whose members typically have a complex, large, multicellular body; eat other organisms; and are mobile.

animal model Any disease that occurs in a non-human animal and has parallels to a similar disease of humans. Studied by medical researchers in hopes that findings may apply to human disease.

anion A negatively charged ion.

annelids Members of the phylum Annelida (segmented worms). Distinguished by a segmented body and a coelom that functions as a hydrostatic skeleton. Annelids belong to the lophotrochozoan branch of the protostomes.

annual Referring to a plant whose life cycle normally lasts only one growing season—less than one year. Compare with **perennial**.

anoxygenic Referring to any process or reaction that does not produce oxygen. Photosynthesis in purple sulfur and purple nonsulfur bacteria, which does not involve photosystem II, is anoxygenic. Compare with **oxygenic**.

antagonistic muscle group A set of two or more muscles that reextend one another by transmitting their forces via the skeleton.

antenna (plural: antennae) A long appendage of the head that is used to touch or smell.

antenna complex Part of a photosystem, containing an array of chlorophyll molecules and accessory pigments, that receives energy from light and directs the energy to a central reaction center during photosynthesis.

anterior Toward an animal's head and away from its tail. The opposite of posterior.

anterior pituitary The part of the pituitary gland containing endocrine cells that produce and release a variety of peptide hormones in response to other hormones from the hypothalamus. Compare with **posterior pituitary**.

anther The pollen-producing structure at the end of a stamen in flowering plants (angiosperms).

antheridium (plural: antheridia) The sperm-producing structure in most land plants except angiosperms.

anthropoids One of the two major lineages of primates, including apes, humans, and all monkeys. Compare with **prosimians**.

antibiotic Any substance, such as penicillin, that can kill or inhibit the growth of bacteria.

antibody A protein produced by B cells that can bind to a specific part of an antigen, tagging it for removal by the immune system. All monomeric forms of antibodies consist of two light chains and two heavy chains, which vary between different antibodies. Also called *immunoglobulin*.

anticodon The sequence of three bases (a triplet) in a transfer RNA molecule that can bind to an mRNA codon with a complementary sequence.

antidiuretic hormone (ADH) A peptide hormone, secreted from the posterior pituitary gland, that stimulates water retention by the kidney. Also called *vasopressin*.

antigen Any foreign molecule, often a protein, that can stimulate an innate or adaptive response by the immune system.

antigen presentation Process by which small peptides, derived from ingested particulate antigens (e.g., bacteria) or intracellular antigens (e.g., viruses in infected cell) are complexed with MHC proteins and transported to the cell surface, where they are displayed and can be recognized by T cells.

antiparallel Describing the opposite orientation of nucleic acid strands that are hydrogen bonded to one another, with one strand running in the $5' \rightarrow 3'$ direction and the other in the $3' \rightarrow 5'$ direction.

antiporter A carrier protein that allows an ion to diffuse down an electrochemical gradient, using the energy of that process to transport a different substance in the opposite direction *against* its concentration gradient. Compare with **symporter**.

antiviral Any drug or other agent that can interfere with the transmission or replication of viruses.

aorta In terrestrial vertebrates, the major artery carrying oxygenated blood away from the heart.

aphotic zone Deep water receiving no sunlight. Compare with **photic zone**.

apical Toward the top. In plants, at the tip of a branch. In animals, on the side of an epithelial layer

body plan The b[...]
including the nu[...]
body segments, a[...]

bog A freshwate[...]
water flow, result[...]
acidic conditions[...]

Bohr shift The ri[...]
hemoglobin equil[...]
decreasing pH. It[...]
likely to release o[...]
exercising muscle[...]

bone A type of v[...]
ing of living cells[...]
extracellular matr[...]
($CaPO_4$) and sma[...]
($CaCO_3$) and pro[...]

bone marrow Th[...]
large bones; cont[...]
blood cells and le[...]

Bowman's capsu[...]
shaped portion o[...]
erulus in the vert[...]

brain A large ma[...]
head region of an[...]
mation processin[...]
ganglion.

brain stem The r[...]
tebrate brain, con[...]
responsible for au[...]
heart rate, respira[...]

braincase See cr[...]

branch (1) A par[...]
sents populations[...]
a plant's shoot sys[...]

brassinosteroids[...]
found in plants; s[...]

bronchiole Any [...]
lungs that carry a[...]

bronchus (plural[...]
pair of large tube[...]
lung.

bryophyte See n[...]

budding Asexual[...]
growth from the [...]
pendent individu[...]
invertebrates.

buffer A substan[...]
mize changes in t[...]
base is added.

bulbourethral gla[...]
small pair of glan[...]
secrete an alkalin[...]
lubricates the tip [...]
the urethra durin[...]
Cowper's glands.

bulk flow The di[...]
due to pressure di[...]
ter through plant[...]
of blood in anima[...]

bulk-phase endoc[...]
extracellular fluid[...]
brane to form sm[...]
considered to be a[...]

that faces the environment and not other body tissues. Compare with **basal**.

apical–basal axis The shoot-to-root axis of a plant.

apical bud A bud at the tip of a stem or branch, where growth occurs to lengthen the stem or branch.

apical dominance Inhibition of lateral bud growth by the apical meristem at the tip of a plant branch.

apical meristem A group of undifferentiated plant cells, at the tip of a shoot or root, that is responsible for primary growth. Compare with **cambium**.

apodeme Any of the chitinous ingrowths of the exoskeleton to which muscles attach.

apomixis The formation of mature seeds without fertilization occurring; a type of asexual reproduction.

apoplast In plants, the region outside plasma membranes consisting of the porous cell walls and the intervening extracellular air space. Compare with **symplast**.

apoptosis Series of tightly controlled changes that lead to the self-destruction of a cell. Occurs frequently during embryological development and as part of the immune response to remove infected or cancerous cells. Also called *programmed cell death.*

appendix A blind sac (having only one opening) that extends from the cecum in some mammals.

aquaporin A type of channel protein through which water can move by osmosis across a plasma membrane.

aquifer An underground layer of porous rock, sand, or gravel that is saturated with water.

***ara* operon** A set of three genes in *E. coli* that are transcribed into a single mRNA and required for metabolism of the sugar arabinose. Transcription of the *ara* operon is controlled by the AraC regulatory protein.

araC The regulatory gene (written as *araC*) or regulatory protein (when written as AraC) of the *E. coli ara* operon.

arbuscular mycorrhizal fungi (AMF) Fungi from the Glomeromycota lineage whose hyphae enter the root cells of their host plants. Also called *endomycorrhizal fungi.*

Archaea One of the three taxonomic domains of life, consisting of unicellular prokaryotes distinguished by cell walls made of certain polysaccharides not found in bacterial or eukaryotic cell walls, plasma membranes composed of unique isoprene-containing phospholipids, and ribosomes and RNA polymerase similar to those of eukaryotes. Compare with **Bacteria** and **Eukarya**.

archegonium (plural: archegonia) The egg-producing structure in most land plants except angiosperms.

arteriole Any of the many tiny vessels that carry blood from arteries to capillaries.

arteriosclerosis Hardening and loss of elasticity of arteries.

artery Any thick-walled blood vessel that carries blood (oxygenated or not) under relatively high pressure away from the heart to organs throughout the body. Compare with **vein**.

arthropods Members of the phylum Arthropoda. Distinguished by a segmented body; a hard, jointed exoskeleton; paired, jointed appendages; and an extensive body cavity called a hemocoel. Arthropods belong to the ecdysozoan branch of the protostomes.

articulation A movable point of contact between two rigid components of a skeleton, such as between bones of a vertebrate endoskeleton or between segments of cuticle in an arthropod exoskeleton. See **joint**.

artificial selection Deliberate manipulation by humans, as in animal and plant breeding, of the genetic composition of a population by allowing only individuals with desirable traits to reproduce.

ascus (plural: asci) Specialized spore-producing cell found at the ends of hyphae in "sac fungi" (Ascomycota).

asexual reproduction Any form of reproduction where offspring inherit DNA from only one parent. Includes binary fission, budding, and parthenogenesis. Compare with **sexual reproduction**.

astral microtubules Mitotic and meiotic microtubules that have arisen from the two spindle poles and interact with proteins on the plasma membrane.

asymmetric competition Ecological competition between two species in which one species suffers a much greater fitness decline than the other. Compare with **symmetric competition**.

atomic number The number of protons in the nucleus of an atom, giving the atom its identity as a particular chemical element.

atomic weight The average mass of an element that is based on the relative proportions of all the naturally occurring isotopes.

ATP synthase A large membrane-bound protein complex that uses the energy of protons flowing through it to synthesize ATP.

ATP See **adenosine triphosphate**.

atrioventricular (AV) node A region of the heart between the right atrium and right ventricle where electrical signals from the atrium are slowed briefly before spreading to the ventricle. This delay allows the ventricle to fill with blood before contracting. Compare with **sinoatrial (SA) node**.

atrium (plural: atria) A thin-walled chamber of the heart that receives blood from veins and pumps it to a neighboring chamber (the ventricle).

autocrine Relating to a chemical signal that affects the same cell that produced and released it.

autoimmunity A pathological condition in which the immune system attacks self cells or tissues of an individual's own body.

autonomic nervous system The part of the peripheral nervous system that controls internal organs and involuntary processes, such as stomach contraction, hormone release, and heart rate. Includes parasympathetic and sympathetic nerves. Compare with **somatic nervous system**.

autophagy The process by which damaged organelles are surrounded by a membrane and delivered to a lysosome to be recycled.

autopolyploidy (adjective: autopolyploid) The state of having more than two full sets of chromosomes

(polyploidy) due to a mutation that doubled the chromosome number. All the chromosomes come from the same species. Compare with **allopolyploidy**.

autosomal inheritance The inheritance patterns that occur when genes are located on autosomes rather than on sex chromosomes.

autosome Any chromosome other than a sex chromosome (i.e,. any chromosome other than the X or Y in mammals).

autotroph Any organism that can synthesize reduced organic compounds from simple inorganic sources such as CO_2 or CH_4. Most plants and some bacteria and archaea are autotrophs. Also called *primary producer*. Compare with **heterotroph**.

auxin Indoleacetic acid (IAA), a plant hormone that stimulates phototropism and cell elongation.

axillary bud A bud that forms at a node and may develop into a lateral (side) branch. Also called *lateral bud*.

axon A long projection of a neuron that can propagate an action potential.

axon hillock The site in a neuron where an axon joins the cell body and where action potentials are first triggered.

axoneme A structure found in eukaryotic cilia and flagella and responsible for their motion; composed of two central microtubules surrounded by nine doublet microtubules (9 + 2 arrangement).

B cell A type of lymphocyte that matures in the bone marrow and, with T cells, is responsible for adaptive immunity. Produces antibodies and also functions in antigen presentation. Also called *B lymphocyte*.

B-cell receptor (BCR) An immunoglobulin protein embedded in the plasma membrane of mature B cells and to which antigens bind. Apart from the transmembrane domain, it is identical in structure to antibodies.

BAC library A collection of all the sequences found in the genome of a species, inserted into **bacterial artificial chromosomes (BACs)**.

background extinction The average rate of low-level extinction that has occurred continuously throughout much of evolutionary history. Compare with **mass extinction**.

Bacteria One of the three taxonomic domains of life, consisting of unicellular prokaryotes distinguished by cell walls composed largely of peptidoglycan, plasma membranes similar to those of eukaryotic cells, and ribosomes and RNA polymerase that differ from those in archaeans or eukaryotes. Compare with **Archaea** and **Eukarya**.

bacterial artificial chromosome (BAC) An artificial version of a bacterial chromosome that can be used as a cloning vector to produce many copies of large DNA fragments.

bacteriophage Any virus that infects bacteria.

baculum A bone inside the penis; usually present in mammals with a penis that lacks erectile tissue.

balancing selection A mode of natural selection in which no single allele is favored in all populations of a species at all times. Instead, there is a balance among alleles in terms of fitness and frequency.

ball-and-stick ...
where atoms are ...
to indicate the a...
are shown as ro...
the correct geo...

bar coding The ...
quences to iden...

bark The prote...
composed of co...
ary phloem.

baroreceptors ...
of the heart and ...
changes in bloo...
responses by th...

basal Toward t...
or at the base o...
animals, on the ...
underlying bod...

basal body Th...
cilia and flagell...
triplets of micro...
establishes the ...
identical with a...

basal lamina A...
trix that underl...
and connects it...

**basal metabolic...
sumption by an...
environment. F...
amount of oxyg...

**basal transcrip...
eukaryotic cells...
initiate transcri...
transcription f...

base Any com...
up electrons du...
hydrogen ions ...

basidium (plu...
producing cell ...
members of the...

**basilar membr...
bottom portion ...
cochlea.

basolateral To...
mals, the side o...
body tissues an...

**Batesian mimic...
harmless or pa...
ous or poisono...
mimicry.

beak A structu...
associated with ...
pods, and som...

behavior Any ...
sponse to a sti...

**behavioral eco...
respond to par...
their environm...

beneficial In g...
allele, or trait t...
fitness.

benign tumor ...
pears due to u...

cast A type of fossil, formed when the decay of a body part leaves a void that is then filled with minerals that later harden.

catabolic pathway Any set of chemical reactions that breaks down large, complex molecules into smaller ones, releasing energy in the process. Compare with **anabolic pathway**.

catalysis (verb: catalyze) Acceleration of the rate of a chemical reaction due to a decrease in the free energy of the transition state, called the activation energy.

catalyst Any substance that increases the rate of a chemical reaction without itself undergoing any permanent chemical change.

catecholamines A class of small compounds, derived from the amino acid tyrosine, that are used as hormones or neurotransmitters. Include epinephrine, norepinephrine, and dopamine.

cation A positively charged ion.

cation exchange The release (displacement) of cations, such as magnesium and calcium from soil particles, by protons in acidic soil water. The released cations are available for uptake by plants.

CD4 A membrane protein on the surface of some T cells in humans. CD4$^+$ T cells can give rise to helper T cells.

CD8 A membrane protein on the surface of some T cells in humans. CD8$^+$ T cells can give rise to cytotoxic T cells.

Cdk See **cyclin-dependent kinase**.

cDNA See **complementary DNA**.

cDNA library A set of cDNAs from a particular cell type or stage of development. Each cDNA is carried by a plasmid or other cloning vector and can be separated from other cDNAs. Compare with **genomic library**.

cecum A blind sac between the small intestine and the colon. Is enlarged in some species (e.g., rabbits) that use it as a fermentation vat for digestion of cellulose.

cell A highly organized compartment bounded by a thin, flexible structure (plasma membrane) and containing concentrated chemicals in an aqueous (watery) solution. The basic structural and functional unit of all organisms.

cell body The part of a neuron that contains the nucleus; where incoming signals are integrated. Also called the *soma*.

cell crawling A form of cellular movement involving actin filaments in which the cell produces bulges (pseudopodia) that stick to the substrate and pull the cell forward. Also called *amoeboid motion*.

cell cycle Ordered sequence of events in which a eukaryotic cell replicates its chromosomes, evenly partitions the chromosomes to two daughter cells, and then undergoes division of the cytoplasm.

cell-cycle checkpoint Any of several points in the cell cycle at which progression of a cell through the cycle can be regulated.

cell division Creation of new cells by division of preexisting cells.

cell-mediated (immune) response The type of immune response that involves generation of cytotoxic

T cells from CD8$^+$ T cells. Defends against pathogen-infected cells, cancer cells, and transplanted cells. Compare with **humoral (immune) response**.

cell membrane See **plasma membrane**.

cell plate A flattened sac-like structure formed in the middle of a dividing plant cell from Golgi-derived vesicles containing cell wall material; ultimately divides the cytoplasm into two separate cells.

cell sap An aqueous solution found in the vacuoles of plant cells.

cell theory The theory that all organisms are made of cells and that all cells come from preexisting cells.

cell wall A fibrous layer found outside the plasma membrane of most bacteria and archaea and many eukaryotes.

cellular respiration A common pathway for production of ATP, involving transfer of electrons from compounds with high potential energy through an electron transport chain and ultimately to an electron acceptor (often oxygen).

cellulase An enzyme that can break down cellulose by catalyzing hydrolysis of the glycosidic linkages between the glucose residues.

cellulose A structural polysaccharide composed of glucose monomers joined by β-1,4-glycosidic linkages. Found in the cell wall of algae, plants, and some bacteria and fungi.

Cenozoic era The most recent interval of geologic time, beginning 65.5 million years ago, during which mammals became the dominant vertebrates and angiosperms became the dominant plants.

central dogma The scheme for information flow in the cell: DNA → RNA → protein.

central nervous system (CNS) Large numbers of neurons aggregated into clusters called ganglia in bilaterian animals. In vertebrates, the central nervous system consists of the brain and spinal cord. Compare with **nerve net** and **peripheral nervous system (PNS)**.

centriole One of two small cylindrical structures found together within the centrosome near the nucleus of a eukaryotic cell (not found in plants). Consists of microtubule triplets and is structurally identical with a basal body.

centromere Constricted region of a replicated chromosome where the two sister chromatids are joined and the kinetochore is located.

centrosome Structure in animal and fungal cells, containing two centrioles, that serves as a microtubule organizing center for the cell's cytoskeleton and for the spindle apparatus during cell division.

cephalization The formation in animals of a distinct anterior region (the head) where sense organs and a mouth are clustered.

cephalochordates One of the three major chordate lineages (Cephalochordata), comprising small, mobile organisms that live in marine sands and suspension feed; also called *lancelets* or *amphioxus*. Compare with **urochordates** and **vertebrates**.

cephalopods A lineage of mollusks including the squid, octopuses, and nautiluses. Distinguished by large brains, excellent vision, tentacles, and a reduced or absent shell.

cerebellum Posterior section of the vertebrate brain; involved in coordination of complex muscle movements, such as those required for locomotion and maintaining balance.

cerebrum The most anterior section of the vertebrate brain. Divided into left and right hemispheres and four lobes: frontal lobe, involved in complex decision making (in humans); occipital lobe, receives and interprets visual information; parietal lobe, involved in integrating sensory and motor functions; and temporal lobe, functions in memory, speech (in humans), and interpreting auditory information.

cervix The bottom portion of the uterus, containing a canal that leads to the vagina.

chaetae (singular: chaeta) Bristle-like extensions found in some annelids.

channel protein A protein that forms a pore in a cell membrane. The structure of most channels allows them to admit just one or a few types of ions or molecules. Compare with **carrier protein**.

character Any genetic, morphological, physiological, or behavioral characteristic of an organism to be studied.

character displacement The evolutionary tendency for the traits of similar species that occupy overlapping ranges to change in a way that reduces interspecific competition.

chelicerae A pair of clawlike appendages found around the mouth of certain arthropods called chelicerates (spiders, mites, and allies).

chemical bond An attractive force binding two atoms together. Covalent bonds, ionic bonds, and hydrogen bonds are types of chemical bonds.

chemical energy The potential energy stored in covalent bonds between atoms.

chemical equilibrium A dynamic but stable state of a reversible chemical reaction in which the forward reaction and reverse reactions proceed at the same rate, so that the concentrations of reactants and products remain constant.

chemical evolution A The theory that simple chemical compounds in the early atmosphere and ocean combined via chemical reactions to form larger, more complex substances, eventually leading to the origin of life and the start of biological evolution.

chemical reaction Any process in which one compound or element is combined with others or is broken down; involves the making and/or breaking of chemical bonds.

chemiosmosis An energetic coupling mechanism whereby energy stored in an electrochemical proton gradient is used to drive an energy-requiring process such as production of ATP.

chemokine Any of a subset of cytokines that acts as a chemical signal attracting leukocytes to a site of tissue injury or infection.

chemolithotroph An organism (bacteria or archaea) that produces ATP by oxidizing inorganic molecules with high potential energy such as ammonia (NH_3) or methane (CH_4). Also called *lithotroph*. Compare with **chemoorganotroph**.

chemoorganotroph An organism that produces ATP by oxidizing organic molecules with high potential energy such as sugars. Also called *organotroph*. Compare with **chemolithotroph**.

chemoreception A sensory system in which receptors are activated in response to the binding of chemicals.

chemoreceptor A sensory cell or organ specialized for detection of specific molecules or classes of molecules.

chiasma (plural: chiasmata) The X-shaped structure formed during meiosis by crossing over between non-sister chromatids in a pair of homologous chromosomes.

chitin A structural polysaccharide composed of *N*-acetyl-glucosamine (NAG) monomers joined end to end by β-1,4-glycosidic linkages. Found in cell walls of fungi and many algae, and in external skeletons of insects and crustaceans.

chitons A lineage of marine mollusks that have a protective shell formed of eight calcium carbonate plates.

chlorophyll Any of several closely related green pigments, found in chloroplasts, that absorb light during photosynthesis.

chloroplast A chlorophyll-containing organelle, bounded by a double membrane, in which photosynthesis occurs; found in plants and photosynthetic protists. Also the location of starch, amino acid, fatty acid, purine, and pyrimidine synthesis.

choanocyte A specialized, flagellated feeding cell found in choanoflagellates (protists that are the closest living relatives of animals) and sponges (the most ancient animal phylum).

cholecystokinin A peptide hormone secreted by cells in the lining of the small intestine. Stimulates the secretion of digestive enzymes from the pancreas and of bile from the liver and gallbladder.

chordate Any member of the phylum Chordata. Chordates are deuterostomes distinguished by a dorsal hollow nerve cord, pharyngeal gill slits, a notochord, and a post-anal tail. Includes vertebrates, cephalochordata, and urochordata.

chromatid One of the two identical double-stranded DNAs composing a replicated chromosome that is connected at the centromere to the other strand.

chromatin The complex of DNA and proteins, mainly histones, that compose eukaryotic chromosomes. Can be highly compact (heterochromatin) or loosely coiled (euchromatin).

chromatin remodeling The process by which the DNA in chromatin is unwound from its associated proteins to allow transcription or replication. May involve chemical modification of histone proteins or reshaping of the chromatin by large multiprotein complexes in an ATP-requiring process.

chromatin remodeling complexes Sets of enzymes that use energy from ATP hydrolysis shift nucleosomes on DNA to expose regulatory sequences to transcription factors.

chromosome theory of inheritance The principle that genes are located on chromosomes and that patterns of inheritance are determined by the behavior of chromosomes during meiosis.

chromosome Gene-carrying structure consisting of a single long molecule of double-stranded DNA and

associated proteins (e.g., histones). Most prokaryotic cells contain a single, circular chromosome; eukaryotic cells contain multiple noncircular (linear) chromosomes located in the nucleus.

cilium (plural: cilia) One of many short, filamentous projections of some eukaryotic cells, containing a core of microtubules. Used to move the cell as well as to circulate fluid or particles around the surface of a stationary cell. See **axoneme**.

circadian clock An internal mechanism found in most organisms that regulates many body processes (sleep–wake cycles, hormonal patterns, etc.) in a roughly 24-hour cycle.

circulatory system The system responsible for moving oxygen, carbon dioxide, and other materials (hormones, nutrients, wastes) within the animal body.

cisternae (singular: cisterna) Flattened, membrane-bound compartments that make up the Golgi apparatus.

cisternal maturation The process of cargo movement through the Golgi apparatus by residing in cisternae that mature from *cis* to *trans* via the import and export of different Golgi enzymes.

citric acid cycle A series of eight chemical reactions that start with citrate (deprotonated citric acid) and ends with oxaloacetate, which reacts with acetyl CoA to form citrate—forming a cycle that is part of the pathway that oxidizes glucose to CO_2. Also known as the *Krebs cycle* or *tricarboxylic acid* (*TCA*) *cycle*.

clade See **monophyletic group**.

cladistic approach A method for constructing a phylogenetic tree that is based on identifying the unique traits (shared, derived characters, called synapomorphies) of each monophyletic group.

Class I MHC protein A type of MHC protein that is present on the plasma membrane of virtually all nucleated cells and functions in presenting antigen to CD8[+] T cells.

Class II MHC protein A type of MHC protein that is present only on the plasma membrane of certain cells in the immune response, such as dendritic cells, macrophages, and B cells. It functions in presenting epitopes of antigens to CD4[+] T cells.

cleavage In animal development, the series of rapid mitotic cell divisions, with little cell growth, that produces successively smaller cells (blastomeres) and transforms a zygote into a multicellular blastula.

cleavage furrow A pinching in of the plasma membrane that occurs as cytokinesis begins in animal cells and deepens until the cytoplasm is divided into two daughter cells.

climate The prevailing, long-term weather conditions in a particular region.

climax community The stable, final community that develops from ecological succession.

clitoris A rod of erectile tissue in the external genitalia of female mammals. Is formed from the same embryonic tissue as the male penis and has a similar function in sexual arousal.

cloaca In a few mammals and many nonmammalian vertebrates, a body cavity opening to the outside and used by both the excretory and reproductive systems.

clonal selection theory The dominant explanation of the generation of an adaptive immune response. According to the theory, the immune system retains a vast pool of inactive lymphocytes, each with a unique receptor for a unique epitope. Lymphocytes that encounter their complementary epitopes are stimulated to divide (selected and cloned), producing daughter cells that combat infection and confer immunity.

clone (1) An individual that is genetically identical to another individual. (2) A lineage of genetically identical individuals or cells. (3) As a verb, to make one or more genetic replicas of a cell or individual.

cloning vector A plasmid or other agent used to transfer recombinant genes into cultured host cells. Also called simply *vector*.

closed circulatory system A circulatory system in which the circulating fluid (blood) is confined to blood vessels and flows in a continuous circuit. Compare with **open circulatory system**.

cnidocyte A specialized stinging cell found in cnidarians (e.g., jellyfish, corals, and anemones) and used in capturing prey.

co-receptor Any membrane protein that acts with some other membrane protein in a cell interaction or cell response.

cochlea The organ of hearing in the inner ear of mammals, birds, and crocodilians. A coiled, fluid-filled tube containing specialized pressure-sensing neurons (hair cells) that detect sounds of different pitches.

coding strand See **non-template strand**.

codominance An inheritance pattern in which heterozygotes exhibit both of the traits seen in each type of homozygous individual.

codon A sequence of three nucleotides in DNA or RNA that codes for an amino acid or a start or stop signal for protein synthesis.

coefficient of relatedness (*r*) A measure of how closely two individuals are related. Calculated as the probability that an allele in two individuals is inherited from the same ancestor.

coelom An internal, usually fluid-filled body cavity that is completely or partially lined with mesoderm.

coelomate An animal that has a true coelom, completely lined with mesoderm. Compare with **acoelomate** and **pseudocoelomate**.

coenocytic Containing many nuclei and a continuous cytoplasm through a filamentous body, without the body being divided into distinct cells. Some fungi are coenocytic.

coenzyme A small organic molecule that is a required cofactor for an enzyme-catalyzed reaction. Often donates or receives electrons or functional groups during the reaction.

coenzyme A (CoA) A molecule that is required for many cellular reactions and that is often transiently linked to other molecules, such as acetyl groups (see **acetyl CoA**).

coenzyme Q A nonprotein molecule that shuttles electrons between membrane-bound complexes in the mitochondrial electron transport chain. Also called **ubiquinone** or *Q*.

coevolution A pattern of evolution in which two interacting species reciprocally influence each other's adaptations over time.

coevolutionary arms race A series of adaptations and counter-adaptations observed in species that interact closely over time and affect each other's fitness.

cofactor An inorganic ion, such as a metal ion, that is required for an enzyme to function normally. May be bound tightly to an enzyme or associate with it transiently during catalysis.

cohesion The tendency of certain like molecules (e.g., water molecules) to cling together due to attractive forces. Compare with **adhesion**.

cohesion-tension theory The theory that water movement upward through plant vascular tissues is due to loss of water from leaves (transpiration), which pulls a cohesive column of water upward.

cohort A group of individuals that are the same age and can be followed through time.

coleoptile A modified leaf that covers and protects the stems and leaves of grass seedlings.

collagen A fibrous, pliable, cable-like glycoprotein that is a major component of the extracellular matrix of animal cells. Various subtypes differ in their tissue distribution, some of which are assembled into large fibrils in the extracellular space.

collecting duct In the vertebrate kidney, a large straight tube that receives filtrate from the distal tubules of several nephrons. Involved in the regulated reabsorption of water.

collenchyma cell In plants, an elongated cell with cell walls thickened at the corners that provides support to growing plant parts; usually found in strands along leaf veins and stalks. Compare with **parenchyma cell** and **sclerenchyma cell**.

colon The portion of the large intestine where feces are formed by compaction of wastes and reabsorption of water.

colony An assemblage of individuals. May refer to an assemblage of semi-independent cells or to a breeding population of multicellular organisms.

commensalism (adjective: commensal) A symbiotic relationship in which one organism (the commensal) benefits and the other (the host) is not harmed. Compare with **mutualism** and **parasitism**.

communication In ecology, any process in which a signal from one individual modifies the behavior of another individual.

community All of the species that interact with each other in a certain area.

companion cell In plants, a cell in the phloem that is connected via many plasmodesmata to adjacent sieve-tube elements. Companion cells provide materials to maintain sieve-tube elements and function in the loading and unloading of sugars into sieve-tube elements.

compass orientation A type of navigation in which movement occurs in a specific direction.

competition In ecology, the interaction of two species or two individuals trying to use the same limited resource (e.g., water, food, living space). May occur between individuals of the same species (intraspecific competition) or different species (interspecific competition).

competitive exclusion principle The principle that two species cannot coexist in the same ecological niche in the same area because one species will outcompete the other.

competitive inhibition Inhibition of an enzyme's ability to catalyze a chemical reaction via binding of a nonreactant molecule that competes with the substrate(s) for access to the active site.

complement system A set of proteins that circulate in the bloodstream and can destroy bacteria by forming holes in the bacterial plasma membrane.

complementary base pairing The association between specific nitrogenous bases of nucleic acids stabilized by hydrogen bonding. Adenine pairs only with thymine (in DNA) or uracil (in RNA), and guanine pairs only with cytosine.

complementary DNA (cDNA) DNA produced in the laboratory using an RNA transcript as a template and reverse transcriptase; corresponds to a gene but lacks introns. Also produced naturally by retroviruses.

complementary strand A newly synthesized strand of RNA or DNA that has a base sequence complementary to that of the template strand.

complete digestive tract A digestive tract with two openings, usually called a mouth and an anus.

complete metamorphosis See **holometabolous metamorphosis**.

compound eye An eye formed of many independent light-sensing columns (ommatidia); occurs in arthropods. Compare with **simple eye**.

concentration gradient Difference across space (e.g., across a membrane) in the concentration of a dissolved substance.

condensation reaction A chemical reaction in which two molecules are joined covalently with the removal of an $-OH$ from one and an $-H$ from another to form water. Also called a *dehydration reaction*. Compare with **hydrolysis**.

conduction (1) Direct transfer of heat between two objects that are in physical contact. Compare with **convection**. (2) Transmission of an electrical impulse along the axon of a nerve cell.

cone cell A photoreceptor cell with a cone-shaped outer portion that is particularly sensitive to bright light of a certain color. Also called simply *cone*. Compare with **rod cell**.

connective tissue An animal tissue consisting of scattered cells in a liquid, jellylike, or solid extracellular matrix. Includes bone, cartilage, tendons, ligaments, and blood.

conservation biology The effort to study, preserve, and restore threatened genetic diversity, populations, communities, and ecosystems.

constant (C) region The invariant amino acid sequence in polypeptides that are used to make antibodies, B-cell receptors, and T-cell receptors. Apart from antibody class types (IgG, IgM, etc.), this region remains constant within an individual. Compare with **variable (V) region**.

constitutive Always occurring; always present. Commonly used to describe enzymes and other proteins that are synthesized continuously or mutants in which one or more genetic loci are constantly expressed due to defects in gene control.

constitutive defense A defensive trait that is always manifested even in the absence of a predator or pathogen. Also called *standing defense*. Compare with **inducible defenses**.

constitutive mutant An abnormal (mutated) strain that produces a product at all times, instead of under certain conditions only.

consumer See **heterotroph**.

consumption Predation or herbivory.

continental shelf The portion of a geologic plate that extends from a continent under seawater.

continuous strand See **leading strand**.

contraception Any of several methods to prevent pregnancy.

control In a scientific experiment, a group of organisms or samples that do not receive the experimental treatment but are otherwise identical to the group that does.

convection Transfer of heat by movement of large volumes of a gas or liquid. Compare with **conduction**.

convergent evolution The independent evolution of similar traits in distantly related organisms due to adaptation to similar environments and a similar way of life.

cooperative binding The tendency of the protein subunits of hemoglobin to affect each other's oxygen binding such that each bound oxygen molecule increases the likelihood of further oxygen binding.

coprophagy The eating of feces.

coral reef A large assemblage of colonial marine corals that usually serves as shallow-water, sunlit habitat for many other species as well.

co-receptor Any membrane protein that acts with some other membrane protein in a cell interaction or cell response.

core enzyme The enzyme responsible for catalysis in a multipart holoenzyme.

cork cambium One of two types of cylindrical meristem, consisting of a ring of undifferentiated plant cells found just under the cork layer of woody stems and roots; produces new cork cells on its outer side. Compare with **vascular cambium**.

cork cell A cell in the protective outermost layer of a woody stem and root that produces and accumulates waxes that make the cell less permeable to water and gases.

corm A rounded, thick underground stem that can produce new plants via asexual reproduction.

cornea The transparent sheet of connective tissue at the very front of the eye in vertebrates and some other animals. Protects the eye and helps focus light.

corolla All of the petals of a flower.

corona The cluster of cilia at the anterior end of a rotifer; in many species it facilitates suspension feeding.

corpus callosum A thick band of neurons that connects the two hemispheres of the cerebrum in the mammalian brain.

corpus luteum A yellowish structure that secretes progesterone in an ovary. Is formed from a follicle that has recently ovulated.

cortex (1) In animals, the outermost region of an organ, such as the kidney or adrenal gland. (2) In plants, a layer of ground tissue found outside the vascular bundles of roots and outside the pith of a stem.

corticotropin-releasing hormone (CRH) A peptide hormone, produced and secreted by the hypothalamus, that stimulates the anterior pituitary to release ACTH.

cortisol A steroid hormone, produced and secreted by the adrenal cortex, that increases blood glucose and prepares the body for stress. The major glucocorticoid hormone in some mammals. Also called *hydrocortisone*.

cost–benefit analysis Decisions or analyses that weigh the fitness costs and benefits of a particular action.

cotransporter A transmembrane protein that facilitates diffusion of an ion down its previously established electrochemical gradient and uses the energy of that process to transport some other substance, in the same or opposite direction, *against* its concentration gradient. Also called *secondary active transporter*. See **antiporter** and **symporter**.

cotyledon The first leaf, or seed leaf, of a plant embryo. Used for storing and digesting nutrients and/or for early photosynthesis.

countercurrent exchanger In animals, any anatomical arrangement that allows the maximum transfer of heat or a soluble substance from one fluid to another. The two fluids must be flowing in opposite directions and have a heat or concentration gradient between them.

covalent bond A type of chemical bond in which two atoms share one or more pairs of electrons. Compare with **hydrogen bond** and **ionic bond**.

cranium A bony, cartilaginous, or fibrous case that encloses and protects the brain of vertebrates. Forms part of the skull. Also called *braincase*.

crassulacean acid metabolism (CAM) A variant type of photosynthesis in which CO_2 is fixed and stored in organic acids at night when stomata are open and then released to feed the Calvin cycle during the day when stomata are closed. Helps reduce water loss and CO_2 loss by photorespiration.

cristae (singular: crista) Sac-like invaginations of the inner membrane of a mitochondrion. Location of the electron transport chain and ATP synthase.

Cro-Magnon A prehistoric European population of modern humans (*Homo sapiens*) known from fossils, paintings, sculptures, and other artifacts.

crop A storage organ in the digestive system of certain vertebrates.

cross A mating between two individuals that is used for genetic analysis.

cross-pollination Pollination of a flower by pollen from another individual, rather than by self-fertilization. Also called *crossing*.

cross-talk Interactions among signaling pathways, triggered by different signals, that modify a cellular response.

crossing over The exchange of segments of non-sister chromatids between a pair of homologous chromosomes that occurs during meiosis I.

crosstalk Interactions among signaling pathways that modify a cellular response.

crustaceans A lineage of arthropods that includes shrimp, lobster, and crabs. Many have a carapace (a platelike portion of the exoskeleton covering the cephalothorax) and mandibles for biting or chewing.

cryptic species A species that cannot be distinguished from similar species by easily identifiable morphological traits.

culture In cell biology, a collection of cells or a tissue growing under controlled conditions, usually in suspension or on the surface of a dish of solid growth medium.

Cushing's disease A human endocrine disorder caused by loss of feedback inhibition of cortisol on ACTH secretion. Characterized by high ACTH and cortisol levels and wasting of body protein reserves.

cuticle A protective coating secreted by the outermost layer of cells of an animal or a plant; often functioning to reduce evaporative water loss.

cyanobacteria A lineage of photosynthetic bacteria formerly known as blue-green algae. Likely the first life-forms to carry out oxygenic photosynthesis.

cyclic AMP (cAMP) Cyclic adenosine monophosphate; a small molecule, derived from ATP, that is widely used by cells in signal transduction and transcriptional control.

cyclic electron flow Path of electrons in which excited electrons of photosystem I are transferred back to plastoquinone (PQ), the start of the electron transport chain normally associated with photosystem II. Instead of reducing $NADP^+$ to make NADPH, the electron energy is used to make ATP via photophosphorylation. Compare with **noncyclic electron flow**.

cyclin One of several regulatory proteins whose concentrations fluctuate cyclically throughout the cell cycle. Involved in the control of the cell cycle via cyclin-dependent kinases.

cyclin-dependent kinase (Cdk) Any of several related protein kinases that are functional only when bound to a cyclin and are activated by other modifications. Involved in control of the cell cycle.

cytochrome *c* (cyt *c*) A soluble protein that shuttles electrons between membrane-bound complexes in the mitochondrial electron transport chain.

cytokine Any of a diverse group of signaling proteins, secreted largely by cells of the immune system, whose effects include stimulating leukocyte production, recruiting cells to the site of infection, tissue repair, and fever. Generally function to regulate the type, intensity, and duration of an immune response.

cytokinesis Division of the cytoplasm to form two daughter cells. Typically occurs immediately after division of the nucleus by mitosis or meiosis.

cytokinins A class of plant hormones that stimulate cell division and retard aging.

cytoplasm All the contents of a cell, excluding the nucleus, bounded by the plasma membrane.

cytoplasmic determinant A regulatory transcription factor or signaling molecule that is distributed unevenly in the cytoplasm of the egg and that directs early pattern formation in an embryo.

cytoplasmic streaming The directed flow of cytosol and organelles that facilitates distribution of materials within some large plant and fungal cells. Occurs along actin filaments and is powered by myosin.

cytoskeleton In eukaryotic cells, a network of protein fibers in the cytoplasm that are involved in cell shape, support, locomotion, and transport of materials within the cell. Prokaryotic cells have a similar but much less extensive network of fibers.

cytosol The fluid portion of the cytoplasm, excluding the contents of membrane-enclosed organelles.

cytotoxic T cell A type of $CD8^+$ effector T cell that induces apoptosis in infected and cancerous cells. Recognizes target cells via interactions with complementary class I MHC–peptide complexes. Also called *cytotoxic T lymphocyte (CTL)* and *killer T cell*. Compare with **helper T cell**.

dalton (Da) A unit of mass equal to 1/12 the mass of one carbon-12 atom; about the mass of 1 proton or 1 neutron.

daughter strand The strand of DNA that is newly replicated from an existing template strand of DNA.

day-neutral plant A plant whose flowering time is not affected by the relative length of day and night (the photoperiod). Compare with **long-day** and **short-day plant**.

dead space Air passages that are not involved in gas exchange with the blood; examples are the trachea and bronchi.

deciduous Describing a plant that sheds leaves or other structures at regular intervals (e.g., each autumn).

decomposer See **detritivore**.

decomposer food chain An ecological network of detritus, decomposers that eat detritus, and predators and parasites of the decomposers.

deep sequencing A method to learn the types of mRNAs or DNA sequences present in cells, and their relative amounts, involving the preparation and sequencing of cDNA libraries.

definitive host The host species in which a parasite reproduces sexually. Compare with **intermediate host**.

dehydration reaction See **condensation reaction**.

deleterious In genetics, referring to any mutation, allele, or trait that reduces an individual's fitness.

deletion In genetics, refers to the loss of part of a chromosome.

demography The study of factors that determine the size and structure of populations through time.

dendrite A short extension from a neuron's cell body that receives signals from other neurons.

dendritic cell A type of leukocyte that ingests and digests foreign antigens, moves to a lymph node, and presents the antigens' epitopes, in the context of MHC proteins on its membrane, to $CD4^+$ and $CD8^+$ T cells.

dense connective tissue A type of connective tissue, distinguished by having an extracellular matrix dominated by collagen fibers. Found in tendons and ligaments.

density dependent In population ecology, referring to any characteristic that varies depending on population density.

density independent In population ecology, referring to any characteristic that does not vary with population density.

deoxyribonucleic acid (DNA) A nucleic acid composed of deoxyribonucleotides that carries the genetic information of a cell. Generally occurs as two intertwined strands, but these can be separated. See also **double helix**.

deoxyribonucleoside triphosphate (dNTP) A monomer used by DNA polymerase to polymerize DNA. Consists of the sugar deoxyribose, a base (A, T, G, or C), and three phosphate groups.

deoxyribonucleotide See **nucleotide**.

depolarization A change in membrane potential from its resting negative state to a less negative or a positive state; a normal phase in an action potential. Compare with **hyperpolarization**.

depolarization Change in membrane potential from its resting negative state to a less negative or to a positive state; a normal phase in an action potential. Compare with **hyperpolarization**.

deposit feeder An animal that eats its way through a food-containing substrate.

derived trait A trait that is clearly homologous with a trait found in an ancestor of a particular group, but that has a new form.

dermal tissue system The tissue forming the outer layer of a plant; also called *epidermis*.

descent with modification The phrase used by Darwin to describe his hypothesis of evolution by natural selection.

desmosome A type of cell–cell attachment structure, consisting of cadherin proteins, that is anchored to intermediate filaments. Serves to link the cytoskeletons of adjacent animal cells and form strong cell–cell attachments throughout a tissue. Compare with **gap junction** and **tight junction**.

detergent A type of small amphipathic molecule used to solubilize hydrophobic molecules in aqueous solution.

determination In development, the commitment of a cell to a particular differentiated fate. Once a cell is fully determined, it can differentiate only into a particular cell type (e.g., liver cell, brain cell).

detritivore An organism whose diet consists mainly of dead organic matter (detritus). Various bacteria, fungi, protists, and animals are detritivores. Also called *decomposer*.

detritus A layer of dead organic matter that accumulates at ground level or on seafloors and lake bottoms.

deuterostomes A major lineage of bilaterian animals that share a pattern of embryological development, including formation of the anus earlier than the mouth, and formation of the coelom by pinching off of layers of mesoderm from the gut. Includes echinoderms and chordates. Compare with **protostomes**.

developmental homology A similarity in embryonic form, or in the fate of embryonic tissues, that is due to inheritance from a common ancestor.

diabetes mellitus A disease caused by defects in insulin production (type 1) or in the response of cells to insulin (type 2). Characterized by abnormally high blood glucose levels and huge amounts of glucose-containing urine.

diaphragm An elastic, sheetlike structure. In mammals, the muscular sheet of tissue that separates the chest and abdominal cavities. It contracts and moves downward during inhalation, expanding the chest cavity.

diastole The portion of the cardiac cycle during which the atria or ventricles of the heart are relaxed. Compare with **systole**.

diastolic blood pressure The force exerted by blood against artery walls during relaxation of the heart's left ventricle. Compare with **systolic blood pressure**.

dicot Any flowering plant (angiosperm) that has two cotyledons (embryonic leaves) upon germination. The dicots do not form a monophyletic group. Also called *dicotyledonous plant*. Compare with **eudicot** and **monocot**.

dideoxy sequencing A laboratory technique for determining the exact nucleotide sequence of DNA. Relies on the use of dideoxynucleoside triphosphates (ddNTPs), which terminate DNA replication.

diencephalon The part of the mammalian brain that relays sensory information to the cerebellum and functions in maintaining homeostasis.

differential centrifugation Procedure for separating cellular components according to their size and density by spinning a cell homogenate in a series of centrifuge runs. After each run, the supernatant is removed from the deposited material (pellet) and spun again at progressively higher speeds.

differential gene expression Expression of different sets of genes in cells with the same genome. Responsible for creating different cell types.

differentiation The process by which any unspecialized cell becomes a distinct specialized cell type (e.g., liver cell, brain cell), usually by changes in gene expression. Also called *cell differentiation*.

diffusion Spontaneous movement of a substance from one region to another, often with a net movement from a region of high concentration to one of low concentration (i.e., down a concentration gradient).

digestion The physical and chemical breakdown of food into molecules that can be absorbed into the body of an animal.

digestive tract The long tube that begins at the mouth and ends at the anus. Also called *alimentary canal* or *gastrointestinal (GI) tract*.

dihybrid cross A mating between two parents that are heterozygous for two different genes.

dikaryotic Describing a cell or fungal mycelium having two haploid nuclei that are genetically distinct.

dimer An association of two molecules that may be identical (homodimer) or different (heterodimer).

dioecious Describing an angiosperm species that has male and female reproductive structures on separate plants. Compare with **monoecious**.

diploblast (adjective: diploblastic) An animal whose body develops from two basic embryonic cell layers or tissues—ectoderm and endoderm. Compare with **triploblast**.

diploid (1) Having two sets of chromosomes (2n). (2) A cell or an individual organism with two sets of chromosomes, one set inherited from the mother and one set from the father. Compare with **haploid**.

direct sequencing A technique for identifying and studying microorganisms that cannot be grown in culture. Involves detecting and amplifying copies of certain specific genes in the microorganisms' DNA, sequencing these genes, and then comparing the sequences with the known sequences from other organisms.

directional selection A mode of natural selection that favors one extreme phenotype with the result that the average phenotype of a population changes in one direction. Generally reduces overall genetic variation in a population. Compare with **disruptive selection** and **stabilizing selection**.

disaccharide A carbohydrate consisting of two monosaccharides (sugar residues) linked together.

discontinuous strand See **lagging strand**.

discrete trait An inherited trait that exhibits distinct phenotypes rather than the continuous variation characteristic of a quantitative trait such as body height.

dispersal The movement of individuals from their place of origin (birth, hatching) to a new location.

disruptive selection A mode of natural selection that favors extreme phenotypes at both ends of the range of phenotypic variation. Maintains overall genetic variation in a population. Compare with **stabilizing selection** and **directional selection**.

distal tubule In the vertebrate kidney, the convoluted portion of a nephron into which filtrate moves from the loop of Henle. Involved in the regulated reabsorption of sodium and water. Compare with **proximal tubule**.

disturbance In ecology, any strong, short-lived disruption to a community that changes the distribution of living and/or nonliving resources.

disturbance regime The characteristic disturbances that affect a given ecological community.

disulfide bond A covalent bond between two sulfur atoms, typically in the side chains of certain amino acids (e.g., cysteine). Often contributes to tertiary and quaternary levels of protein structure.

DNA See **deoxyribonucleic acid**.

DNA cloning Any of several techniques for producing many identical copies of a particular gene or other DNA sequence.

DNA fingerprinting Any of several methods for identifying individuals by unique features of their genomes. Commonly involves using PCR to produce many copies of certain short tandem repeats (microsatellites) and then analyzing their lengths.

DNA helicase An enzyme that breaks hydrogen bonds between nucleotides of DNA, "unzipping" a double-stranded DNA molecule.

DNA library See **cDNA library** and **genomic library**.

DNA ligase An enzyme that joins pieces of DNA by catalyzing the formation of a phosphodiester bond between the pieces.

DNA methylation The addition of a methyl group ($-CH_3$) to a DNA molecule.

DNA methyltransferase A class of eukaryotic enzymes that add a methyl group to cytosines in DNA. Methylation of DNA leads to chromatin condensation and is an important means of regulating gene expression in eukaryotes.

DNA microarray A set of single-stranded DNA fragments, representing thousands of different genes that are permanently fixed to a small glass slide. Can be used to determine which genes are expressed in different cell types, under different conditions, or at different developmental stages.

DNA polymerase Any enzyme that catalyzes synthesis of DNA from deoxyribonucleotide triphosphates (dNTPs).

domain (1) A taxonomic category, based on similarities in basic cellular biochemistry, above the kingdom level. The three recognized domains are Bacteria, Archaea, and Eukarya. (2) A section of a protein that has a distinctive tertiary structure and function.

dominant Referring to an allele that determines the same phenotype when it is present in homozygous or heterozygous form.. Compare with **recessive**.

dormancy A temporary state of greatly reduced metabolic activity and growth in plants or plant parts (e.g., seeds, spores, bulbs, and buds).

dorsal Toward an animal's back and away from its belly. The opposite of ventral.

dorsal hollow nerve chord See **nerve chord**.

double fertilization An unusual form of reproduction seen in flowering plants, in which one sperm cell fuses with an egg to form a zygote and the other sperm cell fuses with two polar nuclei to form the triploid endosperm.

double helix The secondary structure of DNA, consisting of two antiparallel DNA strands wound around each other.

Down syndrome A human developmental disorder caused by trisomy of chromosome 21.

downstream In genetics, the direction in which RNA polymerase moves along a DNA strand. Compare with **upstream**.

duplication In genetics, refers to an additional copy of part of a chromosome.

dynein A class of motor proteins that uses the chemical energy of ATP to "walk" toward the minus end of a microtubule. Dyneins are responsible for bending of cilia and flagella, play a role in chromosome movement during mitosis, and can transport vesicles and organelles.

early endosome A small transient organelle that is formed by the accumulation of vesicles from receptor-mediated endocytosis and is an early stage in the formation of a lysosome.

ecdysone An insect hormone that triggers either molting (to a larger larval form) or metamorphosis (to the adult form), depending on the level of juvenile hormone.

ecdysozoans A major lineage of protostomes (Ecdysozoa) that grow by shedding their external skeletons (molting) and expanding their bodies. Includes arthropods, nematodes, and other groups. Compare with **lophotrochozoans**.

echinoderms A major lineage of deuterostomes (Echinodermata) distinguished by adult bodies with five-sided radial symmetry, a water vascular system, and tube feet. Includes sea urchins, sand dollars, and sea stars.

echolocation The use of echoes from vocalizations to obtain information about locations of objects in the environment.

ecological selection Also known as environmental selection. A type of natural selection that favors individuals with heritable traits that enhance their ability to survive and reproduce in a certain physical and/or biological environment, excluding their ability to obtain a mate. Compare with **sexual selection**.

ecology The study of how organisms interact with each other and with their surrounding environment.

ecosystem All the organisms that live in a geographic area, together with the nonliving (abiotic) components that affect or exchange materials with the organisms; a community and its physical environment.

ecosystem diversity The variety of biotic components in a region along with abiotic components, such as soil, water, and nutrients.

ecosystem function The sum of biological and chemical processes that are characteristic of a given ecosystem—such as primary production, nitrogen cycling, and carbon storage.

ecosystem services All of the benefits that humans derive, directly or indirectly, from ecosystem functions.

ecotourism Tourism that is based on observing wildlife or experiencing other aspects of natural areas.

ectoderm The outermost of the three basic cell layers (germ layers) in most animal embryos; gives rise to the outer covering and nervous system. Compare with **endoderm** and **mesoderm**.

ectomycorrhizal fungi (EMF) Fungi whose hyphae form a dense network that covers their host plant's roots but do not enter the root cells.

ectoparasite A parasite that lives on the outer surface of the host's body.

ectotherm An animal that gains most of its body heat from external sources as opposed to metabolic processes. Compare with **endotherm**.

effector Any cell, organ, or structure with which an animal can respond to external or internal stimuli. Usually functions, along with a sensor and integrator, as part of a homeostatic system.

efferent division The part of the nervous system that carries commands from the central nervous system to the body. Consists primarily of motor neurons.

egg A mature female gamete and any associated external layers (such as a shell). Larger and less mobile than the male gamete. In animals, also called *ovum*.

ejaculation The release of semen from the copulatory organ of a male animal.

ejaculatory duct A short duct through which sperm move during ejaculation; connects the vas deferens to the urethra.

electric current A flow of electrical charge past a point. Also called *current*.

electrical potential Potential energy created by a separation of electrical charges between two points. Also called *voltage*.

electrocardiogram (EKG) A recording of the electrical activity of the heart, as measured through electrodes on the skin.

electrochemical gradient The combined effect of an ion's concentration gradient and electrical (charge) gradient across a membrane that affects the diffusion of ions across the membrane.

electrogenic fish Any of various kinds of fishes having specialized electric organs that emit a current into the water to detect objects.

electrolyte Any compound that dissociates into ions when dissolved in water. In nutrition, any of the major ions necessary for normal cell function.

electromagnetic spectrum The entire range of wavelengths of radiation extending from short wavelengths (high energy) to long wavelengths (low energy). Includes gamma rays, X-rays, ultraviolet, visible light, infrared, microwaves, and radio waves (from short to long wavelengths).

electron acceptor A reactant that gains an electron and is reduced in a reduction–oxidation reaction.

electron carrier Any molecule that readily accepts electrons from and donates electrons to other molecules. Protons may be transferred with the electrons in the form of hydrogen atoms.

electron donor A reactant that loses an electron and is oxidized in a reduction–oxidation reaction.

electron shell A group of orbitals of electrons with similar energies. Electron shells are arranged in roughly concentric layers around the nucleus of an atom, and electrons in outer shells have more energy than those in inner shells. Electrons in the outermost shell, the valence shell, often are involved in chemical bonding.

electron transport chain (ETC) Any set of membrane-bound protein complexes and mobile electron carriers involved in a coordinated series of redox reactions in which the potential energy of electrons is successively decreased and used to pump protons from one side of a membrane to the other.

electronegativity A measure of the ability of an atom to attract electrons toward itself from an atom to which it is bonded.

electroreception A sensory system in which receptors are activated by electric fields.

electroreceptor A sensory cell or organ specialized to detect electric fields.

element A substance, consisting of atoms with a specific number of protons. Elements preserve their identity in chemical reactions.

elongation (1) The process by which RNA lengthens during transcription. (2) The process by which a polypeptide chain lengthens during translation.

elongation factors Proteins involved in the elongation phase of translation, assisting ribosomes in the synthesis of the growing peptide chain.

embryo A young, developing organism; the stage after fertilization and zygote formation.

embryo sac The female gametophyte in flowering plants.

embryogenesis The production of an embryo from a zygote. Embryogenesis is an early event in development of animals and plants.

Embryophyta An increasingly popular name for the lineage called land plants; reflects their retention of a fertilized egg.

embryophyte A plant that nourishes its embryos inside its own body. All land plants are embryophytes.

emergent property A property that stems from the interaction of simpler elements and that is impossible to predict from the study of individual elements.

emergent vegetation Any plants in an aquatic habitat that extend above the surface of the water.

emerging disease Any infectious disease, often a viral disease, that suddenly afflicts significant numbers of humans for the first time; often due to changes in the host species for a pathogen or to radical changes in the genetic material of the pathogen.

emigration The migration of individuals away from one population to other populations. Compare with **immigration**.

emulsification (verb: emulsify) The dispersion of fat into an aqueous solution. Usually requires the aid of an amphipathic substance such as a detergent or bile salts, which can break large fat globules into microscopic fat droplets.

endangered species A species whose numbers have decreased so much that it is in danger of extinction throughout all or part of its range.

endemic species A species that lives in one geographic area and nowhere else.

endergonic Referring to a chemical reaction that requires an input of energy to occur and for which the Gibbs free-energy change (ΔG) is greater than zero. Compare with **exergonic**.

endocrine Relating to a chemical signal (hormone) that is released into the bloodstream by a producing cell and acts on a distant target cell.

endocrine disruptor An exogenous chemical that interferes with normal hormonal signaling.

endocrine gland A gland that secretes hormones directly into the bloodstream or interstitial fluid instead of into ducts. Compare with **exocrine gland**.

endocrine system All of the glands and tissues that produce and secrete hormones into the bloodstream.

endocytosis General term for any pinching off of the plasma membrane that results in the uptake of material from outside the cell. Includes phagocytosis, pinocytosis, and receptor-mediated endocytosis. Compare with **exocytosis**.

endoderm The innermost of the three basic cell layers (germ layers) in most animal embryos; gives rise to the digestive tract and organs that connect to it (liver, lungs, etc.). Compare with **ectoderm** and **mesoderm**.

endodermis In plant roots, a cylindrical layer of cells that separates the cortex from the vascular tissue and location of the Casparian strip.

endomembrane system A system of organelles in eukaryotic cells that synthesizes, processes, transports, and recycles proteins and lipids. Includes the endoplasmic reticulum (ER), Golgi apparatus, and lysosomes.

endomycorrhizal fungi See **arbuscular mycorrhizal fungi (AMF)**.

endoparasite A parasite that lives inside the host's body.

endophyte (adjective: endophytic) A fungus that lives inside the aboveground parts of a plant in a symbiotic relationship. Compare with **epiphyte**.

endoplasmic reticulum (ER) A network of interconnected membranous sacs and tubules found inside eukaryotic cells. See **rough** and **smooth endoplasmic reticulum**.

endoskeleton Bony and/or cartilaginous structures within the body that provide support. Examples are the spicules of sponges, the plates in echinoderms, and the bony skeleton of vertebrates. Compare with **exoskeleton**.

endosperm A triploid ($3n$) tissue in the seed of a flowering plant (angiosperm) that serves as food for the plant embryo. Functionally analogous to the yolk in some animal eggs.

endosymbiont An organism that lives in a symbiotic relationship inside the body of its host.

endosymbiosis An association between organisms of two different species in which one lives inside the cell or cells of the other.

endosymbiosis theory The theory that mitochondria and chloroplasts evolved from prokaryotes that were engulfed by host cells and took up a symbiotic existence within those cells, a process termed primary endosymbiosis. In some eukaryotes, chloroplasts may have originated by secondary endosymbiosis; that is, when a cell engulfed a chloroplast-containing protist and retained its chloroplasts.

endotherm An animal whose primary source of body heat is internally generated. Compare with **ectotherm**.

endothermic Referring to a chemical reaction that absorbs heat. Compare with **exothermic**.

energetic coupling In cellular metabolism, the mechanism by which energy released from an exergonic reaction (commonly, hydrolysis of ATP) is used to drive an endergonic reaction.

energy The capacity to do work or to supply heat. May be stored (potential energy) or available in the form of motion (kinetic energy).

enhancer A regulatory sequence in eukaryotic DNA that may be located far from the gene it controls or within introns of the gene. Binding of specific proteins to an enhancer enhances the transcription of certain genes.

enrichment culture A method of detecting and obtaining cells with specific characteristics by placing a sample, containing many types of cells, under a specific set of conditions (e.g., temperature, salt concentration, available nutrients) and isolating those cells that grow rapidly in response.

enthalpy (*H*) A quantitative measure of the amount of potential energy, or heat content, of a system plus the pressure and volume it exerts on its surroundings.

entropy (*S*) A quantitative measure of the amount of disorder of any system, such as a group of molecules.

envelope (viral) A membrane that encloses the capsids of some viruses. Normally includes specialized proteins that attach to host-cell surfaces.

environmental sequencing See **metagenomics**.

enzyme A protein catalyst used by living organisms to speed up and control biological reactions.

epicotyl In some embryonic plants, a portion of the embryonic stem that extends above the cotyledons.

epidemic The spread of an infectious disease throughout a population in a short time period. Compare with **pandemic**.

epidermis The outermost layer of cells of any multicellular organism.

epididymis A coiled tube wrapped around each testis in reptiles, birds, and mammals. The site of the final stages of sperm maturation.

epigenetic inheritance Pattern of inheritance involving differences in phenotype that are not due to differences in the nucleotide sequence of genes.

epinephrine A catecholamine hormone, produced and secreted by the adrenal medulla, that triggers rapid responses related to the fight-or-flight response. Also called *adrenaline*.

epiphyte (adjective: epiphytic) A nonparasitic plant that grows on the trunks or branches of other plants and is not rooted in soil.

epithelial tissues See **epithelium**.

epithelium (plural: epithelia) An animal tissue consisting of sheetlike layers of tightly packed cells that line an organ, a gland, a duct, or a body surface. Also called *epithelial tissue*.

epitope A small region of a particular antigen to which an antibody, B-cell receptor, or T-cell receptor binds.

equilibrium potential The membrane potential at which there is no net movement of a particular ion into or out of a cell.

ER signal sequence A short amino acid sequence that marks a polypeptide for transport to the endoplasmic reticulum, where synthesis of the polypeptide chain is completed and the signal sequence removed. See **signal recognition particle**.

erythropoietin (EPO) A peptide hormone, released by the kidney in response to low blood-oxygen levels, that stimulates the bone marrow to produce more red blood cells.

esophagus The muscular tube that connects the mouth to the stomach.

essential amino acid Any amino acid that an animal cannot synthesize and must obtain from the diet. May refer specifically to one of the eight essential amino acids of adult humans: isoleucine, leucine, lysine, methionine, phenylalanine, threonine, tryptophan, and valine.

essential nutrient Any chemical element, ion, or compound that is required for normal growth, reproduction, and maintenance of a living organism and that cannot be synthesized by the organism.

EST See **expressed sequence tag**.

ester linkage The covalent bond formed by a condensation reaction between a carboxyl group and a hydroxyl group. Ester linkages join fatty acids to glycerol to form a fat or phospholipid.

estradiol The major estrogen produced by the ovaries of female mammals and many other vertebrates. Stimulates development of the female reproductive tract, growth of ovarian follicles, and growth of breast tissue in mammals.

estrogens A class of steroid hormones, including estradiol, estrone, and estriol, that generally promote female-like traits. Secreted by the gonads, fat tissue, and some other organs.

estrous cycle A female reproductive cycle in which the uterine lining is reabsorbed rather than shed in the absence of pregnancy and the female is sexually receptive only briefly during mid-cycle (estrus). It is seen in all mammals except Old World monkeys and apes (including humans). Compare with **menstrual cycle**.

ethylene A gaseous plant hormone associated with senescence that induces fruits to ripen and flowers to fade.

eudicot A member of a monophyletic group (lineage) of angiosperms that includes complex flowering plants and trees (e.g., roses, daisies, maples). All eudicots have two cotyledons, but not all dicots are members of this lineage. Compare with **dicot** and **monocot**.

Eukarya One of the three taxonomic domains of life, consisting of unicellular organisms (most protists, yeasts) and multicellular organisms (fungi, plants, animals) distinguished by a membrane-bound cell nucleus, numerous organelles, and an extensive cytoskeleton. Compare with **Archaea** and **Bacteria**.

eukaryote A member of the domain Eukarya; an organism whose cells contain a nucleus, numerous membrane-bound organelles, and an extensive cytoskeleton. May be unicellular or multicellular. Compare with **prokaryote**.

eusociality A complex social structure in which workers sacrifice most or all of their direct reproduction to help rear the queen's offspring. Common in insects such as ants, bees, wasps, and termites.

eutherians A lineage of mammals (Eutheria) whose young develop in the uterus and are not housed in an abdominal pouch. Also called *placental mammals*.

evaporation The energy-absorbing phase change from a liquid state to a gaseous state. Many organisms evaporate water as a means of heat loss.

evo-devo Popular term for evolutionary developmental biology, a research field focused on how changes in developmentally important genes have led to the evolution of new phenotypes.

evolution (1) The theory that all organisms on Earth are related by common ancestry and that they have changed over time, and continue to change, via natural selection and other processes. (2) Any change in the genetic characteristics of a population over time, especially a change in allele frequencies.

ex situ conservation Preserving species outside of natural areas (e.g., in zoos, aquaria, or botanical gardens).

excitable membrane A plasma membrane that is capable of generating an action potential. Neurons, muscle cells, and some other cells have excitable membranes.

excitatory postsynaptic potential (EPSP) A change in membrane potential, usually depolarization, at a neuron dendrite that makes an action potential more likely.

exergonic Referring to a chemical reaction that can occur spontaneously, releasing heat and/or increasing entropy, and for which the Gibbs free-energy change (ΔG) is less than zero. Compare with **endergonic**.

exocrine gland A gland that secretes some substance through a duct into a space other than the circulatory system, such as the digestive tract or the skin surface. Compare with **endocrine gland**.

exocytosis Secretion of intracellular molecules (e.g., hormones, collagen), contained within membrane-bound vesicles, to the outside of the cell by fusion of vesicles to the plasma membrane. Compare with **endocytosis**.

exon A transcribed region of a eukaryotic gene or region of a primary transcript that is retained in the mature RNA. Except for 5′ and 3′ UTRs, mRNA exons code for amino acids. Compare with **intron**.

exoskeleton A hard covering secreted on the outside of the body, used for body support, protection, and muscle attachment. Prominent in arthropods. Compare with **endoskeleton**.

exothermic Referring to a chemical reaction that releases heat. Compare with **endothermic**.

exotic species A nonnative species that is introduced into a new area. Exotic species often are competitors, pathogens, or predators of native species.

expansins A class of plant proteins that break hydrogen bonds between components in the primary cell wall to allow it to expand for cell growth.

exponential population growth The accelerating increase in the size of a population that occurs when the growth rate is constant and density independent. Compare with **logistic population growth**.

expressed sequence tag (EST) A portion of a transcribed gene (synthesized from an mRNA in a cell), used to find the physical location of that gene in the genome.

extant species A species that is living today.

extensor A muscle that pulls two bones farther apart from each other, increasing the angle of the joint, as in the extension of a limb or the spine. Compare with **flexor**.

extinct species A species that has died out.

extracellular digestion Digestion that takes place outside of an organism, as occurs in many fungi that make and secrete digestive enzymes.

extracellular matrix (ECM) A complex meshwork in which animal cells are embedded, consisting of proteins (e.g., collagen, proteoglycan, laminin) and polysaccharides produced by the cells.

extremophile A bacterium or archaean that thrives in an "extreme" environment (e.g., high-salt, high-temperature, low-temperature, or low-pressure).

F_1 generation First filial generation. The first generation of offspring produced from a mating (i.e., the offspring of the parental generation).

facilitated diffusion Passive movement of a substance across a membrane with the assistance of transmembrane carrier proteins or channel proteins.

facilitation In ecological succession, the phenomenon in which early-arriving species make conditions more favorable for later-arriving species. Compare with **inhibition** and **tolerance**.

facultative anaerobe Any organism that can survive and reproduce by performing aerobic respiration when oxygen is available or fermentation when it is not.

FAD/FADH$_2$ Oxidized and reduced forms, respectively, of flavin adenine dinucleotide. A nonprotein electron carrier that functions in the citric acid cycle and oxidative phosphorylation.

fallopian tube A narrow tube connecting the uterus to the ovary in humans, through which the egg travels after ovulation. Site of fertilization and cleavage. In nonhuman animals, called *oviduct*.

fast muscle fiber Type of skeletal muscle fiber that is white in color, generates ATP by glycolysis, and contracts rapidly but fatigues easily. Also called *fast glycolytic*, or *Type IIb*, *fiber*.

fat A lipid consisting of three fatty acid molecules joined by ester linkages to a glycerol molecule. Also called *triacylglycerol* or *triglyceride*.

fatty acid A lipid consisting of a hydrocarbon chain bonded at one end to a carboxyl group. Used by many organisms to store chemical energy; a major component of animal and plant fats and phospholipids.

fauna All the animal species characteristic of a particular region, period, or environment.

feather A specialized skin outgrowth, composed of β-keratin, present in all birds as well as in some non-avian dinosaurs. Used for flight, insulation, display, and other purposes.

feces The waste products of digestion.

fecundity The average number of female offspring produced by a single female in the course of her lifetime.

feedback inhibition A type of control in which high concentrations of the product of a metabolic pathway inhibit one of the enzymes early in the pathway. A form of negative feedback.

fermentation Any of several metabolic pathways that regenerate oxidizing agents, such as NAD$^+$, by transferring electrons to a final electron acceptor in the absence of an electron transport chain. Allows pathways such as glycolysis to continue to make ATP.

ferredoxin In photosynthetic organisms, an iron- and sulfur-containing protein in the electron transport chain of photosystem I. Can transfer electrons to the enzyme NADP$^+$ reductase, which catalyzes formation of NADPH.

fertility The average number of surviving children that each woman has during her lifetime.

fertilization Fusion of the nuclei of two haploid gametes to form a zygote with a diploid nucleus.

fertilization envelope A physical barrier that forms around a fertilized egg in many animals. The fertilization envelope prevents fertilization by more than one sperm (polyspermy).

fetal alcohol syndrome (FAS) A condition marked by hyperactivity, severe learning disabilities, and depression. Thought to be caused by exposure of an

individual to high blood alcohol concentrations during embryonic development.

fetus In live-bearing animals, the unborn offspring after the embryonic stage. It usually is developed sufficiently to be recognizable as belonging to a certain species.

fiber In plants, a type of elongated sclerenchyma cell that provides support to vascular tissue. Compare with **sclereid**.

Fick's law of diffusion A mathematical relationship that describes the rates of diffusion of gases.

fight-or-flight response Rapid physiological changes that prepare the body for emergencies. Includes increased heart rate, increased blood pressure, and decreased digestion.

filament Any thin, threadlike structure, particularly (1) the threadlike extensions of a fish's gills or (2) part of a stamen: the slender stalk that bears the anthers in a flower.

filter feeder See **suspension feeder**.

filtrate Any fluid produced by filtration, in particular the fluid ("pre-urine") in the Malpighian tubules of insects and the nephrons of vertebrate kidneys.

filtration A process of removing large components from a fluid by forcing it through a filter. Occurs in a renal corpuscle of the vertebrate kidney, allowing water and small solutes to pass from the blood into the nephron.

fimbria (plural: fimbriae) A long, needlelike projection from the cell membrane of prokaryotes that is involved in attachment to nonliving surfaces or other cells.

finite rate of increase (λ) The rate of increase of a population over a given period of time. Calculated as the ending population size divided by the starting population size. Compare with **intrinsic rate of increase**.

first law of thermodynamics The principle of physics that energy is conserved in any process. Energy can be transferred and converted into different forms, but it cannot be created or destroyed.

fission (1) A form of asexual reproduction in which a prokaryotic cell divides to produce two genetically similar daughter cells by a process similar to mitosis of eukaryotic cells. Also called *binary fission*. (2) A form of asexual reproduction in which an animal splits into two or more individuals of approximately equal size; common among invertebrates.

fitness The ability of an individual to produce viable offspring relative to others of the same species.

fitness trade-off See **trade-off**.

fixed action pattern (FAP) Highly stereotyped behavior pattern that occurs in a certain invariant way in a certain species. A form of innate behavior.

flaccid Limp as a result of low internal (turgor) pressure (e.g., a wilted plant leaf). Compare with **turgid**.

flagellum (plural: flagella) A long, cellular projection that undulates (in eukaryotes) or rotates (in prokaryotes) to move the cell through an aqueous environment. See **axoneme**.

flatworms Members of the phylum Platyhelminthes. Distinguished by a broad, flat, unsegmented body

that lacks a coelom. Flatworms belong to the lophotrochozoan branch of the protostomes.

flavin adenine dinucleotide See **FAD/FADH₂**.

flexor A muscle that pulls two bones closer together, decreasing the joint angle, as in the flexing of a limb or the spine. Compare with **extensor**.

floral meristem A group of undifferentiated plant stem cells that can give rise to the four organs making up a flower.

florigen In plants, a protein hormone that is synthesized in leaves and transported to the shoot apical meristem, where it stimulates flowering.

flower In angiosperms, the part of a plant that contains reproductive structures. Typically includes a calyx, a corolla, and one or more stamens and/or carpels. See **perfect** and **imperfect flower**.

fluid connective tissue A type of connective tissue, distinguished by having a liquid extracellular matrix; includes blood.

fluid feeder Any animal that feeds by sucking or mopping up liquids such as nectar, plant sap, or blood.

fluid-mosaic model The widely accepted hypothesis that the plasma membrane and organelle membranes consist of proteins embedded in a fluid phospholipid bilayer.

fluorescence The spontaneous emission of light from an excited electron falling back to its normal (ground) state.

follicle In a mammalian ovary, a sac of supportive cells containing an egg cell.

follicle-stimulating hormone (FSH) A peptide hormone, produced and secreted by the anterior pituitary; it stimulates (in females) growth of eggs and follicles in the ovaries or (in males) sperm production in the testes.

follicular phase In a menstrual cycle, the first major phase, during which follicles grow and estradiol levels increase; ends with ovulation.

food Any nutrient-containing material that can be consumed and digested by animals.

food chain A relatively simple pathway of energy flow through a few species, each at a different trophic level, in an ecosystem. Might include, for example, a primary producer, a primary consumer, a secondary consumer, and a decomposer. A subset of a **food web**.

food web The complex network of interactions among species in an ecosystem formed by the transfer of energy and nutrients among trophic levels. Consists of many food chains.

foot One of the three main parts of the mollusk body; a muscular appendage, used for movements such as crawling and/or burrowing into sediment.

foraging Searching for food.

forebrain One of the three main regions of the vertebrate brain; includes the cerebrum, thalamus, and hypothalamus. Compare with **hindbrain** and **midbrain**.

fossil Any physical trace of an organism that existed in the past. Includes tracks, burrows, fossilized bones, casts, and so on.

fossil record All of the fossils that have been found anywhere on Earth and that have been formally described in the scientific literature.

founder effect A change in allele frequencies that often occurs when a new population is established from a small group of individuals (founder event) due to sampling error (i.e., the small group is not a representative sample of the source population).

fovea In the vertebrate eye, a portion of the retina where incoming light is focused; contains a high proportion of cone cells.

frameshift mutation The addition or deletion of a nucleotide in a coding sequence that shifts the reading frame of the mRNA.

free energy The energy of a system that can be converted into work. It may be measured only through the change in free energy in a reaction. See **Gibbs free-energy change**.

free radicals Any substance containing one or more atoms with an unpaired electron. Unstable and highly reactive.

frequency The number of wave crests per second traveling past a stationary point. Determines the pitch of sound and the color of light.

frequency-dependent selection A pattern of selection in which certain alleles are favored only when they are rare; a form of balancing selection.

fronds The large leaves of ferns.

frontal lobe In the vertebrate brain, one of the four major areas in the cerebrum.

fruit In flowering plants (angiosperms), a mature, ripened plant ovary (or group of ovaries), along with the seeds it contains and any adjacent fused parts; often functions in seed dispersal. See **aggregate**, **multiple**, and **simple fruit**.

fruiting body A structure formed in some prokaryotes, fungi, and protists for spore dispersal; usually consists of a base, a stalk, and a mass of spores at the top.

functional genomics The study of how a genome works; that is, when and where specific genes are expressed and how their products interact to produce a functional organism.

functional group A small group of atoms bonded together in a precise configuration and exhibiting particular chemical properties that it imparts to any organic molecule in which it occurs.

fundamental niche The total theoretical range of environmental conditions that a species can tolerate. Compare with **realized niche**.

fungi A lineage of eukaryotes that typically have a filamentous body (mycelium) and obtain nutrients by absorption.

fungicide Any substance that can kill fungi or slow their growth.

G protein Any of various proteins that are activated by binding to guanosine triphosphate (GTP) and inactivated when GTP is hydrolyzed to GDP. In G-protein-coupled receptors, signal binding directly triggers the activation of a G protein, leading to production of a second messenger or initiation of a phosphorylation cascade.

G₁ phase The phase of the cell cycle that constitutes the first part of interphase before DNA synthesis (S phase).

G₂ phase The phase of the cell cycle between synthesis of DNA (S phase) and mitosis (M phase); the last part of interphase.

gallbladder A small pouch that stores bile from the liver and releases it as needed into the small intestine during digestion of fats.

gametangium (plural: gametangia) (1) The gamete-forming structure found in all land plants except angiosperms. Contains a sperm-producing antheridium and an egg-producing archegonium. (2) The gamete-forming structure of some chytrid fungi.

gamete A haploid reproductive cell that can fuse with another haploid cell to form a zygote. Most multicellular eukaryotes have two distinct forms of gametes: egg cells (ova) and sperm cells.

gametogenesis The production of gametes (eggs or sperm).

gametophyte In organisms undergoing alternation of generations, the multicellular haploid form that arises from a single haploid spore and produces gametes. Compare with **sporophyte**.

ganglion (plural: ganglia) A mass of neurons in a centralized nervous system.

ganglion cell In the retina, a type of neuron whose axons form the optic nerves.

gap junction A type of cell–cell attachment structure that directly connects the cytosolic components of adjacent animal cells, allowing passage of water, ions, and small molecules between the cells. Compare with **desmosome** and **tight junction**.

gastrin A hormone produced by cells in the stomach lining in response to the arrival of food or to a neural signal from the brain. Stimulates other stomach cells to release hydrochloric acid.

gastropods A lineage of mollusks distinguished by a large, muscular foot and a unique feeding structure, the radula. Include slugs and snails.

gastrulation The process of coordinated cell movements, including the moving of some cells from the outer surface of the embryo to the interior, that results in the formation of three germ layers (endoderm, mesoderm, and ectoderm) and the axes of the embryo.

gated channel A channel protein that opens and closes in response to a specific stimulus, such as the binding of a particular molecule or a change in voltage across the membrane.

gemma (plural: gemmae) A small reproductive structure that is produced asexually in some plants during the gametophyte phase and can grow into a mature gametophyte; most common in non-vascular plants, particularly liverworts and club mosses, and in ferns

gene A section of DNA (or RNA, for some viruses) that encodes information for building one or more related polypeptides or functional RNA molecules along with the regulatory sequences required for its transcription.

gene duplication The formation of an additional copy of a gene, typically by misalignment of chromosomes during crossing over. Thought to be an important evolutionary process in creating new genes.

gene expression The set of processes, including transcription and translation, that convert information in DNA into aproduct of a gene, most commonly a protein.

gene family A set of genetic loci whose DNA sequences are extremely similar. Thought to have arisen by duplication of a single ancestral gene and subsequent mutations in the duplicated sequences.

gene flow The movement of alleles between populations; occurs when individuals leave one population, join another, and breed.

gene pool All the alleles of all the genes in a certain population.

gene therapy The treatment of an inherited disease by introducing a normal form of the gene.

generation The average time between a mother's first offspring and her daughter's first offspring.

genetic bottleneck A reduction in allelic diversity resulting from a sudden reduction in the size of a large population (population bottleneck) due to a random event.

genetic code The set of all codons and their meaning.

genetic correlation A type of evolutionary constraint in which selection on one trait causes a change in another trait as well; may occur when the same gene(s) affect both traits.

genetic diversity The diversity of alleles or genes in a population, species, or group of species.

genetic drift Any change in allele frequencies due to random events. Causes allele frequencies to drift up and down randomly over time, and eventually can lead to the fixation or loss of alleles.

genetic equivalence Having all different cell types of a multicellular individual possess the same genome.

genetic homology Similarity in DNA nucleotide sequences, RNA nucleotide sequences, or amino acid sequences due to inheritance from a common ancestor.

genetic map A list of genes on a chromosome that indicates their position and relative distances from one another. Also called a *linkage map*. Compare with **physical map**.

genetic marker A genetic locus that can be identified and traced in populations by laboratory techniques or by a distinctive visible phenotype.

genetic recombination A change in the combination of alleles on a given chromosome or in a given individual. Also called *recombination*.

genetic screen Any technique that identifies individuals with a particular type of mutation.

genetic variation (1) The number and relative frequency of alleles present in a particular population. (2) The proportion of phenotypic variation in a trait that is due to genetic rather than environmental influences in a certain population in a certain environment.

genetics The study of the inheritance of traits.

genital (plural: genitalia) Any external copulatory organ.

genome All the hereditary information in an organism, including not only genes but also stretches of DNA that do not contain genes.

genome annotation The process of analyzing a genome sequence to identify key features such as genes, regulatory sequences, and splice sites.

genomic library A set of DNA segments representing the entire genome of a particular organism. Each segment is carried by a plasmid or other cloning vector and can be separated from other segments. Compare with **cDNA library**.

genomics The field of study concerned with sequencing, interpreting, and comparing whole genomes from different organisms.

genotype All the alleles of every gene present in a given individual. Often specified only for the alleles of a particular set of genes under study. Compare with **phenotype**.

genus (plural: genera) In Linnaeus' system, a taxonomic category of closely related species. Always italicized and capitalized to indicate that it is a recognized scientific genus.

geologic time scale The sequence of eons, eras, and periods used to describe the geologic history of Earth.

germ cell In animals, any cell that can potentially give rise to gametes. Also called *germ-line cells*.

germ layer In animals, one of the three basic types of tissue formed during gastrulation; gives rise to all other tissues. See **endoderm**, **mesoderm**, and **ectoderm**.

germ line In animals, any of the cells that are capable of giving rise to reproductive cells (sperm or egg). Compare with **germ cell**.

germ theory of disease The theory that infectious diseases are caused by bacteria, viruses, and other microorganisms.

germination The process by which a seed becomes a young plant.

gestation The period of development inside the mother, from implantation to birth, in those species that have live birth.

gibberellins A class of hormones, found in plants and fungi, that stimulate growth. Gibberellic acid (GA) is one of the major gibberellins.

Gibbs free-energy change (ΔG) A measure of the change in enthalpy and entropy that occurs in a given chemical reaction. ΔG is less than 0 for spontaneous reactions and greater than 0 for nonspontaneous reactions.

gill Any organ in aquatic animals that exchanges gases and other dissolved substances between the blood and the surrounding water. Typically, a filamentous outgrowth of a body surface.

gill arch In aquatic vertebrates, a curved region of tissue between the gills. Gills are suspended from the gill arches.

gill filament In fish, any of the many long, thin structures that extend from gill arches into the water and across which gas exchange occurs.

gill lamella (plural: gill lamellae) Any of hundreds to thousands of sheetlike structures, each containing a capillary bed, that make up a gill filament.

gland An organ whose primary function is to secrete some substance, either into the blood (endocrine gland) or into some other space such as the gut or skin (exocrine gland).

glia Collective term for several types of cells in nervous tissue that are not neurons and do not conduct electrical signals but perform other functions, such as providing support, nourishment, or electrical insulation. Also called *glial cells*.

global carbon cycle See **carbon cycle, global**.

global climate change The global sum of all the local changes in temperature and precipitation patterns that accompany global warming (or in some past events, global cooling).

global gene regulation The regulation of multiple bacterial genes that are not part of one operon.

global nitrogen cycle See **nitrogen cycle, global**.

global warming A sustained increase in Earth's average surface temperature.

global water cycle See **water cycle, global**.

glomalin A glycoprotein that is abundant in the hyphae of arbuscular mycorrhizal fungi; when hyphae decay, it is an important component of soil.

glomerulus (plural: glomeruli) (1) In the vertebrate kidney, a ball-like cluster of capillaries, surrounded by Bowman's capsule, at the beginning of a nephron. (2) In the brain, a ball-like cluster of neurons in the olfactory bulb.

glucagon A peptide hormone produced by the pancreas in response to low blood glucose. Raises blood glucose by triggering breakdown of glycogen and stimulating gluconeogenesis. Compare with **insulin**.

glucocorticoids A class of steroid hormones, produced and secreted by the adrenal cortex, that increase blood glucose and prepare the body for stress. Include cortisol and corticosterone. Compare with **mineralocorticoids**.

gluconeogenesis Synthesis of glucose, often from non-carbohydrate sources (e.g., proteins and fatty acids). In plants, used to produce glucose from products of the Calvin cycle. In animals, occurs in the liver in response to low insulin levels and high glucagon levels.

glucose Six-carbon monosaccharide whose oxidation in cellular respiration is the major source of ATP in animal cells.

glyceraldehyde-3-phosphate (G3P) The phosphorylated three-carbon compound formed as the result of carbon fixation in the first step of the Calvin cycle.

glycerol A three-carbon molecule that forms the "backbone" of phospholipids and most fats.

glycogen A highly branched storage polysaccharide composed of glucose monomers joined by α-1,4- and α-1,6-glycosidic linkages. The major form of stored carbohydrate in animals.

glycolipid Any lipid molecule that is covalently bonded to a carbohydrate group.

glycolysis A series of 10 chemical reactions that oxidize glucose to produce pyruvate, NADH, and ATP.

Used by organisms as part of fermentation or cellular respiration.

glycoprotein Any protein with one or more covalently bonded carbohydrates, typically oligosaccharides.

glycosidic linkage The covalent linkage formed by a condensation reaction between two sugar monomers; joins the residues of a polysaccharide.

glycosylation Addition of a carbohydrate group to a molecule.

glyoxysome Specialized type of peroxisome found in plant cells and packed with enzymes for processing the products of photosynthesis.

gnathostomes Animals with jaws. Most vertebrates are gnathostomes.

Golgi apparatus A eukaryotic organelle, consisting of stacks of flattened membranous sacs (cisternae), that functions in processing and sorting proteins and lipids destined to be secreted or directed to other organelles. Also called *Golgi complex*.

gonad An organ, such as a testis or an ovary, that produces reproductive cells.

gonadotropin-releasing hormone (GnRH) A peptide hormone, produced and secreted by the hypothalamus, that stimulates release of follicle-stimulating hormone (FSH) and luteinizing hormone (LH) from the anterior pituitary.

grade In taxonomy, a group of species that share some, but not all, of the descendants of a common ancestor. Also called a *paraphyletic group*.

Gram-negative Describing bacteria that look pink when treated with a Gram stain. These bacteria have a cell wall composed of a thin layer of peptidoglycan and an outer phospholipid layer. Compare with **Gram-positive**.

Gram-positive Describing bacteria that look purple when treated with a Gram stain. These bacteria have cell walls composed of a thick layer of peptidoglycan and no outer phospholipid later. Compare with **Gram-negative**.

Gram stain A dye that distinguishes the two general types of cell walls found in bacteria. Used to routinely classify bacteria as Gram-negative or Gram-positive.

granum (plural: grana) In chloroplasts, a stack of flattened, membrane-bound thylakoid discs where the light reactions of photosynthesis occur.

gravitropism The growth or movement of a plant in a particular direction in response to gravity.

grazing food chain The ecological network of herbivores and the predators and parasites that consume them.

great apes See **hominids**.

green algae A paraphyletic group of photosynthetic organisms that contain chloroplasts similar to those in green plants. Often classified as protists, green algae are the closest living relatives of land plants and form a monophyletic group with them.

greenhouse gas An atmospheric gas that absorbs and reflects infrared radiation, so that heat radiated from Earth is retained in the atmosphere instead of being lost to space.

gross primary productivity In an ecosystem, the total amount of carbon fixed by photosynthesis (or more

rarely, chemosynthesis), including that used for cellular respiration, over a given time period. Compare with **net primary productivity**.

ground meristem The middle layer of a young plant embryo. Gives rise to the ground tissue system.

ground tissue An embryonic tissue layer that gives rise to parenchyma, collenchyma, and sclerenchyma—tissues other than the epidermis and vascular tissue. Also called *ground tissue system*.

groundwater Any water below the land surface.

growth factor Any of a large number of signaling molecules that are secreted by certain cells and that stimulate other cells to grow, divide, or differentiate.

growth hormone (GH) A peptide hormone, produced and secreted by the mammalian anterior pituitary, that promotes lengthening of the long bones in children and muscle growth, tissue repair, and lactation in adults. Also called *somatotropin*.

guanosine triphosphate (GTP) A nucleotide consisting of guanine, a ribose sugar, and three phosphate groups. Can be hydrolyzed to release free energy. Commonly used in RNA synthesis and also functions in signal transduction in association with G proteins.

guard cell One of two specialized, crescent-shaped cells forming the border of a plant stoma. Guard cells can change shape to open or close the stoma. See also **stoma**.

gustation The perception of taste.

guttation Excretion of water droplets from plant leaves; visible in the early morning. Caused by root pressure.

gymnosperm A vascular plant that makes seeds but does not produce flowers. The gymnosperms include five lineages of green plants (cycads, ginkgoes, conifers, redwoods, and gnetophytes). Compare with **angiosperm**.

H⁺-ATPase See **proton pump**.

habitat degradation The reduction of the quality of a habitat.

habitat destruction Human-caused destruction of a natural habitat, replaced by an urban, suburban, or agricultural landscape.

habitat fragmentation The breakup of a large region of a habitat into many smaller regions, separated from others by a different type of habitat.

Hadley cell An atmospheric cycle of large-scale air movement in which warm equatorial air rises, moves north or south, and then descends at approximately 30° N or 30° S latitude.

hair cell A pressure-detecting sensory cell that has tiny "hairs" (stereocilia) jutting from its surface. Found in the inner ear, lateral line system, and ampullae of Lorenzini.

hairpin A secondary structure in RNA consisting of a stable loop formed by hydrogen bonding between purine and pyrimidine bases on the same strand.

Hamilton's rule The proposition that an allele for altruistic behavior will be favored by natural selection only if *Br* C, where *B* = the fitness benefit to the recipient, *C* = the fitness cost to the actor, and *r* = the coefficient of relatedness between recipient and actor.

haploid (1) Having one set of chromosomes (1*n* or *n* for short). (2) A cell or an individual organism with one set of chromosomes. Compare with **diploid**.

haploid number The number of different types of chromosomes in a cell. Symbolized as *n*.

Hardy–Weinberg principle A principle of population genetics stating that genotype frequencies in a large population do not change from generation to generation in the absence of evolutionary processes (e.g., mutation, gene flow, genetic drift, and selection), and nonrandom mating.

haustorium (plural: haustoria) Highly modified stem or root of a parasitic plant. The haustorium penetrates the tissues of a host and absorbs nutrients and water.

hearing The sensation of the wavelike changes in air pressure called sound.

heart A muscular pump that circulates blood throughout the body.

heart murmur A distinctive sound caused by back-flow of blood through a defective heart valve.

heartwood The older xylem in the center of an older stem or root, containing protective compounds and no longer functioning in water transport.

heat Thermal energy that is transferred from an object at higher temperature to one at lower temperature.

heat of vaporization The energy required to vaporize 1 gram of a liquid into a gas.

heat-shock proteins Proteins that facilitate refolding of proteins that have been denatured by heat or other agents.

heavy chain The larger of the two types of polypeptide chains in an antibody or B-cell receptor molecule; composed of a variable (*V*) region, which contributes to the antigen-binding site, and a constant (*C*) region. Differences in heavy-chain constant regions determine the different classes of immunoglobulins (IgA, IgE, etc.). Compare with **light chain**.

helper T cell A CD4$^+$ effector T cell that secretes cytokines and in other ways promotes the activation of other lymphocytes. Activated by interacting with complementary class II MHC–peptide complexes on the surface of antigen-presenting cells such as dendritic cells.

heme A small molecule that binds to each of the four polypeptides that combine to form hemoglobin; contains an iron ion that can bind oxygen.

hemimetabolous metamorphosis A type of metamorphosis in which the animal increases in size from one stage to the next, but does not dramatically change its body form. Also called *incomplete metamorphosis*.

hemocoel A body cavity, present in arthropods and some mollusks, containing a pool of circulatory fluid (hemolymph) bathing the internal organs. Unlike a coelom, a hemocoel is not lined in mesoderm.

hemoglobin An oxygen-binding protein consisting of four polypeptide subunits, each containing an oxygen-binding heme group. The major oxygen carrier in mammalian blood.

hemolymph The circulatory fluid of animals with open circulatory systems (e.g., insects) in which the fluid is not confined to blood vessels.

herbaceous Referring to a plant that is not woody.

herbivore (adjective: herbivorous) An animal that eats primarily plants and rarely or never eats meat. Compare with **carnivore** and **omnivore**.

herbivory The practice of eating plant tissues.

heredity The transmission of traits from parents to offspring via genetic information.

heritable Referring to traits that can be transmitted from one generation to the next.

hermaphrodite An organism that produces both male and female gametes.

heterokaryotic Describing a cell or fungal mycelium containing two or more haploid nuclei that are genetically distinct.

heterospory (adjective: heterosporous) In seed plants, the production of two distinct types of spores: microspores, which become the male gametophyte, and megaspores, which become the female gametophyte. Compare with **homospory**.

heterotherm An animal whose body temperature varies markedly. Compare with **homeotherm**.

heterotroph Any organism that cannot synthesize reduced organic compounds from inorganic sources and that must obtain them from other organisms. Some bacteria, some archaea, and virtually all fungi and animals are heterotrophs. Also called *consumer*. Compare with **autotroph**.

heterozygote advantage A pattern of natural selection that favors heterozygous individuals compared with homozygotes. Tends to maintain genetic variation in a population, thus is a form of balancing selection.

heterozygous Having two different alleles of a gene.

hexose A monosaccharide (simple sugar) containing six carbon atoms.

hibernation An energy-conserving physiological state, marked by a decrease in metabolic rate, body temperature, and activity, that lasts for a prolonged period (weeks to months). Occurs in some animals in response to winter cold and scarcity of food. Compare with **torpor**.

hindbrain One of the three main regions of the vertebrate brain, responsible for balance and sometimes hearing; includes the cerebellum and medulla oblongata. Compare with **forebrain** and **midbrain**.

histamine A molecule released from mast cells during an inflammatory response that, at high concentrations, causes blood vessels to constrict to reduce blood loss from tissue damage.

histone One of several positively charged (basic) proteins associated with DNA in the chromatin of eukaryotic cells.

histone acetyltransferases (HATs) A class of eukaryotic enzymes that loosen chromatin structure by adding acetyl groups to histone proteins.

histone code The hypothesis that specific combinations of chemical modifications of histone proteins contain information that influences chromatin condensation and gene expression.

histone deacetylases (HDACs) A class of eukaryotic enzymes that condense chromatin by removing acetyl groups from histone proteins.

holoenzyme A multipart enzyme consisting of a core enzyme (containing the active site for catalysis) along with other required proteins.

holometabolous metamorphosis A type of metamorphosis in which the animal completely changes its form; includes a distinct larval stage. Also called *complete metamorphosis*.

homeobox A DNA sequence of about 180 base pairs that codes for a DNA binding motif called the homeodomain in the resulting protein. Genes containing a homeobox usually play a role in controlling development of organisms from fruit flies to humans.

homeostasis (adjective: homeostatic) The array of relatively stable chemical and physical conditions in an animal's cells, tissues, and organs. May be achieved by the body's passively matching the conditions of a stable external environment (conformational homeostasis) or by active physiological processes (regulatory homeostasis) triggered by variations in the external or internal environment.

homeotherm An animal that has a constant or relatively constant body temperature. Compare with **heterotherm**.

homeotic mutation A mutation that causes one body part to be substituted for another.

hominids Members of the family Hominidae, which includes humans and extinct related forms, chimpanzees, gorillas, and orangutans. Distinguished by large body size, no tail, and an exceptionally large brain. Also called *great apes*.

hominins Any extinct or living species of bipedal apes, such as *Australopithecus africanus*, *Homo erectus*, and *Homo sapiens*.

homologous See **homology**.

homologous chromosomes In a diploid organism, chromosomes that are similar in size, shape, and gene content. Also called *homologs*.

homology (adjective: homologous) Similarity among organisms of different species due to their inheritance from a common ancestor. Features that exhibit such similarity (e.g., DNA sequences, proteins, body parts) are said to be homologous. Compare with **homoplasy**.

homoplasy (adjective: homoplastic) Similarity among organisms of different species due to reasons other than common ancestry, such as convergent evolution. Features that exhibit such similarity (e.g., the wings of birds and bats) are said to be homoplastic, or convergent. Compare with **homology**.

homospory (adjective: homosporous) In seedless vascular plants, the production of just one type of spore. Compare with **heterospory**.

homozygous Having two identical alleles of a gene.

hormone Any of many different signaling molecules that circulate throughout the plant or animal body and can trigger characteristic responses in distant target cells at very low concentrations.

hormone-response element A specific sequence in DNA to which a steroid hormone–receptor complex can bind and affect gene transcription.

host An individual that has been invaded by an organism such as a parasite or a virus, or that provides habitat or resources to a commensal organism.

host cell A cell that has been invaded by a parasitic organism or a virus and provides an environment that is conducive to the pathogen's growth and reproduction.

***Hox* genes** A class of genes found in several animal phyla, including vertebrates, that are expressed in a distinctive pattern along the anterior–posterior axis in early embryos and control formation of specific structures. *Hox* genes code for transcription factors with a DNA-binding sequence called a homeobox.

human Any member of the genus *Homo*, which includes modern humans (*Homo sapiens*) and several extinct species.

human chorionic gonadotropin (hCG) A glycoprotein hormone produced by a human embryo and placenta from about week 3 to week 14 of pregnancy. Maintains the corpus luteum, which produces hormones that preserve the uterine lining.

Human Genome Project The multinational research project that sequenced the human genome.

human immunodeficiency virus (HIV) A retrovirus that causes acquired immune deficiency syndrome (AIDS) in humans.

humoral (immune) response The type of immune response that is mediated through the production and secretion of antibodies, complement proteins, and other soluble factors that eliminate extracellular pathogens. Compare with **cell-mediated (immune) response**.

humus The decayed organic matter in soils.

hybrid The offspring of parents from two different strains, populations, or species.

hybrid zone A geographic area where interbreeding occurs between two species, sometimes producing fertile hybrid offspring.

hydrocarbon An organic molecule that contains only hydrogen and carbon atoms.

hydrogen bond A weak interaction between two molecules or different parts of the same molecule resulting from the attraction between a hydrogen atom with a partial positive charge and another atom (usually O or N) with a partial negative charge. Compare with **covalent bond** and **ionic bond**.

hydrogen ion (H⁺) A single proton with a charge of 1+; typically, one that is dissolved in solution or that is being transferred from one atom to another in a chemical reaction.

hydrolysis A chemical reaction in which a molecule is split into smaller molecules by reacting with water. In biology, most hydrolysis reactions involve the splitting of polymers into monomers. Compare with **condensation reaction**.

hydrophilic Interacting readily with water. Hydrophilic compounds are typically polar compounds containing partially or fully charged atoms. Compare with **hydrophobic**.

hydrophobic Not readily interacting with water. Hydrophobic compounds are typically nonpolar compounds that lack partially or fully charged atoms. Compare with **hydrophilic**.

hydrophobic interactions Very weak interactions between nonpolar molecules, or nonpolar regions of the same molecule, when exposed to an aqueous solvent. The surrounding water molecules support these interactions by interacting with one another and encapsulating the nonpolar molecules.

hydroponic growth Growth of plants in liquid cultures instead of soil.

hydrostatic skeleton A system of body support involving a body wall in tension surrounding a fluid or soft tissue under compression.

hydroxide ion (OH⁻) An oxygen atom and a hydrogen atom joined by a single covalent bond and carrying a negative charge; formed by dissociation of water.

hygiene hypothesis The claim that immune disorders arise in individuals less likely to have been exposed to pathogens and parasites, especially in early childhood. Provides an explanation for the increased risk of allergies and autoimmune disease in countries with high levels of sanitation.

hyperosmotic Comparative term designating a solution that has a greater solute concentration, and therefore a lower water concentration, than another solution. Compare with **hyposmotic** and **isosmotic**.

hyperpolarization A change in membrane potential from its resting negative state to an even more negative state; a normal phase in an action potential. Compare with **depolarization**.

hypersensitive reaction An intense allergic response by cells that have been sensitized by previous exposure to an allergen.

hypersensitive response In plants, the rapid death of a cell that has been infected by a pathogen, thereby reducing the potential for infection to spread throughout a plant. Compare with **systemic acquired resistance**.

hypertension Abnormally high blood pressure.

hypertonic Comparative term designating a solution that, if inside a cell or vesicle, results in the uptake of water and swelling or even bursting of the membrane-bound structure. This solution has a greater solute concentration than the solution on the other side of the membrane. Used when the solute is unable to pass through the membrane. Compare with **hypotonic** and **isotonic**.

hypha (plural: hyphae) One of the long, branching strands of a fungal mycelium (the mesh-like body of a fungus). Also found in some protists.

hypocotyl The stem of a very young plant; the region between the cotyledon (embryonic leaf) and the radicle (embryonic root).

hyposmotic Comparative term designating a solution that has a lower solute concentration, and therefore a higher water concentration, than another solution. Compare with **hyperosmotic** and **isosmotic**.

hypothalamic–pituitary axis The functional interaction of the hypothalamus and the anterior pituitary gland, which are anatomically distinct but work together to regulate most of the other endocrine glands in the body.

hypothalamus A part of the brain that functions in maintaining the body's internal physiological state by regulating the autonomic nervous system, endocrine system, body temperature, water balance, and appetite.

hypothesis A testable statement that explains a phenomenon or a set of observations.

hypotonic Comparative term designating a solution that, if inside a cell or vesicle, results in the loss of water and shrinkage of the membrane-bound structure. This solution has a lower solute concentration than the solution on the other side of the membrane. Used when the solute is unable to pass through the membrane. Compare with **hypertonic** and **isotonic**.

immigration The migration of individuals into a particular population from other populations. Compare with **emigration**.

immune system The system whose primary function is to defend the host organism against pathogens. Includes several types of cells (e.g., leukocytes). In vertebrates, several organs are also involved where specialized cells develop or reside (e.g., lymph nodes and thymus).

immunity (adjective: immune) State of being protected against infection by disease-causing pathogens.

immunization The conferring of immunity to a particular disease by artificial means.

immunoglobulin (Ig) Any of the class of proteins that are structurally related to antibodies.

immunological memory The ability of the immune system to "remember" an antigen and mount a rapid, effective adaptive immune response to a pathogen encountered years or decades earlier. Based on the formation of memory lymphocytes.

impact hypothesis The hypothesis that a collision between the Earth and an asteroid caused the mass extinction at the K–P boundary, 65 million years ago.

imperfect flower A flower that contains male parts (stamens) *or* female parts (carpels) but not both. Compare with **perfect flower**.

implantation The process by which an embryo buries itself in the uterine or oviductal wall and forms a placenta. Occurs in mammals and some other viviparous vertebrates.

in situ hybridization A technique for detecting specific DNAs and mRNAs in cells and tissues by use of labeled complementary probes. Can be used to determine where and when particular genes are expressed in embryos.

inbreeding Mating between closely related individuals. Increases homozygosity of a population and often leads to a decline in the average fitness via selection (inbreeding depression).

inbreeding depression In inbred offspring, fitness declines due to deleterious recessive alleles that are homozygous, thus exposed to selection.

inclusive fitness The combination of (1) direct production of offspring (direct fitness) and (2) extra production of offspring by relatives in response to help provided by the individual in question (indirect fitness).

incomplete digestive tract A digestive tract that has just one opening.

incomplete dominance An inheritance pattern in which the heterozygote phenotype is in between the homozygote phenotypes.

incomplete metamorphosis See **hemimetabolous metamorphosis**.

independent assortment, principle of The concept that each pair of hereditary elements (alleles of the same gene) segregates (separates) independently of alleles of other genes during meiosis. One of Mendel's two principles of genetics.

indeterminate growth A pattern of growth in which an individual continues to increase its overall body size throughout its life.

induced fit Change in the shape of the active site of an enzyme, as the result of initial weak binding of a substrate, so that it binds substrate more tightly.

inducer A small molecule that triggers transcription of a specific gene, often by binding to and inactivating a repressor protein.

inducible defense A defensive trait that is manifested only in response to the presence of a consumer (predator or herbivore) or pathogen. Compare with **constitutive defense**.

infection thread An invagination of the plasma membrane of a root hair through which beneficial nitrogen-fixing bacteria enter the roots of their host plants (legumes).

inflammatory response An aspect of the innate immune response, seen in most cases of infection or tissue injury, in which the affected tissue becomes swollen, red, warm, and painful.

ingestion The act of bringing food into the digestive tract.

inhibition In ecological succession, the phenomenon in which early-arriving species make conditions less favorable for the establishment of certain later-arriving species. Compare with **facilitation** and **tolerance**.

inhibitory postsynaptic potential (IPSP) A change in membrane potential, usually hyperpolarization, at a neuron dendrite that makes an action potential less likely.

initiation (1) In an enzyme-catalyzed reaction, the stage during which enzymes orient reactants precisely as they bind at specific locations within the enzyme's active site. (2) In DNA transcription, the stage during which RNA polymerase and other proteins assemble at the promoter sequence and open the strands of DNA to start transcription. (3) In translation, the stage during which a complex consisting of initiation factor proteins, a ribosome, an mRNA, and an aminoacyl tRNA corresponding to the start codon is formed.

initiation factors A class of proteins that assist ribosomes in binding to a messenger RNA molecule to begin translation.

innate behavior Behavior that is inherited genetically, does not have to be learned, and is typical of a species.

innate immune response See **innate immunity**.

innate immunity A set of barriers to infection and generic defenses against broad types of pathogens. Produces an immediate response that involves many different leukocytes, which often activate an inflammatory response. Compare with **acquired immunity**.

inner cell mass (ICM) A cluster of cells, in the interior of a mammalian blastocyst, that eventually develop into the embryo. Contrast with **trophoblast**.

inner ear The innermost portion of the mammalian ear, consisting of a fluid-filled system of tubes that includes the cochlea (which receives sound vibrations from the middle ear) and the semicircular canals (which function in balance).

insects A terrestrial lineage of arthropods distinguished by three tagmata (head, thorax, abdomen), a single pair of antennae, and unbranched appendages.

insulin A peptide hormone produced by the pancreas in response to high levels of glucose (or amino acids) in blood. Enables cells to absorb glucose and coordinates synthesis of fats, proteins, and glycogen. Compare with **glucagon**.

integral membrane protein Any membrane protein that spans the entire lipid bilayer. Also called *transmembrane protein*. Compare with **peripheral membrane protein**.

integrated pest management In agriculture or forestry, systems for managing insects or other pests that include carefully controlled applications of toxins, introduction of species that prey on pests, planting schemes that reduce the chance of a severe pest outbreak, and other techniques.

integrator A component of an animal's nervous system that functions as part of a homeostatic system by evaluating sensory information and triggering appropriate responses. See **effector** and **sensor**.

integrin Any of a class of cell-surface proteins that bind to laminins and other proteins in the extracellular matrix, thus holding cells in place.

intercalated disc A type of specialized connection between adjacent heart muscle cells that contains gap junctions, allowing electrical signals to pass between the cells.

intermediate filament A long fiber, about 10 nm in diameter, composed of one of various proteins (e.g., keratins, lamins); one of the three types of cytoskeletal fibers. Used to form networks that help maintain cell shape and hold the nucleus in place. Compare with **actin filament** and **microtubule**.

intermediate host The host species in which a parasite reproduces asexually. Compare with **definitive host**.

intermediate muscle fiber Type of skeletal muscle fiber that is pink in color, generates ATP by both glycolysis and aerobic respiration, and has contractile properties that are intermediate between slow fibers and fast fibers. Also called fast oxidative fiber.

interneuron A neuron that passes signals from one neuron to another. Compare with **motor neuron** and **sensory neuron**.

internode The section of a plant stem between two nodes (sites where leaves attach).

interphase The portion of the cell cycle between one M phase and the next. Includes the G_1 phase, S phase, and G_2 phase.

intersexual selection The sexual selection of an individual of one gender for mating by an individual of the other gender (usually by female choice).

interspecific competition Competition between members of different species for the same limited resource. Compare with **intraspecific competition**.

interstitial fluid The plasma-like fluid found in the region (interstitial space) between cells.

intertidal zone The region between the low-tide and high-tide marks on a seashore.

intrasexual selection Competition among members of one gender for an opportunity to mate (usually male–male competition).

intraspecific competition Competition between members of the same species for the same limited resource. Compare with **interspecific competition**.

intrinsic rate of increase (r_{max}) The rate at which a population will grow under optimal conditions (i.e., when birthrates are as high as possible and death rates are as low as possible). Compare with **finite rate of increase**.

intron A region of a eukaryotic gene that is transcribed into RNA but is later removed. Compare with **exon**.

invasive species An exotic (nonnative) species that, upon introduction to a new area, spreads rapidly and competes successfully with native species.

inversion A mutation in which a segment of a chromosome breaks from the rest of the chromosome, flips, and rejoins in reversed orientation.

invertebrates A paraphyletic group composed of animals without a backbone; includes about 95 percent of all animal species. Compare with **vertebrates**.

involuntary muscle Muscle that contracts in response to stimulation by involuntary (parasympathetic or sympathetic), but not voluntary (somatic), neural stimulation.

ion An atom or a molecule that has lost or gained electrons and thus carries an electric charge, either positive (cation) or negative (anion), respectively.

ion channel A type of channel protein that allows certain ions to diffuse across a plasma membrane down an electrochemical gradient.

ionic bond A chemical bond that is formed when an electron is completely transferred from one atom to another so that the atoms remain associated due to their opposite electric charges. Compare with **covalent bond** and **hydrogen bond**.

iris A ring of pigmented muscle just behind the cornea in the vertebrate eye that contracts or expands to control the amount of light entering the eye through the pupil.

isosmotic Comparative term designating a solution that has the same solute concentration and water concentration as another solution. Compare with **hyperosmotic** and **hyposmotic**.

isotonic Comparative term designating a solution that, if inside a cell or vesicle, results in no net uptake or loss of water and thus no effect on the volume of the membrane-bound structure. This solution has the same solute concentration as the solution on the other side of the membrane. Compare with **hypertonic** and **hypotonic**.

isotope Any of several forms of an element that have the same number of protons but differ in the number of neutrons.

mycelium (plural: mycelia) A mass of underground filaments (hyphae) that form the body of a fungus. Also found in some protists and bacteria.

mycorrhiza (plural: mycorrhizae) A mutualistic association between certain fungi and the roots of most vascular plants, sometimes visible as nodules or nets in or around plant roots.

myelin sheath Multiple layers of myelin, derived from the cell membranes of certain glial cells, wrapped around the axon of a neuron and providing electrical insulation.

myocardial infarction Death of cardiac muscle cells when deprived of oxygen.

myoD A transcription factor that is a master regulator of muscle cell differentiation (short for "*myo*blast determination").

myofibril Long, slender structure composed of contractile proteins organized into repeating units (sarcomeres) in vertebrate heart muscle and skeletal muscle.

myosin Any one of a class of motor proteins that use the chemical energy of ATP to move along actin filaments in muscle contraction, cytokinesis, and vesicle transport.

myriapods A lineage of arthropods with long segmented trunks, each segment bearing one or two pairs of legs. Includes millipedes and centipedes.

NAD$^+$/NADH Oxidized and reduced forms, respectively, of nicotinamide adenine dinucleotide. A nonprotein electron carrier that functions in many of the redox reactions of metabolism.

NADP$^+$/NADPH Oxidized and reduced forms, respectively, of nicotinamide adenine dinucleotide phosphate. A nonprotein electron carrier that is reduced during the light-dependent reactions in photosynthesis and extensively used in biosynthetic reactions.

natural experiment A situation in which a natural change in conditions enables comparisons of groups, rather than a manipulation of conditions by researchers.

natural selection The process by which individuals with certain heritable traits tend to produce more surviving offspring than do individuals without those traits, often leading to a change in the genetic makeup of the population. A major mechanism of evolution.

nauplius A distinct planktonic larval stage seen in many crustaceans.

Neanderthal A recently extinct European species of hominin, *Homo neanderthalensis*, closely related to but distinct from modern humans.

nectar The sugary fluid produced by flowers to attract and reward pollinating animals.

nectary A nectar-producing structure in a flower.

negative control Of genes, when a regulatory protein shuts down expression by binding to DNA on or near the gene.

negative feedback A self-limiting, corrective response in which a deviation in some variable (e.g., concentration of some compound) triggers responses aimed at returning the variable to normal. Represents a means of maintaining homeostasis. Compare with **positive feedback**.

negative pressure ventilation Ventilation of the lungs by expanding the rib cage so as to "pull" air into the lungs. Compare with **positive pressure ventilation**.

negative-sense RNA virus An ssRNA virus whose genome contains sequences complementary to those in the mRNA required to produce viral proteins. Compare with **ambisense virus** and **positive-sense virus**.

nematodes See **roundworms**.

nephron One of many tiny tubules inside the kidney that function in the formation of urine.

neritic zone Shallow marine waters beyond the intertidal zone, extending down to about 200 meters, where the continental shelf ends.

nerve A long, tough strand of nervous tissue, typically containing thousands of axons, wrapped in connective tissue; carries impulses between the central nervous system and some other part of the body.

nerve cord In chordate animals, a hollow bundle of nerves extending from the brain along the dorsal (back) side of the animal, with cerebrospinal fluid inside a central channel. One of the defining features of chordates.

nerve net A nervous system in which neurons are diffuse instead of being clustered into large ganglia or tracts; found in cnidarians and ctenophores.

nervous tissue An animal tissue consisting of nerve cells (neurons) and various supporting cells.

net primary productivity (NPP) In an ecosystem, the total amount of carbon fixed by photosynthesis over a given time period minus the amount oxidized during cellular respiration. Compare with **gross primary productivity**.

net reproductive rate (R_0) The growth rate of a population per generation; equivalent to the average number of female offspring that each female produces over her lifetime.

neural tube A folded tube of ectoderm that forms along the dorsal side of a young vertebrate embryo; gives rise to the brain and spinal cord.

neuroendocrine Referring to nerve cells (neurons) that release hormones into the blood or to such hormones themselves.

neurogenesis The birth of new neurons from central nervous system stem cells.

neurohormones Hormones produced by neurons.

neuron A cell that is specialized for the transmission of nerve impulses. Typically has dendrites, a cell body, and a long axon that forms synapses with other neurons. Also called *nerve cell*.

neurosecretory cell A nerve cell (neuron) that produces and secretes hormones into the bloodstream. Principally found in the hypothalamus. Also called *neuroendocrine cell*.

neurotoxin Any substance that specifically destroys or blocks the normal functioning of neurons.

neurotransmitter A molecule that transmits signals from one neuron to another or from a neuron to a muscle or gland. Examples are acetylcholine, dopamine, serotonin, and norepinephrine.

neutral In genetics, referring to any mutation or mutant allele that has no effect on an individual's fitness.

neutrophil A type of leukocyte, capable of moving through body tissues, that engulfs and digests pathogens and other foreign particles; also secretes various compounds that attack bacteria and fungi.

niche The range of resources that a species can use and the range of conditions that it can tolerate. More broadly, the role that species plays in its ecosystem.

niche differentiation The evolutionary change in resource use by competing species that occurs as the result of character displacement.

nicotinamide adenine dinucleotide See **NAD$^+$/NADH**.

nitrogen cycle, global The movement of nitrogen among terrestrial ecosystems, the oceans, and the atmosphere.

nitrogen fixation The incorporation of atmospheric nitrogen (N_2) into ammonia (NH_3), which can be used to make many organic compounds. Occurs in only a few lineages of bacteria and archaea.

nociceptor A sensory cell or organ specialized to detect tissue damage, usually producing the sensation of pain.

Nod factors Molecules produced by nitrogen-fixing bacteria that help them recognize and bind to roots of legumes.

node (1) In animals, any small thickening (e.g., a lymph node). (2) In plants, the part of a stem where leaves or leaf buds are attached. (3) In a phylogenetic tree, the point where two branches diverge, representing the point in time when an ancestral group split into two or more descendant groups. Also called *fork*.

node of Ranvier One of the unmyelinated sections that occurs periodically along a neuron's axon and serves as a site where an action potential can be regenerated.

nodule Globular structure on roots of legume plants that contain symbiotic nitrogen-fixing bacteria.

noncyclic electron flow Path of electron flow in which electrons pass from photosystem II, through an electron transport chain, to photosystem I, and ultimately to NADP$^+$ during the light-dependent reactions of photosynthesis. See also **Z scheme**.

nondisjunction An error that can occur during meiosis or mitosis, in which one daughter cell receives two copies of a particular chromosome and the other daughter cell receives none.

nonpolar covalent bond A covalent bond in which electrons are equally shared between two atoms of the same or similar electronegativity. Compare with **polar covalent bond**.

nonsense mutation A point mutation (change in a single base pair) that converts an amino-acid-specifying codon into a stop codon.

non-sister chromatids The chromatids of a particular type of chromosome (after replication) with respect to the chromatids of its homologous chromosome. Crossing over occurs between non-sister chromatids. Compare with **sister chromatids**.

non-template strand The strand of DNA that is not transcribed during synthesis of RNA. Its sequence corresponds to that of the mRNA produced from the other strand. Also called *coding strand*.

non-vascular plants A paraphyletic group of land plants that lack vascular tissue and reproduce using spores. The non-vascular plants include three lineages of green plants (liverworts, mosses, and hornworts). These lineages are sometimes called *bryophytes*.

norepinephrine A catecholamine used as a neurotransmitter in the sympathetic nervous system. Also is produced by the adrenal medulla and functions as a hormone that triggers rapid responses relating to the fight-or-flight response.

notochord A supportive but flexible rod that occurs in the back of a chordate embryo, ventral to the developing spinal cord. Replaced by vertebrae in most adult vertebrates. A defining feature of chordates.

nuclear envelope The double-layered membrane enclosing the nucleus of a eukaryotic cell.

nuclear lamina A lattice-like sheet of fibrous nuclear lamins, which are one type of intermediate filament. Lines the inner membrane of the nuclear envelope, stiffening the envelope and helping to organize the chromosomes.

nuclear lamins Intermediate filaments that make up the nuclear lamina layer—a lattice-like layer inside the nuclear envelope that stiffens the structure.

nuclear localization signal (NLS) A short amino acid sequence that marks a protein for delivery to the nucleus.

nuclear pore An opening in the nuclear envelope that connects the inside of the nucleus with the cytoplasm and through which molecules such as mRNA and some proteins can pass.

nuclear pore complex A large complex of dozens of proteins lining a nuclear pore, defining its shape and regulating transport through the pore.

nuclease Any enzyme that can break down RNA or DNA molecules.

nucleic acid A macromolecule composed of nucleotide monomers. Generally used by cells to store or transmit hereditary information. Includes ribonucleic acid and deoxyribonucleic acid.

nucleoid In prokaryotic cells, a dense, centrally located region that contains DNA but is not surrounded by a membrane.

nucleolus In eukaryotic cells, a specialized structure in the nucleus where ribosomal RNA processing occurs and ribosomal subunits are assembled.

nucleosome A repeating, bead-like unit of eukaryotic chromatin, consisting of about 200 nucleotides of DNA wrapped twice around eight histone proteins.

nucleotide excision repair The process of removing a damaged region in one strand of DNA and replacing it with the correct sequence using the undamaged strand as a template.

nucleotide A molecule consisting of a five-carbon sugar (ribose or deoxyribose), a phosphate group, and one of several nitrogen-containing bases. DNA and RNA are polymers of nucleotides containing deoxyribose (deoxyribonucleotides) and ribose (ribonucleotides), respectively. Equivalent to a nucleoside plus one phosphate group.

nucleus (1) The center of an atom, containing protons and neutrons. (2) In eukaryotic cells, the large organelle containing the chromosomes and surrounded by a double membrane. (3) A discrete clump of neuron cell bodies in the brain, usually sharing a distinct function.

null allele See **knock-out allele**.

null hypothesis A hypothesis that specifies what the results of an experiment will be if the main hypothesis being tested is wrong. Often states that there will be no difference between experimental groups.

nutrient Any substance that an organism requires for normal growth, maintenance, or reproduction.

occipital lobe In the vertebrate brain, one of the four major areas in the cerebrum.

oceanic zone The waters of the open ocean beyond the continental shelf.

odorant Any volatile molecule that conveys information about food or the environment.

oil An unsaturated fat that is liquid at room temperature.

Okazaki fragment Short segment of DNA produced during replication of the lagging strand template. Many Okazaki fragments make up the lagging strand in newly synthesized DNA.

olfaction The perception of odors.

olfactory bulb A bulb-shaped projection of the brain just above the nose. Receives and interprets odor information from the nose.

oligodendrocyte A type of glial cell that wraps around axons of some neurons in the central nervous system, forming a myelin sheath that provides electrical insulation. Compare with **Schwann cell**.

oligopeptide A chain composed of fewer than 50 amino acids linked together by peptide bonds. Often referred to simply as *peptide*.

oligosaccharide A linear or branched polymer consisting of less than 50 monosaccharides joined by glycosidic linkages. Compare with **monosaccharide** and **polysaccharide**.

ommatidium (plural: ommatidia) A light-sensing column in an arthropod's compound eye.

omnivore (adjective: omnivorous) An animal whose diet regularly includes both meat and plants. Compare with **carnivore** and **herbivore**.

oncogene Any gene whose protein product stimulates cell division at all times and thus promotes cancer development. Often is a mutated form of a gene involved in regulating the cell cycle. See **proto-oncogene**.

one-gene, one-enzyme hypothesis The hypothesis that each gene is responsible for making one enzyme. This hypothesis has expanded to include genes that produce proteins other than enzymes or that produce RNAs as final products.

oogenesis The production of egg cells (ova).

oogonium (plural: oogonia) In an ovary, any of the diploid cells that can divide by mitosis to create primary oocytes (which can undergo meiosis) and more oogonia.

open circulatory system A circulatory system in which the circulating fluid (hemolymph) is not confined to blood vessels. Compare with **closed circulatory system**.

open reading frame (ORF) Any DNA sequence, ranging in length from several hundred to thousands of base pairs long, that is flanked by a start codon and a stop codon. ORFs identified by computer analysis of DNA may be functional genes, especially if they have other features characteristic of genes (e.g., promoter sequence).

operator In prokaryotic DNA, a binding site for a repressor protein; located near the start of an operon.

operculum The stiff flap of tissue that covers the gills of teleost fishes.

operon A region of prokaryotic DNA that codes for a series of functionally related genes and is transcribed from a single promoter into one mRNA.

opsin A transmembrane protein that is covalently linked to retinal, the light-detecting pigment in rod and cone cells.

optic nerve A bundle of neurons that runs from the eye to the brain.

optimal foraging The concept that animals forage in a way that maximizes the amount of usable energy they take in, given the costs of finding and ingesting their food and the risk of being eaten while they're at it.

orbital The region of space around an atomic nucleus in which an electron is present most of the time.

organ A group of tissues organized into a functional and structural unit.

organ system Groups of tissues and organs that work together to perform a function.

organelle Any discrete, membrane-bound structure within a cell (e.g., mitochondrion) that has a characteristic structure and function.

organic For a compound, containing carbon and hydrogen and usually containing carbon–carbon bonds. Organic compounds are widely used by living organisms.

organism Any living entity that contains one or more cells.

organogenesis A stage of embryonic development that follows gastrulation and that creates organs from the three germ layers.

origin of replication The site on a chromosome at which DNA replication begins.

osmoconformer An animal that does not actively regulate the osmolarity of its tissues but conforms to the osmolarity of the surrounding environment.

osmolarity The concentration of dissolved substances in a solution, measured in osmoles per liter.

osmoregulation The process by which a living organism controls the concentration of water and salts in its body.

osmoregulator An animal that actively regulates the osmolarity of its tissues.

osmosis Diffusion of water across a selectively permeable membrane from a region of low solute concentration (high water concentration) to a region high solute concentration (low water concentra

ossicles, ear In mammals, three bones found middle ear that function in transferring and ing sound from the outer ear to the inner metabolism, which led to the evolution of proteins and the first life-form.

rod cell A photoreceptor cell with a rod-shaped outer portion that is particularly sensitive to dim light but not used to distinguish colors. Also called simply *rod*. Compare with **cone cell**.

root (1) An underground part of a plant that anchors the plant and absorbs water and nutrients. (2) The most ancestral branch in a phylogenetic tree.

root apical meristem (RAM) A group of undifferentiated plant stem cells at the tip of a plant root that can differentiate into mature root tissue.

root cap A small group of cells that covers and protects the root apical meristem. Senses gravity and determines the direction of root growth.

root hair A long, thin outgrowth of the epidermal cells of plant roots, providing increased surface area for absorption of water and nutrients.

root pressure Positive pressure of xylem sap in the vascular tissue of roots. Generated during the night as a result of the accumulation of ions from the soil and subsequent osmotic movement of water into the xylem.

root system The belowground part of a plant.

rotifer Member of the phylum Rotifera. Distinguished by a cluster of cilia, called a corona, used in suspension feeding in marine and freshwater environments. Rotifers belong to the lophotrochozoan branch of the protostomes.

rough endoplasmic reticulum (rough ER) The portion of the endoplasmic reticulum that is dotted with ribosomes. Involved in synthesis of plasma membrane proteins, secreted proteins, and proteins localized to the ER, Golgi apparatus, and lysosomes. Compare with **smooth endoplasmic reticulum**.

roundworms Members of the phylum Nematoda. Distinguished by an unsegmented body with a pseudocoelom and no appendages. Roundworms belong to the ecdysozoan branch of the protostomes. Also called *nematodes*.

rubisco The enzyme that catalyzes the first step of the Calvin cycle during photosynthesis: the addition of a molecule of CO_2 to ribulose bisphosphate. See also **carbon fixation**.

ruminant Member of a group of hoofed mammals (e.g., cattle, sheep, deer) that have a four-chambered stomach specialized for digestion of plant cellulose. Ruminants regurgitate cud, a mixture of partially digested food and cellulose-digesting bacteria, from the largest chamber (the rumen) for further chewing.

salinity The proportion of solutes dissolved in water in natural environments, often designated in grams of solute per kilogram of water (cited as parts per thousand).

salivary amylase An enzyme that is produced by the salivary glands and that can break down starch by catalyzing hydrolysis of the glycosidic linkages between the glucose residues.

salivary gland A type of vertebrate gland that secretes saliva (a mixture of water, mucus-forming glycoproteins, and digestive enzymes) into the mouth.

sampling error The selection of a nonrepresentative sample from some larger population, due to chance.

saprophyte An organism that feeds primarily on dead plant material.

sapwood The younger xylem in the outer layer of wood of a stem or root, functioning primarily in water transport.

sarcomere The repeating contractile unit of a skeletal muscle cell; the portion of a myofibril located between adjacent Z disks.

sarcoplasmic reticulum Sheets of smooth endoplasmic reticulum in a muscle cell. Contains high concentrations of calcium, which can be released into the cytoplasm to trigger contraction.

saturated Referring to lipids in which all the carbon-carbon bonds are single bonds. Such compounds have relatively high melting points. Compare with **unsaturated**.

scanning electron microscope (SEM) A microscope that produces images of the surfaces of objects by reflecting electrons from a specimen coated with a layer of metal atoms. Compare with **transmission electron microscope**.

scarify To scrape, rasp, cut, or otherwise damage the coat of a seed. Necessary in some species to trigger germination.

Schwann cell A type of glial cell that wraps around axons of some neurons outside the brain and spinal cord, forming a myelin sheath that provides electrical insulation. Compare with **oligodendrocyte**.

scientific name The unique, two-part name given to each species, with a genus name followed by a species name—as in *Homo sapiens*. Scientific names are always italicized, and are also known as Latin names.

sclereid In plants, a relatively short type of sclerenchyma cell that usually functions in protection, such as in seed coats and nutshells. Compare with **fiber**.

sclerenchyma cell In plants, a cell that has a thick secondary cell wall and provides support; typically contains the tough structural polymer lignin and usually is dead at maturity. Includes fibers and sclereids. Compare with **collenchyma cell** and **parenchyma cell**.

scrotum A sac of skin containing the testes and suspended just outside the abdominal body cavity of many male mammals.

second law of thermodynamics The principle of physics that the entropy of the universe or any closed system always increases.

second-male advantage The reproductive advantage, in some species, of a male who mates with a female last, after other males have mated with her.

second messenger A nonprotein signaling molecule produced or activated inside a cell in response to stimulation at the cell surface. Commonly used to relay the message of a hormone or other extracellular signaling molecule.

secondary active transport Transport of an ion or molecule in a defined direction that is often against its electrochemical gradient, in company with an ion or molecule being transported along its electrochemical gradient. Also called *cotransport*.

secondary cell wall The thickened inner layer of a plant cell wall formed by certain cells as they mature and have stopped growing; in water-conducting cells, contains lignin. Provides support or protection.

secondary consumer A carnivore; an organism that eats herbivores. Compare with **primary consumer**.

secondary growth In plants, an increase in the width of stems and roots due to the activity of cambium. Compare with **primary growth**.

secondary immune response The adaptive immune response to a pathogen that the immune system has encountered before. Normally much faster and more efficient than the primary response, due to immunological memory. Compare with **primary immune response**.

secondary metabolites Molecules that are closely related to compounds in key synthetic pathways and that often function in defense.

secondary oocyte A cell produced by meiosis I of a primary oocyte in the ovary. If fertilized, will complete meiosis II to produce an ootid (which develops into an ovum) and a polar body.

secondary spermatocyte A cell produced by meiosis I of a primary spermatocyte in the testis. Can undergo meiosis II to produce spermatids.

secondary structure In proteins, localized folding of a polypeptide chain into regular structures (i.e., alpha-helix and beta-pleated sheet) stabilized by hydrogen bonding between atoms of the peptide backbone. In nucleic acids, elements of structure (e.g., helices and hairpins) stabilized by hydrogen bonding and hydrophobic interactions between complementary bases. Compare with **primary**, **tertiary**, and **quaternary structures**.

secondary succession Gradual colonization of a habitat after an environmental disturbance (e.g., fire, windstorm, logging) that removes some or all previous organisms but leaves the soil intact. Compare with **primary succession**.

secretin A peptide hormone produced by cells in the small intestine in response to the arrival of food from the stomach. Stimulates secretion of bicarbonate (HCO_3^-) from the pancreas.

sedimentary rock A type of rock formed by gradual accumulation of sediment, particularly sand and mud, as in riverbeds and on the ocean floor. Most fossils are found in sedimentary rocks.

seed A plant reproductive structure consisting of an embryo, associated nutritive tissue (endosperm), and an outer protective layer (seed coat). In angiosperms, develops from the fertilized ovule of a flower.

seed bank A repository where seeds, representing many different varieties of domestic crops or other species, are preserved.

seed coat A protective layer around a seed that encases both the embryo and the endosperm.

segment A well-defined, repeated region of the body along the anterior–posterior body axis, containing structures similar to other nearby segments.

segmentation Division of the body or a part of it into a series of similar structures; exemplified by the body segments of insects and worms and by the somites of vertebrates.

segmentation genes A group of genes that control the formation and patterning of body segmentation in embryonic development. Includes maternal genes, gap genes, pair-rule genes, and segment polarity genes.

segregation, principle of The concept that each pair of hereditary elements (alleles of the same gene) separate from each other during meiosis. One of Mendel's two principles of genetics.

selective adhesion The tendency of cells of one tissue type to adhere to other cells of the same type.

selective permeability The property of a membrane that allows some substances to diffuse across it much more readily than other substances.

selectively permeable membrane Any membrane across which some solutes can move more readily than others.

self-fertilization The fusion of two gametes from the same individual to form offspring. Also called *selfing*.

self molecule A molecule that is synthesized by an organism and is a normal part of its cells and/or body; as opposed to nonself, or foreign, molecules.

semen The combination of sperm and accessory fluids that is released by male mammals and reptiles during ejaculation.

semiconservative replication The way DNA replicates, in which each strand of an existing DNA molecule serves as a template to create a new complementary DNA strand. It is called semiconservative because each newly replicated DNA molecule conserves one of the parental strands and contains another, newly replicated strand.

seminal vesicle In male mammals, either of a pair of reproductive glands that secrete a sugar-containing fluid into semen, which provides energy for sperm movement.

senescence The genetically programmed, active process of aging.

sensor Any cell, organ, or structure with which an animal can sense some aspect of the external or internal environment. Usually functions, along with an integrator and effector, as part of a homeostatic system.

sensory neuron A nerve cell that carries signals from sensory receptors to the central nervous system. Compare with **interneuron** and **motor neuron**.

sepal One of the protective leaflike organs enclosing a flower bud and later part of the outermost portion of the flower.

septum (plural: septa) Any wall-like structure. In fungi, septa divide the filaments (hyphae) of mycelia into cell-like compartments.

serotonin A neurotransmitter involved in many brain functions, including sleep, pleasure, and mood.

serum The liquid that remains when cells and clot material are removed from clotted blood. Contains water, dissolved gases, growth factors, nutrients, and other soluble substances. Compare with **plasma**.

sessile Permanently attached to a substrate; not capable of moving to another location.

set point A normal or target value for a regulated internal variable, such as body heat or blood pH.

sex chromosome Chromosomes that differ in shape or in number in males and females. For example, the X and Y chromosomes of many animals. Compare with **autosome**.

sex-linked inheritance Inheritance patterns observed in genes carried on sex chromosomes. In this case, females and males have different numbers of alleles of a gene. Often creates situations in which a trait appears more often in one sex. Also called *sex-linkage*.

sexual dimorphism Any trait that differs between males and females.

sexual reproduction Any form of reproduction in which genes from two parents are combined via fusion of gametes, producing offspring that are genetically distinct from both parents. Compare with **asexual reproduction**.

sexual selection A type of natural selection that favors individuals with traits that increase their ability to obtain mates. Acts more strongly on males than females. (Compare with **ecological selection**.)

shell A hard, protective outer structure.

Shine–Dalgarno sequence See **ribosome binding sequence**.

shoot In a plant embryo, the combination of hypocotyl and cotyledons, which will become the aboveground portions of the plant.

shoot apical meristem (SAM) A group of undifferentiated plant stem cells at the tip of a plant stem that can differentiate into mature shoot tissues.

shoot system The aboveground part of a plant comprising stems, leaves, and flowers (in angiosperms).

short-day plant A plant that blooms in response to long nights (usually in late summer or fall in the Northern Hemisphere). Compare with **day-neutral** and **long-day plant**.

short tandem repeats (STRs) Relatively short DNA sequences that are repeated, one after another, down the length of a chromosome. See **microsatellite**.

shotgun sequencing A method of sequencing genomes that is based on breaking the genome into small pieces, sequencing each piece separately, and then figuring out how the pieces are connected.

sieve plate In plants, a pore-containing structure at each end of a sieve-tube element in phloem.

sieve-tube element In plants, an elongated sugar-conducting cell in phloem that lacks nuclei and has sieve plates at both ends, allowing sap to flow to adjacent cells.

sigma A bacterial protein that associates with the core RNA polymerase to allow recognition of promoters.

signal In behavioral ecology, any information-containing behavior or characteristic.

signal receptor Any cellular protein that binds to a particular signaling molecule (e.g., a hormone or neurotransmitter) and triggers a response by the cell. Receptors for lipid-insoluble signals are transmembrane proteins in the plasma membrane; those for many lipid-soluble signals (e.g., steroid hormones) are located inside the cell.

signal recognition particle (SRP) An RNA–protein complex that binds to the ER signal sequence in a polypeptide as it emerges from a ribosome and transports the ribosome–polypeptide complex to the ER membrane, where synthesis of the polypeptide is completed.

signal transduction The process by which a stimulus (e.g., a hormone, a neurotransmitter, or sensory information) outside a cell is converted into an intracellular signal required for a cellular response. Usually involves a specific sequence of molecular events, or signal transduction pathway, that may lead to amplification of the signal.

signal transduction cascade See **phosphorylation cascade**.

silencer A regulatory sequence in eukaryotic DNA to which repressor proteins can bind, inhibiting gene transcription.

silent mutation A point mutation that changes the sequence of a codon without changing the amino acid that is specified.

simple eye An eye with only one light-collecting apparatus (e.g., one lens), as in vertebrates and cephalopods. Compare with **compound eye**.

simple fruit A fruit (e.g., apricot) that develops from a single flower that has a single carpel or several fused carpels. Compare with **aggregate** and **multiple fruit**.

simple sequence repeat See **microsatellite**.

SINEs (short interspersed nuclear elements) The second most abundant class of transposable elements in human genomes; can create copies of itself and insert them elsewhere in the genome. Compare with **LINEs**.

single nucleotide polymorphism (SNP) A site on a chromosome where individuals in a population have different nucleotides. Can be used as a genetic marker to help track the inheritance of nearby genes.

single-strand DNA-binding proteins (SSBP) A protein that attaches to separated strands of DNA during replication or transcription, preventing them from re-forming a double helix.

sink Any tissue, site, or location where an element or a molecule is consumed or taken out of circulation (e.g., in plants, a tissue where sugar exits the phloem). Compare with **source**.

sinoatrial (SA) node In the right atrium of the vertebrate heart, a cluster of cardiac muscle cells that initiates the heartbeat and determines the heart rate. Compare with **atrioventricular (AV) node**.

siphon A tubelike appendage of many mollusks, often used for feeding or propulsion.

sister chromatids The paired strands of a recently replicated chromosome, which are connected at the centromere and eventually separate during anaphase of mitosis and meiosis II. Compare with **non-sister chromatids**.

sister species Closely related species that occupy adjacent branches in a phylogenetic tree.

skeletal muscle The muscle tissue attached to the bones of the vertebrate skeleton. Consists of long, unbranched muscle fibers with a characteristic striped (striated) appearance; controlled voluntarily. Compare with **cardiac** and **smooth muscle**.

sliding-filament model The hypothesis that thin (actin) filaments and thick (myosin) filaments slide past each other, thereby shortening the sarcomere. Shortening of all the sarcomeres in a myofibril results in contraction of the entire myofibril.

slow muscle fiber Type of skeletal muscle fiber that is red in color due to the abundance of myoglobin, generates ATP by oxidative phosphorylation, and contracts slowly but does not fatigue easily. Also called *slow oxidative*, or *Type I, fiber*.

small intestine The portion of the digestive tract between the stomach and the large intestine. The site of the final stages of digestion and of most nutrient absorption.

small nuclear ribonucleoproteins See **snRNPs**.

smooth endoplasmic reticulum (smooth ER) The portion of the endoplasmic reticulum that does not have ribosomes attached to it. Involved in synthesis and secretion of lipids. Compare with **rough endoplasmic reticulum**.

smooth muscle The unstriated muscle tissue that lines the intestine, blood vessels, and some other organs. Consists of tapered, unbranched cells that can sustain long contractions. Not voluntarily controlled. Compare with **cardiac** and **skeletal muscle**.

snRNPs (small nuclear ribonucleoproteins) Complexes of proteins and small RNA molecules that function as components of spliceosomes during splicing (removal of introns from pre-mRNAs).

sodium–potassium pump A transmembrane protein that uses the energy of ATP to move sodium ions out of the cell and potassium ions in. Also called Na^+/K^+-ATPase.

soil organic matter Organic (carbon-containing) compounds found in soil.

solute Any substance that is dissolved in a liquid.

solute potential (ψ_S) A component of the potential energy of water caused by a difference in solute concentrations at two locations. Can be zero (pure water) or negative. Compare with **pressure potential (ψ_P)**.

solution A liquid containing one or more dissolved solids or gases in a homogeneous mixture.

solvent Any liquid in which one or more solids or gases can dissolve.

soma See **cell body**.

somatic cell Any type of cell in a multicellular organism except eggs, sperm, and their precursor cells. Also called *body cells*.

somatic hypermutation Mutation that occurs in the variable regions of immunoglobulin genes when B cells are first activated and in memory cells, resulting in novel variation in the receptors that bind to antigens.

somatic nervous system The part of the peripheral nervous system (outside the brain and spinal cord) that controls skeletal muscles and is under voluntary control. Compare with **autonomic nervous system**.

somatostatin A hormone secreted by the pancreas and hypothalamus that inhibits the release of several other hormones.

somite A block of mesoderm that occurs in pairs along both sides of the developing neural tube in a vertebrate embryo. Gives rise to muscle, vertebrae, ribs, and the dermis of the skin.

sori In ferns, a cluster of spore-producing structures (sporangia) usually found on the underside of fronds.

source Any tissue, site, or location where a substance is produced or enters circulation (e.g., in plants, the tissue where sugar enters the phloem). Compare with **sink**.

space-filling model A representation of a molecule where atoms are shown as balls—color-coded and scaled to indicate the atom's identify—attached to each other in the correct geometry.

speciation The evolution of two or more distinct species from a single ancestral species.

species An evolutionarily independent population or group of populations. Generally distinct from other species in appearance, behavior, habitat, ecology, genetic characteristics, and so on.

species–area relationship The mathematical relationship between the area of a certain habitat and the number of species that it can support.

species diversity The variety and relative abundance of the species present in a given ecological community.

species richness The number of species present in a given ecological community.

specific heat The amount of energy required to raise the temperature of 1 gram of a substance by 1°C; a measure of the capacity of a substance to absorb energy.

sperm A mature male gamete; smaller and more mobile than the female gamete.

sperm competition Competition between the sperm of different males to fertilize eggs inside the same female.

spermatid An immature sperm cell.

spermatogenesis The production of sperm. Occurs continuously in a testis.

spermatogonium (plural: spermatogonia) Any of the diploid cells in a testis that can give rise to primary spermatocytes.

spermatophore A gelatinous package containing sperm cells that is produced by males of species that have internal fertilization without copulation.

sphincter A muscular valve that can close off a tube, as in a blood vessel or a part of the digestive tract.

spicule Stiff spike of silica or calcium carbonate that provides structural support in the body of many sponges.

spindle apparatus The array of microtubules responsible for moving chromosomes during mitosis and meiosis; includes kinetochore microtubules, polar microtubules, and astral microtubules.

spines In plants, modified leaves that are stiff and sharp and that function in defense.

spiracle In insects, a small opening that connects air-filled tracheae to the external environment, allowing for gas exchange.

spleen A dark red organ, found near the stomach of most vertebrates, that filters blood, stores extra red blood cells in case of emergency, and plays a role in immunity.

spliceosome In eukaryotes, a large, complex assembly of snRNPs (small nuclear ribonucleoproteins) that catalyzes removal of introns from primary RNA transcripts.

splicing The process by which introns are removed from primary RNA transcripts and the remaining exons are connected together.

sporangium (plural: sporangia) A spore-producing structure found in seed plants, some protists, and some fungi (e.g., chytrids).

spore (1) In bacteria, a dormant form that generally is resistant to extreme conditions. (2) In eukaryotes, a single haploid cell produced by mitosis or meiosis (not by fusion of gametes) that is capable of developing into an adult organism.

sporophyte In organisms undergoing alternation of generations, the multicellular diploid form that arises from two fused gametes and produces haploid spores. Compare with **gametophyte**.

sporopollenin A watertight material that encases spores and pollen of modern land plants.

stabilizing selection A mode of natural selection that favors phenotypes near the middle of the range of phenotypic variation. Reduces overall genetic variation in a population. Compare with **disruptive selection** and **directional selection**.

stamen The male reproductive structure of a flower. Consists of an anther, in which pollen grains are produced, and a filament, which supports the anther. Compare with **carpel**.

standing defense See **constitutive defense**.

stapes The last of three small bones (ossicles) in the middle ear of vertebrates. Receives vibrations from the tympanic membrane and by vibrating against the oval window passes them to the cochlea.

starch A mixture of two storage polysaccharides, amylose and amylopectin, both formed from α-glucose monomers. Amylopectin is branched, and amylose is unbranched. The major form of stored carbohydrate in plants.

start codon The AUG triplet in mRNA at which protein synthesis begins; codes for the amino acid methionine.

statocyst A sensory organ of many arthropods that detects the animal's orientation in space (e.g., whether the animal is flipped upside down).

statolith A tiny stone or dense particle found in specialized gravity-sensing organs in some animals such as lobsters, and in gravity-sensing tissues of plants.

statolith hypothesis The hypothesis that amyloplasts (dense, starch-storing plant organelles) serve as statoliths in gravity detection by plants.

stem cell Any relatively undifferentiated cell that can divide to produce a daughter cell that remains a stem cell and a daughter cell that can differentiate into specific cell types.

stems Vertical, aboveground structures that make up the shoot system of plants.

stereocilium (plural: stereocilia) One of many stiff outgrowths from the surface of a hair cell that are involved in detection of sound by terrestrial vertebrates or of waterborne vibrations by fishes.

steroid A class of lipid with a characteristic four-ring hydrocarbon structure.

sticky ends The short, single-stranded ends of a DNA molecule cut by a restriction endonuclease. Tend to form hydrogen bonds with other sticky ends that have complementary sequences.

stigma The sticky tip at the end of a flower carpel to which pollen grains adhere.

stolon A modified stem that runs horizontally over the soil surface and produces new plants at the nodes (a form of asexual reproduction). Compare with **rhizome**.

stoma (plural: stomata) Generally, a pore or opening. In plants, a microscopic pore, surrounded by specialized cells that open the pore, on the surface of a leaf or stem through which gas exchange occurs. See also **guard cells**.

stomach A tough, muscular pouch in the vertebrate digestive tract between the esophagus and small intestine. Physically breaks up food and begins digestion of proteins.

stop codon Any of three mRNA triplets (UAG, UGA, or UAA) that cause termination of protein synthesis. Also called a *termination codon*.

strain The lowest, most specific level of taxonomy that refers to a population of individuals that are genetically very similar or identical.

striated muscle Muscle tissue containing protein filaments organized into repeating structures that give the cells and tissues a banded appearance.

stroma The fluid matrix of a chloroplast in which the thylakoids are embedded. Site where the Calvin cycle reactions occur.

structural formula A two-dimensional notation in which the chemical symbols for the constituent atoms are joined by straight lines representing single (–), double (=), or triple (≡) covalent bonds. Compare with **molecular formula**.

structural homology Similarities in adult organismal structures (e.g., limbs, shells, flowers) that are due to inheritance from a common ancestor.

style The slender stalk of a flower carpel connecting the stigma and the ovary.

suberin Waxy substance found in the cell walls of cork tissue and in the Casparian strip of endodermal cells.

subspecies A population that has distinctive traits and some genetic differences relative to other populations of the same species but that is not distinct enough to be classified as a separate species.

substrate (1) A reactant that interacts with a catalyst, such as an enzyme or ribozyme, in a chemical reaction. (2) A surface on which a cell or organism sits.

substrate-level phosphorylation Production of ATP or GTP by the transfer of a phosphate group from an intermediate substrate directly to ADP or GDP. Occurs in glycolysis and in the citric acid cycle.

succession In ecology, the gradual colonization of a habitat after an environmental disturbance (e.g., fire, flood), usually by a series of species. See **primary** and **secondary succession**.

sucrose A disaccharide formed from glucose and fructose. One of the two main products of photosynthesis.

sugar Synonymous with carbohydrate, though usually used in an informal sense to refer to small carbohydrates (monosaccharides and disaccharides).

summation The additive effect of different postsynaptic potentials on a nerve or muscle cell, such that several subthreshold stimulations can cause or inhibit an action potential.

supporting connective tissue A type of connective tissue distinguished by having a firm extracellular matrix. Includes bone and cartilage.

surface metabolism model Hypothetical explanation for chemical evolution whereby small molecules reacted with one another through catalytic activity associated with a surface, such as the mineral deposits found in deep-sea hydrothermal vents.

surface tension The cohesive force that causes molecules at the surface of a liquid to stick together, thereby resisting deformation of the liquid's surface and minimizing its surface area.

survivorship On average, the proportion of offspring that survive to a particular age.

survivorship curve A graph depicting the percentage of a population that survives to different ages.

suspension feeder Any organism that obtains food by filtering small particles or small organisms out of water or air. Also called *filter feeder*.

sustainability The planned use of environmental resources at a rate no faster than the rate at which they are naturally replaced.

sustainable agriculture Agricultural techniques that are designed to maintain long-term soil quality and productivity.

swamp A wetland that has a steady rate of water flow and is dominated by trees and shrubs.

swim bladder A gas-filled organ of many ray-finned fishes that regulates buoyancy.

symbiosis (adjective: symbiotic) Any close and prolonged physical relationship between individuals of two different species. See **commensalism**, **mutualism**, and **parasitism**.

symmetric competition Ecological competition between two species in which both suffer similar declines in fitness. Compare with **asymmetric competition**.

sympathetic nervous system The part of the autonomic nervous system that stimulates fight-or-flight responses, such as increased heart rate, increased blood pressure, and decreased digestion. Compare with **parasympathetic nervous system**.

sympatric speciation The divergence of populations living within the same geographic area into different species as the result of their genetic (not physical) isolation. Compare with **allopatric speciation**.

sympatry Condition in which two or more populations live in the same geographic area, or close enough to permit interbreeding. Compare with **allopatry**.

symplast In plants, the space inside the plasma membranes. The symplast of adjacent cells is often connected through plasmodesmata. Compare with **apoplast**.

symporter A cotransport protein that allows an ion to diffuse down an electrochemical gradient, using the energy of that process to transport a different substance in the same direction *against* its concentration gradient. Compare with **antiporter**.

synapomorphy A shared, derived trait found in two or more taxa that is present in their most recent common ancestor but is missing in more distant ancestors. Useful for inferring evolutionary relationships.

synapse The interface between two neurons or between a neuron and an effector cell.

synapsis The physical pairing of two homologous chromosomes during prophase I of meiosis. Crossing over is observed during synapsis.

synaptic cleft The space between two communicating nerve cells (or between a neuron and effector cell) at a synapse, across which neurotransmitters diffuse.

synaptic plasticity Long-term changes in the responsiveness or physical structure of a synapse that can occur after particular stimulation patterns. Thought to be the basis of learning and memory.

synaptic vesicle A small neurotransmitter-containing vesicle at the end of an axon that releases neurotransmitter into the synaptic cleft by exocytosis.

synaptonemal complex A network of proteins that holds non-sister chromatids together during synapsis in meiosis I.

synthesis (S) phase The phase of the cell cycle during which DNA is synthesized and chromosomes are replicated.

system A defined set of interacting chemical components under observation.

systemic acquired resistance (SAR) A slow, widespread response of plants to a localized infection that protects healthy tissue from invasion by pathogens. Compare with **hypersensitive response**.

systemic circulation The part of the circulatory system that sends oxygen-rich blood from the lungs out to the rest of the body. It is separate from the pulmonary circulation in mammals and birds.

systemin A peptide hormone, produced by plant cells damaged by herbivores, that initiates a protective response in undamaged cells.

systems biology The study of the structure of networks and how interactions between individual network components such as genes or proteins can lead to emergent biological properties.

systole The portion of the cardiac cycle during which the heart muscles are contracting. Compare with **diastole**.

systolic blood pressure The force exerted by blood against artery walls during contraction of the heart's left ventricle. Compare with **diastolic blood pressure**.

T cell A type of lymphocyte that matures in the thymus and, with B cells, is responsible for adaptive immunity. Involved in activation of B cells (CD4$^+$ helper T cells) and destruction of infected cells (CD8$^+$ cytotoxic T cells). Also called *T lymphocytes*.

T-cell receptor (TCR) A type of transmembrane protein found on T cells that can bind to antigens displayed on the surfaces of other cells. Composed of two polypeptides, called the alpha chain and beta chain, that consist of variable and constant regions. See **antigen presentation**.

T tubule Any of the membranous tubes that extend into the interior of muscle cells, propagating action potentials throughout the cell and triggering release of calcium from the sarcoplasmic reticulum.

tagmata (singular: tagma) Prominent body regions in arthropods, such as the head, thorax, and abdomen in insects.

taiga A vast forest biome throughout subarctic regions, consisting primarily of short coniferous trees. Characterized by intensely cold winters, short summers, and high annual variation in temperature.

taproot A large, vertical main root of a plant's root system.

taste bud A sensory structure, found chiefly in the mammalian tongue, containing spindle-shaped cells that respond to chemical stimuli.

TATA-binding protein (TBP) A protein that binds to the TATA box in eukaryotic promoters and is a component of the basal transcription complex.

TATA box A short DNA sequence in many eukaryotic promoters about 30 base pairs upstream from the transcription start site.

taxon (plural: taxa) Any named group of organisms at any level of a classification system.

taxonomy The branch of biology concerned with the classification and naming of organisms.

tectorial membrane A membrane, located in the vertebrate cochlea, that takes part in the transduction of sound by bending the stereocilia of hair cells in response to sonic vibrations.

telomerase An enzyme that adds DNA to the ends of chromosomes (telomeres) by catalyzing DNA synthesis from an RNA template that is part of the enzyme.

telomere The end of a linear chromosome that contains a repeated sequence of DNA.

telophase The final stage in mitosis or meiosis during which daughter chromosomes (homologous chromosomes in meiosis I) have separated and new nuclear envelopes begin to form around each set of chromosomes.

temperate Having a climate with pronounced annual fluctuations in temperature (i.e., warm summers and cold winters) but typically neither as hot as the tropics nor as cold as the poles.

temperature A measurement of thermal energy present in an object or substance, reflecting how much the constituent molecules are moving.

template strand An original nucleic acid strand used to make a new, complementary copy based on hydrogen bonding between nitrogeneous bases.

temporal lobe In the vertebrate brain, one of the four major areas in the cerebrum.

tendon A band of tough, fibrous connective tissue that connects a muscle to a bone.

tentacle A long, thin, muscular appendage typically used for feeling and feeding. Occurs in different forms in diverse animals such as cephalopod mollusks and sea anemones.

termination (1) In enzyme-catalyzed reactions, the final stage in which the enzyme returns to its original conformation and products are released. (2) In transcription, the dissociation of RNA polymerase from DNA. (3) In translation, the dissociation of a ribosome from mRNA when it reaches a stop codon.

territory An area that is actively defended by an animal from others of its species and that provides exclusive or semi-exclusive use of its resources by the owner.

tertiary consumers In a food chain or food web, organisms that feed on secondary consumers. Compare with **primary consumer** and **secondary consumer**.

tertiary structure The overall three-dimensional shape of a single polypeptide chain, resulting from multiple interactions among the amino acid side chains and the peptide backbone. In single-stranded nucleic acids, the three-dimensional shape is formed by hydrogen bonding and hydrophobic interactions between complementary bases. Compare with **primary**, **secondary**, and **quaternary structure**.

testcross The breeding of an individual that expresses a dominant phenotype but has an unknown genotype with an individual having only recessive alleles for the traits of interest. Used to order to infer the unknown genotype from observation of the phenotypes seen in offspring.

testis (plural: testes) The sperm-producing organ of a male animal.

testosterone A steroid hormone, produced and secreted by the testes, that stimulates sperm production and various male traits and reproductive behaviors.

tetrad The structure formed by synapsed homologous chromosomes during prophase of meiosis I. Also known as a *bivalent*.

tetrapod Any member of the lineage that includes all vertebrates with two pairs of limbs (amphibians, mammals, and reptiles, including birds).

texture A quality of soil, resulting from the relative abundance of different-sized particles.

theory An explanation for a broad class of phenomena that is supported by a wide body of evidence. A theory serves as a framework for the development of new hypotheses.

thermal energy The kinetic energy of molecular motion.

thermocline A steep gradient (cline) in environmental temperature, such as occurs in a thermally stratified lake or ocean.

thermophile A bacterium or archaean that thrives in very hot environments.

thermoreception A sensory system in which receptors are activated by changes in heat energy.

thermoreceptor A sensory cell or an organ specialized for detection of changes in temperature.

thermoregulation Regulation of body temperature.

thick filament A filament composed of bundles of the motor protein myosin; anchored to the center of the sarcomere. Compare with **thin filament**.

thigmotropism Growth or movement of an organism in response to contact with a solid object.

thin filament A filament composed of two coiled chains of actin and associated regulatory proteins; anchored at the Z disk of the sarcomere. Compare with **thick filament**.

thorax A region of the body; in insects, one of the three prominent body regions, along with the head and abdomen, called tagmata.

thorn A modified plant stem shaped as a sharp, protective structure. Helps protect a plant against feeding by herbivores.

threshold potential The membrane potential that will trigger an action potential in a neuron or other excitable cell. Also called simply *threshold*.

thylakoid A membrane-bound network of flattened sac-like structures inside a plant chloroplast that functions in converting light energy to chemical energy. A stack of thylakoid discs is a granum.

thymus An organ, located in the anterior chest or neck of vertebrates, in which immature T cells that originated in the bone marrow undergo maturation.

thyroid gland A gland in the neck that releases thyroid hormone (which increases metabolic rate) and calcitonin (which lowers blood calcium).

thyroid hormones Either of two hormones, triiodothyronine (T_3) or thyroxine (T_4), produced by the thyroid gland. See **triiodothyronine** and **thyroxine**.

thyroid-stimulating hormone (TSH) A peptide hormone, produced and secreted by the anterior pituitary, that stimulates release of thyroid hormones from the thyroid gland.

thyroxine (T_4) A lipid-soluble hormone, derived from the amino acid tyrosine, containing four iodine atoms and produced and secreted by the thyroid gland. Acts primarily to increase cellular metabolism. In mammals, T_4 is converted to the more active hormone triiodothyronine (T_3) in the liver.

Ti plasmid A plasmid carried by *Agrobacterium* (a bacterium that infects plants) that can integrate into a plant cell's chromosomes and induce formation of a gall.

tight junction A type of cell–cell attachment structure that links the plasma membranes of adjacent animal cells, forming a barrier that restricts movement of substances in the space between the cells. Most abundant in epithelia (e.g., the intestinal lining). Compare with **desmosome** and **gap junction**.

tip The end of a branch on a phylogenetic tree. Represents a specific species or larger taxon that has not (yet) produced descendants—either a group living today or a group that ended in extinction. Also called *terminal node*.

tissue A group of cells that function as a unit, such as muscle tissue in an animal or xylem tissue in a plant.

tolerance In ecological succession, the phenomenon in which early-arriving species do not affect the probability that subsequent species will become established. Compare with **facilitation** and **inhibition**.

tonoplast The membrane surrounding a plant vacuole.

tool-kit genes A set of key developmental genes that establishes the body plan of animals and plants; present at the origin of the multicellular lineages and elaborated upon over evolutionary time by a process of duplication and divergence. Includes *Hox* genes.

top-down control The hypothesis that population size is limited by predators or herbivores (consumers).

topoisomerase An enzyme that prevents the twisting of DNA ahead of the advancing replication fork by cutting the DNA, allowing it to unwind, and rejoining it.

torpor An energy-conserving physiological state, marked by a decrease in metabolic rate, body temperature, and activity, that lasts for a short period (overnight to a few days or weeks). Occurs in some small mammals when the ambient temperature drops significantly. Compare with **hibernation**.

totipotent Capable of dividing and developing to form a complete, mature organism.

toxin A poison produced by a living organism, such as a plant, animal, or microorganism.

trachea (plural: tracheae) (1) In insects, any of the small air-filled tubes that extend throughout the body and function in gas exchange. (2) In terrestrial vertebrates, the airway connecting the larynx to the bronchi. Also called *windpipe*.

tracheid In vascular plants, a long, thin, water-conducting cell that has pits where its lignin-containing secondary cell wall is absent, allowing water movement between adjacent cells. Compare with **vessel element**.

trade-off In evolutionary biology, an inescapable compromise between two traits that cannot be optimized simultaneously. Also called *fitness trade-off*.

trait Any observable characteristic of an individual.

transcription The process that uses a DNA template to produce a complementary RNA.

transcription factor General term for a protein that binds to a DNA regulatory sequence to influence transcription. It includes both regulatory and basal transcription factors.

transcriptional activator A eukaryotic regulatory transcription factor that binds to regulatory DNA sequences in enhancers or promoter-proximal elements to promote the initiation of transcription.

transcriptional control Regulation of gene expression by various mechanisms that change the rate at which genes are transcribed to form messenger RNA. In negative control, binding of a regulatory protein to DNA represses transcription; in positive control, binding of a regulatory protein to DNA promotes transcription.

transcriptome The complete set of genes transcribed in a particular cell.

transduction The conversion of information from one mode to another. For example, the process by which a stimulus outside a cell is converted into a response by the cell.

transfer RNA (tRNA) An RNA molecule that has an anticodon at one end and an amino acid attachment site at the other. Each tRNA carries a specific amino acid and binds to the corresponding codon in messenger RNA during translation.

transformation (1) Incorporation of external DNA into a cell. Occurs naturally in some bacteria; can be induced in the laboratory. (2) Conversion of a normal mammalian cell to one that divides uncontrollably.

transgenic A plant or animal whose genome contains DNA introduced from another individual, often from a different species.

transition state A high-energy intermediate state of the reactants during a chemical reaction that must be achieved for the reaction to proceed. Compare with **activation energy**.

transitional feature A trait that is intermediate between a condition observed in ancestral (older) species and the condition observed in derived (younger) species.

translation The process by which a polypeptide (a string of amino acids joined by peptide bonds) is synthesized from information in codons of messenger RNA.

translational control Regulation of gene expression by various mechanisms that alter the life span of messenger RNA or the efficiency of translation.

translocation (1) In plants, the movement of sugars and other organic nutrients through the phloem by bulk flow. (2) A type of mutation in which a piece of a chromosome moves to a nonhomologous chromosome. (3) The movement of a ribosome down a messenger RNA during translation.

transmembrane protein See **integral membrane protein**.

transmission The passage or transfer of (1) a disease from one individual to another or (2) electrical impulses from one neuron to another.

transpiration Loss of water vapor from aboveground plant parts. Occurs primarily through stomata.

transporter See **carrier protein**.

transposable elements Any of several kinds of DNA sequences that are capable of moving themselves, or copies of themselves, to other locations in the genome. Include LINEs and SINEs.

tree of life The phylogenetic tree that includes all organisms.

trichome A hairlike appendage that grows from epidermal cells in the shoot system of some plants. Trichomes exhibit a variety of shapes, sizes, and functions depending on species.

triiodothyronine (T$_3$) A lipid-soluble hormone, derived from the amino acid tyrosine, containing three iodine atoms and produced and secreted by the thyroid gland. Acts primarily to increase cellular metabolism. In mammals, T$_3$ has a stronger effect than does the related hormone thyroxine (T$_4$).

triose A monosaccharide (simple sugar) containing three carbon atoms.

triplet code A code in which a "word" of three letters encodes one piece of information. The genetic code is a triplet code because a codon is three nucleotides long and encodes one amino acid.

triploblast (adjective: triploblastic) An animal whose body develops from three basic embryonic cell layers or tissues: ectoderm, mesoderm, and endoderm. Compare with **diploblast**.

trisomy The state of having three copies of one particular type of chromosome in an otherwise diploid cell.

tRNA See **transfer RNA**.

trochophore A larva with a ring of cilia around its middle that is found in some lophotrochozoans.

trophic cascade A series of changes in the abundance of species in a food web, usually caused by the addition or removal of a key predator.

trophic level A feeding level in an ecosystem.

trophoblast The exterior of a blastocyst (the structure that results from cleavage in embryonic development of mammals).

tropomyosin A regulatory protein present in thin (actin) filaments that blocks the myosin-binding sites on these filaments in resting muscles, thereby preventing muscle contraction.

troponin A regulatory protein, present in thin (actin) filaments, that can move tropomyosin off the myosin-binding sites on these filaments, thereby triggering muscle contraction. Activated by high intracellular calcium.

true navigation The type of navigation by which an animal can reach a specific point on Earth's surface. Also called *map orientation*.

trypsin A protein-digesting enzyme present in the small intestine that activates several other protein-digesting enzymes.

tube foot One of the many small, mobile, fluid-filled extensions of the water vascular system of echinoderms; the part extending outside the body is called a podium, while the bulb within the body is the ampulla. Used in locomotion, feeding, and respiration.

tuber A modified plant rhizome that functions in storage of carbohydrates.

tuberculosis A disease of the lungs caused by infection with the bacterium *Mycobacterium tuberculosis*.

tumor A mass of cells formed by uncontrolled cell division. Can be benign or malignant.

tumor suppressor A protein (e.g., p53 or Rb) that prevents cell division, such as when the cell has DNA damage. Mutant genes that code for tumor suppressors are associated with cancer.

tundra The treeless biome in polar and alpine regions, characterized by short, slow-growing vegetation, permafrost, and a climate of long, intensely cold winters and very short summers.

turbidity Cloudiness of water caused by sediments and/or microscopic organisms.

turgid Swollen and firm as a result of high internal pressure (e.g., a plant cell containing enough water for the cytoplasm to press against the cell wall). Compare with **flaccid**.

turgor pressure The outward pressure exerted by the fluid contents of a living plant cell against its cell wall.

turnover In lake ecology, the complete mixing of upper and lower layers of water of different temperatures; occurs each spring and fall in temperate-zone lakes.

tympanic membrane The membrane separating the middle ear from the outer ear in terrestrial vertebrates, or similar structures in insects. Also called the *eardrum*.

ubiquinone See **coenzyme Q**.

ulcer A hole in an epithelial layer, exposing the underlying tissues to damage.

ultimate causation In biology, the reason that a trait or phenomenon is thought to have evolved; the adaptive advantage of that trait. Also called *ultimate explanation*. Compare with **proximate causation**.

umami The taste of glutamate, responsible for the "meaty" taste of most proteins and of monosodium glutamate.

umbilical cord The cord that connects a developing mammalian embryo or fetus to the placenta and through which the embryo or fetus receives oxygen and nutrients.

unequal crossover An error in crossing over during meiosis I in which the two non-sister chromatids match up at different sites. Results in gene duplication in one chromatid and gene loss in the other.

unsaturated Referring to lipids in which at least one carbon-carbon bond is a double bond. Double bonds produce kinks in hydrocarbon chains and decrease the compound's melting point. Compare with **saturated**.

upstream In genetics, opposite to the direction in which RNA polymerase moves along a DNA strand. Compare with **downstream**.

urea The major nitrogenous waste of mammals, adult amphibains, and cartilaginous fishes. Compare with **ammonia** and **uric acid**.

ureter In vertebrates, a tube that transports urine from one kidney to the bladder.

urethra The tube that drains urine from the bladder to the outside environment. In male vertebrates, also used for passage of semen during ejaculation.

uric acid A whitish excretory product of birds, reptiles, and terrestrial arthropods. Used to remove from the body excess nitrogen derived from the breakdown of amino acids. Compare with **ammonia** and **urea**.

urochordates One of the three major chordate lineages (Urochordata), comprising sessile or floating, filter-feeding animals that have a polysaccharide covering (tunic) and two siphons through which water enters and exits; include tunicates and salps. Compare with **cephalochordates** and **vertebrates**.

uterus The organ in which developing embryos are housed in mammals and some other viviparous vertebrates.

vaccination Artificially producing immunological memory against a pathogen by using isolated antigens or altered versions of the pathogen to stimulate an adaptive immune response in the absence of disease.

vaccine A preparation designed to stimulate an immune response against a particular pathogen without causing illness. Vaccines consist of inactivated (killed) pathogens, live but weakened (attenuated) pathogens, or parts of a pathogen (subunit vaccine).

vacuole A large organelle in plant and fungal cells that usually is used for bulk storage of water, pigments, oils, or other substances. Some vacuoles contain enzymes and have a digestive function similar to lysosomes in animal cells.

vagina The birth canal of female mammals; a muscular tube that extends from the uterus through the pelvis to the exterior.

valence The number of unpaired electrons in the outermost electron shell of an atom; often determines how many covalent bonds the atom can form.

valence electron An electron in the outermost electron shell, the valence shell, of an atom. Valence electrons tend to be involved in chemical bonding.

valence shell The outermost electron shell of an atom.

valve In circulatory systems, any of the flaps of tissue that prevent backward flow of blood, particularly in veins and in the heart.

van der Waals interactions A weak electrical attraction between two nonpolar molecules that have been brought together through hydrophobic interactions. Often contributes to tertiary and quaternary structures in proteins.

variable number tandem repeat See **minisatellite**.

variable (V) region The amino acid sequence that changes in polypeptides used to make antibodies, B-cell receptors, and T-cell receptors, This portion of the protein is highly variable within an individual and forms the epitope-binding site. Compare with **constant (C) region**.

vas deferens (plural: vasa deferentia) A muscular tube that stores and transports semen from the epididymis to the ejaculatory duct. Also called the *ductus deferens*.

vasa recta In the vertebrate kidney, a network of blood vessels that runs alongside the loop of Henle of a nephron. Functions in reabsorption of water and solutes from the filtrate.

vascular bundle In a plant stem, a cluster of xylem and phloem strands that run the length of the stem.

vascular cambium One of two types of cylindrical meristem, consisting of a ring of undifferentiated plant cells in the stem and root of woody plants; produces secondary xylem (wood) and secondary phloem. Compare with **cork cambium**.

vascular tissue In plants, tissue that transports water, nutrients, and sugars. Made up of the complex tissues xylem and phloem, each of which contains several cell types. Also called *vascular tissue system*.

vector A biting insect or other organism that transfers pathogens from one species to another. See also **cloning vector**.

vegetative development The phase of plant development that involves growth and the the production of all plant structures except the flower.

vein Any blood vessel that carries blood (oxygenated or not) under relatively low pressure from the tissues toward the heart. Compare with **artery**.

veliger A distinctive type of larva, found in mollusks.

vena cava (plural: venae cavae) Either of two large veins that return oxygen-poor blood to the heart.

ventral Toward an animal's belly and away from its back. The opposite of dorsal.

ventricle (1) A thick-walled chamber of the heart that receives blood from an atrium and pumps it to the body or to the lungs. (2) Any of several small fluid-filled chambers in the vertebrate brain.

venule Any of the body's many small veins (blood vessels that return blood to the heart).

vertebrae (singular: vertebra) The cartilaginous or bony elements that form the backbones of vertebrate animals.

vertebrates One of the three major chordate lineages (Vertebrata), comprising animals with a dorsal column of cartilaginous or bony structures (vertebrae) and a skull enclosing the brain. Includes fishes, amphibians, mammals, and reptiles (including birds). Compare with **cephalochordates** and **urochordates**.

vessel element In vascular plants, a short, wide, water-conducting cell that has gaps through both the primary and secondary cell walls, allowing unimpeded passage of water between adjacent cells. Compare with **tracheid**.

vestigial trait A reduced or incompletely developed structure that has no function or reduced function, but is clearly similar to functioning organs or structures in closely related species.

vicariance The physical splitting of a population into smaller, isolated populations by a geographic barrier.

villi (singular: villus) Small, fingerlike projections (1) of the lining of the small intestine or (2) of the fetal portion of the placenta adjacent to maternal arteries. Function to increase the surface area available for absorption of nutrients and gas exchange.

virion The infectious extracellular particle that is produced from a viral infection; used for transmitting the virus between hosts. It consists of a DNA or RNA genome enclosed within a protein shell (capsid) that may be further enveloped in a phospholipid bilayer. Compare with **virus**.

virulence (adjective: virulent) Referring to the ability of pathogens to cause severe disease in susceptible hosts.

virus An obligate, intracellular parasite that is acellular, but uses host-cell biosynthetic machinery to replicate. Compare with **virion**.

visceral mass One of the three main parts of the mollusk body; contains most of the internal organs and external gill.

visible light The range of wavelengths of electromagnetic radiation that humans can see, from about 400 to 700 nanometers.

vitamin Any of various organic micronutrients that usually function as coenzymes.

vitelline envelope A fibrous sheet of glycoproteins that surrounds mature egg cells in many vertebrates. Surrounded by a thick gelatinous matrix (the jelly layer) in some aquatic species. In mammals, called the *zona pellucida*.

viviparous Producing live young (instead of eggs) that develop within the body of the mother before birth. Compare with **oviparous** and **ovoviviparous**.

volt (V) A unit of electrical potential (voltage).

voltage Potential energy created by a separation of electric charges between two points. Also called *electrical potential*.

voltage clamping A technique for imposing a constant membrane potential on a cell. Widely used to investigate ion channels.

voltage-gated channel An ion channel that opens or closes in response to changes in membrane voltage. Compare with **ligand-gated channel**.

voluntary muscle Muscle that contracts in response to stimulation by voluntary (somatic), but not involuntary (parasympathetic or sympathetic), neural stimulation.

vomeronasal organ A paired sensory organ, located in the nasal region, containing chemoreceptors that bind odorants and pheromones.

wall pressure The inward pressure exerted by a cell wall against the fluid contents of a living plant cell.

Wallace line A line in the Indonesian region that demarcates two areas, each of which is characterized by a distinct set of animal species.

water potential (ψ) The potential energy of water in a certain environment compared with the potential energy of pure water at room temperature and atmospheric pressure. In living organisms, ψ equals the solute potential ($ψ_S$) plus the pressure potential ($ψ_P$).

water-potential gradient A difference in water potential in one region compared with that in another region. Determines the direction that water moves, always from regions of higher water potential to regions of lower water potential.

water table The upper limit of the underground layer of soil that is saturated with water.

water vascular system In echinoderms, a system of fluid-filled tubes and chambers that functions as a hydrostatic skeleton.

watershed The area drained by a single stream or river.

Watson–Crick pairing See **complementary base pairing.**

wavelength The distance between two successive crests in any regular wave, such as light waves, sound waves, or waves in water.

wax A class of lipid with extremely long, saturated hydrocarbon tails. Harder and less greasy than fats.

weather The specific short-term atmospheric conditions of temperature, moisture, sunlight, and wind in a certain area.

weathering The gradual wearing down of large rocks by rain, running water, temperature changes, and wind; one of the processes that transform rocks into soil.

weed Any plant that is adapted for growth in disturbed soils.

white blood cell Any of several types of blood cells, including neutrophils, macrophages, and lymphocytes, that circulate in blood and lymph and function in defense against pathogens.

wild type The most common phenotype seen in a wild population.

wildlife corridor Strips of wildlife habitat connecting populations that otherwise would be isolated by human-made development.

wilt To lose turgor pressure in a plant tissue.

wobble hypothesis The hypothesis that some tRNA molecules can pair with more than one mRNA codon by tolerating some non-standard base pairing in the third base, so long as the first and second bases are correctly matched.

wood Xylem resulting from secondary growth; forms strong supporting material. Also called *secondary xylem.*

worm An animal with a long, thin, tubelike body lacking limbs.

Woronin body A dense organelle in certain fungi that plugs pores in damaged septa to prevent leakage of cytoplasm.

X-linked inheritance Inheritance patterns for genes located on the mammalian X chromosome. Also called *X-linkage.*

X-ray crystallography A technique for determining the three-dimensional structure of large molecules, including proteins and nucleic acids, by analysis of the diffraction patterns produced by X-rays beamed at crystals of the molecule.

xenoestrogens Foreign chemicals that bind to estrogen receptors or otherwise induce estrogen-like effects.

xeroderma pigmentosum (XP) A human disease characterized by extreme sensitivity to ultraviolet light. Caused by an autosomal recessive allele that inactivates the nucleotide excision DNA repair system.

xylem A plant vascular tissue that conducts water and ions; contains tracheids and/or vessel elements.

Primary xylem develops from the procambium of apical meristems; secondary xylem, or wood, from the vascular cambium. Compare with **phloem.**

Y-linked inheritance Inheritance patterns for genes located on the mammalian Y chromosome. Also called *Y-linkage.*

yeast Any fungus growing as a single-celled form. Also, a specific lineage of Ascomycota.

yolk The nutrient-rich cytoplasm inside an egg cell; used as food for the growing embryo.

Z disk The structure that forms each end of a sarcomere. Contains a protein that binds tightly to actin, thereby anchoring thin filaments.

Z scheme Model for changes in the potential energy of electrons as they pass from photosystem II to photosystem I and ultimately to NADP$^+$ during the light-dependent reactions of photosynthesis. See also **noncyclic electron flow.**

zero population growth (ZPG) A state of stable population size due to fertility staying at the replacement rate for at least one generation.

zona pellucida The gelatinous layer around a mammalian egg cell. In other vertebrates, called the *vitelline envelope.*

zone of (cellular) division In plant roots, a group of apical meristematic cells just behind the root cap where cells are actively dividing.

zone of (cellular) elongation In plant roots, a group of young cells, derived from primary meristem tissues and located behind the apical meristem, that are increasing in length.

zone of (cellular) maturation In plant roots, a group of plant cells, located several millimeters behind the root cap, that are differentiating into mature tissues.

zygosporangium (plural: zygosporangia) The distinctive spore-producing structure in fungi that are members of the Zygomycota.

zygote The cell formed by the union of two gametes; a fertilized egg.

Credits

Photo Credits

and the art overlays do not appear in original paper.) **22.5** Christiane Nüsslein-Volhard, Max Planck Institute of Developmental Biology, Tubingen, Germany **22.6** Wolfgang Driever, University of Freiburg, Germany **22.7ac** Jim Langeland, Stephen Paddock, and Sean Carroll, University of Wisconsin at Madison **22.7b** Stephen J. Small, New York University **22.8L** Thomas Deerinck, NCMIR/Photo Researchers, Inc. **22.8M** David Scharf/Science Source/Photo Researchers, Inc. **22.8R** Eye of Science/Photo Researchers, Inc. **22.11ab** Reprinted from "The Origin of Form" by Sean B. Carroll, *Natural History*, November 2005. Burke, A. C. 2000. *Hox* Genes and the Global Patterning of the Somitic Mesoderm. In *Somitogenesis*. C. Ordahl (ed.) *Current Topics in Developmental Biology*, Vol. 47. Academic Press.

Chapter 23 Opener Dr. Yorgos Nikas/Photo Researchers, Inc. **23.2a** MIA Tegner/Science Photo Library/Photo Researchers, Inc. **23.3a(1–3)** Michael Whitaker/Science Photo Library/Photo Researchers, Inc. **23.3b(1–3)** Victor D. Vacquier, Scripps Institution of Oceanography, University of California at San Diego **23.6** David Shook, University of Virginia **23.8ab** Kathryn W. Tosney, University of Miami

Chapter 24 Opener Bill Stevenson/AGE Fotostock **24.5** Ed Reschke/Peter Arnold Images/Getty Images **24.7ab** Reprinted by permission from *Nature*, Vol. 424, July 24, 2003, p. 439, by Kim et al. ©2004 Macmillan Publishers Ltd. Photograph supplied by Dr. Neelima Sinah. **24.9a–d** John L. Bowman, University of California at Davis

Chapter 25 Opener R.B. Taylor/Photo Researchers, Inc. **25.2L** Gary Crabbe/AGE Fotostock **25.2RTMB** Michael Quinn, Grand Canyon NPS **25.3a** Gerd Weitbrecht for Landesbildungsserver Baden-Wurttemberg (www.schule-bw.de)/Gerd Weitbrecht **25.3b** angelhell/iStockphoto **25.5L** Januario Dolores **25.5R** BSIP/Photo Researchers, Inc. **25.6a1** AGE fotostock/SuperStock **25.6a2** George D. Lepp/Photo Researchers, Inc. **25.6a3** David Hosking/FLPA **25.6a4** Stefan Huwiler/AGE fotostock **25.8a** Photo by Michael K. Richardson. Reproduced by permission from *Anatomy and Embryology* 305, Fig. 7. Copyright © Springer-Verlag GmbH & Co KG, Heidelberg, Germany. **25.8b** Photo by Professor R. O'Rahilly. Reproduced by permission from *Anatomy and Embryology* 305, Fig.8. Copyright © Springer-Verlag GmbH & Co KG, Heidelberg, Germany. **25.8c** From: Richardson, M. K., et al. *Science*. Vol. 280, p. 983c, Issue #5366, May 15, 1998. Embryo from Professor R. O'Rahilly, National Museum of Health and Medicine/Armed Forces Institute of Pathology. **25.10** Walter J. Gehring **25.12LT** PetStockBoys/Alamy **25.12LB** Arco Images GmbH/Alamy **25.12RT** Arco Images GmbH/Alamy **25.12RB** Juniors Bildarchiv/Alamy **25.15** Alison Wright/ National Geographic Stock **25.15inset** D. & S. Tollerton/AGE fotostock **25.16** Kim Quillin **25.18a** American Association for the Advancement of Science **25.18b** The Royal Society of London **25.T2-1** dibrova/Shutterstock **25.T2-2** James H. Robinson/Photo Researchers, Inc. **25.T2-3** Erlend Haarberg/NHPA/Photoshot **25.T2-4** D. & S. Tollerton/AGE fotostock

Chapter 26 Opener REUTERS/Chaiwat Subprasom **26.8** Cyril Laubscher/Dorling Kindersley **26.9** Otorohanga Kiwi House, New Zealand **26.10** Robert Rattner **26.11a** Steve Mackay Photography/Flickr/Getty Images **26.11b** Phil Savoie/Nature Picture Library **26.13** Andre Karwath **26.16** Mark Conlin/V&W/Image Quest Marine **26.18** Alasdair Thomson/iStockphoto

Chapter 27 Opener Jason Rick/Loren H. Rieseberg, Indiana University **27.10B** Cubo Images/SuperStock **27.10TL** Michael Lustbader/Photo Researchers, Inc. **27.10TR** Stephen Lynn Clancy **27.11TMB** H. Douglas Pratt/National Geographic Image Collection

Chapter 28 Opener Indiana9 fossils and L & M Collectibles **28.5a** GEUS **28.5b** imv/iStockphoto **28.5c** Mike Viney/The Virtual Petrified Wood Museum **28.5d** Mike Viney/The Virtual Petrified Wood Museum **28.7a–f** Ron Blakey, Colorado Plateau Geosystems, Inc. **28.9TL** Dr. Gerald D. Carr, PhD **28.9BL** Forest & Kim Starr **28.9R** Forest & Kim Starr **28.10a** Eladio Fernandez **28.10b** Jonathan B. Losos, Washington University in St. Louis **28.11bBL** Verisimilus, Used under a Creative Commons license. http://creativecommons.org/licenses/by-sa/3.0/ **28.11bBR** James Gehling/South Australian Museum **28.11bTL** Chip Clark, NMNH, Smithsonian Institution **28.11bTR** Ed Reschke/Getty **28.12** Jun-Yuan Chen **28.T4a** National Park Service **28.T4b** Glen A. Izett, USGS-NASA Langley **28.T4c** Peter Schulte/American Association for the Advancement of Science **28.T4d** NASA

Chapter 29 Opener John Hammond **29.2** Doc Searls **29.2inset** University of Maryland Biotechnology Institute **29.3** Rodney M. Donlan and Janice Carr, CDC **29.4** Accent Alaska/Alamy **29.5** Reprinted with permission from S.V. Liu, et al., "Thermophilic Fe(III)-reducing bacteria from the deep subsurface." *Science* 277: 1106–1109 (1997), Fig. 2. ©1997 AAAS. **29.8aT** Tom Deerinck **29.8aB** Reprinted with permission from H.N. Schulz, et al., "Dense populations of a giant sulfur bacterium in Namibian shelf sediments." *Science* 284: 493–495, Fig. 1b, August 4, 1999. ©1999 AAAS. **29.8bTB** SciMAT/Photo Researchers, Inc. **29.8cT** Linda M. Stannard, University of Cape Town/Photo Researchers, Inc. **29.8cB** Dr. Ralf Wagner **29.9a** R. Hubert, Microbiology Program, Dept. of Animal Science, Iowa State University **29.11** Bruce J. Russell/Biomedia Associates **29.15** Andrew Syred/Photo Researchers, Inc. **29.16** David Dalton **29.17** S. Amano, S. Miyadoh and T. Shomura/The Society for Actinomycetes Japan **29.18** Bill Schwartz, Public Health Image Library (PHIL), CDC **29.19** Jordan L. Cocchiaro, Yadunanda Kumar, Elizabeth R.

Fischer, Ted Hackstadt, and Raphael H. Valdivia, *PNAS*. **29.20a** Yves V. Brun **29.20b** Dr. George L. Barron **29.21** Martin Koenneke, MARUM, University of Bremen **29.22** Dr. Kenneth M. Stedman/Portland State University **29.22inset** Eye of Science/Photo Researchers, Inc. **29.23inset** Prof. Dr. Karl O. Stetter and Prof. Dr. Reinhard Rachel, University of Regensburg, Germany **29.23** MARUM, University of Bremen

Chapter 30 Opener TIM LAMAN/National Geographic Stock **30.2L** Scenics & Science/Alamy **30.2M** Mirko Zanni/WaterFrame/AGE fotostock **30.2R** Francois Gohier/Photo Researchers, Inc. **30.4** Pete Atkinson/NHPA/Photoshot **30.4inset** Woods Hole Oceanographic Institution Biology **30.6** Biophoto Associates/Photo Researchers, Inc. **30.11a** Steve Gschmeissner/Photo Researchers, Inc. **30.11b** Biophoto Associates/Photo Researchers **30.11c** Andrew Syred/Photo Researchers, Inc. **30.12a** Biophoto Associates/Photo Researchers, Inc. **30.12b** Photoshot Holdings Ltd./Bruce Coleman **30.15** Eric V. Grave/Photo Researchers, Inc. **30.19** Jeffrey Lepore/Photo Researchers, Inc. **30.20** Eye of Science/Photo Researchers, Inc. **30.21** Tai-Soon Young/Chungbuk National University College of Medicine **30.22** David J. Patterson **30.23** Francis Abbott/Nature Picture Library **30.24** @Colomban de Vargas, SBRoscoff **30.25** David J. Patterson **30.26** Inra **30.27** Vern B. Carruthers, Dept. of Microbiology & Immunology, U. of Michigan Medical School **30.28** Andrew Syred/SPL/Photo Researchers, Inc. **30.29** Linden Gledhill Photography **30.30** Kelvin Aitken/V&W/Image Quest Marine

Chapter 31 Opener RF Company/Alamy **31.1** USDA/NRCS **31.2** Hugh Iltis, The Doebley Lab, University of Wisconsin-Madison **31.3L** Linda Graham, University of Wisconsin-Madison **31.3R** Lee W. Wilcox **31.4a** Lee W. Wilcox **31.4b** sacco/iStockphoto **31.4c** Zinatova Olga/Shutterstock **31.7** Lee W. Wilcox **31.10b** BlueRidgeKitties, used under a creative commons license: http://creativecommons.org/licenses/by-nc-sa/2.0/deed.en **31.10LR** Jon Monroe, James Madison University **31.15a** Lee W. Wilcox **31.15bLR** Lee W. Wilcox **31.17** Patrick J. Lynch/Photo Researchers, Inc. **31.20a** Larry Deack **31.20b** Quest786/Shutterstock **31.20c** John Severns, Severnjc **31.22a** Maxim Khytra/Shutterstock **31.22b** Andreas Nilsson/Shutterstock **31.24T.1&2** Dr. Keith Wheeler/Photo Researchers, Inc. **31.24T.3** jonphoto/iStockphoto **31.24T.4** Noodle Snacks **31.24B.1** Garry DeLong/Photo Researchers, Inc. **31.24B.2** Dr. Keith Wheeler/Photo Researchers, Inc. **31.24B.3** ATELIER Simon/iStockphoto **31.24B.4** Shutterstock **31.26a** Doug Allan/Nature Picture Library **31.26b** Roland Birke/Peter Arnold Images/Getty **31.26c** Lee W. Wilcox **31.27** Andrew J. Martinez/Photo Researchers, Inc. **31.28** Linda Graham **31.29** Center for Aquatic and Invasive Plants **31.30** Adrian Davies/Nature Picture Library **31.31a** Wildlife GmbH/Alamy **31.31b** Photononstop/SuperStock **31.32** Lee W. Wilcox **31.33** Lee W. Wilcox **31.34** Gordon Dickson **31.35a** Greg Vaughn/Alamy **31.35b** Lee W. Wilcox **31.36L** F Hecker/AGE fotostock **31.36R** SPL/Photo Researchers, Inc. **31.37a** Martin Land/Science Photo Library/Photo Researchers, Inc. **31.37b** brigg/iStockphoto **31.38** Maxfocus/iStockphoto **31.39** Peter Chadwick/Dorling Kindersley **31.40ab** Lee W. Wilcox **31.41** Geoff Renner/Robert Harding **31.42a** Brian Johnston **31.42b1&2** Lee W. Wilcox

Chapter 32 Opener Michael P. Gadomski/Photo Researchers, Inc. **32.1a** Inga Spence/Photo Researchers, Inc. **32.1b** EGON/Alamy **32.2a** David Read, Department of Animal and Plant Sciences, University of Sheffield **32.2b** Mycorrhizal Applications, Inc. **32.3** Paul W. Romaine **32.4a** Eye of Science/Photo Researchers, Inc. **32.4b** Emmanuel LATTES/Alamy **32.5aR** Susumu Nishinaga/Photo Researchers, Inc. **32.5bL** Dr. George L. Barron **32.5bR** Biophoto Associates/Photo Researchers, Inc. **32.5c** Dr. Patrick C. Hickey, NIPHT Limited & Prof. Nick D. Read, University of Edinburgh **32.6a** Melvin Fuller/The Mycological Society of America **32.6b** Ed Reschke/ Peter Arnold/Getty **32.6c** Biophoto Associates/Photo Researchers, Inc. **32.6d** Nino Santamaria **32.9ab** Mark Brundrett **32.11** Scott Camazine/Photo Researchers, Inc. **32.14** Courtesy of Dr. J.I. Ronny Larsson, Professor of Zoology, and *The European Journal of Protistology*, "The Continuation of Protistologica." 17, 1981, 407–422. **32.15** Thomas J. Volk **32.16** Gregory G. Dimijian/Photo Researchers, Inc. **32.17** Dr. Jim Deacon **32.18** David Slater/DanitaDelimont.com "Danita Delimont Photography"/Newscom **32.19b** Daniel Mosquin **32.20** N. Allen and G. Barron, University of Guelph, Guelph, Ontario **32.21** Adrian Hepworth/NHPA/Photoshot

Chapter 33 Opener Bill Curtsinger/National Geographic Image Collection **33.4** Walker England/Photo Researchers, Inc. **33.5a** Steve Gschmeissner/Photo Researchers, Inc. **33.5b** Susumu Nishinaga/Photo Researchers, Inc. **33.6** www.deepseawaters.com **33.7** From "Origins of bilateral symmetry: *Hox* and *dpp* expression in a sea anemone." JR Finnerty, et al. *Science*. 2004 May 28; 304(5675): 1335–1337. Epub 2004 May 6. Fig. 2i. **33.T2-1** Perennou Nuridsany/Photo Researchers, Inc. **33.T2-2** Barry Mansell/Nature Picture Library **33.T2-3** Pascal Goetgheluck/Photo Researchers, Inc. **33.T2-4** Joze Maucec/Shutterstock **33.T3-1** Juniors Bildarchiv/Alamy **33.T3-2** Olga Khoroshunova/Fotolia **33.T3-3** Len Rue, Jr./Photo Researchers, Inc. **33.T4-1** Sanamyan/Alamy **33.T4-2** Todd Husman **33.T4-3** Heinz-jürgen Landshoe/AGE Fotostock **33.T4-4** Four Oaks/Shutterstock **33.T5-1** Francesco Tomasinelli/Photo Researchers, Inc. **33.T5-2** Michael Zysman/Shutterstock **33.T5-3** D P Wilson/Frank Lane Picture Agency **33.T5-4** Lawson Wood/Specialist Stock **33.T5-5** Jeff Rotman/Photo Researchers, Inc.

33.13 Reproduced by permission from Grace Panganiban, et al., "The origin and evolution of animal appendages." *PNAS* 94: 5162–5166 (May 1997), p. 5163, Figs. 1c and 1f. ©1997 National Academy of Sciences, USA **33.T6-1** William J. Harrigan/Alamy **33.T6-2** David Doubilet/Getty **33.T6-3** Horia Bogdan/Shutterstock **33.15** Masa Ushioda/Alamy **33.16** Gregory G Dimijian/Photo Researchers, Inc. **33.19** Kelvin Aitken/V&W/Image Quest Marine

Chapter 34 Opener Barbara Strnadova/Photo Researchers, Inc. **34.4a** Peter Parks/Image Quest Marine **34.4b** Carmel McDougall, University of Oxford **34.6a** Robert Yin/Corbis Premium RF/Alamy **34.6b** Alexander Semenov/SPL/Photo Researchers, Inc. **34.6c** Solvin Zankl/Nature Picture Library **34.8** Jane Burton/Nature Picture Library **34.9** Adrian Costea/Fotolia **34.9inset** Josef Ramsauer and Prof. Dr. Robert Patzner, Salzburg University, Austria **34.11** Image Quest Marine **34.12** Antonio Guillen Oterino **34.13a** Doug Anderson **34.13b** Medical-on-Line/Alamy **34.13c** Oliver Meckes/Photo Researchers, Inc. **34.15a** JW Alker/Imagebroker/Alamy **34.15b** Jasper Nance **34.15c** Hans-Joachim Schneider/Alamy **34.16** Image Quest Marine **34.17** Roger Steen/Image Quest Marine **34.18** Kjell B. Sandved/Photo Researchers, Inc. **34.19** Mark Conlin/VWPICS/AGE fotostock **34.20a** Dave King/Dorling Kindersley **34.20b** Michael Mules/Alamy **34.21** Eric Isselée/Shutterstock **34.24a** Nigel Cattlin/Alamy **34.24b** Kim Taylor/Nature Picture Library **34.24inset** Frank Greenaway/Dorling Kindersley **34.25a** Dr. Morley Read/Shutterstock **34.25b** Steve Gschmeissner/Photo Researchers, Inc. **34.26a** C. Hough/Custom Medical Stock Photo, Inc. **34.26b** Sinclair Stammers/Photo Researchers, Inc. **34.27** Mark Smith/Photo Researchers, Inc. **34.T1-1** Meul/ARCO/Nature Picture Library **34.T1-2** OGphoto/iStockphoto **34.T1-3** Arlindo71/iStockphoto **34.T1-4** Trounes **34.T1-5** Yxowert/Shutterstock **34.T1-6** Marc Dubus/Fotolia **34.T1-7** Sergey Toronto/Fotolia **34.29** Formiktopus/Shutterstock **34.31a** Holger Mette/Fotolia **34.31b** Peter Waters/Shutterstock

Chapter 35 Opener Fritz Wilhelm **35.2a** D P Wilson/FLPA **35.2b** Juri Vinokurov/Fotolia **35.3b** Jeff Rotman/Nature Picture Library **35.4a** Marevision/AGE fotostock **35.4b** Gary Bell/Oceanwide Images **35.6** Alan James/Nature Picture Library **35.7a** Audra Reese Kirkland **35.7b** Glow Images **35.9** Heather Angel/Natural Visions/Alamy **35.10a** Borut Furlan/WaterFrame/AGE fotostock **35.10b** Laurence Madin/Woods Hole Oceanographic Institution **35.15** Rudie Kuiter/Oceanwide Images **35.19** Radius Images/Alamy **35.23a** Tom McHugh/Photo Researchers, Inc. **35.23b** James L Amos/National Geographic Image Collection **35.23binset** A Hartl/Blickwinkel/AGE fotostock **35.24a** Fred McConnaughey/Photo Researchers **35.24b** Image Quest Marine **35.25** Oceans Image/Photoshot **35.26** Arnaz Mehta **35.27** Jane Burton/Nature Picture Library **35.28a** Tom McHugh/Photo Researchers, Inc. **35.28b** Nigel Pavitt/John Warburton-Lee Photography/Alamy **35.29** Frank Lukasseck/Photographer's Choice/Getty **35.30** Thomas Dressler/AGE fotostock **35.31** George Grall/National Geographic Image Collection **35.32** Life on White/iStockphoto **35.33** Larry Lynch/iStockphoto **35.34** Alamy Creativity/Alamy **35.35a** Holger Ehlers/Alamy **35.35b** J & C Sohns/AGE fotostock

Chapter 36 Opener Jed Fuhrman, University of Southern California **36.2** Science Source/Photo Researchers, Inc. **36.4** David M. Phillips/Photo Researchers, Inc. **36.5a** Science Source/Photo Researchers, Inc. **36.5b** Biophoto Associates/Photo Researchers, Inc. **36.5c** Oliver Meckes/E.O.S./Max-Planck-Institut-Tubingen/Photo Researchers, Inc. **36.5d** Eye of Science/Photo Researchers, Inc. **36.12a** Eye of Science/Photo Researchers, Inc. **36.12b** Biozentrum, University of Basel/Photo Researchers, Inc. **36.17** Jean Roy, Centers for Disease Control and Prevention (CDC) **36.18** Infectious Diseases of Livestock. J.A.W. Coetzer and R.C. Tustin, University of Pretoria **36.19** David Parker/SPL/Photo Researchers, Inc. **36.20** Lowell Georgia/Photo Researchers, Inc. **36.21** David Gray/REUTERS

Chapter 37 Opener Bruno Morandi/Hemis/Alamy **37.T1-1** photofun/Shutterstock **37.T1-2** Matt Meadows/Science Photo Library/Alamy **37.T1-3** rukhsanaphotography/Fotolia **37.T2-1** Geoff Dan/Dorling Kindersley **37.T2-2** Dorling Kindersley **37.T2-3** Lee W. Wilcox **37.T2-4** Nigel Cattlin/Photo Researchers, Inc. **37.T2-5** Pete Oxford/Nature Picture Library **37.6ab** Lee W. Wilcox **37.6cd** David T. Webb, Botany, University of Hawaii **37.7a–c** Lee W. Wilcox **37.7d** Nigel Cattlin/Alamy **37.8** Eric Schrader/Pearson Education **37.T3-1** eAlisa/Shutterstock **37.T3-2** Ivan Kmit/Alamy Images **37.T3-3** Colin Carson/Getty Images **37.T3-4** Filip Fuxa/Shutterstock **37.T3-5** Torsten Brehm/Nature Picture Library **37.9a** Ed Reschke/Peter Arnold/Getty **37.9b** M. I. Walker/Photo Researchers, Inc. **37.12a** Dr. Keith Wheeler/Photo Researchers, Inc. **37.12b** Garry DeLong/Photo Researchers, Inc. **37.14ab** Ed Reschke/Peter Arnold/Getty **37.15** Andrew Syred/Science Photo Library/Photo Researchers, Inc. **37.16a** John Durham/Photo Researchers, Inc. **37.16b** Lee W. Wilcox **37.17LR** Lee W. Wilcox **37.18a–c** Lee W. Wilcox **37.19a** Paul Schulte **37.19b** Garry DeLong/Photo Researchers, Inc. **37.20a** Power and Syred/Photo Researchers, Inc. **37.20b** Mike Clayton **37.20c** Andrew Syred/Photo Researchers, Inc. **37.21L** Jim Bidlack **37.21R** Lee W. Wilcox **37.22ab** Garry DeLong/Photo Researchers, Inc. **37.23** Pascal Goetgheluck/Photo Researchers, Inc. **37.24a** Lee W. Wilcox **37.24b** John Durham/SPL/Photo Researchers, Inc. **37.24c** Adam Hart-Davis/Science Photo Library/Photo Researchers, Inc.

Chapter 38 Opener imagebroker/Alamy **38.4** David Cook/blueshiftstudios/Alamy **38.6LR** Lee W. Wilcox **38.8** Bruce Peters **38.10TB** SPL/Photo Researchers, Inc. **38.13** Lee W. Wilcox **38.18** Martin H. Zimmermann, Harvard Forest, Harvard University

Chapter 39 Opener Angelo Cavalli/Getty Images **39.2a–c** Nigel Cattlin/Photo Researchers, Inc. **39.3** Emanuel Epstein, University of California at Davis **39.6a** Elaine Moczygemba, no 87-286, The UT Institute of Texan Cultures **39.6b** Washington State Dept of Transportation **39.9** Ed Reschke/Peter Arnold/Getty Images **39.13** Photograph courtesy of Dr. Eduardo Blumwald. Reprinted with permission from Apse et al., *Science* 285: 1256–1258, Fig. 3 (1999). ©2004 AAAS. **39.14bL** Andrew Syred/SPL/Photo Researchers, Inc. **39.14bR** E.H. Newcomb & S.R. Tandon, *Science* 212:1394 (1981)/Biological Photo Service, Terraphotographics **39.14T** Hugh Spencer/Photo Researchers, Inc. **39.16a** Ed Reschke/Peter Arnold/Getty Images **39.16b** Florapix/Alamy **39.17L** Frank Greenaway/Dorling Kindersley **39.17R** Carol A. Wilson and Clyde L. Calvin, Dorling Kindersley **39.18** Noah Elhardt

Chapter 40 Opener Lee W. Wilcox **40.3ab** Malcolm B. Wilkins, University of Glasgow, Glasgow, Scotland, U.K. **40.8TB** Malcolm B. Wilkins, University of Glasgow, Glasgow, Scotland, U.K. **40.11** From A. K. Sarkar, et al., 2007. "Conserved factors regulate signalling in *Arabidopsis thaliana* shoot and root stem cell organizers." *Nature* 446: 811–814. ©2007, Rights Managed by Nature Publishing Group. **40.14** E. Wassim Chehab and Janet Braam **40.15** Rafael Campillo/AGE fotostock **40.16L** I Weber/Blickwinkel/AGE fotostock **40.16R** blickwinkel/Alamy **40.17ab** Lee W. Wilcox **40.19** Malcolm B. Wilkins, University of Glasgow, Glasgow, Scotland, U.K. **40.23** Professor Adel A. Kader, University of California at Davis **40.25** Jon Monroe **40.28** Nigel Cattlin/Holt Studios International/Photo Researchers, Inc.

Chapter 41 Opener Walter Siegmund **41.3a** Dan Suzio/Photo Researchers, Inc. **41.3bc** Jerome Wexler/Photo Researchers, Inc. **41.4bT** Bernie Kasper **41.4bM** Rod Planck/Photo Researchers, Inc. **41.4bB** Rafael Ramirez Lee/Shutterstock **41.5ab** Leonard Lessin/Photo Researchers, Inc. **41.6a** Justin Voight/iStockphoto.com **41.6bTB** Lee W. Wilcox **41.9L** Frans Lanting/FLPA **41.9R** Nick Garbutt/Nature Picture Library **41.14aL** Brian Johnston **41.14aR** Pablo Galán Cela/AGE Fotostock **41.14bL** Rafael Campillo/AGE fotostock **41.14bR** Laurie Campbell/NHPA/Photoshot **41.14cL** Nicholas and Sherry Lu Aldridg/FLPA **41.14cR** James Hardy/Getty Images **41.15** Lee W. Wilcox

Chapter 42 Opener Gaston Piccinetti/AGE Fotostock **42.1ab** Dave Funk **42.3** Reproduced by permission from K. Petren, et al. "A phylogeny of Darwin's finches based on microsatellite DNA length variation." *Proceedings of the Royal Society of London*, Series B, 266: 321–329 (1999), p. 327, Fig. 3. ©1999 Royal Society of London. **42.T1-1** Educational Images/Custom Medical Stock Photo, Inc. **42.T1-2** Peter Arnold/Alamy **42.T1-3&4** Nina Zanetti/Pearson Education **42.T1-5** Ed Reschke/Getty Images **42.4** Deco Images II/Alamy Images **42.5ab** Manfred Kage/Photo Reserachers, Inc. **42.5c** Biophoto Associates/Photo Researchers, Inc. **42.6** Ed Reschke/Peter Arnold/Getty Images **42.8** Otmar Winterleitner/Dreamstime **42.12a** G. Kruitwagen, H.P.M. Geurts and E.S. Pierson/Radboud University Nijmegen (vcbio.science.ru.nl/en) **42.12b** Innerspace Imaging/SPL/Photo Researchers, Inc. **42.12c** P.M. Motta, A. Caggiati, G. Macchiarelli/Science Photo Library **42.14** Derrick Hamrick/imagebroker.net/AGE fotostock **42.15bLR** Masato Ono, Tamagawa University **42.15a** Scott Camazine/Photo Researchers, Inc. **42.17** From "Thermoregulation in the Mouths of Feeding Gray Whales." J. E. Heyning, J. G. Mead. *Science* 7 November 1997. Vol. 278, No. 5340, pp. 1138–1140, Figure 1B. Image provided by the Natural History Museum of Los Angeles, County, From specimen: LACM 92044.

Chapter 43 Opener Steve Winter/National Geographic/Getty Images **43.13** Image taken by H. Wartenberg © H. Jastrow from Dr. Jastrow's Electron Microscopic Atlas (http://www.drjastrow.de)

Chapter 44 Opener Gerard Lacz/AGE fotostock **44.2a** Wallace63, Used under a Creative Commons license, http://creativecommons.org/licenses/by-sa/3.0/deed.en **44.2b** Dasypeltis Scabra/Nature Picture Library **44.3** Michael K. Oliver, malawicichlids.com **44.4LMR** Reproduced by permission from K.F. Liem, "Evolutionary strategies and morphological innovations: cichlid pharyngeal jaws." *Systematic Zoology* 22: 425–441 (1974), p. 436, Figs. 15A and 15B. ©1974 by the Society of Systematic Biologists. Image courtesy of Karel F. Liem, Harvard University. **44.12** Biophoto Associates/Photo Researchers, Inc.

Chapter 45 Opener Tony Freeman/PhotoEdit **45.4a** imagebroker.net/SuperStock **45.4b** David Quillin **45.8T** Jake Socha **45.8BLR** Jake Socha **45.20a** Lennart Nilsson/Scanpix **45.20b** Ed Reschke/Peter Arnold Images/Getty Images

Chapter 46 Opener Tom Barrick, Chris Clark, SGHMS/Photo Researchers, Inc. **46.9** Don W. Fawcett/Photo Researchers, Inc. **46.11** Thomas Deerinck, NCMIR/Photo Researchers, Inc. **46.20** The National Library of Medicine **46.22.1** Allstar Picture Library/Alamy **46.22.2** Kathy Hutchins/Newscom **46.22.3** Péter Gudella/Shutterstock **46.22.4** MARKA/Alamy **46.22.5** WDG Photo/Shutterstock

Illustration and Text Credits

The three-dimensional structure of the first EGF-like module of human factor IX: Comparison with EGF and TGF-alpha. *Protein Science* 1: 81–90. **3.12a** PDB ID: 1D1L. P. B. Rupert, A. K. Mollah, M. C. Mossing, et al. 2000. The structural basis for enhanced stability and reduced DNA binding seen in engineered second-generation Cro monomers and dimers. *Journal of Molecular Biology* 296: 1079–1090. **3.12b** PDB ID: 2DN2. S.Y. Park, T. Yokoyama, N. Shibayama, et al. 2006. 1.25 Å resolution crystal structures of human hemoglobin in the oxy, deoxy and carbonmonoxy forms. *Journal of Molecular Biology* 360: 690–701. **3.14a** PDB ID: 1DMO. M. Zhang, T. Tanaka, and M. Ikura. 1995. Calcium-induced conformational transition revealed by the solution structure of apo calmodulin. *Nature Structural Biology* 2: 758–767. **3.14b** PDB ID: 3CLN. Y. S. Babu, C. E. Bugg, and W. J. Cook. 1988. Structure of calmodulin refined at 2.2 A resolution. *Journal of Molecular Biology* 204: 191–204. **3.16** PDB ID: 2PTC. M. Marquart, J. Walter, J. Deisenhofer, et al. 1983. The geometry of the reactive site and of the peptide groups in trypsin, trypsinogen and its complexes with inhibitors. *Acta Crystallographica* Section B39: 480–490.

Chapter 4 **T4-1** PDB ID: 1EHZ. H. Shi and P. B. Moore. 2000. The crystal structure of yeast phenylalanine tRNA at 1.93 Å resolution: A classic structure revisited. *RNA* 6: 1091–1105. **4.11** PDB ID: 1X8W. F. Guo, A. R. Gooding, and T. R. Cech. 2004. Structure of the *Tetrahymena* ribozyme: Base triple sandwich and metal ion at the active site. *Molecular Cell* 16: 351–362.

Chapter 5 **5.6** H. M. Florman, K. B. Bechtol, and P. M. Wassarman. 1984. Enzymatic dissection of the functions of the mouse egg's receptor for sperm. *Developmental Biology* 106: 243–255. Also H. M. Florman and P. M. Wassarman. 1985. O-linked oligosaccharides of mouse egg ZP3 account for its sperm receptor activity. *Cell* 41: 313–324; J. D. Bleil and P. M. Wassarman. 1988. Galactose at the nonreducing terminus of O-linked oligosaccharides of mouse egg zona pellucida glycoprotein ZP3 is essential for the glycoprotein's sperm receptor activity. *PNAS* 85: 6778–6782.

Chapter 6 **Opener** CHARMM-GUI Archive—Library of Pure Lipid Bilayer (www.charmm-gui.org/?doc=archive&lib=lipid_pure), POPE Bilayer Library (pope_n256.pdb). Reference: S. Jo, T. Kim, and W. Im. 2007. Automated builder and database of protein/membrane complexes for molecular dynamics simulations. *PLoS ONE* 2 (9): e880. **6.10** Data: J. de Gier, J. G. Mandersloot, and L. L. van Deenen. 1968. Lipid composition and permeability of liposomes. *Biochimica et Biophysica Acta* 150: 666–675. **6.21** Data: C. E. Bear, F. Duguay, A. L. Naismith, et al. 1992. Purification and functional reconstitution of the cystic fibrosis transmembrane conductance regulator (CFTR). *Cell* 68: 809–818. **6.22** PDB ID: 2ZZ9. K. Tani, T. Mitsuma, Y. Hiroaki, et al. 2009. Mechanism of aquaporin-4's fast and highly selective water conduction and proton exclusion. *Journal of Molecular Biology* 389: 694–706. **6.23** PDB ID: 1K4C. Y. Zhou, J. H. Morais-Cabral, A. Kaufman, et al. 2001. Chemistry of ion coordination and hydration revealed by a K^+ channel–Fab complex at 2.0 Å resolution. *Nature* 414: 43–48. **6.23** PDB ID: 3FB7. L. G. Cuello, V. Jogini, D. M. Cortes, et al. Open KcsA potassium channel in the presence of Rb^+ ion. (To be published.) **7.18** A. D. Mills, R. A. Laskey, P. Black, et al. 1980. An acidic protein which assembles nucleosomes in vitro is the most abundant protein in *Xenopus* oocyte nuclei. *Journal of Molecular Biology* 139: 561–568. Also C. Dingwall, S. V. Sharnick, and R. A. Laskey. 1982. A polypeptide domain that specifies migration of nucleoplasmin into the nucleus. *Cell* 30: 449–458.

Chapter 7 **7.5** Reprinted with kind permission from Springer Science+Business Media B.V. from David S. Goodsell, *The Machinery of Life*. 2nd ed., 2009.

Chapter 8 **8.8** PDB ID: 1Q18. V. V. Lunin, Y. Li, J. D. Schrag, et al. 2004. Crystal structures of *Escherichia coli* ATP-dependent glucokinase and its complex with glucose. *Journal of Bacteriology* 186: 6915–6927. **8.8** PDB ID: 2Q2R. A. T. Cordeiro, A. J. Caceres, D. Vertommen, et al. 2007. The crystal structure of *Trypanosoma cruzi* glucokinase reveals features determining oligomerization and anomer specificity of hexose-phosphorylating enzymes. *Journal of Molecular Biology* 372: 1215–1226. **8.14a&b** Data: N. N. Nawani, B. P. Kapadnis, A. D. Das, et al. 2002. Purification and characterization of a thermophilic and acidophilic chitinase from Microbispora sp. V2. *Journal of Applied Microbiology* 93: 865–975, Figs. 7, 8a. Also N. N. Nawani and B. P. Kapadnis. 2001. One-step purification of chitinase from *Serratia marcescens* NK1, a soil isolate. *Journal of Applied Microbiology* 90: 803–808, Figs. 3, 4. **8.15** PDB ID: 2ERK. B. J. Canagarajah, A. Khokhlatchev, M. H. Cobb, et al. 1997. Activation mechanism of the MAP kinase ERK2 by dual phosphorylation. *Cell* (Cambridge, MA) 90: 859–869. **8.15** PDB ID 3ERK. Z. Wang, B. J. Canagarajah, J. C. Boehm, et al. 1998. Structural basis of inhibitor selectivity in MAP kinases. *Structure* 6: 1117–1128.

Chapter 9 **9.7** PDB ID: 4PFK. P. R. Evans and P. J. Hudson. 1981. Phosphofructokinase: Structure and control. *Philosophical Transactions R. Soc. Lond. B: Biol. Sci.* 293: 53–62. **9.13** Data: X. Li, R. K. Dash, R. K. Pradhan, et al. 2010. A database of thermodynamic quantities for the reactions of glycolysis and the tricarboxylic acid cycle. *Journal of Physical Chemistry B* 114: 16068–16082, Table 4. **9.14** Data: D.F . Wilson, M. Erecinska, and P. L. Dutton. 1974. Thermodynamic relationships in mitochondrial oxidative phosphorylation. *Annual Review of Biophysics and Bioengineering* 3: 203–230, Tables 1, 3. Also V. D. Sled, N. I. Rudnitzky, Y. Hatefi, et al. 1994. Thermodynamic analysis of flavin in mitochondrial NADH: Ubiquinone oxidoreductase (complex I). *Biochemistry* 33: 10069–10075.

9.17 E. Racker and W. Stoeckenius. 1974. Reconstitution of purple membrane vesicles catalyzing light-driven proton uptake and adenosine triphosphate formation. *Journal of Biological Chemistry* 249: 662–663.

Chapter 10 **10.6** T. W. Engelmann. 1882. Oxygen excretion from plant cells in a microspectrum. *Botanische Zeitung* 40: 419–426. **10.7** Reprinted with kind permission from Springer Science+Business Media B.V. from The Photosynthetic Process. In *Concepts in Photobiology: Photosynthesis and Photomorphogenesis*, edited by G. S. Singhal, G. Renger, S. K. Sopory et al. Dordrecht: Kluwer Academic; co-published with Narosa Publishing House (New Delhi), pp. 11–51, Fig. 5. **10.11** R. Govindjee, Govindjee, and G. Hoch. 1964. Emerson enhancement effect in chloroplast reactions. *Plant Physiology* 39: 10–14. **10.18** A. A. Benson, J. A. Bassham, M. Calvin, et al. 1950. The path of carbon in photosynthesis. V. Paper chromatography and radioautography of the products. *Journal of the American Chemistry Society* 72: 1710–1718. **10.20a** PDB ID: 1RCX. T. C. Taylor and I. Andersson. 1997. The structure of the complex between rubisco and its natural substrate ribulose 1,5-bisphosphate. *Journal of Molecular Biology* 265: 432–444.

Chapter 11 **11.4** Reprinted with kind permission from Springer Science+Business Media B.V. from David S. Goodsell, *The Machinery of Life*. 2nd ed., 2009. **11.7b** B. Alberts, A. Johnson, J. Lewis, et al. 2002. *Molecular Biology of the Cell*. 4th ed., Fig. 19.5, p. 1069. **11.9** K. Hatta and M. Takeichi. 1986. Expression of N-cadherin adhesion molecules associated with early morphogenetic events in chick development. *Nature* 320: 447–449. Also M. Takeichi. 1988. The cadherins: Cell–cell adhesion molecules controlling animal morphogenesis. *Development* 102: 639–655.

Chapter 12 **12.6** G. J. Gorbsky, P. J. Sammack, and G. G. Borisey. 1987. Chromosomes move poleward during anaphase along stationary microtubules that coordinately dissemble from their kinetochore ends. *Journal of Cell Biology* 104: 9–18. **12.9** Based on J. L. Ptacin, S. F. Lee, E. C. Garner, et al. 2010. A spindle-like apparatus guides bacterial chromosome segregation. *Nature Cell Biology* 12: 791–798, Fig. 5. **12.10** Y. Masui and C. L. Markert. 1971. Cytoplasmic control of nuclear behavior during meiotic maturation of frog oocytes. *Journal of Experimental Zoology* 177: 129–145. **12.13** Data: the website of the National Cancer Institute (www.cancer.gov), Common Cancer Types, November 2010.

Chapter 13 **13.13** Data: National Down Syndrome Society, 2012. **13.15** L. T. Morran, O. G. Schmidt, I. A. Gelarden, et al. 2011. Running with the red queen: Host-parasite coevolution selects for biparental sex. *Science* 333: 216–218.

Chapter 14 **14.3** G. Mendel. 1866. Versuche über Pflanzen-hybriden. *Verhandlungen des naturforschenden Vereines in Brünn* 4: 3–47. English translation available from Electronic Scholarly Publishing (www.esp.org). **T14.2** Data: G. Mendel. 1866. Versuche über Pflanzen-hybriden. *Verhandlungen des naturforschenden Vereines in Brünn* 4: 3–47. **14.13** Data: T. H. Morgan. 1911. An attempt to analyze the constitution of the chromosomes on the basis of sex-limited inheritance in *Drosophila*. *Journal of Experimental Zoology* 11: 365–414. **14.20** Data: Pan American Health Organization/WHO. 2004. *Epidemiological Bulletin* 25: 9–13, Graph 1.

Chapter 15 **15.2** A. D. Hershey and M. Chase. 1952. Independent functions of viral protein and nucleic acid in growth of bacteriophage. *Journal of General Physiology* 36: 39–56. **15.5** Adapted by permission of Dr. Matthew Meselson after M. Meselson and F. W. Stahl. 1958. The replication of DNA in *Escherichia coli*. *PNAS* 44: 671–682, Fig. 6. **15.14** Data: R. C. Allsopp, H. Vaziri, C. Patterson, et al. 1992. Telomere length predicts replicative capacity of human fibroblasts. *PNAS* 89: 10114–10118. **15.18a** Data: J. E. Cleaver. 1970. DNA repair and radiation sensitivity in human (xeroderma pigmentosum) cells. *International Journal of Radiation Biology* 18: 557–565, Fig. 3. **15.18b** Data: J .E. Cleaver. 1972. Xeroderma pigmentosum: Variants with normal DNA repair and normal sensitivity to ultraviolet light. *Journal of Investigative Dermatology* 58: 124–128, Fig. 1. **15.UN2** Graph adapted by permission of the Radiation Research Society from P. Howard-Flanders and R. P. Boyce. 1966. DNA repair and genetic recombination: Studies on mutants of *Escherichia coli* defective in these processes. *Radiation Research Supplement* 6: 156–184, Fig. 8.

Chapter 16 **16.2** Data: A. M. Srb and N. H. Horowitz. 1944. The ornithine cycle in *Neurospora* and its genetic control. *Journal of Biological Chemistry* 154: 129–139.

Chapter 17 **17.2a** PDB ID: 3IYD. B. P. Hudson, J. Quispe, S. Lara-Gonzalez, et al. 2009. Three-dimensional EM structure of an intact activator-dependent transcription initiation complex. *PNAS* 106: 19830–19835. **17.11** Based on M. B. Hoagland, M. L. Stephenson, J. F. Scott, et al. 1958. A soluble ribonucleic acid intermediate in protein synthesis. *Journal of Biological Chemistry* 231: 241–257, Fig. 6. **17.13** PDB ID: 1ZJW: I. Gruic-Sovulj, N. Uter, T. Bullock, et al. 2005. tRNA-dependent aminoacyl-adenylate hydrolysis by a nonediting class I aminoacyl-tRNA synthetase. *Journal of Biological Chemistry* 280: 23978–23986. **17.14b** PDB IDs: 3FIK, 3FIH. E. Villa, J. Sengupta, L. G. Trabuco, et al. 2009. Ribosome-induced changes in elongation factor Tu conformation control GTP hydrolysis. *PNAS* 106: 1063–1068.

Chapter 18 **18.2** Data: A. B. Pardee, F. Jacob, and J. Monod. 1959. The genetic control and cytoplasmic expression of "inducibility" in the synthesis of ß-glactosidase by *E. coli*. *Journal of Molecular Biology* 1: 165–178.

Chapter 19 Opener PDB ID: 1ZBB. T. Schalch, S. Duda, D. F. Sargent, et al. 2005. X-ray structure of a tetranucleosome and its implications for the chromatin fibre. *Nature* 436: 138–141. **19.5** I. Sandovici, N. H. Smith, and M. D. Nitert. 2011. Maternal diet and aging alter the epigenetic control of a promoter–enhancer interaction at the *Hnf4a* gene in rat pancreatic islets. *PNAS* 108: 5449–5454. **19.7b** PDB ID: 1MDY. P. C. Ma, M. A. Rould, H. Weintraub, et al. 1994. Crystal structure of MyoD bHLH domain–DNA complex: Perspectives on DNA recognition and implications for transcriptional activation. *Cell* (Cambridge, MA) 77: 451–459.

Chapter 21 21.1 Data: European Nucleotide Archive/EMBL–Bank Release Notes. Release 110, December 2011 (www.ebi.ac.uk/embl/). **21.4a&b** Data: (a) Y. Hou and S. Lin. 2009. *PLoS ONE* 4 (9): e6978, Supplemental Table S1. (b) KEGG: *Kyoto Encyclopedia of Genes and Genomes*, KEGG Organisms: Complete genomes (www. genome.jp/kegg/). **21.8** Reproduced by permission of GENDIA from www.paternity. be/information_EN.html#identitytest, Examples 1 and 2. **21.10** Based on G. T. Ryan. 2005. Synergy between sequence and size in large-scale genomics. *Nature Reviews Genetics* 6, Box 3, p. 702. **21.13** V. Pancaldi, O. S. Sarac, C. Rallis, et al. 2012. Predicting the fission yeast protein interaction network. *G3: Genes, Genomes, Genetics* 2 (4): 453–567, Fig. 5.

Chapter 22 22.10a&b Reproduced by permission from Macmillan Publishers Ltd after J. C. Pearson, D. Lemons, and W. McGinnis. 2005. Modulating *Hox* gene functions during animal body patterning. *Nature Reviews Genetics* 6: 893–904, Fig. 1.

Chapter 23 23.12 H. Weintraub, S. J. Tapscott, R. L. Davis, et al. 1989. Activation of muscle-specific genes in pigment, nerve, fat, liver, and fibroblast cell lines by forced expression of MyoD. *PNAS* 86: 5434–5438.

Chapter 25 25.4 Based on E. B. Daeschler et al. 2006. A Devonian tetrapod-like fish and the evolution of the tetrapod body plan. *Nature* 440: 757–763, Fig. 6; P. E. Ahlberg and J. A. Clack. 2006. A firm step from water to land. *Nature* 440: 747–749, Fig. 1; N. H. Shubin et al. 2006. The pectoral fin of *Tiktaalik roseae* and the origin of the tetrapod limb. *Nature* 440: 764–771, Fig. 4; M. Hildebrand and G. Goslow. 2001. *Analysis of Vertebrate Structures*, 5th ed. John Wiley and Sons, Inc. **25.11** *Indohyus* reproduced by permission from Macmillan Publishers Ltd after J. G. M. Thewissen, L. N. Cooper, M. T. Clementz, et al. 2007. Whales originated from aquatic artiodactyls in the Eocene epoch of India. *Nature* 450: 1190–1194, Fig. 5. *Rhodhocetus* reproduced by permission of AAAS after P. D. Gingerich, M. ul Haq, I. S. Zalmout, et al. 2001. Origin of whales from early Artiodactyls: Hands and feet of Eocene Protocetidae from Pakistan. *Science* 293: 2239–2242, Fig. 3. (http://www.sciencemag .org/content/293/5538/2239.abstract). *Dorudon* reproduced by permission after P. D. Gingerich, M. ul Haq, W. Von Koenigswald, et al. 2009. New protocetid whale from the middle Eocene of Pakistan: Birth on land, precocial development, and sexual dimorphism. *PLoS ONE* 4 (2): e4366, Fig. 1B. New data for Dorudon from M. D. Uhen. 2004. Form, function, and anatomy of *Dorudon atrox* (Mammalia, Cetacea): An Archaeocete from the Middle to Late Eocene of Egypt. University of Michigan Papesrs on Paleontology No. 34. *Delphinapterus* reproduced by permission of Skulls Unlimited International, Inc. (www.SkullsUnlimited.com). **25.14** Adapted by permission from D. P. Genereux and C. T. Bergstrom. 2005. Evolution in action: Understanding antibiotic resistance. In *Evolutionary Science and Society: Educating a New Generation*, edited by J. Cracraft and R. W. Bybee. Colorado Springs, CO: BSCS, 145–153, Fig 3. Data: Centers for Disease Control, 2004. **25.16** Data: P. T. Boag and P. R. Grant. 1981. Intense natural selection in a population of Darwin's finches (*Geospizinae*) in the Galápagos. *Science New Series* 214 (4516): 82–85, Table 1. **25.17** Body size and beak shape graph reproduced by permission of AAAS from P. R. Grant and B. R. Grant. 2002. *Science* 296:707–711, Fig. 1. Beak size graph reprinted by permission of AAAS from P. R. Grant and B. R. Grant. 2006. Evolution of character displacement in Darwin's finches. *Science* 313: 224–226, Fig. 2. (http:// www.sciencemag.org/content/296/5568/707.abstract; http://www.sciencemag.org/ content/313/5784/224.abstract).

Chapter 26 T26.1 Data: W. C. Boyd. 1950. *Genetics and the Races of Man*. Boston: Little, Brown. Data are from Chapter 8, Blood Groups: Table 27, p. 238. **T26.2** Data: T. Markow, P. H. Hedrick, K. Zuerlein, et al. 1993. HLA polymorphism in the Havasupai: Evidence for balancing selection. *American Journal of Human Genetics* 53: 943–952, Table 3. **26.4** Data: W. E. Johnson, E. P. Onorato, M. E. Roelke, et al. 2010. Genetic restoration of the Florida panther. *Science* 329: 1641–1645. **26.5b** Data: C. R. Brown and M. B. Brown. 1998. Intense natural selection on body size and wing and tail asymmetry in cliff swallows during severe weather. *Evolution* 52: 1461–1475. **26.6b** Data: M. N. Karn, H. Lang-Brown, H. J. MacKenzie, et al. 1951. Birth weight, gestation time and survival in sibs. *Annals of Eugenics* 15: 306–322. **26.7b** Data: T. B. Smith. 1987. Bill size polymorphism and intraspecific niche utilization in an African finch. *Nature* 329: 717–719. **26.8b** Data: J. D. Blount, N. B. Metcalf, T. R. Birkhead, et al. 2003. Carotenoid modulation of immune function and sexual attractiveness in zebra finches. *Science* 300: 125–127, Fig. 1b. **26.10 b&c** Data: B. J. Le Boeuf and R. S. Peterson. 1969. Social status and mating activity in elephant seals. *Science* 163: 91–93. **26.12** Reproduced by permission of Pearson Education, Inc., from S. Freeman and J. Herron. 2004. *Evolutionary Analysis*. 3rd ed., Figs. 6.15a, 6.15c. ©2004. **26.13** W. E. Kerr and S. Wright. 1954. Experimental studies of the distribution of gene frequencies in very small populations of *Drosophilia melanogaster*: I. Forked. *Evolution* 8: 172–177. **26.16b** Data: H. Araki, B. Cooper, and M. S. Blouin. 2009. Carry-over effect of captive breeding reduces reproductive fitness of wild-born descendants in the wild. *Biology Letters* 5 (5): 621–624, Fig. 2b. **26.17** Data: S. F. Elena, V. S. Cooper, and R. E. Lenski. 1996. Punctuated evolution caused by selection of rare beneficial mutations. *Science New Series* 272: 1802–1804. **26.18b** N. Moran and T. Jarvik. 2010. Lateral transfer of genes from fungi underlies carotenoid production in aphids. *Science* 328: 624–627. **27.3a** Based on J. C. Avise and W. S. Nelson. 1989. Molecular genetic relationships of the extinct dusky seaside sparrow. *Science* 243 (4891): 646–648, Fig. 2.

Chapter 27 27.3b J. C. Avise and W. S. Nelson. 1989. Molecular genetic relationships of the extinct dusky seaside sparrow. *Science* 243 (4891): 646–648. **27.5b** Data: N. Knowlton, L. A. Weigt, L. A. Solorzano, et al. 1993. Divergence in proteins, mitochondrial DNA, and reproductive compatibility across the Isthmus of Panama. *Science* 260 (5114): 1629–1632, Fig. 1. **27.7** Adapted by permission of Wiley-Blackwell from H. R. Dambroski, C. Linn Jr., S. H. Berlocher, et al. 2005. The genetic basis for fruit odor discrimination in *Rhagoletis* flies and its significance for sympatric host shifts. *Evolution* 59 (9): 1953–1964, Figs. 1A, 1B. **27.11b** Data: S. E. Rohwer, E. Bermingham, and C. Wood. 2001. Plumage and mitochondrial DNA haplotype variation across a moving hybrid zone. *Evolution* 55: 405–422. **27.12** Data: L. H. Rieseberg, B. Sinervo, C. R. Linder, et al. 1996. Role of gene interactions in hybrid speciation: Evidence from ancient and experimental hybrids. *Science* 272 (5262): 741–745.

Chapter 28 28.4c Data adapted by permission of the publisher from M. A. Nikaido, P. Rooney, and N. Okada. 1999. Phylogenetic relationships among cetartiodactyls based on insertions of short and long interspersed elements: Hippopotamuses are the closest extant relatives of whales. *PNAS* 96: 10261–10266, Figs. 1, 7. ©1999 National Academy of Sciences, USA. **28.6, 28.7, 28.8** Data: International Commission on Stratigraphy, "International Stratigraphic Chart, 2009" (www.stratigraphy.org/ column.php?id=Chart/TimeScale). The data on this site are modified from F. M. Gradstein and J. C. Ogg (eds.). 2004. *A Geologic Time Scale 2004*. Cambridge, UK: Cambridge University Press; and J. G. Ogg, G. Ogg, and F. M. Gradstein. 2008. *The Concise Geologic Time Scale*. Cambridge, UK: Cambridge University Press. **28.9** B. G. Baldwin, D. W. Kyhos, and J. Dvorak. 1990. Chloroplast DNA evolution and adaptive radiation in the Hawaiian Silversword Alliance (*Asteraceae–Madiinae*). *Annals of the Missouri Botanical Garden* 77 (1): 96–109, Fig. 2. **28.10c** J. B Losos, K. I. Warheit, and T. W. Schoener. 1997. Adaptive differentiation following experimental island colonization in Anolis lizards. *Nature* 387: 70–73. **28.13** Data: J. W. Valentine, D. H. Erwin, and D. Jablonski. 1996. Developmental evolution of metazoan body plans: The fossil evidence. *Developmental Biology* 173: 373–381, Fig. 3. Also R. de Rosa et al. 1999. *Hox* genes in brachiopods and priapulids and protostome evolution. *Nature* 399: 772–776; D. Chourrout, F. Delsuc, P. Chourrout, et al. 2006. Minimal *ProtoHox* cluster inferred from bilaterian and cnidarian *Hox* complements. *Nature* 442: 684–687. **28.14** Data: M. J. Benton. 1995. Diversification and extinction in the history of life. *Science* 268: 52–58.

Chapter 29 29.12 Data: D. F. Wilson, M. Erecinska, and P. L. Dutton. 1974. Thermodynamic relationships in mitochondrial oxidative phosphorylation. *Annual Review of Biophysics and Bioengineering* 3: 203–230, Tables 1, 3. **30.1** S. M. Adl, A. G. Simpson, M. A. Farmer, et al. 2005. The new higher level classification of eukaryotes with emphasis on the taxonomy of protists. *Journal of Eukaryotic Microbiology* 52: 399–451. Also N. Arisue, M. Hasegawa, and T. Hashimoto. 2005. Root of the eukaryota tree as inferred from combined maximum likelihood analyses of multiple molecular sequence data. *Society for Molecular Biology and Evolution, Molecular Biology and Evolution* 22: 409–420; V. Hampl, L. Hug, J. W. Leigh, et al. 2009. Phylogenomic analyses support the monophyly of Excavata and resolve relationships among eukaryotic "supergroups." *PNAS* 106 (10): 3859–3864, Figs. 1, 2, 3; J. D. Hackett, H. S. Yoon, S. Li, et al. 2007. Phylogenomic analysis supports the monophyly of cryptophytes and haptophytes and the association of Rhizaria with chromalveolates. *Molecular Biology and Evolution* 24: 1702–1713, Fig. 1; P. Schaap, T. Winckler, M. Nelson, et al. 2006. Molecular phylogeny and evolution of morphology in the social amoebas. *Science* 314: 661–663.

Chapter 30 30.1, 30.7 S. M. Adl, A. G. Simpson, M. A. Farmer, et al. 2005. The new higher level classification of eukaryotes with emphasis on the taxonomy of protists. *Journal of Eukaryotic Microbiology* 52: 399–451. Also N. Arisue, M. Hasegawa, and T. Hashimoto. 2005. Root of the eukaryota tree as inferred from combined maximum likelihood analyses of multiple molecular sequence data. *Society for Molecular Biology and Evolution, Molecular Biology and Evolution* 22: 409–420; V. Hampl, L. Hug, J. W. Leigh, et al. 2009. Phylogenomic analyses support the monophyly of Excavata and resolve relationships among eukaryotic "supergroups." *PNAS* 106 (10): 3859–3864, Figs. 1, 2, 3; J. D. Hackett, H. S. Yoon, S. Li, et al. 2007. Phylogenomic analysis supports the monophyly of cryptophytes and haptophytes and the association of Rhizaria with chromalveolates. *Molecular Biology and Evolution* 24: 1702–1713, Fig. 1; P. Schaap, T. Winckler, M. Nelson, et al. 2006. Molecular phylogeny and evolution of morphology in the social amoebas. *Science* 314: 661–663.

Chapter 31 **31.6, 31.9, 31.23** Y.-L. Qiu, L. Li, B. Wang, et al. 2006. The deepest divergences in land plants inferred from phylogenomic evidence. *PNAS* 103 (42): 15511–15516, Fig. 1. Also K. S. Renzaglia, S. Schuette, R. J. Duff, et al. 2007. Bryophyte phylogeny: Advancing the molecular and morphological frontiers. American Bryological and Lichenological Society, *The Bryologist* 110 (2): 179–213; J.-F. Pombert, C. Otis, C. Lemieux, et al. 2005. The chloroplast genome sequence of the green alga *Pseudendoclonium akinetum* (*Ulvophyceae*) reveals unusual structural features and new insights into the branching order of chlorophyte lineages. *Molecular Biology and Evolution* 22: 1903–1918. **31.21** Based on M. E. Hoballah, T. Gubitz, J. Stuurman, et al. 2007. Single gene-mediated shift in pollinator attraction in Petunia. *Plant Cell* 19: 779–790, Fig. 6g, 6h. **31.25** P. S. Soltis and D. E. Soltis. 2004. The origin and diversification of angiosperms. *American Journal of Botany* 91 (10): 1614–1626, Figs. 1, 2, 3.

Chapter 32 **32.7** S. M. Adl, A. G. Simpson, M. A. Farmer, et al. 2005. The new higher level classification of eukaryotes with emphasis on the taxonomy of protists. *Journal of Eukaryotic Microbiology* 52: 399–451. Also N. Arisue, M. Hasegawa, and T. Hashimoto. 2005. Root of the Eukaryota tree as inferred from combined maximum likelihood analyses of multiple molecular sequence data. *Molecular Biology and Evolution* 22: 409–420; V. Hampl, L. Hug, J. W. Leigh, et al. 2009. Phylogenomic analyses support the monophyly of Excavata and resolve relationships among eukaryotic "supergroups." *PNAS* 106: 3859–3864, Figs. 1, 2, 3; J. D. Hackett, H. S. Yoon, S. Li, et al. 2007. Phylogenomic analysis supports the monophyly of cryptophytes and haptophytes and the association of Rhizaria with chromalveolates. *Molecular Biology and Evolution* 24: 1702–1713, Fig 1; P. Schaap, T. Winckler, M. Nelson, et al. 2006. Molecular phylogeny and evolution of morphology in the social amoebas. *Science* 314: 661–663. **32.8** T. Y. James, F. Kauff, C. L. Schoch, et. al. 2006. Reconstructing the early evolution of fungi using a six-gene phylogeny. *Nature* 443: 818–822, Fig. 1. Also D. J. McLaughlin, D. S. Hibbett, F. Lutzoni, et al. 2009. The search for the fungal tree of life. *Trends in Microbiology* 17: 488–497; D. S. Hibbett, M. Binder, J. F. Bischoff, et al. 2007. A higher-level phylogenetic classification of the Fungi. *Mycological Research* 111: 509–547; H. Wang, X. Zhao, L. Gao, et al. 2009. A fungal phylogeny based on 82 complete genomes using the composition vector method. *BMC Evolutionary Biology* 9: 195. **32.10** After H. Bücking and W. Heyser. 2001. Microautoradiographic localization of phosphate and carbohydrates in mycorrhizal roots of *Populus tremula* x *Populus alba* and the implications for transfer processes in ectmycorrhizal associations. *Tree Physiology* 21: 101–107.

Chapter 33 **33.2** A. M. A. Aguinaldo, J. M. Turbeville, L. S. Linford, et al. 1997. Evidence for a clade of nematodes, arthropods, and other moulting animals. *Nature* 387: 489–493, Figs. 1, 2, 3. Also C. W. Dunn, A. Hejnol, D. Q. Matus, et al. 2008. Broad phylogenomic sampling improves resolution of the animal tree of life. *Nature* 452: 745–750, Figs. 1, 2. **33.7** Based on J. R. Finnerty, K. Pang, P. Burton, et al. 2004. Origins of bilateral symmetry: *Hox* and *Dpp* expression in a sea anemone. *Science* 304: 1335–1337.

Chapter 34 **34.1** A. D. Chapman. 2009. *Numbers of Living Species in Australia and the World*. 2nd ed. Canberra: Australian Biological Resources Study. **34.2** A. M. A. Aguinaldo, J. M. Turbeville, L. S. Linford, et al. 1997. Evidence for a clade of nematodes, arthropods, and other moulting animals. *Nature* 387: 489–493, Figs. 1, 2, 3. Also C. W. Dunn, A. Hejnol, D. Q. Matus, et al. 2008. Broad phylogenomic sampling improves resolution of the animal tree of life. *Nature* 452: 745–750, Figs. 1, 2. **34.28, 34.30** Based on data in J. Regier, J. W. Schultz, A. Zwick, et al. 2010. Arthropod relationships revealed by phylogenomic analysis of nuclear protein-coding sequences. *Nature* 463: 1079–1083. **34.UN1** Based on M. Averof and N. H. Patel. 1997. Crustacean appendage evolution associated with changes in *Hox* gene expression. *Nature* 388: 682–686, Fig. 4.

Chapter 35 **35.1** After C. W. Dunn, A. Hejnol. D. Q. Matus, et al. 2008. Broad phylogenomic sampling improves resolution of the animal tree of life. *Nature* 452: 745–750, Figs. 1, 2. Also F. Delsuc, H. Brinkmann, D. Chourrout, et al. 2006. Tunicates and not cephalochordates are the closest living relatives of vertebrates. *Nature* 439: 965–968, Fig. 1. **35.12** J. E. Blair and S. B. Hedges. 2005. Molecular phylogeny and divergence times of deuterostome animals. *Molecular Biology and Evolution* 22: 2275–2284, Figs. 1, 3, 4. **35.13** Data: American Museum of Natural History; BirdLife International (birdlife.org); FishBase (www.fishbase.org); IUCN 2010. *The IUCN Red List of Threatened Species* (http://www.iucnredlist.org); The Reptile Database (reptile-database.org). **35.16** Based on E. B. Daeschler, N. H. Shubin, F. A. Jenkins, et al. 2006. A Devonian tetrapod-like fish and the evolution of the tetrapod body plan. *Nature* 440: 757–763, Fig. 6. Also N. H. Shubin, E. B. Daeschler, F. A. Jenkins, et al. 2006. The pectoral fin of *Tiktaalik roseae* and the origin of the tetrapod limb. *Nature* 440:764–771, Fig. 4. **35.20** Color plate by Michael DiGiorgio. From Q. Li, K.-Q. Gao, J. Vinther, et al. 2010. Plumage color patterns of an extinct dinosaur. *Science* 327, Fig. 4, p. 1371. Reprinted by permission of the artist, Michael DiGiorgio. **35.22** J. E. Blair and S. B. Hedges. 2005. Molecular phylogeny and divergence times of deuterostome animals. *Molecular Biology and Evolution*, 22: 2275–2284, Figs. 1, 3, 4. Also F. R. Liu, M. M. Miyamoto, N. P. Freire, et al. 2001. Molecular and morphological supertrees for eutherian (placental) mammals. *Science*

291: 1786–1789; A. B. Prasad, M. W. Allard, E. D. Green, et al. 2008. Confirming the phylogeny of mammals by use of large comparative sequence data sets. Oxford University Press, *Molecular Biology and Evolution* 25: 1795–1808. **35.36** A. B. Prasad, M. W. Allard, E. D. Green, et al. 2008. Confirming the phylogeny of mammals by use of large comparative sequence data sets. Oxford University Press, *Molecular Biology and Evolution* 25 (9):1795–1808, Figs. 1, 2, 3. **35.37** Data: D. E. Lieberman. 2009. Palaeoanthropology: *Homo floresiensis* from head to toe. *Nature* 459: 41–42. **T35.1** S. L. Robson and B. Wood. 2008. Hominin life history: Reconstruction and evolution. *Journal of Anatomy* 212: 394–425. **35.39** J. Z. Li, D. M. Absher, H. Tang, et al. 2008. Worldwide human relationships inferred from genome-wide patterns of variation. *Science* 319: 1100–1104, Fig. 1. **35.40** After L. L. Cavalli-Sforza and M. W. Feldman. 2003. The application of molecular genetic approaches to the study of human evolution. *Nature Genetics Supplement* 33: 266–275, Fig. 3; Data: Genographic Project at National Geographic: https://genographic.nationalgeographic.com/genographic/lan/en/atlas.html; dates from August 2009 summary map: http://ngm.nationalgeographic.com/big-idea/02/queens-genes; M. Rasmussen, X. Guo, Y. Wang, et al. 2011. An Aboriginal Australian genome reveals separate human dispersals into Asia. *Science* 334: 94–98.

Chapter 36 **36.2a** Data: E. Arias. 2010. United States life tables, 2006. *National Vital Statistics Reports* 58 (21): 1–40, Table 10. Hyattsville, MD: National Center for Health Statistics. **36.3** Adapted with permission of Annual Reviews from G. Pantaleo and A. S. Fauci. 1996. Immunopathogensis of HIV infection. *Annual Review of Microbiology* 50: 825–854. Fig. 1a, p. 829. **36.7** Data courtesy of the Undergraduate Biotechnology Laboratory, California Polytechnic State University, San Luis Obispo. **36.8** A. G. Dalgleish, P. C. Beverley, P. R. Clapham, et al. 1984. The CD4 (T4) antigen is an essential component of the receptor for the AIDS retrovirus. *Nature* 312: 763–767. Also D. Klatzmann, E. Champagne, S. Chamaret, et al. 1984. T-lymphocyte T4 molecule behaves as the receptor for human retrovirus LAV. *Nature* 312: 767–768. **36.15** F. Gao, E. Bailes, D. L. Robertson, et al. 1999. Origin of HIV-1 in the chimpanzee *Pan troglodytes troglodytes*. *Nature* 397: 436–441, Fig. 2.

Chapter 37 **37.3** Illustration adapted by permission from "Root Systems of Prairie Plants," by Heidi Natura. ©1995 Conservation Research Institute. **37.5** After J. Clausen, D. D. Keck, and M. Hiesey. 1945. Experimental studies on the nature of species II: Plant evolution through amphiploidy and autoploidy, with examples from the *Madiinae*. Publication No. 564. Washington, DC: Carnegie Institute of Washington.

Chapter 38 **38.3** Data: R. G. Cline and W. S. Campbell. 1976. Seasonal and diurnal water relations of selected forest species. *Ecology* 57: 367–373, Fig. 1. **38.12** Adapted by permission of the American Society of Plant Biologists from C. M. Wei, M. T. Tyree, and E. Steudle. 1999. Direct measurement of xylem pressure in leaves of intact maize plants. A test of the cohesion-tension theory taking hydraulic architecture into consideration. *Plant Physiology* 121: 1191–1205, Fig. 6. ©1999 by American Society of Plant Biologists. **38.22** J. R. Gottwald, P. J. Krysan, J. C. Young, et al. 2000. Genetic evidence for the in planta role of phloem-specific plasma membrane sucrose transporters. *PNAS* 97: 13979–13984.

Chapter 39 **39.3** D. I. Arnon and P. R. Stout. 1939. The essentiality of certain elements in minute quantity for plants with special reference to copper. *Plant Physiology* 14: 371–375.

Chapter 40 **40.4** Based on S. Inoue, T. Kinoshita, M. Matsumoto, et al. 2008. Blue light-induced autophosphorylation of phototropin is a primary step for signaling. *PNAS* 105: 5626–5631. **40.5** After C. Darwin and F. Darwin. 1897. *The Power of Movement in Plants*. New York: D. Appleton. **40.10** After A. Lang, M. K. Chailakhyan, and I. A. Frolova. 1977. Promotion and inhibition of flower formation in a day-neutral plant in grafts with a short-day plant and a long-day plant. *PNAS* 74: 2412–2416. **T40.1** H. A. Borthwick, S. B. Hendricks, M. W. Parker, et al. 1952. A reversible photoreaction controlling seed germination. *PNAS* 38: 662–666, Table 1. **40.21** After P. G. Blackman and W. J. Davies. 1985. Root to shoot communication in maize plants of the effect of soil drying. *Journal of Experimental Botany* 36: 39–48.

Chapter 41 **41.10** Y.-L. Qiu, L. Li, B. Wang, et al. 2006. The deepest divergences in land plants inferred from phylogenomic evidence. *PNAS* 103 (42): 15511–15516, Fig. 1. Also K. S. Renzaglia, S. Schuette, R. J. Duff, et al. 2007. Bryophyte phylogeny: Advancing the molecular and morphological frontiers. American Bryological and Lichenological Society, *The Bryologist* 110 (2): 179–213; J.-F. Pombert, C. Otis, C. Lemieux, et al. 2005. The chloroplast genome sequence of the green alga *Pseudendoclonium akinetum* (*Ulvophyceae*) reveals unusual structural features and new insights into the branching order of chlorophyte lineages. *Molecular Biology and Evolution* 22: 1903–1918. **41.17** J. J. Tewksbury and G. P. Nabhan. 2001. Directed deterrence by capsaicin in chilies. *Nature* 412: 403–404.

Chapter 42 **42.2** A. M. Kerr, S. N. Gershman, and S. K. Sakaluk. 2010. Experimentally induced spermatophore production and immune responses reveal a trade-off in crickets. *Behavioral Ecology* 21: 647–654. **42.10** Data: K. Schmidt-Nielsen. 1984. *Scaling: Why Is Animal Size So Important?* Cambridge, UK: Cambridge University Press. **42.11** Adapted by permission of The Company of Biologists from

P. R. Wells and A. W. Pinder. 1996. The respiratory development of Atlantic salmon. II. Partitioning of oxygen uptake among gills, yolk sac and body surfaces. *Journal of Experimental Biology* 199: 2737–2744, Figs. 1A, 3.

Chapter 43 **43.13** Anterolateral view reproduced by permission from Macmillan Publishers Ltd after S. E. Pierce, J. A. Clack, and J. R. Hutchinson. 2012. Three-dimensional limb joint mobility in the early tetrapod *Ichthyostega*. *Nature* 486: Fig. 1, p. 523. **43.14** Adapted by permission of Elsevier from K. J. Ullrich, K. Kramer, and J. W. Boyer. 1961. Present knowledge of the counter-current system in the mammalian kidney. *Progress in Cardiovascular Diseases* 3: 395–431, Figs. 5, 7. ©1961. **43.19** J. R. Davis and D. F. DeNardo. 2007. The urinary bladder as a physiological reservoir that moderates dehydration in a large desert lizard, the Gila monster *Heloderma suspectum*. *Journal of Experimental Biology* 210 (8): 1472–1480. Reprinted with permission from The Company of Biologists.

Chapter 44 **44.14** After E. M. Wright. 1993. The intestinal Na$^+$/glucose cotransporter. *Annual Review of Physiology* 55: 575–589. Also K. Takata, T. Kasahara, M. Kasahar, et al. 1992. Immunohistochemical localization of Na$^+$-dependent glucose transporter in rat jejunum. *Cell and Tissue Research* 267: 3–9. **44.17a** Graph reproduced with permission from The American Diabetes Association from L. O. Schulz, P. H. Bennett, E. Ravussin, et al. 2006. Effects of traditional and Western environments on prevalence of type 2 diabetes in Pima Indians in Mexico and the U.S. *Diabetes Care* 29: 1866–1871, Fig. 1. ©2006 American Diabetes Association. **44.17b** Data: L. O. Schulz, P. H. Bennett, E. Ravussin, et al. 2006. Effects of traditional and Western environments on prevalence of type 2 diabetes in Pima Indians in Mexico and the U.S. *Diabetes Care* 29: 1866–1871, Table 2.

Chapter 45 **45.7** Based on Y. Komai. 1998. Augmented respiration in a flying insect. *Journal of Experimental Biology* 201: 2359–2366, Fig. 7. **45.9** R. A. Berner. 1999. Atmospheric oxygen over Phanerozoic time. *PNAS* 96: 10955–10957. Fig. 2. Copyright ©1999 National Academy of Sciences, USA. Reprinted with permission. **45.12** W. Bretz and K. Schmidt Nielsen. 1971. Bird respiration: Flow patterns in the duck lung. *Journal of Experimental Biology* 54: 103–118. **45.21** Data: E. H. Starling. 1896. On the absorption of fluids from the connective tissue spaces. *Journal of Physiology* 19: 312–326. **45.23** J. E. Blair and S. B. Hedges. 2005. Molecular phylogeny and divergence times of deuterostome animals. *Molecular Biology and Evolution* 22: 2275–2284, Figs. 1, 3, 4.

Chapter 46 **46.10** Data: O. Loewi. 1921. Über humorale Übertragbarkeit der Herznervenwirkung. *Pflügers Archiv European Journal of Physiology* 189: 239–242. **46.22** Data: R. Q. Quiroga, L. Reddy, G. Kreiman, et al. 2005. Invariant visual representation by single neurons in the human brain. *Nature* 435: 1102–1107, Fig. 1B. **46.24a&c** Based on K. C. Martin, A. Casadio, H. Zhu, et al. 1997. Synapse-specific, long-term facilitation of *Aplysia* sensory to motor synapses: A function for local protein synthesis in memory storage. *Cell* 91: 927–938.

Chapter 47 **47.2a&b** Data: J. E. Rose, J. E. Hind, D. J. Anderson, et al. 1971. Some effects of stimulus intensity on response of auditory nerve fibers in the squirrel monkey. *Journal of Neurophysiology* 34: 685–699. **47.9** K. Pohlmann, J. Atema, and T. Breithaupt. 2004. The importance of the lateral line in nocturnal predation of piscivorous catfish. *Journal of Experimental Biology* 207: 2971–2978. **47.15** Based on D. M. Hunt, K. S. Dulai, J. K. Bowmaker, et al. 1995. The chemistry of John Dalton's color blindness. *Science* 267: 984–988, Fig. 3. **47.UN1** Reprinted by permission from Macmillan Publishers Ltd. from M. K. McClintock. 1971. Menstrual synchrony and suppression. *Nature* 229: 244–245, Fig. 1.

Chapter 48 **48.3** PDB ID: 1KWO. D. M. Himmel, S. Gourinath, L. Reshetnikova, et al. 2002. Crystallographic findings on the internally uncoupled and near-rigor states of myosin: Further insights into the mechanics of the motor. *PNAS* 99: 12645–12650. **48.7** After Exercise Information—Interactive Muscle Map (www.askthetrainer.com/exercise-information/). **48.15** Reprinted by permission from Macmillan Publishers Ltd from D. F. Hoyt and C. R. Taylor. 1981. Gait and the energetics of locomotion in horses. *Nature* 292: 239–240, Fig. 2, p. 240. **48.16** S. Vogel. 2003. *Comparative Biomechanics: Life's Physical World*. 2nd ed. Princeton, NJ: Princeton University Press, Fig. 3.5, p. 60. Reprinted by permission of Princeton University Press. **48.17** Reprinted with permission from AAAS from K. Schmidt-Nielsen. 1972. Locomotion: Energy cost of swimming, flying, and running. *Science* 177: 222–228. (http://www.sciencemag.org/content/177/4045/222.extract).

Chapter 49 **49.4** W. M. Bayliss and E. H. Starling. 1902. The mechanism of pancreatic secretion. *Journal of Physiology* 28: 325–353. **49.7** Graph from J. Vanecek. 1998. Cellular mechanisms of melatonin action. *Psychological Reviews* 78: 687–721, Fig. 1, p. 688. Original Data: H. Illnerová, K. Hoffmann, J. Vanácek, et al. 1984. Adjustment of pineal melatonin and N-acetyltransferase rhythms to change from long to short photoperiod in the Djungarian hamster Phodopus sungorus. *Neuroendocrinology* 38: 226–231. **49.9** Data: B. D. King, L. Sokoloff, and R. L. Wechsler. 1952. The effects of l-epinephrine and l-nor-epinephrine upon cerebral circulation and metabolism in man. *Journal of Clinical Investigation* 31: 273–279. Also P. S. Mueller and D. Horwitz. 1962. Plasma free fatty acid and blood glucose responses to analogues of norepinephrine in man. *Journal of Lipid Research* 3: 251–255. **49.11** Reprinted by permission from Macmillan Publishers Ltd. from D. L. Coleman. 2010. A historical perspective on Leptin. *Nature Medicine* 16: 1097–1099, Fig. 2. **49.12a&b** Data: G. V. Upton, A. Corbin, C. C. Mabry, et al. 1973. Evidence for the internal feedback phenomenon in human subjects: Effects of ACTH on plasma CRF. *Acta Endocrinologica* 73: 437–443. **48.13** Anterolateral view reproduced by permission from Macmillan Publishers Ltd after S. E. Pierce, J. A. Clack, and J. R. Hutchinson. 2012. Three-dimensional limb joint mobility in the early tetrapod *Ichthyostega*. *Nature* 486: Fig. 1, p. 523. **49.14** Data: D. Toft and J. Gorski. 1966. A receptor molecule for estrogens: Isolation from the rat uterus and preliminary characterization. *PNAS* 55: 1574–1581. **49.16b** Data: T. W. Rall, E. W. Sutherland, and J. Berthet, et al. 1957. The relationship of epinephrine and glucagon to liver phosphorylase. *Journal of Biological Chemistry* 224: 463–475. **49.17** Reproduced from E. W. Sutherland. 1971. Studies on the mechanism of hormone action. (A lecture delivered in Stockholm, December 11, 1971.) ©1971 Nobel Foundation; E. W. Sutherland. 1972. Studies on the mechanism of hormone action. *Science* 177: 401–408, Fig. 6. Also G. A. Robison, R. W. Butcher, I. Oye, et al. 1965. The effect of epinephrine on adenosine 3′,5′-phosphate levels in the isolated perfused rat heart. *Molecular Pharmacology* 1: 168–177. **49.UN1** Data: T. B. Hayes, A. Collins, M. Lee, et al. 2002. Hermaphroditic, demasculinized frogs after exposure to the herbicide atrazine at low ecologically relevant doses. *PNAS* 99: 5476–5480.

Chapter 50 **50.3a** Based on R. G. Stross and J. C. Hill. 1965. Diapause induction in *Daphnia* requires two stimuli. *Science* 150: 1462–1464, Fig. 3. **50.3b** Data: O. T. Kleiven, P. Larsson, and A. Hobæk. 1992. Sexual reproduction in *Daphnia magna* requires three stimuli. *Oikos* 65: 197–206, Table 4. **50.7** Based on S. C. S. Price, K. A. Dyer, and J. A. Coyne. 1999. Sperm competition between *Drosophila* males involves both displacement and incapacitation. *Nature* 400: 449–452, Fig. 3c. **50.9** R. Shine and M. S. Y. Lee. 1999. A reanalysis of the evolution of viviparity and egg-guarding in squamate reptiles. The Herpetologists' League, Inc., *Herpetologica* 55 (4): 538–549, Figs. 1, 2, 3. **50.10b** Reproduced by permission of Elsevier from C. Hotzy and G. Arnqvist. 2009. Sperm competition favors harmful males in seed beetles. *Current Biology* 19: 404–407, Fig. 2. ©2009. **50.17, 50.18** Data: R. Stricker, R. Eberhart, M.-C. Chevallier, et al. 2006. Establishment of detailed reference values for luteinizing hormone, follicle stimulating hormone, estradiol, and progesterone during different phases of the menstrual cycle on the Abbott ARCHITECT˚ analyzer. *Clinical Chemistry and Laboratory Medicine* 44: 883–887, Tables 1A, 1B. **50.22b** Data: V. W. Swayze II., V. P. Johnson, J. W. Hanson, et al. 1997. Magnetic resonance imaging of brain anomalies in fetal alcohol syndrome. *Pediatrics* 99: 232–240. **50.24** Adapted by permission of the publisher and author from U. Högberg and I. Joelsson. 1985. The decline in maternal mortality in Sweden, 1931–1980. *Acta Obstetricia et Gynecologica Scandinavica* 64: 583–592, Fig. 1. Taylor & Francis Group, http://www.informaworld.com.

Chapter 51 **51.2** Based on B. Lemaitre, E. Nicolas, L. Michaut, et al. 1996. The dorsoventral regulatory gene cassette *spätzle/Toll/cactus* controls the potent antifungal response in *Drosophila* adults. *Cell* 86: 973–983. **51.5a** PDB ID: 1IGT. L. J. Harris, S. B. Larson, K. W. Hasel, et al. 1997. Refined structure of an intact IgG2a monoclonal antibody. *Biochemistry* 36: 1581–1597. **51.5b** PDB ID: 1TCR. K. C. Garcia, M. Degano, R. L. Stanfield, et al. 1996. An aß T cell receptor structure at 2.5 Å and its orientation in the TCR–MHC complex. *Science* 274: 209–219. **51.6** PDB ID: 1HMG. M. Knossow, R. S. Daniels, A. R. Douglas, et al. 1984. Three-dimensional structure of an antigenic mutant of the influenza virus haemagglutinin. *Nature* 311 (5987): 678–680. **51.10** PDB ID: 1DLH. L. J. Stern, J. H. Brown, T. S. Jardetzky, et al. 1994. Crystal structure of the human class II MHC protein HLA–DR1 complexed with an influenza virus peptide. *Nature* 368: 215–221. **51.17** Based on B. Alberts, A. Johnson, J. Lewis, et al. 2002. *Molecular Biology of the Cell*. 4th ed. New York: Garland Science, Taylor & Francis Group. Modified from Fig. 24-10. **52.1** Data: Food and Agricultural Organization of the United Nations.

Chapter 52 **52.4** S. B. Menke, R. N. Fisher, W. Jetz, et al. 2007. Biotic and abiotic controls of Argentine ant invasion success at local and landscape scales. *Ecology* 88 (12): 3164–3173. **52.9** D. M. Olson, E. Dinerstein, E. D. Wikramanayake, et al. 2001. Terrestrial ecoregions of the world: A new map of life on Earth. *BioScience* 51: 933–938. **52.11, 52.12, 52.13, 52.14, 52.15, 52.16** Data: E. Aguado and J. E. Burt. 1999. *Understanding Weather and Climate*. Upper Saddle River, NJ: Prentice Hall. **52.18a** Based on M. B. Saffo. 1987. New light on seaweeds. *BioScience* 37 (9): 654–664, Fig. 1. ©1987 American Institute of Biological Sciences.

Chapter 53 **53.3** After M. Sokolowski. 2001. *Drosophila*: Genetics meets behavior. *Nature Reviews Genetics* 2: 879–890, Fig. 2. **53.4** Z. Abramsky, M. L. Rosenzweig, and A. Subach. 2002. The costs of apprehensive foraging. *Ecology* 85: 1330–1340. **53.5b** Adapted by permission of the author from D. Crews. 1975. Psychobiology of reptilian reproduction. *Science* 189: 1059–1065, Fig. 1. **53.6** Adapted by permission of the author from D. Crews. 1975. Psychobiology of reptilian reproduction. *Science* 189: 1059–1065, Fig. 2. **53.7** Reprinted by permission from Macmillan Publishers Ltd from K. J. Lohmann, C. M. F. Lohmann, L. M. Ehrhart, et al. 2004. Geomagnetic map used in sea-turtle navigation. *Nature* 428: 909. **53.8** C. Egevang, et al. 2010. Tracking of

Arctic terns *Sterna paradisaea* reveals longest animal migration. *PNAS* 107(5): 2078–2081, Fig. 1B. ©2010 National Academy of Sciences, USA. Reprinted with permission. **53.15** Data: J. L. Hoogland. 1983. Nepotism and alarm calling in the black-tailed prairie dog (*Cynomys ludovicianus*). *Animal Behavior* 31: 472–479.

Chapter 54 **T54.1** Data reproduced from H. Strijbosch and R. C. M. Creemers. 1988. Comparative demography of sympatric populations of *Lacerta vivipara* and *Lacerta agilis*. *Oecologia* 76: 20–26. **54.5b** Data: G. F. Gause. 1934. *The Struggle for Existence*. Baltimore, MD: Williams & Wilkins. **54.6a** G. E. Forrester and M. A. Steele. 2000. Variation in the presence and cause of density-dependent mortality in three species of reef fishes. *Ecology* 81 (9): 2416–2427, Fig. 5. Copyright © 2000 by the Ecological Society of America. Reprinted with permission. **54.6b** Data: P. Arcese and J. N. M. Smith. 1988. Effects of population density and supplemental food on reproduction in song sparrows. *Journal of Animal Ecology* 57: 119–136, Fig. 2. **54.9** Data: D. A. MacLulich. 1937. Fluctuations in the numbers of the varying hare (*Lepus americanus*). *University of Toronto Studies Biological Series* 43. **54.10** Graphs reproduced by permission of AAAS from C. J. Krebs, S. Boutin, R. Boonstra, et al. 1995. Impact of food and predation on the snowshoe hare cycle. *Science* 269: 1112–1115, Fig. 3. (http://www.sciencemag.org/content/269/5227/1112). **54.11a** Data: United Nations, Department of Economic and Social Affairs, Population Division. 2011. *World Population Prospects: The 2010 Revision*. **54.11b** Data: United Nations, Department of Economic and Social Affairs, Population Division. 2011. *World Population Prospects: The 2010 Revision*. **54.13** United Nations, Department of Economic and Social Affairs, Population Division. 2011. *World Population Prospects: The 2010 Revision*.

Chapter 55 **55.2a&b, 55.3a&b** G. F. Gause. 1934. *The Struggle for Existence*. Baltimore. MD: Williams & Wilkins. **55.5** Data: J. H. Connell. 1961. The influence of interspecific competition and other factors on the distribution of the barnacle *Chthamalus stellatus*. *Ecology* 42: 710–723. **55.7** P. R. Grant. 1999. *Ecology and Evolution of Darwin's Finches*. ©1986 Princeton University Press. Fig. 15, p. 78. Reprinted by permission of Princeton University Press. **55.10** Adapted by permission of Ecological Society of America from G. H. Leonard, M. D. Bertness, and P. O. Yund. 1999. Crab predation, waterborne cues, and inducible defenses in the blue mussel, *Mytilus edulis*. *Ecology* 80: 1–14. **55.14** Based on J. H. Cushman and T. G. Whitham. 1989. Conditional mutualism in a membracid–ant association: temporal, age-specific, and density-dependent effects. *Ecology* 70 (4): 1040–1047. **55.15** Based on D. G. Jenkins and A. L. Buikema. 1998. Do similar communities develop in similar sites? A test with zooplankton structure and function. *Ecological Monographs* 68: 421–443. **55.16** Adapted with kind permission from Springer Science+Business Media B.V. from R. T. Paine. 1974. Intertidal community structure. *Oecologia* 15: 93–120. ©1974 Springer-Verlag. **55.18b** Data: T. W. Swetnam. 1993. Fire history and climate change in giant sequoia groves. *Science* 262: 885–889. **55.21a** Adapted by permission of Wiley-Blackwell from R. H. MacArthur and E. O. Wilson. 1963. An equilibrium theory of insular zoogeography. *Evolution* 17 (4): 373–387, Fig. 4. **55.21 b & c** Data: R. H. MacArthur and E. O. Wilson. 1963. An equilibrium theory of insular zoogeography. *Evolution* 17 (4): 373–387, Fig. 5. **55.23** Data: W. V. Reid and K. R. Miller. 1989. *Keeping Options Alive: The Scientific Basis for Conserving Biodiversity*. Washington, DC: World Resources Institute. **56.10** Data: H. Leith and R. H. Whittaker (eds.). 1975. *Primary Productivity in the Biosphere. Ecological Studies and Synthesis*, Vol. 14. New York: Springer-Verlag.

Chapter 56 **56.5** Based on D. E. Canfield, A. N. Glazer, and P. G. Falkowski. 2010. The evolution and future of Earth's nitrogen cycle. *Science* 330: 192–196. **56.13** Data: G. E. Likens, F. H. Bormann, N. M. Johnson, et al. 1970. Effects of forest cutting and herbicide treatment on nutrient budgets in the Hubbard Brook watershed-ecosystem. *Ecological Monographs* 40: 23–47. **56.14** T. Oki and S. Kanae. 2006. Global hydrological cycles and world water resources. *Science* 313: 1068–1072. **56.15** U.S Geological Survey: Changes in water levels and storage in the High Plains Aquifer, predevelopment to 2009. Fact Sheet, July 2011, Fig. 1. **56.16** N. Gruber and J. N. Galloway. 2008. An Earth-system perspective of the global nitrogen cycle. *Nature* 451: 293–296. **56.17** Based on D. E. Canfield, A. N. Glazer, and P. G. Falkowski. 2010. The evolution and future of Earth's nitrogen cycle. *Science* 330: 192–196.

56.19 B. Hönisch, A. Ridgwell, D. N. Schmidt, et al. 2012. The geological record of ocean acidification. *Science* 335: 1058–1063. **56.20** Carbon Dioxide Information Analysis Center (http://cdiac.ornl.gov/). **56.22, top** NOAA. **56.22, middle & bottom** T. R. Karl, J. M. Melillo, and T. C. Peterson (eds.). 2009. *Global Climate Change Impacts in the United States*. New York: Cambridge University Press, p. 6. **56.23** Carbon Dioxide Information Analysis Center (http://cdiac.ornl.gov/). **56.24** T. R. Karl, J. M. Melillo, and T. C. Peterson (eds). 2009. *Global Climate Change Impacts in the United States*. New York: Cambridge University Press, p. 34. **56.25** Reproduced by permission of AAAS from M. Zhao and S. W. Running. 2010. Drought-induced reduction in global terrestrial net primary production from 2000 through 2009. *Science* 329: 940–943, Fig. 2. (http://www.sciencemag.org/content/329/5994/940.abstract).

Chapter 57 **57.2** F.-R. Liu, M. M. Miyamoto, N. P. Freire, et al. 2001. Molecular and morphological supertrees for eutherian (placental) mammals. *Science* 291: 1786–1789, Fig. 1. Also A. B. Prasad, M. W. Mallard, E. D. Green, et al. 2008. Confirming the phylogeny of mammals by use of large comparative sequence data sets. Oxford University Press, *Molecular Biology and Evolution* 25 (9):1795–1808; U. Arnason, A. Gullberg, and A. Janke. 2004. Mitogenomic analyses provide new insights into cetacean origin and evolution. Elsevier, *Gene* 333: 27–34, Fig. 1. **57.4a&b** Adapted by permission of Macmillan Publishers Ltd after C. D. L. Orme, R. G. Davies, M. Burgess, et al. 2005. Global hotspots of species richness are not congruent with endemism or threat. *Nature* 436: 1016–1019, Figs. 1a, 1c. **57.5** D. M. Olson and E. Einerstein. 2002. The Global 200: Priority ecoregions for global conservation. *Annals of the Missouri Botanical Garden* 89: 199–224. Fig. 2, p. 201. Reprinted by permission of the Missouri Botanical Garden Press. **57.6** Data: IUCN 2012. *The IUCN Red List of Threatened Species. Version 2012.2*. (http://www.iucnredlist.org). **57.7** Reproduced by permission of the University of California Press Journals and the author from O. Venter et al. 2006. Threats to endangered species in Canada. *BioScience* 56 (11): 903–910, Fig 2. ©2006 American Institute of Biological Sciences. **57.9** Data: W. F. Laurance, S. G. Laurance, L. V. Ferreira, et al. 2000. Conservation: Rainforest fragmentation kills big trees. *Nature* 404: 836. **57.9** Based on W. F. Laurance, S. G. Laurance, L.V . Ferreira, et al. 1997. Biomass collapse in Amazonian forest fragments. *Science* 278: 1117–1118, Fig. 2. **57.10** Adapted by permission of Elsevier from J. M. Diamond. 1975. The island dilemma: Lessons of modern biogeographic studies for the design of natural reserves. *Biological Conservation* 7: 129–146. ©1975. **57.11** Reproduced by permission of AAAS from B. Sinervo, F. Mendez de la Cruz, D. B. Miles, et al. 2010. Erosion of lizard diversity by climate change and altered thermal niches. *Science* 328: 894–899, Fig. 3. (http://www.sciencemag.org/content/328/5980/894.abstract). **57.12** Graphs reproduced by permission of AAAS from D. Tilman, J. Knops, D. Wedin, et al. 1997. The influence of functional diversity and composition on ecosystem processes. *Science* 277: 1300–1302, Fig. 1. (http://www.sciencemag.org/content/277/5330/1300.short). **57.14** Data: W. E. Johnson, D. P. Onorato, M. E. Roelke, et al. 2010. Genetic restoration of the Florida panther. *Science* 329: 1641–1645. **57.15** Reproduced by permission of AAAS from E. I. Damschen, N. M. Haddad, J. L. Orrock, et al. 2006. Corridors increase plant species richness at large scales. *Science* 313: 1284–1286, Fig. 2B. (http://www.sciencemag.org/content/313/5791/1284.abstract). **57.17** Data: H. P. Jones and O. J. Schmitz. 2009. Rapid recovery of damaged ecosystems. *PLoS One* 4 (5): e5653, Fig. 4. **57.UN1, top** Data: A. P. Dobson. 1996. *Conservation and Biodiversity*. Scientific American Library. New York: Freeman. Original source of England maps: Reproduced by permission of the author from D. S. Wilcove, C. H. McLellan, and A. P. Dobson. 1986. Habitat fragmentation in the temperate zone. In *Conservation Biology: The Science of Scarcity and Diversity*, edited by M. E. Soule. Sunderland, MA: Sinauer Associates, pp. 237–256, Fig. 1. **57.UN1, bottom** Data: A. P. Dobson. 1996. *Conservation and Biodiversity*. Scientific American Library. New York: Freeman. Original source of U.S. Maps: W. B. Greeley, Chief, U.S. Forest Service. 1925. Relation of geography to timber supply. *Economic Geography* 1: 1–11.

Appendix B: BioSkills **TB16.1** A. Crowe, C. Dicks, and M. P. Wenderoth. 2008. Biology in bloom: Implementing Bloom's Taxonomy to enhance student learning in Biology. *CBE–Life Sciences Education* 7: 368–381, Table 3.

Index

Boldface page numbers indicate a glossary entry; page numbers followed by an *f* indicate a figure; those followed by *t* indicate a table.

Boldface page numbers indicate a glossary entry; page numbers followed by an *f* indicate a figure; those followed by *t* indicate a table.

Boldface page numbers indicate a glossary entry; page numbers followed by an
f indicate a figure; those followed by *t* indicate a table.

G

G-protein-coupled receptors, 211–14, 216
G proteins, **212**
G_0 state, 229, 233–34
G_1 phase, 221–**22**, 223*f*, 229, 231–34
G_2 phase, 221–**22**, 223*f*, 232
G3P. *See* Glyceraldehyde-3-phosphate (G3P)
Gage, Phineas, 945
Gaits, 986, 987*f*
Galactose, 73–74, 354
Galactoside permease, 340
Galápagos finches, 449–50, 456–59*f*, 494–95, 845–46*f*, 1127
Galen, 973
Galileo, 987
Gallbladder, 887*f*, **894**
Galls, 385–86
Galvani, Luigi, 973
Gametangia, **587**
Gametes, **219**, **238**, **823**. *See also* Eggs; Sperm
 in external fertilization, 1018
 in fertilization, 237–38, 241–42, 420–21
 fungal, 617, 624
 gametogenesis of plant, 433, 438, 588–89
 meiosis and, 219
 in plant reproduction, 823–25
 polyspermy prevention and, 421–23
Gametic barriers and prezygotic isolation, 491*t*
Gametogenesis, **241**, **1016**. *See also* Oogenesis; Spermatogenesis
 animal, 1016–18*f*
 in external fertilization, 1018
 meiosis and, 241
 plant, 433, 438, 588–89
Gametophytes, **567**, **588**, **823**
 in alternation of generations, 588–89
 land plant evolution away from life cycles dominated by, 589–90
 in land plant life cycles, 823–25
 production of, 827–29
 protist, 568–69
 in seed evolution, 592
Gamma-aminobutyric acid (GABA), 941*t*
Gamow, George, 310
Ganglia, **643**
Ganglion cells, **960**, 961*f*
Gap Analysis Program (GAP), 1191
Gap genes, 412, 413*f*, 414*f*
Gap junctions, 205*f*, **207–8**
Gap phases, 221–22, 223*f*
Gaps, cell–cell. *See* Cell–cell gaps
Garden peas, Mendelian genetics and, 257–66. *See also* Mendelian genetics
Gas exchange
 amniotic egg, 693
 amphibian, 700
 animal, 902–3 (*see also* Circulatory systems; Respiratory systems)
 diffusion and skin-breathing in, 852
 fish lungs in, 692–93
 gills in (*see* Gills)
 during human pregnancy, 1032–33
 insect, 869
 in osmoregulation, 862–63
 plant stomata in, 743–44
 pneumatophore roots in, 734, 735*f*
 by protostomes, 659–60
 by salmon, 851–52

terrestrial adaptations for, 659–60
tree bark and, 750
Gases
 chemical evolution and volcanic, 24, 30, 34–35
 greenhouse, 1163, 1192–93
Gastric juice, 890–91
Gastrin, **893**
Gastropods, **662**, 668
Gastrovascular cavities, 886–87*f*
Gastrula, **424**
Gastrulation, **407**, **424**
 in animal embryonic development, 644–45
 cell movement in, 407, 424–25
 definition of body axes in, 425
 formation of germ layers in, 425
 in frog embryo, 424*f*
 protostome vs. deuterostome, 658
Gated channels, **98–99**. *See also* Membrane proteins
Gause, G. F., 1125–26
GDP (guanosine diphosphate), 211, 214
Gel electrophoresis, 378, 894, B:13–B:14
Gemmae, **601**, 603
Gene-by-environment interactions, 273, 277*t*
Gene-by-gene interactions, 273–75, 277*t*
Gene duplication, 398–99, 483, 485
Gene expression, **305–16**, **337**
 bacterial gene regulation and, 336–37 (*see also* Gene regulation, bacterial)
 bacterial vs. eukaryotic, 360–61
 in cell determination, 429
 in cell differentiation, 409–10 (*see also* Differential gene expression)
 central dogma of molecular biology on, 307–9
 eukaryotic gene regulation and (*see* Gene regulation, eukaryotic)
 evolutionary change from changes in, 416
 functional genomics and, 400–402
 genetic code and, 307, 310–12
 linking genotypes and phenotypes in, 308
 molecular biology, DNA, and, 304–5
 mutations of genes and chromosomes and, 313–15
 one-gene, one-enzyme hypothesis on, 305–7
 plant cell–cell signals and, 795–96
 point mutations and, 483
 protein synthesis and, 317–18 (*see also* Protein synthesis)
 RNA as intermediary between genes and proteins in, 307–8
 steroids and, 1007–8*f*
 transcription and translation in, 308
Gene families, **398–99**, 508
Gene flow, **466**, **482**
 in allopatric speciation, 494–95
 effects of, on fitness, 483
 as evolutionary process, 466, 486*t*
 genetic diversity and, 483
 genetic restoration with artificial, 1190
 Hardy–Weinberg principle and, 468
 measuring, 482–83
 reproductive isolation and, 490
 speciation and, 489
 in sympatric speciation, 495
Gene pools, **466–67**
Gene recombination, immune system, 1046, 1047*f*. *See also* Recombinant DNA technology

Gene regulation, bacterial, 336–47
 eukaryotic vs., 348–49, 360–61
 of gene expression, 336–37 (*see also* Gene expression)
 global, 344–45
 identifying regulated genes for, 339–41
 information flow and overview of, 337–39*f*
 negative control of transcription in, 341–44
 positive control of transcription in, 344, 345*f*
Gene regulation, eukaryotic, 348–65
 bacterial gene regulation vs., 348–49, 360–61
 cancer and defects in, 361–63
 chromatin remodeling in, 349–53
 environment, differential gene expression, and, 348–49
 overview of levels of, 349
 post-transcriptional, 356–60
 in transcription initiation, 353–56
 translational and post-translational, 359–60
Gene therapy, 381, **383–85**
Generations, **1103**
Genes, **107**, **238**, **261**. *See also* Chromosomes; DNA (deoxyribonucleic acid); Genomes
 alleles of, 238, 239*f*, 247, 261 (*see also* Alleles)
 in animal diversification, 646
 Cambrian explosion and new, 519–20
 cancer and tumor-suppressor, 361–63
 case study of *HLA* genes in humans, 469–70
 chromosomes, DNA, and, 220
 cloned, in DNA libraries, 372
 for color vision, 964
 control of flower structures by, 439–41
 in control of plant body axes, 435–36
 definitions of, 392
 in determination of leaf shape, 436–38
 developing phylogenetic tree of life from rRNA sequences in, 6–8
 discovery of split eukaryotic, 321–22
 dwarf plants and defective, 808–9
 effects on phenotypes of interactions between, 273–75, 277*t*
 eukaryotic cell nucleus and, 112
 exceptions and extensions to Mendelian rules on, 277*t*
 expression of (*see* Gene expression)
 finding disease-causing, 378–82
 gene families of, 398–99
 in genetic constraints on natural selection, 461–62
 in genetic homologies, 450–51
 genetic information in, 2 (*see also* Genetics)
 genetic maps of (*see* Genetic maps)
 Hershey–Chase experiment on DNA as genetic material in, 285–86
 hormones and, in leptin and energy homeostasis, 1001–3
 identifying, in genome sequences, 392–93
 identifying regulated, with replica plating, 339–41
 inserting, into human cells, 383
 inserting, into plasmids, 370, 371*f*
 lactose metabolism, 340–44
 lateral gene transfer of, 394–95, 399
 linkage and linked, 270–71, 274*f*, 277*t*

loci of, on chromosomes, 266
 in Mendelian genetics, 258*t*
 multiple traits and pleiotropic, 272–73
 mutations of (*see* Mutation[s])
 natural selection of, in Galápagos finches, 458–59*f*
 number of, in genomes, 394–96, 399–400
 in one-gene, one-enzyme hypothesis on gene expression, 305–7
 in particulate inheritance, 261–63
 plant pathogen resistance, 815
 polygenic inheritance and, 275–76
 prokaryotic, 107–8
 redefinition of, 358
 regulation of bacterial (*see* Gene regulation, bacterial)
 regulatory (*see* Regulatory genes)
 RNA as intermediary between proteins and, 307–8
 for taste receptors, 965
 tumor suppressors, 232–33
Genes-in-pieces hypothesis, 321–22
Genetic bottlenecks, **481–82**
Genetic code, **310**
 code words in, 310–11
 cracking of, 311–12
 hypothesis of, by Francis Crick, 307
 in mRNA specification of amino acids, 325–26
 in transfer of amino acids to proteins by tRNAs, 326–28
Genetic constraints, natural selection and, 461–62
Genetic correlation, **461–62**
Genetic counselors, 376
Genetic disorders, human, 97–98, 277–79, 383–85. *See also* Diseases and disorders, human
Genetic divergence
 in allopatric speciation, 494*f*
 interbreeding and reinforcement of, 499–500
 speciation and, 490
 in sympatric speciation, 496*f*
Genetic diversity, **1173**. *See also* Genetic variation
 behavior and, 1083–84
 biodiversity and, 1173–74
 conservation of, 1189–90
 directional selection and, 473
 fungal, 615
 genetic drift, gene flow, mutations, and, 483–84
 habitat fragmentation and loss of, 1182
 in preservation of metapopulations, 1120
 switching of reproductive modes for, 1016
Genetic drift, **465**, **478–82**
 in allopatric speciation, 495
 causes of, in natural populations, 481–82
 effects of, on laboratory populations, 480*f*
 as evolutionary process, 465, 486*t*
 experimental studies of, 479–81
 founder effect and, 481
 genetic bottlenecks and, 481–82
 genetic diversity and, 483
 Hardy–Weinberg principle and, 468
 key points about fitness, genetic variation, and, 479
 loss of biodiversity from, 1180
 simulation studies of, 478–79

Boldface page numbers indicate a glossary entry; page numbers followed by an
f indicate a figure; those followed by t indicate a table.

Boldface page numbers indicate a glossary entry; page numbers followed by an *f* indicate a figure; those followed by *t* indicate a table.

one gene node one polypeptide note.

three H bonds   2 Hydrogen bonds.

DNA (G—C) (A—T)
RNA (G—C) (A—U)

Central dogma

DNA → mRNA → Protein → activated protein

transcription — is that genetic infor transferred from DNA to RNA

translation — n

genotype

phenotype

coding sequence

-35 -10 boot

promoter

Sigma factor holoenzyme

RNA polymerase

mRNA transcription

mRNA transcription

TATA
box

Protor

TBP

NADPH?
NADP+ is found in chloroplast

NADH

without oxygen, we are not able to make ATP, NADH

How cells respond to their environment

Intro to immune System
immune system develop in bone marrow.
B-cells make up antibodies
T-cells help to regulate conductors in the brain
Neutrophil | macrophages. acts in vector

How do the cell response to the environment?
- Each cell has a receptor on its cell surface that recognize
  a specific part of a microbe
- That receptor triggers a signal transduction pathway
- This triggers gene expression transcription, that leads to
  protein synthesis (translation) that allows

DNA   DNA
Double helix. strand is a polymer of Nucleotide
PO₄⁻ group in DNA is negative charge (—). It is made of
5-carbon sugar (deoxy ribose).
Differ in their nitrogenous base (DNA)
     T, C, A, G
- RNA has uracil (U) in place of thymine

Boldface page numbers indicate a glossary entry; page numbers followed by an *f* indicate a figure; those followed by *t* indicate a table.

I:24 INDEX

Boldface page numbers indicate a glossary entry; page numbers followed by an *f* indicate a figure; those followed by *t* indicate a table.

Boldface page numbers indicate a glossary entry; page numbers followed by an *f* indicate a figure; those followed by *t* indicate a table.

Boldface page numbers indicate a glossary entry; page numbers followed by an *f* indicate a figure; those followed by *t* indicate a table.

Boldface page numbers indicate a glossary entry; page numbers followed by an
f indicate a figure; those followed by *t* indicate a table.

Boldface page numbers indicate a glossary entry; page numbers followed by an
f indicate a figure; those followed by *t* indicate a table.

Boldface page numbers indicate a glossary entry; page numbers followed by an *f* indicate a figure; those followed by *t* indicate a table.

Boldface page numbers indicate a glossary entry; page numbers followed by an *f* indicate a figure; those followed by *t* indicate a table.

Boldface page numbers indicate a glossary entry; page numbers followed by an *f* indicate a figure; those followed by *t* indicate a table.

Lab 7

use in defense against virdes

Restriction enzyme — is a protein enzyme found in many microbial species and function in cutting DNA at very specific sequence found in the palidrone

top
5' GAATTC 3'   5' CTTAAG 3'

is attached to
5' (PO₄) 5' 1' base is attached }
G

3'
OH

3'
P base
A
OH

Canvas

Protse record

Spupgone viewer

usy Spup Sane viewer

N'Suria

Nide

how many times does
5371
PGLO = 531 base pair   5371 basepair = 5.371 kb

↓ double strand
                        there is 3 Nide in PGlo plasmid.
Kb = 1000 bp

↓
kilobase pair

NdeI

PGlo   5371   1300
3 fragments — Nde 2   4900   ±471
of NdeI   ————   ————
                          1   13471

Nde ③   Sul

⊕ — nesative charge   larse ⊖ nesative
     larse
     (+)(−)   — 3300 bp
□□□ □

smal snul   1771bp
      1972
                             300 bp
⊕ charse
snd I ⊕ because dna is negatively charge
        charse